Handbook of Experimental Pharmacology

Volume 119

Springer

Berlin
Heidelberg
New York
Barcelona
Budapest
Hong Kong
London
Milan
Paris
Santa Clara
Singapore
Tokyo

Oral Antidiabetics

Contributors

H.J. Ahr, S.L. Ali, J.M. Amatruda, E.M. Bardolph, H. Bischoff,
H.H. Blume, E.-M. Bomhard, E. Brendel, L. Groop, F. Hartig,
A. Hasselblatt, L.S. Hermann, B. Junge, J. Köbberling,
H.P. Krause, J. Kuhlmann, K.H. Langer, H.E. Lebovitz,
P.J. Lefèbvre, M. Matzke, M.L. McCaleb, G. Neugebauer, M. Noel,
P. Ochlich, U. Panten, H.J. Ploschke, H. Plümpe, E. Prugnard,
W. Puls, W. Rebel, A.J. Scheen, H. Schlecker, F.H. Schmidt,
B.S. Schug, E. Schütz, C. Schwanstecher, M. Schwanstecher,
S. Seip, J. Stoltefuss, R.H. Taylor, N.F. Wiernsperger,
W. Wingender, C. Wünsche

Editors

Jochen Kuhlmann and Walter Puls

 Springer

Professor Dr.med. JOCHEN KUHLMANN
Bayer AG
Institut für Klinische Pharmakologie
Aprather Weg
42096 Wuppertal
Germany

Professor Dr.med. WALTER PULS
Krummacher Straße 200
42115 Wuppertal
Germany

With 140 Figures and 85 Tables

ISBN 3-540-58990-2 Springer-Verlag Berlin Heidelberg New York

Library of Congress Cataloging-in-Publication Data. Oral antidiabetics/contributors. H.J. Ahr . . . [et al.]; editors, Jochen Kuhlmann and W. Puls. p. cm. – (Handbook of experimental pharmacology: v. 119) ISBN 3-540-58990-2 (hardcover: alk. paper) 1. Hypoglycemic agents. 2. Non-insulin-dependent diabetes – Chemotherapy. I. Ahr, H.J. II. Kuhlmann, J. (Jochen) III. Puls, Walter. IV. Series. [DNLM: 1. Hypoglycemic Agents – pharmacology. 2. Hypoglycemic Agents – therapeutic use. 3. Diabetes Mellitus, Non-Insulin-Dependent – drug therapy. W1 HA51L v. 119 1996/WK 825 O63 1996] QP905.H3 vol. 119 [RC661.A1] 615′.1s – dc20 [616.4′62061] DNLM/DLC for Library of Congress 96-23697

© Springer-Verlag Berlin Heidelberg 1996
Printed in Germany

Cover design: Springer-Verlag, Design & Production
Typesetting: Best-set Typesetter Ltd., Hong Kong

SPIN: 10126523   27/3136/SPS – 5 4 3 2 1 0 – Printed on acid-free paper

Preface

The prevalence of diabetes continues to increase worldwide. Traditionally, diabetes in its adult form has not been considered a serious life-threatening disease. This attitude needs to be changed because the complications associated with the adult form of diabetes affect almost every organ system. The high morbidity and mortality of non-insulin-dependent diabetes mellitus (NIDDM) suggest that current treatment strategies are unsatisfactory. We therefore face an urgent need for new therapeutic approaches.

When the first *Handbook of Oral Antidiabetics* was edited by H. Maske in 1971, the risks and benefits associated with the use of oral antidiabetics were still under discussion. Nowadays, oral antidiabetics hold a strong position in the long-term treatment of diabetes. Roughly 30%–50% of the patients with diabetes in Europe and the United States are treated with oral antidiabetics, chiefly sulfonylureas. While acknowledging the value of the β-cytotropic sulfonylureas, we also need to recognize important limitations of their use, e.g., in the treatment of obese diabetic patients.

A re-assessment of the association of biguanides, especially of metformin treatment, with lactic acidosis by investigators in Sweden has brought a revival in the use of biguanides. Today, metformin is the only biguanide recommended for the treatment of NIDDM, and it is included in the large United Kingdom Prospective Diabetes Study (UKPDS) concerning long-term benefits. Metformin has recently undergone extensive investigations in the United States and received marketing approval from the Food and Drug Administration (FDA) at the end of 1994. There is also an increasing interest in the effect of metformin on insulin resistance, hyperinsulinemia and other components of the recently recognized "metabolic syndrome" or "syndrome X."

A brand-new approach to managing IDDM and NIDDM was offered in the early 1990s by the introduction of a pseudo tetrasaccharide of microbial origin, which reduces diet-induced hyperglycemia and endogenous insulin secretion by inhibiting intestinal α-glucosidases. It may be of particular interest that repeated administration of a glucosidase inhibitor ameliorated or prevented diabetes-related neuropathy and microvascular complications in preclinical studies.

We are most grateful to the authors for the high quality of their contributions and (in most cases) for their prompt submission of manuscripts. We believe that the wealth of knowledge of oral antidiabetics presented here offers the clinician the widest possible selection for the optimal treatment of this widespread disease. We hope this information will stimulate scientists to improve therapeutic standards by developing agents that may prevent dangerous late complications or the manifestation of NIDDM.

We would like to thank Mrs. Doris Walker from Springer-Verlag for her efficient management and editorial assistance through the entire project. We would also like to thank Mrs. G. Lion for secretarial and administrative help of exceptional quality.

JOCHEN KUHLMANN
WALTER PULS

List of Contributors

AHR, H.J., Bayer AG, PH-PD Toxicology, Geb. 514, 42096 Wuppertal, Germany

ALI, S.L., Zentrallaboratorium Deutscher Apotheker, Ginnheimerstr. 20 65760 Eschborn, Germany

AMATRUDA, J.M., Miles Inc., Pharmaceutical Division, Institute for Metabolic Disorders, 400 Morgan Lane, West Haven, CT 06516-4175, USA

BARDOLPH, E.M., Department of Medicine, Royal Naval Hospital, Haslar, Gosport, Hampshire PO12 2AA, United Kingdom

BISCHOFF, H., Bayer AG, Institut für Herz/Kreislauf- und Arteriosklero-seforschung, Geb. 500, 42096 Wuppertal, Germany

BLUME, H.H., Zentrallaboratorium Deutscher Apotheker, Ginnheimerstr. 20, 65760 Eschborn, Germany

BOMHARD, E.M., Bayer AG, Fachbereich Toxikologie, Aprather Weg, 42096 Wuppertal, Germany

BRENDEL, E., Bayer AG, Institut für Klinische Pharmakologie, Geb. 405, 42096 Wuppertal, Germany

GROOP, L., Malmö General Hospital, Department of Endocrinology, Södra Förstadsg 101, 21401 Malmö, Sweden

HARTIG, F., Boehringer Mannheim GmbH, Sandhofer Str. 116, 68305 Mannheim, Germany

HASSELBLATT, A., Institut für Pharmakologie und Toxikologie der Univer-sität, Robert-Koch-Str. 40, 37075 Göttingen, Germany

HERMANN, L.S., Meda Sverige AB, P.O. Box 138, 40122 Göteborg, Sweden

JUNGE, B., Bayer AG, Central Research, Geb. Q 18, 51368 Leverkusen, Germany

KÖBBERLING, J., Ferdinand-Sauerbruch-Klinikum Wuppertal, Akademisches Lehrkrankenhaus der Heinrich-Heine-Universität Düsseldorf, Arrenberger Str. 20, 42117 Wuppertal, Germany

KRAUSE, H.P., Bayer AG, PH-PD Preclinical Development, Geb. 402, 42096 Wuppertal, Germany

KUHLMANN, J., Bayer AG, Institut für Klinische Pharmakologie, Geschäftsbereich Pharma, Aprather Weg, 42096 Wuppertal, Germany

LANGER, K.H., Hoechst AG, Pharma Entwicklung Pathologie H 811, 65926 Frankfurt, Germany

LEBOVITZ, H.E., Endocrinology and Metabolism/Diabetes, State University of New York, 450 Clarkson Avenue, Box 1205, Brooklyn, NY 11203, USA

LEFÈBVRE, P.J., Centre Hospitalier de Liège, Département de Médecine, Service de Diabétologie, Nutrition et Maladies métaboliques, C.H.U. – Domaine Universitaire du Sart Tilman, B. 35, 4000 Liège 1, Belgium

MATZKE, M., Bayer AG, Central Research, Geb. Q 18, 51368 Leverkusen, Germany

MCCALEB, M.L., Miles Inc., Pharmaceutical Division, Institute for Metabolic Disorders, 400 Morgan Lane, West Haven, CT 06516-4175, USA

NEUGEBAUER, G., Boehringer Mannheim Gmbh, Abt. Klinische Pharmakologie, Sandhofer Str. 116, 68305 Mannheim, Germany

NOEL, M., Groupe LIPHA, Centre de Recherche Chilly-Mazarin, 4, rue de la Division Leclerc, 91380 Chilly Mazarin, France

OCHLICH, P., Boehringer Mannheim GmbH, Sandhofer Str. 116, 68305 Mannheim, Germany

PANTEN, U., Institut für Pharmakologie und Toxikologie der Technischen Universität, Mendelssohnstr. 1, 38106 Braunschweig, Germany

PLOSCHKE, H.J., Bayer AG, PH-R Structural Research, 42096 Wuppertal, Germany

PLÜMPE, H., Kirschbaum Str. 9, 42115 Wuppertal, Germany

PRUGNARD, E., Groupe LIPHA, Centre de Recherche Chilly-Mazarin, 4, rue de la Division Leclerc, 91384 Chilly-Mazarin, France

PULS, W., Krummacher Str. 200, 42115 Wuppertal, Germany

REBEL, W., Boehringer Mannheim GmbH, Sandhofer Str. 116, 68305 Mannheim, Germany

SCHEEN, A.J., Centre Hospitalier de Liège, Département de Médecine, Service de Diabétologie, Nutrition et Maladies métaboliques, C.H.U. – Domaine Universitaire du Sart Tilman, B. 35, 4000 Liège 1, Belgium

SCHLECKER, H., Bayer AG, PH-R Structural Research, 42096 Wuppertal, Germany

SCHMIDT, F.H., Stockacher Str. 1, 68239 Mannheim, Germany

SCHUG, B.S., Zentrallaboratorium Deutscher Apotheker, Ginnheimerstr. 20, 65760 Eschborn, Germany

SCHÜTZ, E., Am Bräunling 12, 65719 Hofheim-Diedenbergen, Germany

SCHWANSTECHER, C., Institut für Pharmakologie und Toxikologie der Technischen Universität, Mendelssohnstr. 1, 38106 Braunschweig, Germany

SCHWANSTECHER, M., Institut für Pharmakologie und Toxikologie der Technischen Universität, Mendelssohnstr. 1, 38106 Braunschweig, Germany

SEIP, S., Bayer AG, PH-R Structural Research, 42096 Wuppertal, Germany

STOLTEFUSS, J., Bayer AG, PH-R-CR Medical Chemistry III, Geb. 460, 42096 Wuppertal, Germany

TAYLOR, R.H., Department of Medicine, Royal Naval Hospital, Haslar, Gosport, Hampshire PO12 2AA, United Kingdom

WIERNSPERGER, N.F., Groupe LIPHA, Strategic Development Division, International Pharmacological Development Department, 34, rue Saint Romain, 69379 Lyon, Cedex 08, France

WINGENDER, W., Bayer AG, Institut für Klinische Pharmakologie, Geb. 429, 42096 Wuppertal, Germany

WÜNSCHE, C., Bayer AG, PH-R Structural Research, 42096 Wuppertal, Germany

Contents

CHAPTER 3

**Non-Pharmacological Management
of Non-Insulin-Dependent Diabetes**
J. Köbberling 43

Contents

Section I: Sulfonylureas

CHAPTER 4

Sulfonylureas and Related Compounds:
Chemistry and Structure-Activity Relationships
H. Plümpe. With 1 Figure   65

CHAPTER 5

Sulfonylureas: Physicochemical Properties, Analytical Methods
of Determination and Bioavailability
S.L. Ali, H.H. Blume, and B.S. Schug. With 15 Figures   73

CHAPTER 6

Mode of Action of Sulfonylureas
U. PANTEN, M. SCHWANSTECHER, and C. SCHWANSTECHER.

CHAPTER 7

Sulfonylureas: Pharmacokinetics in Animal Experiments

CHAPTER 8

Toxicology of Sulfonylureas
F. HARTIG, K.H. LANGER, W. REBEL, F.H. SCHMIDT,
and E. SCHÜTZ 185

Section II: Biguanides

CHAPTER 10

Chemistry and Structure-Activity Relationships of Biguanides

CHAPTER 11

Physicochemical Properties and Analytical Methods
of Determination of Biguanides

CHAPTER 12

Preclinical Pharmacology of Biguanides

CHAPTER 13

Section III: Glucosidase Inhibitors

CHAPTER 15

Chemistry and Structure-Activity Relationships
of Glucosidase Inhibitors
B. JUNGE, M. MATZKE, and J. STOLTEFUSS. With 63 Figures   411

CHAPTER 16

Analytical Methods of Determination of Glucosidase Inhibitors
H.J. PLOSCHKE, H. SCHLECKER, S. SEIP, and C. WÜNSCHE.

CHAPTER 17

Pharmacology of Glucosidase Inhibitors
W. PULS. With 3 Figures

CHAPTER 18

General Pharmacology of Glucosidase Inhibitors

CHAPTER 19

CHAPTER 21

Clinical Pharmacology of Glucosidase Inhibitors
E. BRENDEL and W. WINGENDER. With 4 Figures 611

CHAPTER 24

New Approaches for the Treatment of Diabetes Mellitus
J.M. AMATRUDA and M.L. McCALEB 697

CHAPTER 1
Introduction

J. KUHLMANN

Diabetes is well recognized as a major global health problem. The clinical manifestation and progression of diabetes often vary considerably between countries and commonly between ethnic groups in the same country. Currently diabetes affects an estimate 15.1 million people in North America, 12.6 million in Latin America, 18.5 million in Europe, 6.6 million in the former USSR, 5.3 million in Africa, 51.4 million in Asia, and just under 1 million in Oceania (McCARTY and ZIMMET 1994).

Of the 110.4 million with diabetes in 1994, non-insulin-dependent diabetes mellitus (NIDDM) or type II diabetes constitutes approximately 85%–90% of all cases of diabetes. Annually, between 500 000 and 600 000 Americans are detected with NIDDM (MAZZE 1994). More than 75% of the individuals with diabetes will develop neurological, microvascular, or macrovascular complications (HERMAN et al. 1984; American Diabetes Association 1993; MAZZE et al. 1985). Each year diabetes is a contributing or underlying cause of 350 000 deaths in the United States. The goal of modern treatment must be to prevent the development and progression of complications of diabetes. At the present time physicians employ a graduated treatment beginning with diet and progressing to the prescription of oral antidiabetics such as sulfonylureas and metformin as monotherapy or in combination (GROOP 1992; BAILEY 1992). These have been shown to decrease fasting plasma glucose levels, but postprandial hyperglycemia persists in more than 60% of patients and probably accounts for sustained increases of hemoglobin A_{1c} (HbA_{1c}) levels (UK Prospective Diabetes Study 1991). The development of innovative, reliable approaches to diabetes management, therefore, is necessary.

This second edition of the volume *Oral Antidiabetics* in the *Handbook of Experimental Pharmacology* aims to discuss the new approaches and to update the older aspects of the subject described in the first edition. The clinical pharmacological aspects of the treatment with oral antidiabetics will also be examined here. The overall goal of clinical pharmacology is to provide information which can be utilized for improving the rational prescribing of drugs. Unfortunately, dosage and dosage intervals are often the only drug-related considerations in many clinical trials. It is also unfortunate that pharmacokinetic investigations are purely descriptive and do not direct adequate attention to the question to what extent drug response is influenced by pharmacokinetics. Through an improved understanding of the pharmacodynamic/pharmacokinetic relationships, the optimal therapeutic dose for the individual patient can be predicted.

The second edition opens with the pathophysiology of NIDDM. The present review provides an overview of the pathogenesis and pathophysiology of NIDDM and emphasizes the dynamic interaction between altered insulin secretion and insulin action in the development of impaired glucose homeostasis. Over the last decade, major advances have been made in our understanding of the pathophysiology and molecular biology of the disease, and this knowledge has provided new insights into the development of therapeutic interventions designed to reverse the specific intracellular defects that characterize the diabetic state.

Chapter 3 focuses on dietary management and exercise as basic parts of the non-pharmacological treatment of the disease. Dietary recommendations for patients with diabetes have to consider differing dietary habits in different countries. Dietary advice, therefore, has to be separated into that for patients of westernized and non-westernized countries. Many patients with this disease who comply with dietary advice will show improvement in the major metabolic abnormalities associated with the condition. Unfortunately, most people with non-insulin-dependent diabetes never receive sufficient nutritional education or are not capable of complying with a strict diet regimen.

Sections I and II deal with the two important classes of oral antidiabetics, sulfonylureas and biguanides. The first edition of *Oral Antidiabetics* in the *Handbook of Experimental Pharmacology* dealt with traditional experimental pharmacological areas such as chemistry and structure-activity relationship, analytics, metabolism. pharmacology, and toxicology. These areas still form an important part in this second edition. However, the clinical pharmacological aspects of treatment with oral antidiabetics will also be examined here. The chapter "Mode of Action of Sulfonylureas" concentrates on the discussion of new results of the hypoglycemic effect of sulfonylureas and their analogues resulting from their binding to the high-affinity receptors in various tissues as well as organs and occuring at drug concentrations relevant to antidiabetic therapy. Some implications of the results from animal pharmacokinetics of sulfonylureas for the clinical use are described in the next chapter. As there are different sulfonylurea compounds on the market acting in principle in the same way, the choice of the right drug for the right patient has to take into account the special kinetic properties of the drug selected. Although most sulfonylureas are readily absorbed, the bioavailability of different brands of the same compound may not be identical. The bioavailability of drugs is on the one hand dependent on the physicochemical and pharmacokinetic characteristics of the drug itself, but on the other hand biopharmaceutical properties of the pharmaceutical formulation are also significant.

Because of its importance in diabetes mellitus therapy, glibenclamide has become the object of extensive generic competition in many countries. As shown by BLUME et al. (1985a,b), there are marked differences in the rate and extent of absorption between different glibenclamide products. Due to these deviations the respective preparations cannot be interchanged during treatment without adjusting the dosing schedule to the differing biopharmaceutical properties.

The chapter "Clinical Pharmacology of Sulfonylureas" focuses on pharmacodynamics, pharmacokinetics, and the safety and tolerance of the most important compounds on the market as well as drugs under clinical investigation. In the pharmacodynamic part of this chapter the effects of sulfonylureas on insulin secretion, hepatic insulin clearance, other pancreatic and extrapancreatic effects as well as combination therapy with insulin and sulfonylurea are described in detail. The pharmacokinetic part covers similarities and differences, timing of drug intake, effect of hyperglycemia, and the dose- and concentration-response relationship in general. Furthermore, it deals with the specific pharmacokinetic behavior of the different substances.

Section II provides an update on the second important class of oral antidiabetics, the biguanides. Since the publication of the first edition, claims have been made for the hypoglycemic properties of several new structures, but none has been subject to further development. The biguanide drugs phenformin, buformin, and metformin were introduced in the late 1950s. Their relationship to lactate metabolism was acknowledged early and lactic acidosis became associated with biguanides. Therefore, phenformin and buformin were withdrawn from the market in the late 1970s and the early 1980s. However, metformin was kept on the market, because only a few cases of lactic acidosis were associated with this biguanide. During the last 10 years there has been a renewed interest in metformin, as it was increasingly realized that metformin offers several therapeutic advantages. The chapters pharmacology, toxicology, and clinical pharmacology of this section, therefore, are mainly concentrated on metformin. The clinical pharmacology part also includes a reevaluation of the clinical trials in NIDDM patients. Because lipid metabolism is influenced in a favorable way, metformin is particularly justified in NIDDM patients with obesity or hyperlipidemia (DUNN and PETERS 1995). Lactate metabolism may be affected, rarely leading to lactic acidosis and then only under special circumstances. Risk factors for lactic acidosis include renal insufficiency, serious cardiovascular and respiratory disease, liver disease, alcohol abuse and severe infections (HERMANN and MELANDER 1992).

Following the introduction of the oral antidiabetics sulfonylureas and biguanides in the 1950s, no substantial progress was achieved in the medical treatment of diabetes. With the development and introduction of the α-glucosidase inhibitor acarbose, the therapy of diabetes mellitus was enriched with a new principle based on a delayed intestinal carbohydrate degradation. This is accompanied by a decrease in the postprandial elevation of blood glucose and insulin. In the light of an accumulating body of evidence that persistent elevations of blood glucose are associated with an increased risk of developing the late microvascular complications of diabetes, attention in the diabetological community has been focused anew on the importance of adequate blood glucose control.

Section III gives a comprehensive review of the most important α-glucosidase inhibitors on the market and in research and development. Many of these data are new and not yet published in detail. It opens with chemistry and structure-activity relationships and analytical methods of determination. Two types of inhibitors represented by either acarbose or 1-deoxynojirimycin analogues are described in detail.

In the following chapters the primary effects of acarbose, voglibose, miglitol, and emiglitate are described comprehensively inclusive of their beneficial secondary effects on insulin insensitivity, lipid metabolism, and dangerous diabetes late complications. Furthermore, an overview of the general pharmacology and toxicology results is given. Since the original guideline studies were of questionable relevance for the clinical situation, in addition to the usual guideline studies required by international authorities for registration, several highly sophisticated special studies have been initiated to clarify the effects that were observed in the original long-term toxicological study on rats. The results are discussed in the toxicological chapter.

Until now only two α-glucosidase inhibitors have reached the market (acarbose in 1990 and voglibose in 1994). A preliminary clinical evaluation of the experiences with this new therapeutic principle has been added following the chapters that deals with pharmacokinetics and clinical pharmacology. The evidence from the clinical evaluation of α-glucosidase inhibitors, much of which is based on acarbose, shows that they can be effective therapeutic agents in both type I and, especially, in type II diabetes.

Chapter 23 focuses on further oral antidiabetics in research and development. The experience with well-established oral antidiabetics has shown that, so far, the ultimate goal of antidiabetic drug therapy, i.e., the prevention of diabetic complications, could not be met satisfactorily. Therefore, a high medical need exists for new therapeutic strategies and approaches to restore euglycemia or to directly prevent diabetic complications. This chapter provides an overview of new pharmacological approaches that were studied in the past decade. Special consideration is given to new agents which stimulate and modulate insulin secretion, suppress hepatic glucose production, enhance peripheral insulin sensitivity, or mimic the action of insulin.

Chapter 24 is devoted to new approaches for the treatment of diabetes mellitus. It concentrates on future prevention strategies of NIDDM and IDDM, future treatments of obesity, and drugs that will decrease insulin resistance or improve insulin sensitivity. Furthermore, the value of alternative insulin delivery devices that include encapsulated islets and an artificial pancreas and the possibility of gene therapy are critically evaluated.

It is our hope that this book will enhance our understanding of the oral antidiabetic agents and their properties to ameliorate, arrest, reverse, or even prevent this damaging chronic disease. We hope that it will be a stimulus for new ideas in this challenging and exciting field of research in favor of diabetic patients.

References

American Diabetes Association (ADA) (1993) Diabetes 1993 vital statistics. American Diabetes Association, Alexandria

Bailey CJ (1992) Biguanides and NIDDM. Diabetes Care 15:755–772

Blume H, Förster H, Stenzhorn G, Askali F (1985a) Zur Bioverfügbarkeit und pharmakodynamischen Aktivität hendelsüblicher Glibenclamid-Fertigarzneimittel, 2nd end. Pharm Ztg 130:1070–1078

Blume H, Ali SL, Stenzhorn G, Stüber W, Siewert M (1985b) Zur Bioverfügbarkeit und pharmakodynamischen Aktivität handelsüblicher Glibenclamid-Fertigarzneimittel, 3rd end. Pharm Ztg 130:2605–2610

Dunn CJ, Peters DG (1995) Metformin: a review of its pharmacological properties and therapeutic use in non-insulin-dependent diabetes mellitus. Drugs 49:721–749

Groop LC (1992) Sulfonylureas in NIDDM. Diabetes Care 15:737–754

Herman W, Sinnock P, Brenner E, Brimberry JL, Langford D, Nakashima A, Sepe SJ, Teutsch SM, Mazze RS (1984) An epidemiological model for diabetes mellitus: incidence, prevalence and mortality. Diabetes Care 7:367–371

Hermann LS, Melander A (1992) Biguanides: basic aspects and clinical uses. In: Alberti KGM, Defronzo RA, Keen H, Zimmet P (eds) International textbook of diabetes mellitus. Wiley, Chichester vol 28, pp 773–795

Mazze RS (1994) A systems approach to diabetes care. Diabetes Care 17:5–11

Mazze RS, Sinnock P, Deeb L, Brimberry J (1985) An epidemiological model for diabetes mellitus in the United States: five major complications. Diabetes Res Clin Pract 1:185–191

McCarty D, Zimmet P (1994) Diabetes 1994 to 2010: global estimates and projections. International Diabetes Institute, Melbourne

UK Prospective Diabetes Study (UKPDS) (1991) VIII. Study design, progress and performance. Diabetologia 34:877–890

CHAPTER 2
Pathophysiology of Type 2 Diabetes Mellitus

A.J. SCHEEN and P.J. LEFÈBVRE

A. Introduction

Non-insulin-dependent diabetes mellitus (NIDDM), also called type 2 diabetes mellitus, is a common metabolic disorder that afflicts 2%–5% of the adult population of most Western countries, with, however, wide international variation (KING et al. 1993). Furthermore, NIDDM is often undiagnosed even in the Western countries (HORTULANUS-BECK et al. 1990), and numerous data indicate that undiagnosed NIDDM is not a benign condition (HARRIS 1993). Certainly, NIDDM is a leading cause of disability and death in developed and developing nations (SONGER 1992).

Over the last decade, major advances have been made in our understanding of the pathophysiology (review in SCHEEN et al. 1986; FAJANS 1986; REAVEN 1990; DeFRONZO et al. 1992; FELBER et al. 1993a) and molecular biology (review in BELL 1991; GRANNER and O'BRIEN 1992; MATSCHINSKY et al. 1993) of the disease, and this knowledge has provided new insights into the development of therapeutic interventions designed to reverse the specific intracellular defects that characterize the diabetic state (BAILEY and FLATT 1990; LEFÈBVRE and SCHEEN 1992; BRESSLER and JOHNSON 1992; SCHEEN and LEFÈBVRE 1993b).

NIDDM is a heterogeneous condition caused by genetic and environmental factors. The present review provides an overview of the pathogenesis and pathophysiology of NIDDM and emphasizes the dynamic interaction between defects in insulin secretion and insulin action in the development of impaired glucose homeostasis (Fig. 1).

I. Heterogeneous Disease

Several lines of evidence indicate that heterogeneity exists within the NIDDM phenotype. These include both genetic and pathophysiological arguments. The former especially include the facts that more than 60 different genetic syndromes are associated with impaired glucose tolerance (VADHEIM and ROTTER 1992), and that within MODY (maturity-onset diabetes of the youth) pedigrees, linkage has been found for selected genetic markers in some families but not others (PERUMTT et al. 1992). The latter include the common observations of physiological heterogeneity in patterns of obesity and insulin

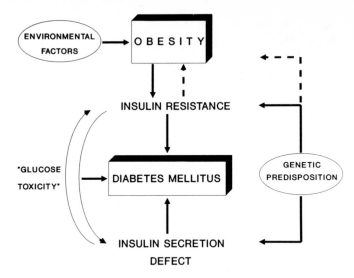

Fig. 1. Effects of genetic predisposition and environmental factors on the combined defect in insulin secretion and insulin action leading to non-insulin-dependent diabetes mellitus

response (Scheen et al. 1986; Fajans 1986; Banerji and Lebovitz 1989; Arner et al. 1991; Scheen et al. 1992b).

II. Genetic Background

NIDDM is known to display a strong familial aggregation. Studies of identical twins with NIDDM have clearly shown that the disease is genetic, although the pattern of inheritance has defied classification (Barnett et al. 1981). The challenge for the end of this century is to identify the genetic factors and specific mutations that cause NIDDM (Bell 1991; Cox et al. 1992). Available studies do indicate that at least some of the genes must act through inherited tendencies to insulin resistance, and to decreased capacity for insulin secretion (see below). The dramatic familial aggregation of NIDDM means that screening of relatives will identify, in the near future, many individuals at earlier stages of the disease process. Several excellent reviews on the genetics of NIDDM have been published recently (Permutt 1990; Serjeantson and Zimmet 1991; Hitman and McCarthy 1991; Vadheim and Rotter 1992; Hamman 1992; Granner and O'Brien 1992).

III. Environmental Factors

One of the major epidemiological arguments for the role of environmental factors in the etiology of NIDDM has been the rapid increase in the prevalence and incidence of NIDDM in populations undergoing rapid westernization, such as in the south Pacific inhabitants and in the American Indians (O'DEA 1992). The westernization transition is usually accompanied by increases in obesity, decreases in physical activity, alterations in dietary intake toward more calories, fat and less complex carbohydrates, and, often, urbanization. These observations may be explained by a "thrifty genotype" which continued to be strongly selected for until this most recent lifestyle change, because it conferred survival advantage under conditions when food shortages were an inevitable part of life (HALES and BARKER 1992; O'DEA 1992). Numerous studies have now become available to support the role of diet and physical inactivity in NIDDM risk (review in HAMMAN 1992; KRISKA and BENNETT 1992). The specific role played by obesity in the pathophysiology of NIDDM will be discussed in detail at the end of this review (Fig. 1).

B. Abnormalities of the Glucose-Insulin Feedback Loop

Plasma glucose concentrations are normally maintained within a fairly narrow range despite wide fluctuations in the body's supply (e.g. meals) and demand (e.g. exercise) for nutrients (Fig. 2) (GERICH 1993). After an overnight fast, insulin-independent tissues, the brain (=50%) and splanchnic organs (=25%), account for most of the total-body glucose disposal. Insulin-dependent tissues, adipose tissue and primarily skeletal muscles, are re-

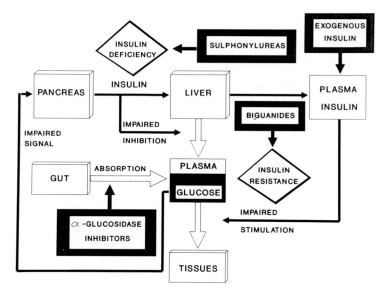

Fig. 2. Abnormalities in the "glucose-insulin" feedback loop in non-insulin-dependent diabetes mellitus. Defects in insulin secretion and in insulin action play a key role. The most important sites of action of the classical pharmacological treatments are illustrated

sponsible for the remaining 25% of glucose utilization. This basal glucose uptake (around 2 mg/kg/min) is precisely matched by the release of glucose from the liver. After a meal, this delicate balance between tissue glucose uptake and hepatic glucose production (HGP) is disrupted and the maintenance of normal glucose homeostasis is dependent upon three processes that occur in a coordinated and tightly integrated fashion. In response to hyperglycaemia, pancreatic insulin secretion is stimulated and the combination of hyperinsulinaemia plus hyperglycaemia promotes glucose uptake (by splanchnic and peripheral, primarily muscle, tissues) and suppresses HGP. It follows, therefore, that defects at the level of the B-cell, muscle and/or liver can lead to the development of glucose intolerance or overt diabetes mellitus (DeFRONZO 1988).

In NIDDM patients with established hyperglycaemia, basal HGP is usually increased while glucose uptake by the tissues is also increased due to mass action. The results obtained during an oral glucose tolerance test (OGTT) or after a mixed meal (review in DeFRONZO et al. 1992; DINNEEN et al. 1992) indicate that both impaired suppression of HGP and decreased peripheral tissue (muscle) glucose uptake contribute approximately equally to the glucose intolerance of NIDDM, while diminished splanchnic glucose uptake is not an important contributory factor. All these abnormalities present in NIDDM basically result from an imbalance between insulin sensitivity and insulin secretion.

I. Insulin Secretion

B-cell function in NIDDM has been the subject of intense investigation for several decades (review in WARD et al. 1986; FAJANS 1986; CERASI 1988; LEAHY 1990; PORTE 1991; WOLFFENBUTTEL and VAN HAEFTEN 1993). Table 1 summarizes the most important abnormalities in insulin secretion described in NIDDM.

Table 1. Most important in vivo abnormalities in insulin secretion of NIDDM patients

1. Abnormal proportion of proinsulin
2. Selective loss of glucose-induced insulin secretion
3. Major defect in first-phase insulin response
4. Loss of rapid insulin oscillations
5. Partial dissociation between oscillatory patterns of insulin secretion and glucose levels
6. Decreased glucose potentiation slope of insulin secretion by other secretagogues
7. Impaired insulin feedback inhibition

1. How to Measure Insulin Secretion?

Standard techniques for evaluating insulin secretion in human subjects involve the measurement of peripheral concentrations of insulin and C peptide, which are secreted in equimolar amounts by the B cells of the islets of Langerhans of the pancreas. However, these techniques do not always provide accurate quantitative estimates of insulin secretion, have important limitations and may lead to several pitfalls (POLONSKY et al. 1986a; FERNER and ALBERTI 1986; TEMPLE et al. 1992; GROOP et al. 1993b).

a) Immunoreactive Insulin

The development of the radioimmunoassay technique satisfied the need for a sensitive and apparently specific measurement of plasma insulin, and allowed a quantitative evaluation of the pancreatic insulin response to various stimuli (YALOW and BERSON 1960). However, insulin undergoes a large and variable hepatic extraction and has a variable peripheral clearance (review in FERRANNINI and COBELLI 1987; CASTILLO et al. 1994b). In fact, hyperinsulinaemia that occurs in compensation for increasing insulin resistance may result not only from an enhanced insulin secretion but also from a decreased insulin clearance, probably due to a reduction of its hepatic extraction (BONORA et al. 1984; HAFFNER et al. 1992).

b) C Peptide

In contrast to insulin, C peptide is not significantly extracted by the liver and thus may be a good alternative for the quantitatation of insulin secretion (POLONSKY et al. 1986a). However, peripheral C peptide levels do not change in proportion to the rate of insulin secretion, since C peptide has a longer half-life and is distributed in more than a single kinetic compartment. Nevertheless, even in non-steady-state conditions, the rate of secretion of insulin can be accurately estimated form peripheral C peptide concentrations by using an open two-compartmental mathematical model (POLONSKY et al. 1986b). Such an approach has been used to study insulin secretion, in a precise quantitative and qualitative manner, in both obesity (POLONSKY et al. 1988a) and NIDDM (POLONSKY et al. 1988b).

c) "True" Insulin

Recently, it has been suggested that NIDDM individuals may be more insulinopenic than previously appreciated because of the presence of high circulating levels of proinsulin and 32–33 split proinsulin, which are biologically much less active than insulin but cross-react substantially with insulin in routine radioimmunoassays (TEMPLE et al. 1990). Until highly specific and sensitive two-sided immunoradiometric assays (IRMAs) based on monoclonal antibodies are more widely available and appropriate standardization is achieved, controversy regarding insulin secretion characteristics of NIDDM patients will remain (Temple et al. 1992).

2. Abnormal Plasma Concentrations

a) Basal State

The fasting plasma insulin concentrations have invariably been found to be normal or increased in NIDDM. However, it is difficult to compare the insulin levels of different individuals unless they are matched for the steady state plasma glucose concentration, and body adiposity. When such matching is performed, it is obvious that there is a marked insulin secretory defect in patients with NIDDM (Porte 1991). An attractive hypothesis has been presented according to which the basal insulin level in NIDDM is the result of B-cell adaptation to hyperglycaemia, the degree of fasting hyperglycaemia in turn being a function of the adaptative capacity of the B cell, the feedback regulatory system thus reaching equilibrium (Turner et al. 1988).

The relationship between fasting glucose and insulin levels is complex, resembling an inverted "U" curve (DeFronzo et al. 1992). As the fasting plasma glucose levels increase from 80 to 140 mg/dl, there is a progressive rise in fasting plasma insulin, which represents a compensatory response by the pancreas to offset the deterioration in glucose metabolism (see below). When the fasting glucose exceeds 140 mg/dl, insulin secretion drops off precipitously, which has important pathophysiological implications: it is at this point that HGP increases in absolute terms and begins to make a major contribution to fasting hyperglycaemia (DeFronzo et al. 1992).

b) Glucose Stimuli

Glucose-stimulated insulin secretion has variably been reported to be normal, increased or decreased in NIDDM subjects. It must be emphasized, however, that even a normal or increased plasma insulin response is deficient when viewed relative to the level of hyperglycaemia. Numerous studies (review in DeFronzo et al. 1992), including our own (Scheen et al. 1991), have shown that the relationship between the insulin response during an OGTT and the severity of diabetes is complex and resembles that seen in the fasting state, i.e. an inverted U-shaped curve. This pattern has been called the Starling's curve of the pancreas, suggesting an initial compensation of the B cell in response to moderate hyperglycaemia followed by a late decompensation when hyperglycaemia progresses (DeFronzo 1988).

c) Non-glucose Stimuli

In NIDDM patients with defective response to glucose, it has been shown that various other insulin secretagogues, e.g. isoproterenol, secretin, arginine and glucagon, are still able to stimulate insulin secretion by B cells (review in ROBERTSON 1989). These observations support the concept of defective glucose recognition as a specific abnormal entity in type 2 diabetic patients (MALAISSE 1988; ROBERTSON 1989).

3. Abnormal Kinetics Response

a) First-Phase Insulin Response

Insulin secretion is biphasic, with an early burst of insulin release within the first minutes after intravenous administration of glucose, followed by a progressively increasing phase of insulin secretion that persists as long as the hyperglycaemic stimulus is present. The first-phase insulin secretory response is characteristically lost in NIDDM (BRUNZEL et al. 1976), and may be one of the earliest detectable abnormalities in patients who will later develop NIDDM (review in CERASI 1988). The loss of the first phase of insulin secretion may have important pathogenic consequences, because this early burst of insulin release plays an important role in priming the target tissues, especially the liver, that are responsible for the maintenance of normal post-prandial glucose homeostasis (MITRAKOU et al. 1992).

In contrast, second-phase insulin release is still sensitive to glucose and therefore partially maintained in patients with compensated NIDDM (fasting plasma glucose levels below 200 mg/dl).

b) Pulsatile Secretion

Insulin is secreted in a pulsatile manner, associating both rapid pulses (12- to 15-min period) and slower ultradian oscillations (90- to 120-min period). Some evidence suggests that rapid pulsatile insulin delivery is more effective in inhibiting HGP than continuous administration and that oscillatory insulin delivery is more effective in promoting peripheral glucose disposal (review in LEFÈBVRE et al. 1987). Loss of regular pulsatile insulin secretion has been shown in NIDDM patients as well as in normoglycaemic first-degree relatives (O'RAHILLY et al. 1988), which suggests that this may be an early, possibly genetic, defect in the natural history of NIDDM. The ultradian oscillations are also perturbed in NIDDM patients: the pulse amplitude is significantly reduced and their relationship to meals is distorted (POLONSKY et al. 1988b). Moreover, a partial dissociation between the oscillatory patterns of insulin secretion and glucose levels has been observed in NIDDM, which could represent a sensitive quantitative marker of the breakdown of the insulin-glucose feedback loop in diabetes (STURIS et al. 1992).

c) Potentiation Slope

Support for a major loss of islet functional capacity in NIDDM comes from studies of the non-glucose regulation of insulin secretion. It has been shown that the effect of various insulin secretagogues (isoproterenol, arginine, glucagon) on insulin release can be markedly enhanced by increasing the plasma glucose level (review in WARD et al. 1986 and PORTE 1991). This phenomenon has been termed "glucose potentiation of insulin secretion". Comparison of the dose-response relationships of the glucose potentiation of insulin secretion indicated that maximal insulin release was markedly decreased in the NIDDM patients; however, after correction for the difference in maximal release, the relative sensitivity of the B cell to glucose proved to be normal (WARD et al. 1986). Such a finding supports the concept that the impaired capacity of the B cell to respond to glucose potentiation constitutes a major islet abnormality in patients with NIDDM (PORTE 1991).

4. Varia

The existence of insulin feedback inhibition is a controversial issue. However, numerous studies have demonstrated that insulin inhibits its own secretion in normal humans. Such inhibition has been shown to be normal in obese subjects, suggesting that insulin resistance of obesity does not extend to the B cells of the islets of Langerhans (SCHEEN et al. 1992a). In NIDDM patients, however, such a pancreatic resistance has been reported with an impaired inhibition of insulin secretion by insulin itself when compared with normal subjects (GARVEY et al. 1985).

II. Insulin Sensitivity

Even if insulin exerts numerous different effects, so far insulin sensitivity has been considered mainly in the context of glucose metabolism, especially at the liver and muscle sited, the two most important organs, besides pancreas, involved in the pathogenesis of NIDDM (DEFRONZO et al. 1983; SCHEEN et al. 1986; and DEFRONZO 1988; GERICH 1988; BECK-NIELSEN and HOTHER-NIELSEN 1988; DEFRONZO et al. 1992; SCHEEN and LEFÈBVRE 1992; GROOP et al. 1993b; SCHEEN et al. 1994). Furthermore, while understanding "insulin resistance", it has been shown that both glucose resistance and insulin resistance are required to model NIDDM in humans (RUDENSKI et al. 1991). Table 2 summarizes the most important abnormalities in insulin action described in NIDDM.

1. Hepatic Glucose Production

Increased glucose production by the liver represents a major factor contributing to hyperglycaemia in both the postabsorptive and the postprandial state in NIDDM (DEFRONZO and FERRANNINI 1987; FERRANNINI and GROOP 1989). Glucose is released by the liver as a result of both breakdown of glycogen stores (glycogenolysis) and de novo synthesis of glucose from nonglucose precursors such as lactate, alanine or glycerol (gluconeogenesis).

Table 2. Most important in vivo abnormalities in insulin action of NIDDM patients

---

1. Hepatic glucose production
   a) Increased in the basal state
   b) Submaximal suppression during insulin clamp
   c) Insufficient inhibition after OGTT or after a meal
2. Splanchnic glucose uptake
   a) No major alterations
3. Peripheral (muscle) glucose uptake
   a) Increased in the basal state ("mass action")
   b) Poor stimulation during insulin clamp
   c) Insufficient stimulation after OGTT or after a meal
   d) Defect in glucose storage more than in glucose oxidation
4. Varia
   a) Resistance to insulin suppression of plasma NEFA
   b) Increased glucagon secretion
   c) Reduced glucose-mediated glucose uptake

---

a) Fasting State

In NIDDM subjects with moderate fasting hyperglycaemia, a consistent increase in basal HGP of about 0.5 mg/kg/min has been demonstrated (review in DeFronzo et al. 1992). Moreover, the increase in basal HGP is closely correlated with the degree of fasting hyperglycaemia. Even if some methodological problems may have somewhat overestimated basal glucose production in type 2 diabetes (Hother-Nielsen and Beck-Nielsen 1990), these results indicate that, in NIDDM subjects, excessive HGP is an important factor in the development of fasting hyperglycaemia above 140 mg/dl. On the other hand, basal plasma insulin levels are frequently increased in those NIDDM subjects. Because hyperinsulinaemia is a powerful inhibitor of HGP, it is clear that hepatic insulin resistance to insulin is present in the postabsorptive state and contributes to the excessive output of glucose by the liver (DeFronzo and Ferrannini 1987; Ferrannini and Groop 1989).

b) Hyperinsulinaemic Euglycaemic Clamp

Apparently conflicting results have appeared concerning the suppression of HGP by insulin in NIDDM, in conditions of euglycaemia and steady hyperinsulinaemia as performed by the clamp technique (review in DeFronzo et al. 1992). This controversy may be explained by various methodological problems (review in Bergman et al. 1992; Scheen et al. 1994). Most recent studies that have used insulin concentrations that produce submaximal suppression of HGP have clearly demonstrated hepatic insulin resistance in NIDDM (Campbell et al. 1988; Butler et al. 1990; Scheen and Lefèbvre 1992).

c) Oral Glucose Tolerance Test

Several studies used the double-tracer isotope technique to examine the disposition of an oral glucose load in NIDDM, and the concomitant inhibition of HGP (review in DeFronzo et al. 1992). All studies concluded that the combined effect of hyperglycaemia plus hyperinsulinaemia to suppress HGP after glucose ingestion is impaired in NIDDM, and that this difference accounts for at least one-third of the defect in total-body glucose homeostasis in NIDDM (review in DeFronzo et al. 1992; Dinneen et al. 1992). Thus, the results obtained with the OGTT emphasize the important contribution of the liver to the impairment of glucose homeostasis in NIDDM (Mitrakou et al. 1992).

2. Splanchnic Glucose Uptake

a) Fasting State

Using the hepatic vein catheterization technique, it has been shown that, in the postabsorptive state, there is a *net* release of glucose from the splanchnic area in both the control and NIDDM subjects, reflecting glucose production by the liver (DeFronzo et al. 1985). Simultraneously, splanchnic glucose uptake is quantitatively small (around 0.5 mg/kg/min) and similar in NIDDM and normal subjects.

b) Hyperinsulinaemic Euglycaemic Clamp

During a euglycaemic insulin clamp, HGP is almost completely inhibited and there is a small net uptake of glucose by the splanchnic area that averages about 0.5 mg/kg/min after 2 h of sustained hyperinsulinaemia, thus a value virtually identical to that observed in the basal state. Importantly, there is no difference between NIDDM and control subjects in the amount of glucose that is taken up by the splanchnic tissues at any time during the insulin-clamp study (DeFronzo et al. 1985).

c) Oral Glucose Tolerance Test

After glucose ingestion, both the oral route of administration and the resultant hyperglycaemia conspire to enhance splanchnic (hepatic) glucose uptake (DeFronzo and Ferrannini 1987). When compared to normal subjects, NIDDM patients have been shown to have normal, slightly decreased or increased splanchnic glucose uptake (review in DeFronzo et al. 1992). Thus, splanchnic glucose uptake does not appear to contribute significantly to the impairment in oral glucose tolerance in NIDDM.

3. Peripheral (Muscle) Glucose Uptake

Glucose taken up by cells has basically three main fates: storage as glycogen (or lipid), oxidation to CO_2 or conversion to lactate (or alanine) that is subsequently released into the circulation. The relative proportion in which glucose undergoes these fates varies between tissues and is dependent upon the degree of fasting, the hormonal milieu and the presence of alternative substrates (GERICH 1993).

a) Fasting State

In the postabsorptive state, about 80% of glucose uptake is independent of insulin and occurs in the brain, splanchnic tissues and red cells (GERICH 1991). Muscle is the insulin-sensitive tissue responsible for the majority of insulin-mediated glucose disposal, but it takes little glucose in the fasting state, because plasma insulin levels are low (GERICH 1991).

By simultaneous femoral artery/vein catheterization to quantitate leg glucose exchange, it has been shown that leg glucose uptake is slightly but significantly increased in NIDDM subjects (DEFRONZO et al. 1985). This accelerated rate of tissue glucose uptake is due to the mass action effect of hyperglycaemia that passively drives glucose into cells. However, the metabolic fate of the glucose that is taken up is not normal as a disproportionate amount is converted to lactate, which is subsequently released and can serve as a substrate to drive gluconeogenesis by the liver (DEFRONZO et al. 1985; CAPALDO et al. 1990).

b) Hyperinsulinaemic Euglycaemic Clamp

Insulin-clamp studies, combined with leg (DEFRONZO et al. 1985) or forearm (CAPALDO et al. 1990) exchange balance measurements, have shown that the absolute rate of insulin-mediated glucose uptake is reduced by approximately 50% and that the onset of insulin action is delayed for 30–60 min in NIDDM subjects. By extrapolating from local to whole-body glucose disposal, one can account that >90% of the impairment in insulin-mediated whole-body glucose disposal is due to impaired muscle glucose uptake. These observations provide conclusive evidence that the primary site of insulin resistance during the euglycaemic insulin-clamp studies performed in overweight NIDDM subjects resides in peripheral tissues and primary skeletal muscle (DEFRONZO et al. 1992).

The combination of the insulin-glucose clamp with indirect calorimetry allows the fraction of glucose which is oxidized to be measured and, by subtracting it from the total amount of infused glucose, the fraction of glucose which is not oxidized and thus considered stored as glycogen. This approach has clearly demonstrated that obese subjects and patients with NIDDM have a defective insulin-stimulated glucose oxidation, and an even more marked defective glucose storage, when compared to lean non-diabetic controls (FELBER et al. 1988); a role for enhanced lipid oxidation in these abnormalities in glucose metabolism has also been demonstrated (review in FELBER et al. 1988, 1993a,b) (see below).

Tissue biopsies at various times during the course of the glucose clamp can also be used to relate a cellular biochemical event to whole-body metabolism (Mandarino 1989). As an example, intracellular defects in glucose metabolism in obese patients with NIDDM have been recently studied in detail by combining euglycaemic insulin clamps, leg balance technique, leg indirect calorimetry and leg muscle biopsies to determine whether alterations in muscle pyruvate dehydrogenase or glycogen synthase activities could explain defects in glucose oxidation or storage (Kelley et al. 1992). The findings confirmed that decreased muscle glucose storage during hyperinsulinaemia is the largest defect in glucose metabolism of *obese* diabetic patients, but also revealed a major defect in glucose oxidation. In contrast, insulin action may be intact in the muscle tissue of *lean* NIDDM patients (Kelley et al. 1993).

Using ^{13}C-NMR spectroscopy for the direct measurement of the rates of muscle glycogen synthesis in response to physiological increments in plasma glucose and insulin concentrations, Shulman et al. (1990) confirmed that, under hyperglycaemic-hyperinsulinaemic conditions, muscle glycogen synthesis is the major pathway of glucose disposal in man and that defective muscle glycogen synthesis plays a predominant role in impairing glucose metabolism in patients with NIDDM.

c) Oral Glucose Tolerance Test

In conditions where glycaemia and insulinaemia are not experimentally restricted as during the clamp procedure, the resistance to glucose storage could be overcome first by a rise in insulinaemia which is present in obese subjects with normal glucose tolerance, and then by a rise in both glycaemia and insulinaemia when impaired glucose tolerance occurs (DeFronzo et al. 1992). Thus, the compensatory response of increased glycaemia and insulinaemia allows the subjects to dispose their dietary carbohydrates normally in the absence of diabetes, in spite of the increased resistance to glucose storage and oxidation (see below). When these compensatory responses are overcome, diabetes occurs: glucose which cannot be stored remains in the glucose space (review in Felber et al. 1988, 1993a).

4. Adipose Tissue Lipolysis

Ambient plasma non-esterified fatty acid (NEFA) concentrations are increased in patients with NIDDM when they have significant fasting hyperglycaemia. Furthermore, there is a defect in the ability of insulin to decrease plasma NEFA levels in diabetic patients, and this resistance to insulin suppression of plasma NEFA concentration is highly correlated with the resistance to insulin stimulation of glucose uptake (review in Reaven 1990). These observations evidence multiple sites of insulin resistance in NIDDM (Groop et al. 1989).

5. Increased Glucagon Secretion

Plasma glucagon concentrations are elevated in NIDDM subjects in the postabsorptive state and are not appropriately suppressed following food ingestion, despite increases in both glucose and insulin (review in LEFÈBVRE 1991; LEFÈBVRE et al. 1991). The mechanism of this hyperglucagonaemia is still unclear. It might result from A-cell insulin resistance and/or A-cell desensitization by prolonged hyperglycaemia and/or loss of paracrine inhibition as a consequence of the loss of the normal intraislet oscillatory secretory pattern of insulin suggested in NIDDM. The precise role of hyperglucagonaemia in the pathophysiology of NIDDM, however, remains a matter of controversy (review in LEFÈBVRE 1991; LEFÈBVRE et al. 1991). Nevertheless, glucagon strongly stimulates liver glucose output, and elevated glucagon levels contribute significantly to the elevated rates of basal HGP in NIDDM. Therefore, searching for compounds that inhibit glucagon secretion or antagonize the effects of this hormone may represent an attractive way for improving management of NIDDM (review in LEFÈBVRE 1991 and LEFÈBVRE et al. 1991).

6. Varia

Glucose inhibits glucose release from the liver and increases glucose uptake independently of insulin. These mechanisms seem to be intact in subjects with NIDDM when studied in an insulin-free system (BARON et al. 1985). On the other hand, chronic hyperglycaemia may reduce glucose-mediated glucose uptake in peripheral tissues, especially skeletal muscle, when insulin is present (BARON et al. 1985). This may be attributable to both insulin resistance and "glucose resistance" (RUDENSKI et al. 1991).

C. Possible Causal Defects

I. Insulin Secretion

Considerable progress has been made during recent years in the knowledge of the physiology and the pathophysiology of insulin secretion (review in RASMUSSEN et al. 1990). However, given the number of variables that could contribute to a disordered insulin secretory response, it is obvious that a precise definition of the sequence of events that leads to this disorder in a given NIDDM patient remains, in most cases, an elusive goal (CERASI 1988; LEAHY 1990; PORTE 1991) (Fig. 3).

1. B-Cell Number

In Type 2 diabetes, the mass of B cells may be increased, normal or moderately decreased, and various morphological abnormalities are observed, though not systematically, in the islets of these patients. From these data, it

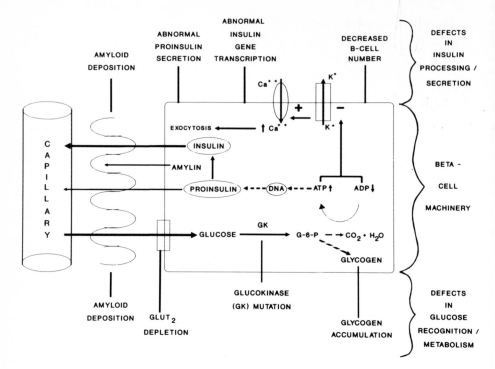

Fig. 3. Main defects likely to contribute to the abnormal insulin secretion of the B cell in response to glucose in non-insulin-dependent diabetes mellitus

can be inferred that the diabetic symptomatology does not result from an overt decrease of the B-cell mass, but rather from abnormal B-cell function. It has been suggested that the abnormal proportion of proinsulin secreted in NIDDM might be a clue to the etiology of islet B-cell dysfunction (Porte and Kahn 1989). It is still unknown which morphological abnormality (if any) is related to the disease (review in Rahier 1988).

2. Insulin Gene

A basic genetic defect in the insulin gene has been suspected to explain the disturbance in insulin secretion in NIDDM, but there is little evidence to support this hypothesis (review in Permutt and Elbein 1990). A few families have been identified in whom a point mutation in the insulin gene has led to the development of mild hyperinsulinaemic diabetes that is similar, in some respects, to NIDDM. However, such mutations appear to be rare and are unlikely to account for most patients with NIDDM (Steiner et al. 1990).

Another interesting topic involves the genes that activate insulin-gene transcription. Because 5'-flanking region *cis*-regulatory elements have been described, it has been postulated that abnormalities of gene products that activate the promoter region of the insulin gene may be responsible for the development of hypoinsulinaemia in NIDDM (review in DeFronzo et al. 1992).

3. Amylin

Recently, considerable excitement has been generated by "islet amyloid polypeptide" (IAPP), or amylin, a 37-amino acid peptide which is structurally similar to the calcitonin gene-related peptide. Deposition of amylin is more frequent in NIDDM patients than in controls, but considerable variation in the occurrence of islet amyloid has been reported, with no significant relationship with body mass index, fasting blood glucose or age. Histologically, amyloid is seen between the islet capillaries and the endocrine cells. The physiological role of amylin is far from clear. The involvement in the intracellular processing of insulin, regulation of enzymatic activity and paracrine modulation of insulin secretion have been suggested as the main actions of the peptide (Johnson et al. 1989; Porte and Kahn 1989; Clark 1992; Westermark et al. 1992). Amylin appears to be synthesized, stored and cosecreted with insulin by the B cells. Therefore, amylin deposits in B cells may be secondary to chronic hypersecretion of insulin, and the amylin deposit may in turn be responsible for the secondary secretory dysfunction of B cells (Johnson et al. 1989).

It has also been recently demonstrated that amylin is detectable in the peripheral circulation, and that pharmacological amounts of amylin are able to induce peripheral insulin resistance in rats in vivo. However, the functional significance of peripheral amylin concentration in man remains to be determined (Clark 1992; Westermark et al. 1992).

4. GLUT 2

Before glucose leads to insulin release, it is taken up by the B cells and subsequently metabolized. Glucose enters the B cells through a membrane-bound facilitative transporter that is designed as GLUT 2 (glucose transporter 2). A reduction of the amount of GLUT 2 transporters, by whatever cause, may play a causal role in the defect in insulin response to glucose and in the development of hyperglycaemia in rodents, and perhaps also in humans. Evidence is now accumulating that this glucose transport is reduced as a consequence of down-regulation of the normal B-cell GLUT 2 transporters by hyperglycaemia (Unger 1991). However, several other findings have obscured the role the reduced amount of GLUT 2 may play in the secretory dysfunction of NIDDM patients (review in Leahy et al. 1992).

5. Glucokinase

Glucokinase is an enzyme that catalyses the formation of glucose-6-phosphate from glucose. It is the major rate-limiting step in glycolysis, and may act as a primary part of the glucose sensor or glucoreceptor in the B cell (Malaisse 1988; Matschinsky 1990; Randle 1993; Matschinsky et al. 1993). Abnormalities in the regulation of the biosynthesis or degradation of glucokinase in the B cells have been proposed to be involved in the pathogenesis of insulin secretion disturbances in NIDDM (review in Matschinsky 1990; Permutt et al. 1992; Matschinsky et al. 1993). Mutations of the glucokinase gene locus have been recorded in some families with MODY (Velho et al. 1992; Cox et al. 1992; Froguel et al. 1993) and in one pedigree thought to have classical Type 2 diabetes, but no other mutations have been yet identified in the common variety of the disease (review in Randle 1993). However, there are many other potential sites for mutations which include various promoter regions and splicing sites, and all have not been explored yet (review in Randle 1993).

6. Accumulation of B-Cell Glycogen

Under normal conditions, B cells contain minuscule amounts of glycogen. In contrast, B cells exposed to hyperglycaemia accumulate large glycogen stores. Malaisse (1988) argued that glycogen accumulation may explain a number of the insulin secretion defects that occur with chronic hyperglycaemia, such as in NIDDM patients. An important question that remains unanswered is what role, if any, stored glycogen plays in the suppressed glucose-induced insulin secretion (Leahy et al. 1992).

7. Varia

Several hypotheses in addition to those already discussed have been suggested to explain the insulin secretory defects in NIDDM: mitochondrial defect, defective hydrolysis of membrane inositol phospholipids, etc. (review in Leahy et al. 1992).

Various pharmacological studies have suggested a potential role of prostaglandins (review in Giugliano et al. 1988), increased α-adrenergic tone (review in Robertson 1989) and opioid peptides (metenkephalin and β-endorphin (review in Lefèbvre et al. 1992) in the alterations of the secretory pattern of insulin in NIDDM. These observations remain, however, to be confirmed and further studies are required before developing new compounds able to modulate these mechanisms (Wolffenbuttel and Van Haeften 1993).

Finally, physiological studies have shown that the incretin effect of both gastric inhibitory peptide (GIP) and glucagon-like peptide-1 (GLP-1) is reduced in NIDDM patients. This abnormality does not result from decreased circulating levels of these hormones but, rather, from a diminished sensitivity of the B cells to their incretin action. Importantly, pharmacological doses of GLP-1 can stimulate insulin secretion in NIDDM, and agonists of its receptor have been suggested as possible therapeutic agents in NIDDM (review in Thorens and Waeber 1993).

II. Insulin Sensitivity

Insulin action results from a complex sequence of extracellular and intracellular events (review in MOLLER and FLIER 1991; MOLLER 1993). Prereceptor, receptor and postreceptor defects all may contribute to the insulin resistance of NIDDM individuals (Fig. 4).

1. Insulin Pre-receptor Level

a) Capillary Diffusion

For insulin to reach its target tissues, the hormone must first diffuse across the vascular endothelium, and a defect in insulin diffusion has been suggested as a potential rate-limiting step for insulin action (BERGMAN et al. 1992). Such "capillary insulin resistance" may be present in extrahepatic tissues characterized by unfenestrated vessels, but not in liver, where there are large fenestrations present in hepatic sinusoids. The presence of insulin receptors on capillaries (KING and JOHNSON 1985) introduces the possibility that defective transport of insulin out of the vascular space could impair the ability of insulin to stimulate glucose uptake. Measurement by microdialysis of the insulin concentration in subcutaneous interstitial fluid in humans recently confirmed that the transcapillary route might be a rate-limiting step for insulin action, at least in adipose tissue (JANSSON et al. 1993). Although an interesting hypothesis, it remains to be proved whether insulin diffusion

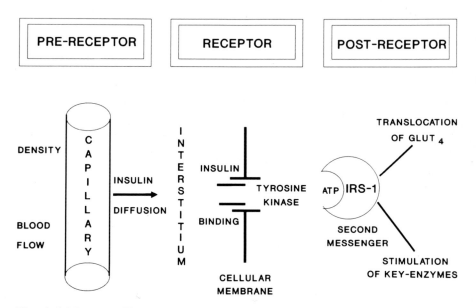

Fig. 4. Main steps likely to contribute to the defect in insulin action in skeletal muscle in non-insulin-dependent diabetes mellitus

across the capillary endothelium significantly contributes to the defect in insulin action in skeletal muscle seen in NIDDM patients (Lillioja and Bogardus 1988; Bergman et al. 1992).

b) Blood Flow

Insulin may also influence glucose uptake by increasing blood flow to insulin-sensitive tissues. It has been claimed that reduced muscle blood flow is an important factor in the development of insulin resistance in skeletal muscles from obese NIDDM patients due to a reduced number of open capillaries (Laakso et al. 1992). This abnormality may play a role in the activation rate of intracellular systems due to a prolonged diffusion distance. However, this haemodynamic perturbation cannot explain specific intracellular defects in muscle metabolism, such as alterations in enzyme activities (Kelley et al. 1992). Nevertheless, these issues require further study since if insufficient insulin-induced stimulation of blood flow indeed contributes to impaired glucose uptake, then agents that facilitate the blood flow response to insulin may improve insulin action in individuals with NIDDM (review in Ganrot 1993).

2. Insulin Receptor Level

Insulin resistance at this step may result from a specific defect in insulin binding per se and/or an abnormality in the insulin receptor function.

a) Insulin Binding to Its Receptor

Two approaches have been used to examine the contribution of alterations in insulin binding to the insulin resistance in NIDDM. The first approach relies on information derived from in vivo insulin dose-response studies where, owing to the spare receptor theory, a right shift of the curve with a preservation of a normal maximal response indicates a specific receptor-binding defect (Kahn 1978). Such a defect is mainly observed in obese patients with mild carbohydrate intolerance, while a decrease in maximal response is frequently seen in patients with overt NIDDM, which implies the presence of postreceptor defects (see below) (Kolterman et al. 1981; Groop et al. 1989).

An alternative approach is to directly measure in vitro insulin binding and/or signal transmission in tissues obtained by venipuncture or biopsy. Numerous studies, but not all, have demonstrated that insulin binding to erythrocytes, monocytes, skin fibroblasts and adipocytes from NIDDM patients is reduced by a mean of 20%–30% (Olefsky 1976). However, some caution should be employed in interpreting the results of these studies. From the quantitative standpoint, the two major organs for insulin action in vivo are the liver and muscle, and very few studies have examined insulin binding to these tissues in humans (OLEFSKY 1988; CARO et al. 1989). The reduction in insulin binding is due to a decrease in the number of insulin receptors without alteration in insulin-receptor affinity (OLEFSKY 1976). As this defect is mainly observed in individuals with mild hyperglycaemia who are usually hyperinsulinaemic, it is likely that the decrease in insulin binding results from a downregulation of insulin-receptor number by sustained hyperinsulinaemia (OLEFSKY 1988).

b) Insulin Receptor Function

In addition to a decrease in the number of cell surface receptors, various defects in insulin-receptor internalization and processing have been described (CARO et al. 1989; HÄRING and MEHNERT 1993). Because of the potentially important role of tyrosine kinase in the insulin-signalling mechanism (KAHN and FOLLI 1993), tyrosine kinase activity has been examined in various cell types, and most investigators found a severe reduction in tyrosine kinase activity in NIDDM patients (review in CARO et al. 1989; HÄRING and MEHNERT 1993). The cause of this defect remains unclear. However, several studies have found that the kinase defect in receptors from NIDDM subjects can be reversed, implying that it is a secondary rather than a primary abnormality, maybe related to chronic hyperglycaemia or some other aspects of the abnormal metabolic state (review in CARO et al. 1989; HARING and MEHNERT 1993).

Following the detection of various mutations in the insulin receptor gene in patients with genetic syndromes of extreme insulin resistance (review in MOLLER and FLIER 1991; MOLLER 1993), it was hoped that NIDDM patients might also have insulin receptor mutations that had a more subtle effect on receptor function. However, the insulin receptor gene is normal in patients with NIDDM (KUSARI et al. 1991). Interestingly, it has been recently shown that the relative expression of the two known subtypes of the insulin receptor, generated by differential splicing, may be altered in NIDDM (review in HARING and MEHNERT 1993). However, the bulk of the evidence suggests that the insulin receptor does not play a major role in the insulin resistance of the most common form of NIDDM.

3. Insulin Post-receptor Level

In the most general sense, postbinding defects in insulin action can be explained by one of three metabolic disturbances: impaired generation of insulin's second messenger, diminished glucose transport into cell, and a postglucose transport abnormality in some critical enzymatic step involved in glucose utilization (GARVEY 1989; HARING and MEHNERT 1993). On the other hand, abnormal regulation of gluconeogenesis and/or glycogenolysis should also be considered to explain increased hepatic glucose release in NIDDM (DEFRONZO et al. 1992).

a) Insulin Second Messenger

Despite a huge amount of studies, the precise identification of the second messenger for insulin action remains elusive (SALTIEL 1990). However, phosphorylation/dephosphorylation of key intracellular proteins appears to be an important signalling mechanism that couples insulin binding to the intracellular action of insulin (KAHN and FOLLI 1993). Following receptor autophosphorylation, endogenous protein substrates are phosphorylated on tyrosine residues and, insofar as these endogenous substrates serve as signalling molecules, defective substrate phosphorylation could be a cause of cellular insulin resistance. The best studied of the endogenous insulin receptor protein substrates is pp185 or IRS-1. A substantial defect and an amino acid polymorphism (ALMIND et al. 1993) have been reported in Type 2 diabetes, but the importance of these abnormalities in the pathogenesis of the cellular insulin resistance in NIDDM remains to be determined (review in HÄRING and MEHNERT 1993).

b) Glucose Transport

After the generation of the second messenger for insulin action, glucose transport is activated. This effect of insulin is brought about by the translocation of a large intracellular pool of glucose transporters, from the microsomes to the plasma membrane. GLUT 4 is the insulin-regulatable transporter found in the insulin-sensitive tissues, i.e. muscle and adipocytes (review in KLIP and PAQUET 1990).

Glucose-transport activity in NIDDM has been studied extensively (review in GARVEY and KOLTERMAN 1988; GARVEY 1989, 1992) and has been found to be decreased in both adipocytes and muscle. Pretranslational events specifically suppress gene expression of the GLUT 4 transporter isoform in adipocytes, which appears to be the key mechanism of insulin resistance in this target tissue (GARVEY 1992). In contrast, in skeletal muscle, total cellular content of GLUT 4 protein and mRNA is normal in NIDDM (ERIKSSON et al. 1992); thus, insulin resistance in muscle may involve impaired GLUT 4 function or translocation and not transporter depletion per se (GARVEY 1992). Moreover, this abnormality may be only present in hyperglycaemic patients. Indeed, in normoglycaemic relatives of patients with NIDDM, the presence of a specific defect in glucose storage compared to a near normal glucose oxidation suggests that glucose transport is not impaired in the earliest stages of development of NIDDM (ERIKSSON et al. 1989). Rather, the results suggest a specific enzymatic defect in the glycogen synthetic pathway.

c) Enzymatic Defects in Glucose Metabolism

Glycogen Synthesis. Glycogen synthase is the key enzyme in the regulation of glycogen synthesis in the skeletal muscle. This enzyme is sensitive to insulin, but in NIDDM patients it has been shown to be markedly resistant to insulin stimulation when measured at euglycaemia (THORNBURN et al. 1991; BECK-NIELSEN et al. 1992). Interestingly, it has been shown that the reduced glycogen synthase activity of individuals with NIDDM correlates well with the defects both in non-oxidative glucose disposal and in muscle glycogen formation (BOGARDUS et al. 1984). A reduced glycogen synthase activity has also been found in normoglycaemic first-relatives of NIDDM patients, indicating that this abnormality precedes development of hyperglycaemia in subjects prone to develop NIDDM (VAAG et al. 1992; SCHALIN-JANTTI et al. 1992). Therefore, this defect may be of primary genetic origin. However, it does not appear to be a defect in the enzyme itself, but rather a defect in the covalent activation of the enzyme system (BECK-NIELSEN et al. 1992). Indeed, glycogen synthase is resistant to insulin but may be activated allosterically by glucose-6-phosphate. This means that the defect in insulin activation can be compensated for by increased intracellular concentrations of glucose-6-phosphate, which could be obtained with both hyperinsulinaemia and hyperglycaemia (BECK-NIELSEN et al. 1992) (see Sect. E.II below).

Studies in the Pima Indians have suggested that impaired non-oxidative glucose storage may be the site of a genetic defect (LILLIOJA and BOGARDUS 1988). Since it was not possible to detect any genetic variants of the entire coding sequence of the glycogen synthase gene, except in one single individual (KUSARI et al. 1991), it was concluded that abnormal pretranslational regulation of the gene may contribute to impaired glycogen synthesis of muscle in NIDDM (VESTERGAARD et al. 1991). A polymorphism of the glycogen synthase gene in an intron, outside the coding region of the gene, has been recently identified in Finnish subjects with NIDDM (GROOP et al. 1993a). This study appears to be the first one to identify a strong association of a candidate gene that is coupled with appropriate metabolic changes consistent with findings in insulin-resistant subjects.

Glucose Oxidation. Glucose oxidation represents the other major fate of glucose disposal (KELLEY et al. 1992). A key regulator of glucose oxidation is pyruvate dehydrogenase, an enzyme whose activity is regulated by insulin. The ability of insulin to stimulate this enzyme was found to be impaired in both adipocytes and muscle from NIDDM patients (review in DEFRONZO et al. 1992). Glucose flux through this pathway also has been shown to be impaired in patients with NIDDM (GOLAY et al. 1988), particularly in obese subjects with high circulating NEFA levels. Importantly, glucose oxidation cannot be normalized even if total glucose flux into the cells is normalized by both hyperglycaemia and hyperinsulinaemia in NIDDM patients (DELPRATO et al. 1993).

d) Abnormal Regulation of Hepatic Glucose Production

*Gluconeogenesis.*Increased gluconeogenesis is the main factor responsible for increased hepatic glucose production (HGP) in the postabsorptive state in NIDDM (Consoli et al. 1989), and preliminary data suggest that increased gluconeogenesis might also be the main factor responsible for impaired suppression of hepatic glucose output after carbohydrate ingestion (Consoli 1992). There are three primary sites at which gluconeogenesis may be regulated: supply of substrates, hepatic substrate extraction and intra-hepatic substrate conversion into glucose (DeFronzo and Ferrannini 1987; Ferrannini and Groop 1989). Although there is considerable evidence that the production of gluconeogenic precursors is increased in NIDDM (especially lactate) (Consoli et al. 1990), several studies suggest that an intrahepatic mechanism must also contribute to the increased glucose pro-duction in this disorder (Consoli 1992).

Glycogenolysis. Increased glycogenolysis does not seem to significantly con-tribute to the increased basal HGP usually observed in NIDDM patients, which almost completely results from enhanced gluconeogenesis (Consoli et al. 1989; review in Consoli 1992). However, this has proven to be a difficult question to address.

4. Varia

a) Muscle Fibre Type

It has been suggested that, besides skeletal muscle capillary density, muscle fibre type may also contribute to the insulin resistance in NIDDM indivi-duals (Lillioja et al. 1987). As already said, the diffusion distance from capillary to muscle cells may be inversely related to the kinetics of insulin action (Lillioja and Bogardus 1988). Furthermore, type 1 muscular fibres are very sensitive to insulin while type 2B fast-twitch fibres are insulin resistant. The proportion of both types of fibres appears to be, at least in part, genetically determined. However, few studies are available yet on fibre typing in skeletal muscles from NIDDM patients.

b) Reduced Intracellular Magnesium Content

As reviewed by our group (Paolisso et al. 1990), low plasma and erythro-cyte magnesium are frequently found in patients with NIDDM. It has been suggested that low intracellular magnesium levels result from both urinary losses and insulin resistance. Furthermore, there are some indications that a reduced intracellular magnesium content might contribute to the impaired insulin response and insulin action that are seen in NIDDM (Paolisso et al. 1988). Recent observations suggest that chronic magnesium supplementa-tion can contribute to an improvement in both islet B-cell response and insulin action in NIDDM (review in Lefèbvre et al. 1994).

D. Role of Associated Obesity

Obesity has been recognized as associated with type 2 diabetes for a very long time (review in VAGUE and VAGUE 1988; PEDERSEN 1989; BRAY 1990; BJÖRNTORP 1990; FELBER et al. 1993a). Several models are possible for the relationship of NIDMM and obesity: (1) obesity precedes NIDDM and is in the etiological pathway; (2) a similar genetic predisposition leads to both obesity and NIDDM; or (3) a similar genetic defect predisposes to both, but additional gene defects (and/or environmental insults) are required to differentiate the NIDDM or obese phenotype. Besides fat mass per se, body fat distribution is now recognized as an important variable in the pathogenesis of insulin resistance and NIDDM. Clearly, android obesity, also known as central, abdominal or upper-body segment obesity, is associated with a higher risk of developing NIDDM (VAGUE and VAGUE 1988; KISSEBAH and PEIRIS 1989; BJÖRNTORP 1988, 1990).

I. Impact of Obesity on Insulin Secretion

Hyperinsulinaemia seen in obesity could be due to an enhanced insulin secretion, an impaired insulin clearance or a combination of both factors (BONORA et al. 1984; SCHEEN et al. 1991). Recent studies analysing the kinetics of plasma C peptide levels demonstrated that obesity is clearly associated with an increase in insulin secretion rate (POLONSKY et al. 1988a). The insulin production rate does not appear to be influenced by the site of body fat localization (KISSEBAH and PEIRIS 1989), but clearly increases as the severity of obesity increases (POLONSKY et al. 1988a). On the other hand, numerous studies, using various techniques to estimate insulin metabolism in vivo (review in CASTILLO et al. 1994b), have also demonstrated that the metabolic clearance rate of insulin is decreased in obesity (BONORA et al. 1984; SCHEEN et al. 1991; CASTILLO et al. 1994a), at least in subjects with fasting hyperinsulinaemia (POLONSKY et al. 1988a). Insulin clearance has been found to be negatively correlated with body mass index (CASTILLO et al. 1994a), and to be more severely reduced in the presence of android obesity with predominant intra-abdominal fat localization (BJÖRNTORP 1988; KISSEBAH and PEIRIS 1989). Thus, while the increased insulin secretion rate in obesity is likely to be the predominant cause of hyperinsulinaemia, a contribution of impaired insulin clearance, maybe due to a reduced hepatic extraction, should also be considered, at least in some patients.

II. Impact of Obesity on Insulin Sensitivity

Obesity clearly diminishes the insulin sensitivity of both hepatic and peripheral tissues in patients with NIDDM (SCHEEN et al. 1992b; CAMPBELL and CARLSON 1993). Both the degree of obesity (LILLIOJA and BOGARDUS 1988; CAMPBELL and CARLSON 1993) and the intra-abdominal body fat distribution

(Lillioja and Bogardus 1988; Björntorp 1988; Kissebah and Peiris 1989)
play a significant role in the increased insulin resistance of obese subjects.
Biophysical changes in muscle such as decreased capillary density, type 1
oxidative fibres and muscle hypertrophy of skeletal muscles may explain the
effects of obesity to produce a reduction in insulin-induced glucose uptake
and the abnormal kinetics of insulin action found in the obese subjects
(Lillioja et al. 1987; Lillioja and Bogardus 1988; Pedersen 1989).
Alternatively, metabolic abnormalities secondary to the increased lipolysis,
especially when triglycerides are stored in the abdominal region with fat cell
hypertrophy, may also play a central role in the increased insulin resistance
characterizing obesity (Vague and Vague 1988; Björntorp 1988; Kissebah
and Peiris 1989; Felber et al. 1993b) (see below). The exacerbation of the
insulin resistance of NIDDM by obesity explains why weight reduction is an
effective treatment for the obese NIDDM patient (Felber et al. 1993a).

III. Lipid Oxidation and Insulin Resistance

Randle and colleagues initially proposed the concept of substrate competition, whereby increased NEFA oxidation restrains glucose oxidation in
muscle by altering the redox potential of the cell and inhibiting several key
enzymatic steps within the glycolytic cascade (Randle et al. 1963). It was
subsequently shown that these events impair not only glucose oxidation
(direct inhibition of pyruvate dehydrogenase and Krebs cycle) but also
glycogen formation (secondary to decreased glucose transport). Increased
NEFA oxidation also augments HGP, most likely secondary to a stimulation
of gluconeogenesis. This could be explained by an enhanced flux of three
carbon precursors and by an increased source of energy to drive gluconeogenesis (Felber et al. 1993a and 1993b).

Therefore, from a theoretical standpoint, an elevated rate of NEFA
oxidation can reproduce all of the major intracellular defects and abnormalities in glucose metabolism that have been described in NIDDM patients
(Felber et al. 1993a and 1993b). While it is probable that excess fat metabolism plays some role in the insulin resistance of NIDDM, the magnitude
of that role, however, remains to be defined. New drugs which could inhibit
either lipolysis or NEFA oxidation are currently under investigation for the
treatment of obese NIDDM subjects (Bailey and Flat 1990; Foley 1992;
Scheen and Lefèbvre 1993b).

IV. From Obesity to Type 2 Diabetes

Obesity has to be considered as a major risk factor for NIDDM. It is known
that diabetes generally appears many years after the beginning of obesity
(Felber et al. 1993a). The first stage, impaired glucose tolerance, seems to
be primarily due to a reduction in insulin action; then the pancreatic B cell is
still able to secrete more insulin, thus contributing to peripheral hyperinsu-

linism in most obese subjects. The moderately elevated plasma glucose levels and the markedly increased plasma insulin concentrations may initiate a cycle leading to B-cell unresponsiveness and to greater cellular insulin resistance, thus resulting in diabetes mellitus (see below). This hypothesis suggests that the development of NIDDM in obese subjects may be a two-step process involving insulin resistance followed by B-cell failure (SAAD et al. 1991; DeFRONZO et al. 1992; SCHEEN 1992; SCHEEN et al. 1992b; SCHEEN and LEFÈBVRE 1993a; FELBER et al. 1993a) (see below).

E. Dynamic Interaction between Insulin Action and Insulin Secretion

Subjects with NIDDM are characterized by both tissue insulin resistance and impaired insulin secretion. The development of NIDDM requires the presence of these two fundamental defects, which disrupt the delicate balance by which insulin-target tissues communicate with the B cells and vice versa (Fig. 2). Numerous observations underscore the important interplay between insulin resistance and insulin secretion (DeFRONZO 1988; DeFRONZO et al. 1992; LILLIOJA and BOGARDUS 1988; TURNER et al. 1988; SCHEEN et al. 1991). Both abnormalities must be looked at in concert and the relative importance of these two factors can be estimated with the aid of a theoretical mathematical model of glucose metabolism (TURNER et al. 1988; RUDENSKI et al. 1991). Insulin resistance alone is, in most instances, insufficient to cause overt glucose intolerance. To observe the development of frank diabetes mellitus, a defect in insulin secretion must be superimposed on insulin resistance.

I. What Is the Primary Defect?

Despite extensive efforts, it is currently not known which defect, of insulin action or of insulin secretion, comes first in the most "common" form of NIDDM (Fig. 5) (GROOP et al. 1993b). Here below are listed the main arguments in favour of one or the other hypothesis.

Insulin resistance is a universal finding in patients with established NIDDM (DeFRONZO 1988; GERICH 1988; REAVEN 1988; SCHEEN and LEFÈBVRE 1992), and is also present in patients at risk of developing the disease (MARTIN et al. 1992), such as normoglucose-tolerant first-degree relatives of NIDDM subjects and offspring of two diabetic parents (ERIKSSON et al. 1989). Such results provide conclusive evidence that insulin resistance is an inherited defect that may initiate the diabetic condition. Prospective studies have demonstrated that insulin resistance and hyperinsulinaemia precede the development of impaired glucose tolerance and predict the later development of NIDDM (review in LILLIOJA and BOGARDUS 1988; DeFRONZO et al. 1992; MARTIN et al. 1992). All these observations have led investigators

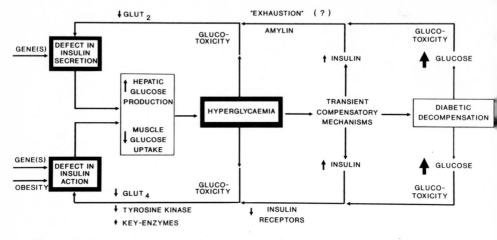

Fig. 5. Vicious circle perpetuating and aggravating non-insulin-dependent diabetes, whatever the priming defect in insulin secretion or in insulin action

to propose that insulin resistance is the initiating event in the development of NIDDM (SAAD et al. 1991; DEFRONZO et al. 1992; MARTIN et al. 1992; FELBER et al. 1993a) (Fig. 5).

An alternative hypothesis suggests that a defect in insulin secretion is the earliest event in the development of NIDDM (TURNER et al. 1988; CERASI 1988; LEAHY 1990; PORTE 1991). Some recent data show that NIDDM subjects have lower insulin concentrations than previously appreciated (TEMPLE et al. 1990, 1992). Furthermore, abnormalities in both the amount and pattern of insulin secretion have been demonstrated in normoglycaemic relatives of patients with NIDDM, indicating that B-cell dysfunction may be the primary defect (O'RAHILLY et al. 1986). A plausible sequence of events, starting with a defect in insulin secretion and leading to the emergence of insulin resistance, can also be described on the basis of the information currently available (Fig. 5).

Thus, while most investigators agree that insulin secretion and insulin action must both be impaired for the development of overt diabetes, there is no consensus as to which comes first. Owing to increasing evidence of heterogeneity (BANERJI and LEBOVITZ 1989; ARNER et al. 1991), it is highly probable that each defect will prove to be primary in some forms of NIDDM.

II. Transient Compensatory Mechanisms

Transient compensatory mechanisms, which all concur to minimize hyperglycaemia, can be observed at the various stages of insulin resistance and glucose intolerance.

In (genetically) insulin-resistant subjects with normal or mildly impaired glucose tolerance, the decrease in glucose oxidation and, more specifically, in glucose storage is corrected by an increase in insulin concentrations, resulting from both enhanced insulin secretion and reduced insulin clearance (HAFFNER et al. 1992). Several studies have shown that hyperinsulinaemia may increase the intrinsic glycogen synthase activity as well as its sensitivity to glucose-6-phosphate (BECK-NIELSEN et al. 1992). Hyperinsulinaemia as a compensatory mechanism to maintain glucose homeostasis despite insulin resistance may, however, play a role in the development of associated metabolic disorders such as hypertension and dyslipidaemia. This association of risk factors for cardiovascular disease has been designed as "syndrome X", and suggests that NIDDM may not be a distinct clinical disorder, but part of a wider metabolic syadrome (REAVEN 1988; LEFÈBVRE 1993; DEFRONZO and FERRANNINI 1991; ESCHWÈGE et al. 1991).

In the diabetic state, the most important compensatory factor appears to be hyperglycaemia. Hyperglycaemia pushes glucose inside the cells by mass action and, because it cannot be metabolized at a normal rate, free glucose and glucose-6-phosphate accumulate, which result in allosteric activation of glycogen synthase (review in BECK-NIELSEN et al. 1992). Glycogenesis is thus improved and may reach a normal rate in the full compensated state. This compensation is essential because a normal glycogen content is obligatory for the muscles to maintain the normal strength. If insulin values are too low, the compensation may not be complete and plasma glucose levels continue to rise, leading to a final decompensation (Fig. 5).

III. Role of Glucose Toxicity

Many studies that used both in vitro and in vivo techniques have provided impressive evidence that chronic hyperglycaemia leads to the development of insulin resistance, both through a downregulation of the glucose-transport system and from an acquired defect in the post-glucose transport steps involved in insulin action (review in DEFRONZO et al. 1992). On the other hand, substantial data have been generated, mainly in animal models, to support the concept that chronic hyperglycaemia also causes the loss of glucorecognition and so the B-cell dysfunction characterizing NIDDM. Many hypotheses are being investigated that span most of the major intracellular steps for glucose-induced insulin secretion, includig abnormalities in glucose transport, storage, metabolism/oxidation and the second messengers (review in LEAHY et al. 1992). Thus, hyperglycaemia must be viewed not only as a manifestation of diabetes mellitus but also as a self-perpetuating and aggravating cause of the diabetic condition (UNGER and GRUNDY 1985), a phenomenon known as "glucotoxicity" (ROSSETTI et al. 1990; YKI-JÄRVINEN 1992) (Figs. 1, 5).

IV. Vicious Circle Leading to Severe Hyperglycaemia

Once it develops, either from insulin deficiency or from insulin resistance, hyperglycaemia will exacerbate both defects, thereby closing a pathological feedback loop (Fig. 5). There appears to be a counterproductive interplay between B-cell inadequacy and insulin resistance: a primary B-cell defect could result in hypoinsulinaemia and subsequent postreceptor insulin resistance, while insulin resistance and hyperglycaemia could exhaust B cells and make them unresponsive to glucose. A vicious circle can therefore be envisioned in which B-cell function and insulin resistance both deteriorate with time (Beck-Nielsen and Hother-Nielsen 1988; DeFronzo et al. 1992). Even though our understanding of the mechanisms is incomplete, this counterproductive interplay and the concept of glucose toxicity have important clinical and therapeutic implications. Vigorous and early treatment of NIDDM with such approaches as diet, regular physical activity, oral agents or insulin should lead to both improved B-cell function and reduction of insulin resistance (review in Rossetti et al. 1990; Yki-Järvinen 1992).

F. Conclusions

NIDDM, or type 2 diabetes, remains a popular topic in biomedical science because of its high prevalence, severe morbidity and undefined pathophysiological aspects. It can be defined as a heterogeneous condition caused by genetic and environmental factors which is expressed late in life. It has been postulated that various environmental factors (physical inactivity, and increased fat and caloric intake) operate on susceptible genotype(s) (probably the glucokinase gene in most subjects with MODY; another still unknown gene in the most common insulin-resistant form of NIDDM) to promote the development of NIDDM.

The development of NIDDM requires the presence of two fundamental defects, i.e. insulin resistance and impaired insulin secretion, which disrupt the delicate balance by which insulin target tissues communicate with the B cells and vice versa. Thus, NIDDM is a heterogeneous disorder characterized by impaired insulin secretion, diminished peripheral insulin action, and increased hepatic glucose production, all defects being present in variable proportions in different individuals. Even though we are not yet able to pinpoint the primary defect in most NIDDM patients, there appears to be a counterproductive interplay between B-cell inadequacy and insulin resistance, which leads to a vicious circle that perpetuates and aggravates the metabolic disorder. On clinical and epidemiological grounds, it would appear that insulin resistance is most commonly an early defect in the sequence of events leading to NIDDM, particularly in overweight people. Once the pancreas cannot secrete enough insulin, due to genetic or acquired defects of the B cells, glucose tolerance deteriorates rapidly and a frank diabetic state ensues. Emphasis should be placed upon developing an approach to the therapy of NIDDM which is firmly based upon recent knowledge of the metabolic defects characteristic of the disease.

References

Almind K, Bjorbaek C, Vestergaard H, Hansen T, Echwald S, Pedersen O (1993) Amino acid polymorphism in insulin receptor substrate-1 in non-insulin-dependent diabetes mellitus. Lancet 342:828–832

Arner P, Pollare T, Lithell H (1991) Different aetiologies of Type 2 (non-insulin-dependent) diabetes mellitus in obese and non-obese subjects. Diabetologia 34:483–487

Bailey CJ, Flatt PR (1990) New antidiabetic drugs. Smith-Gordon, London

Banerji MA, Lebovitz HE (1989) Insulin-sensitive and insulin-resistant variants in NIDDM. Diabetes 38:784–792

Barnett AH, Eff C, Leslie RDG, Pyke DA (1981) Diabetes in identical twins: a study of 200 pairs. Diabetologia 20:87–93

Baron AD, Kolterman OG, Bell J, Mandarino LJ, Olefsky JM (1985) Rates of noninsulin-mediated glucose uptake are elevated in type II diabetic subjects. J Clin Invest 76: 1782–1788

Beck-Nielsen H, Hother-Nielsen O (1988) Insulin resistance. Diabetes Ann 4: 565–591

Beck-Nielsen H, Vaag A, Damsbo P, Handberg A, Hother Nielsen O, Henriksen JE, Thye-Ronn P (1992) Insulin resistance in skeletal muscles in patients with NIDDM. Diabetes Care 15:418–429

Bell GI (1991) Lilly lecture 1990. Molecular defects in diabetes mellitus. Diabetes 40:413–422

Bergman RN, Steil GM, Bradley DC, Watanabe RM (1992) Modeling of insulin action in vivo. Annu Rev Physiol 54:861–883

Björntorp P (1988) Abdominal obesity and the development of noninsulin-dependent diabetes mellitus. Diabetes Metab Rev 4:615–622

Björntorp P (1990) Obesity and diabetes. In: Alberti KGMM, Krall LP (eds) The diabetes annual/5. Elsevier, Amsterdam, pp 373–395

Bogardus C, Lillioja A, Stone K, Mott D (1984) Correlation of muscle glycogen synthase activity and in vivo insulin action in man. J Clin Invest 73:1185–1190

Bonora E, Zavaroni I, Bruschi F, Alpi O, Pezzarossa A, Guerra L, Dall'Aglio E, Coscelli C, Butturini U (1984) Peripheral hyperinsulinemia of simple obesity: pancreatic hypersecretion or impaired insulin metabolism? J Clin Endocrinol Metab 59:1121–1127

Bray GA (1990) Obesity and diabetes. Acta Diabetol Lat 27:81–88

Bressler R, Johnson D (1992) New pharmacological approaches to therapy of NIDDM. Diabetes Care 15:792–805

Brunzell JD, Robertson RP, Lerner RL, Hazzard WR, Ensinck JW, Bierman EL, Porte D (1976) Relationships between fasting plasma glucose levels and insulin secretion during intravenous glucose tolerance tests. J Clin Endocrinol Metab 46:222–229

Butler PC, Kryshak EJ, Schwenk WF, Haymond MW, Rizza RA (1990) Hepatic and extrahepatic responses to insulin in NIDDM and nondiabetic humans: assessment in absence of artifact introduced by tritiated nonglucose contaminants. Diabetes 39:217–225

Campbell PJ, Carlson MG (1993) Impact of obesity on insulin action in NIDDM. Diabetes 42:405–410

Campbell PJ, Mandarino LJ, Gerich JE (1988) Quantification of the relative impairment in actions of insulin on hepatic glucose production and peripheral glucose uptake in non-insulin-dependent diabetes mellitus. Metabolism 37:15–21

Capaldo B, Napoli R, Di Bonito P, Albano G, Sacca L (1990) Glucose and gluconeogenic substrate exchange by the forearm skeletal muscle in hyperglycemic and insulin-treated type II diabetic patients. J Clin Endocrinol Metab 71: 1220–1223

Caro JF, Dohm LG, Pories WJ, Sinha MK (1989) Cellular alterations in liver, skeletal muscle, and adipose tissue responsible for insulin resistance in obesity and type 2 diabetes. Diabetes Metab Rev 5:665–689

Castillo MJ, Scheen AJ, Jandrain B, Lefèbvre PJ (1994a) Relationships between metabolic clearance rate of insulin and body mass index in a female population ranging from anorexia nervosa to severe obesity. Int J Obes 18:47–53

Castillo MJ, Scheen AJ, Letiexhe MR, Lefèbvre PJ (1994b) How to measure insulin clearance. Diabetes Metab Rev 10:119–150

Cerasi E (1988) Insulin secretion in diabetes mellitus. In: Lefèbvre PJ, Pipeleers DG (eds) The pathology of the endocrine pancreas in diabetes. Springer, Berlin Heidelberg New York, pp 191–218

Clark A (1992) Islet amyloid: an enigma of type 2 diabetes. Diabetes Metab Rev 8:117–132

Consoli A (1992) Role of liver in pathophysiology of NIDDM. Diabetes Care 15:430–441

Consoli A, Nurjhan N, Capani F, Gerich J (1989) Predominant role of gluconeogenesis in increased hepatic glucose production in NIDDM. Diabetes 38: 550–557

Consoli A, Nurjhan N, Reilly JJ Jr, Bier DM, Gerich JE (1990) Mechanism of increased gluconeogenesis in non-insulin dependent diabetes mellitus: role of alterations in systemic, hepatic and muscle lactate and alanine metabolism. J Clin Invest 86:2034–2045

Cox NJ, Xiang K-S, Fajans SS, Bell GI (1992) Mapping diabetes-susceptibility genes: lessons learned from search for DNA marker for maturity-onset-diabetes of the young. Diabetes 41:401–407

DeFronzo RA (1988) The triumvirate: B-cell, muscle, liver. A collusion responsible for NIDDM. Diabetes 37:667–687

DeFronzo RA, Ferrannini E (1987) Regulation of hepatic glucose metabolism in humans. Diabetes Metab Rev 3:415–459

DeFronzo RA, Ferrannini E (1991) Insulin resistance: a multifaceted syndrome responsible for NIDDM, obesity, hypertension, dyslipidemia, and atherosclerotic cardiovascular disease. Diabetes Care 14:173–194

DeFronzo RA, Ferrannini E, Koivisto V (1983) New concepts in the pathogenesis and treatment of noninsulin-dependent diabetes mellitus. Am J Med 75:52–81

DeFronzo R, Gunnarsson R, Bjorkman O, Olsson M, Wahren J (1985) Effects of insulin on peripheral and splanchnic glucose metabolism in non-insulin dependent (type II) diabetes mellitus. J Clin Invest 76:146–155

DeFronzo RA, Bonadonna RC, Ferrannini E (1992) Pathogenesis of NIDDM: a balanced overview. Diabetes Care 15:318–368

DelPrato S, Bonadonna RC, Bonora E, Guli G, Solini A, Shank M, DeFronzo RA (1993) Characterization of cellular defects of insulin action in type 2 (non-insulin-dependent) diabetes mellitus. J Clin Invest 91:484–494

Dinneen S, Gerich J, Rizza R (1992) Carbohydrate metabolism in non-insulin-dependent diabetes mellitus. N Engl J Med 327:707–713

Eriksson J, Franssila-Kallunki A, Ekstrand A, Saloranta C, Widen E, Schalin C, Groop L (1989) Early metabolic defects in persons at risk for non-insulin-dependent diabetes mellitus. N Engl J Med 321:337–343

Eriksson J, Koranyi L, Bourey R, Schalin-Jantti C, Widen E, Mueckler M, Permutt AM, Groop LC (1992) Insulin resistance in type 2 (non-insulin-dependent) diabetic patients and their relatives is not associated with a defect in the expression of the insulin-responsive glucose transporter (GLUT-4) gene in human skeletal muscle. Diabetologia 35:143–147

Eschwège E, Mehnert H, Reaven G, Vigneri R (eds) (1991) NIDD today. Diabete Metab 17:75–254

Fajans SS (1986) Heterogeneity of insulin secretion in type II diabetes. Diabetes Metab Rev 2:347–361

Felber JP, Golay A, Felley C, Jéquier E (1988) Regulation of glucose storage in obesity and diabetes: metabolic aspects. Diabetes Metab Rev 4:691–700

Felber JP, Acheson KJ, Tappy L (1993a) From obesity to diabetes. Wiley, Chichester

Felber JP, Haesler E, Jéquier E (1993b) Metabolic origin of insulin resistance in obesity with and without Type 2 (non-insulin-dependent) diabetes mellitus. Diabetologia 36:1221–1229

Ferner RE, Alberti KGMM (1986) Why is there still disagreement over insulin secretion in non-insulin-dependent diabetes? Diabet Med 3:13–17

Ferrannini E, Cobelli C (1987) The kinetics of insulin in man. Diabetes Metab Rev 3:335–363, 365–397

Ferrannini E, Groop LC (1989) Hepatic glucose production in insulin-resistant states. Diabetes Metab Rev 5:711–725

Foley JE (1992) Rationale and application of fatty acid oxidation inhibitors in treatment of diabetes mellitus. Diabetes Care 15:773–784

Froguel Ph, Zouali H, Vionnet N, Velho G, Vaxillaire M, Sun F, Lesage S, Stoffel M, Takeda J, Passa Ph, Permutt MA, Beckmann JS, Bell GI, Cohen D (1993) Familial hyperglycemia due to mutations in glucokinase. Definition of a subtype of diabetes mellitus. N Engl J Med 328:697–702

Ganrot PO (1993) Insulin resistance syndrome: possible key role of blood flow in resting muscle. Diabetologia 36:876–879

Garvey WT (1989) Insulin resistance and noninsulin-dependent diabetes mellitus: which horse is pulling the cart? Diabetes Metab Rev 5:727–742

Garvey WT (1992) Glucose transport and NIDDM. Diabetes Care 15:396–417

Garvey WT, Kolterman OG (1988) Correlation of in vivo and in vitro actions of insulin in obesity and noninsulin-dependent diabetes mellitus: role of the glucose transport system. Diabetes Metab Rev 4:543–569

Garvey TW, Revers RR, Kolterman OG, Rubenstein AH, Olefsky JM (1985) Modulation of insulin secretion by insulin and glucose in type II diabetes mellitus. J Clin Endocrinol Metab 60:559–568

Gerich JE (1988) Role of insulin resistance in the pathogenesis of Type 2 (non-insulin-dependent) diabetes mellitus. Bailliere's Clin Endocrinol Metab 2: 307–326

Gerich JE (1991) Is muscle the major site of insulin resistance in Type 2 (non-insulin-dependent) diabetes mellitus? Diabetologia 34:607–610

Gerich JE (1993) Control of glycaemia. Bailliere's Clin Endocrinol Metab 7:551–586

Giugliano D, Torella R, Scheen AJ, Lefèbvre PJ, D'Onofrio F (1988) Prostaglandines, sécrétion d'insuline et diabète sucré. Diabete Metab 14:721–727

Golay A, Felber JP, Jéquier E, DeFronzo RA, Ferrannini E (1988) Metabolic basis of obesity and noninsulin-dependent diabetes mellitus. Diabetes Metab Rev 4:727–747

Granner DK, O'Brien RM (1992) Molecular physiology and genetics of NIDDM. Importance of metabolic staging. Diabetes Care 15:369–395

Groop LC, Bonadonna RC, DelPrato S, Ratheiser K, Zyck K, Ferrannini E, DeFronzo RA (1989) Glucose and free fatty acid metabolism in non-insulin-dependent diabetes mellitus: evidence for multiple sites of insulin resistance. J Clin Invest 84:205–213

Groop LC, Kankuri M, Schalin-Jantti C, Ekstrand A, Nikula-Ijas P, Widen E, Kuismanen E, Eriksson J, Franssila-Kallunki A, Saloranta C, Koskimies S (1993a) Association between polymorphism of the glycogen synthase gene and non-insulin-dependent diabetes mellitus. N Engl J Med 328:10–14

Groop LC, Widen E, Ferrannini E (1993b) Insulin resistance and insulin deficiency in the pathogenesis of Type 2 (non-insulin-dependent) diabetes mellitus: errors of metabolism or of methods? Diabetologia 36:1326–1331

Haffner SM, Stern MP, Watanabe RN, Bergman RN (1992) Relationship of insulin clearance and secretion to insulin sensitivity in non-diabetic Mexican Americans. Eur J Clin Invest 22:147–153

Hales CN, Barker DJP (1992) Type 2 (non-insulin-dependent) diabetes mellitus: The thrifty phenotype hypothesis. Diabetologia 35:595–601

Hamman RF (1992) Genetic and environmental determinants of non-insulin-dependent diabetes mellitus (NIDDM). Diabetes Metab Rev 8:287–338

Häring HU, Mehnert H (1993) Pathogenesis of Type 2 (non-insulin-dependent) diabetes mellitus: candidates for a signal transmitter defect causing insulin resistance of the skeletal muscle. Diabetologia 36:176–182

Harris MI (1993) Undiagnosed NIDDM: clinical and public health issues. Diabetes Care 16:642–655

Hitman GA, McCarthy MI (1991) Genetics of non-insulin dependent diabetes mellitus. Bailliere's Clin Endocrinol Metab 5:455–476

Hortulanus-Beck D, Lefèbvre PJ, Jeanjean MF (1990) Le diabète dans la province belge de Luxembourg: Fréquence, importance de l'épreuve de surcharge glucosée orale et d'une glycémie à jeun discrètement accrue. Diabete Metab 16:311–317

Hother-Nielsen O, Beck-Nielsen H (1990) On the determination of basal glucose production rate in patients with type 2 (non-insulin-dependent) diabetes mellitus using primed-continuous 3-^{3}H-glucose infusion. Diabetologia 33:603–610

Jansson P-AE, Fowelin JP, Von Schenck HP, Smith UP, Lönnroth PN (1993) Measurement by microdialysis of the insulin concentration in subcutaneous interstitial fluid. Importance of the endothelial barrier for insulin. Diabetes 42:1469–1473

Johnson KH, O'Brien TD, Betsholtz C, Westermark P (1989) Islet amyloid, islet-amyloid polypeptide, and diabetes mellitus. N Engl J Med 321:513–518

Kahn CR (1978) In:sulin resistance, insulin sensitivity, and insulin unresponsiveness: a necessary distinction. Metabolism 27:1893–1902

Kahn CR, Folli F (1993) Molecular determinants of insulin action. Horm Res [Suppl 3]:93–101

Karam JH, Grodsky GM, Forsham PH (1963) Excessive insulin response to glucose in obese subjects as measured by immunochemical assay. Diabetes 12:197–204

Kelley DE, Mokan M, Mandarino LJ (1992) Intracellular defects in glucose metabolism in obese patients with NIDDM. Diabetes 41:698–706

Kelley DE, Mokan M, Mandarino LJ (1993) Metabolic pathways of glucose in skeletal muscle of lean NIDDM. Diabetes Care 16:1158–1166

King GL, Johnson SM (1985) Receptor-mediated transport of insulin across endothelial cells. Science 227:1583–1586

King H, Rewers M, WHO Ad Hoc Diabetes Reporting Group (1993) Global estimates for prevalence of diabetes mellitus and impaired glucose tolerance. Diabetes Care 16:157–177

Kissebah AH, Peiris AN (1989) Biology of regional body fat distribution: relationship to non-insulin-dependent diabetes mellitus. Diabetes Metab Rev 5:83–109

Klip A, Paquet MR (1990) Glucose transport and glucose transporters in muscle and their metabolic regulation. Diabetes Care 13:228–243

Kolterman O, Gray R, Griffin J, Burstein P, Insel J, Scarlett J, Olefsky J (1981) Receptor and postreceptor defects contribute to the insulin resistance of noninsulin-dependent diabetes mellitus. J Clin Invest 68:957–969

Kriska AM, Bennett PH (1991) An epidemiological perspective of the relationship between physical activity and NIDDM: from activity assessment to intervention. Diabetes Metab Rev 8:355–372

Kusari J, Verma US, Buse JB, Henry RR, Olefsky JM (1991) Analysis of the gene sequences of the insulin receptor and the insulin-sensitive glucose transporter (GLUT-4) in patients with common-type non-insulin-dependent diabetes mellitus. J Clin Invest 88:1323–1330

Laakso M, Edelman SV, Hoit B, Brechtel G, Baron AD (1992) Impaired insulin-mediated skeletal muscle blood flow in patients with NIDDM. Diabetes 41:1076–1083

Leahy JL (1990) Natural history of B-cell dysfunction in NIDDM. Diabetes Care 13:992–1010

Leahy JL, Bonner-Weir S, Weir GC (1992) B-cell dysfunction induced by chronic hyperglycemia. Current ideas on mechanism of impaired glucose-induced insulin secretion. Diabetes Care 15:442–455

Lefèbvre PJ (1991) Diabetes. Abnormal secretion of glucagon. In: Samols E (ed) The endocrine pancreas. Raven, New York, pp 191–205

Lefèbvre PJ (1993) Syndrome X. Diab Nutr Metab 6:61–65

Lefèbvre PJ, Scheen AJ (1992) Management of non-insulin-dependent diabetes mellitus. Drugs 44 [Suppl 3]:29–38

Lefèbvre PJ, Paolisso G, Scheen AJ, Henquin JC (1987) Pulsatility of insulin and glucagon release: Physiological significance and pharmacological implications. Diabetologia 30:443–452

Lefèbvre PJ, Paolisso G, Scheen A (1991) The role of glucagon in non-insulin-dependent (type 2) diabetes mellitus. In: Sakamoto N, Angel A, Hotta H (eds) New directions in research and clinical works for obesity and diabetes mellitus. Excerpta Medica, Amsterdam, pp 25–29

Lefèbvre PJ, Scheen AJ, Paolisso G, Giugliano D (1992) Opioid peptides and carbohydrate metabolism. In: Negri M, Lotti G, Grossman A (eds) Clinical perspectives of endogenous opioids production. Wiley, London, pp 255–273

Lefèbvre PJ, Paolisso G, Scheen AJ (1994) Magnésium et métabolisme glucidique. Thérapie 49:1–7

Lillioja S, Young AA, Cutler CL, Ivy JL, Abbott GH, Zawadzki JK, Yki-Jarvinen H, Christin L, Secomb TW, Bogardus C (1987) Skeletal muscle capillary density and fiber type are possible determinants of in vivo insulin resistance in man. J Clin Invest 80:415–424

Lillioja S, Bogardus C (1988) Obesity and insulin resistance: lessons learned from the Pima indians. Diabetes Metab Rev 4:517–540

Malaisse WJ (1988) Possible sites for deficient glucose recognition in islet cells. In: Lefèbvre PJ, Pipeleers DG (eds) The pathology of the endocrine pancreas in diabetes. Springer, Berlin Heidelberg New York, pp 219–232

Mandarino LJ (1989) Regulation of skeletal muscle pyruvate dehydrogenase and glycogen synthase in man. Diabetes Metab Rev 5:475–486

Martin BC, Warram JH, Krolewski AS, Bergman RN, Soeldner JS, Kahn CR (1992) Role of glucose and insulin resistance in development of type 2 diabetes mellitus: results of a 25-year follow-up study. Lancet 340:925–929

Matschinsky FM (1990) Glucokinase as glucose sensor and metabolic signal generator in pancreatic beta-cells and hepatocytes. Diabetes 39:647–652

Matschinsky F, Liang Y, Kesavan P, Wang L, Froguel Ph, Velho G, Cohen D, Permutt MA, Tanizawa Y, Jetton TL, Niswender K, Magnuson MA (1993) Glucokinase as pancreatic B cell glucose sensor and diabetes gene. J Clin Invest 92:2092–2098

Mitrakou A, Kelley D, Mokan M, Veneman T, Pangburn T, Reilly J, Gerich J (1992) Role of reduced suppression of glucose production and diminished early insulin release in impaired glucose tolerance. N Engl J Med 326:22–29

Moller DE (1993) Insulin resistance. Wiley, Chichester

Moller DE, Flier JS (1991) Insulin resistance. Mechanisms, syndromes, and implications. N Engl J Med 325:938–948

O'Dea K (1992) Obesity and diabetes in "the land of milk and honey". Diabetes Metab Rev 8:373–388

Olefsky JM (1976) The insulin receptor: its role in insulin resistance in obesity and diabetes. Diabetes 25:1154–1165

Olefsky JM, Garvey WT, Henry RR, Brillon D, Matthaei S, Freidenberg GR (1988) Cellular mechanisms of insulin resistance in non-insulin-dependent (type II) diabetes. Am J Med 85 [Suppl 5A]: 86–105

O'Rahilly SP, Nugent Z, Rudenski AS, Hosker JP, Burnett MA, Darling P, Turner RC (1986) Beta-cell dysfunction, rather than insulin insensitivity, is the primary defect in familial type 2 diabetes. Lancet II: 360–363

O'Rahilly S, Turner RC, Matthews DR (1988) Impaired pulsatile secretion of insulin in relatives of patients with non-insulin dependent diabetes. N Engl J Med 318:1225–1230

Paolisso G, Sgambato S, Giugliano D, Torella R, Varricchio M, Scheen AJ, D'Onofrio F, Lefèbvre PJ (1988) Impaired insulin-induced erythrocyte magnesium accumulation is correlated to impaired insulin-mediated glucose disposal in type 2 (non-insulin-dependent) diabetic patients. Diabetologia 31:910–915

Paolisso G, Scheen AJ, D'Onofrio F, Lefèbvre PJ (1990) Magnesium and glucose homeostasis. Diabetologia 33:511–514

Pedersen O (1989) The impact of obesity on the pathogenesis of non-insulin-dependent diabetes mellitus: a review of current hypotheses. Diabetes Metab Rev 5:495–509

Permutt MA, Elbein SC (1990) Insulin gene in diabetes: analysis through RFLP. Diabetes Care 13:364–372

Permutt MA, Chiu KC, Tanizawa Y (1992) Glucokinase and NIDDM: a candidate gene that paid off. Diabetes 41:1367–1372

Polonsky K, Frank B, Pugh W, Addis A, Karrison T, Meier P, Tager H, Rubenstein A (1986a) The limitations to and valid use of C-peptide as a marker of the secretion of insulin. Diabetes 35:379–386

Polonsky KS, Licinio-Paixao J, Given BD, Pugh W, Rue P, Galloway J, Karrison T, Frank B (1986b) Use of biosynthetic human C-peptide in the measurement of insulin secretion rates in normal volunteers and type I diabetic patients. J Clin Invest 77:98–105

Polonsky KS, Given BD, Hirsch L, Shapiro EJ, Tillil H, Beebe C, Galloway JA, Frank BH, Karrison T, Van Cauter E (1988a) Quantitative study of insulin secretion and clearance in normal and obese subjects. J Clin Invest 81:435–441

Polonsky KS, Given BD, Hirsch LJ, Tillil H, Shapiro ET (1988b) Abnormal patterns of insulin secretion in non-insulin-dependent diabetes mellitus. N Engl J Med 318:1231–1239

Porte D Jr (1991) B-cells in type II diabetes mellitus. Diabetes 40:166–180

Porte D Jr, Kahn SE (1989) Hyperproinsulinemia and amyloid in NIDDM – clues to etiology of islet B-cell dysfunction? Diabetes 38:1333–1336

Rahier J (1988) The diabetic pancreas: a pathologist's view. In: Lefèbvre PJ, Pipeleers DG (eds) The pathology of the endocrine pancreas in diabetes. Springer, Berlin Heidelberg New York, pp 17–40

Randle PJ (1993) Glucokinase and candidate genes for Type 2 (non-insulin-dependent) diabetes mellitus. Diabetologia 36:269–275

Randle PJ, Garland PB, Hales CN, Newsholme EA (1963) The glucose fatty acid cycle: its role in insulin sensitivity and the metabolic disturbances of diabetes mellitus. Lancet I:785–789

Rasmussen H, Zawalich KC, Ganesan S, Calle R, Zawalich WS (1990) Physiology and pathophysiology of insulin secretion. Diabetes Care 13:655–666

Reaven GM (1988) Role of insulin resistance in human disease. Diabetes 37:1595–1607

Reaven GM (1990) Non-insulin-dependent diabetes (NIDDM): speculations on etiology. In Alberti KGMM, Krall LP (eds) The diabetes annual/5. Elsevier, Amsterdam, pp 51–71

Robertson RP (1989) Type II diabetes, glucose "non-sense", and islet desensitization. Diabetes 38:1501–1505

Rossetti L, Giaccari A, DeFronzo RA (1990) Glucose toxicity. Diabetes Care 13:610–630.

Rudenski AS, Matthews DR, Levy JC, Turner RC (1991) Understanding "insulin resistance": both glucose resistance and insulin resistance are required to model human diabetes. Metabolism 40:908–917

Saad MF, Knowler WC, Pettitt DJ, Nelson RG, Charles MA, Bennett PH (1991) A two-step model for development of non-insulin-dependent diabetes. Am J Med 90:229–235

Saltiel AR (1990) Second messengers of insulin action. Diabetes Care 13:244–256

Schalin-Jantti C, Harkonen M, Groop LC (1992) Impaired activation of glycogen synthase activity in skeletal muscle from obese patients with and without Type 2 (non-insulin-dependent) diabetes mellitus. Diabetologia 34:239–245

Scheen AJ (1992) From obesity to type 2 diabetes. Acta Clin Belg 47 [Suppl 14]: 30–36

Scheen AJ, Lefèbvre PJ (1992) Assessment of insulin sensitivity in vivo. Application to the study of type 2 diabetes. Horm Res 38:19–27

Scheen AJ, Lefèbvre PJ (1993a) De l'obésité au diabète ou de l'insulino-résistance au déficit insulino-sécrétoire. Rev Franç Endocrinol Clin 34:297–310

Scheen AJ, Lefèbvre PJ (1993b) Pharmacological treatment for the obese diabetic patient. Diabete Metab 19:547–559

Scheen AJ, Nemery A, Luyckx AS, Lefèbvre PJ (1986) Etiologie et physiopathologie des diabètes sucrés. Encycl Méd Chir. Paris, Glandes-Nutrition, 10366 C10

Scheen AJ, Paquot N, Lefèbvre PJ (1991) La cellule B dans le diabète de type II: coupable ou victime? In: Journées Annuelles de Diabétologie de l'Hôtel-Dieu. Flammarion Médecine-Sciences, Paris, pp 153–169

Scheen AJ, Castillo M, Paolisso G, Jandrain B, Lefèbvre PJ (1992a) Normal feed-back inhibition of insulin secretion by insulin but reduced metabolic clearance rate of the hormone in obese subjects. In: Ailhaud G, Guy-Grand B, Lafontan M, Ricquier D (eds) Obesity in Europe 91. Libbey, London, pp 201–205

Scheen AJ, Paolisso G, Castillo M, Jandrain B, Paquot N, Lefèbvre PJ (1992b) Insulin secretion and action in various populations with type 2 (non-insulin-dependent) diabetes mellitus (Abstract). Diabetologia 35 [Suppl 1]: A87

Scheen AJ, Paquot N, Castillo MJ, Lefèbvre PJ (1994) How to measure insulin action. Diabetes Metab Rev 10:151–188

Serjeantson SW, Zimmet P (1991) Genetics of non-insulin dependent diabetes mellitus in 1990. Bailliere's Clin Endocrinol Metab 5:477–493

Shulman GI, Rothman DL, Jue T, Stein P, DeFronzo RA, Shulman RG (1990) Quantitation of muscle glycogen synthesis in normal subjects and subjects with non-insulin-dependent diabetes by ^{13}C nuclear magnetic resonance spectros-copy. N Engl J Med 322:223–228

Songer TJ (1992) The economic costs of NIDDM. Diabetes Metab Rev 8:389–404

Steiner DF, Tager HS, Chan SJ, Nanjo T, Sanke T, Rubenstein AH (1990) Lessons learned from molecular biology of insulin-gene mutations. Diabetes Care 13: 600–607

Sturis J, Polonsky KS, Shapiro ET, Blackman JD, O'Meara NM, Van Cauter E (1992) Abnormalities in the ultradian oscillations of insulin secretion and glu-cose levels in type 2 (non-insulin-dependent) diabetic patients. Diabetologia 35:681–689

Temple RC, Clark PMS, Nagi DK, Schneider AE, Yudkin JS, Hales CN (1990) Radioimmunoassay may overestimate insulin in non-insulin-dependent diabe-tics. Clin Endocrinol (Oxf) 32:689–693

Temple RC, Clark PMS, Hales CN (1992) Measurement of insulin secretion in type 2 diabetes: problems and pitfalls. Diabetic Med 9:503–512

Thorens B, Waeber G (1993) Glucagon-like peptide-1 and the control of insulin secretion in the normal state and in NIDDM. Diabetes 42:1219–1225

Thornburn AW, Gumbiner B, Bulacan F, Brechtel G, Henry RR (1991) Multiple defects in muscle glycogen synthase activity contribute to reduced glycogen synthesis in non-insulin dependent diabetes mellitus. J Clin Invest 87:489–495

Turner RC, Matthews DR, Clark A, O'Rahilly S, Rudenski AS, Levy J (1988) Pathogenesis of NIDDM – a disease of deficient insulin secretion. Bailliere's Clin Endocrinol Metab 2:327–342

Unger RH (1991) Diabetic hyperglycemia: link to impaired glucose transport in pancreatic B-cells. Science 251:1200–1205

Unger RH, Grundy S (1985) Hyperglycaemia as an inducer as well as a consequence of impaired islet cell function and insulin resistance: implications for the management of diabetes. Diabetologia 28:119–121

Vaag A, Henriksen JE, Beck-Nielsen H (1992) Decreased insulin activation of glycogen synthase in skeletal muscles in young nonobese Caucasian first-degree relatives of patients with non-insulin-dependent diabetes mellitus. J Clin Invest 89:782–788

Vadheim CM, Rotter JI (1992) Genetics of diabetes mellitus. In: Alberti KGMM, DeFronzo RA, Keen H, Zimmet P (eds) International textbook of diabetes mellitus. Wiley, Chichester, pp 31–98

Vague J, Vague Ph (1988) Obesity and diabetes. In: Alberti KGMM, Krall LP (eds) The diabetes annual/4. Elsevier Science, Amsterdam, pp 311–338

Velho G, Froguel Ph, Clément K, Pueyo ME, Rakotoambinina B, Zouali H, Passa Ph, Cohen D, Robert JJ (1992) Primary pancreatic beta-cell secretory defect caused by mutations in glucokinase gene in kindreds of maturity onset diabetes of the young. Lancet 340:444–448

Vestergaard H, Bjorbaek C, Andersen PH, Bak JF, Pedersen O (1991) Impaired expression of glycogen synthase mRNA in skeletal muscle of NIDDM patients. Diabetes 40:1740–1745

Ward WK, Beard JC, Porte D Jr (1986) Clinical aspects of islet B-cell function in non-insulin-dependent diabetes mellitus. Diabetes Metab Rev 2:297–313

Westermark P, Johnson KH, O'Brien TD, Betsholtz C (1992) Islet amyloid polypeptide – a novel controversy in diabetes research. Diabetologia 35:297–303

Wolffenbuttel BHR, Van Haeften TW (1993) Non-insulin dependent diabetes mellitus: defects in insulin secretion. Eur J Clin Invest 23:69–79

Yalow RS, Berson SA (1960) Plasma insulin concentrations in nondiabetic and early diabetic subjects. Determination by a new sensitive immunoasssay technique. Diabetes 9:254–260

Yki-Järvinen H (1992) Glucose toxicity. Endocr Metab Rev 15:415–431

CHAPTER 3

Non-Pharmacological Management
of Non-Insulin-Dependent Diabetes

J. Köbberling

A. Dietary Management

I. General Basis for Recommendations

There is broad consensus that energy intake and diet are major causes of obesity and thereby main contributors to non-insulin-dependent diabetes (NIDDM). Today's diet, which is rich in fats and relatively low in carbohydrates and fibres, the so-called westernized diet, is highly associated with the occurrence of NIDDM. It is therefore concluded that dietary modifications are the mainstay of treatment for NIDDM. These modifications, concentrating on fat restriction to reduce energy intake and body weight, should be the starting point of long-term treatment of NIDDM. Many patients with this disease who comply with dietary advice will show improvement in the major metabolic abnormalities associated with the condition. The main goal will be to obviate the need for oral agents and insulin. But even for those who cannot be managed without drug therapy, attention to dietary advice may modify blood lipids and will thereby reduce the risk for coronary heart disease. Even in insulin-dependent diabetes mellitus (IDDM), the role of diet is not only to help minimize the short-term fluctuations of blood glucose but also to reduce the risk of long-term complications by helping to achieve optimum glycaemic control and satisfactory levels of blood lipids. Dietary advice for patients with NIDDM and IDDM is rather similar and the major principles resemble those for the general population, especially in populations with a high risk of cardiovascular disease.

In contrast to often-heard opinion, especially in lay people, there is no need for diabetic patients to be given meals that differ from those eaten by the rest of the family. Dietary recommendations for people with diabetes have to consider differing dietary habits in different countries. The broadly accepted recommendations of the Diabetes and Nutrition Study Group of the European Association for the Study of Diabetes (1988) and the rather similar recommendations of the American Diabetes Association (1987) refer to European, North American and other "westernized" countries.

II. Total Energy Intake

Dietary advice according to energy intake needs to be separated into that for patients who are overweight and that for NIDDM patients of normal weight.

1. Overweight NIDDM Patients

The majority of patients with NIDDM are overweight. For these patients the dietary advice should mainly focus on energy restriction (HANSEN 1988). Successful weight reduction is particularly important in these patients and will result in a fall in blood glucose levels, thereby reducing the risk for several conditions associated with diabetes and obesity. A major goal is the reduction of triglycerides and very low density lipoproteins (VLDL's, which are rich in triglycerides. These levels are elevated in most patients with NIDDM and obesity. There are data suggesting that elevated VLDL levels are more important as risk factors for cardiovascular diseases in diabetic patients than in the non-diabetic population. Weight reduction will also reduce elevated levels of blood pressure and thereby will also reduce the risk for cardiovascular disease.

The primary goal in reducing the energy intake is a reduction in plasma glucose concentration, which is often seen even before energy restriction has resulted in an appreciable weight loss. The mechanism by which weight loss improves metabolic control is the reduction of insulin resistance This may lead to an improvement in B-cell function and a fall in hepatic glucose production.

In order to achieve satisfactory results, it is important that patients be given realistic targets for weight reduction. For patients with higher degrees of obesity it may be useful to provide intermediate goals. The patients should know that it may take a long time to achieve the target. In general, a sustained weight loss of 1–2 kg/month is satisfactory. This goal can be achieved by a reduction of at least 500 kcal below that required for average weight maintenance. Whenever possible, oral hypoglycaemic therapy should not be started until the patient has had a fair trial of weight reduction. Oral hypoglycaemic agents and especially insulin will often inhibit weight reduction. Despite satisfactory glycaemic control, they may well lead to even more weight gain. Patients on insulin should be advised to reduce the daily dose while reducing their weight, and they can often appreciably reduce their insulin requirement and sometimes even come off insulin if they are able to lose weight.

With mildly overweight people it might be appropriate not to prescribe a precise energy intake but rather to recommend avoidance of energy-dense food. In these cases dietary advice will focus on the recommendation of appropriate food types. For more overweight people or for those unable to lose weight, it is advisable to prescribe a diet based on calorie counting. Very often this has to be combined with some behavioural modification techniques.

There is some controversy as to whether it is helpful to use very low calorie diets of less than 500 kcal/day (FITZ et al. 1983). Diets of this kind are not infrequently used by people without special medical advice, but it is very questionable whether these diets are really helpful for diabetic patients (DURNIN 1987). The very low calorie diets were introduced in the 1960. It soon became clear that minerals and trace elements had to be added. Later it was learned that at least certain essential amino acids should be supplied. Without these supplementations, a number of deaths were reported and formulations were withdrawn (SOURS et al. 1988). If in individual cases it seems advisable to use a very low calorie diet, this should be not less than 400 calories/day and it should be enriched with essential micronutritients and appropriate proteins. Various formula diets of this kind are on the market. In any case, diets of this kind should not be used for longer periods than 4 weeks. During treatment with a very low calorie diet, any type of hypoglycaemic agent should be withdrawn and patients should be warned of the possibility of hypoglycaemia. There is no question that these diets are effective in reducing blood glucose and improving glycaemic control. Also triglycerides and sometimes lipoprotein levels are improved. The degree of improvement correlates with the degree of weight loss and with the initial degree of abnormality. It should be kipt in mind that most studies of this kind were performed under metabolic ward conditions. These studies do not provide information about long-term benefits. Some studies indicating a long-term benefit in a considerable proportion of patients have been performed under rather unusual conditions, since the patients were seen more frequently than is possible with ordinary outpatients. Most authors are convinced that the long-term results with very low calorie diets are no better than those with moderate weight reduction. Another concern of the use of very low calorie diets is the possibility that part of the weight reduction is due to a loss of lean body mass, which is regained during refeeding.

Long-term weight reduction and maintenance can only be achieved by an adjustment of eating habits. There is no evidence today that very low calorie diets will promote this more effectively than following conventional dietary advice. As long as controlled long-term studies determining the role of very low calorie diets are not available, it seems advisable to restrict diets of this kind to exceptional cases under metabolic ward conditions. Exceptions may be patients with morbid obesity and impaired function of other organs besides diabetes who may need rapid weight reduction, which can best be achieved by a very low calorie diet.

2. Normal Weight NIDDM Patients

Less than 20%, in most societies less than 10%, of all patients with NIDDM, are not overweight. These patients certainly will not require a low-calorie diet but rather normal energy intake. It has to be considered that, after initiation of insulin therapy or even after addition of oral hypoglycaemic agents, even in non-overweight patients a weight gain may be observed

Patients should be aware of this problem and if a tendency of this kind can be observed they should learn to reduce their daily calorie intake in a way similar to primarily overweight patients with NIDDM

III. Carbohydrates

For many years the recommendation has been to use diets low in carbohydrate for people with diabetes. This recommendation, however, was unnecessary and. even undesirale (MANN 1989). The consequence of low carbohydrate content was usually a higher content of saturated fatty acids. Modern dietary recommendations for diabetics suggest that the amount of carbohydrate should be not restricted or even increased. The total daily energy supply should consist of 50%–60% carbohydrates. Since the average carbohydrate content in most Western diets is only around 40%, people need to increase their carbohydrate intake in order to fulfil the recommendations of the above-mentioned European and North American societies. Again, the recommendation of 50%–60% carbohydrate does not apply specifically to diabetics but to the general population. A "modern" diet, sometimes termed a "prudent" diet, has a much higher carbohydrate content than the average diet during recent decades and even than the diet most people usually eat today.

1. Dietary Fibres

It is recommended that the diet should contain approximately 20 g/1000 kcal or, on the average, about 40 g/day dietary febre. This recommendation is based on many studies from recent years. A diet high in fibre usually leads to a reduction in the insulin requirement (KIEHM et al. 1976). Dietary "fibre" is a rather complex description. Fibre may be subdivided into two broad classes, the so-called soluble fibres, which include gums, gels, mucilages and pectic substances, and insoluble fibres, including lignin, cellulose and hemicellulose. Whereas insoluble fibres mainly increase faecal bulk without major metabolic effects, soluble fibres are broken down in the large bowel. They produce gas and short-chain fatty acids and they may influence carbohydrate and lipid metabolism.

There are studies (JENKINS et al. 1977) indicating that the addition of guar, pectin or other fibre to the normal diet may improve glycaemic control. They may even lower cholesterol levels. The mechanism cannot only be a reduction of postprandial glycaemia by slowing absorption since fasting glucose levels are also lowered (ARO et al. 1981). Recently it has become clear that the fibre needs to be incorporated into foods in order to achieve its beneficial effect. Also fibre is only beneficial if it is part of a high-carbohydrate diet (VINIK and JENKINS 1988). There is a definite dose response effect and different types of fibre seem to have different potencies. These observations have led to preparations of fibre which are offered as additional treatment for patients with NIDDM being treated with diet alone, with oral hypoglycaemic agents or even with insulin therapy (PETERSON and MANN 1985). The addition of fibre in tablets or similar preparations should be regarded as pharmacological rather than dietary treatment. The likelihood of side effects is low but long-term studies are still required to establish whether the benefit is definitely higher than any risk of side effect.

Unfortunately, most studies on fibre are performed with fibre supplementation rather than with dietary fibre. It is certainly more difficult to measure fibre content in ordinary foods and to perform controlled studies. Some studies, however, have indicated that the use of foods high in soluble fibre such as legumes, lentils, some fruits, oats and barley can lead to an improvement in the glycaemic control and also to an improvement in some lipoprotein parameters. This effect can also be achieved in patients under treatment with oral hypoglycaemic agents (LOUSLEY et al. 1984). The beneficial effect of some special carbohydrate preparations, e.g., pasta, is not always explained by a high content of dietary fibre (VINIK 1988). The influence of dietary carbohydrate and fibre on the metabolic control of diabetes seems to be independent.

Since, as mentioned above, the addition of fibre without increasing the carbohydrate content of the diet seems not to be beneficial, the most appropriate recommendation will be to increase foods with a high content of carbohydrate and fibre (SIMPSON et al. 1982). This recommendation will also prevent patients from using carbohydrates without fibre, mainly simple sugars, which have no advantageous effect on the disease.

Although, in the strict sense, there are no appropriate long-term studies which clearly show te beneficial effect of diets high in carbohydrate and fibre, there is a far-reaching acceptance that the present data of short-term studies, experimental studies and epidemiological observations provide enough evidence to justify the recommendation to adhere to this diet (McCULLOCH et al. 1985). People should be encouraged to use fibre-rich food in a sufficient amount for at least 50% of total calorie intake to be achieved by the carbohydrates. This will then lead to a fibre consumption of about 40 g/day (GEEKIE et al. 1986).

NUTTAL (1993) in a recent survey pointed out that several studies have been conducted in which a high-carbohydrate diet has been reported to reduce the plasma glucose concentration. In these diets, however, the emphasis has been on foods with a high fibre content. In general, they were not well controlled, and several confounding variables such as weight loss, decreased food energy intake, different food sources with potential for differences in starch digestibility, and decreased dietary fat content were present.

Thus, it has not been possible to determine whether dietary fibre played a significant role. The results of the studies in which specific, defined fibre was added to the diet would suggest that the naturally occurring fibre in foods is likely to play a minor role.

2. Simple Sugars

The use of natural carbohydrate-like foods will lead to a considerable intake of simple sugars. The same applies to sugars in milk. The addition of sucrose as sweetener, however, has traditionally been restricted or forbidden for diabetics (FRANZ 1993). At least for type II diabetics, the basis for such recommendations is rather small. Some studies performed to clarify this question have clearly shown that monosaccharides and disaccharides in limited amounts do not result in a deterioration of glycaemic control or of elevated lipid levels. PETERSON et al. (1986) have recently performed a study in which 45 g sucrose was given to replace an isocaloric amount of bread and potatoes. No deleterious effects were observed. On the basis of this result and those of other studies, we need to adjust our recommendations. Limited quantities of sugar, e.g., 30 g/day, as sweetener should not be precluded, provided that the total energy content is not increased and that foods high in fibre are not replaced. Some people are concerned about possible side effects of artificial sweeteners and would rather see their diabetics use sucrose in limited amounts as sweeteners.

3. Glycaemic Index of Various Carbohydrates

For a long time it was thought that not all carbohydrate-containing foods led to elevations in blood glucose in the same manner. This led to the introduction of the so-called glycaemic index. JENKINS et al. (1988) have classified most carbohydrate-containing foods according to their glycaemic index and have suggested that this should be an appropriate means of identifying the optimum foods. The glycaemic index is defined as the ratio of the incremental blood glucose area after food to the corresponding area after a glucose load or after white bread containing an equal quantity of carbohydrate. The glycaemic index is usually expressed as a percentage and foods with a low index are regarded as the most appropriate carbohydrate-containing foods for diabetics.

This so-called glycaemic index is rather suggestive (THORBURN et al. 1986). It pretends that there are exact figures and that the value of various carbohydrates can be calculated. There are, however, severe objections to the glycaemic index approach. The main objections are as follows (GANNON and NUTTALL 1987): There are large individual variations in response, there is a lack of agreement among centres, there is a lack of difference between mixed meals including foods of different glycaemic indices and there is a lack of studies showing long-term benefits of foods with a lower index. Even the mathematical approach is different among different researchers.

This criticism does not indicate that there is no difference between carbohydrates according to their effect on blood glucose. In general, foods containing higher amounts of fibre have lower glycaemic indexes. Carbohydrates such as legumes, pasta and grains, e.g., oats and barley, cracked wheat and others, are associated with improved glycaemic control and a reduction in LDL cholesterol. The glycaemic index may be used as a rough means for ranking starchy foods. It has to be pointed out that the glycaemic index is not only a characteristic of the food itself but also of the processing and composition of foods. This can differ a lot from country to country or even within countries and can thereby influence the glycaemic index to a large extent.

IV. Dietary Fat

There are very few specific recommendations as to the quantity or quality of fat intake in diabetics. Most recommendations are very close to those given for the general population.

Fat is the most energy rich of all energy-providing nutrients, and whenever a reduction in weight is recommended a reduction of fat consumption is advisable. In addition, a reduction in saturated fat intake is associated with reduced levels of LDL cholesterol, which is an important predictor of cardiovascular diseases in NIDDM. Epidemiological evidence suggests that people with diabetes as well as non-diabetics in populations consuming a low-fat diet have a reduced incidence of morbidity and mortality from cardiovascular diseases. It has been suggested that the advice of a reduction in fat consumption should be restricted to younger people and to those with higher levels of plasma cholesterol. It seems difficult, however, to identify an age beyond which lipid lowering is no longer beneficial. It is therefore recommended that, in people with diabetes, saturated fat should be reduced to 8% or less of total energy and that total fat should be no more than 30% of total energy. Mono-unsaturated and polyunsaturated fatty acids are equally energy rich, but their use should still only partly be restricted or not increased. Mono-unsaturated fatty acids have an LDL-lowering effect and, in contrast to other dietary modifications, are not associated with a reduction in HDL cholesterol (GRUNDY 1987; GARG et al. 1987). Also the increased intakes of polyunsaturated fatty acids help to lower LDL cholesterol. These observations have led to the European recommendations that total fat intake may be higher than 30% provided that mono-unsaturated and polyunsaturated fatty acids and not saturated fatty acids are increased and as long as the total energy intake does not exceed the optimum amount (LEWIS-BARNED et al. 1987).

There has been considerable interest in the effect of special polyunsaturated fatty acids, e.g., the so-called eicosapentaenoic acid. This fatty acid is derived from fish and appears to reduce platelet aggregation by increasing the production fo thromboxane A_2 and its stable metabolite thromboxane B_2. If this holds true, these fatty acids may reduce the risk of thrombosis and coronary heart disease (KROMHOUT et al. 1985). Fish oil in capsules has been offered for special treatment of patients with NIDDM, but there is considerable uncertainty about the optimum quantities. Variable effects on LDL cholesterol preclude recommendations concerning the use of pharmacological preparations of these fatty acids (HAINES et al. 1968). An increased consumption of fish can be recommended to diabetics as well as to healthy people. Fish consumption helps to reduce the intake of saturated fatty acids and in some epidemiological studies it has been associated with a reduced risk for cardiovascular disease.

A reduction of saturated fat will lead to reductions of dietary cholesterol. Although a restriction of dietary cholesterol is not a specific recommendation for diabetics, this general guideline will certainly also apply to patients with NIDDM.

V. Dietary Protein

Dietary protein was traditionally regarded as being less important in dietary recommendations for diabetics. When diets low in carbohydrate were prescribed, the protein intake was usually relatively high, up to 20% of the total energy intake. This is much more than is necessary for a normal balance of nitrates and other nutritional factors. In recent years it has become clear that a high-protein diet even has an adverse effect for diabetics and that total protein should not exceed the requirement of 0.8–1.0 g/kg per day.

Several studies have been published clearly showing that a protein restriction in diabetic patients with established nephropathy has a positive effect (EVANOFF et al. 1987). Urinary protein excretion usually falls, blood pressure often normalizes and in a proportion of patients even the renal function improves. Not only creatinine clearance, which itself is influenced by protein intake, is improved on a a low-protein diet but also the clearance measured by isotopic techniques (WALKER et al. 1989). A restriction of protein to 0.8 g/kg per day led to a decline of the elevated glomerular filtration rate from 0.61 to 0.14 ml/min per month. Urinary albumin excretion falls strikingly and independently of blood pressure.

According to several studies, a so-called microalbuminuria may be used to define a "prenephrotic" diabetic kidney disease. An intervention at this stage by reducing protein intake leads to a significant amelioration of the increased glomerular filtration rate (MOGENSEN and CHRISTENSEN 1984; VIBERTI and KEEN 1984). The benefit of the low-protein diet on renal, function was independent of glycaemic control, arterial blood pressure and renal vascular resistance.

These observations have led to considerations as to whether even earlier stages of diabetic nephropathy or even normal patients may benefit from a low-protein diet. There are short-term studies which favour a positive influence of this diet on various renal parameters.

The difficult question arises as to whether one should recommend a low-protein diet to patients with NIDDM in general. Patients without signs or symptoms of nephropathy usually feel well and it might therefore be rather difficult for those patients to adhere to a low-protein diet. It seems likely that animal protein is more harmful than protein derived from vegetables. It might be easier to recommend and to accept a diet low in animal protein than a diet low in total protein. Unfortunately, there are no good studies which would confirm the theory that a reduction of animal protein alone would slow down the progression of renal disease in patients with NIDDM. It is therefore not yet possible to make definite recommendations as to the quality of dietary protein in diabetics.

VI. Other Dietary Factors

1. Sodium

It is well established that sodium intake corresponds to the tendency for blood pressure to increase. Especially in patients with mild or moderate hypertension, a restriction of sodium intake has a substantial blood pressure lowering effect. According to recent studies, a reduction of sodium intake and weight reduction should be the first line of therapy in such patients. Even patients with higher degrees of hypertension who need hypotensive medication should restrict their dietary sodium intake to less than 3 g/day as an adjunct to other types of therapy (DODSON and PACY 1984; PACY et al. 1986).

Restriction of sodium intake for hypertension is a general recommendation to all patients prone to elevated blood pressure. A sodium intake of not more than 6 g/day is part of the so-called prudent diet, which should be recommended to everybody. All diabetic patients should be advised to restrict their sodium intake to not more than 6 g/day, while those with hypertension should try to lower their sodium intake even further.

2. Alcohol

Alcohol in excessive amounts is noxious to diabetics as well as to other people. The general warning regarding alcohol intake also applies to patients with NIDDM. But even alcohol in doses usually regarded as "socially acceptable" may be harmful to diabetics. Patients with hypertriglyceridaemia should eliminate alcohol as far as possible. Also patients who are overweight and those with hypertension should try to avoid alcohol as much as possible. Patients on insulin or antidiabetic agents are prone to hypoglycaemic effects, especially when alcohol is consumed without food. So-called diabetic beer is carbohydrate replete and sometimes has an even higher content of alcohol. It is not advantageous for diabetics, and is possibly disadvantageous.

3. Vitamins, Minerals and Trace Elements

The ordinary Western diet and even most specialized diets contain sufficient amounts of vitamins, minerals and trace elements. There is no reason to assume that the requirement for these dietary factors is higher in diabetics than in other people. Supplements are therefore usually not required.

VII. Dietary Specialities for Diabetics

1. Sweeteners

Dietetic sweeteners may be subdivided into nutritive and non-nutritive ones (Warshaw 1990). Saccharin and cyclamate are the best-known non-nutritive sweeteners. They are useful for any type of sweet drink, pudding and other sweet fruit. The peptide aspartame is also included in this group and has a sweetness so intensive that only minute quantities are required for a sweetening effect (Rolls 1991). Whenever an energy restriction is wanted without a loss of palatability, the non-nutritive sweeteners are useful and may be recommended. There has been some argument as to their safety but this question has since been resolved. Earlier reports on a possible induction of cancer were based on doses several times higher those recommended for appropriate use in diet.

Nutritive sweeteners may be used as sugars instead of sucrose. Sorbitol and fructose have less effect on blood glucose than glucose or sucrose, and both drugs have been used for many years by people with diabetes. No severe metabolic side effects have been observed. Fructose usually leads to much less hypertriglyceridaemia than glucose. However, monitoring of triglycerides is still recommended when this monosaccharide is used for longer periods (Bantle et al. 1992). Doctors and patients should keep in mind that nutritive sweeteners contain equal amounts of energy to sucrose. It is usually recommended that the daily intake should be less than 50 g of all nutritive sweeteners together. This includes possible surrogate sugars in so-called diabetic foods. If the intake is spread over the day, the gastrointestinal symptoms are minor and diarrhoea will not usually occur.

2. Diabetic Foods

In contrast to often heard opinion, diabetic foods with surrogate sugars are not a recommended part of the diabetic diet. This applies especially to patients with NIDDM, who should lose weight by consuming a low-calorie diet. Most diabetic foods contain nutritive sweeteners and are therefore not at all calorie restricted. The "low-calorie" or "light" sugar-free drinks usually contain non-nutritive sweeteners. They are useful for most patients with NIDDM, especially for those trying to lose weight.

VIII. Problems and Techniques of Dietary Advice

Dietary advice is not just a matter of information. The prescription of special diets for diabetics leads to a major change in lifestyle. Physicians are mostly not competent enough to be the sole providers of such advice. For appropriate advice they need the help of experienced dietitians or of a trained nurse specialized in diet and diabetes. Even then diabetic patients usually need numerous visits to fully understand and accept the need for changing their lifestyle and their eating habits. They have to learn new

Table 1. Key educational points about lower-calorie and dietetic foods (WARSHAW 1993)

1. All sugars, including fructose and sugar alcohols, contain approximately 4 cal/g.
2. So-called natural sugars (e.g., honey, brown sugar and turbinado sugar) contain 4 cal/g and are no healthier than regular sugars.
3. Consuming excess quantities of sugar alcohols (sorbitol, xylitol and mannitol) can lead to gastrointestinal distress or have a laxative effect.
4. The terms "fat free", "no sugar added" or "reduced sugar" do not mean that the food does not contain sugars, carbohydrates and/or calories.
5. The terms "fat free", "light" and "reduced fat" may indicate that a carbohydrate-based fat replacer was used. Although calories from fat may be reduced, calories from carbohydrates may be increased. Effects on blood glucose levels will differ.
6. For teaching purposes, divide lower-calorie foods into two groups: those with calories and those with minimal or no calories. Count the caloric group in exchanges, calories or grams of carbohydrate or fat. Reasonable amounts of minimal/no-calorie food can be used freely in small amounts.
7. High-sugar foods are often high-fat foods. Along with sugar content, check the fat, saturated fat, cholesterol and calories from fat.
8. Timing the consumption of foods containing sugar is important. Eating sweets within a meal may blunt glycaemic increase, compared with eating sweets alone. Macronutrient and fibre content will also affect glycaemic increase.
9. Clients can observe the effects of sugars on their blood glucose by self-monitoring.

approaches to meal planning and to master the many diabetes and nutrition subjects. New techniques of food preparation need to be learned. Cookery demonstrations are very helpful, and it may be recommended that other family members take part in these demonstrations. A charismatic teacher is needed who gives the instruction with enthusiasm. Many educational aids are offered such as flip charts, postcards and plastic devices for various foods.

The education should not be too complicated and sophisticated. WARSHAM (1993) has listed some simple and important educational points. These key educational points about lower-calorie and dietetic foods are listed in Table 1.

Unfortunately, most people with diabetes, especially those with non-insulin-dependent diabetes, never receive nutritional education. The main reason is the lack of physicians referring patients to dietitians. The doctor's involvement in nutrition education and their emphasis on the importance of nutrition in their advice to patients, however, is critical for improving diabetes control.

IX. Dietary Recommendations in "Non-Western" Societies

1. India and Southeast Asia

In India, as in many countries of the southeast Asian region, the diet is mostly based on rice along with a variety of fresh vegetables. Animal protein is derived from milk, meat, fish, poultry and eggs. However, taking the population as a whole, consumption of meat, fish and eggs is uncommon even among those who are non-vegetarian. Besides a small quantity of milk, yogurt is consumed in larger amounts in most parts of India (RAHEJA 1988).

In general, traditional Indian and Asian food is much closer to recent recommendations for a diet for diabetics than Western food. Not only the low animal protein and low content of saturated fat and the high content of unsaturated fat from vegetables or fish oil, but also the limited energy content of Indian food is a distinct advantage with special relevance to the management of diabetes mellitus (RAHEJA 1987). The average energy consumption has increased during the last decade; however, the mean daily calorie intake is around only 2400 kcal.

In previous decades many diabetologists in Asia tried to change the traditional eating habits of their patients and to bring them into line with "Western" habits and the norms of low-carbohydrate diets. The situation has since completely changed. At present many diabetologists in the Western world are recommending a diet which resembles the traditional Asian high-carbohydrate diet, often described as an "oriental" diet.

India has some sociocultural characteriatics which have implications for diet and food consumption. In most oriental societies, to give and share food is a key part of social interaction. A joint family system operates and meal times are an opportunity for family reunion. Instead of counselling the patient, the family has to be trained by health care providers in terms of methods of cooking and consumption of fat. Especially in India, castes and religious beliefs have to be considered since Hindus and Sikhs do not eat beef and Muslims do not eat pork. Also fasting is traditionally performed in various religions of this subcontinent.

Special foods for diabetics are not part of the strategy of national programmes for diabetes care in India. Foods of this kind are not available and the industry is not encouraged to develop and market such products. In general, the situation in India may, in various respects, be regarded as a model for a population-wide effort to improve the nutritional usage and food consumption in diabetics.

2. Japan

The situation in Japan is little different from that in India. The average daily energy intake is low – 2200 kcal/day compared to 3000 kcal in most Western societies (KONIAHI 1983). Fat consumption is also less than half of that of the Western population, with a high ratio of polyunsaturated fat compared to saturated fat. Earlier attempts towards "westernization" have been recognized as a movement in the wrong direction. The Japanese diabetes society has published a list of dietary modifications for diabetic treatment (Japan Diabetes Society 1985). It contains a list of six main types of food, giving the average amount of proteins, lipids and carbohydrates. Well-balanced meals are also discribed individually.

3. China

Also in China diabetic patients were formerly given low-carbohydrate diets, which resulted in drastic changes in the usual diet pattern. In the early 1960s a larger group of patients was first treated with a high-carbohydrate diet and hypoglycaemic drugs, and patients improved dramatically. From that time on all diabetics were advised to stay on their traditional diet.

The characteristics of the Chinese diet are as follows: high levels of complex carbohydrates, mainly semipolished cereals, high dietary fibre, enough protein from both animal and vegetable sources, low fat and low dietary cholesterol. This diet is fully appropriate for non-insulin-dependent diabetics (Du and Wang 1990). The much lower incidence of macrovascular complications among Chinese diabetics is probably a consequence of these dietary habits (Chi 1988).

4. Africa

Nutritional data from African countries are rather heterogeneous both between and within countries. The mean energy supply is 87% of the requirement and more than 70% of the energy is supplied by carbohydrates. Malnutrition is therefore extremely common in many countries of this continent. On the other hand, increasing urbanization in Africa is causing food and health problems resulting from overcrowding, low earnings and poor sanitation. In general the nutritional situation in towns is worse than in rural regions.

Nevertheless, obesity is a growing concern among the privileged and sedentary workers in some cities. Obesity has been reported to be more common among diabetic patients than in non-diabetic urban populations [20]. A major problem is the high alcohol intake. In some rural communities the energy intake through local beer is up to 29% of the total energy intake (Tanner and Lukmanji 1987).

A major problem for adequate dietetic treatment in many African countries is the low educational status with the inability to receive any dietary instructions. In general, the instructions given are as follows: reduced starch, sugar and energy intake to lose weight or prevent weight gain, equal distribution of starch-containing foods throughout the day, high fibre and low fat intake. All these recommendations are especially valid for patients with NIDDM.

A major problem throughout Africa is the absence of adequate dietetic, educational or nursing personal. Dietetic recommendations are therefore usually restricted to more or less general and practical guidelines, written in the patient's native tongue.

B. Exercise

It has long been traditional in diabetology that exercise is recommended as part of the non-pharmacological treatment of the disease. Exercise is often delineated as one of the main cornerstones in the treatment of this disease (SMITH 1983). Shortly after insulin was introduced into the treatment of diabetic patients, it was observed that the amount of insulin needed to control glucose levels was reduced after vigorous exercise. Over a period of several decades, it became dogmatic opinion that diabetics should be instructed to increase their physical activity (JOSLIN 1959). Even today many doctors would not dare to doubt that exercise is an important factor in preenting or treating diabetes. What are the reasons for this strong belief?

I. Insulin-Dependent Diabetes

Shortly after the discovery of insulin it was observed that hypoglycaemia occurred after exercise if the amount of insulin given was kept constant. Exercise was therefore regarded as acting synergistically with insulin. The observation of the ability of exercise to decrease circulating blood glucose levels in insulin-treated diabetic patients gave rise to the recommendation to use physical activity in addition to insulin and diet in the treatment of IDDM. It was believed that exercise could play the part of a third important element besides diet and insulin in normalizing glycaemia and maintaining good metabolic control. According to KEMMER and BERGER (1991), there are three reasons to regard the prescription of physical activity as a means to improve glycaemic control in type 1 diabetic patients as obsolete:

1. Relatively precise concepts on the complexity of fuel fluxes and their regulation during physical activity in normal and diabetic individuals have been developed, and leave no doubt that the use of physical exercise as a therapeutic tool meets almost insurmountable problems in clinical practice.
2. Numerous studies have shown that exercise programmes are not helpful when used to improve metabolic control on a long-term basis.
3. Glycaemic control may very well be optimized and near normoglycaemia be maintained by intensive treatment and teaching programmes per se.

II. Non-Insulin-Dependent Diabetes

1. Short-Term Effects

In NIDDM the hyperglycaemia is mainly due to insulin resistance and it is associated with a variety of cardiovascular risk factors such as hypertension and hyperlipoproteinaemia. These patients are usually obese and their lifestyle is characterized by reduced physical activity. The majority of the patients are older than 60 years, and macrovascular diseases are more important for their prognosis than the microvascular disease characteristic of younger people with diabetes. As for IDDM, the primary therapeutic goal is a normalization of glycaemia. This is achieved by a variety of strategies such as a low-energy diet, oral antidiabetic drugs and insulin. The disease is probably heterogeneous in origin, but in all types the basic defect seems to be a resistance to insulin with a relative deficiency of insulin secretion. Increasing insulin sensitivity is therefore a major aim for any type of treatment. Since physical exercise usually leads to an increase in insulin sensitivity, a training programme must be regarded as a rational approach to the treatment of this type of diabetes.

Only a few studies have examined the effect of acute exercise on glucoregulatory hormones. In NIDDM patients treated with sulphonylurea and/or diet, exercise on a cycle ergometer for 45 min did not suppress circulating insulin levels in spite of a fall of blood glucose by about 50 mg/dl (MINUK et al. 1980). Other authors, on the other hand, found a slight fall in insulin levels during prolonged exercise in hyperinsulinaemic NIDDM patients (KEMMER et al. 1987). The suppression of insulin secretion was much lower than could be expected from normal patients. The failure of exercise to suppress insulin secretion may be a consequence of the prevailing hyperglycaemia or of a defective control of insulin secretion. In patients treated with sulphonylurea, even complications such as exercise-induced hypoglycaemia must be taken into account after vigorous exercise (KOIVISTO and DEFRONZO 1984). On the other hand, no evidence has been provided that any difference exists in hormonal and metabolic response to physical exercise between healthy subjects and NIDDM patients treated with diet alone who have good metabolic control.

In contrast to short-term exercise, training by long-term exercise may have a beneficial effect on glucose tolerance and insulin sensitivity. Studies in obese non-diabetic subjects and in trained athletes have clearly demonstrated an association between physical training and increased insulin sensitivity (CÜPPERS et al. 1982). Also the insulin resistance found in obese subjects with normal glucose tolerance could be reversed by a moderate physical conditioning programme (DEFRONZO et al. 1987). According to various studies, the skeletal muscle is the major site of increased insulin sensitivity after training in rats (HAINES et al. 1985). Studies in man have revealed that at low or high insulin concentrations trained subjects had a markedly increased insulin-stimulated glucose uptake. Their hepatic glucose production, on the other hand, was much lower than in controls. The increased insulin-stimulated glucose utilization after physical training is therefore a consequence of both an increased peripheral and hepatic sensitivity to insulin. The mechanism of this effect has not been fully elucidated. It may be mediated by several factors. Most probably exercise may enhance insulin sensitivity through increased production of glucose transporters (WAKE et al. 1991).

2. Long-Term Effects

In spite of the observations mentioned above and the theoretical anticipation of a beneficial effect of physical training programmes, no unequivocal evidence of the effectiveness of such an approach has been provided so far on the basis of prospective controlled trials. Clinical studies have demonstrated a slight improvement of glucose tolerance after a 6-month training programme (REITMAN et al. 1984; RUDERMAN et al. 1979; SALTIN et al. 1979; SCHNEIDER et al. 1984; TROVATI et al. 1984). This improvement of glucose tolerance is mainly observed in intravenous tolerance tests, not, or to a lower degree, after an oral glucose load. Training programmes were also associated with a decrease in fasting glucose levels and an improvement in haemoglobin A_{1c} (BOGARDUS et al. 1984; KROTKIEWSKI et al. 1985). These studies are suggestive of a beneficial effect of training programmes on overall metabolic control and insulin sensitivity, but they fail to provide evidence for a beneficial effect on oral glucose tolerance and clinical outcome.

The critical evaluation of all these investigations during a conference of the National Institute of Health (National Institutes of Health 1987) gave rise to some sceptical conclusions as to the general beneficial effect of physical training programmes as part of the treatment of NIDDM. In addition, a feasibility study (SKARFORS et al. 1987) over 2 years on physical training as treatment for NIDDM in elderly men has thrown light on a number of problems which had been encountered to a certain extent in previous investigations. These problems include difficulties in separating dietary from exercise effects, heterogeneity of the initial hormonal and metabolic status of the patient, appearance or aggravation of coronary heart disease, deterioration of diabetes and the apparent lack of motivation of patients to participate in physical training over a period lasting for more than 3–6 months.

It needs to be considered that the therapeutic benefit of physical training (if there is one) must be balanced against the cardiovascular risk in elderly patients which is associated with training programmes. Before patients are allocated to training programmes, they have to be screened specifically for cardiovascular risk factors. Most of the elderly NIDDM patients have not been physically active for many years and they are rather vulnerable to all kinds of injury. Also the risk of exercise-induced hypoglycaemia has to be considered in patients treated with sulphonylureas or insulin. Taking all these factors into account, SKARFORS et al. (1987) came to the conclusion that the feasibility and efficacy of physical training as a long-term treatment for the majority of diabetic patients older than 60 years has to be doubted. Only a relatively healthy hyperinsulinaemic group of patients below the age of 60 years may show a positive effect of physical activity upon glucose tolerance.

III. Recommendations

Physical exercise, training and all kinds of sports have become part of the recreational and social activities in modern societies, and many people regard physical activity as an important source of psychosomatic and psychological wellbeing. It is difficult to prove that physical activity really has a positive effect on longevity, but there is no doubt that exercise is important in promoting physical and mental health and should therefore be encouraged from the medical point of view. One of the main goals in treating diabetic patients is to enable them to live as far as possible in the same way as non-diabetics. In this respect, it has to be regarded as progress that with modern diabetes therapy there is no problem about allowing diabetics take part in physical exercise. Physicians should encourage their diabetic patients to exercise and participate in training programmes in the same way as they encourage all their other patients. There is no need to restrict physical activity in diabetics, but, on the other hand, there is no need to argue in favour of physical exercise only by virtue of the fact that a patient is diabetic.

References

American Diabetes Association (1987) Nutritional recommendations and principles for individuals with diabetes mellitus. Diabetes Care 10:126–132

Aro A, Uusitupa M, Voutilainen E (1981) Improved diabetic control and hypocholesterolaemic effect induced by long term supplementation with guar in type 2 diabetics. Diabetologia 21:29–33

Bantle JP, Swanson JE, Thomas W, Laine DC (1992) Metabolic effects of dietary fructose in diabetic subjects. Diabetes Care 15:1468–1476

Bogardus C, Ravussin E, Robbins DC, Wolfe RR, Horton ED, Sims EAH (1984) Effects of physical training and diet therapy on carbohydrate metabolism in patients with glucose intolerance and non-insulin-dependent diabetes mellitus. Diabetes 33:311–318

Chi ZS (1988) Study on the cardiovascular complications of diabetes mellitus in Beijing and Tianjin. In: Mimura G, et al. (eds) Diabetes mellitus in East Asia. Experta Medica, Amsterdam, pp 23–28

Cüppers HJ, Erdmann D, Schubert H, Berchtold P, Berger M (1982) Glucose tolerance, serum insulin and serum lipids in athletes. In: Berger M, Christacopoulos P, Wahren J (eds) Diabetes and exercise. Huber, Bern, pp 155–165

DeFronzo RA, Sherwin RS, Kraemer N (1987) Effect of physical training on insulin action in obesity. Diabetes 36:1379–1385

Diabetes and Nutrition Study Group of the European Association for the Study of Diabetes (1988) Nutritional recommendations for individuals with diabetes mellitus. Diabet Nutr Metab 1:145–149

Dodson PM, Pacy PJ (1984) A controlled trial of high fibre, low fat and low sodium diet for mild hypertension in type 2 (non-insulin-dependent) diabetic patients. Diabetologia 27:522

Du SF, Wang H (1990) Plasma glucose and insulin responses to mixed meals. In: Mimura G, et al. (eds) Recent trends of diabetes in East Asia. Experta Medica, Amsterdam, pp 167–168

Durnin JVGA (1987) Microdiets: love them or leave them? Br Med J 294:1565

Evanoff GV, Thompson CS, Brown J, Weinman EJ (1987) The effect of dietary protein restriction on the progression of diabetic nephropathy: a 12-month follow up. Arch Intern Med 147:492

Fitz JD, Sperling EM, Fein HG (1983) A hypocaloric high-protein diet as primary treatment for adults with obesity related diabetes. Diabetes Care 6:328

Franz MJ (1993) Avoiding sugar: does research support traditional beliefs? Diabetes Educ 19:144–150

Gannon MC, Nuttall FQ (1987) Factors effecting interpretation of postprandial glucose and insulin areas. Diabetes Care 10:754–763

Garg A, Bonanome A, Grundy SM (1987) Comparison of high monounsaturated fat and high carbohydrate diets for NIDDM. Diabetes 36 [Suupl I]:11A

Geekie MA, Porteous J, Hockaday TDR, Mann JI (1986) Acceptability of high-fibre diets in diabetic patients. Diabet Med 3:52–55

Grundy SM (1987) Monounsaturated fatty acids, plasma cholesterol and coronary heart disease. Am J Clin Nutr 45:1168

Haines AP, Sanders TA, Imeson JD, et al. (1985) Effects of fish oil supplement on platelet function, haemostatic variables, and albuminuria in insulin dependent diabetics. Thromb Res 43:643

Hansen BC (1988) Dietary care for obese diabetic subjects. Diabetes Care 11(2):183–186

James DE, Kraegen EW, Chisholm DJ (1985) Effects of exercise training on in vivo insulin action in individual tissues of the rat. J Clin Invest 76:657–666

Japan Diabetes Society (1985) Food exchange lists for diabetic treatment, 4th edn. Kyowakikaku-Tsushin Bunkodo, Tokyo

Jenkins DJA, Leeds AR, Gassull MA, Cochet B, Alberti KGMM (1977) Decrease in postprandial insulin and glucose concentrations for guar and pectin. Ann Intern Med 86:20–23

Jenkins DJA, Wolever TMS, Jenkins AL (1988) Starchy foods and glycemic index. Diabetes Care 11:149–159

Joslin EP (1959) The treatment of diabetes mellitus. In: Joslin EP, Root HF, White P, Marble A (eds) Treatment of diabetes mellitus, 10th edn. Lea and Febiger, Philadelphia, pp 243–300

Kemmer FW, Berger M: Exercise (1992) In: Alberti K-G, DeFronzo RA, Keen H, Zimmet P (eds) International textbook of diabetes mellitus. Wiley, Chichester

Kemmer FW, Tacken M, Berger M (1987) On the mechanism of exercise induced hypoglycemia during sulfonylurea treatment. Diabetes 36:1178–1187

Kiehm TG, Anderson JW, Ward K (1976) Beneficial effects of a high carbohydrate, high fibre diet on hyperglycemic men. Am J Clin Nutr 29: 895–899

Koivisto V, DeFronzo RA (1984) Exercise in the treatment of type II diabetes. Acta Endocrinol 107 [Suppl 1]:107–111

Konishi K (1983) Japanese cooking for health and fitnes. Gakken, Tokyo

Kromhout D, Bosschieter EB, Coulander C, de L (1985) The inverse relation between fish oil consumption and 20-year mortality from coronary heart disease. N Engl J Med 312:1205–1209

Krotkiewski M, Lönnroth P, Mandroukas K, et al. (1985) The effects of physical training on insulin secretion and effectiveness and on glucose metabolism in obesity and type 2 (non-insulin-dependent) diabetes. Diabetes 28:881–890

Lewis-Barned NJ, Mann JI, Carter RC, Hockaday TDR (1987) Modified fat diet can achieve cholesterol lowering in diabetic patients over 10 years. Diabetologia 30:549A

Lousley SE, Jones DB, Slaughter P, Carter RD, Jelfs R, Mann JI (1984) High Carbohydrate/high fibre diets in poorly controlled diabetes. Diabet Med 1:21–25

Mann JI (1989) Diet and diabetes. Diabetologia 18:89

McCulloch DK, Mitchell RD, Ambler J, Tattersall RB (1985) A prospective comparison of "conventional" and high carbohydrate/high fibre/low fat diets in adults with established type 1 (insulin-dependent) diabetes. Diabetologia 28:208

McLarty DG, Pollitt C, Swai ABM (1990) Diabetes in Africa. Diabet Med 7:670–684

Minuk HL, Hanna AK, Marliss EB, Vranic M, Zinman B (1980) Metabolic response to moderate exercise in obese man during prolonged fasting. Am J Physiol 238:E322–E329

Mogensen CE, Christensen CK (1984) Predicting diabetic nephropathy in insulin dependent patients. New Engl J Med 311:89–93

National Institutes of Health (1987) Consensus development conference on diet and exercise in non-insulin-dependent diabetes mellitus. Diabetes Care 10:639–644

Nuttal FO (1993) Dietary fiber in the management of diabetes. Diabetes 42:503–508

Pacy PJ, Dodson PM, Fletcher RF (1986) Effect of high carbohydrate, low sodium and low fat diet in type 2 diabetics with moderate hypertension. Int J Obes 10:43

Peterson DB, Mann JI (1985) Guar: pharmalogical fibre or food fibre? Diabet Med 2:345–347

Peterson DB, Lambert J, Gerring S, et al. (1986) Sucrose in the diabetic diet – just another carbohydrate? Diabet Med 2:345

Raheja BS (1987) Diabetes-associated domplications and traditional Indian diet. Diabetes Care 10:382–383

Raheja BS (1988) Indian diet – diabetes and its complications. Bull Int Diab Fed XXXIII:14–17

Reitman JS, Vasquez B, Klimes I, Nagulesparan M (1984) Improvement of glucose homeostasis after exercise training in non-insulin-dependent diabetes. Diabetes Care 7:434–441

Riccardi G, Rivellese A, Pacioni D, Genovese D, Mastranzo P, Mancini M (1984) Separate influence of dietary carbohydrate and fibre on the metabolic control in diabetes. Diabetologia 26:116–121

Rolls BJ (1991) Effects of intense sweeteners on hunger, food intake, and body weight: a review. Am J Clin Nutr 53:872–878

Ruderman NB, Ganda OP, Johansen K (1979) The effects of physical training on glucose tolerance and plasma lipids in maturity onset diabetes. Diabetes 28 [Suppl 1]:89–92

Saltin B, Lindgarde F, Housten M, Horlin R, Nygaard E, Gad P (1979) Physical training and glucose tolerance in middle-aged men with chemical diabetes. Diabetes 28 [Suppl 1]:30–32

Schneider SH, Amorosa LF, Khachadurian AK, Ruderman NB (1984) Studies on the mechanism of improved glucose control during regular exercise in type-2 (non-insulin-dependent) diabetes. Diabetologia 26:355–360

Simpson HCR, Carter RD, Lousley S, Mann JI (1982) Digestible carbohydrate – an independent effect on diabetic control in type 2 Diabetologia 23:235–239

Skarfors ET, Wegener TA, Lithell H, Selinus I (1987) Physical training as treatment for type 2 (non-insulin-dependent) diabetes in elderly men. A feasibility study over 2 years. Diabetologia 30:930–933

Smith T (1983) Exercise: cult or cure-all? Br Med J 286:1637–1639

Sours HE, Fratelli VP, Brand CD (1988) Sudden death associated with very low calorie weight reduction regimes. Am J Clin Nutr 34:453

Tanner M, Lukmanji Z (1987) Food consumption patterns in a rural Tanzanian community (Kikwawilla Village, Kilombero District, Morogoro region) during lean and post harvest season. Acta Trop 44:229–244

Thorburn AW, Brand JC, Truswell AS (1986) The glycaemic index of foods. Med J Aust 144:580–582

Trovati M, Carta Q, Cavalot F, et al. (1984) Influence of physical training on blood glucose control, glucose tolerance, insulin secretion and insulin action in non-insulin-dependent diabetic patients. Diabetes Care 7:416–420

Viberti GC, Keen H (1984) The patterns of proteinuria in diabetes mellitus: relevance to pathogenesis and prevention of diabetic nephropathy. Diabetes 33:686–692

Vinik AI (1988) Report of the American Diabetes Association's Task Force on Nutrition: introduction. Diabetes Care 11:127–128

Vinik AI, Jenkins DJA (1988) Dietary fibre in the management of diabetes. Diabetes Care 11:160–173

Vranic M, Berger M (1979) Exercise and diabetes mellitus. Diabetes 28:147–167

Wake SA, Sowden JA, Storlien LH, James DE, Clark PW, Shine J, Chisholm DJ, Kraegen EW (1991) Effects of exercise training and dietary manipulation on insulin-regulatable glucose-transporter mRNA in rat muscle. Diabetes 40:275–279

Walker JD, Dodds RA, Murrells TJ, et al. (1989) Restriction of dietary protein and the progression of renal failure in diabetic patients. Lancet II:1411–1415

Warshaw HS (1990) Alternative sweeteners: past, present, pending and potential. Diabetes Spectrum 3:335–343

Warshaw HS (1993) Today's dietetic foods: definitions, ingredients, recommendations for use, and education issues. Clin Diabetes 11:136–142

Section I
Sulfonylureas

CHAPTER 4

Sulfonylureas and Related Compounds:
Chemistry and Structure-Activity Relationships

H. PLÜMPE

A. Introduction

Over the past 40 years, thousands of sulfonamide derivatives have been synthesized and biologically tested for their hypoglycemic activity. A comprehensive review of the molecular variations and optimization attempts described in the literature up to 1967 (AUMÜLLER and HEERDT 1971) highlighted the fact that active substances suitable for human use can be classified as belonging to the compound groups sulfonylureas and sulfonylsemicarbazides with the general formula shown in Fig. 1 and sulfonylaminopyrimidines (-CO-NH-R^2 replaced by a substituted pyrimidine ring).

In principle, nothing has changed to this day, even though numerous other derivatives have since been synthesized and tested and some of these have been marketed. Which elements of chemical structure have proved essential can be clearly identified from the total of 15 marketed active substances from the group of insulin secretion-stimulating sulfonamides (most of which are still in use today) and from a few nonmarketed but interesting related compounds.

The optimization criteria, as defined at an early stage (HAACK 1958; RUSCHIG et al. 1958), are as follows:

1. Efficacy at the lowest possible dose
2. Tolerability, especially in respect of long-term use
3. Pharmacokinetics (duration of action, metabolism)

If the average daily dose in man is taken as the assessment criterion, the formulae presented in Table 1 can readily be used to trace how optimization attempts based on variation of chemical structure have progressed.

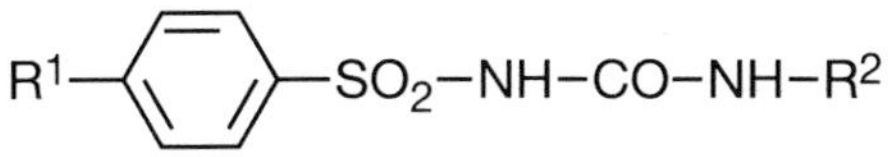

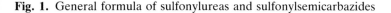

$$R^1-\langle\ \rangle-SO_2-NH-CO-NH-R^2$$

Fig. 1. General formula of sulfonylureas and sulfonylsemicarbazides

B. Sulfonylureas (Table 1a)

During clinical testing of N-(4-aminobenzolsulfonyl)-N'-butyl-urea for antibacterial activity, severe hypoglycemic reactions were observed in some of the patients (Franke and Fuchs 1955). During further structure-activity studies in the laboratories of C.F. Boehringer Mannheim, this compound proved to be the best out of a number of similar blood sugar lowering derivatives (Haack 1958). So it was launched as an oral antidiabetic (INN: carbutamide) in early 1956.

At around the same time a Hoechst study group found that, out of a large number of synthesized non-antibacterial benzolsulfonamide derivatives, N-(4-methyl-benzolsulfonyl)-N'-butyl-sulfonylurea (INN: tolbutamide) had very favorable hypoglycemic properties (Ruschig et al. 1958). This compound was also launched in 1956.

The research work by C.F. Boehringer Mannheim and Hoechst had shown that good efficacy requires a methyl, amino or alkoxy group or (with reservations about tolerability) a chlorine atom as the substituent R^1, whereas R^2 should be a straight, branched or cyclic hydrocarbon residue with three to seven carbon atoms. Compounds with R^1 in position 4 on he benzolsulfonyl residue are uniformly more effective than those with R^1 in position 2 or 3. These findings have subsequently been confirmed and extended (Schröder et al. 1982).

I. Variation of R^1

Whereas 500–2000 mg is required as the daily dose with carbutamide and tolbutamide and 500–1500 mg with tolcyclamide (Ruschig et al. 1958), chlorpropamide (Ruschig et al. 1958) is roughly four times more effective. In comparison with tolcyclamide, acetohexamide (Marshall et al. 1963) is nearly twice as effective owing to the substitution of the methyl group by a methylcarbonyl residue.

A major advance towards lower dosages was made by extending the R^1 residue in the form of an acylaminoalkylene group, in which -CH_2-CH_2- proved to be the optimal alkylene chain. The first and, to date, the best-known member of this "second generation of oral sulfonamide antidiabetics" is glibenclamide (Aumüller et al. 1966); in man it is around 100–200 times more potent than tolbutamide.

The 2-methoxy-5-chlorbenzyol residue in glibenclamide can be replaced by other suitable acyl groups while maintaining the high level of activity, as seen from the chemical structure of glisentide (Morell 1974), glipizide (Ambrogi et al. 1971) and glisolamide (Plümpe et al. 1974). Replacement of the acylamino group by 7-methoxy-2-tetrahydro-isoquinolyl-1,3-dione in gliquidone (Kopitar and Koss 1975) leads to a 20–30 times greater potency than that of tolbutamide.

Table 1. Marketed active substances: chemical structures and INNs of sulfonylureas, sulfonylsemicarbazides and sulfonylamino-pyrimidines

Chemical structure	INN	Daily dosage (mg)	Ref. (dosage)
a) Sulfonylureas			
H_2N—〇—SO_2—NH—CO—NH—C_4H_9	Carbutamide	500–2000	HITZEL, WEYER (1987)
H_3C—〇—SO_2—NH—CO—NH—C_4H_9	Tolbutamide	500–2000	HITZEL, WEYER (1987)
Cl—〇—SO_2—NH—CO—NH—C_3H_7	Chlorpropamide	125–500	HITZEL, WEYER (1987)
H_3C—〇—SO_2—NH—CO—NH—〇	Tolcyclamide	500–1500	ARZNEIBÜRO (1993)
H_3C—CO—〇—SO_2—NH—CO—NH—〇	Acetohexamide	250–1500	HITZEL, WEYER (1987)
H_3C—〇—SO_2—NH—CO—NH— [bicyclic structure]	Glibornuride	12.5–75	HITZEL, WEYER (1987)
Cl—[ring]—CO—NH—CH_2—CH_2—〇—SO_2—NH—CO—NH—〇 O—CH_3	Glibenclamide	2.5–15	HITZEL, WEYER (1987)
H_3C—[ring]—CO—NH—CH_2—CH_2—〇—SO_2—NH—CO—NH—〇 O—CH_3	Glisentide	2.5–20	REYNOLDS (1993)
H_3C—[pyrazine]—CO—NH—CH_2—CH_2—〇—SO_2—NH—CO—NH—〇	Glipizide	2.5–15	HITZEL, WEYER (1987)
H_3C—[isoxazole]—CO—NH—CH_2—CH_2—〇—SO_2—NH—CO—NH—〇	Glisolamide	5–20	REYNOLDS (1993)
H_3C—O—[ring]—N—CH_2—CH_2—〇—SO_2—NH—CO—NH—〇	Gliquidone	15–120	HITZEL, WEYER (1987)

Table 1. *Continued*

Chemical structure	INN	Daily dosage (mg)	Ref. (dosage)
b) Sulfonylsemicarbazides			
	Tolazamide	100–250	Arzneibüro (1993)
	Gliclazide	80–250	Hitzel, Weyer (1987)
	Glisoxepide	2–16	Hitzel, Weyer (1987)
c) Sulfonylaminopyrimidines			
	Glymidine	500–2000	Hitzel, Weyer (1987)

II. Variation of R^2

Replacement of the butyl group in tolbutamide or the cyclohexyl group in tolcyclamide by a bulky substituent such as 2-hydroxy-3-bornyl can lead to a considerable increase in activity by a factor of 30–40 in the case of glibornuride (LORCH et al. 1972).

C. Sulfonylsemicarbazides (Table 1b)

Replacing the butyl group in tolbutamide by a 1-hexamethylenimino group produces the sulfonylsemicarbazide derivative tolazamide (WRIGHT and WILLETTE 1962), with a five- to eightfold increase in efficacy. Gliclazide (DUHAULT et al. 1972), in which a 3-azabicyclooctyl residue is introduced instead of the 1-hexamethylenimino group, is within roughly the same dose range. Replacement of the cyclohexyl residue in glisolamide by the 1-hexamethylenimino group leads to the comparably potent glisoxepide (PLÜMPE et al. 1974).

D. Sulfonylaminopyrimidines (Table 1c)

Replacing the carbamide group -CO-NH-R^2 in the general formula (Fig. 1) by pyrimidine produces sulfonylaminopyrimidines, which lower blood sugar by the same mechanism as corresponding sulfonylureas and sulfonylsemicarbazides. Glymidine (GUTSCHE et al. 1964) as a sodium salt was the only member of this class of compounds on the market although derivatives have been described that match the efficacy of glibenclamide.

E. Various Non-marketed Compounds of Interest (Table 2)

Clinical studies were reported (LADIK et al. 1988) with glimepiride (WEYER and HITZEL 1988), a compound structurally related to glibenclamide, but about five times more active in humans. As further clinical investigations showed favorable results, the product is now being prepared for worldwide registration.

The sulfonylaminopyrimidines glidanile (GUTSCHE et al. 1974) and gliflumide (RUFER et al. 1974) also display certain structural similarities to glibenclamide. In these compounds the "extended" R^1 substituent comprises an acetic acid amide residue instead of the aroylaminoethylene group in glibenclamide. Both compounds are effective in humans at low dosages of 10–15 mg and 5 mg, respectively. Gliflumide, as the S-enantiomer, is experimentally far more effective than the corresponding R-enantiomer (SCHRÖDER et al. 1982).

Gliamilide is an interesting structural variant (SARGES et al. 1976). Whereas the R^1 substituent is present in a similar form in second-generation

Table 2. Various nonmarketed compounds of interest

Chemical structure	INN	Daily dosage (mg)	Ref. (dosage)
	Glimepiride	2	LADIK et al. (1988)
	Glidanile	10–15	SCHROEDER et al. (1982)
	Gliflumide	5	SCHROEDER et al. (1982)
	Gliamilide	3 × 50–3 × 75	HITZEL, WEYER (1987)

sulfonylureas and the R^2 residue is related to the bulky group in glibornuride, a 4-piperidinyl-1-sulfonyl group is found in the middle section of the gliamilide molecule instead of the 4-benzenesulfonyl group. Strictly speaking, therefore, gliamilide is not a sulfonylurea in the usual sense but it does have the same insulin-stimulating properties together with substantial, prolonged activity in humans at a daily dosage of 3×50–3×75 mg (RYAN et al. 1975).

F. Conclusions

Within the classes of hypoglycemic sulfonylureas, sulfonylsemicarbazides and sulfonylaminopyrimidines, 15 new chemical entities emerged as commercial products from many thousands of compounds synthesized and tested. After relatively simple substituents in the "first-generation" sulfonylureas and sulfonylsemicarbazides had led to effective compounds, introduction of an acylaminoalkylene group R^1 brought about a major advance to even more potent "second-generation" compounds.

Variations of R^2 effect additional minor improvements. Both substituents R^1 and R^2 contribute to the rate of activity by binding to appropriate receptor sites as can be shown from experimental studies with simple sulfonylurea derivatives such as tolbutamide as well as from studies with meglitinide, a partial structure of glibenclamide, in which the sulfonylurea moiety is replaced by a carboxyl group (see Chap. 6, this volume). With sulfonylaminopyrimidines and some other nonmarketed structural variances of sulfonylureas, similar structure-activity relationships can be shown.

References

Ambrogi V, Bloch K, Daturi S, Griggi P, Logemann W, Parenti MA, Rabini T, Tommasini R (1971) New antidiabetic drugs. Part I. Arzneimittelforsch 21:200–204

Arzneibüro der Bundesvereinigung Deutscher Apothekerverbände (1993) Pharmazeutische Stoffliste – list of pharmaceutical substances. Werbe- und Vertriebsgesellschaft Deutscher Apotheker mbH, Frankfurt

Aumüller W, Heerdt R (1971) Sulfonylharnstoffderivate und verwandte Verbindungen als blutzuckersenkende Substanzen. In: Maske H (ed) Handbuch der experimentellen Pharmakologie, vol XXIX. Springer, Berlin Heidelberg New York, pp 1–249

Aumüller W, Bänder A, Heerdt R, Muth K, Pfaff W, Schmidt FH, Weber H, Weyer R (1966) Ein neues hochwirksames Antidiabetikum. Arzneimittelforsch 16:1640–1641

Duhault J, Boulanger M, Tisserand F, Beregi L (1972) The pharmacology of S 1702, a new highly effective oral antidiabetic drug with unusual properties. Part I: pharmacological and hypoglycaemic activity, studies in different animal species. Arzneimittelforsch 22:1682–1685

Franke H, Fuchs J (1955) Ein neues antidiabetisches Prinzip. Ergebnisse klinischer Untersuchungen. Dtsch Med Wochenschr 80:1449–1452

Gutsche K, Harwart A, Horstmann H, Priewe H, Raspè G, Schraufstätter E, Wirtz S, Wörffel U (1964) Sulfonamidopyrimidine, eine neue Gruppe blutzuckersenkender Verbindungen. Arzneimittelforsch 14:373–376

Gutsche K, Schröder E, Rufer C, Loge O (1974) Neue blutzuckersenkende Benzol-sulfonamidopyrimidine: N-substituierte 4-[N-(2-Pyrimidinyl)-sulfamoyl]-phenyl-essigsäureamide. Arzneimittelforsch 24:1028–1039

Haack E (1958) Sulfanilyl- und Sulfonyl-carbaminsäure-Derivate und ihre blutzuckersenkende Wirkung. Arzneimittelforsch 8:444–448

Hitzel V, Weyer R (1987) Orale Antidiabetika. In: Kleemann A, Lindner E, Engel J (eds) Arzneimittel: Fortschritte 1972–1985. VCH, Weinheim, pp 818–821

Kopitar Z, Koss FW (1975) Pharmakokinetisches Verhalten von Gliquidone (AR-DF26), einem neuen Sulfonylharnstoff. Arzneimittelforsch 25:1933–1938

Ladik T, Lotz N, Rupp P, Mehnert H (1988) Efficiency of the new sulfonylurea glimepiride in the treatment of type 2 diabetes. Diabetes Res Clin Pract 5 [Suppl 1]: abst. POS-003-257

Lorch E, Gey KF, Sommer P (1972) Glibornurid, ein neues hochwirksames Anti-diabetikum. Arzneimittelforsch 22:2154–2163

Marshall FJ, Sigal MV, Sullivan HR, Cesnik C, Root MA (1963) Further studies on N-arylsulfonyl-N'-alkylureas. J Med Chem 6:60–63

Morell J (1974) Glipentide: a new hypoglycaemic agent. Biochem Pharmacol 23: 2922–2924

Plümpe H, Horstmann H, Puls W (1974) Isoxazolcarboxamidoalkylbenzolsulfonyl-harnstoffe, -semicarbazide und -aminopyrimidine sowie damit verwandte Ver-bindungen und ihre blutzuckersenkende Wirkung. Arzneimittelforsch 24:363–374

Reynolds JEF (ed) (1993) Martindale. The extra pharmacopoeia, 30th edn. Phar-maceutical Press, London

Rufer C, Biere H, Ahrens H, Schröder E (1974) Blood glucose lowering sulfona-mides with asymmetric carbon atoms 1. J Med Chem 17:708–715

Ruschig H, Korger G, Aumüller W, Wagner H, Weyer R, Bänder A, Scholz J (1958) Neue peroral wirksame blutzuckersenkende Substanzen. Arzneimittel-forsch 8:448–454

Ryan JR, McMahon FG, Jain AK (1975) Early clinical evaluation of the sulfamylurea Gliamilide. Clin Pharmacol Ther 17:243

Sarges R, Kuhla DE, Wiedermann HE, Mayhew DA (1976) Sulfamylurea hypo-glycemic agents: 6. high-potency derivatives. J Med Chem 19:695–709

Schröder E, Rufer C, Schmiechen R (1982) Pharmazeutische Chemie. Thieme, Stuttgart, pp 525–530

Weyer R, Hitzel V (1988) Acylureidoalkylphenylsulfonylureas with blood glucose lowering efficacy. Arzneimittelforsch 38:1079–1080

Wright JB, Willette RE (1962) Antidiabetic agents. N^4-arylsulfonylsemicarbazides. J Med Pharm Chem 5:815

CHAPTER 5

Sulfonylureas: Physico-chemical Properties, Analytical Methods of Determination and Bioavailability

S.L. ALI, H.H. BLUME, AND B.S. SCHUG

A. Introduction

Oral antidiabetic drugs are widely used in the treatment of diabetes mellitus and are effective in controlling blood glucose levels. These drugs are of several distinct types with different modes of action and include sulfonylureas, biguanides and α-glucosidase inhibitors. The sulfonylurea antidiabetic drugs include acetohexamide, acetylcarbutamide, carboxytolbutamide, carbutamide, chlorpropamide, glibenclamide, glibornuride, gliclazide, glimepiride, glipizide, gliquidone, glisoxepide, metahexamide, tolazamide and tolbutamide. The structural formulas of some of the important antidiabetic agents are given in Fig. 1.

B. Physical Properties

I. Description

Carbutamide is a white or almost white fine crystalline powder without smell. Chlorpropamide is a white, crystalline, odorless, tasteless powder. Glibenclamide is a white crystalline odorless powder. Gliclazide is a crystalline solid, glipizide is a white powder and gliquidone is a white or slightly yellow crystalline substance. Glimepiride is a white odorless crytalline powder (HOECHST 1994, personal communication). Glisoxepide is a white or slightly yellow crystalline odorless powder (BAYER 1994, personal communication). Tolazamide is a crytalline powder and tolbutamide is an almost odorless, white, crystalline powder.

II. Melting Point

Table 1 shows the melting points of a number of sulfonylureas.

III. Solubility

The solubility of the sulfonylurea compounds is influenced by the lipophilic character of the phenyl group and its substituents and by the hydrophilic properties of the SO_2-NH-CO-N group. The aliphatic substituents of the

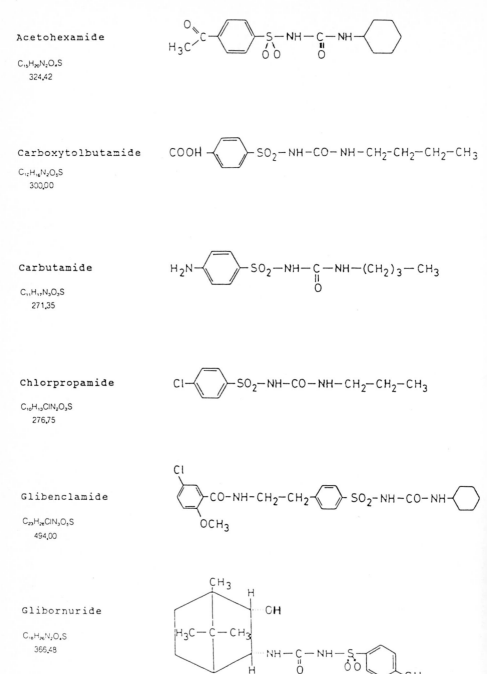

Acetohexamide

$C_{15}H_{20}N_2O_4S$
324,42

Carboxytolbutamide

$C_{12}H_{16}N_2O_5S$
300,00

Carbutamide

$C_{11}H_{17}N_3O_3S$
271,35

Chlorpropamide

$C_{10}H_{13}ClN_2O_3S$
276,75

Glibenclamide

$C_{23}H_{28}ClN_3O_5S$
494,00

Glibornuride

$C_{18}H_{26}N_2O_4S$
366,48

Fig. 1. Structural formulas of important antidiabetic agents

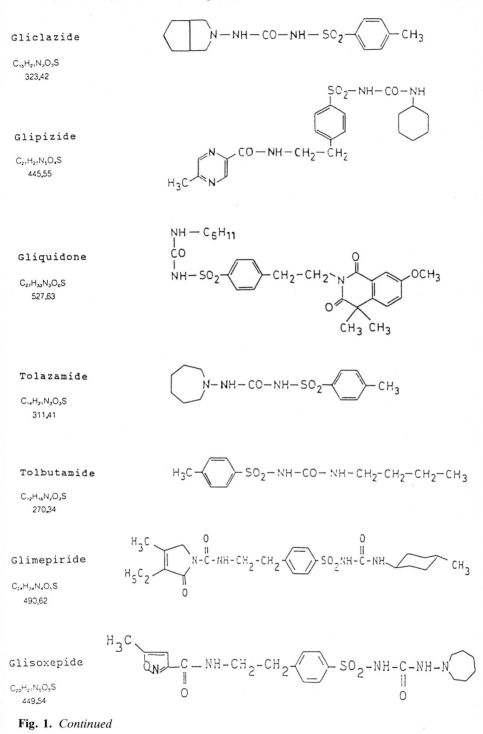

Gliclazide

C₁₅H₂₁N₃O₃S
323,42

Glipizide

C₂₁H₂₇N₅O₄S
445,55

Gliquidone

C₂₇H₃₃N₃O₆S
527,63

Tolazamide

C₁₄H₂₁N₃O₃S
311,41

Tolbutamide

C₁₂H₁₈N₂O₃S
270,34

Glimepiride

C₂₄H₃₄N₄O₅S
490,62

Glisoxepide

C₂₀H₂₇N₅O₅S
449,54

Fig. 1. *Continued*

Table 1. Melting points of a number of sulfonylureas

Substance	Melting point (°C)	Reference
Acetohexamide	183–185	McMahon et al. (1965)
	188–190	Merck Index (1983, p. 9)
Acetylcarbutamide	187–189	Haller and Strauzenberg (1959)
Carboxytolbutamide	211	Mesnard (1960)
Carbutamide	139–142	Deutscher Areneimittel-Codex (1986, p. C-085)
Chlorpropamide	About 128	Pharmaceutical Codex XI (1979, p. 186)
	127–129	Merck Index (1983, p. 309)
Glibenclamide	169–174	Deutsches Arzneibuch (1991)
	About 173	Pharmaceutical Codex XI (1979, p. 391)
Glibornuride	186–189	Clark (1986, p. 639)
Gliclazide	About 181	Clark (1986, p. 640)
Glimepiride	About 207	Hoechst (1994, personal communication)
Glipizide	About 205	Clark (1986, p. 640)
Gliquidone	About 178	Clark (1986, p. 641)
Glisoxepide	192–198	Bayer (1994, personal communication)
Metahexamide	151–152	Mesnard (1960)
Phenbutamide	130–132	Merck Index (1983, p. 1038)
Tolazamide	161–169 with decomposition	Clark (1986, p. 1029)
Tolbutamide	126–130	USP (1990, p. 1386)
	About 129	Pharmaceutical Codex XI (1979, p. 950)

phenyl rest enhance the lipophilic character (Häussler and Pechtold 1969). On the other hand, the NH_2 or COOH groups increase the solubility of the compounds in water. In aqueous solutions the hydrogen atom of the SO_2-NH group reacts with weak bases such as ammonia to form water-soluble salts. The formation of salts is also favored by the COOH and NH_2 substituents of the phenyl moiety. The solubility of the sulfonylureas is very good in strong acidic medium, decreases with increasing pH (poorly soluble in weak acidic medium) and increases again in the alkaline milieu of higher pH values.

1. Acetohexamide

Acetohexamide is practically insoluble in water and ether, slightly soluble in alcohol and chloroform, and soluble in pyridine and dilute solutions of alkali hydroxides (Clark 1986).

2. Acetylcarbutamide

Acetylcarbutamide is almost insoluble in water. Its solubility at 37°C in water varies from 44 mg in 100 g at pH 5 to 20000 mg in 100 g at pH 8 (Achelis and Hardebeck 1965).

3. Carbutamide

Carbutamide is almost insoluble in water, slightly soluble in dilute alkalis and acids under salt formation and almost insoluble in ether, chloroform and methylene chloride. Its solubility at 37°C in water varies from 70 mg in 100 g at pH 5 to 9570 mg in 100 g at pH 8 (Deutscher Arzneimittel-Codex 1986; The Merck Index 1983).

4. Chlorpropamide

Chlorpropamide is practically insoluble in water, dissolves in aqueous solutions of alkali hydroxides, and is soluble at 20°C in 12 parts of alcohol, 200 parts of ether, 9 parts of chloroform and 5 parts of acetone (British Pharmacopeia 1988, vol. I , p. 132; The Pharmaceutical Codex XI, 1979, p. 188). Its solubility at 25°C in water varies between 0.1 and 0.2 mg/ml at pH 2.2 to over 5 mg/ml at pH 6.5 (TOOLAN and WAGNER 1959).

5. Glibenclamide

Glibenclamide is very slightly soluble in water and in ether, and soluble at 20°C in 330 parts of alcohol, 36 parts of chloroform and 250 parts of methanol. It is soluble in dilute alkaline hydroxide solutions (British Pharmacopeia 1988, vol. I, p. 269; Pharmaceutical Codex XI, 1979, p. 391)

6. Glimepiride

Glimepiride is practically insoluble in water, soluble in dimethyl formamide, sparingly soluble in dichlormethane and slightly soluble in methanol, acetonitrile, acetone and diluted alkalis (HOECHST 1994, personal communication).

7. Glipizide

Glipizide is practically insoluble in water and ethanol, sparingly soluble in acetone, and soluble in chloroform and in a dilute solution of alkali hydroxides (CLARK 1986, p. 640).

8. Gliquidone

Gliquidone is practically insoluble in water, slightly soluble in ethanol and methanol, and soluble in acetone and chloroform (CLARK 1986, p. 641).

9. Glisoxepide

Glisoxepide is practically insoluble in water and petroleum benzene, almost insoluble in methanol, ethanol, ethylacetate, and sparingly soluble in acetone. It is soluble in dilute ammonia under salt formation (BAYER 1994, personal communication).

10. Metahexamide

Metahexamide is soluble in water between 0.20 mg/ml at pH 2.0 and 5.0 mg/ml at pH 7.0 (MORGENSTEIN and GARRETT 1959).

11. Tolazamide

Tolazamide is very slightly soluble in water, slightly soluble in ethanol, soluble in acetone and freely soluble in chloroform (CLARK 1986, p. 1029).

12. Tolbutamide

Tolbutamide is practically insoluble in water, soluble at 20°C in 10 parts of alcohol and 3 parts of acetone and chloroform and slightly soluble in ether. It is soluble in sodium hydroxide solution and in dilute mineral acids (The Pharmaceutical Codex XI, 1979, p. 950). The pH dependence of the solubility of tolbutamide and carboxytolbutamide at 37.5°C is illustrated in Fig. 2 (from FORIST and CHULSKI 1958).

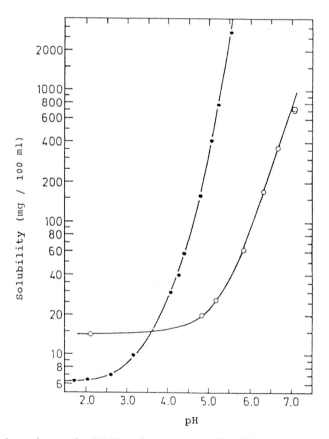

Fig. 2. pH dependence of solubility of tolbutamide (O—O) and carboxytolbutamide (●—●). (FORIST and CHULSKI 1958)

IV. Dissociation Constants

The dissociation constants for different hypoglycemic agents are presented in Table 2.

V. Crystal Shape and Structure

1. Acetohexamide

The X-ray diffraction pattern of acetohexamide conforms to the pattern given in the National Formulary XIII (1970). No polymorphs have been observed and documented.

Table 2. Dissociation constants of sulfonylureas

Substance	pK_a at 20°C	Reference
Carboxytolbutamide	3.54 (37.5°C)	Forist and Chulski (1958)
Carbutamide	6.0	Hager, vol 7 (1993); Martindale 29 (1989, p. 387)
Chlorpropamide	5.0	Martindale 29 (1989, p. 387)
Glibenclamide	5.3	Merck Index (1983, p. 642)
Gliclazide	5.8	Clark (1986, p. 640)
Glimepiride	6.2	Hoechst (1994, personal communication)
Glisoxepide	6.3	Bayer (1994, personal communication)
Metahexamide	5.2	Morgenstern and Garrett (1959)
Tolazamide	3.5, 5.7	Clark (1986, p. 1029)
Tolbutamide	5.3 (25°C)	Pharmaceutical Codex XI (1979, p. 950)

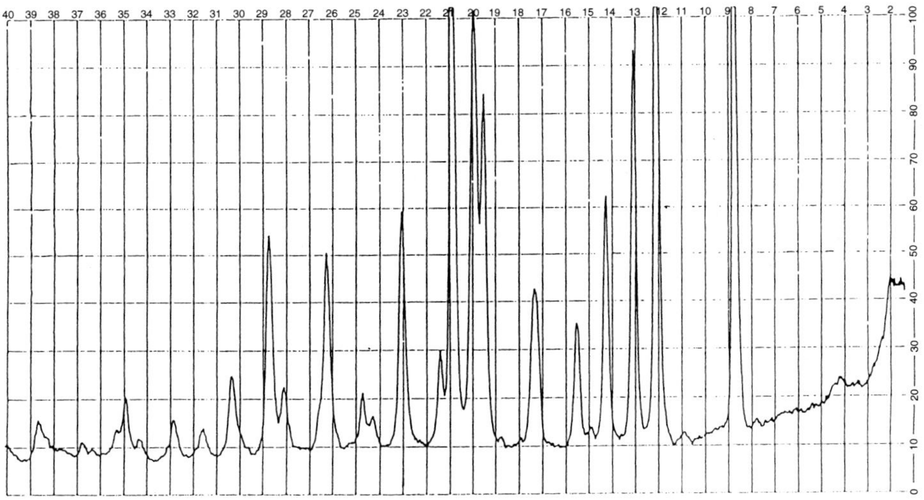

Fig. 3. X-ray diffraction pattern of tolbutamide. (Beyer and Jensen 1974)

2. Glimepiride

Glimepiride is a fine powder consisting of cubic crystals (HOECHST 1994, personal communication).

3. Tolbutamide

Tolbutamide exists in two polymorphic forms which have been characterized by infrared (IR), X-ray diffraction and differential thermal analysis. Tolbu-

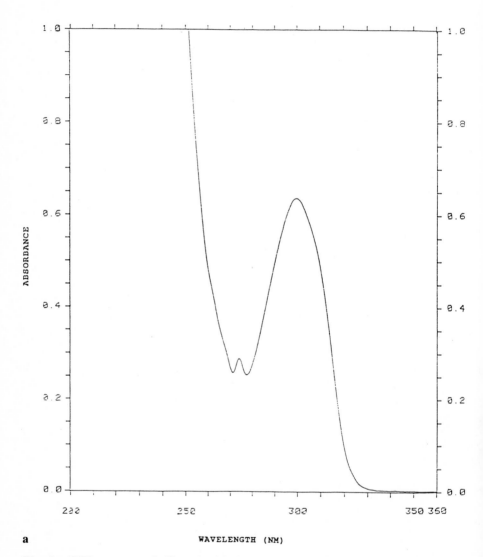

a

Fig. 4. a UV spectrum of glibenclamide in methanol. **b** UV spectrum of tolbutamide in methanol

tamide crystallizes in the orthorhombic space group pnzia (IBRAHIM and AL-BADR 1984). The structure was resolved by the Patterson search method and refined to an *R* factor of 0.083 for 615 visually measured reflections. The structure is stabilized by hydrogen bonding between the polar group and Van der Waals interactions between the nonpolar groups. Figure 3 gives the X-ray diffraction pattern of tolbutamide. The spectrum was obtained with a General Electric XRD-5 Diffractometer using CuKα 1, 50 Kvp and 16 MA (BEYER and JENSEN 1974).

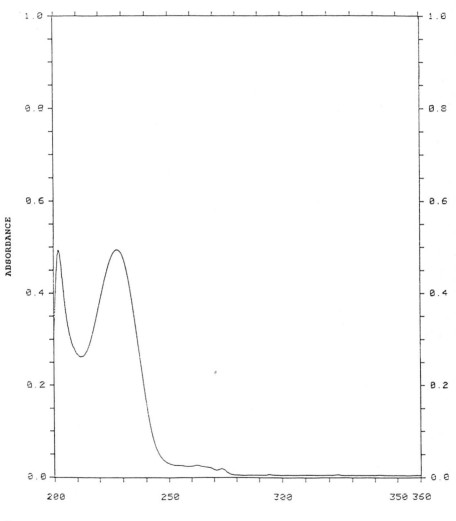

b   WAVELENGTH (NM)

Fig. 4. *Continued*

C. Ultraviolet Spectrum

The UV spectra of the major hypoglycemic agents glibenclamide and tolbutamide in methanol are presented in Fig. 4. Data for UV spectra in different solvent mediums for a number of sulfonylureas are given in Table 3 (DIBBERN 1978; HÄUSSLER and PECHTHOLD 1969; CLARK 1986; BRITISH PHARMACOPEIA 1988; HOECHST 1994).

D. Infrared Spectrum

The IR spectra of the sulfonylureas (KBr disk) are reproduced in Fig. 5a–i (DIBBERN 1978). The IR spectra may differ after recrystallization from alcohols due to the formation of different crystal forms and polymorphism. For identification purposes, it is suggested that the spectrum is compared with the spectrum of a reference substance. The spectral data are given in Table 4.

Table 3. Ultraviolet spectral data of sulfonylureas

Name of compound	Solvent	Absorption maxima (nm)	$E_{1cm}^{1\%}$	ε
Acetohexamide	95% ethanol	247 and 284	–	–
Acetylcarbutamide	Methanol, ethanol	273	–	
Carboxytolbutamide	Methanol, ethanol	237, 280	–	
Carbutamide	Methanol	268	740	20 080
	0.1 N HCl	265	163	4 420
	0.1 N NaOH	254	635	17 230
Chlorpropamide	Methanol	231, 265, 276	575, 20.4, 15.7	15 900, 560, 430
	0.1 N HCl	232	598	16 500
	0.1 N NaOH	228	472	13 100
Glibenclamide	Methanol	274 and 300	28.2, 62.4	1 390, 3080
	0.1 N NaOH	274 and 301	24.1, 54.3	1 190, 2680
	Methanolic 0.1 N HCl	300	63	
Glibornuride	Ethanol	264, 257, 275	17	–
Gliclazide	Aqueous acid	230	440	–
	Aqueous alkali	263	20	–
Glimepiride	Methanol	226		25 700, 759
		227		
Glipizide	Aqueous acid	276	231	–
Gliquidone	Methanol	311	50	–
	Aqueous alkali	276	–	–
Glisoxepide	Methanol	229	510	
Metahexamide	Methanol, ethanol	219, 244, 307	–	
Phenbutamide	Methanol, ethanol	219, 254–273	–	
Tolazamide	Methanol	257, 263, 268, 275	21	–
Tolbutamide	Methanol	228, 263	496, 22.2	13 410, 600
	0.1 N HCl	228	520	14 060

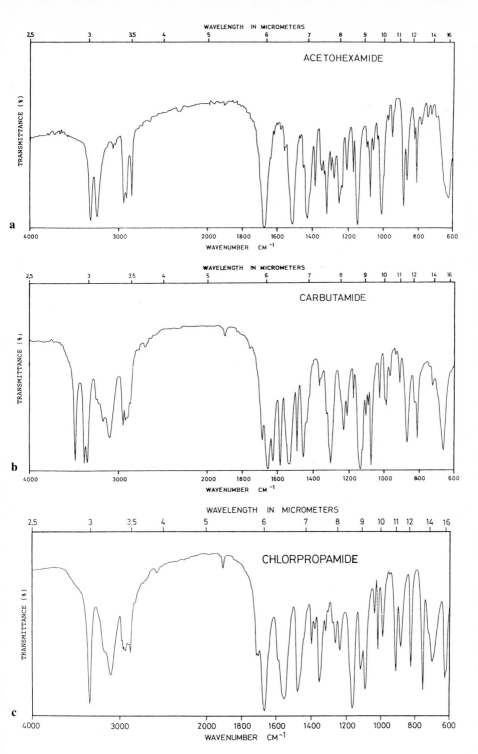

Fig. 5a–i. Infrared spectra of the sulfonylureas. (DIBBERN 1978)

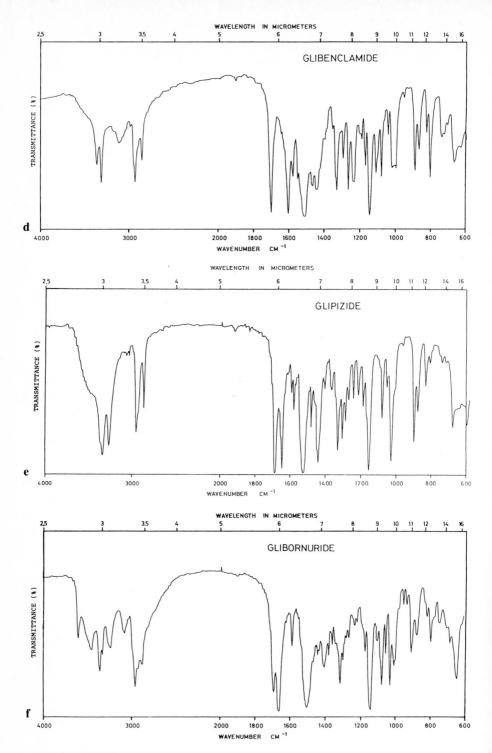

Fig. 5. *Continued*

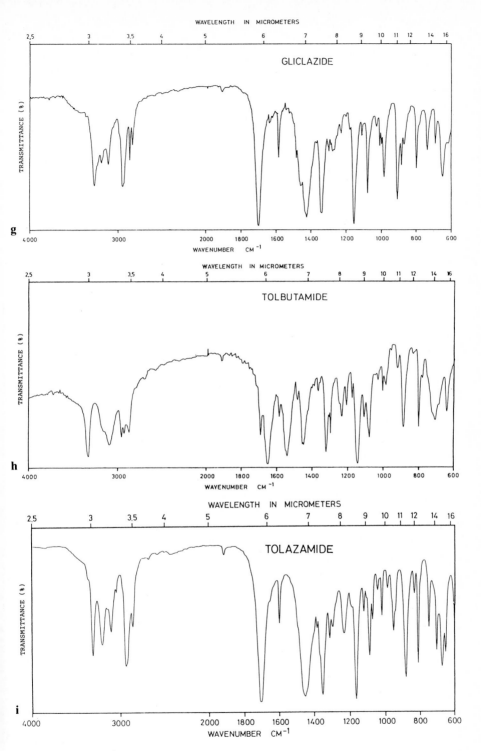

Fig. 5. *Continued*

Table 4. Infrared spectral data (KBr disk) with major peaks for sulfonylureas

Name of compound	Bands with wave numbers (cm^{-1})	Reference
Acetohexamide	1681, 1531, 1264, 1165, 1031, 905 (Nujol mull)	Clark (1986, p. 314)
Carbutamide	1661, 1635, 1599, 1310, 1147, 1089, 872, 678	Hager (1993)
Chlorpropamide	1661, 1553, 1159, 1086, 909, 757	Clark (1986, p. 461)
Glibenclamide	3363, 3313, 1724, 1613, 1515, 1471, 1333, 1163	Pharmaceutical Codex (1979, p. 1041)
Glibornuride	1710, 1682, 1520, 1170, 1100, 1050	Clark (1986, p. 639)
Gliclazide	1707, 1162, 1089, 997, 920, 667	Clark (1986, p. 640)
Glimepiride	3370, 3290, 1710, 1670	Hoechst (1994, personal communication)
Glipizide	1690, 1650, 1528, 1159, 1032, 900	Clark (1986, p. 640)
Gliquidone	1700, 1652, 1530, 1295, 1285, 1160	Clark (1986, p. 641)
Glisoxepide	3400, 3220, 2930, 1696, 1679, 1600, 1554, 1455, 1162, 894, 696	Bayer (1994, personal communication)
Tolazamide	1694, 1176, 1086, 884, 819, 675	Clark (1986, p. 1029)
Tolbutamide	1658, 1552, 1157, 1090, 905, 668	Clark (1986, p. 1030)

E. Nuclear Magnetic Resonance Spectrum

The NMR spectra of the compounds acetohexamide, chlorpropamide, glibenclamide, glisoxepide and tolbutamide are presented in Fig. 6a–c. The compounds were dissolved in deuterated solvents such as chloroform or dimethyl sulfoxide and spectra were taken with tetramethylsilane as the internal standard. The NMR proton spectral assignments for these compounds are given in Table 5.

F. Mass Spectrum

The mass spectral fragmentation pattern in EI (electron impact) mode with the observed m/e (mass/charge) values is presented in Table 6. In addition, the mass spectrum of tolbutamide (Ibrahim and Al-Badr 1984) is shown in Fig. 7. When using the conventional electron impact or chemical ionization techniques, glimepiride tends to form thermal degradation products during the evoporation process in the ion source of the mass spectrometer, resulting in weak and unreliable molecular ion peaks. Therefore, the fast atom bombardment method (FAB) was applied with 3-nitro-benzyl alcohol as the

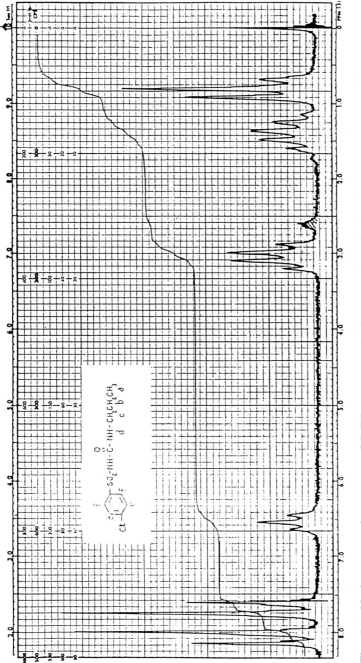

Fig. 6. a Nuclear magnetic resonance (NMR) spectrum of chlorpropamide, (Sadtler Standard Spectra 1970). **b** Nuclear magnetic resonance (NMR) spectrum of glibenclamide (TAKLA 1981). **c** Nuclear magnetic resonance (NMR) spectrum of tolbutamide (BEYER and JENSEN 1974)

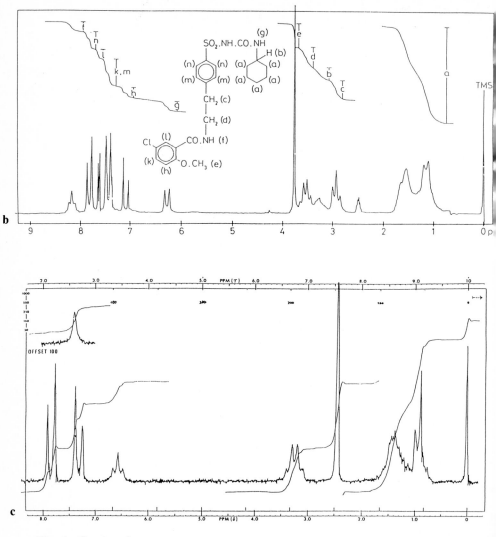

Fig. 6. *Continued*

liquid matrix for glimepiride, which gives reproducible mass spectra (HOECHST 1994, personal communication).

G. Colour and Identification Reactions

Specific color reactions for the identification of sulfonylurea compounds are rare. Color reactions for aromatic amines after diazotization and coupling are used which are also common for other agents having these group, such

Table 5. Nuclear magnetic resonance spectral assignments for sulfonylureas

Name of compound	Spectral assignment for chemical shift (ppm)	Multiplicity	Reference
Acetohexamide	1 to 2, methylene of cyclohexyl group	Multiplet	
	2.5, methyl of CH_3-CO	Singlet	SHAFER (1972)
	7 to 8, aromatic protons	Multiplet	
Chlorpropamide	0.81, methyl	Triplet	Sadtler Standard Spectra (1970)
	1.38 to 3.02, methylene	Multiplet	
	6.53, imine	Triplet	
	7.68 to 8.04, aromatic protons	Doublets	
Glibenclamide	1 to 2, methylene of cyclohexyl	Broad doublets	
	3 to 3.7, methylene	Triplets	HAJDU (1969)
	3.8, methyl of OCH_3	Singlet	
	6.3, imine of NH-CO-NH	Doublet	
	7 to 8, aromatic protons	Doublets	
Glisoxepide	1.55, methylene	Singlet	BAYER (1994, personal communication)
	2.45, methyl	Doublets	
	2.86 to 3.45, methylene	Complex	
	6.50 methine	Doublet	
	7.43 to 7.86, aromatic	Doublet	
	8.38 to 8.80, amide	Triplet	
Tolbutamide	0.88, methyl	Distorted triplet	
	1.40 to 3.23, methylene	Broad multiplet and quartet	IBRAHIM, AL-BADR (1984)
	2.43, methyl of aromatic ring	Singlet	
	6.57, imine (NH-CH_2)	Triplet	
	7.33 to 7.83, aromatic protons	Doublets	

as the sulfonamides. Primary amines also condense with aldehydes such as
p-dimethylaminobenzaldehyde or salicylaldehyde under the formation of
colored Schiff's bases. This reaction could also be used for the identification
of sulfonylurea compounds. The primary amines could be liberated by
heating the sulfonylurea compound, for example, butylamine from carbutamide and tolbutamide, and detected through reactions with color formation.
The treatment with 2,4-dinitrochlorbenzene or 2,4-dinitrofluorbenzene leads
to yellow-colored 2,4-dinitroanilides (HÄUSSLER and PECHTOLD 1969).
Other colored reactions are described under the individual sulfonylurea
compounds.

Table 6. Values of *m/e* for principal peaks in descending order of intensity for sulfonylureas

Name of compound	*m/e*	Reference
Acetohexamide	210, 56, 43, 184, 211, 75, 99, 76	CLARK (1986, p. 314)
Chlorpropamide	111, 175, 75, 85, 30, 276, 127, 113	CLARK (1986, p. 461)
Glibenclamide	394, 368, 352, 288, 198, 169, 125, 99, 82	TAKLA (1981)
Glibornuride	91, 155, 197, 65, 84, 39, 95, 41	CLARK (1986, p. 639)
Glimepiride	126, 152, 181, 352, 491	HOECHST (1994, personal communication)
Glipizide	150, 121, 56, 93, 39, 151, 66, 94	CLARK (1986, p. 640)
Glisoxepide	335, 306, 228, 210, 167, 139, 114, 110, 96, 43	BAYER (1994, personal communication)
Tolazamide	91, 155, 114, 65, 197, 42, 41, 85	CLARK (1986, p. 1029)
Tolbutamide	91, 108, 155, 65, 197, 39, 30, 107	CLARK (1986, p. 1030)

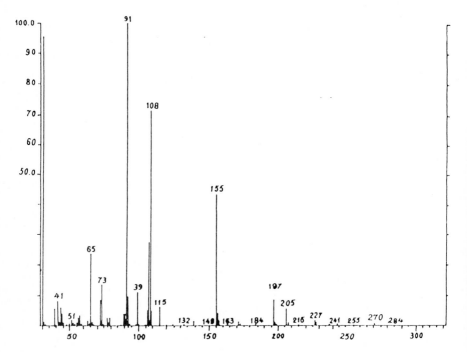

Fig. 7. Mass spectrum of tolbutamide determined in the EI mode by direct sample insertion. (IBRAHIM and AL-BADR 1984)

I. Acetohexamide, Glibornuride, Gliquidone

The substance (50 mg) is dissolved in 1 ml ethanol, one drop 1% cobalt nitrate solution in ethanol is added followed by $10 \mu l$ pyrrolidine and the mixture is well shaken. A violet color is produced (CLARK 1986).

II. Carbutamide

Carbutamide 50 mg is dissolved in 2 ml methanol acidified with 7% HCl, and then 0.2 ml 10% $NaNO_2$ solution is added. After 1–2 min, 1 ml naphthol solution (100 mg in 3 ml 15% NaOH dissolved and filled up with water to 100 ml) is added. An intense orange-red coloration is produced. When the substance is heated with resorcin and concentrated H_2SO_4, a deep-red coloration is produced, which changes into a yellow-green fluorescence after addition of water and an alkali (LUCATELLI 1956). The solution of 50 mg compound in 2 ml methanol is acidified with 7% HCl, 0.2 ml 10% sodium nitrite is added and after 1–2 min 1 ml 5% naphthol 2 solution (3.4% in NaOH) is added and mixed. An intense orange-red coloration is produced (Deutscher Arzneimittel-Codex 1986).

III. Chlorpropamide

About 100 mg chlorpropamide is boiled with 8 ml H_2SO_4 (50% m/m) under a reflux condenser for 30 min. The mixture is cooled and filtered; the residue melts at about 143°C after recrystallization from water. A positive test for chloride is obtained when the substance is ignited with anhydrous sodium carbonate, cooled, extracted with water, filtered and acidified with dilute HNO_3 (British Pharmacopeia 1988, p. 132). The substance (50 mg) is dissolved in 1 ml ethanol, one drop of 1% cobalt nitrate in ethanol is added followed by $10 \mu l$ pyrrolidine and the mixture is well shaken. A violet color is produced (CLARK 1986).

IV. Glibenclamide

Glibenclamide 20 mg is dissolved in 2 ml 96% H_2SO_4. The solution is colorless and shows blue fluorescence in UV light with a wavelength of 365 nm. After the addition of 100 mg chloralhydrate, an intensive yellow color develops after 5 min, which turns brown after 20 min. Positive tests for chloride and sulfate are obtained in an aqueous extract of the residue obtained after igniting 200 mg glibenclamide with 250 mg anhydrous sodium carbonate and 250 mg potassium carbonate (Deutsches Arzneibuch 1991).

V. Tolbutamide

Tolbutamide 200 mg is boiled under a reflux condenser for 30 min with 8 ml $5 M H_2SO_4$ and allowed to cool. The melting point of the resulting crystals, after recrystallization from hot water and drying at 100–105°C, is between 135° and 140°C (Deutsches Arzneibuch 1991).

H. Stability and Degradation

Sulfonylureas are known to undergo a breakdown under various experimental conditions (BOTTARI et al. 1972). Although a hydrolytic process is most frequently observed in such decompositions (VOGT 1959), cases of thermal dissociation have also been reported (BOTTARI et al. 1970). The acid hydrolysis of carbutamide, chlopropamide and tolbutamide has been investigated by VOGT (1959) and HÄUSSLER and HAJDU (1962). The carbutamide and chlorpropamide content of some o/w creams for topical use showed a significant decrease when these drugs were dissolved at 70°–80°C in the oil phase of the emulsions. No such decrease occurred when the sulfonylureas were incorporated into the bases at room temperature. The stabilities of several sulfonylureas are given below.

1. Acetohexamide

Acetohexamide is stable under normal storage conditions. A measurable degradation through spectrophotometric analysis at 249 nm could be detected only at temperatures above 80°C. Irradiation under a Hanovia lamp for 1 and 3 h showed about 25% and 50% decomposition, respectively, by spectrophotometric assay. As a degradation product, p-acetylbenzenesulfonamide was detected through thin-layer chromatography (TLC) (SHAFER 1972).

II. Chlorpropamide

4-Chlorobenzene sulfonamide and 1,3-dipropylurea are the products detected from chlorpropamide. The amounts of these related substances in chlorpropamide are limited to 0.02% each according to the British Pharmacopeia (1988 vol. 1, p. 107). The stability of chlorpropamide urea during the fusion process has been evaluated. Urea was found to decompose to biuret and chlorpropamide to p-chlorbenzene sulfonamide (FORD et al. 1979).

III. Glibenclamide

There are no reports that glibenclamide shows instability under normal storage conditions. It has been postulated in a study that an initial protonation is probably the rate-determining step in the hydrolysis of sulfonylureas. Thus in the case of glibenclamide the hydrolysis products would be a benzenesulfonamide derivative and cyclohexylamine (TAKLA 1981). The benzenesulfonamide derivatives are generally present as impurities in glibenclamide or glibenclamide tablets. The amounts of these impurities as degradation products in glibenclamide are maxima of 0.008% and 0.004%, respectively, according to the British Pharmacopeia (1988).

IV. Glimpiride

Glimpiride has a good stability profile, with stability-specific quality characteristics that remain largely unchanged after storage under normal and accelerated conditions. Glimepiride exhibits no sensitivity to light. It should be stored in tight containers and at room temperature. When stored at room temperature and under accelerated conditions the glimepiride content in tablets remains largely unchanged. A slight increase in the degradation product 4-[2-(3-ethyl-4-methyl-2-oxo-3-pyrroline-1-carbomaxide)-ethyl]-benzenesulfonamide as a function of temperature and time may be observed. There is no significant increase in other by-products. Exposure to light does not impair the stability of the active ingredient (HOECHST 1994, personal communication).

V. Glisoxepide

Glisoxepide is stable under normal storage conditions. Thermal decomposition starts at 170°C, with decomposition products limited to a maximum of 5% (BAYER 1994, personal communication).

VI. Tolbutamide

Tolbutamide cannot be acetylated due to the absence of the p-pamino group. The p-methyl group, however, renders tolbutamide susceptible to oxidation. Thermal decomposition of tolbutamide has been reported with the formation of p-tolylsulfonamide. The hydrolysis of tolbutamide in acid and alkaline environments has been investigated. A quantitative dissociation of tolbutamide to p-toluene sulfonamide and n-butyl isocyanate is reported to take place in an inert solvent at 160°–180°C. A significant degradation occurred in some o/w creams when the drug was dissolved in the oil phase of the emulsion at 70°–80°C. No such loss took place at room temperature. The instability of tolbutamide in the oil phase of the emulsion was attributed to the components containing hydroxyl groups (BEYER and JENSEN 1974).

The dissociation of tolbutamide has been studied at 80°C in 12 primary alcohols and polyethylene glycol 400. Tolbutamide dissociated to give butylamine and p-toluene sulfonyl isocyanate. Other authors (BOTTARI et al. 1972) studying the reaction products of tolbutamide and other N-substituted sulfonylureas in alcohols, water and amines concluded that dissociation rather than solvolysis was the most likely mechanism by which sulfonylureas undergo breakdown at relatively low temperatures. p-Toluenesulfonamide as a related substance in tolbutamide is present at up to 0.015% according to the BRITISH PHARMACOPEIA (1988, vol. I, p. 577).

I. Methods of Analysis

Quantitative determination of hypoglycemic agents in bulk drug or in drug formulations involves titrimetric, UV-spectrophotometric, colorimetric or in some cases spectrofluorometric procedures. Highly sensitive and specific methods such as high-performance liquid chromatography (HPLC), gas-liquid chromatography (GLC) and thin-layer chromatography (TLC) are applied for the determination of these compounds in complicated and biological matrix systems.

I. Titrimetry

The substances with the aromatic primary amino group are titrated with sodium nitrite in HCl medium at a temperature of $0°-5°C$ and the end point may be determined potentiometrically (LUCATELLI 1956; BRÄUNIGER and DUDA 1958). Sulfonylureas can be hydrolyzed in 65% H_2SO_4, made alkaline, and the amines liberated separated by steam distillation and titrated with an acid (VOGT 1959). The acidic reaction of the SO_2-NH-CO group is mostly utilized to titrate in aqueous medium with an alkali using a color indicator (DORFMÜLLER 1957; MESNARD and CROCKETT 1960; WARTMAN-HAFNER and BÜCHI 1965). Acidimetric titration in a non-aqueous system is also possible. The titration is performed in solvents such as dimethylformamide, acetone or pyridine with $0.1 N$ sodium methylate solution using mostly thymol blue as an indicator (BRÄUNIGER and DUDA 1958; THEODORESCU et al. 1966; WARTMAN-HAFNER and BÜCHI 1965). The acidic SO_2-NH-CO group forms highly insoluble salts with silver nitrate or mercuric acetate. Argentometric, complexometric and bromometric methods of determination of sulfonylureas have also been reported in the literature (WARTMANN-HAFNER and BÜCHI 1965; POPA and VOICU 1962; BRÄUNIGER and DUDA 1958). The titrimetric procedures for a number of sulfonylureas are given below.

1. Acetohexamide

Acetohexamide 300 mg is dissolved in 40 ml dimethylformamide, five drops thymol blue indicator are added and titrated with $0.1 N$ sodium methoxide to a blue end point. A blank determination is also performed. Each milliliter of $0.1 N$ sodium methoxide is equivalent to 32.44 mg acetohexamide (The United States Pharmacopeia [USP] 1990, p. 23).

2. Carbutamide

Carbutamide 250 mg is dissolved in 30 ml previously neutralized 96% ethanol and titrated with $0.1 N$ NaOH with phenolphthalein as indicator; 1 ml $0.1 N$ NaOH is equivalent to 27.13 mg carbutamide (HAGER 1993).

3. Chlorpropamide

Chlorpropamide 500 mg is dissolved in 50 ml previously neutralized 96% ethanol, 25 ml water is added and the solution is titrated with 0.1 N NaOH using phenolphthalein as indicator; 1 ml 0.1 N NaOH is equivalent to 27.67 mg chlorpropamide (British Pharmacopeia 1988, vol. I, p. 133). A mercurimetric determination of chlorpropamide in tablets by back titration with NH_4SCN has also been reported (EL-BARDICY et al. 1988).

4. Glibenclamide

Glibenclamide 500 mg is dissolved in 100 ml hot previously neutralized 96% ethanol and titrated with 0.1 N NaOH using phenolphthalein as indicator and protecting against exposure to atmospheric CO_2; 1 ml 0.1 N NaOH is equivalent to 49.40 mg glibenclamide (British Pharmacopeia 1988, vol. I, p. 269). Tetramethylurea has been used as solvent for the titration of glibenclamide with 0.1 N lithium methoxide in a mixture of benzene-methanol, where the end point was determined potentiometrically or by using 0.2% azoviolet in toluene as visual indicator (AGARWAL and WALASH 1972).

5. Glipizide

Glipizide 400 mg is dissolved in 50 ml dimethylformamide and the solution is titrated with 0.1 M lithium methoxide using quinaldine red solution as indicator. A blank has to be determined; 1 ml 0.1 M lithium methoxide is equivalent to 44.55 mg glipizide (British Pharmacopeia 1988, vol. I, p. 270).

6. Glisoxepide

Glisoxepide 450 mg is dissolved in 50 ml dimethylformamide and the solution titrated with 0.1 N-tetrabutyl ammonium hydroxide (TBAOH) solution, where the end point is determined potentiometrically; 1 ml 0.1 N TBAOH solution is equivalent to 44.95 mg glisoxepide (BAYER 1994, personal communication).

7. Tolazamide

Tolazamide 500 mg is dissolved in 20 ml butanone under gentle heating. The solution is cooled, 30 ml ethanol 96% is added and the solution titrated with 0.1 N NaOH using phenolphthalein solution as indicator; 1 ml 0.1 N NaOH is equivalent to 31.14 mg tolazamide (British Pharmacopeia 1988, vol. I, p. 577).

8. Tolbutamide

Tolbutamide 250 mg is dissolved in 40 ml previously neutralized ethanol 96%, 20 ml water is added and the solution is titrated with 0.1 N sodium hydroxide using phenolphthalein as indicator; 1 ml of 0.1 N NaOH is equivalent to 27.03 mg tolbutamide (Deutsches ARZNEIBUCH 1991). Tolbutamide can also be titrated in non-aqueous media: 50–150 mg is dissolved in 10 ml anhydrous acetone or pyridine and titrated with 0.1 N sodium methoxide to the phenolphthalein end point (KRACMAROVA and KRACMAR 1958).

II. Ultraviolet Spectrophotometry

Ultraviolet spectrophotometric methods can be used for quantitative determination of sulfonylureas, most of which have well-defined absorption maxima in alcoholic, acidic or alkaline mediums. Compounds with primary amino groups in the phenyl rest such as carbutamide and metahexamide can be determined quantitatively around 260 nm. UV methods are mostly applied for drug dosage forms where a small amount of substance is to be determined in a large matrix, for example, in tablets. MESNARD and CROCKETT (1960) have given details of UV methods. UV methods are mentioned for several sulfonylureas below.

1. Acetohexamide

Acetohexamide and the metabolite hydroxyhexamide can be determined simultaneously in serum after extraction with chloroform from acidic solution at 247 and 228 nm (SMITH et al. 1965).

2. Carbutamide

Carbutamide can be determined in absolute ethanol at 269 nm and in 0.1 N NaOH at 255 nm (VITALY and PANCRAZIO 1956).

3. Chlorpropamide

Chlorpropamide is extracted from tablets with methanol acidified with 0.1 N HCl and the absorbance of the resulting solution measured at 232 nm (British Pharmacopeia 1988, vol. II, p. 919).

4. Glibenclamide

In tablets glibenclamide is assayed by a spectrophotometric procedure after extraction with 0.1 N methanolic HCl and measurement of absorbance at 300 nm (British Pharmacopeia 1988, vol. II, p. 949).

5. Glipizide

Glipizide is extracted from tablets with sufficient methanol while heating gently on a water bath, cooled, diluted further with methanol and the absorbance of the resulting solution determined at 274 nm (British Pharmacopeia 1988, vol. II, p. 950).

6. Tolbutamide

Tolbutamide tablets are extracted with chloroform and the absorption is determined at 263 nm (USP 1970). Tolbutamide and carboxytolbutamide have been determined together by measuring the absorption of ethanolic solution at 228 and 237 nm, respectively (Stowers 1958). In the dissolution test of tolbutamide tablets in a dissolution medium of pH 7.4 phosphate buffer, the wavelength of maximum absorption at about 226 nm is used (USP 1990, p. 1386). Spectrophotometric methods for the quantitation of tolbutamide without interference from the tablet excipients between 250 and 270 nm have been reported. Tolbutamide was also determined in the presence of thiamine hydrochloride and pyridoxine hydrochloride by measuring the difference in absorbance at 274 and 276 nm in 95% ethanol (Abdel-Hady et al. 1979a). Springler and Kaiser (1956) determined tolbutamide in serum after lyophilization, extraction with acidified ethyl acetate, evaporation of the solvent to dryness and finally redissolution in methanol. Absorbances at 228 and 280 nm were used to quantitate tolbutamide.

III. Colorimetric Methods

Colorimetric methods for the determination of sulfonylureas are more selective than spectrophotometric methods. The color reactions give a certain amount of specificity to assay methods. Compounds with primary aromatic amino groups are measured as azo dyes. This method has been applied for the determination of carbutamide (Mesnard and Crockett 1960; West and Johnson 1959). All sulfonylureas can be determined through the color reaction of the liberated primary amines with dinitrofluorbenzene in isoamyl acetate. Tolbutamide (Mesnard and Crockett 1960), carboxytolbutamide (Nelson et al. 1960), carbutamide (Pignard 1958) and acetohexamide (Maha et al. 1962) have been determined with this method. The amines liberated from sulfonylureas have been determined also by reacting them with naphthoquinine sulfonate, p-nitrophenyl-diazonium chloride and picric acid (Mesnard and Crockett 1960). Colorimetric assay methods for carbutamide, glibenclamide and tolbutamide are given below.

1. Carbutamide

Carbutamide reacts with alloxan in the presence of Ca, Cd, Cu, Mn, Ni and Co, and the absorption of the colored solution is measured at 480 nm (Zorya et al. 1989). Chloroacridine reacts with carbutamide to give a colored solution, with an absorption 434 nm (Gaidukevich and Sidom 1977).

2. Glibenclamide

Glibenclamide is heated in amyl acetate with 2,4-dinitrofluorobenzene to 150°C for 5 min and the absorbance of the resulting solution is measured at 380 nm (Hajdu et al. 1969).

3. Tolbutamide

The reaction between tolbutamide, 2-naphthol, sodium nitrite and concentrated sulfuric acid gives a red color. The method is reported to be
applicable over the concentration range of $50\,\mu g$ to $10\,mg/ml$ (McDonald
and Sawinsky 1958). Dorfmüller (1957) reported the reaction of tolbutamide in alkaline media with diacetylmonoxime and N-phenyl anthranilic
acid followed by acidification, heat and the addition of sodium persulfate
and sodium acetate to form a blue color. A procedure has been reported
for the determination of tolbutamide in blood following extraction with
ethylene chloride at pH 5.0. After nitration, diazotization and coupling with
N-(1-naphthyl) ethylene diamine, the azo dye produced is measured at
547 nm (Kern 1963). Tolbutamide in dosage forms is determined after
extraction with isopropanol and treatment with a solution of acetaldehyde
and p-chloranil in dioxane and heating the mixture. The absorption of this
solution is then measured at 645 nm against blank reagent (Emmanuel and
Fernandez 1989).

IV. Fluorimetry

Fluorimetric methods of determination for carbutamide (Iskender and
Orak 1986), glibornuride (Becker 1977), glibenclamide (Hajdu et al. 1969)
and tolbutamide (Namigohar et al. 1980) have been reported.

V. Miscellaneous Methods

Near infrared (NIR) and Fourier transformation (FT)-Raman spectroscopy
have been used for the qualitative analysis of chlorpropamide polymorphism.
NIR-FT-Raman spectra of the A and B polymorphs of chlorpropamide were
obtained using an Nd-Yag laser operating at $1.046\,\mu m$ with a power of
200 mW (Tudor et al. 1991). A proton magnetic resonance spectrometric
method has been used to determine chlorpropamide in tablets. Chlorpropamide is extracted with deuterated acetone containing maleic acid. The
signals at 0.8 and 7.8 ppm are integrated and compared with that of maleic
acid at 6.3 ppm (El-Khateeb 1987). Assay of tolbutamide in tablets has
been performed by differential scanning colorimetry (Ni and Tong 1988).

J. Chromatographic Methods

Innumerable methods of separation of sulfonylureas by chromatographic
methods have been reported in the literature. It is beyond the scope of this
book to cite all of them. Only some of the various chromatographic techniques such as paper chromatography, thin-layer chromatography, gas
liquid chromatography-mass spectrometry and high-performance liquid
chromatography are described below.

I. Paper Chromatography

Methods of paper chromatographic separation of carbutamide, acetyl carbutamide, chlorpropamide, tolbutamide and its metabolites are presented in Table 7, which lists the sulfonylurea, quality of the paper used, chromatographic technique, mobile phase, Rf value and method of detection with references.

II. Thin-Layer Chromatography

Thin-layer chromatography (TLC) is a most widely used chromatograpic technique for separation, identification and analysis of pharmaceutical substances. The analysis of sulfonylureas through TLC has also been reported in the literature. Only some methods including those of the pharmacopeias are included here. Table 8 gives the salient features of the TLC methods for the different sulfonylureas.

III. Gas-Liquid Chromatography and Mass Spectrometry

Gas-liquid chomatography (GLC) offers rapid, sensitive and specific quantitative assay procedures for the sulfonylureas. Efficient GLC separation columns, sensitive detectors such as the flame ionization detector (FID) and electron capture detector (ECD) and automatic integrators for evaluation of chromatographic peaks have added reliability and reproducibility to GLC methods. Assay requirements involving stability testing of the bulk drug or of sulfonylureas in drug formulation or quantitation of metabolites are met in many instances by GLC methods.

1. Acetohexamide

FRICKIE (1972) described a GLC assay for acetohexamide using Dexil 300 as the liquid phase with FID. KLEBER et al. (1977) developed a sensitive and specific GLC assay for acetohexamide and its metabolite hydroxyhexamide in plasma and urine with GLC on a 0.5% polyethylene glycol 20 M on a gaschrome Q column and FID.

2. Chlorpropamide

SABIH and SABIH (1970) reported a GLC assay for chlorpropamide in tablets, urine and blood using a 5% DC-200 gaschrome Q column and FID. Derivatization was carried out with dimethyl sulfate prior to GLC determination. The determination of chlorpropamide in biological fluids has also been reported in the literature (THOMAS and JUDY 1972; AGGARWAL and SUNSHINE 1974; TAYLOR 1972; FINKLE et al. 1971; MATIN and ROWLAND 1973a). MIDHA et al. (1976) determined chlorpropamide in plasma after extraction of the drug with toluene and treatment of the dried residue with ethereal diazo-

Table 7. Paper chromatography of sulfonylureas

Sulfonylurea	Paper	Mobile phase	Rf value	Detection	Reference
Carbutamide	Schleicher & Schüll 2043 bm, round filter	n-Butanol + water + dimethylformamide $42 + 8 + 4$	1.0	2% alcoholic solution of dimethyl aminocinnamic aldehyde	BRÄUNINGER (1962)
Carbutamide + acetylcarbutamide	Whatman No. 1, ascending at 30–31°C	n-Butanol + ammonia + water $40 + 10 + 50$	0.41–0.43, carbutamide 0.57–0.59, acetylcarbutamide	HCl vapors + Ehrlich's reagent	CORNTER et al. (1957)
Carbutamide, chlorpropamide, tolbutamide	Whatman No. 1, descending at 30–31°C	n-Butanol saturated with ammonia	0.47, carbutamide 0.81, tolbutamide 0.73, chlorpropamide	0.1% concentrated HCl in isoamylacetate heating to 105°C + 0.1% 2,4-dinitrofluorobenzene	CHAKRABARTI (1962)
Carbutamide, chlorpropamide, tolbutamide	Whatman No. 1, ascending at 30–31°C	Ethanol + ammonia $6 + 1$	0.71, carbutamide 0.87, tolbutamide 0.88, chlorpropamide	2% phenylhydrazine in benzene, heating to 195°C, spraying with 1 + 1 mixture of 10% aqueous nickel sulfate + ammonia	HENTRICH (1963)
Tolbutamide and carboxytolbutamide	613 Eaton-Dikeman impregnated with formamide + methanol 1 + 1, ascending	n-Butanol + water + piperidine $81 + 19 + 2$	0.87, tolbutamide 0.46, carboxytolbutamide	0.1% concentrated HCl in isoamylacetate, heating to 105°C, spraying with 0.2% ninhydrin solution	MILLER et al. (1957) MOHNIKE et al. (1958)

Table 8. Thin-layer chromatography methods for sulfonylureas

Sulfonylurea	Stationary phase	Mobile phase	Rf value	Detection	Reference
Carbutamide Chlorpropamide Tolbutamide	Silica gel GF$_{254}$	Butanol + chloroform + methanol + 25% ammonia, 40 + 15 + 15 + 15	–	Ehrlich's reagent	NEIDLEIN et al. (1965)
Carbutamide	Silica gel GF$_{254}$	Chloroform + methanol, 9 + 1	–	Ehrlich's reagent	KIGER and KIGER (1966)
Chlorpropamide Carbutamide Tolbutamide	Silica gel GF$_{254}$	Chloroform + n-butanol + diethylamine, 7 + 7 + 1	0.47, chlorpropamide 0.27, carbutamide 0.57, tolbutamide	UV$_{254}$	REISCH et al. (1964)
Chlorpropamide Tolbutamide Acetohexamide	Silica gel GF$_{254}$	Acetone + benzene + water, 65 + 30 + 5	0.59, chlorpropamide 0.75, tolbutamide 0.52, acetohexamide	Ninhydrin reagent, Ninhydrin + 5% vanillin in H$_2$SO$_4$	MACEK (1972)
Chlorpropamide	Silica gel GF$_{254}$	Methylene chloride + methanol cyclohexane + ammonium hydroxide, 100 + 50 + 30 + 10	–	UV$_{254}$	USP (1990, p. 1386)
Chlorpropamide	Silica gel GF$_{254}$	Chloroform + methanol + cyclohexane + 13.5 M ammonia, 100 + 50 + 30 + 11.5		Heating to 110°C %5 KMnO$_4$, KI + starch solution	British Pharmacopeia (1988, p. 133)

S.L. Ali et al.

Table 8. *Continued*

Sulfonylurea	Stationary phase	Mobile phase	Rf value	Detection	Reference
Glibenclamide	Silica gel GF$_{254}$	Chloroform + cyclohexane + ethanol 96% + glacial acetic acid, 45 + 45 + 5 + 5	–	UV$_{254}$	Deutsches Arzneibuch 10 (1991); British Pharmacopeia (1988)
Glibenclamide	Silica gel GF$_{254}$	Chloroform + ethanol, 50 + 3		UV$_{254}$	ZHU and BAI (1988)
Glisoxepide	Silica gel GF$_{254}$	Chloroform + glacial acetic acid, 99 + 1	0.7	UV$_{254}$	BAYER (1994, personal communication)
Tolazamide	Silica gel G	Chloroform + methanol + cyclohexane + 13.5 M ammonia, 200 + 100 + 60 + 23	–	Heating to 110°C 5% KMnO$_4$, KI + starch solution	British Pharmacopeia (1988, p. 577)
Tolbutamide	Silica gel GF$_{254}$	Chloroform + glacial acetic acid, 99 + 1	0.85	Xanthydrol solution	MENZER et al. (1971)
Tolbutamide	Silica gel G	Chloroform + methanol + anhydrous formic acid, 90 + 8 + 2		Heating to 110°C 5% KMnO$_4$, KI + starch solution	British Pharmacopeia (1988, p. 577)

methane to form the methyl derivative of chorpropamide. The derivative was analyzed on 5% OV-25 on a Chromosorb W 1.83-m column and FID. Operating conditions were 200°C for the injector, 220°C for the column and 285°C detector temperatures, 60 ml/min nitrogen as the carrier gas and tolbutamide as an internal standard. The detector response was linear over a range of 0.20–25 μg/ml and the limit of detection was given as 0.05 μg/ml. Careful choice of the injection port temperature of 200°C minimized the breakdown of the N-methyl derivative of chlorpropamide.

3. Glibenclamide

A procedure employing a column packed with 5% OV-17 on Chromosorb G-AW-DMCS has been used for the determination of glibenclamide in plasma with an Ni63 electron capture detector and tolbutamide as an internal standard. The method involves derivatization of glibenclamide by heating with 2,4-dinitrofluorobenzene in amyl acetate at 130°C for 1 h. The lowest detectable amount of glibenclamide was 100 pg (CASTOLDI and TOFANETTI 1979).

4. Tolazamide

GLC procedures for tolazamide in biological fluids have been reported on a 0.5% Carbowax 20 M on Chromosorb G glass column with FID. The experimental conditions were injection temperature 236°C, column temperature 190°C and detector temperature 220°C. The thermal fragment of tolazamide was p-toluene sulfonamide. The limit of detection was 0.7 μg/ml tolazamide in plasma (WICKRAMASINGHE and SHAW 1971).

5. Tolbutamide

The GLC method reported by MIDHA et al. (1976) for chlorpropamide has also been used for the determination of tolbutamide. After derivatization the retention time of N-methyltolbutamide was 5.9 min, showing a good resolution from the N-methyl derivative of chlorpropamide, which is used here as an internal standard. The detector response was linear over the concentration range of 0.20–25.00 μg/ml tolbutamide. The method was highly specific for tolbutamide, since known metabolites such as p-tolylbenzene-sulfonamide did not interfere in the assay. SABIH and SABIH (1970) have reported a method for the determination of tolbutamide in blood, urine and tablets. A 5% DC 200 on Gaschrome Q column at 210°C, 330°C injection temperature and a FID temperature of 320°C was used. The substance was converted to the methyl derivative prior to GLC determination.

MATIN and ROWLAND (1973a and 1973b) described GLC methods for the determination of tolbutamide and chlorpropamide and the simultaneous analysis of tolbutamide and its metabolites in biological fluids on 3% OV 17 and Chromosorb W, AW, DMCS with Ni 63 EC detection. KNIGHT and

MATIN (1974) worked out a method involving extraction of the sulfonylurea from biological fluids, followed by methylation with diazomethane and subsequent mass spectrometric determination. The chemical ionization-mass spectrometric method of KNIGHT and MATIN (1974) is specific for the determination of tolbutamide and metabolites. AGGARWAL and SUNSHINE (1974) developed a method for the analysis of sulfonylureas by GLC and mass spectrometric determination. BRASELTON et al. (1976) developed a rapid sensitive and specific method for chlorpropamide and tolbutamide. Methyl derivatives of both substances were prepared with diazomethane, which were analyzed by GLC along with mass spectrometric detection. BRASELTON et al. (1976) described the mass spectral and GLC characteristics of the N-methyl and N-methyl trifluorbutyryl derivatives of tolbutamide, chlorpropamide, tolazamide and carboxyltolbutamide. The derivatized compounds were thermally stable as shown by mass spectrometry and exhibited excellent GLC properties on 3% OV-1, OV-17 and SP-2401 phases.

IV. High-Performance Liquid Chromatography

Thermal degradation of compounds occurring in many GLC procedures has given a large impetus to HPLC methods. Labile compounds or those requiring derivatization in GLC due to structural characteristics (OH, NH moieties) can be analyzed by HPLC without derivatization. Assay requirements involving stability testing of the bulk drug or its formulations and determination of drug substances with metabolites in bioligical fluids are successfully met by HPLC methods. HPLC has advanced during the past decade on account of the enormous progress which has been made in high-efficiency columns, the septumless injection system and sensitive detectors to evolve into to a highly sophisticated and widely used chromatographic technique for the analysis of drug substances in diverse matrixes.

Sulfonylureas have been extensively analyzed by HPLC procedures. It is beyond the scope of this chapter to cite all the available literature. A number of well-established HPLC methods are reported here.

1. Acetohexamide, Chlorpropamide, Glibornuride, Gliclazide, Tolazamide

A quantitative HPLC method for the determination of chlorpropamide, tolbutamide and their respective hydrolysis products in solid dosage forms has been developed. This uses a Lichrosorb Si-60 250 × 3.2-mm ID column, a 10-μm mobile phase consisting of 4% absolute ethanol, 8% tetrahydrofurane and 0.06% acetic acid in n-hexane and a 254-nm fixed wavelength UV detector. The method is stability-indicating and can be used to determine the sulfonamide hydrolysis product and the intact drug in the presence of minor degradation products (ROBERTSON et al. 1979). HPLC methods for the quantitation of sulfonylureas in pharmaceutical (BEYER 1972) and for the determination of chlorpropamide in tablet formulations (MOLINS et al. 1975) have been reported. Acetohexamide, chlorpropamide, tolazamide and tolbutamide were separated and analyzed by HPLC on a hydrocarbon polymer (HCP) column with a mobile phase of $0.01 M$ monobasic sodium citrate containing 15% methanol (V/V) with UV detection at 254 nm (Beyer 1972). Chlorpropamide was determined in biological fluids with tolbutamide as internal standard on a 10μm reversed-phase C_{18} Bondapak column using a mobile phase of 17% acetonitrile in $0.05 M$ aqueous ammonium formate and at a UV detection at 254 nm. The linear concentration range for chlorpropamide was from 0.5 to $40 \mu g/ml$. The procedure was also suitable for tolbutamide using chlorpropamide as the internal standard, and sensitive to less than $5 \mu g$ tolbutamide/ml plasma. The HPLC procedure has been applied for the determination of both drugs in single-dose pharmacokinetic studies (Sved et al. 1976).

Chlorpropamide has been analyzed in bulk drug and tablets on a reversed-phase C18 25 cm $\times$ 4.6-mm column with acetonitrile-dilute glacial acetic acid (1 in 100), 50:50 mobile phase at 240 nm UV detection and a flow rate of 1.5 ml/min (Usp 1990). In a collaborative study, chlorpropamide in tablet dosage forms was determined on a Zobrax ODS, $5-6 \mu$m, 25 cm $\times$ 4.6-mm column using a mobile phase of 1% acetic acid-acetonitrile, 13:12, with a flow rate of 1.5 ml/min and detection at 240 nm. The response was rectilinear from 0.2 to $2 \mu g$ chlorpropamide. For a synthetic tablet, a mean recovery of 99.2% was obtained with a coefficient of variation of 1.41%. Propylamine and dipropylurea did not interfere (Everett 1986). Chlorpropamide in pharmaceutical dosage forms is determined on a μ-Bondapack C18, 15 cm $\times$ 3.9-nm column using a mobile phase of methanol-water-acetic acid, 25:74:1, with a flow rate of 1.5 ml/min and at a detection wavelength of 254 nm. The calibration graph was rectilinear from 2 to $10 \mu g$, with a detection limit of $1 \mu g$ (Sadana and Gaonkar 1989).

The separation and determination of chlorpropamide, tolbutamide, glibornuride, glibenclamide, glipizide and tolazamide in plasma have been done on a Spherisorb-ODS, $5-\mu$m, 25 cm $\times$ 4.5-mm HPLC column at 40°C with 0.4% H_3PO_4-acetonitrile, 1:3, mobile phase with a flow rate of 1.2 ml/min and a detection wavelength of 360 nm. The sulfonylureas were extracted with ether from acidified plasma samples, the ether phase evaporated to dryness and the residue dissolved in 0.5 ml butyl acetate containing 2 mg/ml fluorodinitrobenzene and derivatized at 120°C for 1 h. The sulfonylurea derivatives were then separated and quantitated by the HPLC method (Starkey et al. 1988).

Acetohexamide, chlorpropamide, gliclazide and tolbutamide were extracted from serum with chloroform, derivatized with 7-fluoro-4-nitro-1, 4-benzimidazole and separated on a Finepak SIL C18 10-μm, 25 cm $\times$ 4.6-mm HPLC column with an aqueous 60% acetonitrile mobile phase, a flow rate of 1 ml/min and fluorimetric detection using excitation at 470 nm and emission at 534 nm. Calibration graphs were rectilinear for 0.1 to 20 ppm gliclazide and 0.5–100 ppm of other drugs (Igaki et al. 1989).

A screening test detecting sulfonylureas in plasma through HPLC has been reported. The drugs were extracted with ether from acidified plasma, the ether layer evaporated and the residue reconstituted in the mobile phase. A Versapack C18 10-μm, 25 cm × 4.1-mm column was used with a mobile phase of acetonitrile–10 mM H_3PO_4, 1:1, a flow rate of 1 ml/min and detected at 230 nm. The procedure gave semiquantitative results for chlorpropamide, glibenclamide, gliclazide, glipizide, tolbutamide and tolazamide. Detection limits were between 0.1 and 1.0 μg/ml (SHENFIELD et al. 1990). HPLC determinations with UV detection for glibornuride (HARZER 1980), gliclazide (KIMURA et al. 1980) and tolazamide (WELLING et al. 1982) have also been reported.

2. Carbutamide

Carbutamide has been analyzed on a Zobrax SAX ion-exchange column with a phosphate buffer, pH 6.5, along with 2 M cyclodextrin + methanol, 85:15, mobile phase using tolazamide as an internal standard at 269 nm UV detection wavelength (EL-GIZAWY 1991).

3. Glibenclamide, Glipizide

HPLC was recommended by BEYER (1972) for the determination of glibenclamide in tablets. HPLC determination of glibenclamide in biological fluids is carried out on a reversed-phase Lichrosorb RP-8 column using a 50-mM $NH_4H_2PO_4$-acetonitrile (1:1) mobile phase and UV detection at 228 nm. The lower glibenclamide detection limit was about 20 ng/ml serum. Major metabolites did not interfere (ADAMS and KRUEGER 1979).

Another procedure developed for glipizide has been also found to be applicable to glibenclamide. A C18 μ-Bondapak column 30% 0.01 M phosphate buffer, pH 3.5, in 70% methanol and glibornuride as internal standard was used (WÄHLEN-BOLL and MELANDER 1979). Glibenclamide has also been determined in tablets by HPLC (WU and YANG 1987; BLUME et al. 1984). After extraction of powdered tablets with 0.1 N methanolic HCl, the determination was caarried on a Nucleosil C18, 7-μm, 4 × 120-mm column with a mobile phase of 0.05 M ammonium sulfate solution (adjusted to pH 3 with 3N H_2SO_4) and acetonitrile, 55:45. The flow rate was 2.0 ml/min and the UV detection wavelength was 230 nm (BLUME et al. 1984). The same HPLC method was also applied for the determination of glibenclamide in plasma samples (MÖLLER et al. 1983). The calibration curve was linear between 10 and 400 ng/ml.

A selective and sensitive HPLC method for determination of intact glibenclamide in human plasma and urine has been developed. A Spherisorb ODS reversed-phase column 5-μm, 250 × 4.5-mm ID, mobile-phase acetonitrile-phosphate buffer 0.01 M, pH 3,5 50:50, flow rate 1.6 ml/min, UV detection wavelength 225 nm and glibornuride as internal standard was used. The response was linear between 10 and 1000 ng/ml and the detection limit

was 5–10 ng/ml plasma or urine. No interferences from metabolits or endogenous constituents could be noted. The utility of the method was demonstrated by analyzing glibenclamide in samples from diabetic subjects on
therapeutic doses of the drug (EMILSSON et al. 1986).

Glipizide and glibenclamide have been determined in tablets on a μ-
Bondapack phenyl 30 cm × 3.9-mm column with acetonitrile-aqueous
20 mM ammonium acetate (6:10 or 3:7) as a mobile phase with hydrocortisone as an internal standard. The flow rate was 2 ml/min and the UV
detection wavelength was 232 nm. Calibration graphs were rectilinear from
0.2 to 1.0 μg/ml glipizide and from 0.6 to 1.4 μg/ml glibenclamide (DAS
GUPTA 1986). Glibenclamide was determined in human plasma after extraction with methylene chloride from phosphate-buffered sample (pH 4.0).
After evaporation of solvent, the residue was dissolved in the mobile phase
0.05 M ammonium sulfate-acetonitrile, 29:21, and analyzed on a Microsorb
C18, 5-μm, 15 cm × 4.6-mm column with a flow rate of 1.4 ml/min at a
detection wavelength of 229 nm. The limit of detection was 5 ng/ml and
recovery for 25 and 100 ng was 92.4% and 97.2%, respectively (OTHMAN
et al. 1988).

Glibenclamide in serum has been determined with fulfenamic acid as
internal standard on a C8 Spherisorb column with aqueous 45% acetonitrile
(pH 3.7–3.8) as mobile phase, 2 ml/min flow rate and 230 nm detection
wavelength (ABDEL-HAMID et al. 1989). Glibenclamide and its two major
metabolites in human serum and urine have been analyzed on a Chrompak
Chromsep C8, 3-μm, 10 cm × 4.6-mm column with acetonitrile-0.038 M
phosphate buffer (pH 7.5), 27:73 mobile phase, 0.7–1.0 ml/min flow rate
and at a detection wavelength of 203 nm (RYDBERG et al. 1991). Determination of glibenclamide in human plasma by liquid chromatography and fluorescence detection has also been reported. The analysis is carried on an
Ultrasphere ODS 5-μm, 15 cm × 4.6-mm column along with a 1.5-cm RP 16
guard column with a mobile phase of water-acetonitrile–70% HClO$_4$ –
0.25% methanolic tetramethyl ammonium hydroxide, 1000:1000:1:1, using
a flow rate of 1.2 ml/min. Fluorescence detection was done at 300 mm
excitation and 360 nm emission wavelengths. The detection limit was
10 ng/ml (GUPTA 1989).

4. Glimepiride

Depending on the possible by-products in glimepiride to be determined, the
following chromatographic conditions may be used in an isocratic system
with UV detection at 228 nm: Method 1, by-product cis-glimepiride, Lichrosorb Diol column (CGC glass cartridge, Merck) 150 × 3 mm or comparable
column, mobile phase a mixture of anhydrous acetic acid, 2-propanol and
hexane. Method 2, other by-products, Superspher 100 RP 18 column, particle size 4 μm (stainless steel cartridge, Merck), 250 × 4-mm or comparable
column and with a mobile phase of a mixture of acetonitrile and water

containing phosphate buffer of pH 3. Assay may be performed by method 2 (HOECHST 1994, personal communication).

5. Glisoxepide

Glisoxepide and its different decomposition products have been determined by HPLC on a Nucleosil ODS 5-μm, 125-mm column using a mobile phase of acetonitrile + phosphoric acid 85% (2 ml in 1 l), 28:72, at a wavelength of 250 nm, a flow rate of 1 ml/min and a column temperature of 40°C (BAYER 1994, personal communication).

6. Tolbutamide

Tolbutamide has been determined in serum on a 1-m ETH Permaphase column using a mobile phase of 30% methanol and 70% 0.01 M monobasic sodium citrate aqueous solution with UV detection at 254 nm. The HPLC procedure was capable of measuring 2 μg tolbutamide. Metabolites of tolbutamide did not interfere in the assay (WEBER 1976).

Tolbutamide has been determined as bulk drug and as tablets on a 5- to 10-μm 30 cm × 4.0-mm porous silica column with a mixture of n-hexane, water-saturated hexane, tetrahydrofurane, alcohol and glacial acetic acid, 475:475:20:15:9, as mobile phase, tolazamide as internal standard and UV detection at 254 nm (USP 1990). Simultaneous determination of tolbutamide and its hydroxy and carboxy metabolites in plasma and urine has been performed on a Spherisorb CN, 15 cm × 4.6-mm column using a mobile phase of methanol-acetonitrile-0.1 M tetrabutyl ammonium hydrogen sulfate-water 20:20:9:151, adjusted to pH 4.0, and a detection wavelength of 237 nm. There was a linear relationship in the range of 0.1–100 μg/ml tolbutamide and 0.1–10 μg/ml of its metabolites in plasma and 0.5–200 μg/ml of all other compounds in urine. The detection limit in all instances was 0.1 μg/ml (KEAL et al. 1986).

There are further reports of the HPLC analysis of tolbutamide and its metabolites using the photodiode assay detector (CSILLAG et al. 1989) and UV detection (ST. HILARIE and BELANGER 1989; ARCELLONI et al. 1990; RAGHOW and MEYER 1981; HILL and CHAMBERLAIN 1978). Tolbutamide has been determined in plasma on a ODS-SIL-X-1 column with a 22% aqueous acetonitrile mobile phase at a detection wavelength of 223 nm using chlorpropamide as an internal standard. The calibration graph was rectilinear between 25 and 500 μg/ml (PECORARI et al. 1981).

K. Bioavailability

Bioavailability of drugs is, on the one hand, dependent on the physico-chemical and pharmacokinetic characteristics of the drug itself; on the other hand, biopharmaceutical properties of the pharmaceutical formulation are also of significant importance. Accordingly, pharmaceutical quality has to be determined in vitro and in vivo in order to characterize a formulation.

Sulfonylureas are normally administered in the form of immediate-release tablets, since short t_{max} values are desirable in order to prevent postprandial hypoglycemia. Thus, modified-release preparations are not appropriate in this respect.

In vitro, tablets are tested for identity, purity, potency, content uniformity and in vitro dissolution by use of respective pharmacopeial methods. First of all these in vitro tests are used as a quality control tool in order to describe homogeneity as well as consistency of production batches. However, predictive information concerning bioavailability in vivo also needs to be derived especially from in vitro dissolution data. For this purpose in vitro/in vivo correlations need to be established in accordance with USP level "A", "B" or "C" methods, which are described in the USP 1990, Chap. 1088. Nowadays a "mapping" approach is also under discussion for controlled-release dosage forms. A detailed description of the "state of the art" was given by SKELLY and SHIN (1993).

Generally in vitro dissolution tests may be used as a surrogate for in vivo bioavailability testing in order to ensure bioequivalence from batch to batch, after changes in the site of manufacture, the production process or so-called minor changes (e.g., regarding dyes, preservatives) of the formulation, although only preliminary definitions of such "minor" changes are available (SKELLY et al. 1993). For a conclusive valuation of the consequences of the respective changes again in vitro/in vivo correlations form a proper basis.

Concerning biopharmaceutical quality testing in vivo, plasma concentration versus time curves of sulfonylureas are commonly measured after single-dose administrations. From these profiles areas under the curves (AUC) ascertained either until the last measured plasma concentration or after extrapolation to infinity with the help of calculated elimination half-lifes, maximum peak plasma concentrations (C_{max}) and time to reach peak levels have to be determined preferably by model-free methods. Additionally, it may be useful in certain cases also to evaluate the half-value duration (HVD), which is normally used to characterize the degree of retardation, but especially in the case of slowly absorbed or slowly eliminated drugs it may be an important parameter. Because of the multiple dosing of sulfonylureas in therapy, steady-state studies may complete the characterization of the biopharmaceutical characteristics of a formulation in vivo. In this regard, special parameters which characterize steady-state situations, such as the peak-trough fluctuation (PTF), need to be determined. A detailed description of the mathematical and biometric methods is given elsewhere (STEINIJANS and HAUSCHKE 1990; STEINIJANS et al. 1992; CPMP-Note for Guidance 1991; APV Guideline).

In order to test the in vivo quality of generic products, bioequivalence has to be assessed in comparison with an appropriate reference formulation.

Two drugs are considered bioequivalent if their respective rate of absorption, characterized by C_{max} and t_{max}, and extent of bioavailability, characterized by AUC, do not differ by more than a given maximum deviation. According to current knowledge, bioequivalence can be assessed when confidence intervals do not exceed the preset limits of 80 and 125 for AUC. Based on therapeutic experience, limits of 70 and 143 are often accepted for C_{max} (BLUME et al. 1988). A detailed overview concerning statistical analysis is given elsewhere (STEINIJANS and HAUSCHKE 1990; STEINIJANS et al. 1992).

Clinical trials showing a clear relationship between drug concentrations in blood and therapeutically relevant drug effects are generally lacking (FERNER and CHAPLIN 1987). However, for certain drugs, respective associations have been established. Accordingly, a correlation between bioavailability, insulin concentration and resulting blood glucose levels has been demonstrated for glibenclamide (Fig. 8) (BLUME et al. 1985b; BLUME et al. 1987). In conclusion, bioavailability is generally regarded as a relevant parameter with regard to biopharmaceutical quality, and thus its determination is an important part of the characterization of a formulation. Thus, evaluation of bioavailability and assessment of bioequivalence are generally requested for antidiabetic drugs by the regulatory authorities.

Furthermore, bioavailability is not only influenced by the physico-chemical properties of the drug compound and the biopharmaceutical quality of the formulation but also by subject- and situation-dependent

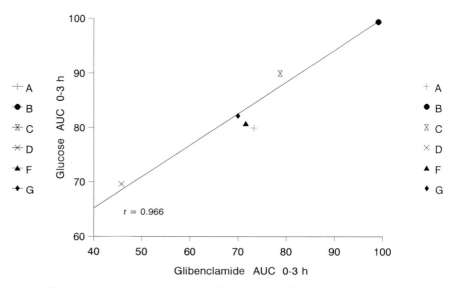

Fig. 8. Correlation between bioavailability data of glibenclamide and reduction of blood glucose levels for six different formulations: Mean differences in AUC_{0-3h} values of glucose with and without (placebo phase) administration of glibenclamide tablets (3.5 mg) in comparison with mean AUC_{0-3h} values of glibenclamide after oral administration of the respective formulation to seven healthy volunteers. (From BLUME et al. 1985a)

factors, e.g., food intake, circadian pharmacokinetics and metabolic pheno-type. Moreover, in the case of diabetic patients bioavailability can also be influenced by the disease state itself or by concomitant diseases (GWILT et al. 1991). Frequently, disorders of gastrointestinal motility are associated with diabetes mellitus: the most common symptom is constipation followed by gastroparesis diabeticorum (which occurs along with retarded gastric emptying due to a relative inability to increase intragastric pressure) and diarrhea (FELDMAN and SCHILLER 1983). IKEGAMI et al. (1986) observed that in some patients with non-insulin-dependent diabetes mellitus absorption of, e.g., glibenclamide, was (reproducibly) delayed. A correlation to impaired autonomic nerve function could be demonstrated, which causes delayed gastric emptying and/or abnormal gastric acid secretion. In general the influence of gastric pH especially upon the rate of absorption has to be considered because sulfonylureas are weak acids. Nevertheless in vivo re-sults concerning the influence of gastric pH are sometimes contradictory especially regarding the concomitant administration of H_2-antagonists.

For certain sulfonylureas, detailed knowledge of pharmacokinetic be-havior and bioavailability exists. For example, this is the case for some drugs of the "first generation" such as tolbutamide and chlorpropamide as well as for "second-generation" drugs such as glibenclamide (= glyburide) and glipizide. On the other hand, less useful information has been published concerning, e.g., acetohexamide, gliquidone, gliclazide and tolazamide. All sulfonylureas mentioned above show small distribution volumes and exten-sive plasma protein binding, and most of them have no relevant first-pass metabolism.

1. Acetohexamide

Acetohexamide is a sulfonylurea of the "first generation" and like, e.g., chlorpropamide, it is only of very limited use nowadays. Absolute bio-availability of acetohexamide is not known; however, an almost complete absorption of the compound is claimed (FERNER and CHAPLIN 1987). One of its metabolites, hydroxyhexamide, contributes significantly to the clinical effect (GALLOWAY et al. 1967) and therefore its plasma concentrations should be considered in bioavailability studies (Fig. 9). Furthermore, this metabolic step is stereoselective, and consequently enantioselective deter-mination may be necessary (IMAMURA et al. 1985). Peak plasma concentra-tions of acetohexamide are reached between 1 and 2 h after administration of immediate release formulations, and C_{max} values of hydroxyhexamide are recorded 2–5 h p.a. Elimination half-life of the metabolite was determined to be longer (3.7–6.4 h) than that of the parent compound (0.8–2.4 h).

2. Chlorpropamide

Chlorpropamide is a drug with a long elimination half-life (24–48 h), and accumulation can result after daily oral administration (FERNER and CHAPLIN 1987). Accordingly, the timing of the daily dose is of less importance. Large

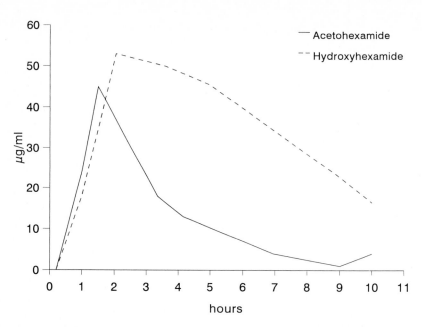

Fig. 9. Mean plasma concentrations of acetohexamide and its active metabolite hydroxyhexamide after oral administration of an immediate-release formulation containing 1 g of the parent compound. (From GALLOWAY et al. 1967)

interindividual variations have been detected in steady-state plasma concentrations. Extent of plasma protein binding is less relevant than for other sulfonylureas. The main metabolites 2-hydroxychlorpropamide, 3-hydroxy-chlorpropamide and p-chlorobenzene sulfonylurea are more rapidly eliminated than the parent compound and are probably of negligible activity. Therefore the metabolites should not be analyzed in comparative bioavailability studies. Chlorpropamide is partly eliminated unchanged via kidney, and renal elimination can be influenced by a change of urine pH (NEUVONEN et al. 1987). When the drug was given after breakfast, a (slight) reduction of peak plasma concentrations was observed while extent of bioavailability remained unchanged (SARTOR et al. 1980b). Moreover, bioavailability was lightly increased by magnesium hydroxide in the early phase after drug intake (KIVISTÖ and NEUVONEN 1992), while only a very slight reduction of bioavailability was observed after coadministration of sucralfate (LETENDRE et al. 1986).

Comparative bioavailability studies with chlorpropamide preparations published in Italy and the United Kingdom have shown pronounced differences between the products (Fig. 10) (MONRO and WELLING 1974; TAYLOR et al. 1977). However, probably because of the reduced therapeutic importance of the compound, no actual data are available.

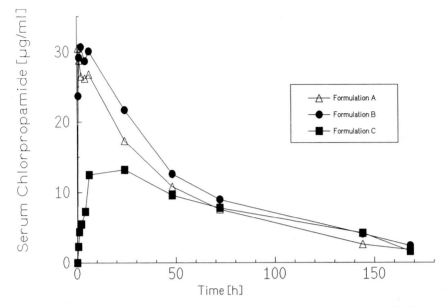

Fig. 10. Comparative bioavailability study of three marketed brands of chlorpropamide: mean serum concentrations of chlorpropamide in six healthy volunteers following single-dose administration of three different formulations (A, B, and C) each containing 250 mg chlorpropamide. (From MONRO and WELLING 1974)

3. Glibenclamide (Glyburide)

In European countries including Germany, glibenclamide is still the most frequently used oral antidiabetic drug (RATZMANN and THOELK 1992). With regard to the in vitro characterization of biopharmaceutical properties of glibenclamide tablets and their correlation to in vivo behavior, CHALK et al. (1986) were unable to prove a clear correlation between in vitro dissolution behavior tested by the British Pharmaco peial method and in vivo data, whereas BLUME et al. (1985c) demonstrated an in vitro/in vivo correlation between bioavailability data and dissolution data obtained with the USP paddle method for several glibenclamide formulations on the German market (Fig. 11).

After oral administration, rate of absorption may differ significantly between individuals; maximum plasma concentrations are reached 1.5–4 h after administration of tablets and 0.5–2 h after administration of oral solutions (IKEGAMI et al. 1985), depending on the biopharmaceutical characteristics of the solid oral dosage forms and the administration conditions (fasted or fed state). Rapid as well as delayed or even biphasic absorption has been observed not only in diabetic patients but also in elderly, nondiabetic volunteers.

As with other sulfonylureas, glibenclamide is a drug with a small volume

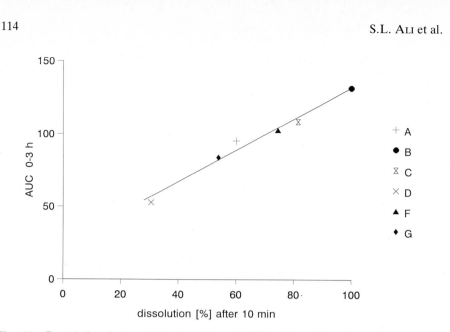

Fig. 11. Correlation between in vivo bioavailability data of glibenclamide and in vitro dissolution of six different formulations from the German market: AUC_{0-3h} values of glibenclamide after oral administration of six different formulations to seven healthy volunteers in comparison to in vitro dissolution (%) of the respective formulation after 10 min. (From BLUME et al. 1985c)

of distribution but extensive binding to plasma albumin (PEARSON 1985; FELDMAN 1985). Glibenclamide is rapidly and almost completely metabolized, resulting in at least two hydroxylated metabolites which are considered to be therapeutically active. Thus, a determination of metabolites of glibenclamide may be meaningful in bioavailability studies. Although a short elimination half-life of 1–2 h has been reported, glibenclamide is considered as "long-acting," so that prolonged hypoglycemias can be a problem. It has been hypothesized in the literature that this long action may be due to the existence of a "deep" compartment (BALANT et al. 1977); another possible explanation is the accumulation within the pancreatic islets to concentrations in excess of those in the extracellular water space. Elimination can be influenced by severe but not by moderate renal impairment in contrast to some "first-generation" oral antidiabetics (PEARSON et al. 1986).

Because genetic differences between subjects concerning glibenclamide metabolism are not known (PEART et al. 1989), phenotyping of participants in bioavailability studies is not regarded as necessary.

As glibenclamide is poorly soluble, its bioavailability, especially rate of absorption, is highly dependent on the physical properties of the raw material. Thus, introduction of micronized formulations resulted in a higher rate and extent of bioavailability (HAUPT et al. 1984; SCHEEN et al. 1987). This was also confirmed for some generic products by in vitro dissolution testing (BLUME et al. 1984) and bioavailability studies (BLUME et al. 1985a). Accordingly, dosage could be reduced from 5 to 3.5 mg or, from 2.5 to 1.75 mg/tablet, respectively, resulting in the similar AUC-values.

In contrast to other sulfonylureas, e.g., tolbutamide, glibenclamide follows a uniform elimination pattern in healthy Caucasian male subjects (SPRAUL et al. 1989), and thus varying bioavailability is mainly a result of varying absorption. No significant food effect concerning plasma concentrations of glibenclamide has been detected (PRENDERGAST 1984); not even a delayed peak concentration has been observed (SARTOR et al. 1980a). However, absorption can be influenced by coadministration of antacids, which may result from the weak acidic character of the drug. Extent of bioavailability of nonmicronized glibenclamide increases significantly during the first 4 h after administration together with antacids, whereas bioavailability of micronized glibenclamide is only slightly influenced (KIVISTÖ et al. 1993; ZUCCARO et al. 1989). However, as KUBACKA et al. (1987) found AUC of glibenclamide to be increased after concomitant administration with cimetidine, but not with ranitidine, increased gastric pH seems not to be the most relevant reason for this interaction. Nevertheless influence of pH on rate of glibenclamide absorption is emphasized by the results of BROCKMEIER et al. (1985), who showed that glibenclamide is absorbed to the same extent after direct instillation into stomach, duodenum or ascending colon but with significant differences in rate: absorption was fastest from duodenum. Although a relevant food effect was not assessed, bioavailability of glibenclamide is influenced by dietary fiber (SHIMA et al. 1983). Regarding the influence of food, bioavailability and pharmacodynamic effect have to be differentiated between, because it is well established that time of food intake influences glucose and insulin plasma levels after administration of sulfonylureas (SARTOR et al. 1982).

Because of its importance in the therapy of diabetes mellitus, glibenclamide has become the object of intense generic competition in many countries, particularly in Germany. As the compound has critical bioavailability properties, comparative in vitro dissolution and in vivo bioequivalence studies have been performed (BLUME et al. 1985c; BLUME et al. 1985a). The results of these investigations show marked differences concerning the dissolution behavior (Fig. 12) as well as in rate and extent of bioavailability (Fig. 13) between the products. Due to these differences, the respective preparations are not interchangeable during treatment without consequent adjustment of the dosing schedule to the differing biopharmaceutical properties.

4. Gliclazide

Rate of absorption of gliclazide varies markedly and a bimodal distribution of rates of absorption has been postulated. Moreover, elimination half-life also varies (6–14 h) and consequently gliclazide may be considered as being

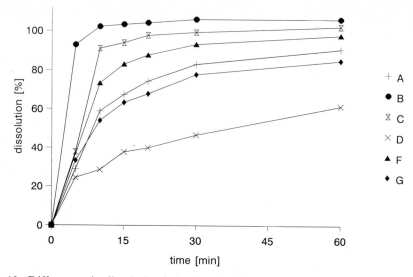

Fig. 12. Differences in dissolution behavior of glibenclamide tablets: in vitro dissolution of six different formulations from the German market in an USP paddle apparatus (900 ml buffer, pH 7.4, 75 U/min. (From BLUME et al. 1985c)

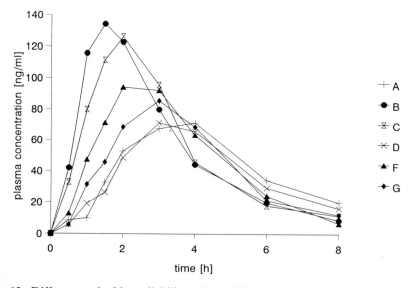

Fig. 13. Differences in bioavailability of six different glibenclamide formulations from the German drug market: mean plasma concentrations of glibenclamide after oral administration of 3.5-mg formulations to seven healthy volunteers. (From BLUME et al. 1985a)

highly variable in bioavailability (FERNER and CHAPLIN 1987). Although gliclazide is extensively metabolized – seven metabolites have been identified in urine – only the parent drug has been determined in plasma (OIDA et al. 1985). Elimination half-life is prolonged in males (compared with females) and in the elderly (IGAKI et al. 1992). However, no significant differences in pharmacokinetics have been found between single- and multiple-dose administrations (KOBAYASHI et al. 1984).

5. Glimepiride

Glimepiride is a new "second-generation" sulfonylurea. Regarding bioavailability and pharmacokinetics, little information is available from studies with tablets containing micronized glimepiride (HOECHST 1994, personal communication). In a single-dose crossover study, an absolute oral bioavailability of 107% was determined based on AUC of the parent compound. From the AUC of the metabolites, an absolute bioavailability of 109% was calculated. After administration of tablets containing between 1.0 mg and 8.0 mg glimepiride, dose linearity was observed concerning recovery of its metabolites in urine, AUCs and serum peak concentrations. Urinary excretion of the metabolites was nearly complete (97%). When the drug was administered following a standardized breakfast and an 8 h-fast, respectively, the overall pharmacokinetics of glimepiride showed no significant differences between C_{max}, t_{max} and AUC. Thus, absorption of glimepiride is not influenced by food intake.

6. Glipizide

Glipizide is a very common hypoglycemic drug in the United States, and a detailed review of its pharmacokinetics was given by LEBOVITZ (1985). After fasted administration the drug is rapidly and almost completely absorbed (GROOP et al. 1985), resulting in maximum plasma concentrations after 1.5 h (tablets) and 0.5 h (solution), respectively. Absolute oral bioavailability has been documented to be about 100% (WÅHLIN-BOLL et al. 1982). Probably caused by its more rapid and extensive absorption, glipizide exhibits a more powerful stimulation of insulin secretion than glibenclamide (GROOP et al. 1985). Glipizide undergoes extensive hepatic metabolism, mainly by hydroxylation of the cyclohexane ring, resulting in metabolites of little or no hypoglycemic activity. Consequently, these biotransformation products do not require determination in bioavailability studies. Elimination half-life is between 1 and 4 h after single-dose administration and 2–5 h in steady state after long-term treatment. For glipizide an enterohepatic recirculation is discussed (FERNER and CHAPLIN 1987), which may influence the assessment of bioavailability. On the other hand, pharmacokinetic data are not influenced by age, diabetes and multiple dosing (KRADJAN et al. 1989).

In contrast to glibenclamide, food intake influences bioavailability of glipizide formulations: the absorption was delayed (Fig. 14) when the drug was given together with food (PRENDERGAST 1984; WÅHLIN-BOLL et al. 1980).

Antacids such as sodium bicarbonate may cause a delay in absorption without influencing total AUC and C_{max} values (KIVISTÖ and NEUVONEN 1991). On the other hand, AUC of glipizide was significantly increased in patients treated with both cimetidine and ranitidine (FEELY et al. 1993). Cimetidine is well known as a potent inhibitor of hepatic metabolism in contrast to ranitidine; thus primarily the influence of increased gastric pH caused by both components has to be taken into consideration. It has been shown that hyperglycemia in healthy volunteers will delay the absorption of glipizide significantly, possibly due to impairment of gastric motility and/or gastric emptying as a result of high glucose plasma concentrations (GROOP et al. 1989).

7. Gliquidone

Little information is available on the bioavailability of gliquidone. Absorption is postulated to be fast and complete (KOPITAR and WILLIM 1975). The parent drug undergoes extensive hepatic demethylation and hydroxylation (KOPITAR and WILLIM 1975). One of the hydroxylated metabolites still exhibits hypoglycemic activity; however, because of its very low concentrations in plasma only a small contribution to the antidiabetic effect can be expected. In bioavailability studies relatively long measurements are necessary because of the biphasic elimination of the compound and its long terminal elimination half-life ($t_{1/2\alpha}$, 1 h, $t_{1/2\beta}$, 24 h).

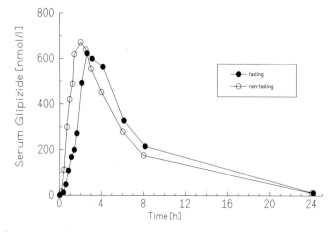

Fig. 14. Influence of food intake on the bioavailability of glipizide: mean serum concentrations of glipizide in nine healthy volunteers following ingestion of 5 mg in fasting and nonfasting (standardized breakfast) state. The delay of about 40 min in the time to reach peak plasma concentration was significant ($P < 0.05$). (From WÅHLIN-BOLL et al. 1980)

8. Glisoxepide

Glisoxepide is rapidly and almost completely absorbed after oral administration of immediate-release tablets; absolute bioavailability was assumed to be nearly 100% (SPECK et al. 1974). After oral administration of ^{3}H-glisoxepide to healthy volunteers, maximum plasma concentrations were determined after 2–5 h. According to SPECK et al. (1974), rate of absorption was slightly faster in diabetic patients, but no correlation to gastric pH and age, respectively, could be observed.

Metabolic pattern of glisoxepide in human is not completely known. However, it can be assumed that metabolites do not contribute significantly to clinical efficacy of the compound. Accordingly, the analytical determination of the parent compound can be considered sufficient in bioavailability studies. According to SCHWARZKOPF and KEWITZ (1977), plasma elimination half-life and extent of bioavailability of the parent compound was unchanged in patients with renal failure.

9. Tolazamide

After oral administration of immediate-release tablets, tolazamide is absorbed slowly with a lag time of 45 min, and maximum plasma concentrations were determined after 3.3 h (FERNER and CHAPLIN 1987). Absorption is even slower in diabetic patients with asymptomatic neuropathy compared with healthy volunteers. Absolute bioavailability is not known. Extensive hepatic hydroxylation takes place and the hydroxymetabolite exhibits therapeutic activity; however, the extent of its contribution to the clinical effect has not been clearly evaluated so far. In vitro dissolution experiments (USP 1990 method for uncoated tablets) and investigating mean disintegration times of different formulations failed to be predictive for rate and extent of bioavailability of the products (WELLING et al. 1982). An in vitro/in vivo correlation has therefore not been discovered so far for tolazamide products.

10. Tolbutamide

Absolute bioavailability of tolbutamide, a "first-generation" sulfonyl urea, is not known (FERNER and CHAPLIN 1987). Tolbutamide is metabolized almost completely and its main metabolites, hydroxytolbutamide and carboxytolbutamide, have little or no activity (BACK and ORME 1989). Thus, these biotransformation products should not necessarily be determined in bioavailability/bioequivalence studies. The first metabolic step is claimed to be genetically polymorphic and both phenylbutazone and cimetidine may act as inhibitors (SCOTT and POFFENBARGER 1979). A trimodal distribution has been postulated; however, the published results are contradictory and have not been explained conclusively (BACK and ORME 1989; VERONESE et al. 1993). A group of poor metabolizers has been identified (CHEN et al. 1993) and it was shown that tolbutamide is metabolized by cytochrome P450 to a highly

varying degree. Thus phenotyping of the participating volunteers with
mephenytoin in order to reduce the variability in bioequivalence studies may
be useful (BACK and ORME 1989).

No significant food effect was observed for tolbutamide by ANTAL et al.
(1982) and SARTOR et al. (1980b). Absorption is markedly reduced by concomitant administration of activated charcoal or antacids (KIVISTÖ and
NEUVONEN 1992). Coadministration of cimetidine increased tolbutamide
AUC and decreased its elimination half-life (BACK et al. 1988). Furthermore
the AUC ratio of carboxytolbutamide/tolbutamide was also decreased by

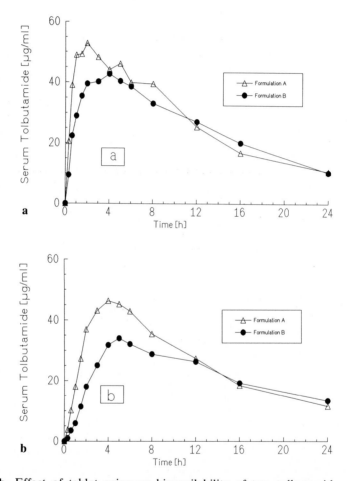

Fig. 15a,b. Effect of tablet aging on bioavailability of two tolbutamide products:
mean serum concentrations of tolbutamide in 16 healthy volunteers following administration of the recently manufactured formulations A and B together with food (**a**)
and the same formulations after short-term storage (98% humidity, 3 days, ambient
temperature) under fasting conditions (**b**); each formulation contained 500 mg tolbutamide. (From OLSON et al. 1985)

cimetidine, whereas ranitidine showed no respective effect (CATE et al. 1986). Consequently, this interaction should primarily be a result of the metabolic interference of cimetidine with the cytochrome P450 system and not of the elevated gastric pH caused by both components. Moreover, it was shown that tablet aging may significantly influence the bioavailability of tolbutamide tablets (OLSON et al. 1985). Plasma concentration versus time profiles were reported to be decreased after storage for 3 days at accelerated conditions (98% relative humidity), when compared with the results of recently manufactured tablets (Fig. 15a,b). However, in this study the freshly manufactured tablets were coadministered with food, whereas administration of the stored tablets was performed after a 9 h-fast.

Acknowledgements. The authors thank Mrs. M. Koellner and Ms. G. Lühr for their very efficient help in preparing this chapter. Mrs. E. Will kindly assisted in drawing the figures. Mrs. M. Elbert took great pains in typing the manuscript.

References

Abdel-Hady EM, Belal SF, El-Walily AM, Abdine H (1979a) Spectrophotometric determination of tolbutamide in tablets. J Assoc Off Anal Chem 62:533–537

Abdel-Hady EM, Belal SF, El-Walily AM, Abdine H (1979b) Spectrophotometric determination of tolbutamide, thiamine hydrochloride and pyridoxine hydrochloride in combination products. J Pharm Sci 68:739–741

Abdel-Hamid M, Suleiman MS, El-Sayed YM, Najib NM, Hasan MM (1989) Rapid HPLC assay of glibenclamide in serum. J Clin Pharm Ther 14:181–188

Achelis JD, Hardebeck K (1965) Über eine blutzuckersenkende Substanz. Dtsch Med Wschr 80:1452–1455

Adams WJ, Krueger DS (1979) Specific and sensitive HPLC determination of glyburide. J Pharm Sci 68:1138–1140

Agarwal SP, Walash MJ (1972) Indian J Pharm 34:109

Aggarwal V, Sunshine I (1974) Determination of sulphonylureas and metabolites by pyrolysis-GC. Clin Chem 20:200

Antal EJ, Gillespie WR, Phillips JP, Albert KS (1982) The effect of food on the bioavailability and pharmacodynamics of tolbutamide in diabetic patients. Eur J Clin Pharmacol 22:459–462

APV Guideline (1987) Studies on bioavailability and bioequivalence. Drugs made in Germany 30:161–166

Arcelloni C et al. (1990) Glibenclamide and tolbutamide in human serum rapid measurement of the free fraction. J Liq Chromatogr 13:175–189

Back DJ, Orme MLE (1989) Genetic factors influencing the metabolism of tolbutamide. Pharmacol Ther 44:147–155

Back DJ, Tjia J, Mönig H, Ohnhaus EE, Park BK (1988) Selective inhibition of drug oxidation after simultaneous administration of two probe drugs, antipyrine and tolbutamide. Eur J Clin Pharmacol 34:157–163

Balant L, Zahnd GR, Weber F, Fabre J (1977) Behaviour of glibenclamide on repeated administration to diabetic patients. Eur J Clin Pharmacol 11:19–25

Becker R (1977) Fluorimetrische Bestimmung von Glibornuride in Plasma und Serum. Arzneimittel-forschung 27:102–105

Beyer WF (1972) Quantitative liquid chromatography of sulphonylureas in pharmaceutical products. Anal Chem 44:1312–1314

Beyer WF, Jensen EH (1974) Tolbutamide. In: Florey K (ed) Analytical profiles of drug substances, vol 3. Academic, New York

Blume H, Ali SL, Siewert M (1984) Zur pharmazeutischen Qualität von glibenclamid-haltigen Fertigarzneimitteln. Pharm Ztg 129:983–989

Blume H, Ali SL, Stenzhorn G, Stüber W, Siewert M (1985a) Zur Bioverfügbarkeit und pharmakodynamischen Aktivität handelsüblicher Glibenclamid-Fertigarzneimittel; 3. Mitteilung. Pharm Ztg 130:2605–2610

Blume H, Förster H, Stenzhorn G, Askali F (1985b) Zur Bioverfügbarkeit und pharmakodynamischen Aktivität handelsüblicher Glibenclamid-Fertigarzneimittel; 2. Mitteilung. Pharm Ztg 130:1070–1078

Blume H, Stenzhorn G, Ali SL (1985c) Zur Bioverfügbarkeit und pharmakodynamischen Aktivität handelsüblicher Glibenclamid-Fertigarzneimittel; 1. Mitteilung. Pharm Ztg 130:1062–1069

Blume H, Walter-Sack I, Ali SL, Siewert M, Stenzhorn G, Nowak N, Weber E (1987) Untersuchungen zur therapeutischen Relevanz der Bioäquivalenz und zur Chargenhomogenität glibenclamidhaltiger Fertigarzneimittel. Pharm Ztg 132:2352–2362

Blume H, Kübel-Thiel K, Reutter B, Siewert M, Stenzhorn G (1988) Nifedipin: Monographie zur Prüfung der Bioverfügbarkeit/Bioäquivalenz von schnell-freisetzenden Zubereitungen. Pharm Ztg 133:389–393

Bottari F, Mannelli M, Saettone MF (1970) Thermal dissociation of sulphonylureas I. J Pharm Sci 59:1663–1666

Bottari F, Giannaccini B, Mannipieri E, Saettone MF (1972) Thermal dissociation of sulphonylureas II. J Pharm Sci 61:602–606

Braselton WE, Bransome ED, Ashline HC, Stewart JT, Honigberg IL (1976) Gas chromatographic and mass spectral properties of sulphonylurea n-methyl-n-perfluoroacyl-derivatives. Anal Chem 48:1386–1394

Bräuniger H, Duda H (1958) Beitrag zur Analytik des Sulfanilyl-n-butylcarbamid. Pharm Ztg 97:305–310

Bräuninger H, Moede F (1962) Paperchromatographie der Sulfonamide und Sulfonamide-N-glycoside. Pharma Zh 101:383–389

British Pharmacopeia (1988) Her Majesty's Stationery Office, London

Brockmeier D, Grigoleit H-G, Leonhardt H (1985) Absorption of glibenclamide from different sites of the gastro-intestinal tract. Eur J Clin Pharmacol 29:193–197

Castoldi D, Tofanetti O (1979) Gas-chromatographic determination of glibenclamide in plasma. Clin Chim Acta 93:195

Cate EW, Roger JF, Powell JR (1986) Inhibition of tolbutamide elimination by cimetidine but not ranitidine. J Clin Pharmacol 26:372–377

Chakrabarti JK (1962) Detection of hypoglycemic sulphonylureas on paper chromatograms. J Chromatogr 8:414–416

Chalk JB, Patterson M, Smith MT, Eadie MJ (1986) Correlations between in vitro dissolution, in vivo bioavailability and hypoglycaemic effects of oral glibenclamide. Eur J Clin Pharmacol 31:177–182

Chen L-S, Yasumori T, Yamazoe Y, Kato R (1993) Hepatic microsomal tolbutamide hydroxylation in Japanese: in vitro evidence for rapid and slow metabolizers. Pharmacogenetics 3:77–85

Clark's Isolation and Identification of Drugs (1986) 2nd edn. Pharmaceutical Press, London

Courter M et al. (1957) Congres Soc Pharm France 9e. Clermont-Ferrand, pp 267–277

CPMP Note for Guidance (1991) Investigation of bioavailability and bioequivalence. CPMP working party on efficacy of medicinal products

Csillag K, Vereczkey L, Gachalyi B (1989) Simple HPLC method for the determination of tolbutamide and its metabolites in human plasma and urine using photodiode-assay detection. J Chromatogr 490:355–363

Das Gupta V (1986) Quantitation of glipizide and glyburide (glibenclamide) in tablets using HPLC. J Liq Chromatogr 16:3607–3615

Deutsches Arzneibuch (1991) 10th edn, vols I, II, III. Deutscher Apotheker-Verlag Govi-Verlag, Stuttgart

Deutscher Arzneimittel-Codex (1986) Vols I, II. Govi-Verlag Deutscher Apotheker-Verlag, Frankfurt

Dibbern HW (1978) UV- und IR-Spektren wichtiger pharmazeutischer Wirkstoffe, vol II. Oberschwäbische Verlagsanstalt, Ravensburg

Dorfmüller T (1957) Das Ärztl Laboratorium 3:8–18

El-Bardicy MG, Khateeb SZ, Assaad HN, Ahmad AS (1988) Mercurimetric determination of chlorpropamide by back titration. Indian J Pharm Sci 56:171–172

El-Gizawy S (1991) Bulletin of Faculty of Science, Assint University, Egypt, vol 20, p 65

El-Khateeb SZ, Assaad HN, Ellaithy MM, Ahmad AS (1987) Determination of chlorpropamide and its tablets by PMR spectrometry. Spectroscopic Letters 20:89–96

Emilsson H, Sjöberg S, Svedner M, Christenson I (1986) High-performance liquid-chromatographic determination of glibenclamide in human plasma and urine. J Chromatogr 383:93–102

Emmanuel J, Fernandes NN (1989) Colorimetric estimation of tolbutamide and its dosage forms. Indian Drugs 26:189–190

Everett RL (1986) Liquid-chromatographic determination of chlorpropamide in tablet dosage forms: collaborative studies. J Assoc Off Anal Chem 69:519–521

Feely J, Collins WCJ, Cullen M, El Debani AH, Macwalter RS, Peden NR, Stevenson IH (1993) Potentiation of the hypoglycaemic response to glipizide in diabetic patients by histamine H_2-receptor antagonists. Br J Clin Pharmacol 35:321–323

Feldman JM (1985) Glyburide: a second-generation sulphonylurea hypoglycemic agent. Pharmacotherapy 5:43–62

Feldman M, Schiller LR (1983) Disorders of gastrointestinal motility associated with diabetes mellitus. Ann Intern Med 98:378–384

Ferner RE, Chaplin S (1987) The relationship between the pharmacokinetics and pharmacodynamic effects of oral hypoglycaemic drugs. Clin Pharmacokinet 12:379–401

Finkle BS, Cherry EJ, Taylor DM (1971) J Chromatogr Sci 9:393

Ford JL, Stewart AF, Rubinstein MH (1979) The assay and stability of chlorpropamide in solid dispersion with urea. J Pharm Pharmacol 31:726–729

Forist AA (1959) Determination of methexamide in human plasma. Ann Acad Sci 82:496–501

Forist AA, Chulski T (1958) pH-solubility relationship for 1-butyl-3-p-tolysulphony-lurea and its metabolite. Metabolism 5:807–812

Fricke FL (1972) Analysis for drugs in pharmaceuticals, using simple extractions and semi-automated GLC, J Assoc Off Anal Chem 55:1162

Gaidukevich AN, Sidom MB (1977) Farm Zh (Kiew) 4:70–73

Galloway JA, McMahon RE, Culp HW, marshall F J, Youn E C (1967) Metabolism, blood levels and rate of excretion of acetohexamide in human subjects. Diabetes 16:118–127

Groop L, Wåhlin-Boll E, Groop P-H, Töttermann K-J, Melander A, Tolppanen E-M, Fyhrqvist F (1985) Pharmacokinetics and metabolic effects of glibenclamide and glipizide in type 2 diabetics. J Clin Pharmacol 28:697–704

Groop L, Luzi L, DeFronzo RA, Melander A (1989) Hyperglycaemia and absorption of sulphonylurea drugs. Lancet 15:129–130

Gupta RN (1989) Determination of glyburide (glibenclamide) in human plasma by liquid chromatography with fluorescence detection. J Liq Chromatogr 12:1471

Gwilt PR, Nahhas RR, Tracewell WG (1991) The effects of diabetes mellitus on pharmacokinetics and pharmacodynamics in humans. Clin Pharmacokinet 20:477–490

Hager's Handbuch der pharmazeutischen Praxis (1993) 5th edn, vol 7, Stoffe A dis D, Springer, Berlin Heidelberg New York

Hajdu P, Kohler KF, Schmidt FH, Speingler H (1969) Phys-chemische und analytische untersuchungen an HB 419. Arzneimittelforschung 19:1381–1386

Haller H, Strauzenberg SE (1959) Perorale Diabetestherapie. Thieme, Leipzig

Harzer K (1980) Nachweis von Glibornurid im Serum durch HPLC mit umgekehrten Phasen. J Chromatogr 183:115–117

Haupt E, Putschky F, Zoltobrocki M, Schöffling K (1984) Pharmakodynamik und Pharmakokinetik zweier Glibenclamid-Zubereitungen beim Typ-II-Diabetes. Dtsch Med Wschr 109:210–213

Häussler A, Hajdu P (1962) Die Spaltung von N-[4 Methylbenzolsulphonyl]-N'-butylharnstoff (Rastinon Hoechst) in alkalischer Lösung. Arch Pharm 295: 471–474

Häussler A, Pechtold F (1969) Die Analytik der Sulphonylharnstoffe, Handbuch der experimentellen Pharmakologic, vol XXIV, edn. Diuretika. Springer, Berlin Heidelberg New York, p 251–290

Hentrich K (1963) Zur papierchromatographischen Unterscheidung von Sulfonamid-Diuretika und Sulfonamiden anderer Indikationsgebiete. Pharmazie 18:405–409

Hill HM, Chamberlain J (1978) Determination of oral anti-diabetic agents in human body fluids using HPLC. J Chromatogr 149:349–358

Ibrahim SE, Al-Badr AA (1984) Tolbutamide. In: Florey K (ed) Analytical profiles of drug substances, vol 13. Academic, New York

Igaki A, Kobayashi K, Kimura M, Sakoguchi T, Matsuoka A (1989) Determination of serum sulphonylureas by HPLC with fluorimetric detection. J Chromatogr 493:222–229

Igaki A, Kobayashi K, Kimura M, Sakoguchi T, Matsuoka A (1992) Influence of blood proteins on biomedical analysis. XII. Effects of glycation on gliclazide (oral hypoglycemic drug) binding with serum albumin in diabetics. Chem Pharm Bull (Tokyo) 40:255–257

Ikegami H, Shima K, Tanaka A, Tahara Y, Hirota M, Kumahara Y (1985) Pharmacokinetics of glibenclamide: heterogeneity in its absorption. Med J Osaka Univ 35:55–61

Ikegami H, Shima K, Tanaka A, Tahara Y, Hirota M, Kumahara Y (1986) Interindividual variation in the absorption of glibenclamide in man. Acta Endocrinol (Copenh) 111:528–532

Imamura Y, Kojima Y, Ichibagase H (1985) Effect of simultaneous administration of drugs on absorption and excretion. Chem Pharm Bull (Tokyo) 33:1281–1284

Iskender G, Orak F (1986) Istanbul University EC Zacihk Fak Mecur 22:9

Keal J, Stockely C, Somogyi A (1986) Simultaneous determination of tolbutamide and its hydroxy-and carboxy-metabolites in plasma and urine by HPLC. J Chromatogr 378:237–241

Kern W (1963) Chemical microdetermination of phenyl and tolylsulphonylurea derivatives in blood. Anal Chem 35:50–53

Kiger JL, Kiger JG (1966) Ann Pharm Fr 24:593

Kimura M, Kobayashi K, Hata M, Matsuoka A, Kitamura H, Kimura Y (1980) Determination of gliclazide in human serum by HPLC using an anion-exchange resin. J Chromatogr 183:467–473

Kivistö KT, Neuvonen PJ (1991) Differential effects of sodium bicarbonate and aluminium hydroxide on the absorption and activity of glipizide. Eur J Clin Pharmacol 40:383–386

Kivistö KT, Neuvonen PJ (1992) Effect of magnesium hydroxide on the absorption and efficacy of tolbutamide and chlorpropamide. Eur J Clin Pharmacol 42: 675–680

Kivistö KT, Lehto P, Neuvonen PJ (1993) The effects of different doses of sodium bicarbonate on the absorption and activity of non-micronized glibenclamide. Int J Clin Pharmacol Ther Toxicol 31:236–240

Kleber JW, Galloway JA, Rodda BE (1997) GLC determination of acetohexamide and hydroxyhexamide in biological fluids. J Pharm Sci 66:635–638

Knight JB, Matin SB (1974) Use of solid sampler in chemicall-ionisation mass spectrometry for the determination of compounds unstable under GLC conditions. Analysis for tolbutamide from biological fluid. Anal Lett 7:529

Kobayashi K, Kimura M, Sakoguchi T, Hase A, Matsuoka A, Kaneko S (1984) Pharmacokinetics of gliclazide in healthy and diabetic subjects. J Pharm Sci 73:1684–1687

Kopitar Z, Willim KD (1975) Humanpharmakokinetik und Metabolismus von ^{14}C-markiertem Gliquidone (AR-DF 26). Arzneimittelforschung/Drug Res 25: 1455–1460

Kracmarova J, Kracmar J (1958) Zur Bestimmung von neuen peroralen Antidiabetics Cs. Farm 7:566–569

Kradjan WA, Kobayashi KA, Bauer LA, Horn JR, Opheim KE, Wood F Jr (1989) Glipizide pharmacokinetics: effects of age, diabetes, and multiple dosing. J Clin Pharmacol 29:1121–1127

Kubacka RT, Antal EJ, Juhl RP (1987) The paradoxical effect of cimetidine and ranitidine on glibenclamide pharmacokinetics and pharmacodynamics. Br J Clin Pharmacol 23:743–751

Lebovitz HE (1985) Glipizide: a second-generation sulphonylurea hypoglycemic agent. Pharmacotherapy 5:63–77

Letendre PW, Carlson JD, Seifert RD, Dietz AJ, Dimmit D (1986) Effect of sucralfate on the absorption and pharmacokinetics of chlorpropamide. J Clin pharmacol 26:622–625

Lucattelli I (1956) Nachweis und Bestimmung von N-(Butyl-Carbamyl)-sulfanilamid. Farmaco 11:452–457

Macek K (1972) Pharmaceutical applications of thin-layer and paper-chromatographie, oral antidiabetics. Elsevier, Amsterdam

Maha GE, Kirtley WR, Root MA, Anderson RC (1962) Acetohexamide: preliminary report on a new oral hypoglycemic agent. Diabetes 11:83–91

Martindale (1989) Martindale: the extra pharmacopeia, 29th edn. EF Reynolds Ltd, Pharmaceutical Press, London

Matin SB, Rowland M (1973a) Determination of tolbutamide and chlorpropamide in biological fluids. J Pharm Pharmacol 25:186–187

Matin SB, Rowland M (1973b) Simultaneous determination of tolbutamide and its metabolites in biological fluids. Anal Lett 6:865

McDonald H, Sawinsky V (1958) Texas Rep Biol Med 16:479

McMahon RE, Marshall FJ, Culp HW (1965) The nature of the metabolites of acetohexamide in the rat and in the human. J Pharmacol Exp Ther 149:272–279

Menzer M, Presewowski J, Haug U (1971) Pharmazie 26:738

Mesnard P (1960) Boll Chim Farm 99:818–836

Mesnard P, Crockett R (1960) Les méthodes de dosage des sulfamides hypoglycémiants non amines. Chim Anal 42:346–354

Midha KK, McGilveray, Charette C (1976) GLC determination of plasma levels of intact chlorpropamide or tolbutamide. J Pharm Sci 65:576–579

Miller WL, Krake JJ, Vander Brook MJ, Reineke LM (1957) Studies on the absorption mechanism of action and excretion of tolbutamide in the rat. Ann NY Acad Sci 71:118–124

Mohnike G, Wittenhager G, Langenbeck W (1958) Über das Ausscheidungsprodukt von N(4-Methyl-benzolsulphonyl)-N'-butylharnstoff beim Hund. Naturwissenschaften 45:13

Molins D, Wong CK, Cohen DM, Munnelly KP (1975) HPLC determination of chlorpropamide in tablet formulations. Pharm Sci 64:123–124

Möller H, Ali SL, Gundert-Remy U (1983) Biopharmazeutische, pharmakokinetische und pharmakodynamische Aspekte von Glibenclamid-Zubereitungen. Die Medizinische Welt 34:949–954

Monro AM, Welling PG (1974) The bioavailability in man of marketed brands of chlorpropmaide. Eur J Clin Pharmacol 7:47–49

Morgenstern LL, Garrett ER (1959) Determination of the half-life of metahexamide in normal humans. Ann NY Acad Sci 82:502–507

Namigohar F, Makhani M, Radparvar S (1980) Trav Soc Pharm Montpellier 40:55

National Formulary (1970) XIII edn. American Pharmaceutical Association, Washington

Neidlein R, Klügel G, Lebert U (1965), Dünnschichtchromatographische Trennung einiger oraler Antidiabetica und Sulfonamide. Pharm Ztg 20:651–652

Nelson E, O'Reilly J, Chulski T (1960) Determination of carboxytolbutamide in urine. Clin Chim Acta 5:774–776

Neuvonen PJ, Kärkkäinen S, Lehtovaara R (1987) Pharmacokinetics of chlorpropamide in epileptic patients: effects of enzyme induction and urine pH on chlorpropamide elimination. Eur J Clin Pharmacol 32:297–301

Ni W, Tong Y (1988) Assay of tolbutamide tablets by differential scanning colorimetry. YiYao Gongye 19:132–134

Oida T, Yoshida K, Kagemot A, Sekine Y, Higashijima T (1985) The metabolism of gliclazide in man. Xenobiotica 15:87–96

Olson SC, Ayres JW, Antal EJ, Albert KS (1985) Effect of food and tablet age on relative bioavailability and pharmacodynamics of two tolbutamide products. J Pharm Sci 74:735–740

Othman S, Shareen O, Jalal I, Awidi A, Al-Turk W (1988) Liquid chromatographic determination of glibenclamide in human plasma. J Assoc Off Anal Chem 71:942–944

Pearson JG (1985) Pharmacokinetics of glyburide. Am J Med 79[Suppl 3B]. 67–71

Pearson JG, Antal EJ, Raehl CL, Gorsch HK, Craig WA, Albert KS (1986) Pharmacokinetic disposition of ^{14}C-glyburide in patients with varying renal function. Clin Pharmacol Ther 39:318–324

Peart GF, Boutagy J, Shenfield GM (1989) The metabolism of glyburide in subjects of known debrisoquin phenotype. Clin Pharmacol Ther 45:277–284

Pecorari P, Albasini A Melegari M, Vampa G, Maneti F (1981) An HPLC method for the simultaneous determination of tolbutamide and its metabolites in plasma. Farmaco 36:7

Pignard P (1958) Ann Biol Clin (Paris) 16:471–480

Popa J, Voicu A (1962) Dozarea complexometrica a diabetamidului si cicloralului. Farmacia (Buc) 10:399–402

Prendergast B D (1984) Glyburide and glipizide, second-generation of oral sulphonylurea hypoglycemic agents. Clin Pharm 3:473–485

Prescott LF, Redmann DR (1972) GLC estimation of tolbutamide and chlorpropamide in plasma. J Pharm Pharmacol 24:713–716

Raghow G, Meyer MC (1981) HPLC assay of tolbutamide and carboxytolbutamide in human plasma. J Pharm Sci 70:1166–1167

Ratzmann KP, Thoelk H (1992) Trends in der Verordnung von oralen Antidiabetika. Med Klin 87:8–11

Reisch J, Bornfleth H, Tittel GL (1964) Zur Analytik von Arzneigemischen, 4. Mitteilung: Die Dünnschichtchromatographie einiger als Antidiabetica verwendeter Sulfonamide. Pharm Ztg 2:74–75

Robertson DL, Butterfield AG, Kolasinski H, Lovernig EG, Matsui FF (1979) Stability-indicating HPLC determination of chlorpropamide, tolbutamide, and their respective sulfonamide degradates. J Pharm Sci 68:577–580

Rydberg T, Wählen-Boll E, Melander E (1991) Determination of glibenclamide and its two major metabolites in human serum and urine by column liquid chromatography. J Chromatogr 564:223–233

Sabih K, Sabih K (1970) Gas chromatographic method for determination of tolbutamide and chlorpropamide. J Pharm Sci 59:782–784

Sadana GS, Gaonkar MV (1989) Quantitative HPLC determination of chlorpropamide in pharmaceutical dosage forms. Indian Drugs 26:180–184

Saffar F et al. (1982) Chem Pharm Bull (Tokyo) 3:679

Sartor G, Melander A, Scherstén B, Wåhlin-Boll E (1980a) Serum glibenclamide in diabetic patients, and influence of food on the kinetics and effects of glibenclamide. Diabetologia 18:17–22

Sartor G, Melander A, Scherstén B, Wåhlin-Boll E (1980b) Influence of food and age on the single-dose kinetics and effects of tolbutamide and chlorpropamide. Eur J Clin Pharmacol 17:285–293

Sartor G, Lundquist I, Melander A, Scherstén B, Wåhlin-Boll E (1982) Improved effects of glibenclamide on administration before breakfast. Eur J Clin Pharmacol 21:403–408

Scheen AJ, Jaminet C, Luyckx AS, Lefebvre PJ (1987) Pharmacokinetics and pharmacological properties of two galenical preparations of glibenclamide, HB419 and HB420, in non insulin-dependent (type 2) diabetes. Int J Pharm Ther Toxicol 25:70–76

Schwartzkopff T, Kewitz H (1977) Elimination of glisoxipide in patients with renal failure. Arch Pharmacol 297[Suppl 2]:R61

Scott J, Poffenbarger PL (1979) Pharmacogenetics of tolbutamide metabolism in humans. Diabetes 28:41–51

Shafer CE (1972) Acetohexamide. In: Florey K (ed) Analytical profiles of drug substances, vol 1. Academic, New York

Shenfield GM, Bontagy IS, Webb C (1990) Screening test for detecting sulphonylureas in plasma. Ther Drug Monit 12:393

Shiba T, Kajinuma H, Suzuki K, Hagura R, Kawai A, Katagiri H, Sando H, Shirakawa W, Kosaka K, Kuzuya N (1986) Serum gliclazide concentration in diabetic patients. Diabetes Res Clin Prac 2:301–306

Shima K, Tanaka A, Ikegami H, Tabata M, Sawazaki N, Kumahara Y (1983) Effect of dietary fiber, glucomannan, on absorption of sulphonylurea in man. Horm Metab Res 15:1–3

Skelly JP, Shin GF (1993) In vitro/in vivo correlations in biophamaceutics: scientific and regulatory implications. Eur J Drug Metab Pharmacokinet 18:121–129

Skelly JP, van Buskirk GA, Savello DR, Amidon GL, Arbit HM, Dighe S, Fawzi MB, Gonzalez MA, Malick AW, Malinowski H, Nedich R, Peck GE, Pearce DM, Shah V, Shangraw RF, Schwartz JB, Truelove J (1993) Scala-up of immediate release oral solid dosage forms. J Parenter Sci Technol 47:52–56

Smith DL, Vecchio Th J, Forist AA (1965) Metabolism of antidiabetic sulphonylurea in man. Metabolism 14:229–240

Speck U, Mützel W, Kolb KH, Acksteiner B, Schulze PE (1974) Pharmakokinetik und Metabolitenspektrum von Glisoxepid beim Menschen. Drug Res 24:404–409

Spraul M, Streek A, Nieradzik M, Berger M (1989) Uniform elimination pattern for glibenclamide in healthy caucasian males. Arzneimittelforschung/Drug Res 39:1449–1450

Springler H, Kaiser F (1956) Die Bestimmung von N-(4-Methyl-benzolsulphonyl) N'-butyl-harnstoff in Serum. Arzneimittelforschung 6:760–762

St-Hilaire S, Belanger PM (1989) Simultaneous determinations of tolbutamide and its hydroxy- and carboxy-metabolites in serum and urine: application to pharmacokinetic studies of tolbutamide in the rat. J Pharm Sci 78:763–766

Starkey BJ, Mould GP, Teale ID (1988) Measurement of sulphonylurea drugs in plasma by HPLC. Ann Clin Biochem [Suppl] 25:206–207

Steinijans VW, Hauschke D (1990) Update on the statistical analysis of bioequivalence studies. Int J Clin Pharmacol Ther Toxicol 28:45–50

Steinijans VW, Hauschke D, Jonkman JHG (1992) Controversies in bioequivalence studies. Clin Pharmacokinet 22:247–253

Stowers JM (1958) Pharmacology and mode of action of the sulphonylurea in man. Lancet I:278–283

Sved S, McGilveray IJ, Beaudoin N (1976) Assay of sulphonylureas in human plasma by HPLC. J Pharm Sci 65:1356–1359

Takla PG (1981) Glibenclamide. In: Florey K (ed) Analytical profiles of drug substances, vol 10. Academic, New York
Takla PG, Joshi SR, Mahbouba M (1982) Fluorimetric determination of glibenclamide in plasma using pyrene-1-aldehyde. Analyst 107:1246–1254
Taylor JA (1972) Clin Pharmacol Ther 13:710
Taylor T, Assinder DF, Chasseaud LF, Bradford PM, Burton JS (1977) Plasma concentrations, bioavailability and dissolution of chlorpropamide. Eur J Clin Pharmacol 11:207–212
The Merck Index (1983) An encyclopedia of chemicals, drugs and biologicals, 10th edn. Merck, Rahway
The Pharmaceutical Codex (1979) 11th edn. Pharmaceutical Press, London
The Sadtler Standard Spectra (1970) Philadelphia
The United States Pharmacopeia (USP) (1970) XVIII-th rev, United States Pharmacopeial Convention, Rockville
The United States Pharmacopeia (USP) (1990) 22nd edn. United States Pharmacopeial Convention, Rockville
Theodorescu N, Circoana R, Văzoiu V (1966) Farmacia (Buc) 14:167–171
Thomas RC, Judy RW (1972) J Med Chem 15:964
Toolan TJ, Wagner R Jr. (1959) The physical properties of chlorpropamide and its determination in human serum. Ann NY Acad Sci 74:449–458
Tudor M, Melia CD, Davies MC (1991) The application of NIR FT-Raman spectroscopy to the qualitative analysis of chlorpropamide polymorphism. J Pharm Pharmacol 43[Suppl]:67
Veronese ME, Miners JO, Rees DLP, Birkett DJ (1993) Tolbutamide hydroxylation in humans: lack of bimodality in 106 healthy subjects. Pharmacogenetics 3:86–93
Vitaly M, Pancrazio G (1956) Analisi delle compresse di N_1-sulfanilil-N_2-N-butilurea. Farmaco 11:512
Vogt H (1959) Versuche zur Spaltung einiger substituierter Benzolsulphonylharnstoffe durch Einwirkung vo Säuren. Pharm Ztg 98:651–655
Wåhlin-Boll E, Melander A (1979) HPLC determination of glipizide and some other sulphonylurea drugs in serum. J Chromatogr 164:541–546
Wåhlin-Boll E, Melander A, Sartor G, Scherstén B (1980) Influence of food intake on the absorption and effect of glipizide in diabetics and in healthy subjects. Eur J Clin Pharmacol 18:279–283
Wåhlin-Boll E, Almér L-O, Melander A (1982) Bioavailability, pharmacokinetics and effects of glipizide in type 2 diabetics. Clin Pharmacokinet 7:363–372
Wartmann-Hafner F, Büchi J (1965) Untersuchungen über die Gehaltsbestimmung von Carbutamide. Pharm Acta Helv 40:592–609
Weber DJ (1976) HPLC analysis of tolbutamide in serum. J Pharm Sci 65:1502–1504
Welling PG, Patel RB, Patel UR, Gillespie WR, Craig WA, Albert KS (1982) Bioavailability of tolazamide from tablets: comparison of in vitro and in vivo results. J Pharm Sci 71:1259–1263
West KM, Johnson PC (1959) The comparative pharmacology of tolbutamide, carbutamide, chlorpropamide and metahexamide in man. Metabolism 8:596–605
Wickramasinghe JAF, Shaw SR (1971) GC-determination of tolazamide in plasma. J Pharm Sci 60:1669–1672
Wu J, Yang X (1987) HPLC of glibenclamide in its pharmaceutical preparation. Yiyao Gongye 18:21–23
Zhu P, Bay Y (1988) Quantitative determination of glibenclamide in Xiao-ke pills by TLC-UV spectrophotometry. Yacowa Fenxi Zazhi 8:27–29
Zorya BP et al. (1989) Farmatsiya (Moscow) 38:69–71
Zuccaro P, Pacifici R, Pichini S, Avico U, Federzoni G, Pini LA, Sternieri E (1989) Influence of antacids on the bioavailability of glibenclamide. Drugs Exp Clin Res 15:165–169

CHAPTER 6
Mode of Action of Sulfonylureas

U. Panten, M. Schwanstecher, and C. Schwanstecher

A. Introduction

Both pancreatic and extrapancreatic effects have been suggested to contribute to the therapeutic benefit of sulfonylureas for type II diabetic patients. Clear evidence for stimulation of insulin secretion by sulfonylureas has been presented (Loubatières 1957a,b; Yalow et al. 1960; Bouman and Gaarenstrom 1961; Coore and Randle 1964; Malaisse et al. 1967). On the other hand, sulfonylurea-induced extrapancreatic effects observed in clinical studies may have been consequences of an increase in insulin secretion (Gerich 1989; Melander et al. 1989). This view is not invalidated by the numerous in vitro investigations demonstrating direct effects of sulfonylureas on metabolism of extrapancreatic cells (for a review see Beck-Nielsen et al. 1988). Most of the latter effects occurred at free (non-protein-bound) sulfonylurea concentrations well beyond those measured in the plasma from treated patients (for references see Panten et al. 1989). In the plasma, sulfonylureas are highly protein bound, e.g., 6% of tolbutamide and less than 1% of glibenclamide are free. The free proportions of sulfonylureas easily cross capillary walls (perhaps with the exception of the blood-brain barrier capillaries) for the following reasons: Firstly, due to their high lipid solubility sulfonylureas rapidly penetrate membranes (Panten et al. 1989). Secondly, fenestrated endothelia (e.g., in pancreatic islets) are no barriers for sulfonylureas ($M_r < 500$). Hence, the free drug concentrations at the target cells are similar to the plasma concentrations of the free drugs, at least under steady-state conditions.

This chapter concentrates on the discussion of effects which occur at sulfonylurea concentrations relevant to antidiabetic therapy and result from actions on the specific sulfonylurea receptors. Additional information on sulfonylurea-induced effects is available in several reviews published during the last 20 years (Hellman and Täljedal 1975; Loubatières 1977; Skillman and Feldman 1981; Malaisse et al. 1983; Gylfe et al. 1984; Boyd 1988; De Weille et al. 1989; Siconolfi-Baez et al. 1990; Lebovitz 1990; Ashford 1990; Rajan et al. 1990; Malaisse and Lebrun 1990; Boyd et al. 1991; Ashcroft and Rorsman 1991; Dunne and Petersen 1991; Ashcroft and Ashcroft 1990, 1992; Petit and Loubatières-Mariani 1992; Panten et al. 1992, 1993; Edwards and Weston 1993; Gopalakrishnan et al. 1993; Ashcroft 1994).

B. Actions on Pancreatic β Cells

I. Stimulation of Insulin Secretion

Sulfonylureas directly stimulate insulin release from the B cells in the islets of Langerhans, and this effect does not require the presence of nutrients (e.g., glucose) or other secretagogues (MALAISSE et al. 1967; GORUS et al. 1988). Electrophysiological studies demonstrated that sulfonylureas depolarize the B cells and produce electrical spike activity (DEAN and MATTHEWS 1968; for reviews see MATTHEWS et al. 1973; MATTHEWS 1985; HENQUIN 1987). The depolarization is due to a decrease in the K^+ permeability of the B-cell membrane (HENQUIN 1980) and opens voltage-dependent Ca^{2+} channels. The resultant enhancement of Ca^{2+} influx across the cell membrane (partly reflected in the electrical spike activity) leads to an increase in the cytosolic Ca^{2+} concentration and to subsequent stimulation of insulin release (HENQUIN 1987).

Patch-clamp experiments revealed that the sulfonylurea-induced decrease in K^+ permeability results from inhibition of an ATP-sensitive K^+ channel (K_{ATP} channel) in the B-cell membrane (see Sect. B.II). There is strong evidence that inhibition of the K_{ATP} channel is the sole mechanism by which sulfonylureas enhance insulin secretion: (1) A close correlation between K_{ATP}-channel-inhibiting and insulin-releasing potencies of sulfonylureas exists (ZÜNKLER et al. 1988a; PANTEN et al. 1989). In the latter studies, the potencies were determined in steady states of drug action since the onset of action of second-generation sulfonylureas is slow unless maximally effective drug concentrations are applied. (2) When allowing for differences in the kinetics of secretory responses to first- and second-generation sulfonylureas, the same maximum secretory rates are caused by maximally effective concentrations of tolbutamide, glipizide, glibenclamide and the benzoic acid derivative meglitinide (PANTEN et al. 1989). This result is in favor of a single insulin-releasing mechanism of sulfonylureas. (3) Sulfonylureas do not exert direct effects on delayed rectifying K^+ channels, Ca^{2+}-activated K^+ channels, Ca^{2+} channels and nonselective cation channels, at least in B cells (TRUBE et al. 1986; ASHCROFT and RORSMAN 1991). (4) Enhancement of phosphoinositide hydrolysis by sulfonylureas is a consequence of the elevation of the cytosolic Ca^{2+} concentration in B cells (LAYCHOK 1983; ZAWALICH et al. 1988). (5) The insulin-releasing effect of sulfonylureas is not mediated through cAMP (HELLMAN and TÄLJEDAL 1975; GORUS et al. 1988). (6) There is also no clear evidence for direct effects of sulfonylureas on B-cell metabolism. Sulfonylurea-induced metabolic changes may be consequences of the workload imposed by stimulation of the secretion process (GYLFE et al. 1984; PANTEN and LENZEN 1988). (7) Increase of the sodium content in B cells induced by sulfonylureas and some structurally related compounds seems to result from depolarization in response to closure of K_{ATP} channels (SAHA and HELLMAN 1994) and from side effects of the drugs [the potencies of the drugs to alter the sodium content do not closely correlate with the potencies to release insulin (ALI et al. 1988, 1989)]. (8) Short-term application of sulfonylureas does not enhance insulin biosynthesis (TAYLOR and PARRY 1967; HELLMAN and TÄLJEDAL 1975). (9) Glibenclamide does not promote hyperplasia or hypertrophy of B cells in adult pancreatic islets (GUIOT et al. 1994).

It was suggested that the sulfonylurea glimepiride ($EC_{50}=0.3\,nM$ for K_{ATP}-channel inhibition; SCHWANSTECHER et al. 1994d) controls insulin secretion by a direct effect on B-cell mitochondria (modulation of glucokinase/hexokinase binding to porin; MÜLLER et al. 1994b). This view cannot be readily accepted since insulin-secreting tumor cells (RINm5F) were incubated in the presence of a high ($20\,nM$) glimepiride concentration. Furthermore, a comparison of the potencies of several sulfonylureas for modulation of glucokinase/hexokinase binding on the one hand and for stimulation of insulin secretion on the other hand was not presented.

Tolbutamide ($100\,\mu M$) and glibenclamide ($2\,\mu M$) stimulated insulin release from electropermeabilized B cells in which K_{ATP} channels are inoperative (FLATT et al. 1994). Provided this effect occurs also at low sulfonylurea concentrations, it might lead to the identification of intracellular components mediating specific responses to sulfonylureas.

Glibenclamide concentrations too low to cause prolonged stimulation of insulin secretion in the presence of $5.5\,mM$ glucose were reported to increase insulin gene expression (YAMATO et al. 1993). However, this effect might have resulted from glibenclamide-induced elevation of the cytosolic Ca^{2+} concentration in B cells.

II. Inhibition of the K_{ATP} Channel

In the plasma membrane of rodent or human B cells, an inwardly rectifying K^+ channel has been identified which is inhibited when the B cells are exposed to glucose or other insulin-releasing fuels (ASHCROFT et al. 1984; for a review see ASHCROFT and RORSMAN 1991). The fuel-induced inhibition of the distinct K^+ channel is due to an increase of the ATP and a decrease of the ADP concentration at the cytosolic face of the B-cell membrane. Channel activity is controlled by at least three separate binding sites for cytosolic nucleotides (Fig. 1). (1) Occupation of the binding site for ATP and some related nucleotides causes closure of the K^+ channel both in the absence and presence of Mg^{2+} (ATP-sensitive K^+ channel, K_{ATP} channel) (COOK and HALES 1984). (2) The Mg complexes of ADP (MgADP) and some other nucleoside diphosphates open the K_{ATP} channel via a separate site (FINDLAY 1987; ASHCROFT and RORSMAN 1991). (3) The Mg complex of ATP (MgATP) considerably slows the rapid decline of K_{ATP}-channel activity (channel rundown) observed in excised membrane patches in the absence of ATP (OHNO-SHOSAKU et al. 1987). As rundown is also reduced in Mg^{2+}-free cytosol-like solution (KOZLOWSKI and ASHFORD, 1990), it may involve a

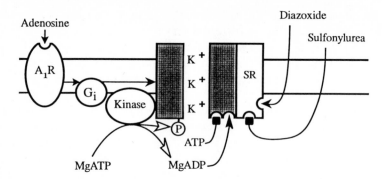

Fig. 1. Simplified model for regulation of the K_{ATP} channel by nucleotides, drugs and adenosine. The pore-forming subunit contains the binding sites for ATP and MgADP. The sulfonylurea receptor (*SR*) is suggested to represent a separate protein containing the binding sites for sulfonylurea and diazoxide. In analogy to the binding site for sulfonylureas (see Sect. B.III), the binding site for diazoxide is suggested to be located at the internal side of the plasma membrane. The protracted reversal of diazoxide-induced inhibition of insulin secretion seems to result from intracellular storage of diazoxide, which is highly lipid-soluble at pH 7.4 (PANTEN et al. 1989). The binding sites for sulfonylureas and diazoxide are coupled by negative allosterism. Channel activation (*filled arrows*) is induced by interaction of adenosine with cardiac and vascular A_1 receptors (A_1R) (KIRSCH et al. 1990; XU and LEE 1994) or by MgADP. The response to MgADP is enhanced by diazoxide, which is ineffective in the absence of MgADP (see Sect. B.II). Channel inhibition (*filled squares*) is induced by ATP. Sulfonylureas shift the competitive interaction between inhibitory and stimulatory nucleotides towards the inhibitory nucleotides. In the absence of any nucleotides, sulfonylureas are only weakly inhibitory (see Sect. B.II). Normal regulation requires phosphorylation (*P*) of the K_{ATP} channel and the sulfonylurea receptor by a tightly associated protein kinase

Mg^{2+}-dependent protein dephosphorylation, and MgATP might act by serving as substrate for one or several protein kinases closely associated with the B-cell membrane.

Sulfonylureas (e.g., tolbutamide or glibenclamide) inhibit and certain K^+-channel openers (e.g., diazoxide or pinacidil) activate the K_{ATP} channel by interaction with separate binding sites in the B-cell plasma membrane (Fig. 1) (STURGESS et al. 1985; TRUBE et al. 1986; SCHWANSTECHER et al. 1992d; for reviews see ASHFORD 1990; ASHCROFT and RORSMAN 1991; DUNNE and PETERSEN 1991; PANTEN et al. 1992; ASHCROFT and ASHCROFT 1992; EDWARDS and WESTON 1993; GOPALAKRISHNAN et al. 1993). Binding studies and patch-clamp experiments suggest that the binding sites for these drugs are not identical with the sites mediating regulation of the K_{ATP} channel by cytosolic nucleotides (SCHWANSTECHER et al. 1992b,d,e). Phosphorylation of the K_{ATP} channel and/or regulatory proteins appears to be involved in the mechanism of action of diazoxide (KOZLOWSKI et al. 1989; DUNNE 1989). Moreover, diazoxide is only effective when the site for stimulatory nucleoside diphosphates is occupied, i.e., diazoxide sensitizes the channel-activating effect of MgADP and related nucleoside diphosphates (SCHWANSTECHER et al. 1992b; LARSSON et al. 1993). The potency of sulfonylureas is much lower in excised membrane patches than in intact B cells, unless both inhibitory and stimulatory nucleotides are present at the cytoplasmic face of the B-cell membrane (ZÜNKLER et al. 1988b; PANTEN et al. 1990; SCHWANSTECHER et al. 1992a). This nucleotide effect reflects a complex interaction between sulfonylureas and cytosolic nucleotides (SCHWANSTECHER et al. 1994a). Firstly, channel-inhibiting nucleotides increase the potency of sulfonylureas. Secondly and most importantly, sulfonylureas shift the competitive interaction between inhibitory and stimulatory nucleotides towards the inhibitory nucleotides.

Occupation of all specific sulfonylurea receptors by ligands appears to cause only partial inhibition of the K_{ATP} channels in the absence of nucleotides (SCHWANSTECHER et al. 1994e). This might indicate that sulfonylureas are partial agonists for their receptor site or that the sulfonylurea receptor and the K_{ATP} channel are not tightly coupled or that a proportion of the K_{ATP} channels is coupled to a receptor state unable to bind ligands. A nonbinding state of the sulfonylurea receptor might be induced by protein phosphorylation (see Sect. B.IV). However, rapid alteration of phosphorylation of the sulfonylurea receptor does not seem to be involved in the mechanism of action of sulfonylureas. Presumably prolonged removal of intracellular Mg^{2+} inhibits protein phosphorylation, but does not prevent strong and rapidly reversible responses to tolbutamide in B cells (SCHWANSTECHER et al. 1994a). The latter finding is at variance with the view of LEE et al. (1994a) that intracellular Mg^{2+} is critically important for the interaction between the sulfonylurea receptor and the K_{ATP} channel in insulin-secreting cells.

Application of trypsin at the cytoplasmic side of the B-cell plasma membrane reduced the sulfonylurea sensitivity of the K_{ATP} channel (PROKS and ASHCROFT 1993; LEE et al. 1994c). This effect of trypsin might reflect altered properties of the sulfonylurea receptor and/or the K_{ATP} channel since trypsin also diminished the responses to channel-activating and channel-inhibiting nucleotides (TRUBE et al. 1989; PROKS and ASHCROFT 1993).

In the resting B cell (presence of $3\,mM$ glucose, $37°C$) 6%–8% of all K_{ATP} channels are still open (PANTEN et al. 1990; SCHWANSTECHER et al. 1992b). The threshold potential for activation of voltage-dependent Ca^{2+} channels is reached when less than 3% of the K_{ATP} channels are open (COOK et al. 1988; PANTEN et al. 1990; SCHWANSTECHER et al. 1992b). These percentages of channel closure are induced by glucose concentrations above $6\,mM$. At glucose concentrations near $6\,mM$, the insulin-releasing potencies of sulfonylureas closely correlate with their K_{ATP}-channel-blocking potencies (presence of cytosolic nucleotides), and the effective free sulfonylurea concentrations are in the range of therapeutic plasma concentrations of the free drugs (PANTEN et al. 1989). In B cells exposed to maximally effective glucose concentrations ($30-40\,mM$), sulfonylureas are still effective (Panten et al. 1986; Henquin 1988). This finding suggests that not all K_{ATP} channels are closed by high glucose concentration due to the rise in cytosolic Ca^{2+} (Henquin 1990a) and subsequent channel activation via Ca^{2+}-dependent protein kinases.

III. Location of the Sulfonylurea Receptor

Patch-clamp experiments with excised membrane patches revealed that sulfonylureas and their analogues (e.g., meglitinide) inhibit the K_{ATP} channel from both sides of the B-cell membrane (Trube et al. 1986; Sturgess et al. 1988). This can be explained by rapid distribution of these drugs within the lipid phase of the cell membrane. Sulfonylureas and their analogues are weak organic acids, and the proportions of their lipophilic undissociated forms decrease with increasing pH. When performing whole-cell clamp experiments and raising the pH value in the extracellular solution at constant total tolbutamide concentration, both the rate of development and the degree of tolbutamide-induced K_{ATP}-channel block was diminished (Zünkler et al. 1989). This finding and additional evidence suggested that the undissociated forms were the effective forms and reached the binding site on the sulfonylurea receptor protein from the lipid phase of the B-cell membrane. However, the possibility was not ruled out that the binding site was located at the cytoplasmic face of the membrane and that the undissociated forms were only required for transport of externally applied sulfonylureas and related compounds across the membrane. We therefore synthesized a benzenesulfonic acid derivative representing the glibenclamide molecule devoid of its cyclohexylurea moiety (Fig. 2). At pH 7.4 the benzenesulfonic acid derivative (Fig. 2) is only present in charged form and is much less lipophilic than the sulfonylureas and sulfonylurea analogues used to far. The dissociation constant (K_D) for binding of the benzenesulfonic acid derivative (Fig. 2) to the sulfonylurea receptor in insulin-secreting cells was similar to the K_D value for tolbutamide (Schwanstecher et al. 1994e). Drug concentrations inhibiting the K_{ATP} channel half-maximally (EC_{50}) were then determined in B cells by use of the patch-clamp technique. When the drugs were applied to the extracellular side of outside-out or the intracellular side of inside-out membrane patches, the ratio of extracellular to intracellular EC_{50} value was 281 for the benzenesulfonic acid derivative (Fig. 2) and 1.2 for tolbutamide (Schwanstecher et al. 1994e). This finding can be readily explained by location of the sulfonylurea receptor site at the cytoplasmic face of the B-cell membrane and a low rate of transport of the benzenesulfonic acid derivative (Fig. 2) across the membrane, but a high rate of tolbutamide transport. In outside-out patch experiments the rate of membrane transport controls the drug concentration effective at the internal membrane side, since the drug molecules diffuse rapidly into the recording pipette, which represents an effectively infinite volume (Penner et al. 1987;

PUSCH and NEHER 1988). As the benzenesulfonic acid derivative (Fig. 2) binds to the sulfonylurea receptor, these patch-clamp experiments clearly indicate that the sulfonylurea receptor site is located at the cytoplasmic face of the B-cell membrane (Fig. 1).

The effects of trypsin on membrane binding of [³H]glibenclamide are consistent with the location of the sulfonylurea receptor site at the cytoplasmic membrane side. Binding of [³H]glibenclamide to microsomes from insulin-secreting tumor cells (CRI-G1) was totally removed by trypsin applied directly to the microsomes, but was not affected by pretreatment of intact CRI-G1 cells with extracellular trypsin (LEE et al. 1994c). These findings alone do not rule out a specific binding site for sulfonylureas lying within the plasma membrane bilayer (e.g., a site on the lipid-exposed regions of a membrane-spanning protein). Moreover, it is unclear why all outside-out vesicles in the microsomal preparation, but not the intact cells, displayed trypsin sensitivity of [³H]glibenclamide binding.

The location of the sulfonylurea receptor site at the internal membrane side can also explain the differences in the kinetics of secretory responses to first- and second-generation sulfonylureas. A steep first peak of insulin release is observed at all effective concentrations of first-generation sulfonylureas ($EC_{50} \geq 1 \mu M$ of free drug; e.g., tolbutamide), but only at supermaximally effective concentrations (i.e., micromolar concentrations) of second-generation sulfonylureas ($EC_{50} = 4 nM$ for free glipizide and $0.5 nM$ for free glibenclamide) (PANTEN et al. 1989). Increase in insulin secretion is slowed down more and more with lowering of the concentrations of glipizide or glibenclamide. These slow kinetics might reflect retardation of increase in drug concentration at the internal membrane side due to drug diffusion into various intracellular pools. However, when micromolar concentrations of sulfonylureas are applied, flux into intracellular pools does not seem to be rapid enough to delay the increase in drug concentration at the internal membrane side. The very slow reversibility of the effects of glibenclamide (LOUBATIÈRES et al. 1969; LOUBATIÈRES 1977; GORUS et al. 1988; PANTEN et al. 1989) seems to result from intracellular accumulation of glibenclamide the lipophilicity of which is exceptionally high (GYLFE et al. 1984; CARPENTIER et al. 1986; PANTEN et al. 1989). Thus, intracellular stores can provide the B-cell sulfonylurea receptor site with glibenclamide even after virtual disappearance of glibenclamide from the extracellular space. This special feature of glibenclamide might be the reason for the rather high incidence of prolonged hypoglycemia seen with glibenclamide (FERNER and CHAPLIN 1987).

IV. Properties of the Binding Sites for Sulfonylureas

The cDNA encoding a novel member of the inward rectifier K^+-channel family (ROMK1/IRK1/GIRK1 family) was isolated from cardiac cells (ASHFORD et al. 1994). The cDNA predicted a 48-kDa polypeptide and expressed a K^+ channel in kidney epithelial cell lines which was inhibited by ATP and activated by MgUDP (MgADP analogue). The mRNA for the new member of the inward rectifier family was detected in a variety of tissues (e.g., brain and heart), but not in insulin-secreting cells. The findings seem to indicate that a pore-forming component of cardiac K_{ATP} channels was cloned. However, the expressed K^+ channel showed no sulfonylurea sensitivity, suggesting that the sulfonylurea receptor represents a regulatory protein, but not a pore-forming component of the K_{ATP} channel (Ashford et al. 1994). It is still unclear whether the sulfonylurea receptor is permanently or only temporarily associated with the K_{ATP} channel (Aguilar-Bryan et al. 1992; Khan et al. 1993).

A high-affinity binding site for sulfonylureas has been found in microsomal membranes from insulinomas, insulin-secreting cell lines and pancreatic islets (Geisen et al. 1985; for reviews see Ashcroft and Ashcroft 1992; Panten et al. 1992; Edwards and Weston 1993; Gopalakrishnan et al. 1993). Correlations between K_{ATP}-channel-blocking and insulin-releasing potencies of sulfonylureas, on the one hand, and binding affinities of sulfonylureas, on the other hand, suggest that the high-affinity site represents the sulfonylurea receptor (Schmid-Antomarchi et al. 1987; Gaines et al. 1988; Panten et al. 1989; Schwanstecher et al. 1994d). Using SDS-polyacrylamide gel electrophoresis of membrane proteins from insulin-secreting tumor cells, specific photoincorporation of [^{3}H]glibenclamide or an ^{125}I-labeled glibenclamide analogue was described into a peptide of 140 kDa (Kramer et al. 1988; Aguilar-Bryan et al. 1990; Nelson et al. 1992; Skeer et al. 1994). Under nondenaturing conditions, using gel filtration chromatography, a sulfonylurea-binding peptide with a molecular weight of 166–182 kDa was identified which appeared to be a subunit of a 250-kDa protein complex (Skeer et al. 1994). A 65-kDa peptide photolabeled by the novel sulfonylurea [^{3}H] glimepiride was suggested to represent a subunit of the B-cell sulfonylurea receptor (Kramer et al. 1994; Müller et al. 1994a). However, this peptide seems to be a low-affinity binding site since 100 μM glibenclamide did not prevent photoincorporation of [^{3}H]glimepiride (31 nM). The 140-kDa peptide was purified 60-fold from insulin-secreting tumor cells (Aguilar-Bryan et al. 1990). Up to now sequence data of this peptide and cloning and expression of the corresponding gene have not been published, and thus it is unknown whether the 140-kDa peptide represents the sulfonylurea receptor of insulin-secreting cells.

In order to provide a probe with increased photoreactivity, a glibenclamide analogue with iodo and azido substituents (^{125}IN$_3$-GA, Fig. 2) was synthesized (Chudziak et al. 1994). This novel probe was specifically photoincorporated into a 160-kDa peptide from membranes of insulin-secreting tumor cells when samples were boiled in an SDS-buffer prior to separation with SDS-polyacrylamide gel electrophoresis (Schwanstecher et al. 1994c). However, omitting the heating step revealed specific labeling of an additional peptide with a molecular weight of 38 kDa. The amount of radioactivity specifically photoincorporated into this peptide was three to fourfold higher than that incorporated into the 160-kDa peptide. Both peptides displayed similar dissociation constants for apparent binding of the sulfonylureas IN_3-GA, glibenclamide, glipizide and tolbutamide. Analysis of photoaffinity labeling of solubilized fractions indicated an almost exclusive specific linkage to the 38-kDa peptide. The data support the view that the 38-kDa peptide represents the sulfonylurea receptor in insulin-secreting cells. This peptide seems to be tightly coupled to a 160-kDa peptide (SCHWANSTECHER et al. 1994c). Both peptides could be components of a protein assembly controlling the opening activity of the K_{ATP} channel.

In membrane preparations from pancreatic islets, MgATP inhibited high-affinity binding of [^{3}H]glibenclamide (SCHWANSTECHER et al. 1991). In the absence of Mg^{2+}, ATP was ineffective. Competition experiments with unlabeled glibenclamide and tolbutamide confirmed this effect of MgATP (SCHWANSTECHER et al. 1991, 1992e), suggesting that MgATP-induced inhibition of binding reflects a general phenomenon being valid for all sulfonylureas. Evidence was presented that the effect of MgATP was not due to a competitive interaction between MgATP and glibenclamide at the same binding site (SCHWANSTECHER et al. 1992e).

In solubilized membranes of insulin-secreting tumor cells (HIT-T 15), the inhibitory effect of MgATP on glibenclamide binding was maintained (SCHWANSTECHER et al. 1992c). MgATP caused both an increase of the K_D value for glibenclamide by 4.4-fold and a reduction of the B_{max} for binding of glibenclamide by 57%. As the effect of MgATP was observed after solubilization of membranes with nonionic (Triton X-100, digitonin) or zwitterionic (CHAPS) detergents, which differ in their efficiency to prevent aggregation of solubilized proteins, the separate binding sites for MgATP and sulfonylureas appear to be located at the same protein or at tightly associated proteins.

In the presence of Mg^{2+}, not only ATP but also ATPγS, ADP, GTP, GTPγS, GDP and GDPβS inhibited binding of glibenclamide (SCHWANSTECHER et al. 1991, 1992c). These effects were not observed in the absence of Mg^{2+}. The responses to MgADP, MgGDP and MgGDPβS were probably indirect and seemed to result from their conversion into the corresponding nucleoside triphosphates by transphosphorylating enzymes in the membrane preparation, The non-hydrolyzable analogues AMP-PNP and GMP-PNP did not alter glibenclamide binding to particulate or solubilized binding sites, neither in the absence nor in the presence of Mg^{2+}. All of these findings suggested that the inhibitory nucleotides acted by providing substrate for one or several protein kinases which regulate the binding properties of the sulfonylurea receptor by phosphorylating or thiophosphorylating the receptor protein or a tightly associated protein (SCHWANSTECHER and RIETZE 1990; SCHWANSTECHER et al. 1990, 1991, 1992c). In support of this view, the half-maximally inhibitory concentrations for MgATP ($12-30\,\mu M$; SCHWANSTECHER et al. 1991, 1992c) were well within the range characteristic for protein kinases (FLOCKHART and CORBIN 1982). Further evidence for a role of protein phosphorylation in the regulation of sulfonylurea binding emerged from the finding that exogenous alkaline phosphatase accelerated the reversal of MgATP-induced inhibition of glibenclamide binding (SCHWANSTECHER et al. 1991, 1992c).

Niki and Ashcroft (1991), too, reported inhibition of glibenclamide binding by MgATP and MgADP in membrane preparations from insulin-secreting tumor cells (HIT-T15). However, the authors concluded that MgATP-induced inhibition of glibenclamide binding might not be mediated by protein phosphorylation but might be due to formation of MgADP in the incubation media. The following observations were made: (1) 1 mM MgADP induced a stronger inhibition of glibenclamide binding than 1 mM MgATP. (2) MgATP concentrations below 1 mM did not inhibit glibenclamide binding. (3) No effect on glibenclamide binding was observed for MgATPγS. (4) Addition of phospho*enol*pyruvate and pyruvate kinase to generate ATP from ADP abolished the inhibitory effect of MgATP on glibenclamide binding. All of these findings are at variance with the results of SCHWANSTECHER et al. (1991, 1992c).

When membrane preparations from HIT cells were exposed to Mg^{2+}-free solutions, ATP, AMP, GTP, GDP, GMP, NADP$^+$, NADPH, NAD$^+$, NADH, cyclic 3′,5′-AMP and cyclic 3′,5′-GMP did not alter glibenclamide binding (NIKI et al. 1989). However, under similar experimental conditions, free ADP (0.5 – 1 mM) markedly inhibited glibenclamide binding (NIKI et al. 1989, 1990). This finding, the inhibition of glibenclamide binding to intact HIT cells by ADP and ADP agarose and the pronounced insulin release from HIT cells induced by extracellular ADP (in the presence of Mg^{2+}), led to the conclusion that the B-cell sulfonylurea receptor is an ADP-binding protein. However, others did not observe an effect of 1 mM free ADP on glibenclamide binding (SCHWANSTECHER et al. 1991, 1992c). Moreover, in the presence of Mg^{2+}, ADP (2 mM) did not markedly stimulate insulin release from perifused pancreatic islets (SCHWANSTECHER et al. 1992e). Finally, inhibition of glibenclamide binding to intact HIT cells by ADP agarose could not be due to competition for binding to a sulfonylurea receptor site located at the internal side of the plasma membrane (see Sect. B. III). Thus, the results of NIKI et al. (1989, 1990; NIKI and ASHCROFT 1991) might reflect a special feature of some subclones of the insulin-secreting tumor-cell line HIT-T15.

In the absence of MgATP, diazoxide only weakly inhibited glibenclamide binding to membranes from insulin-secreting cells, in comparison with the potency of diazoxide for K$_{ATP}$-channel activation in pancreatic B cells (NIKI et al. 1989; PANTEN et al. 1989; SCHWANSTECHER et al. 1991; SCHWANSTECHER et al. 1992b). However, MgATP-induced inhibition of sulfonylurea binding to membranes from insulin-secreting cells was accompanied by enhanced displacement of glibenclamide from its receptor by diazoxide (SCHWANSTECHER et al. 1990, 1991; SCHWANSTECHER and PANTEN 1993b; NIKI and ASHCROFT

1991). In the presence of $100\,\mu M$ MgATP, the EC_{50} value of diazoxide for inhibition of glibenclamide binding ($140\,\mu M$; SCHWANSTECHER et al. 1991) corresponded reasonably well to the diazoxide concentration half-maximally effective on K_{ATP}-channel activity in intact B cells ($43-87\,\mu M$; SCHWANSTECHER et al. 1992b). Thus, binding of diazoxide to its receptor might require protein phosphorylation. This view was supported by the finding that ATP in the absence of Mg^{2+} or nonhydrolyzable ATP analogues did not enhance diazoxide-induced inhibition of glibenclamide binding.

MgATP also enhanced pinacidil-induced displacement of glibenclamide binding (SCHWANSTECHER et al. 1992d), suggesting that the K^+-channel-opening drugs pinacidil and diazoxide exert their effects on insulin-secreting cells by binding to the same site. K^+-channel-opening drugs (cromakalim, minoxidil sulfate) structurally unrelated to pinacidil and diazoxide did not interact with sulfonylurea binding, neither in the absence nor in the presence of MgATP. Pinacidil and diazoxide did not increase the EC_{50} for displacement of [^{3}H]glibenclamide binding by unlabeled glibenclamide in the presence of MgATP (SCHWANSTECHER et al. 1992d; SCHWANSTECHER and PANTEN 1993b). Thus, the MgATP-dependent inhibitory effects of pinacidil and diazoxide cannot be explained by a competitive interaction with sulfonylureas at the same site, but might be due to negative allosterism. The binding sites for these K^+-channel-opening drugs and sulfonylureas appear to be located at the same protein or at tightly associated proteins (SCHWANSTECHER et al. 1992c,d).

Diazoxide was reported to inhibit glibenclamide binding to intact HIT cells (NIKI and ASHCROFT 1991). It is unclear whether this diazoxide effect reflected the same phenomenon observed in microsomes exposed to MgATP (see above) since ATP depletion with deoxyglucose and oligomycin did not increase, but decreased, glibenclamide binding to HIT cells (NIKI and ASHCROFT 1991).

Trinitrobenzene sulfonic acid, which modifies lysine residues, has been shown to inhibit [^{3}H]glibenclamide binding in membranes of the insulin-secreting tumor cell CRI-G1 (LEE et al. 1994b). This could indicate that lysine residues of the sulfonylurea receptor are important for ligand binding. Alternatively, trinitrobenzene sulfonic acid might compete with [^{3}H] glibenclamide for receptor binding.

It was reported that two different high-affinity sites for sulfonylureas exist in the insulin-secreting tumor cell RINm5F, one site preferring glibenclamide and the other one preferring gliquidone (VERSPOHL et al. 1990). However, the binding data shown do not support this interpretation since the specific radioactivities of the applied [^{3}H]glibenclamide and [^{3}H]gliquidone were most different. Low-affinity sites for sulfonylureas were repeatedly found in membranes from insulin-secreting tumor cells (ASHCROFT and ASHCROFT 1992; EDWARDS and WESTON 1993). These sites are probably not involved in the mechanism of insulin-releasing action of sulfonylureas because they were not seen in membranes from pancreatic islets (PANTEN et al. 1989; SCHWANSTECHER et al. 1991). Moreover, the K_D values for binding of sulfonylureas to the low-affinity sites were well beyond the range of therapeutic plasma concentrations of the free sulfonylureas.

V. Structure of Compounds Interacting with the Sulfonylurea Receptor

The sulfonamide 2254RP [2-(p-amino-benzenesulfamido)-5-isopropyl-thiadiazole] is the mother compound of the insulin-releasing agents used in the treatment of type II diabetic patients (LOUBATIÈRES 1957a,b). Compound 2254RP is not a sulfonylurea, but its mode of action is similar to that of hypoglycemic sulfonylureas (HENQUIN 1992). Several other modifications of the basic structure (e.g., formation of sulfonylaminopyrimidines or sul-famylureas) lead to effective compounds (AUMÜLLER and HEERDT 1971). The hypoglycemic and insulin-releasing potency of sulfonylureas depends on the lipophilicity of the substituents R_1 and R_2 (Fig. 2) and is enhanced by a carboxamido structure in R_1 (characteristic of second-generation sulfony-lureas). Unfortunately, structure/function relationships for sulfonylureas and related compounds determined by in vivo studies or by experiments with intact B cells reflect not only the interaction of the drugs with the sulfonylurea receptor but also the requirements of drug transport to the internal side of the B-cell plasma membrane.

A benzenesulfonic acid derivative representing the glibenclamide molecule devoid of its cyclohexylurea moiety (Fig. 2) inhibits the K_{ATP} channel, stimulates insulin secretion and competes with glibenclamide for binding to the sulfonylurea receptor (SCHWANSTECHER et al. 1994e). These findings suggest receptor binding of the anionic forms of hypoglycemic sulfonylureas and their analogues since the benzenesulfonic acid derivative (Fig. 2) is completely dissociated at pH 7.4.

Meglitinide, a hypoglycemic benzoic acid derivative structurally related to glibenclamide (Fig. 2; GEISEN et al. 1978), has been shown to stimulate insulin release by a mechanism of action identical to that of hypoglycemic sulfonylureas (GARRINO et al. 1985). The carboxamido structure and the high lipophilicity of the substituent of benzoic acid ensure sufficient affinity of meglitinide for the sulfonylurea receptor. A further increase in lipophilicity of the substituent leads to benzoic acid derivatives as potent as second-generation sulfonylureas (HENQUIN 1990b; RONNER et al. 1992). Brown and Foubister (1984) proposed that the COOH group of meglitinide and the SO_2NHCO group of sulfonylureas are isosteres and that these acidic groups are required for hypoglycemic activity. The effects of the benzenesulfonic acid derivative (Fig. 2) also demonstrate that the urea group of sulfonylureas is not a prerequisite for interaction of these drugs with their receptor. The urea group is mainly a means of attaching a hydrophobic residue to ben-zenesulfonic acid with conservation of an acidic group. Unlike the urea

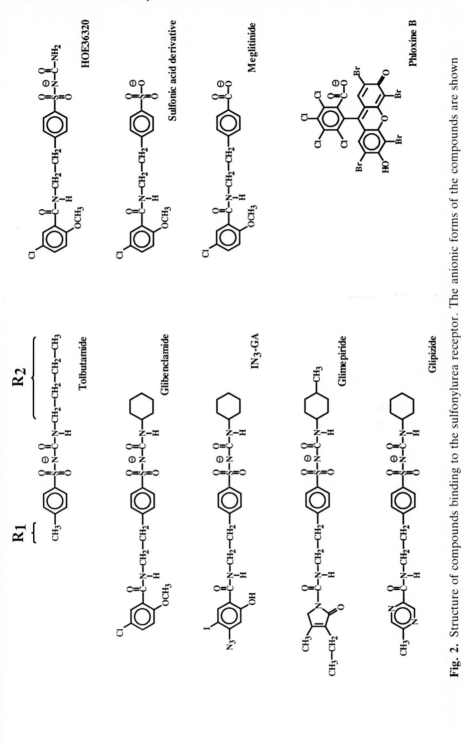

Fig. 2. Structure of compounds binding to the sulfonylurea receptor. The anionic forms of the compounds are shown

group, the hydrophobic cyclohexyl group strongly enhances receptor binding. This is reflected in the finding that the binding affinity of HOE 36320 (Fig. 2) is not much higher than that of the benzenesulfonic acid derivative (Fig. 2), whereas the binding affinity of glibenclamide is higher by several orders of magnitude (Gaines et al. 1988; Schwanstecher et al. 1992c; Schwanstecher et al. 1994e).

Replacement of the COOH group of meglitinide by various nonacidic groups decreased but did not abolish the effects on K^+ permeability and insulin secretion in pancreatic B cells (Henquin et al. 1987). It was concluded that the acidic group is not a prerequisite for the effectiveness of meglitinide. However, the possibility was not ruled out that all effects of the nonacidic derivatives were due to action on sites not representing the sulfonylurea receptor. There are other compounds (e.g., quinine, chlorpromazine, thiopentone) which do not interact with the sulfonylurea receptor, but inhibit the K_{ATP} channel in insulin-secreting cells (Bokvist et al. 1990; Kozlowski and Ashford 1991; Müller et al. 1991; Schwanstecher et al. 1992d).

Fluorescein derivatives (phloxine B, Bengal rose) have been shown to exert both stimulatory and inhibitory effects on the K_{ATP} channels in insulin-secreting tumor cells (De Weille et al. 1992). As fluorescein derivatives are known to interact with the nucleotide-binding sites of various proteins, phloxine B and Bengal rose-induced inhibition of [^{3}H]glibenclamide binding to membranes from Insulin-secreting tumor cells was taken as evidence that the site for stimulatory nucleotides at the K_{ATP} channel is coupled by negative allosterism to the sulfonylurea receptor site. This view is inconsistent with the previous demonstration that the stimulatory nucleotide MgADP per se did not inhibit glibenclamide binding under both phosphorylating and dephosphorylating conditions (Schwanstecher et al. 1991, 1992c,e). Moreover, recent findings suggest that fluorescein derivatives and sulfonylureas compete for binding to the same site (the sulfonylurea receptor site) (Fig. 2; Schwanstecher et al. 1995).

Recently, a novel 3-phenylpropionic acid derivative (KAD-1229) was found to bind to the sulfonylurea receptor of insulin-secreting cells and to have potent K^+-channel-inhibiting and insulin-releasing effects (Ohnota et al. 1994; Mogami et al. 1994). The 3-phenylpropionic acid derivative HB093 and the novel D-phenylalanine derivative A-4166, too, may exert their insulin-releasing effect via binding to the sulfonylurea receptor (Geisen et al. 1978; Hirose et al. 1994; Fujitani and Yada 1994). It is proposed that receptor binding of KAD-1229, HB093 and A-4166 induces a position of the COO^- group relative to the phenyl ring which resembles the position of the SO_2N^- group of sulfonylureas (Fig. 3).

Peptides (endosulfines) displacing [^{3}H]glibenclamide from the cerebral sulfonylurea receptor have been extracted from pancreas and brain and have been shown to stimulate insulin secretion (Virsolvy-Vergine et al. 1992). The role of these peptides in regulating B-cell function is unclear because

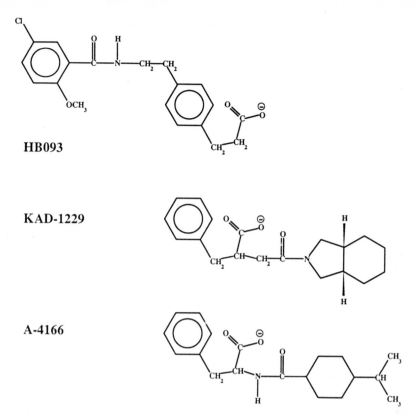

Fig. 3. Structure of insulin-releasing 3-phenylpropionic acid derivatives

the sulfonylurea receptor site is located at the cytoplasmic side of the B-cell plasma membrane (see Sect. B.III).

C. Actions on Non-β Cells in Pancreatic Islets

To our knowledge, unequivocal evidence has not been presented that therapeutic concentrations of free sulfonylureas directly change the function of glucagon-secreting α-cells exposed to normal or high glucose levels (for references see PANTEN et al. 1989; RONNER et al. 1993; RAJAN et al. 1993). K_{ATP} channels were not detected in normal α-cells from pancreatic islets (RORSMAN and HELLMAN 1988). This report contrasts with the demonstration of sulfonylurea-sensitive K_{ATP} channels in glucagon-secreting tumor cells (αTCs) (RONNER et al. 1993; RAJAN et al. 1993). The αTCs also contained typical high-affinity binding sites for sulfonylureas. A ^{125}I-labeled glibenclamide analogue was specifically photoincorporated into two peptides of 140 kDa and 150 kDa (RAJAN et al. 1993). But it is conceivable that α-cells do not express the K_{ATP} channel unless they are transformed.

Sulfonylureas stimulated somatostatin release from pancreas and stomach (CHIBA et al. 1982 and references contained therein). Therefore, K_{ATP} channels might be present in D cells. However, it is unlikely that somatostatin from pancreatic islets produces an antidiabetic effect.

D. Actions on Neurons

By use of single-channel recordings, a sulfonylurea-sensitive K_{ATP} channel with properties characteristic of the K_{ATP} channel in B cells has been identified on demyelinated fibers of frog sciatic nerve and on the somata of neurons in hippocampus, neocortex, substantia nigra pars reticulata (SNr) and caudate nucleus (JONAS et al. 1991; POLITI and ROGAWSKI et al. 1991; OHNO-SHOSAKU and YAMAMOTO 1992; SCHWANSTECHER and PANTEN 1993a, 1994). The striatal K_{ATP} channel which was detected on the somata of GABAergic spiny neurons and the K_{ATP} channel in SNr were closed under physiological conditions, but could be opened by diazoxide or inhibition of ATP production and then showed a tolbutamide sensitivity similar to that of B cells (SCHWANSTECHER and PANTEN 1993a, 1994). ATP-sensitive K^+ channels with biophysical and pharmacological properties quite different from those of the typical K_{ATP} channel in B cells have been found in cortical, hypothalamic, nigral, hippocampal and striatal neurons (ASHFORD et al. 1988, 1990a,b; AMOROSO et al. 1990; SELLERS et al. 1992; TROMBA et al. 1992, 1994; LIN et al. 1993; JIANG et al. 1994).

High-affinity binding sites for sulfonylureas have been found in brain (KAUBISCH et al. 1982; for reviews see ASHCROFT and ASHCROFT 1992; PANTEN et al. 1992; EDWARDS and WESTON 1993; GOPALAKRISHNAN et al. 1993). The SNr contains the highest density of binding sites for sulfonylureas (MOURRE et al. 1990). The binding sites of this area are partly located on the terminals of striatonigral projection neurons (HICKS et al. 1994). Since the cerebral binding sites were partially destroyed by proteolytic or lipolytic treatment (LUPO and BATAILLE 1987), the cerebral sulfonylurea receptor may be a lipoprotein. Photoaffinity labeling by [^{3}H]glibenclamide of a 145- to 150-kDa peptide from pig brain was observed by BERNARDI et al. (1988). This peptide was reported also to contain the binding sites for ATP, MgADP and diazoxide (BERNARDI et al. 1992). Although the 145- to 150-kDa peptide was purified 2500-fold, it is still unknown whether it represents the cerebral sulfonylurea receptor. The novel probe $^{125}IN_3$-GA (Fig. 2) was specifically photoincorporated into a 175-kDa peptide from cerebral cortex microsomes when samples were boiled in an SDS buffer prior to separation with SDS polyacrylamide gel electrophoresis (SCHWANSTECHER et al. 1994b). As observed for membranes from insulin-secreting cells, omitting the heating step revealed preferential specific labeling of an additional peptide with a molecular weight of 38-kDa (SCHWANSTECHER et al. 1994c).

MgATP and MgGTP inhibited high-affinity binding of [^{3}H]glibenclamide to both the particulate and the solubilized sulfonylurea receptor from brain (GOPALAKRISHNAN et al. 1991; SCHWANSTECHER et al. 1992f; NIKI and ASHCROFT 1993). Both nucleotides were ineffective in the absence of Mg^{2+}, suggesting regulation of the binding properties of the cerebral sulfonylurea receptor by protein phosphorylation (SCHWANSTECHER et al. 1992f). In support of this view, the nonhydrolyzable ATP analogue AMP-PNP was ineffective (SCHWANSTECHER et al. 1992f) or only slightly inhibitory (NIKI and ASHCROFT 1993). Inhibitory effects on [^{3}H]glibenclamide binding were also induced by ADP and GDP in an Mg-dependent manner and seemed to result from their conversion into the corresponding nucleoside triphosphates by transphosphorylating enzymes in the preparations (SCHWANSTECHER et al. 1992f). In other studies, the inhibitory effects of ADP and GDP did not require the presence of Mg^{2+} (GOPALAKRISHNAN et al. 1991; NIKI and ASHCROFT 1993). The differences could be due to variations in enzyme activities and/or ion concentrations in the incubations. LEE and ASHFORD (1993) reported that Mg^{2+} reduced high-affinity binding of [^{3}H]glibenclamide to brain microsomes pretreated with EDTA. This finding might reflect formation of inhibitory Mg complexes of nucleotides from endogenous nucleotides. As observed for insulin-secreting cells (see Sect. B.IV), MgATP enhanced both diazoxide-induced and pinacidil-induced inhibiton of [^{3}H]glibenclamide binding to the cerebral high-affinity site (SCHWANSTECHER et al. 1992d,f; NIKI and ASHCROFT 1993).

All of these findings suggest that many neurons contain a sulfonylurea receptor which resembles the pancreatic receptor and is linked to a typical K_{ATP} channel. In neurons, but not in B cells, the levels of creatine kinase and phosphocreatine are high (PANTEN et al. 1986; GHOSH et al. 1991; WALLIMANN et al. 1992). Thus, buffering of high ATP and low ADP concentrations in neurons appears to keep the K_{ATP} channels silent, unless channel opening by transmitters, hormones, drugs or exhaustion of energy metabolism takes place. It has been discussed that hypoxia and ischemia might open K_{ATP} channels and thereby prolong cell survival and that sulfonylureas would antagonize this beneficial effect of K_{ATP} channel activation (JIANG et al. 1992; OHNO-SHOSAKU and YAMAMOTO 1992; HEURTEAUX et al. 1993). Sulfonylureas pass the intact blood-brain barrier with difficulty (KELLNER et al. 1969; KOLB et al. 1974; SUGITA et al. 1982), but prolonged cerebral ischemia is associated with blood-brain barrier disruption (KUROIWA et al. 1988; GUMERLOCK 1989).

E. Actions on Cardiac Cells

The K_{ATP} channel was first described by Noma (1983) in cardiac myocytes. The cardiac channel resembles the B-cell channel with regard to biophysical properties and control by nucleotides (for reviews see ASHCROFT and

ASHCROFT 1990; NICHOLS and LEDERER 1991; TAKANO and NOMA 1993; GOPALAKRISHNAN et al. 1993). When the cytoplasmic side of the cardiac myocyte membrane was exposed to nucleotide-free solution, inhibition of K_{ATP} channels was half-maximal at $0.38-0.94$ mM tolbutamide or $0.5\,\mu M$ glibenclamide (BELLES et al. 1987; VENKATESH et al. 1991). These values suggest that the sulfonylurea sensitivity of the cardiac K_{ATP} channel is low, unless both inhibitory and stimulatory nucleotides are present at the internal membrane side, as described for the K_{ATP} channel in B cells (see Sect. B.II). This view is supported by the EC_{50} values for glibenclamide measured in perforated patch recordings (4 nM; HAMADA et al. 1990) or in whole-cell clamp experiments with 2 mM MgATP in the patch pipette (6 nM; FINDLAY 1992a). In the latter experiments, a proportion of MgATP was most probably transformed into MgADP by intracellular enzymes. Furthermore, the precipitous decline of intracellular ATP which caused rigor contracture was coincident with loss of effectiveness of glibenclamide (FINDLAY 1993). VIRÁG et al. (1993) reported that the K_{ATP}-channel-inhibiting effect of glibenclamide was weaker if, in addition to Mg^{2+} and ATP ($100\,\mu M$), ADP ($100\,\mu M$) was present at the cytoplasmic face of inside-out patches from ventricular myocytes. However, in the presence of Mg^{2+}, ATP and ADP glibenclamide might have been more inhibitory than in the presence of Mg^{2+} and ADP alone (see Sect. B.II). In accordance with findings for B cells, the effectiveness of glibenclamide decreased with increasing extracellular pH (FINDLAY 1992b) and after trypsin treatment of the cytoplasmic membrane side (NICHOLS and LOPATIN 1993; DEUTSCH and WEISS 1994). High-affinity binding sites for sulfonylureas were also identified in heart membranes (FOSSET et al. 1988; for reviews see ASHCROFT and ASHCROFT 1992; EDWARDS and WESTON 1993; GOPALAKRISHNAN et al. 1993). All these observations together are in favor of a great similarity between the sulfonylurea receptors in cardiac myocytes and B cells.

K_{ATP} channels in cardiac myocytes appear to be all closed under physiological conditions since high ATP and low ADP concentrations in the cytosol are buffered by the creatine-phosphocreatine system (for a review see NICHOLS and LEDERER 1991). Depletion of phosphocreatine during hypoxia or ischemia is accompanied by a decrease of the ATP concentration and by increases of the concentrations of ADP and adenosine, with subsequent opening of K_{ATP} channels and an increase in outward K^+ current. Adenosine diffuses out of the cardiac myocytes and contributes to the opening of K_{ATP} channels via interaction with A_1 receptors (KIRSCH et al. 1990). Action potential shortening, cellular K^+ loss and extracellular K^+ accumulation are brought about and facilitate arrhythmias (for reviews see GROSS and AUCHAMPACH 1992; COETZEE 1992; GOPALAKRISHNAN et al. 1993; WILDE and JANSE 1994). On the other hand, survival of cardiac myocytes is prolonged by action potential shortening since reduced Ca^{2+} current reduces contractility and energy consumption. Therefore, sulfonylurea-induced block of K_{ATP} channels during hypoxia or ischemia might inhibit arrhythmias, but might increase tissue damage (for reviews see GROSS and AUCHAMPACH 1992; COETZEE 1992; LYNCH et al. 1992). However, sulfonylureas have also been suspected to produce proarrhythmic effects, mainly during reperfusion (GROSS and AUCHAMPACH 1992; COETZEE 1992; LYNCH et al. 1992).

F. Actions on Smooth Muscle

The characterization of high-affinity binding sites for sulfonylureas in membrane preparations from vascular and nonvascular smooth muscle (KOVACS and NELSON 1991; GOPALAKRISHNAN et al. 1991; ZINI et al. 1991) suggests that these cells contain a typical sulfonylurea receptor. In smooth muscle of portal vein, a K_{ATP} channel with a unitary conductance of 50 pS (in symmetrical high-K^+ solutions) was identified (KAJIOKA et al. 1990, 1991). Cytosolic ATP inhibited this channel half-maximally at $29 \mu M$ when the channel was opened by application of $1 mM$ GDP and $0.1 mM$ pinacidil. This channel was also activated by MgADP (PFRÜNDER et al. 1993). Single-channel recordings revealed that the glibenclamide sensitivity of the K_{ATP} channel in a membrane preparation from aortic smooth muscle was very low in the absence of cytosolic nucleotides (KOVACS and NELSON 1991). Whole-cell clamp experiments did not give a uniform picture. In coronary smooth muscle cells internally dialyzed with nucleotide-free solutions, the ATP-sensitive K^+ current was reduced by glibenclamide with an EC_{50} value of $20 nM$ (XU and LEE 1994). In smooth muscle from portal vein, a similar EC_{50} value ($25 nM$) was estimated for glibenclamide when the cells were dialyzed with solutions containing $10 mM$ GDP, but the potency of glibenclamide was very low if dialysis was performed with nucleotide-free solutions (BEECH et al. 1993). The difference between these studies could be due to differences in efficiency of dialysis. Further studies are required to find out whether the opening activity of K_{ATP} channels in intact vascular smooth muscle is altered by free sulfonylurea concentrations measured in the plasma of treated type II diabetic patients.

G. Actions on Skeletal Muscle

In skeletal muscle a K_{ATP} channel has been detected which displays biophysical properties and nucleotide-dependent regulation characteristic of the K_{ATP} channels in B cells, neurons and cardiac myocytes (SPRUCE et al. 1985, 1987; WOLL et al. 1989; WEIK and NEUMCKE 1989; PARENT and CORONADO 1989; VIVAUDOU et al. 1991; FORESTIER and VIVAUDOU 1993; ALLARD and LAZDUNSKI 1992; for reviews see ASHCROFT and ASHCROFT 1990; DAVIES et al. 1991). As described for neurons and cardiac cells, buffering of high ATP and low ADP concentrations in skeletal muscle appears to keep the K_{ATP} channels silent unless channel opening by hormones, drugs or exhaustion of energy metabolism takes place (DAVIES et al. 1991). Evidence was also found for K_{ATP}-channel activation during muscle fatigue (LIGHT et al. 1994). SPRUCE et al. (1987) and VIVAUDOU et al. (1991) reported a rather low channel-inhibiting potency of ATP, whereas WEIK and NEUMCKE (1989) observed a potency of ATP which was not lower than that for inhibition of the K_{ATP} channel in B cells. High concentrations of tolbutamide or glibenclamide were required for inhibition of the muscular K_{ATP} channel in the absence of cytosolic nucleotides (WOLL et al. 1989; VIVAUDOU et al. 1991; ALLARD and LAZDUNSKI 1992, 1993). To our knowledge, the effects of cytosolic nucleotides on the sulfonylurea sensitivity of the K_{ATP} channel have not been examined in skeletal muscle. However, specific binding of [^{3}H]glibenclamide to membrane preparations from skeletal muscle was of high affinity (K_D = 1.5 nM; GOPALAKRISHNAN et al. 1991). Thus, it is conceivable that the opening activity of the K_{ATP} channel in hypoxic or ischemic skeletal muscle is inhibited by free sulfonylurea concentrations similar to those in the plasma of treated patients.

In contrast to glibenclamide, tolbutamide stimulated hexose uptake in a myocyte cell line at free concentrations found in the plasma of treated patients (COOPER et al. 1990). This effect probably does not contribute to the therapeutic effect of sulfonylureas since there is no clinical evidence that the first-generation sulfonylureas are more effective antidiabetic drugs than the second-generation sulfonylureas.

H. Actions on Miscellaneous Cells

Sulfonylurea-sensitive K_{ATP} channels have also been found in adenohypophysis cells and endothelial cells from aorta and brain microvessels (BERNARDI et al. 1993; JANIGRO et al. 1993). It is unknown whether the function of these K_{ATP} channels is altered during antidiabetic therapy with sulfonylureas.

The same sulfonylurea-sensitive hexose uptake system which was described for a myocyte cell line (see SECT. G) was also demonstrated in adipocytes (MALOFF and LOCKWOOD 1981; MARTZ et al. 1989). As discussed above, effects on this uptake system probably do not contribute to the antidiabetic effect of sulfonylureas. Other in vitro studies reporting extrapancreatic metabolic effects induced by therapeutic concentrations of free sulfonylureas were not found in the literature, an outcome similar to that of a previous search (PANTEN et al. 1989).

I. Conclusions

The hypoglycemic effect of sulfonylureas and their analogues results solely from their binding to the high-affinity receptor site located at the internal side of the B-cell plasma membrane. This receptor site mediates closure of the K_{ATP} channel and thereby stimulation of insulin release. Non-B cell sulfonylurea receptors do not contribute to the therapeutic benefit of sulfonylureas, but might be involved in presumed adverse effects of sulfonylureas in hypoxic or ischemic heart and brain.

Acknowledgements. We are grateful to Ms. C. Stephan for expertly typing the manuscript. Work in the authors' laboratory mentioned in this article was supported by grants from the Deutsche Forschungsgemeinschaft, the Deutsche Diabetes-Stiftung and the Deutsche Diabetes-Gesellschaft.

References

Aguilar-Bryan L, Nelson DA, Vu QA, Humphrey MB, Boyd AE III (1990) Photoaffinity labeling and partial purification of the β-cell sulfonylurea receptor using a novel, biologically active glyburide analog. J Biol Chem 265:8218–8224

Aguilar-Bryan L, Nichols CG, Rajan AS, Parker C, Bryan J (1992) Co-expression of sulfonylurea receptors and K_{ATP} channels in hamster insulinoma tumor (HIT) cells. J Biol Chem 267:14934–14940

Ali L, Wesslén N, Hellman B (1988) Sulphonamide modulation of sodium content in rat pancreatic islets. Eur J Pharmacol 158:257–262

Ali L, Wesslén N, Hellman B (1989) The effect of glibenclamide and its non-sulfonylurea analogue HB 699 on the sodium content of rat pancreatic islets. Exp Clin Endocrinol 93:299–306

Allard B, Lazdunski M (1992) Nucleotide diphosphates activate the ATP-sensitive potassium channel in mouse skeletal muscle. Pflügers Arch 422:185–192

Allard B, Lazdunski M (1993) Pharmacological properties of ATP-sensitive K^+ channels in mammalian skeletal muscle cells. Eur J Pharmacol 236:419–426

Amoroso S, Schmid-Antomarchi H, Fosset M, Lazdunski M (1990) Glucose, sulfonylureas, and neurotransmitter release: role of ATP-sensitive K^+ channels. Science 247:852–854

Ashcroft SJH (1994) The beta-cell sulfonylurea receptor. Diab Nutr Metab 7:149–163

Ashcroft SJH, Ashcroft FM (1990) Properties and functions of ATP-sensitive K-channels. Cell Signal 2:197–214

Ashcroft SJH, Ashcroft FM (1992) The sulfonylurea receptor. Biochim Biophys Acta 1175:45–59

Ashcroft FM, Rorsman P (1991) Electrophysiology of the pancreatic β-cell. Prog Biophys Mol Biol 54:87–143

Ashcroft FM, Harrison DE, Ashcroft SJH (1984) Glucose induces closure of single potassium channels in isolated rat pancreatic β-cells. Nature 312:446–448

Ashford MLJ (1990) Potassium channel and modulation of insulin secretion. In: Cook NS (ed) Potassium channels: structure, classification, function and therapeutic potential. Ellis Horwood, Chichester, pp 300–325

Ashford MLJ, Sturgess NC, Trout NJ, Gardner NJ, Hales CN (1988) Adenosine-5′ triphosphate-sensitive ion channels in neonatal rat cultured central neurons. Pflügers Arch 412:297–304

Ashford MLJ, Boden PR, Treherne JM (1990a) Glucose-induced excitation of hypothalamic neurones is mediated by ATP sensitive K^+ channels. Pflügers Arch 415:479–483

Ashford MLJ, Boden PR, Treherne JM (1990b) Tolbutamide excites rat glucoreceptive ventromedial hypothalamic neurones by indirect inhibition of ATP-K^+ channels. Br J Pharmacol 101:531–540

Ashford MLJ, Bond CT, Blair TA, Adelman JP (1994) Cloning and functional expression of a rat heart K_{ATP} channel. Nature 370:456–459

150   U. Panten et al.

Aumüller W, Heerdt R (1971) Sulfonylharnstoffderivate und verwandte Verbindungen als blutzuckersenkende Substanzen. In: Maske H (ed) Oral wirksame
Antidiabetika. Springer, Berlin Heidelberg New York, pp 1–249 (Handbook of
experimental pharmacology, vol 29)

Beck-Nielsen H, Hother-Nielsen O, Pedersen O (1988) Mechanism of action of
sulphonylureas with special reference to the extrapancreatic effect: an overview.
Diabet Med 5:613–620

Beech DJ, Zhang H, Nakao K, Bolton TB (1993) K channel activation by nucleotide
diphosphates and its inhibition by glibenclamide in vascular smooth muscle cells.
Br J Pharmacol 110:573–582

Belles B, Hescheler J, Trube G (1987) Changes of membrane currents in cardiac
cells induced by long whole-cell recordings and tolbutamide. Pflügers Arch
409:582–588

Bernardi H, Fosset M, Lazdunski M (1988) Characterization, purification, and affinity
labeling of the brain [^{3}H]glibenclamide-binding protein, a putative neuronal
ATP-regulated K$^+$ channel. Proc Natl Acad Sci USA 85:9816–9820

Bernardi H, Fosset M, Lazdunski M (1992) ATP/ADP binding sites are present in
the sulfonylurea binding protein asociated with brain ATP-sensitive K$^+$ channels.
Biochemistry 31:6328–6332

Bernardi H, De Weille JR, Epelbaum J, Mourre C, Amoroso S, Slama A, Fosset M,
Lazdunski M (1993) ATP-modulated K$^+$ channels sensitive to antidiabetic
sulfonylureas are present in adenohypophysis and are involved in growth
hormone release. Proc Natl Acad Sci USA 90:1340–1344

Bokvist K, Rorsman P, Smith PA (1990) Block of ATP-regulated and Ca^{2+}-activated
K$^+$-channels in mouse pancreatic β-cells by external tetraethylammonium and
quinine. J Physiol (Lond) 423:327–343

Bouman PR, Gaarenstroom JH (1961) Stimulation by carbutamide and tolbutamide
of insulin release from rat pancreas in vitro. Metabolism 10:1095–1099

Boyd AE III (1988) Sulfonylurea receptors, ion channels and fruit flies. Diabetes
37:847–850

Boyd AE III, Aguilar-Bryan L, Bryan J, Kunze DL, Moss L, Nelson DA, Rajan
AS, Raef H, Xiang H, Yaney GC (1991) Sulfonylurea signal transduction.
Recent Prog Horm Res 47:299–317

Brown GR, Foubister AJ (1984) Receptor binding sites of hypoglycemic sulfonylureas
and related [(acylamino) alkyl] benzoic acids. J Med Chem 27:79–81

Carpentier J-L, Sawano F, Ravazzola M, Malaisse WJ (1986) Internalization of ^{3}H-
glibenclamide in pancreatic islet cells. Diabetologia 29:259–261

Chiba T, Taminato T, Kadowaki S, Chihara K, Matsukura S, Seino Y, Fujita
T (1982) Tolbutamide stimulates gastric somatostatin release from isolated
perfused rat stomach. Diabetes 31:119–121

Chudziak F, Schwanstecher M, Laatsch H, Panten U (1994) Synthesis of a ^{125}I
labelled azidosubstituted glibenclamide analogue for photoaffinity labelling of
the sulfonylurea receptor. J Labelled Compd Radiopharm 34:675–680

Coetzee WA (1992) ATP-sensitive potassium channels and myocardial ischemia: why
do they open? Cardiovasc Drugs Ther 6:201–208

Cook DL, Hales CN (1984) Intracellular ATP directly blocks K$^+$-channels in
pancreatic B-cells. Nature 311:271–173

Cook DL, Satin LB, Ashford MLJ, Hales CN (1988) ATP-sensitive K$^+$ channels in
pancreatic β-cells. Spare channel hypothesis. Diabetes 37:495–498

Cooper DR, Vila MC, Watson JE, Nair G, Pollet RJ, Standaert M, Farese RV
(1990) Sulfonylurea-stimulated glucose transport association with diacylglycerol-
like activation of protein kinase C in BC$_3$H1 myocytes. Diabetes 39:1399–1407

Coore HG, Randle PJ (1964) Regulation of insulin secretion studied with pieces of
rabbit pancreas incubated in vitro. Biochem J 93:66–78

Davies NW, Standen NB, Stanfield PR (1991) ATP-dependent potassium channels
of muscle cells: their properties, regulation, and possible functions, J Bioenerg
Biomembr 23:509–535

Dean PM, Matthews EK (1968) Electrical activity in pancreatic islet cells. Nature 219:389–390

Deutsch N, Weiss JN (1994) Effects of trypsin on cardiac ATP-sensitive K^+ channels. Am J Physiol 266:H613–H622

De Weille JR, Fosset M, Mourre C, Schmid-Antomarchi H, Bernardi H, Lazdunski M (1989) Pharmacology and regulation of ATP-sensitive K^+ channels. Pflügers Arch 414:S80–S87

De Weille JR, Müller M, Lazdunski M (1992) Activation and inhibition of ATP-sensitive K^+ channels by fluorescein derivatives. J Biol Chem 267:4557–4563

Dunne MJ (1989) Protein phosphorylation is required for diazoxide to open ATP-sensitive potassium channels in insulin (RINm5F) secreting cells. FEBS Lett 250:262–266

Dunne MJ, Petersen OH (1991) Potassium selective ion channels in insulin-secreting cells: physiology, pharmacology and their role in stimulus-secretion coupling. Biochim Biophys Acta 1071:67–82

Edwards G, Weston AH (1993) The pharmacology of ATP-sensitive potassium channels. Annu Rev Pharmacol Toxicol 33:597–637

Ferner RE, Chaplin S (1987) The relationship between the pharmacokinetics and pharmacodynamic effects of oral hypoglycaemic drugs. Clin Pharmacokinet 12:379–401

Findlay I (1987) The effects of magnesium upon adenosine triphosphate-sensitive potassium channels in a rat insulin-secreting cell line. J Physiol (Lond) 391:611–629

Findlay I (1992a) Inhibition of ATP-sensitive K^+ channels in cardiac muscle by the sulphonylurea drug glibenclamide. J Pharmacol Exp Ther 261:540–545

Findlay I (1992b) Effects of pH upon the inhibition by sulphonylurea drugs of ATP-sensitive K^+ channels in cardiac muscle. J Pharmacol Exp Ther 262:71–79

Findlay I (1993) Sulphonylurea drugs no longer inhibit ATP-sensitive K^+ channels during metabolic stress in cardiac muscle. J Pharmacol Exp Ther 266:456–467

Flatt PR, Shibier O, Szecowka J, Berggren P-O (1994) New perspectives on the actions of sulphonylureas and hyperglycaemic sulphonamides on the pancreatic β-cell. Diabete Metab 20:157–162

Flockhart DA, Corbin JD (1982) Regulatory mechanisms in the control of protein kinases. Crit Rev Biochem 12:133–186

Forestier C, Vivaudou M (1993) Modulation by Mg^{2+} and ADP of ATP-sensitive potassium channels in frog skeletal muscle. J Membr Biol 132:87–94

Fosset M, De Weille JR, Green RD, Schmid-Antomarchi H, Lazdunski M (1988) Antidiabetic sulfonylureas control action potential properties in heart cells via high affinity receptors that are linked to ATP-dependent K^+ channels. J Biol Chem 263:7933–7936

Fujitani S, Yada T (1994) A novel D-phenylalanine-derivative hypoglycemic agent A-4166 increases cytosolic free Ca^{2+} in rat pancreatic β-cells by stimulating Ca^{2+} influx. Endocrinology 134:1395–1400

Gaines KL, Hamilton S, Boyd AE III (1988) Characterization of the sulfonylurea receptor on beta cell membranes. J Biol Chem 263:2589–2592

Garrino M-G, Schmeer W, Nenquin M, Meissner HP, Henquin JC (1985) Mechanism of the stimulation of insulin release in vitro by HB 699, a benzoic acid derivative similar to the non-sulphonylurea moiety of glibenclamide. Diabetologia 28:697–703

Geisen K, Hübner M, Hitzel V, Hrstka VE, Pfaff W, Bosies E, Regitz G, Kühnle HF, Schmidt FH, Weyer R (1978) Acylaminoalkyl-substituierte Benzoe- und Phenylalkansäuren mit blutglukose-senkender Wirkung. Arzneimittelforschung 28:1081–1083

Geisen K, Hitzel V, Ökomonopoulos R, Pünter J, Weyer R, Summ HD (1985) Inhibition of ^{3}H-glibenclamide binding to sulfonylurea receptors by oral anti-diabetics. Arzneimittelforschung 35:707–712

Gerich JE (1989) Oral hypoglycemic agents. N Engl J Med 321:1231–1245

Gopalakrishnan M, Johnson DE, Janis RA, Triggle DJ (1991) Characterization of binding of the ATP-sensitive potassium channel ligand, [^{3}H]glyburide, to neuronal and muscle preparations. J Pharmacol Exp Ther 257:1162–1171

Gopalakrishnan M, Janis RA, Triggle DJ (1993) ATP-sensitive K$^+$ channels: pharmacologic properties, regulation, and therapeutic potential. Drug Devel Res 28:95–127

Gorus FK, Schuit FC, In't Veld PA, Gepts W, Pipeleers DG (1988) Interaction of sulfonylureas with pancreatic β-cells. Diabetes 37:1090–1095

Ghosh A, Ronner P, Cheong E, Khalid P, Matschinsky FM (1991) The role of ATP and free ADP in metabolic coupling during fuel-stimulated insulin release from islet β-cells in the isolated perfused rat pancreas. J Biol Chem 266:22887–22892

Gross GJ, Auchampach JA (1992) Role of ATP dependent potassium channels in myocardial ischaemia. Cardiovasc Res 26:1011–1016

Guiot Y, Henquin JC, Rahier J (1994) Effects of glibenclamide on pancreatic β-cell proliferation in vivo. Eur J Pharmacol 261:157–161

Gumerlock MK (1989) Cerebrovascular disease and the blood-brain barrier. In: Neuwelt EA (ed) Implications of the blood-brain barrier and its manipulation, vol 2. Plenum, New York, pp 495–565

Gylfe E, Hellman B, Sehlin J, Täljedal I-B (1984) Interaction of sulfonylurea with the pancreatic B-cell. Experientia 40:1126–1134

Hamada E, Takikawa R, Ito H, Iguchi M, Terano A, Sugimoto T, Kurachi Y (1990) Glibenclamide specifically blocks ATP-sensitive K$^+$ channel current in atrial myocytes of guinea pig heart. Jpn J Pharmacol 54:473–477

Hellman B, Täljedal I-B (1975) Effects of sulfonylurea derivatives on pancreatic β-cells. In: Hasselblatt A, von Bruchhausen F (eds) Insulin II. Springer, Berlin Heidelberg New York, pp 175–194 (Handbook of experimental pharmacology, vol 32/2)

Henquin JC (1980) Tolbutamide stimulation and inhibition of insulin release: studies of the underlying ionic mechanisms in isolated rat islets. Diabetologia 18: 151–160

Henquin JC (1987) Regulation of insulin release by ionic and electrical events in B cells. Horm Res 27:168–178

Henquin JC (1988) ATP-sensitive K$^+$ channels may control glucose-induced electrical activity in pancreatic B-cells. Biochem Biophys Res Commun 156:769–775

Henquin JC (1990a) Glucose-induced electrical activity in β-cells. Feedback control of ATP-sensitive K$^+$ channels by Ca^{2+}. Diabetes 39:1457–1460

Henquin JC (1990b) Established, unsuspected and novel pharmacological insulin secretagogues. In: Bailey CJ, Flatt PR (eds) New antidiabetic drugs. Smith-Gordon, London, pp 93–106

Henquin JC (1992) The fiftieth anniversary of hypoglycaemic sulphonamides. How did the mother compound work? Diabetologia 35:907–912

Henquin JC, Garrino M-G, Nenquin M (1987) Stimulation of insulin release by benzoic acid derivatives related to the non-sulphonylurea moiety of glibenclamide: structural requirements and cellular mechanisms. Eur J Pharmacol 141:243–251

Heurteaux C, Bertaina V, Widmann C, Lazdunski M (1993) K$^+$ channel openers prevent global ischemia-induced expression of c-fos, c-jun, heat shock protein, and amyloid β-protein precursor genes and neuronal death in rat hippocampus. Proc Natl Acad Sci USA 90:9431–9435

Hicks GA, Hudson AL, Henderson G (1994) Localization of high affinity [^{3}H]glibenclamide binding sites within the substantia nigra zona reticulata of the rat brain. Neuroscience 61:285–292

Hirose H, Maruyama H, Ito K, Seto Y, Kido K, Koyama K, Dan K, Saruta T, Kato R (1994) Effects of N-[(trans-4-isopropylcyclohexyl)-carbonyl]-D-phenylalanine (A-4166) on insulin and glucagon secretion in isolated perfused rat pancreas. Pharmacology 48:205–210

Janigro D, West GA, Gordon EL, Winn HR (1993) ATP-sensitive K$^+$ channels in rat aorta and brain microvascular endothelial cells. Am J Physiol 265:C812–C821

Jiang C, Xia Y, Haddad GG (1992) Role of ATP-sensitive K$^+$ channels during anoxia: major differences between rat (newborn and adult) and turtle neurons. J Physiol (Lond) 448:599–612

Jiang C, Sigworth FJ, Haddad GG (1994) Oxygen deprivation activates on ATP-inhibitable K$^+$ channel in substantia nigra neurons. J Neurosci 14:5590–5602

Jonas P, Koh D-S, Kampe K, Hermsteiner M, Vogel W (1991) ATP-sensitive and Ca-activated K channels in vertebrate axons: novel links between metabolism and excitability. Pflügers Arch 418:68–73

Kajioka S, Oike M, Kitamura K (1990) Nicorandil opens a calcium-dependent potassium channel in smooth muscle cells of the rat portal vein. J Pharmacol Exp Ther 254:905–913

Kajioka S, Kitamura K, Kuriyama H (1991) Guanosine diphosphate activates an adenosine 5'-triphosphate-sensitive K$^+$ channel in the rabbit portal vein. J Physiol (Lond) 444:397–418

Kaubisch N, Hammer R, Wollheim C, Renold AE, Offord RE (1982) Specific receptors for sulfonylureas in brain and in a β-cell tumor of the rat. Biochem Pharmacol 31:1171–1174

Kellner H-M, Christ O, Rupp W, Heptner W (1969) Resorption, Verteilung und Ausscheidung nach Gabe von ^{14}C-markiertem HB 419 an Kaninchen, Ratten und Hunde. Arzneimittelforschung 19:1388–1400

Khan RN, Hales CN, Ozanne SE, Adogu AA, Ashford MLJ (1993) Dissociation of K$_{ATP}$ channel and sulphonylurea receptor in the clonal insulin-secreting cell line, CR-D11. Proc R Soc Lond [Biol] 253:225–231

Kirsch GE, Codina J, Birnbaumer L, Brown AM (1990) Coupling of ATP-sensitive K$^+$ channels to A$_1$ receptors by G proteins in rat ventricular myocytes. Am J Physiol 259:H820–H826

Kolb KH, Schulze PE, Speck U, Acksteiner B (1974) Pharmakokinetik von radioaktiv markiertem Glisoxepid beim Tier. Arzneimittelforschung 24:397–403

Kovacs RJ, Nelson MT (1991) ATP-sensitive K$^+$channels from aortic smooth muscle incorporated into planar lipid bilayers. Am J Physiol 261:H604–H609

Kozlowski RZ, Ashford MLJ (1990) ATP-sensitive K$^+$ channel run-down is Mg^{2+} dependent. Proc R Soc [Biol] 240:397–410

Kozlowski RZ, Ashford MLJ (1991) Barbiturates inhibit ATP-K$^+$ channels and voltage-activated currents in CRI-G1 insulin-secreting cells. Br J Pharmacol 103:2021–2029

Kozlowski RZ, Hales CN, Ashford MLJ (1989) Dual effects of diazoxide on ATP-K$^+$ currents recorded from an insulin-secreting cell line. Br J Pharmacol 97:1039–1050

Kramer W, Oekonomopulos R, Pünter J, Summ H-D (1988) Direct photoaffinity labeling of the putative sulfonylurea receptor in rat β-cell tumor membranes by [^{3}H]glibenclamide. FEBS Lett 229:355–359

Kramer W, Müller G, Girbig F, Gutjahr U, Kowalewski S, Hartz D, Summ H-D (1994) Differential interaction of glimepiride and glibenclamide with the β-cell sulfonylurea receptor. II. Photoaffinity labeling of a 65 kDa protein by [^{3}H]glimepiride. Biochim Biophys Acta 1191:278–290

Kuroiwa T, Shibutani M, Okeda R (1988) Blood-brain barrier disruption and exacerbation of ischemic brain edema after restoration of blood flow in experimental focal cerebral ischemia. Acta Neuropathol (Berl) 76:62–70

Larsson O, Ämmälä C, Bokvist K, Fredholm B, Rorsman P (1993) Stimulation of the K$_{ATP}$ channel by ADP and diazoxide requires nucleotide hydrolysis in mouse pancreatic β-cells. J Physiol (Lond) 463:349–365

Laychock SG (1983) Identification and metabolism of polyphosphoinositides in isolated islets of Langerhans. Biochem J 216:101–106

Lebovitz HE (1990) Oral hypoglycemic agents. In: Ellenberg M (ed) Diabetes mellitus "theory and practice." Elsevier, Amsterdam, pp 554–574

Lee K, Ashford MLJ (1993) Mg^{2+} modulates the binding of [^{3}H]glibenclamide to its receptor in rat cerebral cortical membranes. Eur J Pharmacol 247:347–351

Lee K, Ozanne SE, Hales CN, Ashford MLJ (1994a) Mg^{2+}-dependent inhibition of K_{ATP} by sulphonylureas in CRI-G1 insulin-secreting cells. Br J Pharmacol 111:632–640

Lee K, Ozanne SE, Hales CN, Ashford MLJ (1994b) Effects of chemical modification of amino and sulfhydryl groups on K_{ATP} channel function and sulfonylurea binding in CRI-G1 insulin-secreting cells. J Membr Biol 139:167–181

Lee K, Ozanne SE, Rowe ICM, Hales CN, Ashford MLJ (1994c) The effects of trypsin on ATP-sensitive potassium channel properties and sulfonylurea receptors in the CRI-G1 insulin-secreting cell line. Mol Pharmacol 45:176–185

Light PE, Comtois AS, Renaud JM (1994) The effect of glibenclamide on frog skeletal muscle: evidence for K^+_{ATP} channel activation during fatigue. J Physiol (Lond) 475:495–507

Lin Y-J, Greif GJ, Freedman JE (1993) Multiple sulfonylurea-sensitive potassium channels: a novel subtype modulated by dopamine. Mol Pharmacol 44:907–910

Loubatières A (1957a) The hypoglycemic sulfonamides: history and development of the problem from 1942 to 1955. Ann NY Acad Sci 71:4–11

Loubatières A (1957b) The mechanism of action of the hypoglycemic sulfonamides: a concept based on investigations in animals and in human beings. Ann NY Acad Sci 71:192–206

Loubatières A (1977) Effects of sulfonylureas on the pancreas. In: Volk BW, Wellmann KF (eds) The diabetic pancreas. Plenum New York, pp 489–515

Loubatières A, Mariani MM, Ribes G, de Malbosc H, Alric R, Chapal J (1969) Pharmakologische Untersuchungen eines neuen hochwirksamen blutzuckersenkenden Sulfonamids, des Glibenclamid (HB 419). Arzneimittelforschung 19: 1334–1363

Lupo B, Bataille D (1987) A binding site for [^{3}H]glipizide in the rat cerebral cortex. Eur J Pharmacol 140:157–169

Lynch JJ, Sanguinetti MC, Kimura S, Bassett AL (1992) Therapeutic potential of modulating potassium currents in the diseased myocardium. FASEB J 6: 2952–2960

Malaisse WJ, Lebrun P (1990) Mechanisms of sulfonylurea-induced insulin release. Diabetes Care 13:9–17

Malaisse WJ, Malaisse-Lagae F, Mayhew DA, Wright PH (1967) Effects of sulfonylureas upon insulin secretion by the rat's pancreas. In: Butterfield WJH, von Westering W (eds) Tolbutamide after ten years. Excerpta Medica, Amsterdam, pp 49–60

Malaisse WJ, Hubinont C, Lebrun P, Herchuelz A, Couturier E, Deleers M, Malaisse-Lagae F, Sener A (1983) Mode of action of hypoglycaemic sulfonylureas in the pancreatic β-cell: coinciding and conflicting views. In: Serrano-Rios M, Krall LP (eds) Clinical and pharmacological activities of sulfonylurea drugs. Excerpta Medica, Amsterdam, pp 24–38

Maloff BL, Lockwood DH (1981) In vitro effects of a sulfonylurea on insulin action in adipocytes. J Clin Invest 68:85–90

Martz A, Jo I, Jung CY (1989) Sulfonylurea binding to adipocyte membranes and potentiation of insulin-stimulated hexose transport. J Biol Chem 264: 13672–13678

Matthews EK (1985) Electrophysiology of pancreatic islet β-cells. In: Poisner AM, Trifaro JM (eds) the electrophysiology of the secretory cell. Elsevier, Amsterdam, pp 93–112

Matthews EK, Dean PM, Sakamoto Y (1973) Biophysical effects of sulphonylureas on islet cells. In: Acheson GH (ed) Pharmacology and the future of man. Proc 5th Int Congr Pharmacology, San Francisco 1972, vol 3. Karger, Basel, pp 221–229

Melander A, Bitzén P-O, Faber O, Groop L (1989) Sulphonylurea antidiabetic drugs. An update of their clinical pharmacology and rational therapeutic use. Drugs 37:58–72

Mogami H, Shibata H, Nobusawa R, Ohnota H, Satou F, Miyazaki J, Kojima I (1994) Inhibition of ATP-sensitive K^+ channel by a non-sulfonylurea compound KAD-1229 in a pancreatic β-cell line, MIN 6 cell. Eur J Pharmacol 269:293–298

Mourre C, Widmann C, Lazdunski M (1990) Sulfonylurea binding sites associated with ATP-regulated K^+ channels in the central nervous system: autoradiographic analysis of their distribution and ontogenesis, and of their localization in mutant mice cerebellum. Brain Res 519:29–43

Müller M, de Weille JR, Lazdunski M (1991) Chlorpromazine and related phenothiazines inhibit the ATP-sensitive K^+ channel. Eur J Pharmacol 198:101–104

Müller G, Hartz D, Pünter J, Ökonomopulos R, Kramer W (1994a) Differential interaction of glimepiride and glibenclamide with the β-cell sulfonylurea receptor. I. Binding characteristics. Biochim Biophys Acta 1191:267–277

Müller G, Korndörfer A, Kornak U, Malaisse WJ (1994b) Porin proteins in mitochondria from rat pancreatic islet cells and white adipocytes: identification and regulation of hexokinase binding by the sulfonylurea glimepiride. Arch Biochem Biophys 308:8–23

Nelson DA, Aguilar-Bryan L, Bryan J (1992) Specificity of photolabeling of β-cell membrane proteins with an ^{125}I-labeled glyburide analog. J Biol Chem 267: 14928–14933

Nichols CG, Lederer WJ (1991) Adenosine triphosphate-sensitive potassium channels in the cardiovascular system. Am J Physiol 261:H1675–H1686

Nichols CG, Lopatin AN (1993) Trypsin and alpha-chymotrypsin treatment abolishes glibenclamide sensitivity of K_{ATP} channels in rat ventricular myocytes. Pflügers Arch 422:617–619

Niki I, Ashcroft SJH (1991) Possible involvement of protein phosphorylation in the regulation of the sulfonylurea receptor of a pancreatic β-cell line, HIT T15. Biochim Biophys Acta 1133:95–101

Niki I, Ashcroft SJH (1993) Characterization and solubilization of the sulphonylurea receptor in rat brain. Neuropharmacology 32:951–957

Niki I, Kelly RP, Ashcroft SJH, Ashcroft FM (1989) ATP-sensitive K-channels in HIT T15 β-cells studied by patch-clamp methods, ^{86}Rb efflux and glibenclamide binding. Pflügers Arch 415:47–55

Niki I, Nicks JL, Ashcroft SJH (1990) The β-cell glibenclamide receptor is an ADP-binding protein. Biochem J 268:713–718

Noma A (1983) ATP-regulated K^+ channels in cardiac muscle. Nature 305:147–148

Ohno-Shosaku T, Yamamoto C (1992) Identification of an ATP-sensitive K^+ channel in rat cultured cortical neurons. Pflügers Arch 422:260–266

Ohno-Shosaku T, Zünkler BJ, Trube G (1987) Dual effects of ATP on K^+ currents of mouse pancreatic β-cells. Pflügers Arch 408:133–138

Ohnota H, Koizumi T, Tsutsumi N, Kobayashi M, Inoue S, Sato F (1994) Novel rapid and short-acting hypoglycemic agent, a calcium (2 s)-2-benzyl-3-(cis-hexahydro-2-isoindolinyl-carbonyl)propionate (KAD-1229) that acts on the sulfonylurea receptor: comparison of effects between KAD-1229 and gliclazide. J Pharmacol Exper Ther 269:489–495

Panten U, Lenzen S (1988) Alterations in energy metabolism of secretory cells. In: Akkerman J-W N (ed) Energetics of secretion responses, vol 2. CRC Press, Boca Raton, pp 109–123

Panten U, Zünkler BJ, Scheit S, Kirchhoff K, Lenzen S (1986) Regulation of energy metabolism in pancreatic islets by glucose and tolbutamide. Diabetologia 29: 648–654

Panten U, Burgfeld J, Goerke F, Rennicke M, Schwanstecher M, Wallasch A, Zünkler BJ, Lenzen S (1989) Control of insulin secretion by sulfonylureas, meglitinide and diazoxide in relation to their binding to the sulfonylurea receptor in pancreatic islets. Biochem Pharmacol 38:1217–1229

Panten U, Heipel C, Rosenberger F, Scheffer K, Zünkler BJ, Schwanstecher C (1990) Tolbutamide-sensitivity of the adenosine 5'-triphosphate-dependent K^+ channel in mouse pancreatic B-cells. Naunyn Schmiedebergs Arch Pharmacol 342:566–574

Panten U, Schwanstecher M, Schwanstecher C (1992) Pancreatic and extrapancreatic sulfonylurea receptors. Horm Metab Res 24:549–554

Panten U, Schwanstecher C, Schwanstecher M (1993) ATP-sensitive K^+ channel: properties, occurrence, role in regulation of insulin secretion. In: Dickey BF, Birnbaumer L (eds) GTPases in biology II. Springer, Berlin Heidelberg New York, pp 547–559 (Handbook of experimental pharmacology, vol 108/II)

Parent L, Coronado R (1989) Reconstitution of the ATP-sensitive potassium channel of skeletal muscle. J Gen Physiol 94:445–463

Penner R, Pusch M, Neher E (1987) Washout phenomena in dialyzed mast cells allow discrimination of different steps in stimulus-secretion coupling. Biosci Rep 7:313–321

Petit P, Loubatières-Mariani MM (1992) Potassium channels of the insulin-secreting B cell. Fundam Clin Pharmacol 6:123–134

Pfründer D, Anghelescu I, Kreye VAW (1993) Intracellular ADP activates ATP-sensitive K^+ channels in vascular smooth muscle cells of the guinea pig portal vein. Pflügers Arch 423:149–151

Politi DMT, Rogawski MA (1991) Glyburide-sensitive K^+ channels in cultured rat hippocampal neurons: activation by cromakalim and energy-depleting conditions. Mol Pharmacol 40:308–315

Proks P, Ashcroft FM (1993) Modification of K-ATP channels in pancreatic β-cells by trypsin. Pflügers Arch 424:63–72

Pusch M, Neher E (1988) Rates of diffusional exchange between small cells and a measuring patch pipette. Pflügers Arch 411:204–211

Rajan AS, Aguilar-Bryan L, Nelson DA, Yaney GC, Hsu WH, Kunze DL, Boyd AE III (1990) Ion channels and insulin secretion. Diabetes Care 13:340–363

Rajan AS, Aguilar-Bryan L, Nelson DA, Nichols CG, Wechsler SW, Lechago J, Bryan J (1993) Sulfonylyurea receptors and ATP-sensitive K^+ channels in clonal pancreatic α cells. J Biol Chem 268:15221–15228

Ronner P, Hang TL, Kraebber MJ, Higgins TJ (1992) Effect of the hypoglycaemic drug (-)-AZ-DF-265 on ATP-sensitive potassium channels in rat pancreatic β-cells. Br J Pharmacol 106:250–255

Ronner P, Matschinsky FM, Hang TL, Epstein AJ, Buettger C (1993) Sulfonylurea-binding sites and ATP-sensitive K^+ channels in α-TC glucagonoma and β-TC insulinoma cells. Diabetes 42:1760–1772

Rorsman P, Hellman B (1988) Voltage-activated currents in guinea pig pancreatic a_2 cells. Evidence for Ca^{2+}-dependent action potentials. J Gen Physiol 91:223–242

Saha S, Hellman B (1994) Sulfonylureas mimic glucose in stimulating the uptake of Na^+ in pancreatic islets exposed to ouabain. Eur J Pharmacol 258:145–149

Schmid-Antomarchi H, De Weille J, Fosset M, Lazdunski M (1987) The receptor for antidiabetic sulfonylureas controls the activity of the ATP-modulated K^+ channel in insulin-secreting cells. J Biol Chem 262:15840–15844

Schwanstecher C, Panten U (1993a) Tolbutamide- and diazoxide-sensitive K^+ channel in neurons of substantia nigra pars reticulata. Naunyn Schmiedebergs Arch Pharmacol 348:113–117

Schwanstecher M, Panten U (1993b) Protein phosphorylation regulates receptor binding of potassium channel openers in insulin secreting cells and cerebral cortex. Biol Chem Hoppe Seyler 374:151

Schwanstecher C, Panten U (1994) Identification of an ATP-sensitive K^+ channel in spiny neurons of rat caudate nucleus. Pflügers Arch 427:187–189

Schwanstecher M, Rietze I (1990) Hydrolyzable nucleotides inhibit glibenclamide binding in pancreatic islets. Naunyn Schmiedebergs Arch Pharmacol 341:R72

Schwanstecher M, Löser S, Rietze I, Panten U (1990) $Mg^{2+}ATP$ controls gliben-clamide- and diazoxide-binding to their receptor in pancreatic B-cells. Diabeto-logia 33:A78

Schwanstecher M, Löser S, Rietze I, Panten U (1991) Phosphate and thiophosphate group donating adenine and guanine nucleotides inhibit glibenclamide binding to membranes from pancreatic islets. Naunyn Schmiedebergs Arch Pharmacol 343:83–89

Schwanstecher C, Dickel C, Panten U (1992a) Cytosolic nucleotides enhance the tolbutamide sensitivity of the ATP-dependent K^+ channel in mouse pancreatic B cells by their combined actions at inhibitory and stimulatory receptors. Mol Pharmacol 41:480–486

Schwanstecher C, Dickel C, Ebers I, Lins S, Zünkler BJ, Panten U (1992b) Diazoxide-sensitivity of the adenosine 5'-triphosphate dependent K^+ channel in mouse pancreatic β-cells. Br J Pharmacol 107:87–94

Schwanstecher M, Behrends S, Brandt C, Panten U (1992c) The binding properties of the solubilized sulfonylurea receptor from a pancreatic B-cell line are modu-lated by the Mg^{++}-complex of ATP. J Pharmacol Exper Ther 262:495–502

Schwanstecher M, Brandt C, Behrends S, Schaupp U, Panten U (1992d) Effect of MgATP on pinacidil-induced displacement of glibenclamide from the sul-phonylurea receptor in a pancreatic β-cell line and rat cerebral cortex. Br J Pharmacol 106:295–301

Schwanstecher M, Löser S, Brandt Ch, Scheffer K, Rosenberger F, Panten U (1992e) Adenine nucleotide-induced inhibition of binding of sulphonylureas to their receptor in pancreatic islets. Br J Pharmacol 105:531–534

Schwanstecher M, Schaupp U, Löser S, Panten U (1992f) The binding properties of the particulate and solubilized sulfonylurea receptor from cerebral cortex are modulated by the Mg^{2+} complex of ATP. J Neurochem 59:1325–1335

Schwanstecher C, Dickel C, Panten U (1994a) Interaction of tolbutamide and cytosolic nucleotides in controlling the ATP-sensitive K^+ channel in mouse β-cells. Br J Pharmacol 111:302–310

Schwanstecher M, Löser S, Chudziak F, Bachmann C, Panten U (1994b) Photo-affinity labeling of the cerebral sulfonylurea receptor using a novel radioiodinated azidoglibenclamide analogue. J Neurochem 63:698–708

Schwanstecher M, Löser S, Chudziak F, Panten U (1994c) Identification of a 38-kDa high affinity sulfonylurea-bending peptide in insulin-secreting cells and cerebral cortex. J Biol Chem 269:17768–17771

Schwanstecher M, Männer K, Panten U (1994d) Inhibition of K^+ channels and stimulation of insulin secretion by the sulfonylurea, glimepiride, in relation to its membrane binding in pancreatic islets. Pharmacology 49:105–111

Schwanstecher M, Schwanstecher C, Dickel C, Chudziak F, Moshiri A, Panten U (1994e) Location of the sulphonylurea receptor at the cytoplasmic face of the β-cell membrane. Br J Pharmacol 113:903–911

Schwanstecher M, Bachmann C, Löser S, Panten U (1995) Interaction of fluorescein derivatives with sulfonylurea binding in insulin-secreting cells. Pharmacology 50:182–191

Sellers AJ, Boden PR, Ashford MLJ (1992) Lack of effect of potassium channel openers on ATP-modulated potassium channels recorded from rat ventromedial hypothalamic neurones. Br J Pharmacol 107:1068–1074

Siconolfi-Baez L, Banerji MA, Lebovitz HE (1990) Characterization and significance of sulfonylurea receptors. Diabetes Care 13 [Suppl 3]:2–8

Skeer JM, Degano P, Coles B, Potier M, Ashcroft FM, Ashcroft SJH (1994) Determination of the molecular mass of the native beta-cell sulfonylurea receptor. FEBS Lett 338:98–102

Skillman TG, Feldman JM (1981) The pharmacology of sulfonylureas. Am J Med 70:361–372

Spruce AE, Standen NB, Stanfield PR (1985) Voltage-dependent ATP-sensitive potassium channels of skeletal muscle membrane. Nature 316:736–738

Spruce AE, Standen NB, Stanfield PR (1987) Studies of the unitary properties of adenosine-5'-triphosphate-regulated potassium channels of frog skeletal muscle. J Physiol 382:213–236

Sturgess NC, Ashford MLJ, Cook DL, Hales CN (1985) The sulphonylurea receptor may be an ATP-sensitive potassium channel. Lancet 8453:474–475

Sturgess NC, Kozlowski RZ, Carrington CA, Hales CN, Ashford MLJ (1988) Effects of sulphonylureas and diazoxide on insulin secretion and nucleotide-sensitive channels in an insulin-secreting cell line. Br J Pharmacol 95:83–94

Sugita O, Sawada Y, Sugiyama Y, Iga T, Hanano M (1982) Physiologically based pharmacokinetics of drug-drug interaction: a study of tolbutamide-sulfonamide interaction in rats. J Pharmacokinet Biopharm 10:297–316

Takano M, Noma A (1993) The ATP-sensitive K+ channel. Prog Neurobiol 41:21–30

Taylor KW, Parry DG (1967) Tolbutamide and the incorporation of [^{3}H]leucine into insulin in vitro. J Endocrinol 39:457–458

Tromba C, Salvaggio A, Racagni G, Volterra A (1992) Hypoglycemia-activated K+ channels in hippocampal neurons. Neurosci Lett 143:185–189

Tromba C, Salvaggio A, Racagni G, Volterra A (1994) Hippocampal hypoglycaemia-activated K+ channels: single-channel analysis of glucose and voltage dependence. Pflügers Arch 429:58–63

Trube G, Rorsman P, Ohno-Shosaku T (1986) Opposite effects of tolbutamide and diazoxide on the ATP-dependent K+ channel in mouse pancreatic B-cells. Pflügers Arch 407:493–499

Trube G, Hescheler J, Schröter K (1989) Regulation of ATP-dependent K+ channels in pancreatic B-cells. In: Oxford GS, Armstrong CM (eds) Secretion and its control, society of general physiologists series, vol 44. Rockefeller University Press, New York, pp 84–95

Venkatesh N, Lamp ST, Weiss JN (1991) Sulfonylureas, ATP-sensitive K+ channels, and cellular K+ loss during hypoxia, ischemia, and metabolic inhibition in mammalian ventricle. Circ Res 69:623–637

Verspohl EJ, Ammon HPT, Mark M (1990) Evidence for more than one binding site for sulfonylureas in insulin-secreting cells. J Pharm Pharmacol 42:230–235

Virág L, Furukawa T, Hiraoka M (1993) Modulation of the effect of glibenclamide on K_{ATP} channels by ATP and ADP. Mol Cell Biochem 119:209–215

Virsolvy-Vergine A, Leray H, Kuroki S, Lupo B, Dufour M, Bataille D (1992) Endosulfine, an endogenous peptidic ligand for the sulfonylurea receptor: purification and partial characterization from ovine brain. Proc Natl Acad Sci USA 89:6629–6633

Vivaudou MB, Arnoult C, Villaz M (1991) Skeletal muscle ATP-sensitive K+ channels recorded from sarcolemmal blebs of split fibers: ATP inhibition is reduced by magnesium and ADP. J Membr Biol 122:165–175

Wallimann T, Wyss M, Brdiczka D, Nicolay K, Eppenberger HM (1992) Intracellular compartmentation, structure and function of creatine kinase isoenzymes in tissues with high and fluctuating energy demands: the "phosphocreatine circuit" for cellular energy homeostasis. Biochem J 281:21–40

Weik R, Neumcke B (1989) ATP-sensitive potassium channels in adult mouse skeletal muscle: characterization of the ATP-binding site. J Membr Biol 110:217–226

Wilde AAM, Janse MJ (1994) Electrophysiological effects of ATP sensitive potassium channel modulation: implications for arrhythmogenesis. Cardiovasc Res 28:16–24

Woll KH, Lönnendonker U, Neumcke B (1989) ATP-sensitive potassium channels in adult mouse skeletal muscle: different modes of blockage by internal cations, ATP and tolbutamide. Pflügers Arch 414:622–628

Xu X, Lee KS (1994) Characterization of the ATP-inhibited K+current in canine coronary smooth muscle cells. Pflügers Arch 427:110–120

Yalow RS, Black H, Villazon M, Berson SA (1960) Comparison of plasma insulin levels following administration of tolbutamide and glucose. Diabetes 9:356–362

Yamato E, Ikegami H, Tahara Y, Fukuda M, Cha T, Kawaguchi Y, Fujioka Y, Noma Y, Shima K, Ogihara T (1993) Cellular mechanism of glyburide-induced insulin gene expression in isolated rat islets. Biochem Biophys Res Commun 197:957–964

Zawalich WS, Diaz VA, Zawalich KC (1988) Influence of cAMP and calcium on [^{3}H]inositol efflux, inositol phosphate accumulation, and insulin release from isolated rat islets. Diabetes 37:1478–1483

Zini S, Ben-Ari Y, Ashford MLJ (1991) Characterization of sulfonylurea receptors and the action of potassium channel openers on cholinergic neurotransmission in guinea pig isolated small intestine. J Pharmacol Exp Ther 259:566–573

Zünkler BJ, Lenzen S, Männer K, Panten U, Trube G (1988a) Concentration-dependent effects of tolbutamide, meglitinide, glipizide, glibenclamide and diazoxide on ATP-regulated K^+ currents in pancreatic B-cells. Naunyn Schmiedebergs Arch Pharmacol 337:225–230

Zünkler BJ, Lins S, Ohno-Shosaku T, Trube G, Panten U (1988b) Cytosolic ADP enhances the sensitivity to tolbutamide of ATP-dependent K^+ channels from pancreatic B-cells. FEBS Lett 239:241–244

Zünkler BJ, Trube G, Panten U (1989) How do sulfonylureas approach their receptor in the B-cell plasma membrane? Naunyn Schmiedebergs Arch Pharmacol 340:328–332

CHAPTER 7
Sulfonylureas: Pharmacokinetics in Animal Experiments

A. HASSELBLATT

A. Introduction

This chapter reviews the pharmacokinetics of the sulfonylurea compounds in animals. This is a rather restricted area which has so far failed to attract significant research effort. As sulfonylureas were developed to be used in man, more attention has been dedicated to their behavior within the human body than to their fate in animals. Thus the data available from animals have not usually exceeded those required for the decision to be made to use these agents in man. In addition, it is likely that not all the relevant data have been made available to the public.

The analytical methods used in the early days of the sulfonylureas, when the available animal data were mostly obtained, were not as sophisticated as those employed later for human use. Thus the fate of radioactivity after injection of a labeled compound has frequently been taken to represent the real fate of the parent compound in the body. This may have led to erroneous conclusions, since only radioactivity but not the agent itself was localized in the tissues or was measured in the excretions. At a later stage, when plasma and tissue levels of sulfonylureas were measured in man, more specific methods, such as high-pressure liquid chromatography (HPLC), were usually employed. As this chapter is restricted to animal data, only those results on the fate of the drugs in the human body are included which offer information applicable to both animals and human beings.

A second limitation concerns the drugs to be covered. At present there are ten sulfonylurea compounds in clinical use and it is mainly these compounds that are dealt with in the following sections. They include carbutamide, tolbutamide, chlorpropamide, gliquidone, glibornuride, gliclazide, glisoxepide, glipizide, acetohexamide, and glibenclamide. As will be seen, the scope of the available information on each of these individual compounds differs greatly. This may well lead to a generalized view when the documented properties of one member of the group are taken to represent the entire class.

B. Absorption of Sulfonylurea Derivatives After Oral Administration

I. Species Differences in the Rate of Absorption

Sulfonylurea derivatives are acid compounds, forming salts with alkali. They are thus partly dissociated and do not readily penetrate across biological membranes at the physiological pH level by non-ionic diffusion. On the other hand, it may be anticipated that they are precipitated in an acid environment as in gastric juice; thus they are not soluble and for this reason are not absorbed.

In order to achieve optimal intestinal absorption, the sulfonylurea compounds should thus be kept in solution with an alkaline pH in the gut and at the same time be given the opportunity to partly pass into the undissociated and lipid-soluble state at the mucous membrane. It is these non-ionized molecules that are free to move across the lipid membrane of the mucosal cells of the intestinal epithelium by diffusion and thus to pass from the intestine into the blood.

The fact that the sulfonylureas are acidic in nature also serves to explain their fate in the body. Being less dissociated and more lipophilic at an acid pH, they will penetrate a lipid membrane more readily when it lines a more acidic space. The cells of the body are more acid than the extracellular fluid. For this reason sulfonylureas should be present in the extracellular space at a slightly higher concentration than inside the cell.

A high affinity to bind to plasma proteins will protect sulfonylurea compounds from undergoing filtration in the kidney. Protein binding, which is mainly binding to plasma albumin, will prevent sulfonylurea compounds from distributing freely across the body but retain them to some degree within the vascular space. The apparent volume of distribution will thus be smaller than the water space of the body.

As in man, sulfonylurea compounds are readily absorbed from the intestine of all the animal species that have been investigated. This holds for carbutamide (ACHELIS and HARDEBECK 1955) and for tolbutamide (SCHOLZ and BÄNDER 1956). Following oral administration of labeled glisoxepide, similar amounts of radioactivity were excreted into the urine to after intravenous injection of the same dose in the monkey. The bioavailability was thus considered to approach 100% for glisoxepide (KOLB et al. 1974b).

The early compounds carbutamide and tolbutamide had a rather low potency and thus were given at a high dose to obtain a blood glucose lowering response in the dog, rabbit, or rat. Tolbutamide not being soluble in water was less active than its soluble sodium salt. Thus in the rat a dose 100 mg/kg of the substance itself was not active, while the same dose of the sodium salt lowered blood glucose by 40%–50%. In the rabbit 400 mg/kg tolbutamide reduced blood glucose levels by 20%, when given together with 0.5 NaOH by 30% and with 3 N NaOH by 50% (SCHOLZ et al. 1956). It was concluded that in rodents the amount of alkali available in the gut may be limiting to the absorption of those sulfonylurea compounds which have to be applied in a high dose of more than 100 mg. This does not appear to apply to carnivores, such as the dog, and also not for man. Both species seem to have sufficient amounts of alkali available within the gut to transform the sulfonylureas to water-soluble salts.

II. Kinetics of Absorption of Sulfonylureas and Effects of Food and Other Drugs

After having been swallowed, sulfonylureas are insoluble in the acid gastric juice and are therefore not readily taken up within the stomach; moreover, the surface of the gastric mucosa is rather small compared with that of the gut. For this reason the main site of absorption of sulfonylureas is in the intestine. They are more readily absorbed when given together with food, which accelerates gastric emptying and thus the movement of the drug to the large surface of the intestine. Thus the absorption of tolbutamide was more rapid when it was given together with a standard meal. The highest blood levels of radioactivity were obtained within 3 h following an oral dose of 100 mg/kg ^{35}S-labeled tolbutamide (BÄNDER and SCHOLZ 1956). In rats, maximal blood levels were measured within 1 h when tolbutamide (25–100 mg/kg) was given orally. In dogs it took somewhat longer. When the sodium salt of glimidine, which is a pyrimidine derivative, was placed into the duodenum of rats, more than 90% of the dose was absorbed within 4 h (KOLB et al. 1964). The highly active compound glibenclamide was absorbed more slowly from the gut. Following an oral dose of labeled glibenclamide (0.2 mg/kg), the plasma level of radioactivity rose with some delay, taking from 2 to 8 h, depending on the animal species employed, to reach its maximal value (KELLNER et al. 1969). In animal studies gliclazide has been well absorbed by several species, consistently reaching peak plasma levels within 2–3 h (CAMPBELL et al. 1980).

Comparable data obtained in man showed food intake did not affect the total rate of absorption of tolbutamide and chlorpropamide, despite of that the peak concentration of chlorpropamide in the plasma was reduced by the simultaneous intake of food. This did not hold for tolbutamide (SARTOR et al. 1980). In agreement with the animal data, the absorption of tolbutamide was more rapid in diabetic patients when the drug was taken with the meal. About 6% of the dose was lost to absorption when the patients were eating while taking the drug. This small difference was judged to be of no clinical significance (ANTAL et al. 1982). In man, peak plasma concentrations occurred 2.8 h after dosing and plasma half-life of tolbutamide was found to be 8.1–8.3 h. Also in man, eating delayed the absorption of gliplizide in diabetic patients by about 0.5 h without affecting the peak plasma levels, the elimination half-life, or the bioavailability of the drug (WÄHLIN-BOLL et al. 1980).

The effect of dietary fiber on the absorption of glibenclamide has only been studied in man. While plasma levels of the drug reached a peak at 60 min in the controls, the rise in plasma glibenclamide levels was blunted by the dietary fiber glucomannan taken simultaneously (SHIMA et al. 1983). Charcoal taken immediately after the sulfonylurea drugs reduced the absorption of tolbutamide (NEUVONEN et al. 1983) and of chlorpropamide (NEUVONEN and KÄRKKAINEN 1983) by 90%.

III. Dependence of Absorption on Galenic Formulation

Differences in the rate of absorption between tablets of tolbutamide supplied by different manufacturers have been observed in man (RUPP et al. 1975; OLSON et al. 1985). The bioavailability of three marketed brands of chlorpropamide was compared (MONRO and WELLING 1971). Following a single oral dose, one of these gave significantly lower serum levels than the other two. In man, only 45% of an older formulation of glibenclamide was absorbed (RUPP et al. 1969). Incomplete absorption was found to result from larger particle size in this preparation. Two later studies found commercial tablets to give an almost complete absorption. In the first study, 93% was taken up (RUPP et al. 1972), while 80%–100% was absorbed in the second study (PEARSON 1985). Thus the intestinal uptake of glibenclamide almost equals that of glibornuride, which has been found to reach 98% (LORCH et al. 1971). Further studies on the bioavailability of sulfonylurea compounds have been carried out in man are not dealt with in the present chapter, which is devoted to results from animal experimentation.

C. Distribution of Sulfonylurea Derivatives in the Organism

I. Space of Distribution of Sulfonylurea Derivatives

For tolbutamide a volume of distribution of 17.5% of the body weight has been calculated in man. This has been taken to indicate that the drug is dissolved mainly within the extracellular space (STOWERS et al. 1958). Apparently the other sulfonylureas, with the exception of glibenclamide, are distributed in a manner similar to that of tolbutamide. Their volumes of distribution are given in liters per kilogram bodyweight: 0.10–0.15 for tolbutamide, 0.09–0.27 for chlorpropamide, 0.16 for glipizide, and 0.2 for gliclazide (MARCHETTI and NAVALESI 1989). Despite its being bound to plasma proteins to a larger extent than other sulfonylurea drugs, glibenclamide has the highest volume of distribution, which is 0.3 l/kg, and thus twice that of tolbutamide in man. Far higher values have been calculated in animals (see below). This may be taken to indicate that glibenclamide is taken up into the tissues more readily than are the other compounds.

A space of distribution of this size does not necessarily imply that sulfonylurea compounds are in fact free to leave the blood and enter into the extracellular space. All these agents are strongly bound to plasma proteins and are retained in the vascular space by this binding. A high affinity to bind to plasma albumin has been demonstrated for tolbutamide (WISHINSKY et al. 1962), with a fraction bound of 97%. Nevertheless tolbutamide must be able to enter the liver cells, as it is metabolized by the hepatic microsomal enzyme system. Moreover, in the dog following a dose of labeled tolbutamide, some radioactivity was found in the bile (BÄNDER and SCHOLZ 1956). There is no evidence from animal data that tolbutamide or its metabolites accumulate somewhere in the body or penetrate into the cells outside the liver (NISTRUP-MADSEN et al. 1971). Accordingly, the volume of distribution of ^{35}S-tolbutamide was found to be approximately 16%–25% of body weight, which corresponds roughly to the extracellular space (WICK et al. 1956).

Following the injection of labeled gliclazide into rats, whole body autoradiography showed rapid localization of radioactivity in the liver and kidneys (BENAKIS and GLASSON 1980). In man the volume of distribution of gliclazide was found to increase with age, values around 24 l being found in elderly patients (cited by HOLMES et al. 1984). While the level of tolbutamide present in the liver water is about half that of the plasma, carbutamide is better able to penetrate into the liver. Its concentration in the liver water equaled that in the plasma in rats (KUETHER et al. 1956). In other experiments, even an accumulation of carbutamide in the liver was found to occur, resulting in a concentration twice as high as that present in the plasma (STUHLFAUT et al. 1960). The binding of carbutamide to plasma proteins at therapeutic plasma levels was 75% and thus less than that found for tolbutamide (97%) and chlorpropamide (87%) (WISHINSKY et al. 1962). When injected intravenously into rats, chlorpropamide was distributed within the body in a way quite similar to that of tolbutamide, the difference being that chlorpropamide failed to penetrate into the liver cells in measurable amounts in (NISTRUP-MADSEN et al. 1971).

In common with other sulfonylureas, gliclazide is strongly bound to plasma protein. At therapeutic concentrations, binding is higher in humans (93%) and rhesus monkeys (94%) than in beagle dogs (87%) and rabbits (89%). The binding capacity has been found to be reduced by therapeutic concentrations of acetylsalicylic acid, phenylbutazone, and sulfisoxazole. It was not significantly altered by chlorpropamide, warfarin, chlorothiazide, and clofibrate (CAMPBELL et al. 1980). The extent of glibornuride binding to plasma proteins was 97%, which is thus similar to that of other sulfonylureas. Autoradiographic studies, done in mice at intervals of 1 min to 28 h following an injection of labeled glibornuride, showed accumulation of radioactivity to occur in the first 6 h in the kidney, liver, gallbladder, and lungs. At no time was radioactivity found in the CNS (BIGLER et al. 1971, 1972; RENTSCH et al. 1972). It thus seems unlikely that sulfonylureas are capable of crossing the blood-brain barrier in considerable amounts. In agreement with this, no radioactivity was detected in the cerebrospinal fluid following an injection of

[14]C-labeled glisoxepide. In rats the highest radioactivity was found in the liver, kidneys, and blood (Kolb et al. 1974b).

Unlike the older sulfonylureas, the highly potent glibenclamide accumulates in tissues within the body. This is already evident from its high volume of distribution. Thus the space across which glibenclamide distributes was found to be 78% of body weight in the rabbit, 77% in the dog, and 160% in the rat (Heptner et al. 1969). A large part of the given dose must thus have left the extracellular space and been bound within the tissues. Measurements of the radioactivity in different organs revealed that glibenclamide (or its metabolites) accumulated within the liver. The concentration in the liver was calculated to exceed the blood level threefold in the rabbit, sixfold in the dog, and 50-fold in the rat (Kellner et al. 1969). In these animal species, the liver may thus act as a "deep compartment" (Garret 1970) in the sense that it accumulates glibenclamide and releases it slowly again into the blood. It may be anticipated from these data that any effect of glibenclamide on the hepatic metabolism may well outlast its plasma half-life. There is some evidence that this is in fact the case. This can be derived from the kinetics of the elimination of glibenclamide from the blood. Glibenclamide leaves the blood obeying two different kinetics.

In the first phase the capacity to eliminate the drug is considered to determine the rate at which the plasma levels are reduced. This yielded a rather short half-life of 1.4 h in the rat and a more prolonged one of 7.4 h in the rabbit. In the subsequent second phase, a far longer half-life has been identified. It must be assumed that here the capacity of the eliminating process was no longer rate limiting but rather the slow release of the drug from a deep compartment determined the delay in the elimination. This would serve to explain the long half-life found for the second phase, being a whole day in the rabbit and as much as 2 days in the rat (Keller et al. 1969).

The question arose from these animal data to what extent glibenclamide accumulates in the human liver or the islet tissue. Such a localized accumulation might explain the prolonged effect of this drug on liver metabolism or insulin release, giving rise to prolonged and severe hypoglycemias (Hasselblatt 1973). In fact, hypoglycemic events are caused more frequently by glibenclamide than by tolbutamide. The time course of the elimination of glibenclamide from the blood is compatible with the assumption that a "deep compartment" also exists in man. Glibenclamide is released from such a compartment in the human body more readily than in the animal species mentioned. This is testified by the short duration of its second half-life in human blood. Following a more rapid first phase of elimination, where the plasma level decreased by 50% within 1.9 h, the elimination was delayed in the second phase and thus resulted in a prolonged half-life of 9.3 h (Rupp et al. 1972). The fact that glibenclamide accumulates in the tissues is further demonstrated by the rather high volume of distribution in man of 52% of body weight (Rupp et al. 1969).

Of the other sulfonylurea compounds, chlorpropamide is the one most likely also taken up into a "deep compartment." At least a much prolonged second half-life would favor this assumption. In man chlorpropamide was eliminated from the blood at two different rates. In the beginning, the plasma level of chlorpropamide declined by half within 24 h, and in the delayed second phase it took 16 days (JOHNSON et al. 1959).

II. Protein Binding of Sulfonylureas and Interaction with Other Compounds on Plasma Protein-Binding Sites

As mentioned previously, sulfonylurea compounds are bound to plasma proteins and mainly to albumin in the blood. An understanding of the nature of association of the sulfonylureas with serum albumin may be helpful in understanding their pharmacokinetic behavior and the changes they induce in the distribution of other agents. Tolbutamide is bound to 99% over the therapeutic range of $100-150\,\mu g/ml$. The affinity constant of this binding is rather low. The same holds for chlorpropamide (96% bound) and tolazamide (94% bound).

The binding of these derivatives is easily saturated as doubling of the concentration of tolazamide increases the level of free fraction from 6% to 14.6% (CROOKS and BROWN 1974). It may be predicted from these data that plasma protein binding should influence the distribution of these drugs only at lower drug levels. Glibenclamide is more strongly bound as only 0.17% of the drug is free at therapeutic serum levels of less than $1\,\mu g/ml$. A 50-fold increase in its concentration from 1 to $50\,\mu g/ml$ only gives rise to an increase of the free fraction from 0.17% to 0.3%.

Scatchard plots of the binding to human or bovine serum albumin showed two binding sites for tolbutamide, chlorpropamide, and tolazamide. Binding to the first, high-affinity site was strongly dependent on pH and moderately independent of temperature. This was taken as evidence that the drugs are linked to the first site by ionic binding. In contrast, the second association constant is greatly affected by temperature, which is typical of an exothermic reaction between drug and protein. Thus temperature dependence of the binding supports the assumption that it is not based on an ionic interaction alone.

Like all sulfonylurea compounds, glipizide is extensively bound to serum proteins to the extent of 97%–99% (CROOKS and BROWN 1975). The sites of binding differ from those of tolbutamide, chlorpropamide, and acetohexamide (HSU et al. 1974), but resemble those of glibenclamide (CROOKS and BROWN 1975). Unlike the binding of tolbutamide, that of glibenclamide to albumin showed little dependence on pH. Lowering it to 6.4 changed the binding to a small extent compared with pH 7.4. The association constant increased with decreasing temperature, the value of K at 14°C being twice that found at 37°C. Moreover, Scatchard analysis revealed only one population of binding sites for glibenclamide. Thus, according to these results (CROOKS and BROWN 1974), glibenclamide seems to bind to albumin by a mechanism which differs from that of tolbutamide, chlorpropamide, and tolazamide.

The finding that tolbutamide accelerated the removal of the dye sulf-obromophthalein sodium (Bromsulphalein) from the blood, a dye that has been employed to check hepatic function, supported the assumption that tolbutamide may improve liver function. This erroneous conclusion was further substantiated by the finding that in addition the plasma levels of endogenous bilirubin were reduced by tolbutamide treatment. Experiments on rabbits and observations in healthy humans revealed this effect to result merely from tolbutamide displacing both the diagnostic dye and the endogenous bilirubin from plasma protein-binding sites. This enabled Bromsulphalein and bilirubin, which were no longer retained in the blood by protein binding, to leave the circulation and enter the tissues, mainly the liver (HASSELBLATT and HUKUHARA 1964; HASSELBLATT 1965). Tolbutamide binds to albumin with a rather low affinity; the association constant K has been found to be 6.8×10^4 (BUETTNER and PORTWICH 1967). This constant for Bromsulphalein is given as $K = 1.54 \times 10^5$, which indicates a higher affinity for binding to albumin. Nevertheless, tolbutamide proved to be a successful competitor for the protein-binding site, as its concentration was $800 \mu M$ and thus 160 times higher than that of Bromsulphalein, which was only $5 \mu M$ (HASSELBLATT 1971). Thus tolbutamide displaces Bromsulphalein bound to serum proteins from rabbits, rats, and also humans. Only the dog seems to be the exception, as the addition of tolbutamide to canine serum failed to liberate bound Bromsulphalein (HAUFE et al. 1965).

Glisoxepide as a sulfonylurea derivative of high potency, and a member of the so-called second generation, is capable of displacing bound bilirubin or phenprocoumon from bovine serum albumin in a manner similar to that of tolbutamide. Due to its high potency, therapeutic plasma levels are low and thus do not interfere with the binding of bilirubin or phenprocoumon neither in serum nor in diluted albumin solutions. As the bound fraction of glisoxepide is higher in serum (93%) than in a 4% solution of albumin (69%), fractions of serum proteins other than albumin apparently contribute to the binding (SCHLOSSMANN 1974a,b)

The same situation as for Bromsulphalein and tolbutamide will result whenever a highly active drug is retained in the blood by a firm binding to plasma proteins and a second drug has access to the circulation that by virtue of affinity or high concentration successfully competes for the same class of binding site. The displacement of sulfonylureas from bovine or human serum albumin has been investigated a number of times. Competition for binding sites has been shown to occur between tolbutamide and ethylsulfadiazine in vitro (BÜTTNER and PORTWICH 1967). An inhibition of albumin binding of sulfonylurea compounds has been found for sodium salicylate, acetylsalicylate, phenylbutazone, and the sulfonamides sulfadimethoxine, sulfaphenazole, and sulfisoxazole (JUDIS 1972). Similarly, coumarin derivatives displace tolbutamide, acetohexamide, and chlorpropamide from serum albumin in vitro (JUDIS 1973). In man sulfaphenazole and phenylbutazone increased the hypoglycemic response to tolbutamide. This effect was considered to result mainly from an inhibition of the metabolism of tolbutamide. But the presence of these drugs also gave rise to elevated levels of free unbound tolbutamide in the serum. This effect was considered to possibly contribute to a blood sugar lowering action of phenylbutazone and the sulfonamide in patients treated with tolbutamide (CHRISTENSEN et al. 1963).

The binding of three sulfonylurea compounds, tolbutamide, chlorpropamide, and glibenclamide, to bovine and human albumin has been estimated using a dynamic dialysis technique (BROWN and CROOKS 1976). A number of acidic drugs were added and the resulting displacement was measured. The degree of displacement of tolbutamide and chlorpropamide observed in these experiments was greater than anticipated from a competitive binding equation. According to these findings, the displacement of the two sulfonylureas tested by acidic drugs, such as phenylbutazone, sulfaphenazole, and salicylate, may have been noncompetitive in nature. It was concluded that not only competition but rather some distortion of the albumin molecule might have been responsible for the interference with the binding of tolbutamide and chlorpropamide. The displacement of glibenclamide was far less than that calculated by the competitive binding equation. This may have been caused by the different mechanism of binding of glibenclamide, which, unlike the other two sulfonylureas, seems to be bound by non-ionic forces.

D. Accumulation of Sulfonylurea Derivatives in Different Organ Systems

I. Accumulation in the Liver

A characteristic of the sulfonylurea drug glibenclamide appears to be that it is taken up and stored in the liver of various animal species. Thus the drug-related radioactivity in the liver of animals treated with labeled glibenclamide exceeded the plasma levels severalfold. It was elevated threefold in the rabbit, sixfold in the dog, and 50-fold in the rat (KELLNER et al. 1969). Such accumulation might favor direct effects of this sulfonylurea in hepatic metabolism. It is known from in vitro experiments that sulfonylures are capable of inhibiting lipolysis and ketogenesis in the liver (HASSELBLATT 1969). Glibenclamide was not used in these early experiments.

II. Accumulation of Sulfonylurea Derivatives in the Pancreatic Islets

When injected into animals, neither labeled tolbutamide, chlorpropamide, nor glibenclamide induced a special accumulation of radioactivity in the pancreatic gland as a whole. When the uptake by isolated pancreatic islets was measured, glibenclamide proved to be exceptional among the hypoglycemic sulfonylureas in accumulating in B-cell-rich pancreatic islets

(HELLMAN et al. 1984). Glibenclamide differed from tolbutamide, chlorpropamide, glibornuride, and glipizide in not being rapidly bound to reach an
equilibrium, but in accumulating progressively in amounts exceeding by far
the water space. Such progressive accumulation has also been found to
occur inside the pancreatic islet cells. Labeled glibenclamide was taken up
by the endocrine cells, and was found preferentially within the B cells as
judged by autoradiography (CARPENTIER et al. 1986). In accordance with this
slow and progressive binding, the effect of glibenclamide on blood glucose
was slow in onset (HAUPT et al. 1971). Similarly it took some time to
develop its maximal action in promoting the entry of calcium ions into the
cells (HELLMAN et al. 1984).

Within recent years our knowledge of how sulfonylurea compounds
elicit insulin release from the islet cells has greatly improved. One primary
effect of these agents is to bind to the plasma membrane at a specific site,
which has been referred to as the "sulfonylurea receptor." This binding
results in an inhibition of an ATP-sensitive potassium channel (see PANTEN
et al. 1992 for review). By closure of this channel, positively charged potassium ions are prevented from leaving the interior of the islet cell. This
results in a loss of the negative charge inside the cytosol and consequently in
a decline in the membrane potential. As soon as voltage-dependent calcium
channels open and allow calcium ions to enter the cell, the exocytosis of
insulin-containing granules is initiated.

Based on the finding that tolbutamide apparently failed to enter the islet
cell to a measurable extent (HLLMAN et al. 1971), it has been suggested that
sulfonylurea derivatives bind to the plasma membrane and trigger the
secretory response by altering the conformation of the membrane (HELLMAN
and TÄLJEDAL 1975). This site of action was not likely to be located on the
outer surface or the plasma membrane. Polymer-linked sulfonylurea derivatives, which may not enter the cell but could interact with binding sites at
the outer surface of the cell, failed to initiate an insulin release (JOOST and
HASSELBLATT 1977).

The access of sulfonylureas to their site of action was measured in detail
when it became possible to visualize their primary effect on the ATP-sensitive
potassium channel by the patch clamp technique. As the sulfonylureas are
acidic compounds, the rate of their dissociation is reduced in an acid environment at low pH and increased when the pH is elevated to more alkaline
levels. The response of islet cells exposed to sulfonylureas was more rapid
and more pronounced when the pH at the outside of the cell was shifted to
the acid region. This finding suggests that the sulfonylurea molecule enters
the lipid membrane of the islet cell in a lipid-soluble undissociated state to
reach its site of action on the inner surface of or within the membrane
(ZÜNKLER et al. 1989).

The binding of various sulfonylurea compounds to the islet cell membranes has been measured and compared with their effects on insulin release
(PANTEN et al. 1989). The different kinetics of the response to tolbutamide,
meglitinide, glipizide, and glibenclamide was found to depend upon the lipid

solubility of the drug employed. The ability of the sulfonylurea compounds to pass the lipid membrane thus appears to control their penetration into the cells. Allowing for the different kinetics, the same maximal secretory response was obtained by saturating concentrations of the sulfonylureas tested. Moreover, the relative potencies of tolbutamide, meglitinide, glipizide, and glibenclamide corresponded well to their affinity to bind to islet cell membranes. This suggested that the binding site does represent the sulfonylurea receptor.

ATP-sensitive potassium channels have been found to be present in the tissues in addition to the islet cells. Due to high ATP levels, they are probably not normally opened. They may gain importance in pathological situations, however, and in this way the ATP-sensitive potassium channel in the cardiac muscle is supposed to open in severe ischemia. This serves to shorten the action potential, thus allowing less calcium ions to enter the cell. This would reduce the workload imposed on the cardiac muscle. As in the pancreatic islet, the sulfonylureas penetrated into the cardiomyocytes in the non-ionized state. Thus acidosis increased the effective concentration entering the cell, while lowering of pH inside the cell, achieved by acidification of the solution inside the patch pipette, had no effect upon the inhibition of the potassium channels by glibenclamide or tolbutamide (FINDLAY 1992). The high-affinity binding site, considered to represent the sulfonylurea receptor, has also been identified in microsomal membranes from brain tissues (KAUBISCH et al. 1982; GEISEN et al. 1985; BERNARDI et al. 1988). Their physiological significance has not been clarified. Sulfonylureas have not yet been shown to affect brain functions or even to pass the blood-brain barrier.

The binding of sulfonylureas to their binding sites has been found to be modified by the presence of nucleotides. It thus increased in the presence of ADP (NIKI et al. 1989). On the other hand, phosphate-donating magnesium complexes of adenine and guanine nucleotides have been shown to inhibit the binding (SCHWANSTECHER et al. 1991). Corresponding results were obtained with particulate and solubilized sulfonylurea-binding sites from the cerebral cortex. The sulfonylurea site from brain and from islet cells seems to be modified in a similar manner by protein phosphorylation (SCHWANSTECHER et al. 1992).

Further efforts have been directed towards isolation and identification of the protein representing the 'sulfonylurea receptor" and to clarify how it is connected to the ATP-responsive potassium channel. These experiments employing specific photolabels (NELSON et al. 1992; SCHWANSTECHER et al. 1994) are beyond the scope of this review.

E. Elimination of Sulfonylurea Derivatives

Carbutamide, being the oldest member of the sulfonylurea family, is a sulfanilamide and retains some antibacterial potency. Like other sulfonamides, the sulfanilamide moiety is acetylated and the resulting conjugate leaves the body with the urine (ACHELIS and HARDEBECK 1955). In the rat its biological half-life is about 13 h (BARGETON et al. 1965), while in man it was found to take 33 h before half of the drug was removed from the blood (STOWERS et al. 1958).

The methyl group on the benzene ring in the molecule of tolbutamide is hydroxylated by hepatic microsomal enzymes in the rat (TAGG et al. 1967) and in man (DARBY et al. 1972). The resulting hydroxytolbutamide retains some blood glucose lowering potency of the parent compound. In mice it is about half as active as tolbutamide, and in man it is also capable of reducing blood glucose levels (SCHULZ and SCHMIDT 1970a). In humans tolbutamide is transformed further to the corresponding carboxylic acid (WITTENHAGEN and MOHNIKE 1956; LOUIS et al. 1956). In rats and rabbits the first metabolite hydroxytolbutamide constitutes the main excretory product (TAGG et al. 1967).

In man the first hydroxylating step is rate limiting for the elimination of tolbutamide and has been shown to proceed with a half-life of 5.7 h (SCHULZ and SCHMIDT 1970b). Once hydroxytolbutamide has been formed, it is oxidized rapidly to form the carboxylic acid, the half-life of this process being 19 (SCHULZ and SCHMIDT 1970a) or 30 min (NELSON and O'REILLY 1961). The resulting carboxy derivative is devoid of biological activity and leaves the body with the urine.

Of the species tested, the dog is unique in splitting the molecule of tolbutamide. This results in the excretion of the p-toloylsulfonylurea with the urine after the butyl moiety of the tolbutamide molecule has been cleaved off (MOHNIKE et al. 1958). This special metabolic pathway in the dog might explain why tolbutamide interferes with liver function in this species (ELRICK and PURNELL 1957).

A cleavage of the sulfonylurea molecule comparable to that of tolbutamide in the dog has been observed to occur with chlorpropamide in man and other species. Chlorpropamide differs from tolbutamide mainly because the methyl group on the benzene ring has been replaced by chlorine; thus the site for hydroxylation in the tolbutamide molecule is not available in chlorpropamide. The compound is therefore eliminated much more slowly than tolbutamide. It was assumed originally that this molecule passes through the body without being metabolized (JOHNSON et al. 1959). Later, as more specific analytical methods became available, up to 80% of a dose of chlorpropamide was recovered from the urine as metabolites (TAYLOR 1972; BROTHERTON and McMARTIN 1969). These were p-chlorobenzene sulfonylurea and p-chlorobenzene sulfonamide. This shows that the propyl chain had been split off from the molecule. Two additional metabolites have been identified where hydroxylation has occurred in the propyl chain, i.e., a 2-hydroxy- and a 3-hydroxy derivative of chlorpropamide (TAYLOR 1972). Both are still capable of releasing insulin from the pancreatic gland (TAYLOR 1974). It is not known whether the chlorobenzene sulfonylurea, which originates from chlorpropamide by splitting off the propyl moiety, retains a capacity to lower blood glucose levels (BRASSELTON et al. 1977).

As chlorpropamide and its metabolites are weak acids, their renal elimination depends on the urinary pH. In man the excretion via the kidneys can be increased fourfold by raising urinary pH. The renal elimination is lowered to one-tenth when the urine turns acid (KÄRKKAINEN et al. 1983).

The pyridine derivative glymidine undergoes demethylation to yield an alcohol, i.e., a 5β-hydroxyethoxypyrimidin derivative, which is still capable of lowering blood glucose levels but is oxidized further to form a carbonic acid which is inactive (GERHARDS et al. 1964).

The metabolic fate of glibenclamide includes two different pathways. In the rabbit, dog, rat, and man, the cyclohexyl ring is hydroxylated to a metabolite of very little blood glucose lowering activity. The second pathway leads to a splitting of the molecule at three different sites, yielding metabolites of low or no activity (HEPTNER et al. 1969).

Gliclazide is extensively metabolized and less than 20% is excreted in the urine unchanged (CAMPBELL et al. 1980). It is oxidized in the azobicyclo-octyl portion of the molecule to produce hydroxylated or N-oxygenated compounds. In addition, oxidation of the p-toloyl group results in the alcohol, which is further transformed into the corresponding carboxylic acid. These routes of metabolism are common with other sulfonylureas. In plasma, gliclazide represents 90% of all drug-related material. Following an injection of ^{14}C-gliclazide, radioactivity is found in the urine and the feces. The quantities excreted by the two routes were similar for all species studied, namely 60%–70% of the dose was found in the urine and 10%–20% in feces. An enterohepatic circulation comprising 20% of the dose has been demonstrated to occur in the rat (CAMPBELL et al. 1980). The half-life of the disappearance from plasma of gliclazide was 3 h in the rat, rabbit, and monkey; in the dog as in man it is three times longer, i.e., 10 h.

Gliplizide is almost completely metabolized and the products formed are virtually devoid of hypoglycemic activity. Only about 3%–9% of an administered dose is excreted unchanged. The half-life of the parent drug in man is 3–4 h (FUCCELLA et al. 1973). Additional pharmacokinetic studies in man have been published (SCHMIDT et al. 1973; BALANT et al. 1975).

Glibornuride is absorbed almost completely and is distributed in 24.7% of the body weight. Elimination occurs almost exclusively by hepatic metabolism, following a half-life of 8.2 h. The major part of the metabolites formed are excreted into the urine (60%–72%), and a minor portion passes into the bile and feces (23%–33%) (RENTSCH et al. 1972). One metabolite is formed in a way similar to the metabolism of tolbutamide by hydroxylation of the methyl group in the para position of the benzene ring, and subsequent oxidation to the carboxylic acid. Three more metabolites result from hydroxylation of the borneol part of the molecule (LORCH et al. 1971; BIGLER et al. 1971).

Glisoxepide has been injected as a labeled compound. Radioactivity is excreted predominantly with the urine (70%) and the remainder with the feces (30%) by rabbits, baboons, and dogs (KOLB et al. 1974b).

After the initial phase of distribution, radioactivity is eliminated from the blood with a half-life of about 8 h in rabbits, 22 h in dogs, and 1.7 h in baboons. The long half-life in dogs has been taken as resulting from the formation of metabolites with a low volume of distribution. In rats the label was concentrated in the liver and hardly penetrated into the brain, and in dogs and baboons the blood-brain barrier was not crossed, even after daily dosing for 28 days. On the day following such long-term treatment, only 10% more radioactivity was retained in the body of rats and dogs than was measured after a single dose on the previous day. In the baboon, less than 40% was retained after such prolonged treatment. The hexamethyleneimine moiety of glisoxepide was oxidized to CO_2 to 16% of the dose in the rat and 5% in the baboon (Kolb et al. 1974).

Gliquidone is transformed into four metabolites in man. A special characteristic of this sulfonylurea is the fact that only 5% of the drug-related radioactivity is released into the urine, while the main portion is eliminated with the feces. This may be an advantage in patients with impaired renal function. Autoradiography of the whole rat shows that radioactivity is concentrated in liver and spleen within 10 min. Only 3% of the active material is found in the urine of rats. In man the four metabolites are formed from the parent compound by demethylation or hydroxylation (Kopitar and Koss 1975).

Acetohexamide is metabolized by dogs, rats, rabbits, and man to the active compound hydroxyhexamide in the liver (Welles et al. 1961). This blood glucose lowering agent is subsequently eliminated by the kidney and may thus accumulate in renal disease (Cohen et al. 1967).

Glibenclamide, when administered as a single dose, was found unchanged in the liver of dogs and rats after 4 h, while some metabolite was present in the rabbit liver. Only metabolites were released into the bile. The cyclohexyl ring was hydroxylated by all three species. In addition the molecule was split in the dog (Heptner et al. 1969). The liver retained radioactivity for 8 h, as shown by autoradiography. Outside the liver only the gut contained some activity (Hellman et al. 1969). In man biological half-life of glibenclamide is not significantly correlated with renal function as long as a creatinine clearance of 30 ml/min is maintained (Pearson 1985). The elimination of the drug appears to be evenly distributed between the biliary and renal routes. The drug accumulates in the liver and kidney as the excretory organs in the dog, rabbit, and rat (Kellner et al. 1969). The hydroxy derivative formed from glibenclamide has been tested for blood glucose lowering activity in the rat. This metabolite was found to induce hypoglycemia, but it was six to seven times less potent than the parent compound (Samini et al. 1977). Nevertheless this agent, 4-trans-hydroxy-glibenclamide, is more potent than tolbutamide; it could exert some effects if it were allowed to accumulate in renal failure.

F. Influence of Other Drugs on Rate of Metabolism and Excretion of Sulfonylurea Drugs

The time course of the effect of tolbutamide is determined by the capacity of the microsomal hydroxylating enzymes in the liver. These enzymes are subject to induction by phenobarbital, benzpyrene, and tolbutamide itself among other drugs (TAGG et al. 1967). It is in the dog that, unlike other animal species, splits the molecule of tolbutamide, where this sulfonylurea accelerates its own metabolism (REMMER et al. 1964). In the dog it proved to be impossible to maintain the original plasma level for 3 weeks, even if the daily dose was increased from 100 to 300 mg/kg (WELCH 1972).

In man tolbutamide does not induce its own metabolism to an extent similar to that in the dog. Thus the plasma levels of tolbutamide declined in diabetic patients about 20% more rapidly when they had received the drug for some time. Peak plasma levels and the blood glucose lowering effects were not reduced to a clinically significant degree (SÜDHOF et al. 1958).

The microsomal enzymes in the liver of rats can be induced by exposure to phenobarbital. Thus the metabolism of tolbutamide and of glymidine can be accelerated by pretreating the animals with the barbiturate (GERHARDS et al. 1966). Phenobarbital as an enzyme-inducing agent modified the action of chlorpropamide and glibenclamide in rats. It further reduced the activity of hepatic glucose-6-phosphatase in addition to the inhibitory effect of the sulfonylureas on this enzyme. Blood glucose levels were not altered by the addition of phenobarbital (STENGARD et al. 1986).

In man, rifampicin, a potent inductor of hepatic drug-metabolizing enzymes, led to a considerable decrease in the plasma levels of tolbutamide. Following a 4-week course of a tuberculostatic regimen containing rifampicin, the plasma level of tolbutamide decreased by 49%, and the plasma half-life by 43% (SYVÄLAHTI et al. 1974).

A wide variety of drugs was found to inhibit biotransformation of sulfonylureas and thus to give rise to severe hypoglycemic events. Almost all observations pertinent to this most important interaction with the blood glucose lowering agents have been obtained from diabetic patients rather than in laboratory animals. Phenylbutazone is prominent among the drugs retarding the elimination of tolbutamide. In man the half-life of tolbutamide was prolonged from 4.5 h in the control period to 10.5 h after a 1- to 2-week course on phenylbutazone (HANSEN and CHRISTENSEN 1977).

Glymidine is applied at a high dose and thus belongs to the "first generation" of the sulfonylureas. In this respect it is similar to tolbutamide, carbutamide, and chlorpropamide. As these sulfonylureas are present in the blood in higher amounts, the enzymatic pathways transforming them are nearer to being saturated and any interference by a second drug is likely to delay their elimination. Thus phenylbutazone prolongs the plasma half-life of glymidine in man from 4.6 to 12 h, which is quite similar to the effect on the elimination of tolbutamide (HELD et al. 1970). Glibenclamide is approximately 200 times more potent than tolbutamide in man. It is therefore present in the blood at very low levels and phenylbutazone has no effect on its elimination (Schulz et al. 1971). Another sulfonylurea of high potency and classified as belonging to the "second generation" is glibornuride. Again, phenylbutazone is only marginally effective in retarding the elimination of this blood glucose lowering agent. When phenylbutazone was present the plasma half-life of glibornuride was found to be prolonged by merely 10% (Eckhardt et al. 1972). Neither the concentration of glibenclamide in the blood nor its half-life were affected by the second drug phenylbutazone. Despite that the blood glucose lowering effect was markedly enhanced. One possible explanation for this could be competition of both drugs for plasma protein-binding sites. This has been demonstrated to occur on isolated mouse islets by in vitro experiments (Hellman 1974). Addition of albumin to such islets reduced both the uptake of glibenclamide by the islets and its insulin-releasing effect. Phenylbutazone and sulfonamides increased the amount of glibenclamide bound to the islets and also of the insulin released. There was no such effect of the second drugs, when the islets were incubated in the absence of albumin. These results may be taken to indicate that these drugs potentiate the hypoglycemic action of glibenclamide by translocating this agent from plasma protein-binding sites to the pancreatic islets.

G. Possible Implications of the Results from Animal Experiments on the Pharmacokinetics of Sulfonylurea Derivatives for Clinical Applications in Humans

As there are about nine different sulfonylurea drugs on the market which act in principle in the same way but differ in their pharmacokinetics, the choice of agent to be employed has to take into account the kinetic properties of the drug selected.

Although the bioavailability of different brands of the same compound may not be the same, most sulfonylurea drugs are readily absorbed. Chlorpropamide enters the blood slowly, reaching its peak plasma level not before 3–6 h (Marchetti and Navalesi 1989). Due to its long half-life of about 35 h, it takes an entire week for a steady-state plasma level to build up. A long half-life and a slow onset of action are reasons for not choosing chlorpropamide as an antidiabetic agent. In addition, the metabolic fate and the route of excretion may both given rise to additional problems. At least two of the metabolites formed from chlorpropamide have been shown to retain some activity (Taylor 1974). These metabolites and chlorpropamide itself are removed from the body by renal clearance. They are thus apt to accumulate in patients whose renal function is impaired, a finding which is quite usual in elderly diabetics. A relatively mild reduction of the glomerular filtration rate to 43 ml/min has been found to prolong the plasma half-life of chlorpropamide from 35.6 to 86 h (Petitpierre and Fabre 1972).

The kinetic characteristics of chlorpropamide may well explain that this sulfonylurea has caused more severe and prolonged hypoglycemic events than most other compounds employed in the treatment of diabetes (BERGER 1971).

The disadvantage of a long half-life is shared by carbutamide, which in addition has retained antibacterial activity as being a sulfonamide and might thus interfere with the physiological bacterial growth in the gut. Despite a shorter plasma half-life of 9 h, glibenclamide has induced dangerous hypoglycemic reactions in patients more often than tolbutamide or several other sulfonylurea drugs (BERGER 1971). In several animal species, glibenclamide is retained by the liver and the pancreatic islets. This might explain that its blood glucose lowering effect is also prolonged in man (HAUPT et al. 1971). The treatment of diabetic patients with impaired renal function poses special problems. Sulfonylurea drugs have no place in the treatment of patients whose renal function has been reduced to a glomerular filtration rate of 30 ml/min or less (BALANT 1981). In such cases metabolic control should be achieved with insulin alone. In mild to moderate renal impairment with glomerular filtration rates from 30 to 60 ml/min, oral drugs with well-known and favorable pharmacokinetic properties should be selected.

Tolbutamide is one of the oldest and best-known sulfonylurea drugs. It is known to be fully metabolized within the liver to form a carboxy derivative. Although this compound may accumulate in renal failure, it is no longer active, and unchanged tolbutamide itself is not retained. The first hydroxylated metabolite still has some glucose-lowering effect. As it accounts for some 22% of the total amount excreted with the urine (PEART et al. 1987), some of the hydroxy derivative may accumulate in renal failure before being oxidized further. Tolbutamide has an average half-life of 5.7 (NELSON and O'REILLY 1961) or 6.3 h (SCHULZ and SCHMIDT 1970a,b). For this reason, tolbutamide is preferred in old patients and has qualified as being a drug of choice in diabetics with compromised physiological functions including renal clearance. Nevertheless, tolbutamide has given rise to severe hypoglycemic events in patients with reduced renal function (BERGER and SPRING 1970).

Like tolbutamide, glibenclamide is also fully metabolized in the liver. The hydroxylated compounds formed are in part excreted by the kidneys. Again the first metabolite retains some hypoglycemic activity. As long as the glomerular filtration rate is more than 30 ml/min, metabolites do not accumulate and hypoglycemic events due to such accumulation are thus not likely to occur (SCHMIDT et al. 1974). Glibenclamide is long acting and thus not indicated in any case of more severe renal insufficiency.

Glipizide has a potency and metabolic fate similar to that of glibenclamide. Hydroxylated metabolites with little or no hypoglycemic activity are rapidly formed. Biotransformation to polar metabolites does not depend on renal function. Only 5% of the primary drug is excreted unchanged. There was no retention of metabolites even when the glomerular filtration rate was reduced below 30 ml/min (BALANT et al. 1973). Thus a moderate renal insufficiency should not significantly increase accumulation of the active principle as long as the capacity of the liver to metabolize the drug is not seriously impaired.

A prolonged hypoglycemia was precipitated by acetohexamide in patients with renal failure. In one case the hypoglycemia lasted for 8 days. The renal retention of acetohexamide metabolites, which possess full blood glucose lowering activity, is thought to have caused these prolonged reactions (ALEXANDER 1966).

Glibornuride is metabolized totally to less active or inactive compounds. Its metabolism was found not to be changed by moderate renal insufficiency (RAAFLAUB et al. 1978). It has been claimed that gliquidone achieves identical plasma levels in patients with or without renal impairment (KOPITAR 1975). Gliclazide is apparently eliminated within the normal range in patients with a creatinine clearance reduced to 50 ml/min. Glisoxepide is in part excreted by the kidneys unchanged. Some accumulation of the drug has been shown to occur in patients with renal failure (SCHWARTZKOPF and KEWITZ 1977).

A profound knowledge of the pharmacokinetics of sulfonylurea drugs is thus essential for adequate treatment of patients with diabetic nephropathy or reduced renal function. Although most detailed information stems from clinical data obtained in man, animal experiments had to precede human application of these drugs. Their proper use in man has been based on the knowledge of their effects in various animal species and not least on their fate in the body, i.e., on their pharmacokinetics, in animal experiments.

References

Achelis JD, Hardebeck K (1955) Über eine neue blutzuckersenkende Substanz. Dtsch Med Wochenschr 80:1452–1455

Alexander RW (1966) Prolonged hypoglycemia following acetohexamide administration. Diabetes 15:362–364

Antal EJ, Gillespie WR, Phillips JP, Albert KS (1982) The effect of food on the bioavailability and pharmacodynamics of tolbutamide in diabetic patients. Eur J Clin Pharmacol 22:459–462

Balant L (1981) Clinical pharmacokinetics of sulphonylurea hypoglycaemic drugs. Clin Pharmacokinet 6:215–241

Balant L, Zahnd G, Gorgia A, Schwarz R, Fabre J (1973) Pharmacokinetics of glipizide in man: influence of renal insufficiency. Diabetologia 9:331–338

Balant L, Fabre J, Zahnd GR (1975) Comparison of the pharmacokinetics of glipizide and glibenclamide in man. Eur J Clin Pharmacol 8:63–69

Bänder A, Scholz J (1956) Spezielle pharmakologische Untersuchungen mit D860. Dtsch Med Wochenschr 81:889–891

Bänder A, Häussler A, Scholz J (1957) Ergänzende pharmakologische Untersuchungen über Rastinon. Dtsch Med Wochenschr 82:1557–1564

Bargeton D, Roquet J, Rouques A, Chassain A, Bieder A (1965) Action sur la glycemic du rat d'un derivé du theadiazole. Arch Int Pharmacodyn Ther 153:379–404

Benakis A, Glasson B (1980) Metabolic study of ^{14}C-labelled gliclazide in normal rats and in rats with streptozotocin induced diabetes. In: Gliclazide: The Royal Society of Medicine. Int Congr Symp Ser 20:57–69

Berger W (1971) Schwere Hypoglykämiezwischenfälle unter der Behandlung mit Sulfonylharnstoffen. Resultate einer gesamtschweizerischen Umfrage in den Jahren 1968 und 1969. Schweiz Med Wochenschr 101:1013–1022

Berger W, Spring P (1970) Beeinflussung der blutzuckersenkenden Wirkung oraler Antidiabetika durch andere Medikamente und Niereninsuffizienz. Internist (Berl) 11:436–441

Bernardi H, Fosset M, Lazdunski M (1988) Characterisation, purification and affinity labelling of the brain ^{3}H-glibenclamide-binding protein, a putative neuronal ATP-regulated K^{+}-channel. Proc Natl Acad Sci USA 85:9816–9820

Bigler F, Rentsch G, Rieder J (1971) Metabolism and pharmacokinetics of Ro 6-4563. In: Dubach UC, Bückert A (eds) Recent hypoglycemic sulfonylureas. Huber, Bern

Brasselton WE, Bransome ED, Huff TA (1977) Measurement of antidiabetic sulfonylureas in serum by gas chromatography with electron capture detection. Diabetes 26:50–57

Brotherton PM, McMartin C (1969) A study of the metabolic fate of chlorpropamide in man. Clin Pharmacol Ther 10:505–514

Brown KF, Crooks MJ (1976) Displacement of tolbutamide, glibenclamide and chlorpropamide from serum albumin by anionic drugs. Biochem Pharmacol 25:1175–1178

Büttner H, Portwich F (1967) Kompetitionsphänomene bei der Bindung von Pharmaka an Albumin. Klin Wochenschr 45:225–230

Campbell DB, Adriaenssens P, Hopkins YW, Gordon B (1980) Pharmacokinetics and metabolism of gliclazide: a review. In: Gliclazide: The Royal Society of Medicine. Int Congr Symp Ser 20

Carpentier JL, Sawano F, Ravazzola M, Malaisse WJ (1986) Internalization of ^{3}H-glibenclamide in pancreatic islet cells. Diabetologia 29:259–261

Christensen LK, Hansen JM, Kristensen M (1963) Sulphaphenazole-induced hypoglycaemic attacks in tolbutamide-treated patients. Lancet II:1298–1301

Cohen BD, Galloway JA, McMahon RE, Culp HW, Root MA, Henriques KJ (1967) Carbohydrate metabolism in uremia: blood glucose response to sulfonylurea. Am J Med Sci 254:608–618

Crooks MJ, Brown KF (1974) The binding of sulphonylureas to serum albumin. J Pharm Pharmacol 26:304–311

Crooks MJ, Brown KF (1975) Interaction of glipizide with human serum albumin. Biochem Pharmacol 24:298–299

Darby FJ, Grundy RK, Price-Evans DA (1972) Apparent Michaelis constants for the metabolism of (ureyl-^{14}C)-colbutamide by human liver preparations. Biochm Pharmacol 21:407–414

Eckhardt W, Rudolph R, Sauer H, Schubert WR, Undeutsch D (1972) Zur pharmakologischen Interferenz von Glibornurid mit Sulphaphenazol, Phenylbutazon und Phenprocoumon beim Menschen. Arzneimittelforschung 22:2212–2219

Elrick H, Purnell A (1957) The response of kidney, liver and peripheral tissues to tolbutamide. Ann NY Acad Sci 71:38–45

Findlay J (1992) Effect of pH upon the inhibition by sulphonylurea drugs of ATP-sensitive K^{+}-channels in cardiac muscle. J Pharmacol Exper Therap 262:71–79

Fuccella LM, Tamassia V, Valzelli G (1973) Metabolism and kinetics of the hypoglycemic agent glipizide in man. Comparison with glibenclamide. J Clin Pharmacol 13:68–75

Garret ER (1970) The clinical significance of pharmacokinetics. In: Dengler HJ (ed) Pharmacological and clinical significance of pharmacokinetics. Stuttgart, Schattauer, pp 5–21

Geisen K, Hitzel U (1985) Ökonomoupoulos, R., Pünter, J., Weyer, R., and Summ, H.-D.: Inhibition of ³H-glibenclamide binding to sulfonylurea receptors by oral antidiabetics. Arzneimittelforschung 35:707–712

Gerhards E, Gibian H, Kolb KH (1964) Der Stoffwechsel von Glykodiazin beim Menschen. Arzneimittelforschung 14:394–402

Gerhards E, Kolb KH, Schulze PE (1966) Über 2 Benzolsulfonylamino-5(β-methoxy-äthoxy)-pyrimidin (Glykodiazin). V. In vitro und in vivo Versuche zum Einfluß von Phenyläthylbarbitursäure (Luminal) auf den Stoffwechsel und die blutzuckersenkende Wirkung des Glykodiazin. Naunyn Schmiedeberg's Arch Pharmak 231:407–419

Hansen JM, Christensen LK (1977) Drug-interactions with oral sulphonylurea hypoglycaemic drugs. Drugs 13:24–34

Hasselblatt A (1965) Die Verdrängung von Bromsulphalein und von Bilirubin aus dem Blut, eine Nebenwirkung von Tolbutamid. Gastroenterologia (Basel) 104[Suppl]:148–152

Hasselblatt A (1969) Die Hemmung der Ketogenese im Lebergewebe durch Tolbutamid und Glykodiazin in vitro. Naunyn-Schmiedeberg's Arch Pharmak 202:152–164

Hasselblatt A (1971) Interactions of drugs at plasma protein binding sites. In: Toxicological problems of drug combinations. Excerpta Medica Congress Ser 254:89–97

Hasselblatt A (1973) The pharmacology of the blood sugar lowering sulfonylurea drugs. In: Proc 5th Int Congr Pharmacol 1972. Karger, Basel, pp 206–213

Hasselblatt A, Hukuhara T (1964) Die Wirkung von Tolbutamid auf die Elimination von Bromsulphalein aus dem Blut. Klin Wochenschr 42:449–454

Haufe F, Hasselblatt A, Schoepf HJ (1965) Speciesunterschiede in der Wirkung von Tolbutamid und Probenecid auf die Konzentration von injiziertem Bromsulphalein (BSP) im Blut, seine Konjugation in der Leber und die Ausscheidung mit der Galle. Naunyn-Schmiedeberg's Arch Pharmak 250:256

Haupt E, Köbereich W, Beyer J, Schöffling K (1971) Pharmacodynamic aspects of tolbutamide, glibenclamide, glibornuride and glisoxepide. Diabetologia 7:449–454

Held H, Kaminski B, von Oldershausen HF (1970) Die Beeinflussung der Elimination von Glykodiazin durch Leber- und Nierenfunktionsstörungen und durch eine Behandlung mit Phenylbutazon, Phenprocoumon und Doxycyclin. Diabetologia 6:386–391

Hellman B (1974) Potentiating effects of drugs on the binding of glibenclamide to pancreatic beta cells. Metabolism 23:839–846

Hellman B, Täljedal JB (1975) Effects of sulfonylurea derivatives on pancreatic Heffter Heubner Handb Exp Pharm 32 (2):175–194

Hellman B, Sehlin J, Täljedal JB (1984) Glibenclamide is exceptional among hypoglycaemic sulphonylureas in accumulating progressively in β-cell-rich pancreatic islets. Acta Endocrinol (Copenh) 105:385–390

Hellman B, Idahl LA, Tjälve H, Danielsson A, Lernmark A (1969) Beobachtungen zum Wirkungsmechanismus des hypoglykämisch wirksamen Sulfonylharnstoff-Präparates HB 419. Arzneimittelforschung 19:1472–1476

Hellman B, Sehlin J, Täljedal IB (1971) The pancreatic B-cell recognition of insulin secretagogues. Site of action of tolbutamide. Biochem Biophys Res Commun 45:1384–1388

Heptner W, Christ O, Kellner HM, Rupp W (1969) Pharmacokinetics of a new highly effective hypoglycemic sulfonylurea derivative. Acta Diabetol Lat 6[Suppl 1]:105–115

Holmes B, Heal RC, Brogden RM, Speight TM, Avery GS (1984) Gliclazide. Drugs 27:301–327

Hsu PL, Ma JKH, Luzzi LA (1974) Interactions of sulfonylureas with plasma proteins. J Pharmac Sci 63:570–573

Johnson PC, Hermes AR, Driscoll T, West KM (1959) Metabolic fate of chlorpropamide in man. Ann NY Acad Sci 74:459–470
Joost HG, Hasselblatt A (1977) Effects of polymer-linked sulfonylurea derivatives
on insulin release. Naunyn-Schmiedeberg's Arch Pharmacol 297:81–84
Judis J (1972) Binding of sulfonylureas to serum proteins. J Pharm Sci 61:89–93
Judis J (1973) Displacements of sulfonylureas from human serum proteins by
coumarin derivatives and cortical steroids. J Pharm Sci 62:232–237
Kärkkainen S, Vapaatalo H, Neuvonen PJ (1983) Urine pH is important for chlor-
propamide elimination. Diabetes Care 6:313
Kaubisch N, Hammer R, Wollheim C, Renold AE, Offord RE (1987) Specific
receptors for sulfonylureas in brain and in a B-cell tumour of the rat. Biochem
Pharmacol 31:1171–1174
Kellner HM, Christ O, Rupp W, Heptner W (1969) Resorption, Verteilung und
Ausscheidung nach Gabe von ^{14}C-markiertem HB419 an Kaninchen, Ratten
und Hunden. Arzneimittelforschung 19:1388–1400
Kolb KH, Kramer M, Schulze PE (1964) Resorption, Verteilung und Ausscheidung
von radioaktiv markiertem 2-Benzolsulfonamido-5-(β-methoxy-äthoxy)-pyrimidin
(Glykodiazin) im Tierversuch. Arzneimittelforschung 14:385–389
Kolb KH, Mützel W, Speck U, Schulze PE (1974a) Pharmakokinetik und Metabo-
lismus von Pro-Diaban^R. In: Schöffling K, Kroneberg G, Laudahn G (eds) Pro-
Diaban (Glisoxepid). Schattauer, Stuttgart, pp 29–37
Kolb KH, Schulze PE, Speck U, Acksteiner B (1974b) Pharmakokinetik von
radioaktiv markiertem Glisoxepid beim Tier. Arzneimittelforschung 24:397–403
Kopitar Z (1975) Humanpharmakokinetik und Metabolismus von ^{14}C-markiertem
Gliquidon. Arzneimittelforschung 25:1455–1460
Kopitar Z, Koss FW (1975) Pharmakokinetisches Verhalten von Gliquidone (AR-
DF 26), einem neuen Sulfonylharnstoff. Arzneimittelforschung 25:1933–
1983
Kuether CA, Clark MR, Scott EG, Lee HM, Pettinga CW (1956) Lack of effect of
carbutamide on activity of rat liver glucose-6-phosphatase. Proc Soc Exp Biol
(NY) 93:215–217
Lorch E, Gey KF, Bigler F, Rieder J, Rentsch G, Schärer K, Hummler H (1971)
Tierexperimentelle Untersuchungen mit Glibornurid, vol. 1. In: Magyar und L,
Beringer A (eds) Verh. II. Int Donau Sympos. Wiener Med Akad, pp 359–363
Louis H, Fajans SS, Conn JW, Struck WA, Wright JB, Johnson JL (1956) The
structure of a urinary excretion product of 1-butyl-3-p-tolylsulfurea
(orinase). J Am Chem Soc 78:5701–5702
Marchetti P, Navalesi R (1989) Pharmacokinetic-pharmacodynamic relationships of
oral hypoglycemic agents. Clin Pharmacokinet 16:100–128
Mohnike G, Wittenhagen G, Langenbeck W (1958) Über das Ausscheidungsprodukt
von N-(4-methyl-benzolsulfonyl)-N'-butylharnstoff beim Hund. Naturwissenschaften 45:13
Monro AM, Welling PG (1971) The bioavailability in man of marketed brands of
chlorpropamide. Eur J Clin Pharmacol 7:47–49
Nelson DA, Aquilar-Bryan L, Bryan J (1992) Specificity of photolabeling of β-cell
membrane proteins with an ^{125}I-labeled glyburide analog. J Biol Chem 267:
14928–14933
Nelson E, O'Reilly I (1961) Kinetics of carboxytolbutamide excretion following
tolbutamide and carboxytolbutamide administration. J Pharmacol Exper Ther
132:103–109
Neuvonen PJ, Kärkkainen S (1983) Effects of charcoal, sodium bicarbonate, and
ammonium chloride on chlorpropamide kinetics. Clin Pharmacol Ther 33:386–
393
Neuvonen PJ, Kannisto H, Hirvisalo E (1830) Effect of activated charcoal on
absorption of tolbutamide and valproate in man. Eur J Clin Pharmacol 24:243–
246

Niki I, Kelly RP, Ashcroft SJH, Ashcroft FM (1989) ATP-sensitive K-channels in HIT-T15 β-cells studied by patch clamp methods. ^{86}Rb efflux and glibenclamide binding. Pflüger's Arch 415:47–55

Nistrup-Madsen S, Fog-Möller F, Persson I (1971) Distribution of tolbutamide and chlorpropamide after administration to non diabetic rats. Eur J Pharmacol 13:374–380

Olson SC, Ayres EJ, Albert KS (1985) Effect of food and tablet age on relative bioavailability and pharmacodynamics of two tolbutamide products. J Pharm Sci 74:735–739

Panten U, Burgfeld J, Goerke F, Rennicke M, Schwanstecher M, Wallasch A, Zünkler BJ, Lenzen S (1989) Control of insulin secretion by sulfonylureas, meglitinide and diazoxide in relation to their binding to the sulfonylurea receptor in pancreatic islets. Biochem Pharmacol 38:1217–1229

Panten U, Schwanstecher M, Schwanstecher C (1992) Pancreatic and extrapancreatic sulfonylurea receptors. Horm Metab Res 24:549–554

Pearson JC (1985) Pharmacokinetics of glyburide. Am J Med 79 [Suppl3B]:67–71

Peart GF, Boutagy J, Shenfield GM (1987) Lack of relationship between tolbutamide metabolism and debrisoquine oxidation phenotype. Eur J Clin Pharmacol 33: 397–401

Petitpierre B, Fabre J (1972) Effect de l'isuffisance renale sur l'action hypoglycémiante des sulfonylurées. Schweiz Med Wochenschr 102:570–582

Raaflaub J, Baethke R, Sorge F, Meier JM (1978) Pharmacokinetics of glibornuride and its metabolites in patients with renal disease. 7th Int Congr Pharmacol Paris

Remmer H, Siegert M, Mercker HJ (1964) Vermehrung arzneimittelhydroxylierender Fermente durch Tolbutamid. Naunyn-Schmiedeberg's Arch Pharmacol 249:71–84

Rentsch G, Schmidt HAE, Rieder J (1972) Zur Pharmakokinetik von Glibornurid. Arzeimittelforschung 22:2209–2212

Rupp W, Christ O, Heptner W (1969) Resorption, Ausscheidung und Metabolismus nach intravenöser und oraler Gabe von HB419-^{14}C an Menschen. Arzneimittelforschung 19:1428–1434

Rupp W, Christ O, Fulberth W (1972) Untersuchungen zur Bioavailability von Glibenclamid. Arzneimittelforschung 22:471–473

Rupp W, Dibbern HW, Hajclu P, Ross G, Vander EE (1975) Untersuchungen zur Bioaequivalenz von Tolbutamid. Dtsch Med Wochenschr 100:690–695

Samimi H, Loutan L, Balant L, Tillol'Es M, Fabre J (1977) Metabolites des sulfonylurées hypoglycemiantes: experiences avec le glibenclamide chez le rat. Schweiz Med Wochenschr 107:1291–1296

Sartor G, Melander A, Scherstén B, Wählin-Boll E (1980) Influence of food and age on the single-dose kinetics and effects of tolbutamide and chlorpropamide. Eur J Clin pharmacol 17:285–293

Schlossmann K (1974a) Proteinbindung von Glisoxepid und sein Einfluß auf die Proteinbindung von Phenprocoumon. Arzneimittelforschung 24:392–403

Schossmann K (1974b) Wechselwirkungen in der Proteinbindung zwischen Pro-Diaban^R und Phenprocoumon bzw. Bilirubin im Vergleich zu Tolbutamid. & In: O Schöffling K, Kroneberg G, Laudahn G (eds) Schattauer, stuttgart, pp 39–44

Schmidt HAE, Schoog M, Schweer KH, Winkler E (1973) Pharmacokinetics and pharmacodynamics as well as metabolism following orally and intravenously administered ^{14}C-gliplizide, a new antidiabetic. Diabetologia [Suppl 9]:320–330

Schmidt FH, Hrska VE, Heesen D, Schulz O, Schulz E (1974) Plasmaspiegel und Ausscheidung von Glibenclamid bei niereninsuffizienten und leberkranken Patienten im akuten Versuch. Congress of the German Diabetes Society Abstr no 89, 9

Scholz J, Bänder A (1956) Über die orale Behandlung des Diabetes mellitus mit N-(4-Methyl-benzol-sulfonyl)-N'-butyl-harnstoff (D860) Pharmakologie. Dtsch Med Wochenschr 81:825–826

Schulz E, Schmidt FH (1970a) Blutzuckersenkende Wirkung von Hydroxytolbutamid beim Menschen. Klin Wochenschr 48:759–760

Schulz E, Schmidt FH (1970b) Abbauhemmung von Tolbutamid durch Sulfaphenazol beim Menschen. Pharmacologia Clinica 2:150–154

Schulz E, Koch K, Schmidt FH (1971) Pharmakokinetik und Metabolismus von Glibenclamid (HB419) in Gegenwart von Phenylbutazon. Eur J Clin Pharmacol 4:32–37

Schwanstecher M, Löser S, Rietze I, Panten U (1991) Phosphate and thiophosphate group donating adenine and guanine nucleotides inhibit glibenclamide binding to membranes from pancreatic islets. Naunyn-Schmiedeberg's Arch Pharmacol 343:83–89

Schwanstecher M, Schaupp U, Löser S, Panten U (1992) The binding properties of the particulate and solubilized sulfonylurea receptor from cerebral cortex are modulated by the Mg^{2+} complex of ATP. J Neurochem 59:1325–1335

Schwanstecher M, Löser S, Chudziak F, Bachmann C, Panten U (1994) Photoaffinity labeling of the cerebral sulfonylurea receptor using a novel radioiodinated azidoglibenclamide analogue. J Neurochem 63: in press

Schwarzkopf T, and Kewitz H (1977) Elimination of glisoxepide in patients with renal failure. Naunyn-Schmiedeberg's Arch Pharmacol 297:R61

Shima K, Tanaka A, Ikegami H, Tabata M, Sawazaki N, Kumahara Y (1983) Effect of dietary fibre, glucomannan, on absorption of sulfonylurea in man. Horm Metabol Res 15:1–3

Stengard JH, Saarni HU, Knip M, Lahtela JT, Stenbäck F, Sotaniemi EA (1986) Sulphonylurea and glucose metabolism in phenobarbital induced rats. Res Commun Chem Path Pharmacol 54:147

Stowers JM, Mahler RF, Hunter RB (1958) Pharmacology and mode of actions of sulphonylureas in man. Lancet I:278–283

Stuhlfauth K, Mehnert H, Schäfer G, Kallampetsos G (1960) Untersuchungen zum Wirkungsmechanismus der Sulfonylharnstoffe. Klin Wochenschr 38:825–826

Südhof H, Altenburg S, Sander E (1958) Zur Frage der D860-Eliminationsgeschwindigkeit aus dem Serum beim Diabetiker. Klin Wochenschr 36:585

Syvälahti EKG, Pihlajamäki, Iisalo EJ (1974) Rifampicin and drug metabolism. Lancet II:232–233

Tagg J, Yasuda DM, Tanabe M, Mitoma C (1967) Metabolic studies of tolbutamide in the rat. Biochem Pharmacol 16:143–153

Taylor JA (1972) Pharmacokinetics and biotransformation of chlorpropamide in man. Clin Pharmacol Ther 13:710–718

Taylor JA (1974) Pharmacokinetics and biotransformation of chlorpropamide in the rat and dog. Drug Metabolism Disposition Biol Fate Chem 2:221–227

Wählin-Boll E, Melander A, Sartor G, Scherstein B (1980) Influence of food intake on the absorption and effect of glipizide in diabetics and in healthy subjects. Eur J Clin Pharmacol 18:279–283

Welch RM (1972) Altered drug toxicity associated with hepatic enzyme induction. In: de Baker SB and Neuhaus GA (eds) Proc. Europ Soc Drug Toxicity XIII. Excerpta Medica, Amsterdam

Welles JS, Root MA, Anderson RC (1961) Metabolic reduction of 1-(p-acetylbenzenesulfonyl)-3-cyclohexylurea (Acetohexamide) in different species. Proc Soc Exp Biol Med 107:583–585

Wick AN, Britton B, Grabowski, R (1956) The action of a sulfonylurea hypoglycemic agent (orinase) in extrahepatic tissues. Metabolism 5:739–743

Wishinsky H, Glaser FJ, Perkal S (1962) Protein interactions of sulfonylurea compounds. Diabetes 11 [Suppl]:18–28

Wittenhagen G und Mohnike G (1956) Über das Ausscheidungsprodukt von D860. Dtsch Med Wochenschr 81:887–888

Zünkler BJ, Trube G, Panten U (1989) How do sulfonylurea compounds approach their receptor in the B-cell plasma membrane? Naunyn-Schmiedeberg's Arch Pharmacol 340:328–332

CHAPTER 8
Toxicology of Sulfonylureas

F. Hartig, K.H. Langer, W. Rebel, F.H. Schmidt, and E. Schütz

A. Introduction

Since 1958, Hoechst AG, Frankfurt, and Boehringer Mannheim GmbH have synthesized and pharmacologically tested a large number of sulfonylurea analogues which elicit hypoglycemic activity in the milligram range and in some cases at less than 0.1 mg/kg body weight. Glibenclamide (HB419) was chosen from a group of the most potent compounds for joint research in 1964 and introduced onto the market as Daonil by Hoechst and as Euglucon by Boehringer (Hebold et al. 1969a,b; Mitsukami et al. 1969).

The toxicity profile of β-cytotropic antidiabetic drugs including gliben-clamide was comprehensively dealt with by Bänder (1971) in the first edition of this handbook. This review will focus on the toxicological inves-tigations reported since then and on the antidiabetic sulfonylureas which have been developed. It should be noted that the majority of both acute and chronic toxicity studies were conducted before the strict regulations of GLP (good laboratory practice) were introduced into preclinical research. Even some long-term studies do not meet today's guidelines for testing carcinogenicity.

B. Acute Toxicity

The LD_{50} of a number of β-cytotropic compounds administered by different routes to various animal species is shown in Table 1. A single dose of these drugs is slightly toxic in rodents and dogs. Comparison of the sulfonylureas of the "first generation" (carbutamide, tolbutamide, chlorpropamide, acetohexamide, tolazamide, glycodiazine, gliclazide) and the much more potent analogues of the "second generation" (Pfeiffer 1984) clearly shows that the difference between pharmacologically active doses and toxic doses is far greater for glibenclamide, glipizide, gliquidone, glibornuride and glisoxepide than for the aforementioned antidiabetics of the "first generation."

Table 1. LD_{50} values (mg/kg) of β-cytotropic antidiabetics

Ref.	Compound	Route	Mouse	Rat	Guinea pig	Rabbit	Cat	Dog
1	Glyprothiazole (IPTD)	p.o.	4 000	3 940				
2	Carbutamide	p.o.	2 500					
3	Tolbutamide	p.o.	2 500	4 000				
		s.c.	750					
		i.p.	467					
4	Chlorpropamide	p.o.	1 675	2 390				800
		s.c.	780	760				
5	Acetohexamide	p.o.	2 500					
6	Tolazamide	p.o.	>5 000	5 000				
		i.p.	2 239					
7	Glycodiazine (glymidine)	p.o.	5 300	2 850				
		i.v.	1 480	2 000				
8	Glibenclamide	p.o.	>15 000	>15 000	>15 000	>10 000		>10 000
		i.p.	>12 500	>12 500				
		s.c.	>20 000	>20 000				
9	Gliclazide	p.o.	3 000	3 000				
		i.v.	295	382				
		s.c.	1 034	>1 000				
10	Glipizide	p.o.	4 000	4 000	4 000			
		i.p.	3 000	1 200	2 400			
11	Gliquidone	p.o.	>10 000	>10 000		>10 000		
		i.c.	>10 000	>5 000				
		i.p.	>10 000	>3 500				
12	Glibornuride	p.o.	>20 000	>18 000				
		i.p.	1 530	1 360				
		s.c.	20 000	10 800				
13	Glisoxepide	p.o.	>10 000	>10 000		>4 000	>4 000	>2 000
		i.v.	283	196				

IPTD, isopropylthiodiazol.
References: 1, PENHOS 1957; quoted by BÄNDER 1971; 2, ACHELIS and HARDEBECK 1955; 3, SCHOLZ and BÄNDER 1956; 4, SCHNEIDER et al. 1959; 5, MAHA et al. 1962; 6, DULIN et al. 1961; 7, KRAMER et al. 1964; 8, HEBOLD et al. 1969b; 9, DUHAULT et al. 1972; 10, AMBROGI et al. 1971; 11, KAST et al. 1975; 12, KRÖO to PRAGUE 1972; 13, TETTAMONI 1974.

C. Chronic Toxicity

I. Carbutamide

1. Rats

After 18 months of administration of carbutamide as a food additive (1000, 2250, 5000 ppm) to male (n = 30) and female (n = 30) rats, only few alterations were found (G. HEBOLD, H. CZERWEK, F. HARTIG, F.H. SCHMIDT, H.W. TEUTE, 1971, Pathologie und Toxikologie von Carbutamid – Laufzeit 1,5 Jahre an Ratten, unpublished data). Rats of both sexes which had received the high dose exhibited a reduced gain in body weight. In male rats from the middle-dose group, hyperplasia in the bile duct and adenomas and hyperplasia in the thyroid glands were found. An increased weight of the thyroid gland was seen in male and female rats from the middle-dose group. The alterations in the bile ducts and thyroid glands were obviously not dose related. The increased thyroid weight was normalized within 4 weeks after withdrawal of carbutamide.

2. Dogs

Oral administration of 12 mg/kg body weight during 18 months was tolerated by male (n = 4) and female (n = 4) dogs without adverse effects, whereas 30 mg/kg body weight caused elevated activity of alkaline phosphatase in the plasma and increased excretion of bilirubin, urobilinogen and hemoglobin in the urine (G. HEBOLD, H. CZERWEK, F. HARTIG, F.H. SCHMIDT, H.W. TEUTE 1971, Pathologie und Toxikologie von Carbutamid – Laufzeit 1,5 Jahre an Hunden, unpublished data). Alterations were found in the liver: cytosolic protein coalescence and inclusions in the nuclei of liver cells. Higher doses of carbutamide (75 and 185 mg/kg) caused the death of five out of eight and seven out of eight animals, respectively. The dogs exhibited tonic-clonic seizures, diarrhea, salivary flow and vomiting. They were extremely dehydrated, cachectic and had generalized hemorrhages. The surviving dog of the high-dose group was anemic, has an elevated erythrocyte sedimentation rate and an increased number of reticulocytes, retention of Bromosulphthalein (sulfobromophthalein sodium), increased values of alkaline phosphatase, alanine aminotransferase (ALT) and glutamic lactate dehydrogenase (GLDH).

II. Tolbutamide

The carcinogenicity of tolbutamide was tested in long-term bioassays (NCI Carcinogenesis Bioassay 1978). Rats were given 12000 or 24000 ppm and mice of both sexes received 25000 or 50000 ppm for 5 days/week over 78 weeks. Rats and mice were observed for an additional 28 and 24–26 weeks, respectively, after withdrawal of tolbutamide: a variety of tumors were found but there was no significant difference between the drug groups and the control animals of both species and sexes. Thus it was concluded that tolbutamide exerted no carcinogenic effect.

III. Chlorpropamide

1. Rats and Mice

Two long-term experiments with male and female rats have been conducted but not published (G. Hebold, H. Czerwek, F. Hartig, F.H. Schmidt, H.W. Teute 1971, Pathologie und Toxikologie von Chlorpropamid – Laufzeit 1,5 Jahre an Ratten, unpublished data). Chlorpropamide was administered as a feed additive in three different concentrations/study (1000 up to 8000 ppm) during a period of 17 and 18 months, respectively. The daily drug intake was calculated in the range of 48–696 mg/kg body weight.

In the first experiment (18 months), the following alterations were seen in male and female rats: rough coat, skin eczema, a delayed decrease of alkaline phosphatase in the plasma and an increased incidence of hyperplasia in the bile duct. The relative thyroid weight was increased but normalized during a drug-free recovery period. In the second study, the rats had a dose dependently reduced body weight gain, altered spermatogenesis and formation of cataracts. Urinary calculi were found in five rats. These animals displayed epithelial metaplasia, suggesting an early preliminary stage of tumor formation.

A bioassay for carcinogenicity was performed in Fisher 344 rats and BGC 3F1 mice (NCI Carcinogenesis Bioassay 1978). Rats received 3000 and 6000 ppm, respectively; mice received 3317 and 6635 ppm chlorpropamide, respectively. It was concluded that chlorpropamide displayed no carcinogenic effect in mice and rats.

2. Dogs

Oral administration of 45, 90 or 180 mg chlorpropamide/kg body weight during a 1-year experiment brought about a reduced gain of body weight in the high-dose group. With the exception of the low-dose animals, alterations of the skin were seen, e.g., erythema, pustules and necrotic lesions. In the high-dose group, ulcerations were found, predominantly on the limbs. Blood cell counts revealed leukopenia, neutropenia and lymphocytosis. The hemoglobin concentration was dose dependently reduced, and the alkaline phosphatase activity was elevated in the middleand high-dose group. Lipid deposits were found in the myocardium and in the liver, sometimes accompanied by hyaline deposits in enlarged liver cells (J. Scholz, R. Brunk, H. Kief, K. Engelbart 1967, Chronische orale Toxitätsprüfung von Chlorpropamid sm Hund, unpublished data).

IV. Acetohexamide

A review of toxicity studies was presented in the first edition of this handbook by BÄNDER (1971). New contributions are not available except a carcinogenicity study conducted in Fisher 344 rats and BGC 3F1 mice (NCI report 1977a). Acetohexamide was administered as a food additive in concentrations of 10000 or 20000 ppm (rats) and 6359 or 12718 ppm (mice) during a period of 103 weeks. The incidence of leukemic tumors was increased in male rats which had received the low dose. It remains questionable whether acetohexamide caused leukemia since this disease is a common finding in Fisher rats and was not seen in the high-dose group. A higher incidence of lymphomas was found in male mice, but the difference was not statistically significant in comparison to control animals. It was suggested that acetohexamide displays no carcinogenic activity.

V. Tolazamide

A carcinogenicity study was reported in 1977 (NCI report 1977b). Rats and mice received 1000 or 5000 ppm tolazamide in the feed during a period of 103 weeks. A carcinogenic effect was not found.

VI. Glibenclamide

Extensive toxicological investigations were done by HEBOLD et al. (1969a,b), MITSUKAMI et al. (1969) and summarized by BÄNDER (1971) and HEBOLD (1971). This very potent sulfonylurea analogue was very well tolerated by various animal species, even after administration of very high doses. The favorable result of these studies were confirmed by long-term experiments with rats, rabbits and dogs (KRALL 1984).

In feeding experiments with male and female rabbits which ingested 0.19, 0.9 or 4.4 mg glibenclamide/kg body weight during 6 months, a number of hypoglycemic shocks occurred, resulting in clonic convulsions which caused death due to fractures of the vertebrae and legs (G. HEBOLD, H. CZERWEK, F. HARTIG, F.H. SCHMIDT, H. STORK 1969, Pathology and toxicology HB419 (Hoe 33419) in rabbits, unpublished data). In the high-dose female group, the body weight gain was reduced. The number of red blood cells, hemoglobin and hematocrit were decreased in male animals. The hepatic glycogen content and lipid droplets in liver cells were markedly reduced.

VII. Gliclazide

The toxicity of gliclazide was examined in acute, subacute and long-term studies using diabetic animals (YASUBA et al. 1981). An oral administration of 1600 mg gliclazide was lethal in diabetic rabbits, but tolerated by intact rabbits. Administration of 100 mg/kg body weight over 13 weeks had no toxic effects. In feeding experiments over a period of 52 weeks, 0.006% up to 0.06% gliclazide as a food additive did not elicit toxic alterations in mice.

VIII. Glipizide

In rats and dogs, 800 mg/kg per day glipizide administered orally for 1 year caused no toxic alterations (AMBROGI et al. 1971).

IX. Gliquidone

Gliquidone has been tested in subchronic and chronic studies with rats and rabbits. In subacute experiments, oral administration of 100, 200 and 2500 mg/kg (rats) and 10, 200 and 3000 mg/kg (rabbits) (ARDF 26 SE) caused no toxic effects that could be ascribed to the drug (KAST et al. 1975). The pharmacological activity was shown by the reversible degranulation of β cells of the islets of Langerhans. A significantly dose dependent decrease in serum glucose parallels this finding in rabbits only. Oral administration of 10, 100 or 1000 mg/kg body weight in chronic experiments did not cause adverse effects.

X. Glibornuride

Studies in rats, mice, dogs and rabbits have shown this compound to be 50 times more effective than tolbutamide and only slightly less effective than glibenclamide (LORCH et al. 1972). Glibornuride had less effect on free fatty acids in blood plasma than tolbutamide at equipotent dose levels. Concentrations of triglycerides, phospholipids and cholesterol were not significantly affected by glibornuride. There was no evidence of carcinogenic potential (HOFFMANN-LA ROCHE 1985).

XI. Glisoxepide

1. Subchronic and Chronic Toxicity in Rats

In subchronic toxicity studies (13 or 14 weeks), glisoxepide was orally administered to rats in doses of 10, 30, 300 and 1000 mg/kg body weight and given to dogs in doses of 3, 10, 30 and 100 mg/kg body weight. Even after increasing the highest dose after 6 weeks from 300 to 1000 mg in rats and from 30 to 100 mg/kg body weight in dogs, no toxic alterations were found (TETTENBORN 1974). In chronic toxicity experiments over 78 weeks, rats were administered 10, 50, 250 or 1000 mg glisoxepide /kg body weight. With the exception of a reduced body weight gain in the highest dose group, glisoxepide was well tolerated (TETTENBORN 1974).

2. Chronic Toxicity in Dogs

Two studies on beagle dogs have been reported (TETTENBORN 1974). In the first 52-week experiment, dogs were administered 10, 50, or 250 mg glisoxepide/kg body weight per os. Body weight gain, food consumption and water intake were not affected and hematological/clinicochemical assays of the blood and urine showed no drug-related alterations. However, blood glucose concentrations were singificantly reduced 2 h after the first administration of glisoxepide compared with control animals. Oral glucose tolerance tests which were performed in the 10th and 50th weeks of the study showed that the blood glucose concentrations of the glisoxepide-treated dogs increased much more than in control dogs within 30 min after oral glucose load. The postprandial glucose concentration normalized very quickly in those dogs which had received 10 mg glisoxepide/kg, but normalized with a marked delay in dogs which had received higher doses. Individual dogs of all the experimental groups exhibited hypoglycemic cramps and lethal shocks. In order to avoid the death of animals due to the exaggerated pharmacological effects of glisoxepide, a second chronic experiment was conducted with a changed timing of glisoxepide administration (10, 100, 1000 mg/kg), which was performed after an interval of at least 2 h after feeding (TETTENBORN 1974). No deaths occurred in this 52-week study.

Transient adverse effects were seen in one dog of each glisoxepide-treated group beginning in study week 15: these had excessive salivary flow and short-term convulsions and were lying on their side. The symptoms were independent of the actual blood glucose concentration. One month after the start of this experiment, the dogs in the high-dose group showed a slight reduction of packed cell volume, hemoglobin concentration and number of red blood cells. Later on in this study all animals of the high-dose group displayed mild anemia with slight anisocytosis and Howell-Jolly bodies in the red blood cells. Near the end of the experiment the hematological parameters improved. Bone marrow tests in the 51st week showed no differences between experimental and control animals. The weight of the thyroid glands and prostate were significantly increased in the high-dose group. Slight degenerative changes in isolated skeletal muscle fibers with marginal infiltration of mononuclear cells were found in four out of eight dogs of the high-dose group and in two out of eight animals of the middle-dose group. Transient increases in plasma GPT were also reported.

D. Reproduction Toxicology

I. Glibenclamide

ICR-JCL mice were administered 0.2, 2, 20, 200 and 2000 mg/kg per day glibenclamide from day 5 to day 14 of gestation, and Wistar and Sprague-Dawley rats from day 7 to day 16 of pregnancy. The dose of 2000 mg/kg per day represents 10 000-fold the effective hypoglycemic threshold dose in rats. At this dose level, ossification of the cervical vertebra was delayed. Intrauterine deaths did not occur. Both rats and mice displayed vertebral alterations. After administration of the very high dose level of 50 000 ppm (equivalent to a daily dose of 2500 mg/kg body weight) from day 6 of gestation to day 21/22 postpartum, the offspring had deformed lungs and limb bones. These alterations are considered to be perinatal or postnatal effects.

Rabbits were given 0.035, 3.5, and 350 mg glibenclamide/kg body weight from day 7 to day 17 of gestation. The high dose caused premature intrauterine deaths. The surviving offspring showed neither impaired viability nor development (Baeder and Sakaguchi 1969). Earlier reports on the teratogenic effects of glibenclamide were summarized by Bänder (1971).

II. Glisoxepide

The compound was tested in mice (1, 10, 100 mg/kg), rats (1, 10, 100 mg/kg) and rabbits (0.5, 5, 50 mg/kg). No teratogenic effects were detected (Tettenborn 1974).

III. Gliquidone

Gliquidone was given orally to rats (10, 50, 2500 mg/kg) and rabbits (10, 50, 250 mg/kg). In rats no deleterious effect on embryos was detected (Iida et al. 1976). In rabbits distinct embryolethality occurred at dosages of 50 and 250 mg/kg. The resorption rate of fetuses was increased and the number of viable fetuses was reduced. Malformations were seen in 7 out of 144 viable fetuses, namely at D1 = 2.9%, D2 = 7.5% and D3 = 5.7%, while no abnormality was found among 85 control fetuses. An increased rate of variation and malformation, independent of dose, was seen in rabbits only. Therefore a true teratogenic action cannot be excluded by these experiments in rabbits, but in later experiments by another research group a teratogenic action was not seen (Iida et al. 1976).

IV. Glibornuride

Teratogenic effects were not observed in mouse, rat or rabbit after administration of dose levels up to 100 mg/kg body weight (Schärer and Hummler 1971).

E. Mutagenicity

Sulfonylureas had been regarded for a long time to be neither mutagenic nor genotoxic. However, according to Watson et al. (1976), patients who had been given sulfonylureas, and especially chlorpropamide, displayed a higher level of chromatid aberrations and chromosomal exchange aberrations in lymphocyte cultures than nontreated patients. Therefore, carbutamide, chlorpropamide, tolbutamide, tolazamide, azetohexamide, glimidine, gliquidone, glipizide, glibornuride, glibenclamide and glisoxepide were tested in Chinese hamsters and in mice using the sister chromatid exchange test. Only chlorpropamide and tolbutamide displayed a positive reaction, although they were both negative in the Ames test. For chlorpropamide, the micronucleus test was positive in three strains of mice, and for tolbutamide in one strain. The micronucleus test was negative for both compounds in the Chinese hamster and rat (RENNER and MÜNZINGER 1980).

F. Other Toxicological Studies

The long-term UGDP study (University Group Diabetes Program 1970) was discontinued due to an increased rate of untoward cardiovascular deaths in type 2 diabetic patients who had been treated with tolbutamide. A great number of preclinical studies have been performed in vitro and in vivo using various animal species to elucidate the question of whether the cardiovascular deaths were caused by toxic or exaggerated pharmacological effects of tolbutamide or by unknown causes.

I. In Vivo Studies

Intravenous injections of tolbutamide (20, 40, 80 mg/kg body weight) caused a small increase in blood pressure of anesthetized dogs (WALES et al. 1971). This blood pressure increment was more readily demonstrated after a reduction in blood pressure brought about by antihypertensive drugs. The pressure-increasing effect was not associated with release of insulin and was not mediated by catecholamine release. This vascular effect of tolbutamide was confirmed in conscious dogs (LEE et al. 1988). The mean arterial pressure and left ventricular diastolic pressure increased. Cardiac output was decreased while heart rate, d(LVP)/dt, and regional myocardial performance at the left ventricle were not significantly affected. The pressor effect of norepinephrine was enhanced by pretreatment with 45 mg tolbutamide/kg body weight. In an isolated tissue preparation using ring segments of canine femoral arteries, neither tolbutamide nor its major hepatic metabolites caused any smooth muscle contraction. The cardiovascular effects of chronic tolbutamide administration (250 mg/day orally) were examined in mildly diabetic dogs in comparison to untreated diabetics and five intact dogs with and without tolbutamide administration (WU et al. 1977). After 1 year, resting hemodynamic studies in conscious dogs showed that tolbutamide-treated diabetic dogs had a significantly higher left ventricular end-diastolic pressure associated with normal end-diastolic volume compared with untreated diabetic and intact dogs. Enhanced stiffness of mycoardium appeared to be related to interstitial accumulation of periodic acid Schiff staining material, intensified in tolbutamide-treated diabetic dogs by triglyceride accumulation observed on electron microscopy and by chemical analysis. Investigation of anesthetized rabbits showed that the amount of ouabain necessary to produce ventricular fibrillation and ventricular ectopic beats was significantly decreased by 2 h pretreatment with 50, 100 or 200 mg tolbutamide/kg body wt. (HERMAN et al. 1982). Injection of insulin which lowered the blood glucose concentration did not alter the amount of ouabain that produced ventricular fibrillation and ventricular ectopic beats.

The effects of intravenous injection of carbutamide, gliclazide or tolbutamide on the cardiovascular system were examined in rabbits (BALLAGI-PORDANY et al. 1990). These antidiabetics exerted hypertensive and positive inotropic effects, whereas glibenclamide and glipizide failed to produce any alterations. A 28-week study of diabetic rabbits revealed improved metabolic control after administration of tolbutamide but no "tolbutamide-specific" alterations of ECG, heart rate, peripheral and ventricular blood pressure, myocardial muscle and coronary vessels (KLEISS and RECH 1977).

The effect of various doses of glibenclamide on coronary circulation and myocardial function was investigated in open chest pigs (SCHAD et al. 1993). The aortic blood pressure increased dose dependently. Reduction of coronary flow, flow reserve and myocardial contraction was only observed at many times the maximal therapeutically achievable concentrations of glibenclamide. Cardiovascular alterations, explored in in vivo studies, may possibly be caused by feedback mechanisms induced by the desired hypoglycemic effect of β-cytotropic tolbutamide. Most such indirect effect can be excluded by appropriate in vitro investigations.

II. In Vitro Studies

The effect of tolbutamide on contractibility and the concentration of cyclic AMP in ventricular muscle of the beating rat heart was examined in the non-recirculated Langendorff heart (BROWN and BROWN 1977). The increase in contractility caused by tolbutamide was not preceded by or associated with any change in the concentration of cyclic AMP in the ventricular muscle. The findings suggest that in the beating rat heart the positive inotropic effect was not mediated via an increase of cardiac cAMP. Further investigations showed that tolbutamide does not significantly potentiate the effect of catecholamines on cardiac cAMP. In the isolated rat heart subjected to global ischemia under conditions of low flow of the perfusion medium, it was found that short-term tolbutamide application has a protective effect on ischemic myocardium (SCHAFFER et al. 1981). In isolated working rat hearts, perfused with glucose and acetate, tolbutamide led to a dramatic stimulation in glucose utilization and glycolytic flux (KRAMER et al. 1983).

In an isolated tissue preparation using ring segments of canine femoral arteries neither tolbutamide nor its major hepatic metabolites caused any smooth muscle contraction. The glycolytic flux was associated with a rise in lactate production. The most pronounced effect of tolbutamide was the stimulation of glucose oxidation and as a consequence a dramatically increased oxygen consumption. The contribution of glucose to overall ATP production rose from 8% in the absence of tolbutamide to about 30% in the presence of the sulfonylurea. An enhanced glucose utilization of the isolated, glucose perfused heart was also seen in studies on heart of diabetic rats (TAN et al. 1984). The glycogenolysis as a glucose source was dramatically increased in favor of the preferential use of fatty acids as an energy source. Tolbutamide has been found to protect the ischemic myocardium against irreversible mechanic failure of ischemic rats hearts perfused with glucose, acetate and insulin (LAMPSON et al. 1985).

Investigations on rabbit isolated atrial strips, rabbit isolated atria, cat isolated papillary muscle, canine isolated papillary muscle and intact canine heart revealed that tolbutamide exerts species-dependent effects (CURTIS et al. 1975). It was concluded that tolbutamide possesses positive inotropic effects in rabbits and cats but not in the dog. In intact dog heart, the intracoronary administration of tolbutamide did not lead to disturbances in cardiac rhythm, providing evidence that the sulfonylurea does not increase ventricular automaticity. These data do not support the conclusions of previous investigators concerning possible deleterious cardiac effects of tolbutamide. The thyroid hormone stimulation of the Ca^{2+}-transporting ATPase activity in sarcolemma-enriched rabbit myocardial membranes was shown to be inhibited by clinically achievable concentrations of tolbutamide and tolazamide. In contrast to these sulfonylureas, glibenclamide had no effect (WARNIK et al. 1986). It was concluded from the appropriate studies that inhibitory action of certain sulfonylureas on Ca^{2+}-ATPase is mediated by interference of the binding of calmodulin to cardiac membranes. Studies with ATP-sensitive K^+ (KATP) channels of ventricular myocytes isolated from guinea pig hearts showed that tolbutamide and glibenclamide in an un-ionized form are responsible for closure of KATP channels (FINDLAY 1992). It was concluded that extracellular acidification during ischemia will increase the effective concentration of sulfonylureas and may be responsible for cardiovascular disorders. The left ventricular pressure and lactate dehydrogenase (LDH) release of perfused isolated guinea pig hearts, which were subjected to 15 min hypoxia, were increased under the influence of a very high glibenclamide concentration (MCKEAN and BRANZ 1992). No alterations were seen in experiments under normoxic conditions. It was concluded that ATP-sensitive K^+-channels are important in the protection of the myocardium in hypoxia. In accordance with LEBOVITZ and MELANDER (1992), it is suggested, based on toxico-pharmacological experiments, that an increased cardiovascular mortality in the UGDP study due to administration of tolbutamide is unjustified. However, complete mimicking of all type 2 diabetes related pathophysiological alterations and drug interferences seems to be impossible in animal experiments.

G. Conclusions

Fundamental new knowledge concerning toxic effects of β-cytotropic sulfonylureas in preclinical experiments has not been published since the early 1970s, when Bänder reported on this subject in the first edition of this handbook, including the first extremely potent glibenclamide (glyburide). There is some evidence that the more specific-acting sulfonylurea analogues, glibenclamide, gliclazide, glipzide, gliquidone, glibornuride and glisoxepide, have to be administered in higher doses than carbutamide, tolbutamide, chlorpropamide, tolazamide and acetohexamide in relation to the pharmacologically active doses to produce adverse effects in animals. There is reason to believe that this holds true particularly for deleterious alterations of the thyroid gland, liver, kidneys, heart and hematopoietic system.

References

Achelis JD, Hardebeck K (1955) Über eine neue blutzuckersenkende Substanz. Dtsch Med Wochenschr 80:1452–1455

Ambrogi V, Bloch K, Daturi S, Griggi P, Logemann W, Mandelli V, Parenti MA, Rabini T, Usardi MM, Tommasini R (1971) Pharmacological study on a new oral antidiabetic: N-4-beta-(5-methylpryrazine-2-carboxamido)ethylebenzenesulfonyl-n'-cyclohexylurea or K4024. Arzneimittelforschung 21(2):208–215.

Baeder C, Sakaguchi T (1969) Teratologische Untersuchungen mit HB 419, Arzneimittel Forsch 1419–1420

Bänder A (1971) Zur Pharmakologie und Toxikologie der blutzuckersenkenden Sulfonamide. Handbuch der experimentellen Pharmakologie. Springer, Berlin Heidelberg New York, pp 318–401

Ballagi-Pordány G, Koltai M-Z, Aranyi Z, Pogátsa G (1990) Direct cardiovascular effect of hypoglycemic sulphonylurea compounds. Diabetologia 33 [Suppl]:A53

Brown JD, Brown ME (1977) The effect of tolbutamide on contractility and cyclic adenosine 3':5'-monophosphate concentration in the intact beating rat heart. J Pharmacol Exp Ther 200:166–173

Curtis GP, Setchfield J, Lucchesi BR (1975) The cardiac pharmacology of tolbutamide. J Pharmacol Exp Ther 194:264–273

Duhault J, Boulanger M, Tisserand F, Beregi L (1972) The pharmacology of S 1072, a new highly effective oral antidiabetic drug with unusual properties. Arzneimittelforschung (Drug Res) 22:1682–1685

Dulin WE, Oster HL, McMahon FG (1961) A new high potency antidiabetic sulfonylurea [n-(1-hexahydro-1-azepinyl)-N'-p-tolylsulfonylurea]. Proc Soc Exp Biol 107:245–248

Findlay I (1992) Effects of pH upon the inhibition by suphonylurea drugs of ATP-sensitive K$^+$ channels in cardiac muscle. J Pharmacol Exp Ther 262:71–79

Hebold G (1971) Experimentelle, geschlechtsdifferente Beeinflussung der Langerhans'schen Inseln des Pankreas. Habilitationsschrift, University of Heidelberg

Hebold G, Scholz J, Schütz E, Czerwek H, Sagaguchi T, Brunk R, Notdurft H, Kief H, Bäder C, Hartig F (1969a) Experimental investigations of the new sulfonylurea-derivate glibenclamide Hb 419. Horm Metab Res 1:4–10

Hebold G, Scholz J, Schütz E, Czerwek H, Brunk R (1969b) Verträglichkeitsprüfungen von Hb 419 im Tierversuch. Arzneimittelforschung 19:1404–1413

Herman EH, Krop S, Jordan W (1982) Tolbutamide enhancement of ouabain cardiotoxicity in rabbits. Pharmacology 24:111–117

Hoffmann-La Roche (1985) Basic documentation on Glutril. Hoffmann-La Roche, Basel, pp 1–15

Iida H, Kast A, Tsunenari Y (1976) Studies on the teratogenicity of a new sulfonylurea derivative (ARDF 26 SE) in rats and rabbits. Pharmocometrics 11:119–131

Kast A, Tsunenari Y, Honma M, Nishikawa J, Shibata T, Torii M (1975) Toxicological tests of a new sulfonylurea derivate (ARDF 26 SE) in rats, mice and rabbits. Pharmacometrics 10:383–394

Kiso To Rinsho (1972) Clinical report (Yubunsh Co). 6:1925

Kleiss D, Rech M (1977) Langzeituntersuchungen an alloxandiabetischen Kaninchen — Stoff-wechselverhalten, Allgemeinbefinden, Kreislaufuntersuchungen und Organbefunde — unter chronischer Gabe von Tolbutamid und Methoxy-Methyl-Indolcarbonsäure. Dissertation, University of Heidelberg

Krall LP (1984) Glyburide (DiaBeta®): a new second generation hypoglycemic agent. Clin Ther 6(6):746–762

Kramer JH, Lampson WG, Schaffer SW (1983) Effect of tolbutamide on myocadial anergy metabolism. Ann J Physiol 245:H313–H319

Kramer M, Hecht G, Lanecker H, Harwart A, Richter KD, Gloxhuber C (1964) Pharmako-logie des 2-Benzolsulfonamido-5(β-methoxy-äthoxy)-pyrimidins (Glycodiazin), einer neuen blutzuckersenkenden Verbindung. Arzneimittelforschung 14:377–385

Lampson WG, Kramer JH, Schaffer SW (1985) Effect of tolbutamide on myocardial energy metabolism of the ischemic heart. Biochem Pharmacol 34:803–809

Lebovitz HE, Melander A (1992) Sulfonylureas: basic aspects and clinical uses. In: Alberti KGMM, DeFronzo RA, Keen H, Zimmet P (eds) International textbook of diabetes mellitus. Wiley, New York

Lee KC, Wilson RA, Randall DC, Altiere RJ, Kiritsy-Ray JA (1988) An analysis of the haemodynamic effects of tolbutamide in conscious dogs. Clin Exp Pharmacol Physiol 15:379–390

Lorch E, Gey KF, Sommer P (1972) Glibornurid, ein neues hochwirksames Antidiabetikum Pharmakologische und biochemische Vergleichsuntersuchungen an verschiedenen Tierspezies und an tierexperimentellen Modellen. Arzneimittelforschung (Drug Res) 22(12a):2154–2163

Maha GE, Kirtley WR, Root MA, Anderson RC (1962) Acetohexamide, preliminary report on a new oral hypoglycemic agent. Diabetes 11:83–90

McKean, Branz AJ (1992) Influence of ATP-sensitive potassium channel blocker on hypoxia-induced damage of isolated guinea pig heart. Gen Pharmacol 23:921–923

Mitsukami K, Myamoto M, Hayashi S, Kobayashi T, Sukurai M, Sakagushi T (1969) Toxikologische Untersuchungen von N-4-[2-(5-cholor-2-methoxy-benzamido)-acetyl]phenylsulfonyl-N-cyclohexylharnstoff (Hb 419). Arzneimittelforschung 19:1413–1419

NCI report (1977a) Bioassay of acetohexamide for possible carcinogenicity (CAS no 968 81-0)

NCI report (1977b) Bioassay of tolazamide for possible carcinogenicity (CAS no 1156-19-0)

NCI Carcinogenesis Bioassay (1978) Chlorpropamide (NCITR* NCI-CG-TR-45,78); tolbutamide (NCITR* NCI-CG-TR-31,77)

Pfeiffer EF (1984) Are the "second generation" oral hypoglycemic agents really different. Acta Diabetol Lat 21(1):1–32

Renner HW, Münzinger R (1980) Mutagenicity of sulphonylureas. Mutat Res 77:349–355

Schad H, Heimisch W, Maier-Rudolph W, Mendler M (1993) Effect of glibenclamide on myocardial blood flow and function. Eur J Physiol 422 [Suppl 1]:R111

Schaffer SW, Poole CG, Lampson WG, Kramer JH (1981) Effect of tolbutamide on the mechanical function of the isolated rat heart subjected to global ischemia. J Mol Cell Cardiol 13:341–345

Schärer K, Hummler H (1971) Toxicological experiments with drugs of the sulfonylurea type in animals. In: Dubach UC, Bückert A (eds) Recent hypoglycemic sulfonylureas, mechanisms of action and clinical indications. Huber, Bern, pp 163–170

Schneider JA, Salgado ED, Jäger D, Delahunt C (1959) The pharmacology of chlorpropamide. Ann N Y Acad Sci 74:427–442

Scholz J, Bänder A (1956) Über die orale Behandlung des Diabetes mellitus mit N-[4-Methylbenzolsulfonyl]-N'-butylharnstoff (D860), Pharmakologie. Dtsch Med Wochenschr 81:825–826

Tan BH, Wilson GL, Schaffer SW (1984) Effect of tolbutamide on myocardial metabolism and mechanical performance of the diabetic rat. Diabetes 33:1138–1143

Tettenborn D (1974) Zur Toxikologie von Glisoxepid, einem neuen oralen Antidiabetikum. Ergebnisse der Tierversuche. Arzneimittelforschung (Drug Res) 24:409–418

University Group Diabetes Program (1970) A study of the effects of hypoglycemic agents on vascular complications in patients with adult onset. Diabetes 19 [Suppl II]:789–830

Wales JK, Grant AM, Wolff FW (1971) The effect of tolbutamide on blood pressure. J Pharmacol Exp Ther 178:130–140

Warnik PR, Davis FB, Davis PJ, Mylotte KM, Blas SD (1986) Differential activities of tolbutamide, tolazamide, and glyburide in vitro on rabbit myocardial membrane Ca^{2+}-transporting ATPase activity. Diabetes 35:1044–1048

Watson WAF, Petrie JC, Galloway DB, Bullock I, Gilbert JC (1976) In vivo cytogenetic activity of sulphonylurea drugs in man. Mutat Res 38:71–80

Wu CF, Haider B, Ahmed SS, Oldewurtel HA, Lyons MM, Regan TJ (1977) The effects of tolbutamide on the myocardium in experimental diabetes. Circulation 55:200–205

Yasuba M, Matsuoka N, Iida M, Maeda K, Nishiwaki T, Ueda N, Ohnishi K, Tatsumi H, Hashimoto M (1981). Acute, subacute, and long-term toxicity studies of gliclazide in diabetic animals. Yakuri To Chiryo 9(11):4497–4520

CHAPTER 9

Clinical Pharmacology of Sulfonylureas

L. Groop and G. Neugebauer

A. Pharmacodynamics

I. Mode of Action of Sulfonylureas

1. Effects on Insulin Secretion

In vitro studies using the perfused rat pancreas and in vivo studies using the hyperglycemic clamp have demonstrated that sulfonylureas stimulate insulin secretion in a biphasic fashion (Loubatieres 1957; Malaisse et al. 1972; Grodsky et al. 1977; Groop et al. 1987b). The insulinotropic effect of sulfonylureas is augmented by glucose, and sulfonylureas have therefore been proposed to increase B-cell sensitivity to glucose and non-glucose stimuli rather than to increase the synthesis of insulin by the pancreatic B cell (Basabe et al. 1976; Pfeifer et al. 1980; Dunbar and Foá 1974; Grodsky et al. 1977). Sulfonylureas close ATP-dependent potassium channels, which, in turn, results in depolarization of the B cell and influx of calcium (Sturgess et al. 1985; Boyd AE III 1988). The final result is stimulation of insulin secretion. Sulfonylureas bind to receptor-like structures on the B cell, which may be closely linked to or be part of the potassium channels (Schmid-Antomarchi et al. 1987; Gaines et al. 1988; Siconolfi-Baez et al. 1990). The binding capacity of different sulfonylureas reflects their ability to stimulate insulin secretion.

a) First-Phase Insulin Secretion

Glucose stimulates insulin secretion in a biphasic fashion with a rapid first-phase burst (0–10 min) and a more slowly evoking second-phase secretion. Manifest non-insulin-dependent diabetes mellitus (NIDDM) is characterized by loss of first-phase insulin secretion and impairment of the second phase (Perley and Kipnis 1966; Simpson et al. 1968; Porte 1991). In the early stages of NIDDM, only first-phase insulin secretion is impaired. As the transition from impaired glucose tolerance (IGT) to manifest NIDDM is associated with loss of first-phase insulin secretion, restoration of first-phase insulin secretion could be of therapeutic value in the prevention of NIDDM (Groop et al. 1993b). Glucose stimulates first-phase insulin secretion in a log-linear fashion in healthy subjects, but has no effect in NIDDM subjects (Groop et al. 1991a; Fig. 1). In healthy subjects sulfonylureas shift the dose-response curve to the left. Most studies have failed to demonstrate any significant effect of sulfonylurea on first-phase insulin secretion in patients with manifest NIDDM (Shapiro et al. 1989; Groop et al. 1991a; Hosker et al. 1989; Groop et al. 1993a,b). The results from two studies with gliclazide seem at first glance somewhat more promising. However, the authors reported an effect on C peptide but not on insulin concentrations. Secondly, a considerable part of the increase in first-phase insulin concentration was

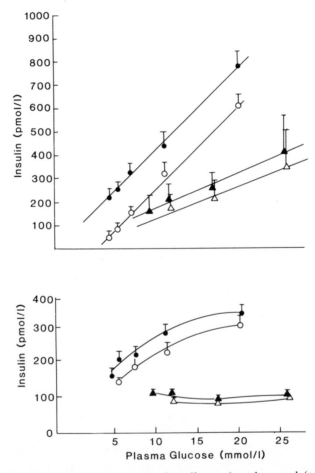

Fig. 1. Dose-response curve relating the first (*bottom*) and second (*top*) phases of plasma insulin responses to ambient plasma glucose concentration in 12 healthy subjects (*circles*) and 6 NIDDM (*triangles*) subjects. *Open circles and triangles* denote values without and *filled circles and triangles* values with prior intake of glipizide. Values are mean ± SEM. (From Groop et al. 1991a)

due to an increase in basal insulin secretion. Third, the patients included had relatively mild diabetes with preserved first-phase insulin secretion. It is possible that sulfonylurea may exert such an effect in patients with mild diabetes and preserved insulin secretion. Indirect support for this view comes from studies in healthy subjects and in patients with mild diabetes (GROOP 1993a; BITZÉN et al. 1988). The lower the fasting blood glucose value, the more sulfonylurea has been shown to enhance early insulin release during an oral glucose load. In contrast to the findings in healthy subjects (GROOP 1993a), i.v. administration of glipizide at the beginning of a test meal to patients with frank diabetes could not restore the early insulin secretion (GROOP 1993b). Importantly, overreplacement of the early inulin peak by i.v. administration of insulin was also unable to achieve this (GROOP 1993b). Postprandial hyperglycemia in patients with frank diabetes, therefore, seems to be due to insulin resistance, which cannot be overcome by early timing of insulin or the insulin secretagogue. In keeping with this, restoration of first-phase insulin secretion by i.v. insulin in patients with IGT did not improve peripheral glucose uptake (WIDÉN 1993). In this regard the total amount of insulin seems more important than timing of insulin.

b) Second-Phase Insulin Secretion

In both normal and NIDDM subjects, second-phase insulin secretion increases linearly with the ambient plasma glucose. Sulfonylureas shift the dose-response curve to the left without changing the slope of the curve (Fig. 1), suggesting that sulfonylureas enhance B-cell responsiveness rather than change B-cell sensitivity to glucose. In practice this means that more insulin is released at every glucose level, but that the effects of glucose and sulfonylurea on second-phase insulin secretion are additive, not synergistic (GROOP et al. 1987b, 1991a). The clinical consequences are that sulfonylureas can be expected to increase insulin secretion and lower blood glucose concentrations to the same extent at high and low blood glucose concentrations.

c) Chronic Effects on Insulin Secretion

The idea of extra pancreatic effects of sulfonylureas originated in the fact that, during chronic treatment, sulfonylureas were shown to lower blood glucose without a concomitant increase in insulin concentrations (REAVEN and DRAY 1967; CHU et al. 1968, DUCKWORTH et al. 1972; BARNES et al. 1974; SHENFIELD et al. 1977; JUDZEWITSCH et al. 1982; GROOP et al. 1987b). The diabetic patients were restudied at a time when plasma glucose concentration was significantly lower than the pretreatment level. Instead, if the plasma glucose concentration is raised to the pretreatment level, the insulin response to glucose and other stimuli is significantly enhanced by sulfonylurea therapy (GROOP et al. 1987b). Recently, BIRKELAND et al. (1994) provided conclusive evidence that insulin secretion is increased after 15 months of sulfonylurea therapy, demonstrating a leftward shift of the dose-response curve relating plasma glucose and insulin concentrations before and after 3 and 15 months of sulfonylurea therapy (Fig. 2). Using the deconvolution method, SHAPIRO et al. (1989) showed that the total amount of insulin secreted during 24 h increased by 25% during glibenclamide therapy, although the insulin secretion rate did not change. Pulsatile insulin secretion was uninfluenced by glibenclamide in that study. Taken together the data

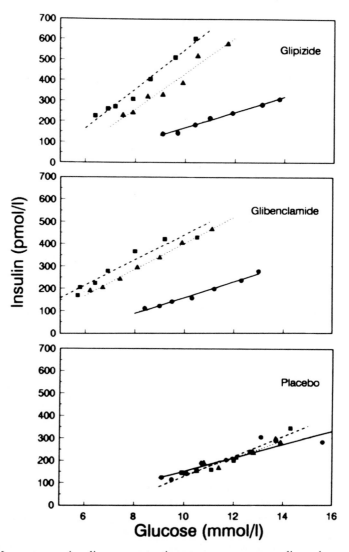

Fig. 2. Mean serum insulin concentrations versus corresponding plasma glucose concentrations at 60–240 min during the test meal before (*circles*) and after 3 (*squares*) and 15 (*triangles*) month of treatment. (From BIRKELAND et al. 1994)

indicate that glucose is a stronger insulin secretagogue than sulfonylurea. Lowering of glucose concentrations by sulfonylurea therapy masks the insulinotropic effect of the compound, which is maintained even during chronic treatment.

II. Hepatic Insulin Clearance

Systemic availability of insulin is dependent not only upon secretion from the pancreas but also on hepatic insulin clearance. There are some data to suggest that sulfonylurea may reduce hepatic insulin extraction and thereby increase peripheral insulin concentrations (MARSHALL et al. 1970; SCHEEN et al. 1988; GROOP et al. 1988; SIMPSON et al. 1990). Sulfonylureas significantly diminish uptake of insulin by the perfused rat liver (MARSHALL et al. 1970). The in vivo data are based upon discrepant peripheral concentrations of insulin and C peptide or upon measurements of clearance of exogenous insulin during insulin clamp studies after administration of sulfonylureas. Given the different half-lives of insulin and C peptide, the ratio of their peripheral concentrations can be used as a measure of hepatic insulin clearance only if the concentrations of both peptides reach baseline levels during follow-up (POLONSKY and RUBENSTEIN 1984). Hepatic insulin clearance also tends to decrease when glucose control is improved by insulin (GARVEY et al. 1985). In contrast, hepatic insulin extraction increase after weight reduction (HENRY et al. 1988). Evidence against a significant effect of sulfonylurea on hepatic insulin clearance was recently provided by WIDÉN (1993). When potential confounding effects of the sulfonylurea on endogenous insulin secretion were excluded by infusion of somatostatin, glibenclamide infusion did not decrease the hepatic extraction of exogenous insulin. It seems thus likely that the major effect of sulfonylurea is to stimulate insulin secretion, not to decrease its clearance.

III. Other Pancreatic Effects

Sulfonylureas have been proposed to reduce glucagon concentrations in vitro and in vivo (SAMOLS et al. 1969; ÖSTENSON et al. 1986; KAJINUMA et al. 1974; TSALIKIAN et al. 1977; PFEIFER et al. 1983), but the presence of such an effect has also been questioned (UNGER et al. 1970; PEK et al. 1972). On the other hand, the effect on glucagon secretion could be the consequence rather than the cause of improved glucose control. The physiologic importance of sulfonylurea-induced reduction in glucagon concentrations seems minor.

At low glucose concentration, glibenclamide was shown to stimulate somatostatin release in the isolated, perfused rat pancreas, whereas the effect at higher glucose concentrations was absent (EFENDIC et al. 1979). The importance of this finding remains uncertain.

IV. Extrapancreatic Effects

Insulin resistance is a characteristic feature of NIDDM, obesity, hypertension and associated conditions, also referred to as syndrome X or the insulin resistance syndrome (Reaven 1988). Impaired activation of glycogen synthase, resulting in impaired glycogen formation in skeletal muscle, represents one of the major defects causing insulin resistance in NIDDM (Eriksson et al. 1989). Insulin resistance has also been ascribed a major role in the sequence of events leading to NIDDM (Martin et al. 1992; Lillioja et al. 1988; Eriksson et al. 1989). Prevention and treatment of NIDDM should therefore include means not only to improve insulin secretion but also to improve insulin sensitivity. Lowering of plasma glucose at unchanged plasma insulin levels was difficult to explain without assuming that sulfonylureas would improve insulin sensitivity (Feldman and Lebovitz 1969b). Several studies have shown improved insulin sensitivity (Groop et al. 1987b; Olefsky and Reaven 1976; Beck-Nielsen et al. 1979; Kolterman et al. 1984; Simonson et al. 1984; Groop et al. 1985) and insulin receptor binding (Olefsky and Reaven 1976; Beck-Nielsen et al. 1979; Kolterman et al. 1984) during chronic sulfonylurea therapy. In the presence of insulin, tolbutamide stimulates glucose transport in cultured muscle cells (Rogers et al. 1987). Similarly, tolazamide potentiates hexose transport in adipose tissue (Maloff and Lockwood 1981). Gliclazide was recently shown to increase insulin-stimulated glycogen synthase activity in skeletal muscle (Johnson et al. 1991). With the notion that impaired glycogen synthesis in skeletal muscle represents an early and consistent finding in NIDDM, therapeutic activation of its key enzyme, glycogen synthase, could be of importance. However, there was no correlation between the degree of improvement in glycemic control and the improvement in response of the glycogen synthase enzyme. A reduction in fasting plasma glucose is always related to a reduction in hepatic glucose production. Consequently, it has been tempting to search for hepatic effects of sulfonylureas (Recant and Fischer 1957; Kolterman 1987). Chronic sulfonylurea therapy reduced hepatic glucose production and thereby fasting plasma glucose in NIDDM patients with only minor changes in the fasting plasma insulin concentration (Kolterman et al. 1984; Best et al. 1982). Glipizide increased significantly the number of insulin receptors on rat hepatocyte plasma membranes (Feinglos and Lebovitz 1978); and sulfonylureas have been ascribed an inhibitory effect on hepatic glucose output from gluconeogenesis and glycogenolysis in the perfused rat liver (Blumenthal and Whitmer 1979; Patel 1986a; Davidson and Sladen 1987). It was proposed that part of this effect could be mediated through inhibition of long-chain fatty acid oxidation in the liver (Patel 1986b). It is, however, difficult to distinguish in vivo between direct or indirect (through small increments in portal insulin concentrations) effects of sulfonylureas on the liver. There are more arguments against than in favor of significant extrapancreatic effects of sulfonylureas:

(1) Sulfonylureas do not reduce plasma glucose in pancreatectomized animals or in patients with type 1 diabetes (MIRSKY et al. 1956; HOUSSAY and PENHOS 1956; GRUNBERGER et al. 1982; RATZMANN et al. 1984), or in NIDDM patients made insulinopenic with somatostatin (WIDÉN 1993). (2) Early studies examining the extrapancreatic effects of sulfonylureas did not appreciate the problem with restudying patients at identical plasma glucose levels. (3) Many of the conclusions about hepatic effects of sulfonylureas were based upon unchanged peripheral insulin concentrations. Given the extreme sensitivity of liver for small changes in the insulin concentration, it is quite possible that small changes in portal insulin concentrations have been overlooked. (4) There have also been difficulties in confirming initial reports of increased insulin binding during sulfonylurea therapy (GRUNBERGER et al. 1982; RATZMANN et al. 1984; DOLAIS-KITABGI et al. 1983). (5) Finally, and probably most important, chronic hyperglycemia can cause both peripheral and hepatic insulin resistance, which can be attenuated by improved glucose control independently of the mode of treatment (YKI-JÄRVINEN 1990; ROSSETTI et al. 1987). ROSSETTI and coworkers (1987b) demonstrated attenuation of insulin resistance when hyperglycemia was reduced by phlorhizin in partially pancreatectomized rats. Given the above evidence, extrapancreatic effects of sulfonylureas are hardly of clinical significance for their antidiabetic effect.

V. Combination of Insulin and Sulfonylurea

Combination therapy with insulin and sulfonylurea was used in the 1950s (FRIEDLANDER 1957; LAZARUS and VOLK 1959). The benefit of adding an insulin secretagogue to insulin was questioned in the 1960s (JACKSON 1969). It was recommended to increase the amount of exogenous insulin. The concept of extrapancreatic effects of sulfonylureas was raised in the 1970s (FELDMAN and LEBOVITZ 1969), and we thought it might be useful to add a compound that could potentiate the effect of insulin (GROOP and HARNO 1979). Since then evidence has accumulated that has led us to reevaluate this concept (GROOP et al. 1990). Although several studies have demonstrated beneficial effects of such combination therapy in NIDDM patients, the average effect has been rather modest and observed in only about half of the patients (GROOP et al. 1990; PETERS and DAVIDSON 1991). However, there are several problems with most studies in this field, including small sample size, selection of patients and simultaneous use of several end points. First, residual B-cell function has been considered as a prerequisite for a beneficial effect of sulfonylurea. Inclusion of patients with impaired B-cell function will therefore attenuate the effect. Second, most studies have used glycemic control as the end point. Nevertheless, the insulin dose has been reduced by about 30% to avoid hypoglycemia. Such studies have, in fact, compared treatments with a smaller insulin dose plus sulfonylurea with a larger insulin dose without sulfonylurea. A recent study also demonstrated less weight

gain and lower insulinemia with the combination of bedtime neutral protamine Hagedorn (NPH) insulin and oral agents (including both sulfonylurea
and metformin) compared with treatment with insulin twice or four times
daily (YKI-JÄRVINEN et al. 1992). In addition, patients with bedtime insulin
in combination with sulfonylurea also gained less weight and had lower
levels of peripheral insulin concentrations than those who received NPH
insulin in the morning in combination with sulfonylurea (YKI-JÄRVINEN et al.
1992). We were unable, however, to find any advantage of bedtime over
morning NPH insulin in combination with sulfonylurea (GROOP et al. 1992).
Instead, in most patients the combination of NPH insulin with sulfonylurea
was not effective enough to reduce blood glucose concentrations to the
treatment goals.

The patient most likely to benefit from combination therapy is slightly
obese, has a relatively short duration of NIDDM, and has preserved B-cell
function. In such a patient, combined-insulin sulfonylurea therapy predominantly stimulates basal insulin secretion, resulting in more effective
suppression of hepatic glucose production and lower fasting plasma glucose.
Therefore, combination therapy generally improves fasting blood glucose
more than postprandial blood glucose concentrations (STENMAN et al. 1988).
The side effects are few, most notably more frequent but mild hypoglycemic
reactions (STENMAN et al. 1988; CASNER 1988).

VI. Rational Use of Sulfonylurea Drugs

Given the importance of insulin resistance in the pathogenesis of NIDDM,
we need agents which can restore insulin sensitivity. On the other hand,
impaired insulin secretion is always present in patients with manifest NIDDM.
The only available agent known to influence insulin sensitivity without
affecting insulin secretion is metformin. However, the overall effect of
metformin on glycemic control seems to be smaller than that of sulfonylureas
(WIDÉN and GROOP 1994). Although a number of "insulin sensitizers" are
under investigation (BAILEY and FLATT 1990), sulfonylureas and metformin
will remain the cornerstones of oral therapy in NIDDM for years.

The use of these agents could be more rational than it has been so far.
We tend to start treatment relatively late, and continue long after the
decision to switch to insulin should have been taken. Sulfonylureas should
be used when they are effective and we should stop sulfonylurea treatment
when they are no longer able to achieve the treatment goal. This requires a
change in the treatment policy. To keep on increasing the drug dose is
usually only an excuse to postpone the decision to start insulin therapy. We
should actively try to normalize blood glucose concentrations in all patients
with NIDDM, including those with the early stages of the disease. Since this
can be achieved with diet alone in less than 30% (WALES 1982; HENRY et al.
1985; UKPDS 7, 1990), we need pharmacotherapy. In patients with signs or
symptoms of the insulin resistance syndrome, i.e., abdominal obesity,

hypertension, lipid disorders, metformin seems at the moment to be the drug of choice. In the majority of patients sulfonylureas will still be needed. How early the agents should be initiated is still an open question. The question whether oral agents such as sulfonylurea and metformin or acarbose can prevent progression from IGT to manifest diabetes deserves much more attention than so far (Sartor et al. 1980d; Jarrett et al. 1979; Papoz et al. 1978). The combination of insulin and sulfonylurea provides some advantages in the treatment of patients failing on sulfonylureas alone, but one should avoid running into a situation of polypharmacy. Therefore, it may be better to stop treatment with sulfonylurea before initiation of insulin therapy and later on add sulfonylurea again if the insulin dose tends to rise or the treatment goals are not achieved. In this case we know what insulin does and what sulfonylurea adds on top of that.

B. Pharmacokinetics

I. Similarities and Differences

The hypoglycemic sulfonylureas are all weak acids with pK_a values between 4.9 and 6.5 and therefore completely ionized at physiologic pH (Balant 1981; Ferner and Chaplin 1987; Marchetti and Navalesi 1989; Gerich 1989; Melander et al. 1989). Like other organic acids they are highly bound to plasma proteins, mainly albumin (Crooks and Brown 1974), which explains their distribution characteristics of small distribution volumes between 0.1 and 0.25 l/kg. All compounds are nearly completely absorbed, the lowest extent being at least 80%. Absorption rates, however, differ between compounds and also depend on formulation (Haupt et al. 1984; Olson et al. 1985). No relevent first-pass effect has been demonstrated with any sulfonylurea.

Whereas the majority have a short to intermediate elimination half-life of 1–10 h, only carbutamide and chlorpropamide exhibit half-lives of 40 h and 42 h, respectively. Hepatic metabolism is the predominant if not exclusive pathway of elimination, whereas the metabolites are excreted mainly with urine. In some agents such as carbutamide, chlorpropamide, glipizide, glisoxepide and tolazamide, renal excretion of the parent compound contributes to elimination of between 10% and 50% and therefore elimination may vary depending on urine pH (Neuvonen and Kärkäinen 1983).

With the exception of three agents, i.e., acetohexamide, glymidine, tolazamide, and also recently suggested glibenclamide, active metabolites generated during biotransformation do not contribute to the hypoglycemic action unless impaired renal function leads to accumulation of mainly less potent active metabolites. Although systemic clearance of sulfonylureas is low with a range of 14–100 ml/min, half-lives are short due to the small apparent distribution volume. An overview on different pharmacokinetic variables from the available literature is presented in Table 1 without reference to individual sources. A detailed description of each sulfonylurea drug is given in the following chapters in this volume. The chemical structures of the different agents are shown in Fig. 3, including one which is not contained in Table 1. Drugs which have nearly exclusive importance in one country only are carbutamide (France), glisoxepide (Germany), glymidine (Japan), tolazamide (United States), glyclopyramide (Japan) and glipentide (Spain). Glibenclamide, gliclazide, glipizide, chlorpropamide and tolbutamide are available in all major markets.

II. Timing of Drug Intake

The acute insulin release in response to meals is impaired in patients with NIDDM. The prolonged hyperglycemia in these patients is a major risk factor for the outcome of this disease. It is therefore highly desirable with any therapeutic regimen to normalize this pathologic reaction. In order to achieve this therapeutic goal, oral hypoglycemic sulfonylureas have to be administered relative to meals according to their pharmacodynamic and pharmacokinetic properties to lower postprandial glucose excursions. In theory, rapid-acting agents with a pharmacokinetic profile independent of food intake should be preferred. The absorption rates vary among the different sulfonylureas, with the highest value being observed probably for glipizide (Wahlin-Boll et al. 1982a), but similar also for gliquidone (Kopitar and Koss 1975a,b), glibenclamide (Haupt et al. 1984; Groop et al. 1985; Neugebauer et al. 1985), tolbutamide (Sartor et al. 1980b), tolazamide (Welling et al. 1982) and lowest for gliclazide (Campbell et al. 1980; Forette et al. 1982) and chlorpropamide (Taylor 1972; Sartor et al. 1980c). The absorption rate has been found to differ between formulations of the same compound, i.e., with tolbutamide (Olson et al. 1985), chlorpropamide (Monro and Welling 1974), glibenclamide (Haupt et al. 1984), tolazamide (Welling et al. 1982), gliquidone (Kopitar and Koss 1975a,b) and glipizide (Haaber et al. 1993), and it is therefore of relevance which formulation is used to achieve improvement of postprandial glycemia. A further complicating factor is the influence of food itself on rate and extent of absorption. Food delays the absorption of chlorpropamide (not the extent) and glipizide (Sartor et al. 1980c; Wahlin-Boll et al. 1980), but not of tolbutamide (Sartor et al. 1980c; Antal et al. 1982). Equivocal findings have been reported for glibenclamide indicating no influence (Sartor et al. 1980a; Coppack et al. 1990), or delayed absorption in type 2 diabetic patients (Grill et al. 1986). In addition, dietary fibers such as glucomannan and guar may have different effects depending on the formulation administered. With nonmicronized formulations they delay absorption, whereas with micronized formulations they do not (Neugebauer et al. 1983; Shima et al. 1983). All these different effects have not been systematically explored with respect to their influence on postprandial glycemia. In general it has been found that

Table 1. Synopsis of pharmacokinetic properties of sulfonylurea drugs in humans

Drug	pK_a	Protein binding (%)	Elimination half-life (h)	Systemic clearance (ml/min)	Distribution volume (l/kg)	Urinary recovery (% dose)	Active metabolites	Absorption (%)
Acetohexamide		90	0.8–2.4 P / 4–6 M	116[a]	0.20[a]	74 M, 1–2 P	1[b]	~100
Carbutamide		50–60	40	8	0.26[a]	50 M, 40 P		95–98
Chlorpropamide	4.9	96	33–50	1.8–3.3	0.11–0.18	50–80 M, 20–50 P	2[c]	100
Glibenclamide/ glyburide	6.5	97–99.8	1.3–15	50–170	0.13–0.20	50 M	2[d]	100
Glibornuride		94–95	9.2	24	0.27	60–72 M	3–4[e]	91
Gliclazide	5.8	85–99	8–17	13–26	0.19–0.26	61–81 M, <1 P		80
Glimepiride		99.5	1.3–3.4	48	0.18[a]	50 M, 0.1–0.4 P	1[f]	100
Glymidine	5.7	89	3–6	30–50	0.15	93 M, 1 P	1[d]	89–90
Gliquidone		99	1.4	90[a]	0.16	5 M, 0.2–0.5 P	1[e]	80–100
Glipentide		98	4	30[a]	0.15			
Glipizide	5.9	98–99.4	2–9	33–52	0.12–0.26	60–77 M, 10 P	?	100
Glisoxepide		93	1.7–3	65[a]	0.20	33 M, 43 P		96
Tolbutamide	5.3	95–98	7–9	12–17	0.10–0.15	80–90 M, 0.1–0.2 P	2[e]	85–100
Tolazamide	5.7	98	3–7		0.16[a]	79 M, 6 P	3[g]	85

P, parent drug, M, metabolites(s).
[a] Estimated value.
[b] 240% activity.
[c] 100% activity in vitro.
[d] 100% activity.
[e] Weak activity.
[f] 80%–100% activity.
[g] 5%, 20% and 70% activity.

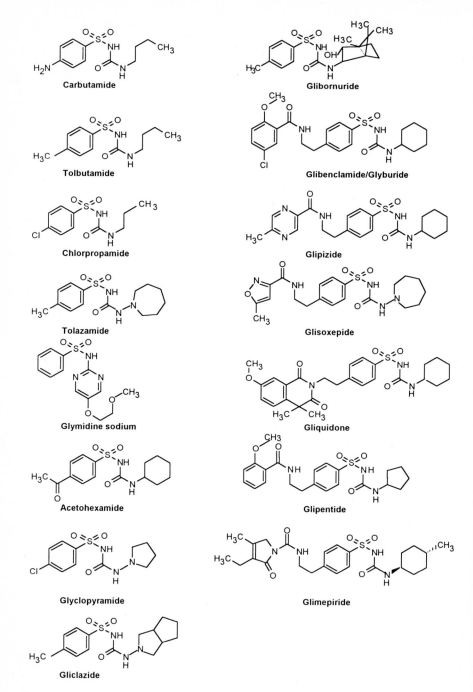

Fig. 3. Structures of sulfonylurea agents in clinical use

the intake of glipizide (SARTOR et al. 1978), glibenclamide (SARTOR et al. 1982) and tolbutamide (SAMANTA et al. 1984) 30 min before a meal leads to lower plasma glucose concentrations than when the drug is taken together with a meal. On the other hand, during chronic treatment in two studies no difference could be found between prior and concomitant drug to meal administration on postprandial glucose (BATCH et al. 1990; FABER et al. 1990). In addition, with compounds such as carbutamide and chlorpropamide, which accumulate extensively, the timing of the dose seems to be irrelevant. Furthermore, in a recent study in patients with manifest NIDDM, the enhanced absorption rate of glipizide from a soft gelatin capsule and even the immediate concentration rise following i.v. administration before a meal produced no greater effect on post-breakfast hyperglycemia and only little difference on hyperinsulinemia than the more slowly absorbed tablet (HAABER et al. 1993; GROOP et al. 1993b). This questions to some extent the importance of formulation differences at least in manifest NIDDM patients, which may be in contrast to those in newly detected NIDDM.

III. Effect of Hyperglycemia

Increasing levels of plasma glucose during a hyperglycemic clamp procedure are able to reduce plasma glipizide concentrations in healthy subjects (GROOP et al. 1989a). This finding was confirmed in NIDDM patients, where a parallel reduction of glipizide concentrations with increasing glucose levels was demonstrated. This effect is most likely related to delayed gastric emptying, since it was shown that hyperglycemia causes concentration-dependent prolongation of gastric transit time in patients with IDDM (FRASER et al. 1990). Since a similar study with glibenclamide confirmed the aforementioned results (HOFFMANN et al., to be published), it can be assumed that not only the presence of diabetic autonomic neuropathy (IKEGAMI et al. 1986) but also hyperglycemia per se can reduce the absorption rate of these agents. It is even conceivable that in some NIDDM patients with primary or secondary failure to drug response this may be attributable to pronounced hyperglycemia.

IV. Dose- and Concentration-Response Relationship

There is no simple way to describe dose- or concentration-response relationships with any sulfonylurea drug, because different effects, namely an increase in insulin secretion (cf. Chap. 1, this volume) or fasting and postprandial plasma glucose concentrations, can be taken as the target parameter. Confounding factors in the relation between dose and response are the amount and type of diet, time of food intake and severity of disease state. Furthermore, long-term continuous drug exposure may lead to a state of hyporesponsiveness, which has so far only been observed for tolazamide (KARAM et al. 1986) and not, for example, for glibenclamide (HOLLIS et al.

1991). There is also evidence that the dose-response curve of sulfonylureas may be bell-shaped, with impairment of metabolic control appearing at very high drug concentrations (Wahlin-Boll et al. 1982b). Contribution of active metabolites to the hypoglycemic action of certain compounds such as acetohexamide (Galloway et al. 1967) or glymidine (Gerhards et al.1964) may obscure the concentration-response relationship.

All these various factors render it rather difficult if not impossible to demonstrate a firm relation between the plasma drug concentration (or dose) and its effect. Under highly standardized conditions as in healthy volunteers (Haupt et al. 1972) or by means of the euglycemic clamp technique (Groop et al. 1991b), this goal may, however, be achievable.

1. First-Phase Insulin Secretion

Only little systematic information is available on dose or concentration dependence on first-phase insulin response. Populations studied under different conditions provide only indirect evidence for this relationship. In a study by Stenman et al. (1993), the pharmacokinetic and pharmacodynamic properties of glipizide were determined in 23 NIDDM patients with doses of 10 mg o.d. and 10 and 20 mg b.i.d. over 3 months. Under this dose schedule, trough concentrations increased from 33 nmol/l (15 ng/ml) to 170 nmol/l (76 ng/ml) and 306 nmol/l (136 ng/ml), respectively Although a true first-phase insulin-response to a test meal was not present in these patients, as a substitute this may be represented by the areas under the time curve from 0 to 90 min, which declined from 126 (10 mg/day) to 118 (20 mg/day) and 101 (40 mg/day), the latter value being significantly different from the first one. This clearly demonstrates that the lowest dose evoked maximal insulinotropic action of a test meal, with concentrations lying approximately between 100 and 250 nmol/l, at the time of food intake. In an acute study with tolbutamide, 250–1000 mg i.v. in seven NIDDM patients Ferner et al. (1990) was similarly unable to find a clear dose-effect relationship on insulin response fasting, after a mixed meal or during a hyperglycemic glucose clamp. This is at variance with earlier results of Haupt et al. (1976) in healthy subjects under fasting conditions, who found a nice log-linear relationship for the maximum insulin response with 5.5–10 mg/kg i.v. tolbutamide. Similar dose-response curves were also obtained for glibenclamide, glisoxepide, gliquidone and glibornuride, but only after i.v. and not oral administration. Pontiroli et al. (1991) confirmed the absence of a significant insulin release with oral glipizide doses of 0.5–2.5 mg.

Further studies have reported on the influence of a variation in pharmacokinetic profiles on insulin response (Helqvist et al. 1991; Pentikäinen et al. 1983; Wahlin-Boll et al. 1982a; Groop et al. 1993a,b; Haaber et al. 1993, Haupt et al. 1984; Neugebauer et al. 1985; Arnqvist et al. 1983; Borthwick et al. 1984; Schieen et al. 1987).

In other studies the effects of single or divided-dose regimens were compared (GINIER et al. 1985; OSTMAN et al. 1981; ALMÉR et al. 1981; HUUPPONEN et al. 1982b; WAHLIN-BOLL et al. 1986; ZILKER and BOTTERMANN 1975). When the daily drug dose was split into two or even three administrations, which consequently led to differing concentration-time profiles, mainly no or only small postbreakfast changes in insulin release were observed, even though the trough concentrations were higher with the split regimen. A greater postbreakfast response was found by the study of ALMÉR et al. (1982) with single-dose administration.

2. Second-Phase Insulin Secretion

Similarly to first-phase insulin release, the relationship of doses or concentrations with total stimulated insulin release or basal insulin concentrations is poorly characterized. In contrast to the result of PONTIROLI et al. (1991), who found no relevant stimulated insulin release in the oral dose range of 0.5–2.5 mg glipizide, an additive effect on insulin response after acute administration was evident from other studies (SCHEEN et al. 1984; AHREN et al. 1986; HAABER et al. 1993; GROOP et al. 1993a,b; CHIASSON et al. 1991; SCHWINGHAMMER et al. 1991 HEINE et al. 1974; PFEIFER et al. 1981; CERASI et al. 1979; PETERSON et al. 1982). Long-term treatment has also provided evidence for continued increase in endogenous insulin secretion and basal insulin levels (JUDZEWITSCH et al. 1982; PEACOCK et al. 1987; SIMONSON et al. 1987; GROOP et al. 1985, 1987a; JENG et al. 1989; BAYNES et al. 1993; WAJCHENBERG et al. 1993). Interestingly, fasting and incremental insulin levels increased after 3 months treatment with gliclazide and returned to baseline levels after 1 year of treatment (WAJCHENBERG et al. 1992). The only study with multiple-dose levels described in the previous section failed to demonstrate a dose-response relationship, because the lowest dose of 10 mg glipizide likewise exhibited peak effect on stimulated total insulin release (STENMAN et al. 1993). Basal insulin levels were not reported in this study.

3. Fasting Blood Glucose

At the time of measurement of fasting plasma glucose, the concentrations of sulfonylureas are mainly low or often below the limit of quantitation, which, however, depends on dose, elimination half-life and dosing interval. MELANDER et al. (1978) determined pre-dose concentrations of tolbutamide (0.5–3 g) and chlorpropamide (125–500 mg) under steady-state conditions, which ranged from 0 to 370 mol/l (0–100 g/ml) and from 0 to 882 mol/l (0–244 g/ml). Concentrations were not correlated with dose or fasting blood glucose. Other investigators demonstrated significant, albeit weak correlation of chlorpropamide trough levels with dose and inverse relationships with fasting glucose levels (BERGMAN et al. 1980; ALMÉR et al. 1982; HUUPPONEN et al. 1982a).

A trend to lower mean fasting glucose was seen with 5 mg b.i.d glipizide treatment accompanied by only slightly higher mean trough drug levels compared with 10 mg o.d. administration, whereas even twice the concentrations after single-dose administration also produced only a trend toward lower glucose compared with the split regimen (Huupponen et al. 1982b). Concerning fasting glucose, similar observations were made by Almér et al. (1982). Wahlin-Boll et al. (1986) were unable to confirm these findings in a larger group of patients receiving 7.5 mg as a single or split dose, although twice as high trough concentrations at steady state were achieved with the latter. Peterson et al. (1982) noted a significant hypoglycemic effect in their study even when drug levels were undetectable in plasma. Despite dose-dependent increases in plasma glipizide concentrations, the differences in home-monitored fasting glucose between the 10-mg, 20-mg and 40-mg doses were too small to reach statistical significance (Stenman et al. 1993). All doses, however, differed significantly from placebo. Increasing the dosage of gliclazide to above 160 mg/day similarly failed to achieve a greater hypoglycemic activity (Shaw et al. 1985).

Also for glibenclamide, a wide variation in steady-state trough levels with no correlation to the dose was reported in patients, using apparently adequate controls (Sartor et al. 1980a; Huupponen et al. 1982a; Matsuda et al. 1983), and furthermore no correlation between plasma gliclazide levels and fasting glucose was deomonstrated after 1 month's treatment (Fagerberg and Gamstedt 1980).

4. Postprandial Blood Glucose

Both acute oral and intravenous administration of sulfonylureas mainly leads to a dose-dependent fall in blood glucose as demonstrated for glisoxepide, glibornuride, glibenclamide, tolbutamide, gliquidone and glipizide (Ganda et al. 1975; Haupt et al. 1972, 1976; Kaubisch et al. 1979; Pontiroli et al. 1991). However, concentration-response relationships are far less clear, since for instance dissimilar concentration profiles after the same intravenous and oral dose of glipizide produced consistently different responses in healthy subjects (Helqvist et al. 1991; Groop et al. 1993a), but indistinguishable responses in NIDDM patients (Groop et al. 1993b; Haaber et al. 1993). This indicates that at least in patients the amount of insulin released appears to be of greater importance for a certain effect than the drug concentration-time profile generated. Furthermore, with a 5-mg dose of glibenclamide, Schwinghammer et al. (1991) demonstrated in young nondiabetic subjects a maximal fractional decrease in response to a glucose tolerance test independent of a wide range of areas under the glibenclamide concentration-time curves during the first 4 h so that the regression slope between the two variables was not significantly different from zero. In elderly nondiabetic subjects, however, a significant slope was obtained, indicating a concentration-response relationship, although for the whole group the mean response was smaller than in the young. Clear differences in sensitivity to and not in pharmacokinetics of glibenclamide were also demonstrated between obese (exhibiting higher sensitivity which may be related to their greater B-cell reserve, i.e., smaller insulin deficiency) and non-obese type 2 diabetics under steady-state conditions, thus revealing differences in concentration-response relationships and consequently the need for different doses for a specified effect (JABER et al. 1993a).

Similarly, glucose tolerance was progressively improved by increasing oral doses of glipizide and normalized by 1 mg glipizide in obese subjects with impaired glucose tolerance but only by 2.5 mg glipizide in obese NIDDM patients (PONTIROLI et al. 1991). Higher doses, i.e., 10–40 mg, however, were only equieffective in reducing the glucose response to a test meal (STENMAN et al. 1993). Similarly, with about double the mean daily concentration of chlorpropamide no greater effect on 24-h glucose profile was obtained (WAHLIN-BOLL et al. 1982b). In the study by COPPACK et al. (1990), glibenclamide was administered with breakfast in 5-, 10- and 20-mg doses to NIDDM patients. Despite interindividual variations in drug absorption, peak concentrations and areas under the concentration-time curve increased dose-dependently over the dose range. By contrast, no significant concentration-response behavior was observed in respect of mean glucose concentrations and AUC (0–24 h), suggesting that the 5-mg dose was already on the top of the dose-response curve in these patients. Other investigators confirmed the achievement of maximum responses with relatively low doses of other sulfonylureas also in acute i.v. studies (HEINE et al. 1974; FERNER et al. 1990; GROOP et al. 1991b). Controversial results were reported regarding time-related changes in sensitivity not explained by changes in pharmacokinetics. Whereas PETERSON et al. (1982) found a more restrained postprandial hyperglycemia with time after glipizide and apparently no effect on stimulated insulin (where, however, the lowered fasting glucose levels were not taken into consideration), either no change or changes in the opposite direction were reported by others (JABER et al. 1991, 1992). Moreover, time-related changes in insulin were not paralleled by corresponding responses of glucose (WAJCHENBERG et al. 1992). A similar equivocal situation exists for studies where the daily dose was either administered as a single or split dose. The higher postdose concentrations with single administration were sometimes associated with a greater effect on breakfast-induced glucose excursions (MATSUDA et al. 1983; WAHLIN-BOLL et al. 1986), but sometimes not (GINIER et al. 1985). Larger studies are needed to resolve this problem.

5. Euglycemic and Hyperglycemic Clamp

In only one study were steady-state plasma concentrations of glibenclamide from 50 to 800 nmol/l (25–395 ng/ml) related to insulin response and glucose disposal under euglycemic and hyperglycemic conditions in healthy subjects (GROOP et al. 1991b). No such study has as yet been performed in diabetic subjects. In agreement with the early results of Haupt et al. (1972), a clear log-linear concentration-response behavior for insulin secretion and glucose uptake was found under euglycemic conditions, although the data were not presented in this way. Hyperglycemia of 11.6 ± 0.2 mM shifted the curves tremendously to the left and appeared to increase the slope, so that the maximum insulin response was obtained at a concentration of 100 nmol/l (49.4 ng/ml).

The following general conclusions may be tentatively drawn from the data presented in the above sections:

1. In healthy subjects with unimpaired insulin secretory response and normoglycemia, the sulfonylurea concentrations are log-linearly related to insulin secretion and glucose disposal. Sensitivity may be decreased in the elderly.
2. Hyperglycemia causes a great shift in the concentration-response curves to the left with maximum effects obtained at far lower concentrations.
3. Sensitivity appears to decrease from healthy subjects to subjects with impaired glucose tolerance, obese and non-obese NIDDM patients in this order.
4. Sulfonylureas operate within a narrow range of concentrations in diabetic patients, which limits the theapeutic response at a certain threshold dose.
5. Fasting glucose levels are generally not related to drug trough concentrations, with the possible exception of chlorpropamide due to its long half-life.

V. Specific Pharmacokinetics

1. Chlorpropamide

a) Absorption

The absorption of chlorpropamide is almost complete (Taylor 1972) and in the absence of any first-pass effect complete absolute bioavailability was confirmed (Huupponen and Lammintausta 1981). Peak concentrations are reached between 3 and 6 h or even later (Taylor 1972; Taylor et al. 1977; Neuvonen et al. 1987) and rate of absorption but not its extent is decreased by food (Sartor et al. 1980c). In vitro dissolution rate of tablets depends on pH, being fast at pH 7.2 but slow at pH 2.0 (Taylor et al. 1977), which may explain differences in absorption rate. Therefore, 850 mg magnesium hydroxide administered simultaneously under fasting conditions increased the rate of chlorpropamide absorption (Kivistö and Neuvonen 1992), but sucralfate was without effect (Letendre et al. 1986). Formulation-related differences in absorption have also been reported (Monro and Welling 1974; Evans et al. 1979), and it has been suggested that this may be dependent on particle size, but not all brands differ in bioavailability (Taylor et al. 1977). Some studies demonstrate both inter- and intrasubject dose linearity (BERGMAN et al. 1980; ALMÉR et al. 1982; HUUPPONEN et al. 1982a; WAHLIN-BOLL et al. 1982b; TAYLOR et al. 1977).

b) Distribution

The distribution volume of chlorpropamide is small, about 0.11–0.18 l/kg (TAYLOR et al. 1977; HUUPPONEN and LAMMINTAUSTA 1981; NEUVONEN et al. 1987), and protein binding is 96% (CROOKS and BROWN 1974; NEUVONEN et al. 1987).

c) Metabolism

Chlorpropamide undergoes extensive metabolism to 2-hydroxychlorpropamide (2-OH-CP), 3-hydroxychlorpropamide (3-OH-CP), *p*-chlorobenzenesulfonylurea (CBSU) and *p*-chlorobenzenesulfonamide (CBSA), a degradation product of CBSU (BROTHERTON et al. 1969; TAYLOR 1972). Eighteen percent (10%–31%) is excreted as unchanged drug in urine, 55% (43%–69%) as 2-OH-CP, 21% (15%–25%) as CBSU and small amounts as 3-OH-CP (2%–3%) and CBSA (1.5%–4%), which accounts for complete recovery. CBSU in serum varies between 8% and 53% of the concentration of the parent compound, but 2-OH-CP hardly exceeds 1%, indicating high renal clearance. Hypoglycemic potencies of hydroxy metabolites appear comparable to chlorpropamide in vitro but not in vivo (THOMAS and JUDY 1972). The pharmacokinetics of chlorpropamide are not related to the debrisoquine polymorphic metabolism, i.e., CYP2D6 appears not to be involved in chlorpropamide hydroxylation (KALLIO et al. 1990).

d) Elimination

Chlorpropamide and its metabolites are completely eliminated through the renal route. The usual renal excretion of unchanged drug of about 20% of the ingested dose can be markedly increased by alkalinization or reduced by acidification of the urine (NEUVONEN and KÄRKKÄINEN 1983; NEUVONEN et al. 1987). At a urine pH below 5.5, the metabolic clearance almost equals the total plasma clearance, whereas at pH 6.5–7 and above the renal clearance represents more than half of the total clearance.

The systemic clearance of the compound is rather low, ranging from 1.8 to 3.3 ml/min (HUUPPONEN and LAMMINTAUSTA 1981; NEUVONEN et al. 1987; TAYLOR et al. 1977). In connection with the small volume of distribution, this leads to a long half-life of considerable variability even with mean values from 33–50 h (SARTOR et al. 1980b; TAYLOR 1972; TAYLOR et al. 1977; NEUVONEN et al. 1987). It has been discussed that enterohepatic circulation may contribute to the long elimination half-life (HUUPPONEN and LAMMINTAUSTA 1981).

e) Influence of Age, Weight and Diseases

In elderly diabetics but not healthy subjects, half-life increases due to both a greater distribution volume probably as a result of diminished protein binding and a reduction in systemic clearance (Sartor et al. 1980c; Arrigoni et al. 1987). It is not clear if a dose adjustment would be advisable. An inverse correlation between body weight and chlorpropamide concentrations has been found (Groop et al. 1984). Whereas no data are available on the pharmacokinetics of chlorpropamide in liver disease, in patients with renal impairment the half-life is prolonged, with a more pronounced effect in patients with predominant tubular damage, and the ratio of metabolites to parent drug in serum is increased (Petitpierre et al. 1972). The relation to the decrease in glomerular filtration remains to be defined.

f) Pharmacokinetic Interactions

A competitive inhibition of or by numerous drugs known to be metabolized through the cytochrome P450 CYP2D6 can be excluded (Kallio et al. 1990). Similarly no pharmacokinetic interactions have been reported for lovastatin, salicylate, cimetidine and enoxacin (Johnson et al. 1990; Richardson et al. 1986; Shah et al. 1985; Logemann 1986). Activated charcoal can reduce the absorption (Neuvonen and Kärkäinen 1983) and chloramphenicol, sulfaphenazole, bishydroxycoumarin or phenylbutazone may inhibit chlorpropamide metabolism (Dalgas et al. 1965; Kristensen and Hansen 1968; Petitpierre et al. 1972). Enzyme inducers such as phenytoin, carbamazepine, phenobarbitone and rifampicin increased the hepatic clearance of chlorpropamide (Self and Morris 1980; Neuvonen et al. 1987). These interactions are more likely to be significant with an acid urine, i.e., when renal chlorpropamide clearance is low.

2. Glibenclamide

a) Absorption

Although glibenclamide is absorbed throughout the gastrointestinal tract to the same extent but at a different rate (Brockmeier et al. 1985) depending on the pH characteristics at the specific site, the extent of absorption is also related to the particle size or formulation used (Rupp et al. 1969, 1975; Sartor et al. 1980b; McEwen et al. 1982; Ayanoglu et al. 1983; Arnqvist et al. 1983; Haupt et al. 1984; Neugebauer et al. 1985; Blume et al. 1985; Karttunen et al. 1985; Chalk et al. 1986; Scheen et al. 1987; Blume et al. 1987; Shaheen et al. 1987; Gramatté et al. 1989; Meyer et al. 1989; Chi et al. 1993). In principle, two different formulations exist on the market, the older one with incomplete bioavailability of about 70% (2.5- and 5-mg dose strengths) and the other with 100% absolute bioavailability (1.75- and 3.5-mg dose strengths) (Arnqvist et al. 1983; Haupt et al. 1984; Neugebauer et al. 1985; KARTTUNEN et al. 1985). For the former, a zero-order absorption process was suggested (INGS et al. 1982; McEWEN et al. 1982). Peak concentrations occur earlier with the new rapidly absorbed form between 0.5 and 2 h vs 1.5–4 h after the classical form. The rate-limiting step in the absorption of the new rapidly dissolved form appears to be gastric emptying (GANLEY et al. 1984). Whereas prior food intake had no influence on rate and extent of absorption of the classical form (SARTOR et al. 1980a; GROOP 1991), a marked delaying effect on in vivo disintegration and dispersion of a novel form and hence on the lag time between dosing and the start of absorption was reported (GANLEY et al. 1984). Therefore one can imagine that diabetic autonomic neuropathy and marked hyperglycemia delay absorption and thus enhance the variability in absorption rate (IKEGAMI et al. 1986; HOFFMANN et al., to be published). Since solubility increases with pH (HAJDÚ et al. 1969), absorption of nonmicronized but not of micronized glibenclamide was augmented by concomitant intake of antacids (ZUCCARO et al. 1989; NEUVONEN and KIVISTÖ 1991). Glibenclamide exhibits dose-linear pharmacokinetic behavior after either intravenous or oral administration (COPPACK et al. 1990; GROOP et al. 1991b).

b) Distribution

The distribution volume of glibenclamide is small and mean values vary by an approximate factor of 1.5 between 0.13 and 0.20 l/kg (ROGERS et al. 1982; NEUGEBAUER et al. 1985; KARTTUNEN et al. 1985; CHALK et al. 1986; SPRAUL et al. 1989). Protein binding is primarily to binding site I of albumin and is from 97% to 99.8% (RUPP et al. 1969; CROOKS and BROWN 1974; SCHWINGHAMMER et al. 1991). Albumin glycation does not alter the binding properties (OLSEN et al. 1992). Controversy is ongoing about the existence of a deep compartment (JÖNSSON et al. 1994), evidence for which has either been found (BALANT et al. 1975, 1977) or not (ROGERS et al. 1982; PEARSON 1985; PEARSON et al. 1986). In this connection the unique property of glibenclamide among the sulfonylureas to progressively accumulate within isolated pancreatic islets and long retention during washout is always mentioned (HELLMAN et al. 1984). Glibenclamide does not cross the normal and diabetic human placenta, suggesting insignificant fetal exposure to maternally administered drug (ELLIOT et al. 1991, 1993).

c) Metabolism

Glibenclamide is completely metabolized in the liver by hydroxylation of the cyclohexyl ring to trans-4-hydroxy-glibenclamide (M1, approximately 73%), cis-3-hydroxyglibenclamide (M2, approximately 19%) and a third unidentified metabolite (approximately 6%) (RUPP et al. 1969). The metabolites are eliminated in nearly equal amounts through biliary and renal routes. Recently, M1 and M2 have been found to exert almost equipotent hypoglycemic effects to glibenclamide itself in healthy subjects (MELANDER et al. 1993) in contrast to earlier findings in animals (Heptner et al. 1969; Balant et al. 1979). Metabolite concentrations measured in patients (Pearson et al. 1986) therefore suggest at least M1 to contribute to the biologic activity. Glibenclamide disposition is unaffected by genetically determined polymorphism related to CYP2D6 or CYP2C18/19 (Spraul et al. 1989; Peart et al. 1989; Dahl-Puustinen et al. 1990).

d) Elimination

The systemic clearance of glibenclamide, which is completely metabolic, has been reported to be between 50 and 170 ml/min (Ings et al. 1982; Rogers et al. 1982; Neugebauer et al. 1985; Karttunen et al. 1985; Chalk et al. 1986; Spraul et al. 1989), mainly near 100 ml/min. The half-life obtained from intravenous data is between 1.2 and 2.5 h (Ings et al. 1982; Rogers et al. 1982; Neugebauer et al. 1985; Spraul et al. 1989; McEwen et al. 1982) and between 1.3 and 15 h after oral administration. Most studies obtained values of 1.5–5 h (Karttunen et al. 1985; Neugebauer et al. 1985; Chalk et al. 1986; Pearson et al. 1986; Schwinghammer et al. 1991). In agreement with these values are half-lives of an initial phase in two studies (Coppack et al. 1990; Jönsson et al. 1994), which exhibit a terminal elimination phase of 9.7–15 h. Two further studies only found terminal phases of 9.9–15.2 h (Schwinghammer et al. 1991; Jaber et al. 1993a). Consistent with the absence of any indication for drug accumulation (Balant et al. 1977; Coppack et al. 1990; Jaber et al. 1991, 1993a) is the notion that the slow terminal phase only insignificantly contributes to the overall drug clearance. Continued prolonged absorption or enterohepatic cycling may partially explain a slow terminal phase (McEwen et al. 1982; Kühnle et al. 1984).

e) Influence of Age, Weight and Diseases

In elderly subjects the half-life was longer at 9.9 h against 4.9 h in younger subjects due to a 52% higher free drug concentration followed by an increase in distribution volume. Clearance was not changed (Schwinghammer et al. 1991). In contrast, no age-related difference in pharmacokinetics was found in type 2 patients. The kinetic profile, however, was analyzed over a time period too short to cover the true elimination phase in this study (Scheen et al. 1989). Clearance and other pharmacokinetic variables are not different in obese type 2 diabetics (Jaber et al. 1993a), which agrees with animal data that the drug primarily partitions into highly perfused organs such as the liver and kidneys and not into fat (Kellner et al. 1969).

Impaired kidney funtion per se should have no effect on the clearance of glibenclamide. This was demonstrated except in one patient with a creatinine clearance of 5 ml/min/1.73 m^2 (Pearson et al. 1986). Renal failure may, however, affect drug metabolism by the liver and thus change the clearance of drugs (Elston et al. 1993). Two studies showed that even end-stage renal failure did not change the clearance of glibenclamide, but that especially metabolite M1, which has a renal clearance of about 100 ml/min, might accumulate (BEHRLE 1980; BRIER and STALKER 1993). M1 and M2 appear at 27%–44% and 8%–18% in urine, respectively (PEARSON et al. 1986; DAHL-PUUSTINEN et al. 1990). As expected with a completely metabolized drug, clearance decreases dramatically in patients with liver cirrhosis by about 60%, which makes dose adjustment necessary (HELLSTERN et al. 1985).

f) Pharmacokinetic Interactions

Enzymes involved in the metabolism of caffeine or aminopyrine seem not to be induced or inhibited by glibenclamide (JUAN et al. 1990). In addition, several studies demonstrate the absence of any interaction between gliben-clamide and pirprofen, acarbose, lisinopril, ibuprofen, ranitidine, trimetho-prim-sulfamethoxazole, vinpocetine, enoxacin, carvedilol and moxonidine (MORRISON et al. 1982; GERARD et al. 1984; DANHOF et al. 1986; KUBACKA et al. 1987; SJOEBERG et al. 1987; GRANDT et al. 1989; GÖBEL et al. 1990; MÜLLER et al. 1993; HARDER et al. 1993). With ranitidine, however, glucose and insulin concentrations are elevated. Whereas verapamil and cimetidine inhibit the metabolism of glibenclamide (SEMPLE et al. 1986; KUBACKA et al. 1987), acetylsalicylic acid and rifampicin may increase the apparent oral clearance (KUBACKA et al. 1986; SELF et al. 1989). Increased nonmicronized glibenclamide absorption is reported for magnesium hydroxide, sodium bicarbonate and antacids (NEUVONEN and KIVISTÖ 1991; KIVISTÖ et al. 1993; ZUCCARO et al. 1989), but not with micronized glibenclamide (NEUVONEN and KIVISTÖ 1991). The absorption rate of nonmicronized but not of micronized glibenclamide is decreased by glucomannan and guar (SHIMA et al. 1983; NEUGEBAUER et al. 1983; UUSITUPA et al. 1990) and may be accelerated by erythromycin (FLEISHAKER and PHILLIPS 1991). Absorption is decreased by the glucosidase inhibitor miglitol (SALVATORE et al. 1990).

3. Gliclazide

a) Absorption

Absorption of gliclazide is not well characterized and amounts to about 80%, but absolute bioavailability is not known (CAMPBELL et al. 1980; KOBAYASHI et al. 1981). Absorption rate is variable with fast (peak concentration with 4 h) and slow absorbers (peak concentration between 4 and 8 h) having been identified. Incomplete absorption of the current market formulation is very likely since its rate and extent could be improved almost 100% by complex formation with betacyclodextrin (CHRYSTIN et al. 1994; WINTERS et al. 1993). Food delays absorption significantly and peak concentrations may also be reduced (BATCH et al. 1990; ISHIBASHI and TAKASHINA 1990). Dose linearity has been studied only insufficiently. The absorption profile appears to be no different between healthy subjects and NIDDM patients (KOBAYASHI et al. 1981, 1984; SHIBA et al. 1986).

b) Distribution

The distribution volume assuming complete absorption ranges between 0.19 and 0.26 l/kg (KOBAYASHI et al. 1981, 1984; FORETTE et al. 1982). Gliclazide is highly bound to albumin by about 95%, with a broad range of 85%–99% (CAMPBELL et al. 1980; KOBAYASHI et al. 1981, 1984). A few drugs displace gliclazide in vitro (FUJII et al. 1983). Interestingly, albumin glycosylation reduces free gliclazide levels in patients (IGAKI et al. 1992).

c) Metabolism

Gliclazide is extensively metabolized to at least eight inactive metabolites by three major metabolic routes: (1) oxidation of the tolyl group to carboxylic acid (20% of the dose in urine), (2) hydroxylation of the azabicyclo-octyl moiety (16% in urine) and (3) glucuronidation, etc. (3%–9% each). Gliclazide represents more than 90% of the total radioactivity found in plasma, and trace levels of individual metabolites with none greater than 1%–2% are observed. Metabolism in Caucasian and Japanese subjects is similar (CAMPBELL et al. 1980; OIDA et al. 1985; LUPO and BATAILLE 1987). Gliclazide does not exhibit genetic polymorphism in debrisoquine-phenotyped subjects (BOUTAGY et al. 1987).

d) Elimination

The proportion of a dose excreted in the urine is 61%–81%, the remainder being eliminated in feces (CAMPBELL et al. 1980; OIDA et al. 1985). Less than 1% of gliclazide is excreted unchanged in urine due to the low renal clearance of 0.5 ml/min. The oral plasma clearance ranges between 13 and 26 ml/min and the terminal elimination half-life between 8 and 17 h, with wide individual variation (CAMPBELL et al. 1980; KOBAYASHI et al. 1981, 1984; FORETTE et al. 1982; OIDA et al. 1985; SHIBA et al. 1986). Because of the variability of different studies, a gender-related difference remains to be established. Steady state is reached after about 3 days dosing with an accumulation ratio of approximately twofold. Repeated dosing appears not to change kinetic parameters, which, however, needs to be confirmed in a larger study population. Gliclazide undergoes enterohepatic recirculation in rats (BENAKIS and GLASSON 1980; MIYAZAKI et al. 1983), and some evidence suggests that this also occurs in humans (CAMPBELL et al. 1980). The relevance of a gliclazide-degrading factor in serum remains unclear (KOBAYASHI et al. 1985).

e) Influence of Age, Weight and Diseases

In the elderly, gliclazide is more slowly absorbed and peak concentrations are reached after 6 h. Half-life is prolonged (20 h) because distribution volume is increased (approximately 0.34 l/kg). Since apparent clearance remains unchanged, no dose adjustment would be necessary (FORETTE et al.

1982). In renal disease the volume of distribution and half-life also increase, whereas the oral clearance at 21–26 ml/min is even higher than in other studies, although lower than in a comparatively healthy volunteer group (CAMPBELL et al. 1986). Therefore no recommendation for dosage reduction is made. The effect of hepatic disease on gliclazide kinetics has not been studied, but in view of the extensive metabolism the dosage should be reduced.

f) Pharmacokinetic Interactions

Dietary fibers such as konjac mannan and guar gum appear to have only little influence on gliclazide absorption (SHIMA et al. 1982). No other specific studies with the aim of evaluating the interaction potential of gliclazide seem to exist.

4. Glipizide

a) Absorption

Glipizide is completely absorbed and also 100% bioavailable, i.e., no first-pass effect is present (SCHMIDT et al. 1973; BALANT et al. 1975; WÄHLIN-BOLL et al. 1982a; PENTIKÄINEN et al. 1983). Peak concentrations are generally reached 1–3 h after oral dosing on an empty stomach, but in "slow" absorbers may even be delayed for up to 12 h (SARTOR et al. 1980b; HUUPPONEN et al. 1982b; PETERSON et al. 1982; KRADJAN et al. 1989; JABER et al. 1992). The rate of absorption depends on the formulation administered, being fastest with solutions or liquid-filled soft gelatin capsules and slower with the commercial tablets (WAHLIN-BOLL et al. 1982a; HELQVIST et al. 1991; HAABER et al. 1993). Food further delays the absorption (SARTOR et al. 1980b; WAHLIN-BOLL et al. 1980) as does hyperglycemia in healthy subjects and NIDDM patients above glucose concentrations of 7 mmol/l (GROOP et al. 1989a, 1991a). Dose linearity is insufficiently supported by kinetic data but probably exists (WAHLIN-BOLL et al. 1982a; HELQVIST et al. 1991).

b) Distribution

The apparent volume of distribution of glipizide is small, ranging between 0.12 and 0.26 l/kg (HUUPPONEN et al. 1982b; PENTIKÄINEN 1983; KRADJAN et al. 1989), which reflects the high amount of binding to albumin of 98%–99.4% (CROOKS and BROWN 1975; KRADJAN et al. 1989). In obese NIDDM patients the distribution volume is no different from that in nonobese patients at 0.20 vs. 0.21 l/kg, respectively (JABER et al. 1993b).

c) Metabolism

Hepatic biotransformation is nearly the exclusive pathway of elimination except for about 10%, which is excreted unchanged with urine (FUCCELLA et al. 1973; Schmidt et al. 1973; Balant et al. 1975; Pentikäinen et al. 1983). The proportion of a dose found in urine is 65%–87%, 9%–12% as 3-*cis*-hydroxy-cyclohexyl derivative, 46%–62% in the form of the 4-*trans*-hydroxy-cyclohexyl derivative, 1%–2% as the *N*-acetyl derivative and about 8% as unextractable and unidentified compounds. The remaining 5%–10% is excreted in feces. Of the total radioactivity in the plasma, 72% accounts for unchanged drug. The kinetics of the metabolites is formation dependent (Pentikäinen et al. 1983). Although the metabolites are reported to be devoid of hypoglycemic activity (Tamassia 1975), in the light of the recent findings with the structurally similar 3- and 4-hydroxy metabolites of glibenclamide (Melander et al. 1993), this should be reevaluated.

d) Elimination

UP to 87% of a dose may be excreted with urine, the remainder with feces (Fuccella et al. 1973), with mean elimination half-lives ranging between 2 and 9 h (Sartor et al. 1980b; Wahlin-Boll et al. 1982a; Huupponen et al. 1982b; Peterson et al. 1982; Pentikäinen et al. 1983; Kradjan et al. 1989; Jaber et al. 1992, 1993b). Values for systemic clearance between 33 and 52 ml/min are attained. With continuous oral treatment the clearance remains constant (Peterson et al. 1982; Kradjan et al. 1989; Jaber et al. 1992, 1993b). The renal clearance, which contributes only about 5% to the total plasma clearance, can be increased by alkalinization of the urine (Pentikäinen et al. 1983). Enterohepatic cycling has been hypothesized but has never been proven (Pentikäinen et al. 1983; Wahlin-Boll 1986).

e) Influence of Age, Weight and Diseases

The free fraction of glipizide is less than 1% in the young and significantly less than this (more than 30% less) in elderly populations. However, no age-related difference in pharmacokinetic variables is seen, although there is a trend toward higher systemic clearance from young healthy subjects to elderly healthy to elderly diabetics (Kobayashi et al. 1988; Kradjan et al. 1989; Jaber et al. 1992). Furthermore, the pharmacokinetics is unchanged in obese NIDDM patients despite the longest half-life of 9 h found in this group after chronic therapy (Jaber et al. 1993b). The only small study in patients with renal impairment demonstrated an unchanged glipizide half-life; however, there was a slow elimination of total metabolites with a half-life greater than 20 h (Balant et al. 1973). From the extensive metabolism of glipizide, one would predict that hepatic dysfunction would impair its metabolism.

f) Pharmacokinetic Interactions

Concomitant intake of nifedipine, guar gum or ethanol does not influence the disposition of glipizide; however, ethanol prolongs the hypoglycemia induced by glipizide (HUUPPONEN et al. 1985; CONNACHER et al. 1987; HARTLING et al. 1987). Several compounds such as indobufen, cimetidine and ranitidine seem to inhibit the metabolic degradation of glipizide, subsequently increasing the average concentrations, i.e., by cimetidine of 23% and ranitidine of 34% (MELANDER and WAHLIN-BOLL 1980, 1981; ELVANDER-STAHL et al. 1984; FEELY et al. 1993). Whereas nothing is known for indobufen, cimetidine has a broad inhibitory spectrum on several P450 enzymes. Inhibition of glipizide metabolism by trimethoprim-sulfamethoxazole is also assumed, but drug concentrations have not been measured (JOHNSON and DOBMEIER 1990). Intake with cholestyramine reduces glipizide absorption by 29% and with activated charcoal by 81%, so that the latter may be taken in acute overdose situations (KIVISTÖ and NEUVONEN 1990). As with other sulfonylureas, absorption of glipizide can be enhanced with antacids such as sodium bicarbonate and magnesium hydroxide, but not aluminum hydroxide (KIVISTÖ and NEUVONEN 1991a,b).

5. Tolbutamide

a) Absorption

The extent of absorption seems to reach 85% or more (THOMAS and IKEDA 1966). The absolute bioavailability has not been determined directly, but from metabolite recovery in urine of different studies with i.v. and oral administration a value of 109% can be calculated (KNODELL et al. 1987; MILLER et al. 1990; VERONESE et al. 1990a; PAGE et al. 1991). Peak concentration is reached at 3–4h after oral administration (SARTOR et al. 1980b,c; ANTAL et al. 1982; KIVISTÖ and NEUVONEN 1992) and the rate but not extent of absorption is decreased by food in healthy subjects, whereas in diabetics only the extent is decreased by 6%, which is considered therapeutically irrelevant. In addition, different formulations exhibit differences in bioavailability (RUPP et al. 1975; OLSON et al. 1985), which seems to be proportional to the apparent specific tolbutamide surface area (SANO et al. 1992). Dose linearity seems to be present, but only partial areas under the serum concentration-time curve have been measured (SARTOR et al. 1980c).

b) Distribution

Under the assumption of complete absorption, the volume of distribution can be estimated to be in a range between 0.10 and 0.15 l/kg (SARTOR et al. 1980b,c; VERONESE et al. 1990a; KIVISTÖ and NEUVONEN 1992; TASSANEEYAKUL et al. 1992). Tolbutamide is about 95% protein-bound and binding decreases with age in parallel with albumin concentration (CROOKS and BROWN 1974; ADIR et al. 1982), but the fraction bound increases with concentration to about 98% (AYANOGLU et al. 1986). Both sites I (warfarin-binding site) and II (diazepam-binding site) of albumin are involved in tolbutamide binding (SJÖHOLM et al. 1979).

c) Metabolism

Tolbutamide is metabolized by microsomal hydroxylation of the tolyl methyl group (OHTB), which is the first and rate-limiting step, followed by further oxidation to a carboxyl derivative (CTB) in the cytosol (THOMAS and IKEDA 1966; PURBA et al. 1987). Early investigations with the tritiated compound revealed that 33% of a dose is metabolized to OHTB and 52% to CTB, the remaining 9% in feces being unknown; subsequently with direct assay of the metabolites in urine, figures of 14% OHTB and 66% CTB were found (VERONESE et al. 1990a). The urinary recovery of both metabolites after i.v. and oral administration ranges from 51% to 96% and 47%–93%, respectively (MILLER et al. 1990; VERONESE et al. 1990a; PAGE et al. 1991). The metabolites have minimal hypoglycemic effects (FELDMAN and LEBOVITZ 1969a). The wide variability of tolbutamide disappearance suggested genetic polymorphic control of the first oxidation step, but this awaits confirmation. In screening programs, slow metabolizers may be detected at a rate of about 1%–2% (SCOTT and POFFENBARGER 1979; PAGE et al. 1991). The enzyme involved in the initial metabolic step seems to be CYP2C9 (VERONESE et al. 1990b, 1991; RELLING et al. 1990; VERONESE 1991; LEEMANN et al. 1993), whereas no association appears to exist with CYP2D6 (debrisoquine) (BOUTAGY and SHENFIELD 1987; PEART et al. 1987) or CYP2C 18/19 (S-mephenytoin) (KNODELL et al. 1987; RELLING et al. 1990).

d) Elimination

Renal excretion is the predominant route of elimination, approaching 80%–90% of a dose in the form of OHTB and CTB, only 0.1%–0.2% being excreted unchanged (KNODELL et al. 1987; VERONESE et al. 1990a). The low oral clearance varies from 12 to 17 ml/min with a mean half-life of 7–9 h, the total range being about 5–16 h (PAGE et al. 1991). Only subjects with half-lives greater than 20 h may be considered slow metabolizers, which needs to be confirmed, however, by determination of oral plasma clearance or urinary metabolic ratio (VERONESE et al. 1990a). The pharmacokinetics of the metabolites is clearly formation dependent (PEART et al. 1987).

e) Influence of Age, Weight and Diseases

The renal excretion of both metabolites of tolbutamide decreases with age, which is probably related to the age-associated decrease of creatinine clearance. However, the ratio of OHTB to the total amount of metabolites also changes as a function of age, suggesting an alteration in the oxidation of the parent drug or in the conversion to CBT (MILLER et al. 1990). SARTOR et al. (1980c) earlier demonstrated higher peak concentrations in elderly, but the reason for this was not elucidated.

Half-life is slightly prolonged in patients with liver cirrhosis, but clearance has not been determined (UEDA et al. 1963). In contrast, tolbutamide clearance is increased in the acute phase of viral hepatitis, but remains unchanged when calculations are based on unbound drug concentrations, indicating no dose adjustment (WILLIAMS et al. 1977). Presumably chronic hypoxemia as occurring in asthma or with chronic respiratory failure appears to increase hepatic clearance, thus decreasing tolbutamide half-life (SOTANIEMI et al. 1971a,b). In patients with renal disease, half-life increased especially when glomerular filtration rate was below 10 ml/min; however, the assay method used is not specific for tolbutamide and therefore decreased renal elimination of the metabolites might have caused these findings (UEDA et al. 1963).

f) Pharmacokinetic Interactions

As already mentioned in a previous chapter, tolbutamide is most likely metabolized by CYP2C9 such as phenytoin, whereas CYP2D6 and CYP2C 18/19 are not involved. Therefore, competitive inhibition may occur, with all drugs metabolized by the same enzyme, namely bishydroxycoumarin (SKOVSTED et al. 1976), chloramphenicol (CHRISTENSEN and SKOVSTED 1969; NATION et al. 1990), cotrimoxazole (WING and MINERS 1985), fuconazole (LAZAR and WILNER 1990; BLUM et al. 1991), ketoconazole (KRISHNAIAH et al. 1994), phenylbutazone (POND et al. 1997), sulfinpyrazone (MINERS et al. 1982) and sulfaphenazole (VERONESE et al. 1990a). Sulfaphenazole, e.g., decreases mean plasma clearance by 80% and increases half-life fivefold.

Cimetidine does not inhibit the metabolism of tolbutamide (DEY et al. 1983; SHAH et al. 1985; STOCKLEY et al. 1986; BACK et al. 1988; ADEBAYO and COKER 1988) unless high doses are administered (CATE et al. 1986; BACK et al. 1988). Atenolol, metoprolol and propanolol (MINERS et al. 1984), dextropropoxyphene (ROBSON et al. 1987), oral contraceptives (PAGE et al. 1991), prednisone (SCHENK 1976), primaquine (BACK et al. 1988) and ranitidine (CATE et al. 1986; ADEBAYO and COKER 1988) appear not to interfere with tolbutamide disposition. Enzyme induction by rifampicin is known to enhance the elimination of tolbutamide (ZILLY et al. 1975; BÜRKL 1977) similarly to ethanol (KATER et al. 1969), but conflicting data regarding phenobarbitone (REDMAN and PRESCOTT 1973; CARULLI et al. 1976) and cigarette smoke exist (UPPAL et al. 1986; PAGE et al. 1991).

Displacement from protein binding by chenodeoxycholic acid and sulfadimethoxine have little influence since unbound tolbutamide levels remain almost constant (GOTO et al. 1985). The absorption rate can be increased by magnesium hydroxide (KIVISTÖ and NEUVONEN 1992) and the amount absorbed reduced by activated charcoal (NEUVONEN et al. 1983).

6. Tolazamide

a) Absorption

At least 85% of a dose of tolazamide is absorbed as judged from urinary excretion of radioactivity (THOMAS et al. 1978). Maximum serum drug concentrations are reached 3.3 h after administration. Rate and extent of absorption are influenced by formulation (Welling et al. 1982). The absolute bioavailability is not known.

b) Distribution

Although not presented, from the data reported a distribution volume assuming complete absorption of 0.16 l/kg can be estimated (Welling et al. 1982). Protein binding, mainly to albumin, is in the range of 98% (Crooks and Brown 1974).

c) Metabolism

Six percent of a dose of tolazamide is excreted unchanged in urine, 9% as hydroxy-methyltolazamide, which has approximately 20% of the hypoglycemic activity of the parent compound as determined in the rat, 14% as carboxy-tolazamide, with 5% activity, 21% as hexadydroazepine-ring-hydroxylated metabolite, with 70% activity, 22% as p-toluenesulfonamide, with no activity, and the remaining 13% not identified (Thomas et al. 1978).

d) Elimination

Eighty-five percent of a dose of tolazamide is recovered in urine and 7% in feces (Thomas et al. 1978). The elimination half-life is approximately 3–7 h (Welling et al. 1982; Jackson and Bressler 1981).

e) Influence of Age

In healthy volunteers aged 40–71 years, no age-related changes in pharmacokinetics have been reported (Wright and Antal 1985).

7. Glibornuride

a) Absorption

The mean absorption of glibornuride is 91% (range 60%–116%), with peak concentrations achieved between 3 and 4 h. No data on absolute bioavailability are available (Rentsch et al. 1972a).

b) Distribution

On average, the distribution volume for total radioactivity is 0.24 l/kg (Rentsch et al. 1972a). Although it is stated that concentrations of total radioactivity and parent compound are practically identical, only 85% of radioactivity in plasma could be attributed to glibornuride (Bigler et al. 1972). With direct determination of glibornuride after intravenous administration, the mean volume of distribution is only slightly higher, i.e., 0.27 l/kg (Stoeckel et al. 1985), and 94%–95% of glibornuride is bound to plasma proteins.

c) Metabolism

Glibornuride is extensively metabolized in man. Six metabolites have been identified, some with relatively moderate activity in animals. About 9% of a dose is oxidized at the methyl group of the tolyl moiety, 4% to the hydroxy and 5% to the carboxy metabolite, the remainder at the methyl and methylene groups of the borneol part, i.e., 45% of a dose. Eleven percent was not extractable or lost during the extraction procedure. The metabolites excreted through the biliary route are not known (BIGLER et al. 1972; RENTSCH et al. 1972a).

d) Elimination

The proportion of the metabolites excreted into the urine is 60%–72%, where no parent compound is detected, and 23%–33% of a dose is recovered from feces within 5 days (RENTSCH et al. 1972a). The range of urinary recovery within 6 days in another study is 54%–94% (ECKHARDT et al. 1972). The mean elimination half-lives of total radioactivity range between 8.2 and 8.7h, which is in agreement with 9.2h for glibornuride itself (STOECKEL et al. 1985). The systemic clearance of the drug is 24ml/min on average. The pharmacokinetics of glibornuride does not change under chronic treatment (DUBACH et al. 1975).

e) Influence of Diseases

In two patients with a creatinine clearance of 5ml/min, the elimination half-life of total radioactivity is prolonged to 30h due to a reduction in renal excretion of the metabolites to 16% (RENTSCH et al. 1972b).

f) Pharmacokinetic Interactions

There is no detectable influence of phenylbutazone (ECKHARDT et al. 1972) and tenoxicam (STOECKEL et al. 1985) on the pharmacokinetics of glibornuride. Phenprocoumon appears to prolong the half-life of total radioactivity after oral glibornuride by about 29%, but renal excretion of metabolites is also increased by 10%. In contrast, sulfaphenazole inhibits the metabolism of glibornuride, which consequently increases the half-life by 34% and reduces renal metabolite excretion by 10% (ECKHARDT et al. 1972), but this is considerably less than determined for tolbutamide.

8. Gliquidone

a) Absorption

The absorption of an oral solution of gliquidone has been estimated to reach 80%, but from the excretory pattern and extent of radioactivity following intravenous and oral administration it may be almost complete with no first-pass effect present (KOPITAR 1975). Peak concentrations are achieved within

1 h after the solution, but between 2 and 3 h after the tablet (Kopitar and Koss 1975a,b; Koss et al. 1976). Food delays the absorption of the tablet only insignificantly (Talaulicar and Willms 1976). The bioavailability of the tablet has not been characterized.

b) Distribution

The distribution volume for total radioactivity of gliquidone is approximately 0.16 l/kg and plasma protein binding 99% and higher (Kopitar 1975; Kopitar and Koss 1975b).

c) Metabolism

The proportion of a dose of gliquidone undergoing mainly 0-demethylation, but also hydroxylation at the cyclohexyl ring in the liver, is 98%–99% four metabolites have been identified, with one hydroxylated metabolite exhibiting some hypoglycemic activity, which is thought not to contribute to the effect of the drug. After both intravenous and oral administration, 46%–47% has been identified as 0-demethyl gliquidone (AR-DF 33), 20%–25% as 0-demethyl-para-(trans)-hydroxy gliquidone (AR-DF 35), 6%–7% as the meta-(cis/trans)-hydroxy metabolite of AR-DF 33 (AL-DF 3), 1.8%–2.1% as para-(trans)-hydroxy gliquidone (AC-DF 1) and 1.2%–1.5% as the parent compound in feces and urine. The remaining 17%–24% was not identified (Kopitar 1975; Kopitar and Koss 1975b).

d) Elimination

Five percent of a dose (0.2%–0.5% unchanged) is eliminated through the kidney in the form of metabolites, the remaining 95% (1% unchanged) through the bile (Kopitar 1975). The elimination half-life of total radioactivity after intravenous administration, at least 80% of which represents the parent compound, is 1.3 h. This corresponds with the first-phase half-life of 1.4 h after oral administration, which is attributed to the fast metabolic and biliary elimination (Koss et al. 1976). The terminal elimination half-life of 16.5 h makes only a very small contribution to the elimination. However, with 30 mg b.i.d., steady state still does not appear to be reached on day 3 of administration. An oral clearance of approximately 90 ml/min clearance has been estimated by Kopitar and Koss (1975a,b).

e) Influence of Diseases

In patients with liver disease, the metabolism of gliquidone is inhibited. This leads to higher and later peak concentrations of total radioactivity and larger areas under the concentration-time curve (Profozic 1976; Talaulicar and Willms 1976; Büchele and Kuhlmann 1978). No data for clearance and elimination half-life have been presented. In one patient with cholestatic jaundice, an extreme increase in concentration and slowing of elimination was demonstrated. With the inhibition of biliary excretion, about 70% of the dose was eliminated in urine. The higher amount of metabolites in plasma normally eliminated through the bile contributes essentially to the concentration increase of total radioactivity and is reflected by increased renal elimination approaching 24% in patients with severe liver disease (BÜCHELE and KUHLMANN 1978). NIDDM patients with impaired renal function (creatinine clearance 2–60 ml/min, mean 24 ml/min) demonstrate no consistent change in pharmacokinetics except diminished renal elimination of metabolites.

f) Pharmacokinetic Interactions

One formal interaction study with acute administration of alcohol demonstrated no effect on gliquidone concentrations, although they were measured for only 5 h (BOTTERMANN et al. 1976).

9. Acetohexamide

The absorption of acetohexamide may be nearly complete, but the absolute bioavailability is not known. About 75% of an oral dose is eliminated in urine and 15% of an intravenous dose in the feces (GALLOWAY et al. 1967). Absorption depends on formulation (NASH et al. 1977). A distribution volume of 0.2 l/kg can be estimated from published data (SMITH et al. 1965; COHEN et al. 1967). Acetohexamide is highly protein bound at about 90%, the main component at approximately 80% being albumin (JUDIS 1972; HSU et al. 1974). Binding of the main active metabolite (-)-hydroxyhexamide to albumin is considerably lower than that of acetohexamide (IMAMURA et al. 1985). Non-enzymatic glycosylation of albumin increases free acetohexamide concentrations by 44% (TSUCHIYA et al. 1984).

From urinary recovery, 47% of a dose can be attributed to metabolism to (-)-hydroxyhexamide, which has 2.4 times the potency of acetohexamide (MCMAHON et al. 1965), 11% to 4-hydroxyacetohexamide, 14% to dihydroxyhexamide and 1%–2% unchanged parent drug (GALLOWAY et al. 1967).

Peak concentrations of acetohexamide occur after 1–2 h and of (−)-hydroxyhexamide after 2–5 h. Serum concentrations of both compounds are in the same range and therefore most of the activity is due to the metabolite. The elimination half-life of the parent compound varies between 0.8 and 2.4 h, and between 4 and 6 h for the active metabolite. An apparent oral clearance of 116 ml/min may be estimated from published data (SMITH et al. 1965; GALLOWAY et al. 1967). Concentrations appear not to be changed during long-term administration (SHELDON et al. 1965).

A prolonged half-life of acetohexamide has been reported in renal disease, but more important is probably the reduction of the renal elimination of the active metabolite (COHEN et al. 1967), although this can be further converted to dihydroxyhexamide.

10. Glisoxepide

It may be assumed that absorption of glisoxepide is almost complete at 96% as judged from fecal (20%) and urinary (76%) recovery of radioactivity after an oral dose compared with urinary recovery (57%) after an intravenous dose (Speck et al. 1974; Schwartzkopff and Kewitz 1977). Peak concentrations are reached about 2 h postdose. The distribution volume after oral administration is small at 0.2 l/kg, and 93% of the drug is bound to plasma proteins (Schlossmann 1974).

The metabolites of glisoxepide in man have not been characterized, about 43% of a dose being found unchanged in the urine. It is not known whether acidification or alkalinization of urine changes renal excretion of glisoxepide. Mean elimination half-lives range from 1.7 to 3 h, whereas total plasma radioactivity declines with a half-life of 25 h (Speck et al. 1974). An estimate of oral clearance gives a value of 65 ml/min.

In two patients with severely impaired kidney function, renal excretion of glisoxepide and of total radioactivity is reduced to about 7% of an intravenous dose with a subsequent increase in glisoxepide half-life to 5 or 9 h. In a medium range of impaired renal function the reduction in glisoxepide excretion may be compensated by metabolic degradation (Schwartzkopff and Kewitz 1977).

11. Glyclopyramide

Although this compound has some importance on the Japanese market, no pharmacokinetic data have been published in the available literature.

12. Glymidine Sodium

Glymidine is the only compound which is administered as the sodium salt and having the urea portion replaced by an aminopyrimidine. Therefore, it lacks immunologic cross-reactivity with other sulfonylureas (Borthwick and Stowers 1979). It is used in Japan. Glymidine is almost completely absorbed, 89%–95% of oral radioactivity being recovered in urine (Gerhards et al. 1964). The absolute bioavailability is not known. The absorption proceeds relatively slowly with peak concentrations after 4–6 h.

Glymidine is virtually completely metabolized in the liver by 0-demethylation as the rate-limiting step followed by oxidation of the generated hydroxyl group to the carboxylic acid. Since a correlation of benzo[a]pyrene hydroxylase activity and clearance of glymidine exists, CYP2A1 and 2 and CYP3A could be involved in demethylation (Held 1980). No conjugated metabolites have been detected in urine. Desmethyl-glymidine has similar hypoglycemic potency to glymidine (Kramer et al. 1964). Approximately 6% of a dose is recovered from feces, the remainder being eliminated in urine with a little less than 1% of a dose unchanged, 19%–37% as the desmethyl metabolite and 56%–74% as carboxylic acid. The distribution volume of glymidine is 0.15 l/kg (HELD 1980) and that of the desmethyl metabolite in one subject 0.12 l/kg (GERHARDS et al. 1964; GERHARDS and KOLB 1965). Eighty percent of glymidine is bound to albumin (total binding 89%), between 5% and 10% to red blood cells, and 5%–10% present in free form. The elimination half-life of glymidine ranges from 3 to 6 h (HELD et al. 1973a). Desmethyl-glymidine is eliminated more rapidly, with a half-life of 3 h by exclusive renal excretion (GERHARDS and KOLB 1965). A systemic clearance of 30–50 ml/min for glymidine can be calculated (HELD 1980). The half-life of glymidine increases to a range between 6 and 27 h in patients with acute or chronic liver disease, indicating a reduction in hepatic clearance (HELD et al. 1973a). Impaired kidney function may affect only the renal elimination of the active metabolite (HELD et al. 1973b).

Severe diarrhea probably reduces the absorption of glymidine (GERHARDS et al. 1964). In hepatic patients with increased serum bilirubin, glymidine is displaced from protein binding, but the consequences are not clear (HELD et al. 1973a). From a possible involvement of CYP2A1, 2 and CYP3A, it may be anticipated that rifampicin, isoniazid and ethambutol increase the clearance of glymidine (HELD 1980). Oxyphenbutazone and marginally phenylbutazone seem to inhibit glymidine metabolism (HELD and SCHEIBLE 1981). Doxycycline and phenprocoumon exhibit inhibitory effects on metabolic degradation of glymidine (HELD et al. 1970).

13. Carbutamide

Carbutamide is the first oral sulfonylurea introduced into therapy and is the only drug in this class with a sulfanilamide moiety relating it to the antibaterial sulfonamides. The drug is still on the market in France.

The absorption of carbutamide appears to be complete, since fecal and urinary recovery approaches 95%–98% (RIDOLFO and KIRTLEY 1956; QUATTRIN et al. 1957; GUGLIELMI and ZUCCONI 1958). The absolute bioavailability is not known and absorption appears to depend on formulation (SAFFAR et al. 1982). Peak concentrations are achieved after 4 h and more (ACHELIS and HARDEBECK 1955).

Carbutamide is acetylated at the amino group. This metabolite is excreted in urine, accounting for 50% or more of a dose, but the behavior in poor and extensive acetylators has not been determined. The distribution volume is said to be twice that of tolbutamide (STOWERS et al. 1958); hence 0.26 l/kg may be estimated. Fifty to 60% of carbutamide is bound to plasma proteins.

Carbutamide and the acetyl metabolite are predominantly eliminated with urine, only 5%–8% being found in the feces. Carbutamide is reabsorbed in the renal tubule to a considerable extent (KLAUS and STRIPECKE 1957). The mean elimination half-life is 40 h with a large individual variability from 6 to 89 h, which probably reflects the existence of poor and extensive metabolizers. A mean systemic clearance of 8 ml/min has been determined.

Approximately 10% of the total drug concentration in plasma is accounted for by the acetyl metabolite. Considerable concentrations of parent compound are detected in bile, so that enterohepatic recycling is very likely (Quattrin et al. 1956).

14. Glipentide

Glipentide has a structure very similar to that of glibenclamide and is used only in Spain. Information on pharmacokinetic data is poor. The elimination half-life of glipentide is about 4 h and the apparent distribution volume 0.15 l/kg. From these data, a clearance of approximately 30 ml/min can be estimated. Protein binding is close to 98% (Rimbau Barreras et al. 1976; Anton Fos et al. 1992; Mis et al. 1992).

15. Drugs Under Clinical Investigation

a) Glimepiride

The new compound glimepiride, developed by Hoechst, is completely absorbed, the absolute bioavailability having been determined as 107% (Badian et al. 1992). Peak concentrations are achieved 2–3 h after administration under fasting conditions.

Glimepiride is metabolized at the methyl group of the cyclohexyl ring to a hydroxymethyl derivate (M1) and further to carboxyglimepiride (M2) (Eckert et al. 1993). Only traces of a few other metabolites exist in urine. M1 exerts hypoglycemic activity in man (Badian et al. 1993a). From the data presented, a distribution volume of 0.18 l/kg can be estimated. Protein binding is 99.5%.

About 50% of a dose is excreted with the urine after both intravenous and oral administration, 33% as M1 and 17% as M2. Only 0.1%–0.4% is eliminated unchanged after an intravenous dose (Badian et al. 1992, 1993b). Mean elimination half-lives of 1.3–3.4 h and a systemic clearance 48 ml/min have been reported. The systemic clearance of M1 is higher with 78 ml/min and mainly renal, i.e., 44 ml/min. M1 exhibits a shorter half-life (1.2 h) than the parent compound (Badian et al. 1993a). In patients with renal disease, the terminal half-life remained unchanged while both oral clearance and distribution volume increased with reduction in creatinine clearance (Rosenkranz et al. 1991). Elimination of both metabolites is reduced. Propranolol appears to reduce the clearance of glimepiride by about 17%, but renal elimination of the metabolites is not changed (Badian et al. 1993b).

C. Safety and Tolerance

I. Hypoglycemia

Prolonged hypoglycemia represents the most common and severe side effect of sulfonylureas, and can lead to permanent neurologic damage and death in elderly people (BERGER 1971; 1985; ASPLUND et al. 1983; CAMPBELL 1985; FERNER and NEIL 1988). Hypoglycemic symptoms can be easily misdiagnosed as a cerebrovascular accident. Twenty percent of patients treated with sulfonylureas in Britain reported at least one episode of symptomatic hypoglycemia during a 6-month period (JENNINGS et al. 1989). In Sweden (ASPLUND et al. 1983; WIHOLM and WESTERHOLM 1984) and Switzerland (BERGER et al. 1986), only about 0.22 severe hypoglycemic episodes/1000 patient-years were reported,'as compared with an incidence of 100/1000 patient-years for insulin (BERGER et al. 1986; GERICH 1988). In the surveys a fatality rate of 3.4%–10% among the hospital-admitted patients has been reported (BERGER et al. 1986; SELTZER 1989).

Most cases of severe and fatal hypoglycemia have been reported with the long-acting sulfonylureas chlorpropamide and glibenclamide, the frequency of severe cases being 0.38, 0.34, 0.16 and 0.07 per 1000 treatment years for glibenclamide, chlorpropamide, glipizide and tolbutamide, respectively (BERGER et al. 1986). From the greater lowering of fasting plasma glucose concentration of glibenclamide, more nocturnal hypoglycemia should be expected than with glipizide (GROOP et al. 1987a,b).

It must be emphasized that almost all severe cases with prolonged hypoglycemia have involved patients over 70 years old (ASPLUND et al. 1983). Other risk factors include alcohol intake, poor nutrition, intercurrent gastrointestinal disease, hepatic disease, impaired renal function and drug interactions (BERGER 1985; SELTZER 1989). Sulfonylureas that cross the placenta may cause severe hypoglycemia in newborns of mothers with diabetes (KEMBALL et al. 1970).

Severe hypoglycemia requires hospital admission for successfull treatment. A bolus of 50% glucose should be given intravenously followed by continuous infusion of 10% or 20% glucose and regular monitoring of blood glucose levels for at least 3 days. If glucose concentrations cannot be maintained at 6–8 mmol/l, hydrocortisone and glucagon administration may help (FERNER and NEIL 1988).

II. Sulfonylurea Failure

In up to 30% of patients initially treated with sulfonylureas, satisfactory control of glycemia is not achieved (STOWERS and BREWSHER 1962; BERNHARD 1965; SINGER and HURWITZ 1967) mainly due to lack of dietary compliance (KOLTERMAN et al. 1984) or more importantly due to deficiency in insulin secretory capacity (MADSBAD et al. 1981; RENDELL et al. 1983; HOSKER et al.

1985). This rate of primary failure can be reduced to about 15% (LEBOVITZ 1983). Secondary failure may occur in about 5% of patients each year who initially responded to therapy (KRALL 1985), but this figure may vary considerably due to the lack of a widely accepted definition and increases with the duration of the disease (HAUPT et al. 1977). The causes of secondary failure have been divided into patient-related factors (diet failure, poor knowledge of the disease, lifestyle, stress, intercurrent illness), disease-related factors (increasing insulin deficiency or resistance) and therapy-related factors (inadequate drug dose, impaired drug absorption in the presence of severe hyperglycemia, concomitant therapy with diabetogenic drugs) (GROOP et al. 1989b). If the reasons for secondary failure are analyzed according to their mechanism, then 43% could be explained by hepatic (26%) and peripheral (17%) insulin resistance and only 13% by impaired B-cell function. However, 44% of the causes remain unknown.

III. Other Adverse Effects

Sulfonylureas are usually well tolerated, and the overall frequency of adverse effects is low, in the range of 2%–5%. Most adverse effects are mild and reversible on withdrawal (SINGER et al. 1961; ANONYMOUS 1971; CERDINO et al. 1975; PANNEKOEK 1975; KRANS 1979; JACKSON and BRESSLER 1981).

Allergic skin reactions are rare and include rashes, pruritus, erythema nodosum, erythema multiforme, exfoliative dermatitis, Steven-Johnson syndrome, purpura and photosensitivity (CLARKE et al. 1974). Hematologic complications such as leukopenia, thrombocytopenia, hemolytic and aplastic anemia, agranulocytosis and bone marrow aplasia are also rare events (MALACARNE et al. 1977; JACKSON and BRESSLER 1981; LEVITT 1987).

Gastrointestinal discomfort is more common, including dyspepsia, nausea and vomiting (O'DONOVAN 1959; EMANUELI et al. 1972; GUNDERSEN et al. 1975). Abnormal liver function tests, jaundice and cholestasis can occur with all agents, but reversible intrahepatic cholestasis is more common with chlorpropamide. Weight gain may occur as a consequence of improved glycemic control (ANONYMOUS 1983).

Chlorpropamide causes a disulfiram-like reaction in predisposed individuals in connection with alcohol intake (PODGAINY and BRESSLER 1968; GROOP et al. 1984). Unique to this compound is also the property to cause hyponatremia and fluid retention by enhancing the effect of the antidiuretic hormone in the distal tubule (WEISSMAN et al. 1971; KADOWAKI et al. 1983; BERGER1985). Conversely, acetohexamide, tolazamide and glibenclamide have a mild diuretic action (MOSES et al. 1973).

There is little, if any, evidence to suggest that sulfonylureas increase cardiovascular mortality in patients with NIDDM as originally suspected with tolbutamide (KLIMT et al. 1970; ANONYMOUS 1976). The validity of the results is questionable (SCHOR 1971; SELTZER 1972; KOLATA 1979; ANONYMOUS 1979), and the results of some other studies rather support the conclusion of improved survival in patients after myocardial infarction (PAASIKIVI and WAHLBERG 1971), reduced progression to manifest diabetes (SARTOR et al. 1980d) and cardiovascular morbidity (KNOWLER et al. 1987).

IV. Interactions

Numerous agents have been reported to influence the hypoglycemic effects of sulfonylureas through pharmacokinetic and/or pharmacodynamic interactions (HANSEN and CHRISTENSEN 1977; JACKSON and BRESSLER 1981; JACKSON 1990; O'BYRNE and FEELY 1990). Pharmacokinetic drug interactions on absorption, protein binding (which are rarely of importance unless metabolism is also inhibited, which almost regularly occurs (JACKSON and BRESSLER 1981), metabolism (inhibition or induction) and elimination are described in the previous chapters on the individual sulfonylurea drugs.

Agents that aggravate the diabetic state by inhibition of insulin secretion or action and other mechanisms include glucocorticoids, estrogens, oral contraceptives, medroxyprogesterone, thiazides and loop diuretics, β-adrenergic agonists and γ-antagonists (VERSCHOOR et al. 1986), calcium channel blockers, thyroid hormones, phenytoin, phenothiazines, indomethacin, azetazolamide, isoniazid, nicotinic acid and diazoxide.

Alcohol, salicylates in higher doses, biguanides, angiotensin-converting enzyme inhibitors (RETT et al. 1988; ARAUZ-PACHECO et al. 1990; BELL 1992; VEYRE et al. 1993), guanethidine and betanidine, monoamine oxidase inhibitors (ROWLAND et al. 1994), β-blockers and sympatholytic agents may increase the risk of hypoglycemia in patients taking sulfonylureas by (a) inhibition of gluconeogenesis, (b) an increase in peripheral glucose uptake and insulin release, (c) antagonism of endogenous hyperglycemia hormones, and (d) other mechanisms. It is noteworthy that differences may exist between sulfonylureas concerning the extent of a certain interaction.

Acknowledgement. The excellent secretarial help of Mrs. E. Grote is greatly acknowledged.

References

Achelis JD, Hardebeck K (1955) Über eine neue blutzuckersenkende Substanz. Dtsch Med Wschr 80:1452–1455
Adebayo GI, Coker HAB (1988) Lack of efficacy of cimetidine and ranitidine as inhibitors of tolbutamide metabolism. Eur J Clin Pharmacol 34:653–656
Adir J, Miller AK, Vestal RE (1982) Effects of total plasma concentration and age on tolbutamide plasma protein binding. Clin Pharmacol Ther 31:488–493
Ahren B, Lundquist J, Scherstén B (1986) Effects of glipizide on various consecutive insulin secretory stimulations in patients with type 2 diabetes. Diabetes Res 3:293–300
Almér LO, Johansson E, Melander A, Wahlin-Boll E (1981) Effects of sulfonylurea on the secretion and disposition of insulin and C-peptide. Acta Med Scand 210 [Suppl]:11–18

Almér LO, Johansson E, Melander A, Wahlin-Boll E (1982) Influence of sulfonylureas on the secretion, disposal and effect of insulin. Eur J Clin Pharmacol 22:27-32

Anonymous (1971) Glibenclamide: a review. Drugs 1:116-140

Anonymous (1976) University group diabetes program: a study of the effects of hypoglycemic agents on vascular complications in patients with adult-onset diabetes. VI. Supplementary report on nonfatal events in patients treated with tolbutamide. Diabetes 25:1129-1153

Anonymous (1979) American Diabetes Association: policy statement: the UGDP controversy. Diabetes Care 2:1-3

Anonymous (1983) UK prospective study of therapies of maturity-onset diabetes I. Effect of diet, sulphonylurea, insulin or biguanide therapy on fasting plasma glucose and body weight over one year. Diabetologia 24:404-411

Antal EG, Gillespie WR, Phillips JP, Albert KS (1982) The effect of food on the bioavailability and pharmacodynamics of tolbutamide in diabetic patients. Eur J Clin Pharmacol 22:459-462

Anton Fos GM, Garcia Domenech R, Perez Gimenez F, Galvez Alvarez J, Garcia March F, Soler Roca RM, Salabert Salvador MT (1992) Estudios de predicción de propiedades farmacocinéticas de hipoglucemiantes orales utilizando relaciones Q SA R An Real Acad Farm 58:551-562

Arauz-Pacheco, Ramirez LC, Rios JM, Raskin P (1990) Hypoglycemia induced by angiotensin-converting enzyme inhibitors in patients with non-insulin-dependent diabetes receiving sulfonylurea therapy. Am J Med 89:811-813

Arnqvist HJ, Karlberg BE, Melander A (1983) Pharmacokinetics and effects of glibenclamide in two formulations, HB419 and HB420, in type 2 diabetics. Ann Clin Res 15 [Suppl 37]:21-25

Arrigoni L, Fundak G, Horn J, Krakgän W, Ellesworth A, Opheim K, Taylor T, Bauer LA (1987) Chlorpropamide pharmacokinetics in young healthy adults and older diabetic patients. Clin Pharm 6:162-164

Asplund K, Wiholm B-E, Lithner F (1983) Glibenclamide-associated hypoglycaemia. A report on 57 cases. Diabetologia 24:412-417

Ayanoglu G, Witte PU, Badian M (1983) Bioavailability and pharmacodynamics of a sustained-release glibenclamide product (Deroctyl) in comparison to a standard tablet formulation (Euglucon, Daonil). Int J Clin Pharmacol Ther Toxicol 21:479-482

Ayanoglu G, Uihlein M, Grigoleit HG (1986) A new aspect of serum protein binding of tolbutamide. Int J Clin Pharmacol Ther Toxicol 24:65-68

Back DJ, Tjia J, Moenig H, Ohnhaus EE, Park BK (1988) Selective inhibition of drug oxidation after simultaneous administration of two probe drugs, antipyrine and tolbutamide. Eur J Clin Pharmacol 34:157-163

Badian M, Korn A, Lehr K-H, Malercyk V, Waldhäusl W (1992) Determination of the absolute bioavailability of glimepiride (HOE 490), a new sylphonylurea. Int J Clin Pharmacol Ther Toxicol 30:481-482

Badian M, Korn A, Lehr K-H, Malercyk V, Waldhäusl W (1993a) Pharmacokinetics and pharmacodynamics after intravenous administration of the hydroxymetabolite (M1) of glimepiride (HOE 490). Arch Pharmacol 347 [Suppl]:R27

Badian M, Korn A, Lehr K-H, Malercyk V, Waldhäusl W (1993b) Pharmacokinetic interaction between propranolol and glimepiride in healthy volunteers. Klin Pharmakol Akt 2:25

Bailey CJ, Flatt PR (1990) New antidiabetic agents. Smith-Gordon, Nishimura

Balant L (1981) Clinical pharmacokinetics of sulphonylurea hypoglycaemic drugs. Clin Pharmacokin 6:215-241

Balant L, Zahnd GR, Gorgia A, Schwarz R, Fabre J (1973) Pharmacokinetics of glipizide in man: influence of renal insufficiency. Diabetologia 9 [Suppl]:331-338

Balant L, Fabre J, Zahnd GR (1975) Comparison of the pharmacokinetics of glipizide and glibenclamide in man. Eur J Clin Pharmacol 8:63-69

Balant L, Zahnd GR, Weber F, Fabre J (1977) Behaviour of glibenclamide on repeated administration to diabetic patients. Eur J Clin Pharmacol 11:19–25

Balant L, Fabre J, Loutan L, Samini H (1979) Does 4-hydroxyglibenclamide show hypoglycaemic activity? Arzneimittelforschung 29:162–163

Barnes AJ, Garbien KJT, Crowley MF, Bloom A (1974) Effect of short and long term chlorpropamide treatment on insulin release and blood-glucose. Lancet II:69–72

Basabe JC, Farina JMS, Chieri RA (1976) Studies on the dynamics and mechanism of glibenclamide-induced insulin secretion. Horm Metab Res 8:413–419

Batch J, Ma A, Bird D, Noblke R, Charles B, Ravenscroft P, Cameron D (1990) The effects of ingestion time of gliclazide in relationship to meals on plasma glucose, insulin and C-peptide levels. Eur J Clin Pharmacol 38:465–467

Baynes C, Elkeles RS, Henderson AD, Richmond W, Johnston DG (1993) The effects of glibenclamide on glucose homeostasis and lipoprotein metabolism in poorly controlled type 2 diabetes. Horm Metab Res 25:96–101

Beck-Nielson H, Pederson O, Lindskov HO (1979) Increased insulin sensitivity and cellular insulin binding in obese diabetics following treatment with glibenclamide. Acta Endocrinol 90:451–462

Behrle M (1980) Untersuchung zur Pharmakokinetik von Glibenclamid bei nierengesunden und niereninsuffizienten Diabetikern. Thesis, Ruprecht-Karl-Universität, Heidelberg

Bell DSH (1992) Hypoglycemia induced by enalapril in patients with insulin resistance and NIDDM. Diabetes Care 15:934–936

Benakis A, Glasson B (1980) Metabolic study of ^{14}C-labelled gliclazide in normal rats and in rats with streptozotocin induced diabetes. In: Keen H et al. (eds) Gliclazide and the treatment of diabetes. Int Congr Symp Series 20. Academic, London, pp 57–69

Berger W (1971) 88 schwere Hypoglykämiezwischenfälle unter der Behandlung mit Sulfonylharnstoffen. Schweiz Med Wschr 71:1013–1022

Berger W (1985) Incidence of severe sideeffects during therapy with sulfonylureas and biguanides. Horm Metab Res 17 [Suppl 15]:111–115

Berger W, Caduff F, Pasquel M, Rump A (1986) Die relative Häufigkeit der schweren Sulfonylharntoff-Hypoglykämie in den letzten 25 Jahren in der Schweiz. Schweiz Med Wschr 116:145–151

Bergman U, Christenson I, Jansson B, Wiholm BE, Ostnam J (1980) Wide variation in serum chlorpropamide concentration in outpatients. Eur J Clin Pharmacol 18:165–169

Bernhard H (1965) Long-term observations on oral hypoglycemic agents in diabetes: the effect of carbutamide and tolbutamide. Diabetes 14:59–70

Best JD, Judzewitch RG, Pfeifer MA, Beard JC, Halter J, Porte D Jr (1982) The effect of chronic sulfonylurea therapy on hepatic glucose production in non-insulin dependent diabetes. Diabetes 31:333–338

Bigler F, Quitt P, Vecchi M, Vetter W (1972) Über den Stoffwechsel von Glibornurid beim Menschen. Arzneimittelforschung 22:2191–2198

Birkeland KI, Furuseth K, Melander A, Mowinckel P, Vaaler S (1994) Long-term randomized placebo-controlled double-blind therapeutic comparison of glipizide and glyburide. Diabetes Care 17:45–49

Bitzén PO, Melander A, Scherstén B, Wahlin-Boll E (1988) The influence of glipizide on early insulin release and glucose disposal before and after dietary regulation in diabetic patients with different degrees of hyperglycemia. Eur J Clin Pharmacol 34:31–37

Blum RA, Wilton JH, Hilligan DM, Gardner MJ, Henry EB, Harrison NJ, Schentag JJ (1991) Effect of fluconazole on the disposition of phenytoin. Clin Pharmacol Ther 49:420–425

Blume H, Stenzhorn G, Ali SL (1985) Zur Bioverfügbarkeit und pharmakodynamischen Aktivität handelsüblicher Glibenclamid-Fertigarzneimittel. 1. Mitteilung: Bioäquivalenzprüfung an gesunden Probanden unter oraler Kohlenhydrabelastung (Bioavailability and pharmacodynamic activity of normal retail glibenclamide finished preparations – 1st communication: bioequivalence test with healthy volunteers under oral carbohydrate load). Pharm Ztg 130:1062–1069

Blume H, Walter-Sack I, Ali SL, Siewert M, Stenzhorn G, Nowak N, Weber E (1987) Untersuchungen zur therapeutischen Relevanz der Bioäquivalenz und zur Chargenhomogenität glibenclamidhaltiger Fertigarzneimittel (Studies for the therapeutic importance of bioequivalence and batch to batch equivalence of glibenclamide tablets) Pharm Ztg 132:2352–2362

Blumenthal SA, Whitmer KR (1979) Hepatic effects of chlorpropamide. Inhibition of glucagon-stimulated gluconeogenesis in perfused livers of fasted rats. Diabetes 28:646–650

Borthwick LJ, Stowers JM (1979) Oral hypoglcaemic agents. Practitioner 222:358–366

Borthwick LJ, Davies IB, Jafri S, Lawrence JR, McEwen J, Pidgen AW (1984) Hypoglycaemic action of glibenclamide in diabetes: pharmacokinetic manipulation of insulin-glucose-glibenclamide relationships. Clin Sci 66:438–448

Bottermann P, Zilker T, Henderkott U, Ermler R (1976) The compatibility of Glurenorm with alcohol. Diab Croat 5:447–465

Boutagy J, Peart GF, Shenfield GM (1987) Gliclazide metabolic clearance in debrisoquine phenotyped subjects: a comparison with tolbutamide. Clin Exp Pharmacol Physiol 79 [Suppl 12]:79

Boutagy PJ, Shenfield GM (1987)Tolbutamide metabolic clearance in subjects of known debrisoquine phenotype. Clin Exp Pharmacol Physiol 88 [Suppl 10]:106–114

Boyd AE III (1988) Sulfonylurea receptors, ion channels and fruit flies. Diabetes 37:847–850

Brier ME, Stalker DJ (1993) Pharmacokinetics and pharmacodynamics of glyburide in dialysis patients. Pharm Res 10 [Suppl]:s384

Brockmeier D, Grigoleit HG, Leonhardt H (1985) Absorption of glibenclamide from different sites of the gastro-intestinal tract. Eur J Clin Pharmacol 29:193–197

Brotherton PM, Grievson P, McMartin C (1969) A study of the metabolic fate of chlorpropamide in man. Clin Pharm Ther 10:505–514

Büchele W, Kuhlmann H (1978) Gliquidon in der antidiabetischen Therapie nieren- und leberinsuffizienter Patienten. Med Welt 29:897–900

Bürkl B (1977) Der Einfluß einer Rifampicinbehandlung auf die Pharmakokinetik von Tolbutamid bei jugendlichen Normalpersonen. Thesis, Julius-Maximilians-Universität, Würzburg

Campbell IW (1985) Metformin and the sulfonylureas: the comparative risk. Horm Metab Res [Suppl] 15:105–111

Campbell DB, Adriaenssens P, Hopkins YW, Gordon B, Williams JRB (1980) Pharmacokinetics and metabolism of gliclazide: a review. In: Keen H et al. (eds) Gliclazide and the treatment of diabetes. Int Congr Symp Series 20. Academic, London, pp 71–82

Campbell DB, Gordon BH, Ings RMJ, Beaufils M, Meynier A, et al. (1986) The effect of renal disease on the pharmacokinetics of gliclazide in diabetic patients. Br J Clin Pharmacol 21:572P–573P

Carulli N, Ponz de Leon M, Mauro E, Moncti F, Ferrarri A (1976) Alteration of drug metabolism in Gilbert's syndrome. Gut 17:581–587

Casner PR (1988) Insulin-glyburide combination therapy for non-insulin dependent diabetes mellitus: a long-term double-blind, placebo-controlled trial. Clin Pharmacol Ther 44:594–603

Cate EW, Rogers JF, Powell JR (1986) Inhibition of tolbutamide elimination by cimetidine but not ranitidine. J Clin Pharm 26:372–377

Cerasi E, Efendic S, Thornqvist C, Luft R (1979) Effect of two sulfonylureas on the dose kinetics of glucose-induced insulin release in normal and diabetic patients. Acta Endocrinol 91:282–293

Cerdino V, Persigli A, Calvert J, Soler J (1975) Clinical evaluation of glipizide: results of a multicenter study in Spain. Rev Iberica Endocrinol 22:43–60

Chalk JB, Patterson M, Smith MH, Eadie MJ (1986) Correlations between in vitro dissolution, in vivo bioavailability and hypoglycaemic effect of oral glibenclamide. Eur J Clin Pharmacol 31:177–182

Chi HD, Jiang WD, Zhu XX, Guo Y, Karras HO (1993) Pharmacokinetics and relative bioavailability of a tablet of micronized glibenclamide in 4 healthy men. Acta Pharmacol Sin 14:193–197

Chiasson J-L, Hamet P, Verdy M (1991) The effect of Diamicron® on the secretion and action of insulin. Diab Res Clin Practice 14 [Suppl 2]:47–52

Christensen LK, Skovsted I (1969) Inhibition of drug metabolism by chloramphenicol. Lancet II:1397–1399

Chrystyn H, Winters C, Bramley PN, Wong V, Burgul R, Losowsky MS (1994) Improved absorption from a gliclazide betacyclodextrin complex formulation. Br J Clin Pharmacol 37:111 P

Chu P-C, Conway MJ, Krouse HA, Goodner CJ (1968)The pattern of response of plasma insulin and glucose to meals and fasting during chlorpropamide therapy. AnnIntern Med 68:757–768

Clarke BF, Campbell IW, Ewing DJ, Beveridge GW, MacDonald MK (1974) Generalized hypersensitivity reaction and visceral arteritis with fatal outcome during glibenclamide therapy. Diabetes 23:739–742

Cohen BD, Galloway JA, McMahon RE, Culp HW, Root MA, Henriques KJ (1967) Carbohydrate metabolism in uremia: blood glucose response to sulfonylurea. Am J Med Sci 254:608–618

Connacher AA, el Debani AH, Isles TE, Stevenson IH (1987) Disposition and hypoglycaemic action of glipizide in diabetic patients given a single dose of nifedipine. Eur J Clin Pharmacol 33:81–83

Cooper AJ, Ashcroft G (1967) Modification of insulin and sulfonylurea hypoglycaemia by monoamine-oxidase inhibitor drugs. Diabetes 16:272–274

Coppack SW, Laut AF, McIntosh SC, Rodgers AV (1990) Pharmacokinetic and pharmacodynamic studies of glibenclamide in non-insulin dependent diabetes mellitus. Br J Clin Pharmacol 29:673–684

Crooks MJ, Brown KF (1974) The binding of sulfonylureas to serum albumin. J Pharm Pharmacol 26:304–311

Crooks MJ, Brown KF (1975) Interaction of glipizide with human serum albumin. Biochem Pharmacol 24:298–299

Dahl-Puustinen ML, Alm C, Bertilsson L, Christenson I, Ostman J, Thunberg E (1990) Lack of relationship between glibenclamidase metabolism and debrisquine or mephenytoin hydroxylation phenotypes. Br J Clin Pharmacol 30:467–480

Dalgas M, Christiansen IB, Kjerulf K (1965) Fenylbutazoninduceret hypoglykaemitifaelde hos klorpropamidbehandlet diabetiker. Ugeskr Laeg 127:834–836

Danhof M, Danhof-Pont MB, Pank S, Bosijinga JK, Breimer DD (1986) Lisinopril in combination with glibenclamide: no pharmacokinetic interaction in healthy volunteers. Acta Pharmacol Toxicol 59 [Suppl V]:174

Davidson MB, Sladen G (1987) Effect of glyburide on glycogen metabolism in cultured rat hepatocytes. Metabolism 36:925–930

Dey NG, Castleden CM, Ward J, Cornhill J, McBurney A (1983) The effect of cimetidine in tolbutamide kinetics. Br J Clin Pharmacol 16:438–440

Dolais-Kitabgy J, Alengrin F, Freychet P (1983) Sulphonylureas in vitro do not alter insulin binding or insulin effect on amino acid transport in rat hepatocytes. Diabetologia 24:441–444

Dubach UC, Korn A, Raaflaub J (1975) Mehrdosenkinetik von Glibornurid beim Menschen. Arzneimittelforschung 25:1967–1969

Duckworth WC, Solomon SS, Kitabchi AE (1972) Effect of chronic sulfonylurea therapy on plasma insulin and proinsulin levels. J Clin Endocrinol Metab 35: 585–591

Dunbar JC, Foá PP (1974) An inhibitory effect of tolbutamide and glibenclamide (glyburide) on the pancreatic islets of normal animals. Diabetologia 10:27–35

Eckert HG, Kellner H-M, Gantz D, Jantz H, Hornke I, von puttkamer G-D (1993) Pharmacokinetics and metabolism of glimepiride. Clin Rep 27:61–92

Eckhardt W, Rudolph R, Sauer H, Schubert WR, Underutsch D (1972) Zur pharmakologischen Interferenz von Glibornurid mit Sulfaphenazol, Phenylbutazon, und Phenprocoumon beim Menschen. Arzneimittelforschung 22:2212–2219

Efendic S, Enzmann F, Nylén A, Uvnäs-Wallensten K, Luft R (1979) Effect of glucose/sulfonylurea interaction on release of insulin, glucagon and somatostatin from isolated perfused rat pancreas. Proc Natl Acad Sci USA 76:5901–5904

Elliot BD, Langer O, Schenker S, Johnson RF (1991) Insignificant transfer of glyburide occurs across the human placenta. Am J Obstet Gynecol 165 [pt 1]:807–812

Elliot BD, Bynum D, Langer O (1993) Glyburide does not cross the diabetic placenta in significant amounts. Am J Obstet Gynecol 168 [pt 2]:360

Elston AC, Bayliss MK, Park GR (1993) Effect of renal failure on drug metabolism by the liver. Br J Anaest 71:282–290

Elvander-Stahl E, Melander A, Wahlin-Boll E (1984) Indobufen interacts with the sulfonylurea, glipizide, but not with the beta-adrenergic receptor antagonists, propranolol and atenolol. Br J Clin Pharmacol 18:773–778

Emanueli A, Molari E, Pirola LC, Caputo G (1972) Glipizide, a new sulfonylurea in the treatment of diabetes mellitus: summary of clinical experience in 1064 cases. Arzneimittelforschung 22:1881–1885

Eriksson J, Franssila-Kallunki A, Ekstrand A, Saloranta C, Widén E, Schalin C, Groop L (1989) Early metabolic defects in persons at increased risk for non-insulin dependent diabetes mellitus. N Engl J Med 321:337–343

Evans M, Glass RC, Mitchard M, Munday BM, Yates R (1979) Bioavailability of chlorpropamide. Br J Clin Pharmacol 7:101–105

Faber OK, Beck-Nielsen H, Binder C, Butzer P, Damsgaard EM, Frölund F, Hjollund E, Lindskov HO, Melander A, Pedersen O, Petersen P, Schewarts-Sorensen N, Wahlin-Boll E (1990) Acute actions of sulfonylurea drugs during long term treatment of NIDDM. Diabetes Care 13 [Suppl 3]:26–31

Fagerberg SE, Gamstedt A (1980) Paired observations between different sulphonylureas in treatment. In: Keen H et al. (eds) Gliclazide and the treatment of diabetes. Int Congr Symp Series 20. Academic, London, pp 143–151

Feely J, Collins WCJ, Cullen M, El Debani AH, MacWalter RS, Peden NR (1993) Potentiation of the hypoglycaemic response to glipizide in diabetic patients by histamine H_2-receptor antagonists. Br J Clin Pharmacol 35:321–323

Feinglos MN, Lebovitz HE (1978) Sulphonylureas increase the number of insulin receptors. Nature 276:184–185

Feldman JM, Lebovitz HE (1969a) Biological activities of tolbutamide and its metabolites. Diabetes 18:529–537

Feldman JM, Lebovitz HE (1969b) Appraisal of the extrapancreatic actions of sulfonylureas. Arch Intern Med 123:314–322

Ferarri C, Frezzati S, Testori GP, Bertazzoni A (1976) Potentation of hypoglycemic response to intravenous tolbutamide by clofibrate. N Engl J Med 294:1184

Ferner RE, Chaplin S (1987) The relationship between the pharmacokinetics and pharmacodynamic effects of oral hypoglycaemic drugs. Clin Pharmacokin 12:379–401

Ferner RE, Neil HAW (1988) Sulphonylureas and hypoglycaemia. Br Med J 296:949–950

Ferner RE, Alberti KGMM, Rawlins MD (1990) The acute effects of tolbutamide during a glucose clamp and after a mixed meal in patients with non-insulin dependent diabetes. Br J Clin Pharmacol 29:638 P

Fleishaker JC, Phillips JP (1991) Evaluation of a potential interaction between erythromycin and glyburide in diabetic volunteers. J Clin Pharmacol 31:259–262

Forette B, Rolland A, Hopkins Y, Gordon B, Campbell B (1982) Gliclazide kinetics in the elderly. In: Alberti KGMM, Ogada T, Aluoch JA, Mngola EN (eds) 11 Congr Intern Diab Found. Excerpta Medica, Amsterdam, pp 8–9

Fraser RJ, Horowitz M, Maddox AF, Harding PE, Chatterton BE, Dent J (1990) Hyperglycemia slows gastric emptying in type I (insulin-dependent) diabetes mellitus. Diabetologia 33:675–680

Friedlander EO (1957) Use of tolbutamide in insulin resistant diabetes: report of a case. N Engl J Med 257:11–14

Fuccella LM, Tamassia V, Valzelli G (1973) Metabolism and kinetics of the hypoglycemic agent glipizide in man: comparison with glibenclamide. J Clin Pharm 13:68–75

Fujii T, Nakamura K, Furukawa H, et al. (1983) Drug interactions of gliclazide and other sulphonylureas in protein binding in vitro and in hypoglycaemic effects in rats. Arzneimittelforschung 33:1535–1537

Gaines KL, Hamilton S, Boyd AE III (1988) Characterization of the sulfonylurea receptor on beta-cell membranes. J Biol Chem 263:2589–2592

Galloway JA, McMahon RE, Culp HW, Marshall FJ, Young EC (1967) Metabolism, blood levels and rate of excretion of acetohexamide in human subjects. Diabetes 16:118–127

Ganda OP, Kahn CB, Soeldner JS, Gleason RE (1975) Dynamics of tolbutamide, glucose and insulin interrelationships following varying doses of intravenous tolbutamide in normal subjects. Diabetes 24:354–361

Ganley JA, Mc Ewen J, Calvert RT, Barker MCJ (1984) The effect of in-vivo dispersion and gastric emptying on glibenclamide absorption from a novel, rapidly dissolving capsule formulation. J Pharm Pharmacol 36:734–739

Garvey TW, Olefsky JM, Hamman RF, Kolterman O (1985) The effect of insulin treatment on insulin secretion and insulin action in type II diabetes mellitus. Diabetes 34:222–234

Gerard J, Lefebvre PJ, Luyckx AS (1984) Glibenclamide pharmacokinetics in acarbose-treated type 2 diabetics. Eur J Clin Pharmacol 27:233–236

Gerhards E, Gibian H, Kolb KH (1964) 2-Benzol-sulfonamide-5-(β-methoxyäthoxy)-pyrimidine (Glycodiazin). Untersuchungen mit einer neuen blutzuckersenkenden Substanz. Arzneim Forsch 14:394–402

Gerhards E, Kolb KH (1965) 2-Benzolsulfonamido-5(β-methozy-äthoxy)-pyrimidin (Glykodiazin). II Der Stoffwechsel von 2-Benzolsulfonamido-5(β-hyd oxyäthoxy)-pyrimidin, einem blutzuckersenkenden Metaboliten des Glykodiazin beim Menschen. Arzneimittelforschung 15:1375–1379

Gerich JE (1988) Glucose counterregulation and its impact on diabetes mellitus. Diabetes 37:1608–1617

Gerich JE (1989) Oral hypoglycaemic agents. N Engl J Med 321:1231–1245

Ginier P, Madan S, Fajardo F, Levin SR (1985) Once-daily use of glyburide. Am J Med 79 [Suppl 3 B]:72–77

Göbel KJ, Moldrzyk D, Vollmer KO, Linn C (1990) Effect of multiple dose enoxacin administration on the bioavailability of glibenclamide. Eur J Drug Metab Pharmacokinet 15 [Suppl]: Abstr 122

Goto S, Watanabe S, Maehara S, Nakata Y, Tateyama T (1985) Displacing effects of chenodexoycholic acid, urodeoxycholic acid and sulfadimethoxine on plasma protein binding of tolbutamide. J Pharmacobiodyn 8:440–447

Gramatté T, Terhaag B, LePEtit G, Richter K, Feller K (1989) In-vivo-Bioverfügbarkeit und in-vitro-Liberation von Glibenclamid aus drei Maninil®-Zubereitungen (In vivo bioavailability and in vitro liberation of three Maninil® preparations of glibenclamide) Z Klin Med 44:183–186

Grandt R, Braun W, Schulz H-U, Lührmann B, Frercks H-J (1989) Glibenclamide steady state plasma levels during concomitant vinpocetine administration in type II diabetic patients. Arzneimittelforschung 39 (II):1451–1454

Grill V, Efendic S, Regitz G (1986) Potentiation by previous nutrients of gliben-clamide-induced insulin release in man. An effect which is counteracted by meal-induced retardation of drug absorption. Scand J Clin Lab Inv 46:527–532

Grodsky GM, Epstein GH, Fanska R, Karam JH (1977) Pancreatic actions of sulfonylureas. Fed Proc 36:2714–2719

Groop L, Harno K (1979) The combination of insulin and sulfonylurea – an approach to improved metabolic control in insulin resistant diabetics. Acta Endocrinol [Suppl] 227:33

Groop L, Eriksson CJP, Huupponen R, Ylikahri R, Pelkonen R (1984) Roles of chlorpropamide, alcohol and acetaldehyde in determining the chlorpropamide-alcohol flush. Diabetologia 26:34–38

Groop L, Wahlin-Boll E, Groop P-H, Tötterman K-J, Melander A, Tolppanen E-M, Fyhrquist F (1985) Pharmacokinetics and metabolic effects of glibenclamide and glipizide in type 2 diabetics. Eur J Clin Pharmacol 28:697–704

Groop L, Groop P-H, Stenman S, Saloranta C, totterman K-J, Fyhrquist F, Melander A (1987a) Comparison of pharmacokinetics, metabolic effects and mechanisms of action of glyburide and glipizide during long-term treatment. Diabetes Care 10:671–678

Groop L, Luzi L, Melander A, Groop P-H, Ratheiser K, Simonson DC, De Fronzo RA (1987b) Different effects of glyburide and glipizide on insulin secretion and hepatic glucose production in normal and NIDDM subjects. Diabetes 36:1320–1328

Groop L, Groop P-H, Stenman S, Saloranta C, Tötterman K-J, Fyhrquist F, Melander A (1988) Do sulfonylureas influence hepatic insulin clearance? Diabetes Care 11:689–690

Groop LC, Luzi L, DeFronzo RA, Melander A (1989a) Hyperglycaemia and absorption of sulfonylurea drugs. Lancet II:129–130

Groop L, Schalin C, Franssial-Kallunki A, Widén E, Ekstrand A, Eriksson J (1989b) Characteristics of non-insulin dependent diabetic patients with secondary failure to oral antidiabetic therapy. Am J Med 87:183–190

Groop LC, Groop P-H, Stenman S (1990) Combined insulin-sulfonylurea therapy in treatment of NIDDM. Diabetes Care 13 [Suppl 3]:47–52

Groop LC, Ratheiser K, Luzi K, Melander A, Simonson DC, Petrides A, Bonadonna RC, Widén E, De Fronzo RA (1991a) Effect of sulfonylurea on glucose stimulated insulin secretion in healthy and non-insulin dependent diabetic subjects: a dose-response study. Acta Diabetol 28:162–168

Groop LC, Barzilai N, Ratheiser K, Luzi L, Wahlin-Boll E, Melander A, De Fronzo RA (1991b) Dose-dependent effects of glyburide on insulin secretion and glucose uptake in humans. Diabetes Care 14:724–727

Groop LC, Widén E, Ekstrand A, Saloranta C, Franssila-Kallunki A, Schalin-Jäntti C, Eriksson JG (1992) Morning or Bedtime NPH Insulin combined with sulf-onylurea in treatment of NIDDM. Diabetes Care

Groop P-H, Melander A, Groop LC (1993a) The relationship between early insulin release and glucose tolerance in healthy subjects. Scand J Clin Lab 53:405–409

Groop P-H, Melander A, Groop LC (1993b) The acute effect of preprandial exogenous and endogenous sulphonylurea-stimulated insulin secretion on post-prandial glucose excursions in patients with type 2 diabetes. Diab Med 10:633–637

Groop LC, Widén E, Ferrannini E (1993c) Insulin deficiency or insulin resistance in the pathogenesis of NIDDM. Inborn errors of metabolism or of methods? Diabetologia 36:1326–1331

Grunberger G, Ryan J, Gorden P (1982) Sulfonylureas do not affect insulin binding or glycemic control in insulin-dependent diabetics. Diabetes 31:890–896

Guglielmi G, Zucconi (1958) Fissazione ed eliminazione negli organi della carbut-amida marcata con S^{35}. Minerva Med 49:1509–1511

Gundersen K, Molony BA, Crim JA, Hearron AE Jr, Maile (1975) Micronase® (glyburide): clinical overview. In: Rifkin H (ed) Micronase® glyburide: pharmacology and clinical evaluation. Excerpta Medica, Amsterdam: pp 254–264

Haaber AB, Groop PH, Blitzén PO, Faber OK, Hansen PM, Helmquist S, Wahlin-Boll E, Groop LC, Melander A (1993) Pharmacokinetics and effects of different formulations of glipizide in patients with non-insulin-dependent diabetes mellitus. Drug Invest 5:114–120

Hajdú P, Kohler KF, Schmidt FH, Springler H (1969) Physikalisch-chemische und analytische Untersuchungen an HB 419. Arzneimittelforschung 19:1381–1386

Hansen JM, Christensen LK (1977) Drug interactions with oral sulphonylurea hypolycaemic drugs. Drugs 13:24–35

Harder S, Merz PG, Rietbrock N (1993) Lack of pharmacokinetic interaction between carvedilol and digitoxin, phenprocoumon or glibenclamide. Cardiovasc Drugs Ther 7 [Suppl 2]:447

Hartling SG, Faber OK, Wegmann M-L, Wahlin-Boll E, Melander A (1987) Interaction of ethanol and glipizide in humans. Diabetes Care 10:683–686

Haupt E, Köberich W, Cordes V, Beyer J, Schöffling K (1972) Untersuchungen zu Dosis-Wirkungs-Relationen verschiedener Sulfonylharnstoffderivate der alten und neuen Generation. Arzneimittelforschung 22:2203–2208

Haupt E, Küllmer KA, Schöffling K (1976) Pharmacodynamics of glurenorm. Diab Croat 5:373–391

Haupt E, Laube F, Loy H, Schöffling K (1977) Secondary failures in modern therapy of diabetes mellitus with blood glucose lowering sulfonamides. Med Klin 72: 1529–1536

Haupt E, Putschky F, Zoltobrocki M, Schöffling K (1984) Pharmacodynamics and pharmacokinetics of 2 glibenclamide preparations in type 2 diabetes: intraindividual double-blind comparison of Euglucon 5 (HB 419) and Euglucon N (HB 420). Dtsch Med Wschr 109:210–213

Heine P, Kewitz H, Schnapperelle U (1974) Dose-response relationship of tolbutamide and glibenclamide in diabetes mellitus. Eur J Clin Pharmacol 7:321–330

Held H (1980) Correlation between the activity of hepatic benzo [a] pyrene hydroxylase activity in human needle biopsies and the clearance of glymidine. Hepato-Gastroenterol 27:266–270

Held H, Scheible G (1981) Interaktion von Phenylbutazon und Oxyphenbutazon mit Glymidin. Arzneimittelforschung 31:1036–1038

Held H, Kaminski B, von Oldershausen HF (1970) Die Beeinflussung der Elimination von Glykodiazin durch Leber- und Nierenfunktionsstörungen und durch eine Behandlung mit Phenylbutazon, Phenprocumarol und Doxycyclin. Diabetologia 6:386–391

Held H, Eisert R, von Oldershausen HF (1973a) Pharmakokinetik von Glymidine (Glykodiazin) und Tolbutamid bei akuten und chronischen Leberschäden. Arzneimittelforschung 23:1801–1807

Held H, Eisert R, von Oldershausen HF(1973b) Über die Pharmakokinetik von Tolbutamid und Glykodiazin bei Leber- und Nierenschäden. Verhandl Dtsch Ges Inn Med 79:1220–1223

Hellmann B, Sehlin J, Täljedal IB (1984) Glibenclamide is exceptional among hypoglycemic sulfonylureas in accumulating progressively in β-cell-rich pancreatic islets. Acta Endocrinol 105:385–390

Hellstern A, Hellenbrecht D, Saller R, Wiest K, Hellstern C (1985) Pilotstudie zur Pharmakokinetik und -dynamik von Glibenclamid bei Patienten mit Lebercirrhose. Klin Wochenschr 63 [Suppl IV]:134

Helqvist S, Hartling SG, Faber OK, Launchbury P, Wahlin-Boll E, Melander A (1991) Pharmacokinetics and effects of glipizide in healthy volunteers. A comparison between conventional tablets and a rapid-release soft gelatin capsule. Drug Invest 3:69–75

Henry RR, Schaeffer L, Olefsky JM (1985) Glycemic effects of intensive caloric restriction and isocaloric refeeding in noninsulin-dependent diabetes mellitus. J Clin Endocrinol Metab 61:917–925

Henry RR, Brechtel G, Griver K (1988) Secretion and hepatic extraction of insulin after weight loss in obese noninsulin dependent diabetes mellitus. J Clin Endocrinol Metab 66:979–986

Heptner W, Kellner H-M, Christ O, Weihrauch D (1969) Metabolismus von HB 419 am Tier. Arzneimittelforschung 19:1400–1404

Hoffmann A, Fischer Y, Gilhar D, Raz I (to be published) Effects of hyperglycemia on glibenclamide absorption in patients with non-insulin dependent diabetes mellitus. Eur J Clin Pharmacol   ·

Hollis CR, McNeill DB, Deatkine D, Allen BT, Feinglos MN (1991) Use of liquid glyburide to assess insulin secretory reserve in type 2 diabetes. Diabetes 40 [Suppl 1]:16A

Hosker JP, Burnett MA, Davies EG, Harris EA, Turner RC (1985) Sulfonylurea therapy doubles β-cell response to glucose in type II diabetic patients. Diabetologia 28:809–814

Hosker JP, Rudenski AS, Burnett MA, Matthews DR, Turner RC (1989) Similar reduction of first- and second-phase B-cell responses at three different glucose levels in type II diabetes and the effect of gliclazide therapy. Metabolism 38:767–772

Houssy BA, Penhos JC (1956) Action of the hypoglycemic sulfonylurea compounds in hypophysectomized, adrenalectomized and depancreatectomized animals. Metabolism 5:727–732

Hsu PL, Ma JKH, Luzzi LA (1974) Interaction of sulfonylureas with plasma proteins. J Pharmac Sci 63:570–573

Huupponen R, Lammintausta R (1981) Chlorpropamide bioavailability and pharmacokinetics. Int J Clin Pharmacol Ther Toxicol 19:331–333

Huupponen R, Viikari J, Saarimaa H (1982a) Chlorpropamide and glibenclamide serum concentrations in hospitalized patients. Ann Clin Res 14:119–122

Huupponen R, Seppälä P, Iisalo E (1982b) Glipizide parmacokinetics and response in diabetics. Int J Clin Pharmacol Ther Toxicol 20:417–422

Huupponen R, Karhuvaara S, Seppala P (1985) Effect of guar gum on glipizide absorption in man. Eur J Clin Pharmacol 28:717–719

Igaki A, Kobayashi K, Kimura M, Sagoguchi T, Matsuoka A (1992) Influence of blood proteins on biomedical analysis. XII. Effects of glycation on gliclazide (oral hypoglycaemic drug) – binding with serum albumin in diabetics. Chem Pharm Bull 40:255–257

Ikegami H, Shima K, Tanaka A, Tahara Y, Hirota M, Kumahara Y (1986) Interindividual variation in the absorption of glibenclamide in man. Acta Endocrinol 111:528–532

Imamura Y, Kojima Y, Ichibagase H (1985) Binding of acetohexamide and its major metabolite, (-)-hydroxyhexamide, to human serum albumin. Chem Pharm Bull 33:1281–1284

Ings RMJ, Lawrence JR, Mc Donald A, Mc Ewen J, Pidgen AW, Robinson JD (1982) Glibenclamide pharmacokinetics in healthy volunteers: evidence for zero-order drug absorption. Br J Clin Pharmacol 13:264P–265P

Ishibashi F, Takashina S (1990) The effect of timing on gliclazide absorption and action. Hiroshima J Med Sci 39:7–9

Jaber LA, Wenzloff NJ, Welshman IR, Antal EJ (1991) The effect of age upon the pharmacokinetics /pharmacodynamics of glyburide after acute and chronic administration to type II diabetics. Diabetes 40 [Suppl 1]:344A

Jaber LA, Ducharm MP, Slaughter RL, Edwards DJ (1992) Influence of age on pharmacokinetics and pharmacodynamics of glipizide in patients with NIDDM. Pharmacotherapy 12:502

Jaber LA, Antal EJ, Slaughter RL, Welshman IR (1993a) The pharmacokinetics and pharmacodynamics of 12 weeks of glyburide therapy in obese diabetics. Eur J Clin Pharmacol 45:459–463

Jaber LA, Ducharme M, Slaughter RL, Edwards DJ (1993b) Influence of obesity on pharmacokinetics and pharmacodynamics of glipizide in patients with NIDDM. Pharmacotherapy 13:276

Jackson WPU (1969) Sulfonylureas in the management of human diabetes. In: Campbell GD (ed) Oral hypoglycemic agents. Academic, New York, pp 135–192

Jackson JE (1990) Sulfonylurea hypoglycaemic agents. Drug Ther 20:39–53

Jackson JE, Bressler R (1981) Clinical pharmacology of sulphonylurea hypoglycaemic agents, parts 1 and 2. Drugs 22:211–245; 295–320

Jarrett RJ, Keen H, Fuller JH, McCartney M (1979) Worsening to diabetes in men with impaired glucose tolerance ("borderline diabetes"). Diabetologia 16:25–30

Jeng C-Y, Hollenbeck CB, Wu M-S, Chen Y-DI, Reaven GM (1989) How does glibenclamide lower plasma glucose concentration in patients with type 2 diabetes? Diab Med 6:303–308

Jennings AM, Wilson RM, Ward JD (1989) Symptomatic hypoglycemia in NIDDM patients treated with oral hypoglycemic agents. Diabetes Care 12:203–208

Johnson JF, Dobmeier ME (1990) Symptomatic hypoglycaemia secondary to a glipizide-trimethoprim/sulfamethoxazole drug interaction. DICP Ann Pharmacother 24:250–251

Johnson BF, La Belle P, Wilson J, Allan J, Zupkis RV, Ronca PD (1990) Effects of lovastatin in diabetic patients treated with chlorpropamide. Clin Pharmacol Ther 48:467–472

Johnson AB, Argyraki M, Thow JC, Broughton D, Miller M, Taylor R (1991) The effect of sulphonylurea therapy on skeletal muscle glycogen synthase activity and insulin secretion in newly presenting type 2 (non-insulin dependent) diabetic patients. Diabetic Medicine 8:243–253

Jönsson A, Rydberg T, Ekberg G, Hallengren B Melander A (1994) Slow elimination of glyburide in NIDDM subjects. Diab Care 17:142–145

Juan D, Molitch ME, Johnson MK, Carlson RF, Antal EG (1990) Unaltered drug metabolizing systems in type II diabetes mellitus before and during glyburide therapy. J Clin Pharmacol 30:943–947

Judis J (1972) Binding of sulfonylureas to serum proteins. J Pharmac Sci 61:89–93

Judzewitsch RG, Pfeifer MA, Best JD, Beard JC, Halter JB, Porte D (1982) Chronic chlorpropamide therapy of non-insulin-dependent diabetes augments basal and stimulated insulin secretion by increasing islet sensitivity to glucose. J Clin Endocrinol Metab 55:321–328

Kadowaki T, Hagura R, Kajinuma H, Kuzuya N, Yoshida S (1983) Chlorpropamide-induced hyponatremia: incidence and risk factors. Diabetes Care 6:468–471

Kajinuma H, Kuzuya TT, Ide T (1974) Effects of hypoglycemic sulfonamides on glucagon and insulin secretion in ducks and dogs. Diabetes 23:412–417

Kallio J, Huupponen R, Pyykkö K (1990) The relationship between debrisoquine oxidation phenotype and the pharmacokinetics of chlorpropamide. Eur J Clin Pharmacol 39:93–95

Karam JH, Sanz N, Salamon E, Nolte MS (1986) Selective unresponsiveness of pancreatic B-cells to acute sulfonylurea stimulation during sulfonylurea therapy in NIDDM. Diabetes 35:1314–1320

Karttunen P, Uusitupa M, Nykänen S, Robinson JD, Sipilä J (1985) The pharmacokinetics of glibenclamide: a single dose comparison of four preparation in human volunteers. Int J Clin Pharmacol Ther Toxicol 23:642–646

Kater RMH, Tobon F, Iker FL (1969) Increased rate of tolbutamide metabolism in alcoholic patients. J Am Med Assd 207:363–365

Kaubisch N, Koss FW, Hammer R (1979) The biochemistry of Glurenorm. In: Pieterse and Klarenbeck (eds) New developments in the oral therapy of diabetes mellitus. Boehringer Ingelheim, Symposium Diabetes Mellitus, Utrecht, pp 37–50

Kellner HM, Christ O, Rupp W, Heptner W (1969) Resorption, Verteilung und Ausscheidung nach Gabe von ^{14}C-markiertem HB 419 an Kaninchen, Ratten und Hunden. Arzneimittelforschung 19:1388–1400

Kemball ML, McIver C, Milner RD, Nourse CH, Schiff D, Tiernan JR (1970) Neonatal hypoglycaemia in infants of diabetic mothers given sulfonylurea drugs in pregnancy. Arch Dis Child 45:696–701

Kivistö KT, Neuvonen PJ (1990) The effect of cholestyramine and activated charcoal on glipizide absorption. Eur J Clin Pharmacol 30:733–736

Kivistö KT, Neuvonen PJ (1991a) Enhancement of absorption and effect of glipizide by magnesium hydroxide. Clin Pharmacol Ther 49:39–43

Kivistö KT, Neuvonen PJ (1991b) Differential effects of sodium bicarbonate and aluminium hydroxide on the absorption and activity of glipizide. Eur J Clin Pharmacol 40:383–386

Kivistö KT, Neuvonen PJ (1992) Effect of magnesium hydroxide on the absorption and efficacy of tolbutamide and chlorpropamide. Eur J Clin Pharmacol 42:675–680

Kivistö KT, Lehto P, Neuvonen PJ (1993) The effects of different doses of sodium bicarbonate on the absorption and activity of non-micronized glibenclamide. Int J Clin Pharmacol Ther Toxicol 31:236–240

Klaus D, Stripecke W (1957) Untersuchungen über die totale und renale Clearance des N_1-sulfanilyl-N_2-n-butyl-carbamids (Nadisan). Z Ges Inn Med 12:289–294

Klimt CR, Knatterud GL, Meinert CL, Prout TE (1970) University Group Diabetes Program: a study of the effects of hypoglycemic agents on vascular complications in patients with adult-onset diabetes. Diabetes 19 [Suppl 2]:747–815

Knodell RG, Hall SD, Wilkinson GR, Guengerich FP (1987) Hepatic metabolism of tolbutamide: characterization of the form of cytochrome P-450 involved in methyl-droxylation and relationship to in vivo disposition. J Pharmacol Exp Ther 241:1112–1119

Knowler WC, Sartor G, Scherstén B (1987) Effects of glucose tolerance and treamtment of abnormal tolerance on mortality in Malmöhus County, Sweden [Abstract]. Diabetologia 30:541A

Kobayashi K, Kimura M, Sakoguchi T, Kitani Y, Mitsuo, et al (1981) Influence of blood proteins on biomedical analysis, III: pharmacokinetics and protein binding of gliclazide. J Pharmacobiodyn 4:436–442

Kobayshi K, Kimura M, Sakoguchi T, Hase A, Matsuoka A, Kaneko S (1984) Pharmacokinetics of gliclazide in healthy and diabetic subjects. J Pharm Sci 73:1684–1687

Kobayashi K, Hase A, Kimura M, Sakoguchi T, Shimosawa M, et al (1985) Influence of blood proteins on biomedical analysis, VIII: attempts at purification of gliclazide-degrading factor in human serum. Life Sci 37:2015–2019

Kobayashi KA, Bauer LA, Horn JA, Opheim K, Wood F, Kradjan WA (1988) Glipizide pharmacokinetics in young and elderly volunteers. Clin Pharm 7:224–228

Kolata GB (1979) Controversy over study of diabetes drugs continues for nearly a decade. Science 203:986–990

Kolterman EG (1987) The impact of sulfonylureas on hepatic glucose metabolism in Type II diabetes. Diabetes Metab Rev 3:399–414

Kolterman EG, Gray RS, Shapiro G, Scarlett JA, Griffin J, Olefsky JM (1984) The acute and chronic effects of sulfonylurea therapy in type II diabetic subjects. Diabetes 33:346–354

Kopitar Z (1975) Humanpharmakokinetik und Metabolismus von ^{14}C-markiertem Gliquidone (ARDF 26). Arzneimittelforschung 24:1455–1460

Kopitar Z, Koss FW (1975a) Pharmacokinetic behaviour of ARDF 26, a new sulfonylurea. Proc Intern Conf Glurenorm 13–14 March, pp 20–29

Kopitar Z, Koss FW (1975b) Pharmakokinetisches Verhalten von Gliquidone (AR-DF 26), einem neuen Sulfonylharnstoff (Zusammenfaseung bisheriger Untersuchungen) Arzneimittelforschung 25:1933–1938

Koss FW, Kopitar Z, Hammer R (1976) The pharmacokinetic profile of Glurenorm. Diab Croat 5:355–371

Kradjan W, Kobayashi KA, Bauer LA, Horn JR, Opheim K, Wood F Jr (1989) Glipizide pharmacokinetics: effects of age, diabetes and multiple dosing. J Clin Pharmacol 29:1121–1127

Krall LP (1985) Oral hypoglycemic agents. In: Marble et al (eds) Joslin's diabetes mellitus. Lea and Febiger, Philadelphia, pp 412–452

Kramer M, Hecht G, Langecker H, Harwart A, Richter KD, Gloxhuber C (1964) Pharmakologie des 2-Benzol-sulfonamido-5(β-methyoxy-äthoxy)-pyrimidins (Glydodiazin), einer neuen blutzuckersenkenden Verbindung. Arzneimittelforschung 14:377–385

Krans H (1979) Insulin, glucagon and oral hypoglycaemic drugs. In: Dukes MNG (ed) Meyler's side effects of drugs: annual 3. Excerpta Medica, Amsterdam, pp 343–353

Krishnaiah YSR, Satyanarayana S, Visweswaram D (1994) Interaction between tolbutamide and ketoconazole in healthy subjects. Br J Clin Pharmacol 37:205–207

Kristensen M, Hansen JM (1968) Accumulation of chlorpropamide caused by dicoumarol. Acta Med Scand 183:83–86

Kubacka RT, Antal EJ, Juhl RP (1986) The effect of aspirin and ibuprofen on glucose metabolism and glyburide pharmacokinetics and pharmacodynamics in normal subjects. Pharm Res 3 [Suppl]:111 S

Kubacka R, Antal EJ, Juhl RP (1987) The paradoxical effect of cimetidine and ranitidine on glibenclamide pharmacokinetics and pharmacodynamics. Br J Clin Pharmacol 23:743–751

Kühnle HF, Hrstka V, Reiter J, Mennicken C, Schmidt FH (1984) Untersuchungen zur Pharmakokinetik und zur biliären Elimination von Glibenclamid. Aktuel Endokrinol Stoffwechsel 5:101

Lazar JD, Wilner KD (1990) Drug interaction with fluconazole. Rev Inf Dis 12 [Suppl 3]:S327–S333

Lazarus SS, Volk BB (1959) Physiological basis of the effectiveness of combined insulin-tolbutamide therapy in stable diabetes. Ann NY Acad Sci 82:590–602

Lebovitz H (1983) Clinical utility of oral hypoglycemic agents in the management of patients with noninsulin-dependent diabetes mellitus. Am J Med 75 [Suppl 5B]:94–99

Leemann T, Transon C, Dayer P (1993) Cytochrome P 450 TB (CYP2C): a major monooxygenase catalyzing diclofenac 4′-hydroxylation in human liver. Life Sci 52:29–34

Letendre PW, Carlson JD, Seifert RD, Dietz AJ, Dimmit D (1986) Effect of sucralfate on the absorption and pharmacokinetics of chlorpropamide. J Clin Pharmacol 26:622–625

Levitt LJ (1987) Chlorpropamide-induced pure white-cell aplasia. Blood 69:394–400

Lillioja S, Mott DM, Howard BV, Bennett PH, Yki-Järvinen H, Freymond D, Nyomba BL, Zurlo F, Swinburn B, Bogardus C (1988) Impaired glucose tolerance as a disorder of insulin action. Longitudinal and cross-sectional studies in Pima Indians. New Engl J Med 318:1217–1225

Logemann C (1986) Interactions of enoxacin with glibenclamide and chlorpropamide. Study of possible inhibition of the microsomal liver enzyme systems by means of the model substance antipyrine. Thesis, University Essen, Germany

Loubatieres A (1957) The hypoglycemic sulfonamides: history and development of the problem from 1942 to 1945. Ann NY Acad Sci 71:4–11

Lupo B, Bataille D (1987) A binding site for (^{3}H) glipizide in the rat cerebral cortex. Eur J Pharmacol 140:157–69

Madsbad S, Krarup T, McNair P, et al. (1981) Practical clinical value of C-peptide response to glucagon stimulation in the choice of treatment in diabetes mellitus. Acta Med Scand 210:153–156

Malacarne P, Castaldi, G, Bertusi M, Zavagli G (1977) Tolbutamide-induced hemolytic anemia. Diabetes 26:156–158

Malaisse WJ, Mahy M, Brisson GR, Malaisse-Lagae F (1972) The simulus-secretion coupling of glucose-induced insulin release. VIII. Combined effects of glucose and sulfonylureas. Eur J Clin Invest 2:85–90

Maloff BL, Lockwood DH (1981) In vitro effects of sulfonylurea on insulin action in adipocytes: potentiation of insulin-stimulated hexose transport. J Clin Invest 68:85–90

Marchetti P, Navalesi R (1989) Pharmacokinetic-pharmacodynamic relationships of oral hypoglycaemic agents. Clin Pharmacokin 16:100–128

Marshall A, Gingereich RL, Wright PH (1970) Hepatic effect of sulfonylurea. Metabolism 19:1046–1052

Martin BC, Warram JH, Krolewski AS, Bergman RN, Soeldner JS, Kahn CR (1992) Role of glucose and insulin resistance in development of Type 2 diabetes mellitus: results of a 25-year follow-up study. Lancet 340:925–929

Matsuda A, Kuzuya T, Sugita Y, Kawashima K (1983) Plasma levels of glibenclamide in diabetic patients during its routine clinical administration determined by a specific radioimmunoassay. Horm Metabol Res 15:425–428

McEwen J, Lawrence JR, Ings RMJ, Pidgen AW, Robinson JD, Walker SE (1982) Characterisation of glibenclamide half-life in man: acute concentration-effect relationships. Clin Sci 63:11P

McMahon RE, Marshall FJ, Culp HW (1965) The nature of the metabolites of acetohexamide in the rat and in the human. J Pharmacol Exp Ther 149:272–279

Melander A, Wahlin-Boll E (1980) Kinetic interaction of glipizide and indoprofen in healthy volunteers. Acta Endocrinol 239 [Suppl]:9–10

Melander A, Wahlin-Boll E (1981) Interaction of glipizide and indoprofen. Eur J Rheumatol Inflamm 4:22–25

Melander A, Sartor G, Wahlin E, Schersten B, Bitzen PO (1978) Serum tolbutamide and chlorpropamide concentrations in patients with diabetes mellitus. Br Med J I:142–144

Melander A, Bitzen P-O, Faber O, Groop L (1989) Sulphonylurea antidiabetic drugs: an update of their clinical pharmacology and rational therapeutic use. Drugs 37:58–72

Melander A, Jönsson A, Rydberg T (1993) Glibenclamide has long half-life and active metabolites in man. Diabetologia 36 [Suppl 1]:A180

Meyer BH, Müller FO, Luns HG, Eckert HG (1989) Bioavailability of three formulations of glibenclamide. S Afr Med J 76:146–147

Miller AK, Adir J, Vestal RE (1990) Excretion of tolbutamide in young and old subjects. Eur J Clin Pharmacol 38:523–524

Miners JO, Foenander T, Wanwimolruk S, Gallus AS, Birkett DJ (1982) The effect of sulphinpyrazone on oxidative drug metabolism in man: inhibition of tolbutamide elimination. Eur J Clin Pharmacol 23:321–326

Miners JO, Wing LMH, Lillywhite KJ, Smith KJ (1984) Failure of "therapeutic" doses of deta-adrenoceptor antagonists to alter the disposition of tolbutamide and lignocaine. Br J Clin Pharmacol 18:853–860

Mirsky JA, Perisutti G, Jinks R (1956) Ineffectiveness of sulfonylureas in alloxan diabetic rats. Proc Soc Exp Biol Med 91:475–477

Mis R, Ramis J, Conte L, Forn J (1992) In-vitro protein binding interaction between a metabolite of triflusal, 2-hydroxy-4-trifluoromethylbenzoic acid and other drugs. J Pharm Pharmacol 44:935–937

Miyazaki H, Fujii T, Yoshida K, Arakawa S, Furukawa H, et al. (1983) Disposition and metabolism of 3-H-gliclazide in rats. Eur J Drug Metab Pharmacokin 8:117–131

Monro AM, Welling PG (1974) The bioavailability in man of marketed brands of chlorpropamide. Eur J Clin Pharmacol 7:47–49

Morrison PJ, Rogers HJ, Spector RG, Bradbrook ID, John VA (1982) Effect of pirprofen on glibenclamide kinetics and response. Br J Clin Pharmacol 14: 123–126

Moses AM, Howanitz J, Miller M (1973) Diuretic action of three sulfonylurea drugs. Ann Intern Med 78:541–544

Müller M, Weimann HJ, Eden G, Weber W, Michaelis K, Dilger C (1993) Steady state investigation of possible pharmacokinetic interactions or moxonidine and glibenclamide. Eur J Drug Metab Pharmacokinet 18:277–283

Nash JF, Galloway JA, Garner AD, Johnson DW, Kleber JW, Rodda BE (1977) In vivo and in vitro availability of acetohexamide from tablets. Can J Pharm Sci 12:59–64

Nation RL, Evans AM, Milne RW (1990) Pharmacokinetic drug interaction with phenytoin. Clin Pharmacokin 18:37–60

Neugebauer G, Akpan W, Abshagen U (1983) Interaktion von Guar mit Glibenclamid und Bezafibrat. Beitr Infusionsther Klin Ernähr 12:40–47

Neugebauer G, Betzien G, Hrstka V, Kaufmann B, Möllendorff E, Abshagen U (1985) Absolute bioavailability and bioequivalence of glibenclamide (Semi-Euglucon®N). Int J Clin Pharmacol Ther Toxicol 23:453–460

Neuvonen PJ, Kärkkäinen S (1983) Effects of charcoal, sodium bicarbonate and ammonium chloride on chlorpropamide kinetics. Clin Pharmacol Ther 33: 386–393

Neuvonen PJ, Kivistö KT (1991) The effects of magnesium hydroxide on the absorption and efficacy of two glibenclamide preparations. Br J Clin Pharmacol 32:215–220

Neuvonen PJ, Kannisto H, Hirvasalo EL (1983) Effect of activated charcoal on absorption of tolbutamide and valproate in man. Eur J Clin Pharmacol 24: 243–246

Neuvonen PJ, Kärkkäinen S, Lehtovaara R (1987) Pharmacokinetics of chlorpropamide in epileptic patients: effects of enzyme induction and urine pH on chlorpropamide elimination. Eur J Clin Pharmacol 32:297–301

O'Byrne S, Feely J (1990) Effects of drugs on glucose tolerance in non-insulin dependent diabetics. I, II. Drugs 40:6–18; 203–219

O'Donovan CJ (1959) Analysis of long-term experience with tolbutamide (orinase) in the management of diabetes. Curr Ther Res 1:69–87

Östenson CC, Nylén A, Grill V, Gutniak M, Efendic S (1986) Sulfonylurea-induced inhibition of glucagon secretion from the perfused rat pancreas: evidence for a direct, nonparacrine effect. Diabetologia 29:861–867

Oida T, Yoshida K, Kagemoto A, Sekine J, Higashijima T (1985) The metabolism of gliclazide in man. Xenobiotica 15:87–96

Olefsky JM, Reaven GM (1976) Effects of sulfonylurea therapy on insulin binding in mononuclear leukocytes of diabetic patients. Am J Med 60:89–95

Olsen KM, Kearns GL, Kemp SF (1992) Glyburide protein binding and the effect of albumin glycation in elderly type II diabetes. Pharmacotherapy 12:264–265

Olson SC, Ayres JW, Antal EJ, Albert KS (1985) Effect of food and tablet age on relative bioavailability and pharmacodynamics of two tolbutamide products. J Pharm Sci 74:735–739

Ostman J, Christenson J, Jansson B, Weiner L (1981) The antidiabetic effect and pharmacokinetic properties of glipizide. Comparison of a single dose with divided dose regime. Acta Med Scand 210:173–180

Paasikivi J, Wahlberg F (1971) Preventive tolbutamide treatment and arterial disease in mild hyperglycemia. Diabetologia 7:323–327

Page MA, Boutagy JS, Shenfield GM (1991) A screening test for slow metabolizers of tolbutamide. Br J Clin Pharmacol 31:649–654

Pannekoek J (1975) Insulin, glucagon and oral hypoglycaemic drugs. In: Dukes MNG (ed) Meyler's side effects of drugs: a survey of unwanted effects of drugs reported in 1972–1975, vol. 8. Excerpta Medica, Amsterdam, pp 904–927

Papoz L, Job D, Eschwege E, Aboulker JP, Cubeau J, Pequignot G, Rathery M, Rosselin G (1978) Effect of oral hypoglycaemic drugs on glucose tolerance and insulin secretion in borderline, diabetic patients. Diabetologia 15:373–580

Patel TB (1986a) Effects of tolbutamide on gluconeogenesis and glycolysis in isolated perfused rat liver. Am J Physiol 250:E82–E86

Patel TB (1986b) Effect of sulfonylureas on hepatic fatty acid oxidation. Am J Physiol 251:E241–E246

Peacock I, Watts R, Selby C, Tattersall B (1987) Serum C-peptide after 6 months on glibenclamide remains higher than during insulin treatment. Diab Res 6:57–59

Pearson JC (1985) Pharmacokinetics of glyburide. Am J Med 79 [Suppl 3B]:67–71

Pearson JG, Antal EJ, Raehi GL, Gorsch HK, Craig WA, et al. (1986) Pharmacokinetic disposition of ^{14}C-glyburide in patients with varying renal function. Clin Pharmacol Ther 39:318–324

Peart GF, Boutagy J, Shenfield GM (1987) Lack of relationship between tolbutamide metabolism and debrisoquine oxidation phenotype. Eur J Clin Pharmacol 33:397–402

Peart GF, Boutagy J, Shenfield GM (1989) The metabolism of glyburide in subjects of known debrisoquine phenotype. Clin Pharmacol Ther 45:277–284

Pek S, Fajans SS, Floyd JC Jr, Knopf RF, Conn JW (1972) Failure of sulfonylureas to suppress plasma glucagon in man. Diabetes 21:216–223

Pentikäinen PJ, Neuvonen PJ, Penttilä A (1983) Pharmacokinetics and pharmacodynamics of glipizide in healthy volunteers. Int J Clin Pharmacol Ther Tox 21:98–107

Perley M, Kipnis DM (1966) Plasma insulin responses to glucose and tolbutamide of normal weight and obese diabetic and nondiabetic subjects. Diabetes 15:867–874

Peters AL, Davidson MB (1991) Insulin plus sulfonylurea agent for treating type 2 diabetes. Ann Intern Med 115:45–53

Peterson CM, Sims RV, Jones RL, Rieders F (1982) Bioavailability of glipizide and its effect on blood glucose and insulin levels in patients with non-insulin-dependent diabetes. Diabetes Care 5:497–500

Petitpierre B, Perrin L, Reidhart M, Herrera A, Fabre J (1972) Behaviour of chlorpropamide in renal insufficiency and under the effect of associated drug therapy. Int J Clin Pharm 6:120–124

Pfeifer MA, Halter JB, Graf R, Porte D Jr (1980) Potentiation of insulin secretion to nonglucose stimuli in normal man by tolbutamide. Diabetes 29:335–340

Pfeifer MA, Halter JB, Beard JC, Porte D (1981) Differential effects of tolbutamide on first and second phase insulin secretion in non insulin dependent diabetes mellitus. J Clin Endocrinol Metab 53:1256–1262

Pfeifer MA, Beard JC, Halter JB, Judzewitsch R, Best JD, Porte D Jr (1983) Suppression of glucagon secretion during a tolbutamide infusion in normal and noninsulin-dependent diabetic subjects. J Clin Endocrinol Metab 56:586–591

Podgainy H, Bressler R (1968) Biochemical basis of the sulfonylurea induced antabus syndrome. Diabetes 17:679–682

Polonsky KH, Rubenstein AH (1984) C-peptide as a measure of the secretion and hepatic extraction of insulin: pitfalls and limitations. Diabetes 33:486–494

Pond SM, Birkett DJ, Wade DN (1977) Mechanism of inhibition of tolbutamide metabolism: phenylbutazone, oxyphenbutazone, sulphaphenazole. Clin Pharmacol Ther 22:573–579

Pontiroli AE, Perfetti MG, Pozza G (1991) Acute effect of glipizide on glucose tolerance in obesity and diabetes mellitus (NIDDM). Eur J Clin Pharmacol 40:23–26

Porte D Jr (1991) β-cells in Type II diabetes. Diabetes 40:166–180

Profozic' V (1976) Pharmacokinetics of Glurenorm in subjects with liver disease. Diab Croat 5:629–640

Purba HS, Back DJ, Orme ML (1987) Tolbutamide 4-hydroxylase activity of human liver microsomes: effect of inhibitors. Br J Clin Pharmacol 24:230–234

Quattrin N, Jacono G, Brancaccio AG (1956) Ricerche sul metabolismo della sulfabutilurea. Minerva Med 47:1777–1780

Ratzmann KP, Schulz B, Heinke P, Besch W (1984) Tolbutamide does not alter insulin requirement in Type 1 (insulin-dependent) diabetes. Diabetologia 27:8–12

Reaven GM (1988) Role of insulin resistance in human disease. Diabetes 37: 1595–1607

Reaven G, Dray J (1967) Effect of chlorpropamide on serum glucose and immunoreactive insulin concentrations in patients with maturity-onset diabetes mellitus. Diabetes 16:487–492

Recant L, Fischer GL (1957) Studies on the chemanism of tolbutamide hypoglycemia in animal and human subjects. Ann NY Acad Sci 71:62–70

Redman DR, Prescott LF (1973) Failure of induction of liver microsomal enzymes by tolbutamide in maturity-onset diabetes. Diabetes 22:210–221

Relling MV, Aoyama T, Gonzalez FJ, Meyer UA (1990) Tolbutamide and mephenytoin hydroxylation by human cytochrome P 450s in the CYP 2 C subfamily. J Pharmacol Exp Ther 252:442–447

Rendell M (1983) C-peptide levels as a criterion in treatment of maturity-onset diabetes. J Clin Endocrinol Metab 57:1198–1206

Rentsch G, Schmidt HAE, Rieder J (1972a) Zur Pharmakokinetik von Glibornurid. Arzneimittel forschung 22:2209–2212

Rentsch G, Forgó I, Dubach UC (1972b) Pharmakokinetik eines neuen Antidiabetikums bei Patienten mit eingeschränkter Nierenfunktion. Schweiz Med Wschr 102:650–653

Rett K, Wicklmayr M, Dietze GJ (1988) Hypoglycemia in hypertensive diabetic patients treated with sufonylureas, biguanides, and captopril. N Engl J Med 319:1609

Richardson T, Foster J, Mawer GE (1986) Enhancement by sodium salicylate of the blood glucose lowering effect of chlorpropamide – drug interaction or summation of similar effects? Br J Clin Pharmacol 22:43–48

Ridolfo AS, Kirtley WR (1956) Clinical experiences with carbutamide, an orally given hypoglycemic agent. J Am Med Assoc 160:1285–1288

Rimbau Barreras V, Uriach Marsal J, Pou Torello JM (1976) Cinetica de la glipentida (UR-661) tras administracion i.v. a voluntarios sanos y diabeticos. Archivos Farmacol Toxicol 2:115–122

Robson RA, Miners JO, Whitehead AG, Birkett DJ (1987) Specificity of the inhibitiory effect of dextropropoxyphene on oxidative drug metabolism in man: effects in theophylline and tolbutamide disposition. Br J Clin Pharmacol 23: 772–775

Rogers BJ, Spector RG, Morrison PG, Bradbrook ID (1982) Pharmacokinetics of intravenous glibenclamide investigated by a higher performance lipid chromatographic assay. Diabetologia 23:37–40

Rogers BJ, Standaert ML, Pollet RJ (1987) Direct effects of sulfonylurea agents on glucose transport in the BBC3H-1 myocyte. Diabetes 36:1292–1296

Rosenkranz B, Malercky V, Lehr K-H, Profozic V, Mrzljak V, Skrabalo Z (1991) Pharmacokinetics of glimepiride in kidney disease. Clin Pharmacol Ther 49:170

Rossetti L, Shulman GI, Zawalich W, DeFronzo RA (1987a) Effect of chronic hyperglycemia on in vivo insulin secretion in partially pancreatectomized rats. J Clin Invest 80:1037–1040

Rossetti L, Smith D, Shulman GI, Papachristou D, DeFronzo RA (1987b) Correction of hyperglycemia with phlorizin normalizes tissue sensitivity to insulin in diabetic rats. J Clin Invest 79:1510–1515

Rowland MJ, Bransome Jr ED, Hendry B (1994) Hypoglycemia caused by selegiline, an antiparkinsonian drug: can such side effects be predicted? J Clin Pharmacol 34:80–85

Rupp W, Christ O, Heptner W (1969) Rosorption, Ausscheidung und Metabolismus nach intravenöser und oraler Gabe von HB 419-14C an Menschen. Arzneim Forsch 19:1428–1434

Rupp W, Dibbern HW, Hajdú P, Ross G, Vander EE (1975) Untersuchungen zur Bioäquivalenz von Tolbutamid. Dtsch Med Wschr 100:690–695

Saffar F, Ogata H, Ejima A (1982) Biopharmaceutical studies on the clinical inequivalence of two carbutamide tablets. Chem Pharm Bull 30:679–683

Salvatore T, Scheen AJ, Ferreira Alves de Magalhaes AC, Jaminet C, Lefebvre PJ (1990) Slight modifications of the pharmacokinetic parameters of glibenclamide after treatment with the alpha-glucosidase inhibitor miglitol in normal subjects. Therapie 45:365

Samanta A, Jones GR, Burden AC, Shakir J (1984) Improved effect of tolbutamide when given before food in patients on long-term therapy. Br J Clin Pharmacol 18:647–648

Samols E, Tyler JM, Mialhe P (1969) Suppression of pancreatic glucagon release by the hypoglycemic sulphonylureas. Lancet I:174–176

Sano A, Kuriki T, Kawashima Y, Takeuchi H, Hino T, Niwa T (1992) Particle design of tolbutamide by spherical crystallization technique. V. Improvement of dissolution and bioavailability of direct compressed tablets using tolbutamide agglomerated crystals. Chem Pharm Bull 40:3030–3035

Sartor G, Scherstén B, Melander A (1978) Effects of glipizide and food intake on the blood levels of glucose and insulin in diabetic patients. Acta Med Scand 203: 211–214

Sartor G, Melander A, Scherstén B, Wahlin-Boll E (1980a) Serum glibenclamide in diabetic patients and influence of food on the kinetics and effects of glibenclamide. Diabetologia 18:17–22

Sartor G, Melander A, Scherstén B, Wahlin-Boll E (1980b) Comparative single-dose kinetics and effects of four sulfonylureas in healthy volunteers. Acta Med Scand 208:301–307

Sartor G, Melander A, Scherstén B, Wahlin-Boll E (1980c) Influence of food and age on the single-dose kinetics and effects of tolbutamide and chlorpropamide. Eur J Clin Pharmacol 17:285–293

Sartor G, Scherstén B, Carlström S, Melander AS, Nordén A, Persson G (1989d) Ten-year follow-up of subjects with impaired glucose tolerance. Prevention of diabetes by tolbutamide and diet regulation. Diabetes 29:41–49

Sartor G, Lundquist I, Melader A, Wahlin-Boll E (1982) Improved effect of glibenclamide on administration before breakfast. Eur J Clin Pharmacol 21:403–408

Scheen AJ, Lefebvre PJ, Luyckx AS (1984) Glipizide increases plasma insulin but not C-peptide level after a standardized breakfast in type 2 diabetic patients. Eur J Clin Pharmacol 26:471–474

Scheen AJ, Jaminet C, Luyckx AS, Lefebvre PJ (1987) Pharmacokinetics and pharmacological properties of two galenical preparations of glibenclamide, HB 419 and HB 420, in non insulin-dependent (type 2) diabetes. Int J Clin Pharmacol Ther Tox 25:70–76

Scheen AJ, Castillo MJ, Lefèbvre PJ (1988) Decreased or increased insulin metabolism after glipizide in Type II diabetes? Diabetes Care 11:687–689

Scheen AJ, Jaminet C, Stassen MP, Fereira Alves de Magalhaes AC, Salvatore T, Gerard J, Lefebvre PJ (1989) Absence of significant age-related alterations of glibenclamide pharmacokinetics: studies in young healthy volunteers, and in middle-aged and elderly type 2 diabetic patients. Eur J Clin Pharmacol 36 [Suppl]:A64

Schenk G (1976) Der Finfluß von Sorbit und Prednison auf die Pharmakokinetik von Tolbutamid bei Lebergesunden. Thesis Julius-Maximilans-Universität, Würzburg

Schlossmann K (1974) Proteinbindung von Glisoxepid und sein Einfluß auf die Proteinbindung von Phenprocoumon. Arzneimittelforschung 24:392–397

Schmid-Antomarchi H, DeWeille J, Fosset M, Lazdunski M (1987) The receptor for the antidiabetic sulfonylureas controls the activity of the ATP-modulated K^+ channel in insulin secreting cells. J Biol Chem 262:15840–15844

Schmidt HAE, Schoog M, Schweer KM, Winkler E (1973) Pharmacokinetics and pharmacodynamics as well as metabolism following orally and intravenously administered C14-glipizide, a new antidiabetic. Diabetologia 9 [Suppl]:320–330

Schor S (1971) The University Group Diabetes Program: a statistician looks at the mortality results. J Am Med Assoc 217:1673–1675

Schwartzkopff T, Kewitz H (1977) Elimination of glisoxepide in patients with renal failure. Arch Pharmacol 297 [Suppl 2]:R61

Schwinghammer TL, Antal EJ, Kubacka RT, Hackimer ME, Johnston JM (1991) Pharmacokinetics and pharmacodynamics of glyburide in young and elderly nondiabetic adults. Clin Pharm 10:532–538

Scott J, Poffenbarger PL (1979) Pharmacogenetics of tolbutamide metabolism in humans. Diabetes 28:41–51

Self TH, Morris T (1980) Interaction of rifampicin and chlorpropamide. Chest 77:800–801

Self TH, Tsin SJ, Bowld WF, Fowler JW (1989) Interaction of rifampicin and glyburide. Chest 96:1443–1444

Seltzer HS (1972) A summary of criticism of the findings and conclusions of the University Group Diabetes Program (UGDP) study. Diabetes 21:976–979

Seltzer HS (1989) Drug-induced hypoglycemia: a review of 1418 cases. Endocrinol Metab Clin North Am 18:163–183

Semple CG, Omile C, Buchanan KD, Beastall GH, Peterson KR (1986) Effect of oral verapamil on glibenclamide stimulated insulin secretion. Br J Clin Pharmacol 22:187–190

Shah GF, Gandhi TP, Patel PR, Patel MR, Gilbert RN, Shridhar PA (1985) Tolbutamide and chlorpropamide kinetics in presence of cimetidine in human volunteers. Indian Drugs 22:455–458

Shaheen O, Othman S, Jalal I, Awidi A, Al-Turk W (1987) Comparison of pharmacokinetics and pharmacodynamics of a conventional and a new rapidly dissolving glibenclamide preparation. Int J Pharm 38:123–131

Shapiro ET, van Cauter E, Tillil H, Given BD, Hirsch L, Beebe C, Rubenstein AH, Polonsky KS (1989) Glyburide enhances the responsiveness of the B-cell to glucose but does not correct the abnormal patterns of insulin secretion in noninsulin-dependent diabetes mellitus. J Clin Endocrinol Metab 69:571–576

Shaw KM, Wheeley MSG, Campbell DB, Ward JP (1985) Home blood glucose monitoring in non-insulin-dependent diabetes: the effect of gliclazide on blood glucose and weight control, a multicentre trial. Diab Med 2:484–490

Sheldon J, Anderson J, Stoner L (1965) Serum concentration and urinary excretion of oral sulfonylurea compounds: relation to diabetic control. Diabetes 14:362–367

Shenfield GM, Logan A, Shirling D, Baird J (1977) Plasma insulin and glucose levels in maturity onset diabetics treated with chlorpropamide. Diabetologia 13:367–371

Shiba T, Kajinuma H, Suzuki K, Hagura R, Kawai A, et al. (1986) Serum gliclazide concentration in diabetic patients. Relationship between gliclazide and serum concentration. Diabetes Res Clin Pract 2:301–306

Shima K, Ikegami H, Tanaka A, Ezaki A, Kumahara Y (1982) Effect of dietary fibre, konjac mannan and guar gum, on absorption of sulphonylurea in man. Nutr Rep Int 26:297–302

Shima K, Tanaka A, Ikegami H, Tabata M, Sawazaki N, et al. (1983) Effect of dietary fiber, glucomannan, on absorption of sulfonylurea in man. Horm Metab Res 15:1–3

Shulman GI, Rothman DL, Jue T, Stein P, DeFronzo RA, Shulmann RG (1990) Quantitation of muscle glycogen synthesis in normal subjects and subjects with non-insulin dependent diabetes by 13C nuclear magnetic resonance spectroscopy. N Engl J Med 322:233–228

Siconolfi-Baez L, Banerji MA, Lebovitz HE (1990) Characterization and significance of sulfonylurea receptors. Diabetes Care 13 [Suppl 3]:2–8

Simonson DC, Ferrannini E, Bevilacqua S, Smith D, Barrett E, Carlson R, DeFronzo RA (1984) Mechanism of improvement in glucose metabolism after chronic glyburide therapy. Diabetes 33:838–845

Simonson DC, Del Prato S, Castellino P, Groop L, De Fronzo RA (1987) Effect of glyburide on glycemic control, insulin requirement, and glucose metabolism in insulin-treated diabetic patients. Diabetes 36:136–146

Simpson RG, Benedetti A, Grodsky GM, Karam JH, Forsham PH (1968) Early phase of insulin release. Diabetes 17:684–692

Simpson HCR, Sturkley R, Stirling CA, Reckless JPD (1990) Combination of insulin with glipizide increases peripheral glucose disposal in secondary failure type 2 diabetic patients. Diabetic Med 7:143–147

Singer DL, Stewart RC, Hurwitz D (1961) Chlorpropamide in patients on high insulin dosage. N Engl J Med 265:823–826

Singer DL, Hurwitz D (1967) Long-term experience with sulfonylureas and placebo. N Engl J Med 277:450–456

Sjoeberg S, Wiholm BE, Gunnarsson R, Emilsson H, Thunberg E, Christenson I, Oestman J (1987) Lack of pharmacokinetic interaction between glibenclamide and trimethoprim-sulphamethoxazole. Diabetic Med 4:245–247

Sjöholm I, Ekman B, Kober A, Ljungstedt-Pahlman I, Seiving B, Sjödin T (1979) Binding of drugs to human serum albumin. XI. The specificity of three binding sites as studied with albumin immobilized in microparticles. Mol Pharmacol 16:767–777

Skovsted I, Kristensen M, Molholm-Hansen J, Siersbaek-Nielsen K (1976) The effect of different oral anticoagulants on diphenylhydantoin (DPH) and tolbutamide metabolism. Acta Med Scand 199:513–515

Smith DL, Vecchio TJ, Forist AA (1965) Biological half-lives of the p-acetylbenzene-sulfonylureas U-18536 and acetohexamide and their metabolites. Metab 14: 229–240

Sotaniemi E, Arvela P, Huhti E (1971a) Increased clearance of tolbutamide from the blood of asthmatic patients. Ann Allerg 29:139–141

Sotaniemi E, Arvela P, Huhti E, Kovisto O (1971b) Half-life of tolbutamide in patients with chronic respiratory failure. Eur J Clin Pharmacol 4:29–31

Speck U, Mützel W, Kolb KH, Acksteiner B, Schulze PE (1974) Pharmakokinetik und Metabolitenspektrum von Glisoxepid beim Menschen. Arzneimittelfor-schung 24:404–409

Spraul M, Streeck A, Nieradzik M, Berger M (1989) Uniform elimination patterns for glibenclamide in healthy caucasian males. Arzneimittelforschung 39: 1449–1450

Stenman S, Groop P-H, Saloranta C, Tötterman K-J, Fyhrquist F, Groop L (1988) Effects of the combination of insulin and glibenclamide in type II (non-insulin dependent) patients with secondary failure to oral hypoglycemic agents. Diabetologia 31:206–213

Stenman S, Melander A, Groop R-H, Groop LC (1993) What is the benefit of increasing the sulfonylurea dose? Ann Intern Med 118:169–172

Stockley C, Keal J, Rolan P, Bochner F, Somogyi A (1986) Lack of inhibition of tolbutamide hydroxylation by cimetidine in man. Eur J Clin Pharmacol 31: 235–237

Stoeckel K, Trueb V, Dubach UC, Heintz RC, Ascalone V, Forgó I, Hennes U (1985) Lack of effect of tenoxicam on glibornuride kinetics and response. Br J Clin Pharmacol 19:249–254

Stowers JM, Mahler RF, Hunter RB (1958) Pharmacology and mode of action of the sulfonylurea in man. Lanet I:278–283

Stowers JM, Brewsher PD (1962) The long-term use of sulfonylureas in diabetes mellitus. Lancet I:122–124
Sturgess NC, Ashford MLJ, Cook DL, Hales CN (1985) The sulphonylurea receptor may be an ATP-sensitive potassium channel. Lancet II:474–475
Talaulicar M, Willms B (1976) Investigations on Glurenorm blood levels. Diab Croat 5:613–622
Tamassia V (1975) The pharmacokinetics and bioavailability of glipizide. Curr Med Res Opinion 3 [Suppl]:20–30
Tassaneeyakul W, Veronese ME, Birkett DJ, Doecke CJ, McManus ME, Sansom LN (1992) Co-regulation of phenytoin and tolbutamide metabolism in humans. Br J Clin Pharmacol 34:494–498
Taylor JA (1972) Pharmacokinetics and biotransformation of chlorpropamide in man. Clin Pharmacol Ther 13:710–718
Taylor T, Assinder DF, Chasseaud LF, Bradford PM, Burton JS (1977) Plasma concentrations, bioavailability and dissolution of chlorpropamide. Eur J Clin Pharmacol 11:207–212
Thomas RC, Ikeda GJ (1966) The metabolic fate of tolbutamide in man and in the rat. J Med Chem 9:507–510
Thomas RC, Judy RW (1972) The metabolic fate of chlorpropamide in man and in the rat. J Med Chem 15:964–968
Thomas RC, Duchamp DJ, Judy RW, Ikeda GJ (1978) Metabolic fate of tolazamide in man and in the rat. J Med Chem 21:725–732
Thorn G (1966) Clinical considerations in the use of corticosteroids. N Engl J Med 274:775–781
Tsalikian E, Dunphy TW, Bohannon NV, Lorenzi M, Gerich JE, Forsham PH, Kane JP, Karam JH (1977) The effect of chronic oral antidiabetic therapy on insulin and glucagon responses to a meal. Diabetes 26:314–321
Tsuchiya S, Sakurai T, Sekiguchi SI (1984) Non enzymatic glycosylation of human serum albumin and its influence on binding capacity of sulfonylureas. Biochem Pharmacol 33:2967–2971
Ueda H, Sakurai T, Ota M, Nakajima A, Kamiti K, et al. (1963) Disappearance rate of tolbutamide in normal subjects and in diabetes mellitus, liver cirrhosis, and renal disease. Diabetes 12:414–419
UKPDS Group (1990) UK prospective diabetes study 7: response of fasting plasma glucose to diet therapy in newly presenting Type II diabetic patients. Metabolism 39:905–912
Unger RH, Aguilar-Parada E, Muller WA, Eisentraut AM (1970) Studies of pancreatic alpha cell function in normal and diabetic subjects. J Clin Invest 49:837–848
Uppal R, Sharma PL, Prakash C, Gupta HR (1986) Tolbutamide kinetics in cigarette smokers in the Indian population. Int J Clin Pharmacol Ther Toxicol 24:82–84
Uusitupa M, Soedervik H, Silvasti M, Karttunen P (1990) Effects of a gel forming dietary fiber, guar gum, on the absorption of glibenclamide and metabolic control and serum lipids in patients with non-insulin-dependent (type 2) diabetes. Int J Clin Pharmacol Ther Toxicol 28:153–157
Veronese ME (1991) Tolbutamide and phenytoin metabolism by human cytochrome P450s in the CYP2C subfamiliy. Clin Exp Pharmacol Physiol [Suppl 18]:64
Veronese ME, Miners JO, Randles D, Gregov D, Birkett DJ (1990a) Validation of the tolbutamide metabolic ratio for population screening with use of sulfaphenazole to produce model phenotype poor metabolizers. Clin Pharmacol Ther 47:403–411
Veronese ME, Doecke CJ, McManus ME, Sansom LN, Miners JO, Birkett DJ (1990b) Relationship between cytochrome P450 dependent tolbutamide and phenytoin hydroxylation in human liver microsomes. Clin Exp Pharmacol Physiol [Suppl 16]:50

Veronese ME, Mackenzie PI, Doecke CJ, McManus ME, Miners JO, Birkett DJ (1991) Tolbutamide and phenytoin hydroxylations by cDNA-expressed human liver cytochrome P450 2C9. Biochem Biophys Res Commun 175:1112–1118

Verschoor L, Wolffenbüttel BHR, Weber RFA (1986) Beta-blockade and carbohydrate metabolism: theoretical aspects and clinical implications. J Cardiovasc Pharmacol 8 [Suppl 11]:592–595

Veyre B, Ginon I, Vial T, Dragol F, Daumont M (1993) Hypoglycémies par interférence entre un inhibiteur de l'enzyme de conversion et un sulfamide hypoglycémiant. Presse Médicale 22:738

Wahlin-Boll E (1986) Kinetics-effect relations of glipizide. Doctoral Thesis, University of Lund, Dalby, Sweden

Wahlin-Boll E, Melander A, Sartor G, Schersten B (1980) Influence of food intake on the absorption and effect of glipizide in diabetics and in healthy subjects. Eur J Clin Pharmacol 18:279–283

Wahlin-Boll E, Almer LO, Melander A (1982a) Bioavailability, pharmacokinetics and effects of glipizide in type 2 diabetics. Clin Pharmacokin 7:363–372

Wahlin-Boll E, Sartor G, Melander A, Schersten B (1982b) Impaired effect of sulfonylurea following increased dosage. Eur J Clin Pharmacol 22:21–25

Wahlin-Boll E, Groop L, Karhumaa S, Groop P-H, Tötterman K-J, Melander A (1986) Therapeutic equivalence of once- and thrice-daily glipizide. Eur J Clin Pharmacol 31:95–99

Wajchenberg BL, Santomauro AT, Gianella-Neto D, Borghi VC, Porrelli RN (1992) Short- and long-term gliclazide effects on pancreatic islet cell function and hepatic insulin extraction in non-insulin-dependent diabetes mellitus. Diab Res Clin Practice 17:89–97

Wajchenberg BL, Santomauro AT, Porrelli RN (1993) Effect of a sulfonylurea (gliclazide) treatment on insulin sensitivity and glucose-mediated glucose disposal in patients with non-insulin-dependent diabetes mellitus (NIDDM). Diab Res Clin Practice 20:147–154

Wales JK (1982) Treatment of Type 2 (non-insulin dependent) diabetic patients with diet alone. Diabetologia 23:240–245

Weissmann PN, Shenkman L, Gregerman RI (1971) Drug-induced inappropriate anti-diuretic hormone activity. N Engl J Med 284:65–71

Welling PG, Patel RB, Patel UR, Gillespic WR, Craig WA, et al. (1982) Bioavailability of tolazamide from tablets: comparison of in vitro and in vivo results. J Pharm Sci 71:1259–1263

Widén E (1993) Does glibenclamide influence the clearance of insulin in patients with type 2 diabetes? Scand J Clin Lab Invest 53:395–403

Widén E, Groop L (1994) Biguanides: metabolic effects and potential use in the treatment of the insulin resistance syndrome. Diabet Annual 8:227–241

Wiholm BE, Westerholm B (1984) Drug utilization and morbidity stastistics for the evaluation of drug safety in Sweden. Acta Med Scand [Suppl] 683:107–117

Williams RL, Blaschke TF, Meflin PJ, Melmar KL, Rowland M (1977) Influence of acute viral hepatitis on disposition and plasma binding of tolbutamide. Clin Pharmacol Ther 21:301–309

Wing LMH, Miners JO (1985) Cotrimoxazole as an inhibitor of oxidative drug metabolism: effects of trimethoprim and sulfamethoxazole separately and combined on tolbutamide disposition. Br J Clin Pharmacol 20:482–485

Winters CS, Chrystyn H, York P, Timmius P, Bramley P, Burgul R (1993) Improved bioperformance of gliclazide on complexation with beta-cyclodextrin. Pharm Res 10 [Suppl]:S263

Wright CE, Antal EJ (1985) The influence of age on the pharmacokinetics of tolazamide. Drug Int Clin Pharm 19:458

Yki-Järvinen H (1990) Acute and chronic effects of hyperglycemia on glucose metabolism. Diabetologia 33:579–585

Yki-Järvinen H, Kauppila M, Kujansuu E, Lahti J, Marjanen T, Niskanen L, Rajala S, Ryysy L, Salo S, Seppälä P, Tulokas T, Viikari J, Karjalainen J, Taskinen MR (1992) Comparison of insulin regimens in patients with non-insulin-dependent diabetes mellitus. N Engl J Med 327:1426–1433

Zilker T, Bottermann P (1975) Blutzuckertagesprofil und Insulinsekretion nach Gabe von Glurenorm in Abhängigkeit von Dosishöhe und Dosierungsintervall. Therapiewoche 25:4809–4818

Zilly W, Breimer DD, Richter E (1975) Induction of drug metabolism in man after rifampicin treatment measured by increased hexobarbital and tolbutamide clearance. Eur J Clin Pharmacol 9:219–227

Zuccaro P, Pacifici R, Pichini S, Avico U, Federzoni G, Pini LA, Sternieri E (1989) Influence of antacids on the bioavailability of glibenclamide. Drugs Exp Clin Res 15:165–169

Section II
Biguanides

CHAPTER 10

Chemistry and Structure-Activity Relationships of Biguanides

E. Prugnard and M. Noel

A. Introduction

I. The First Known Synthesis

In 1878–1879, Rathke obtained a new compound in very low yield via a condensation reaction of thiourea and phosphorus trichloride (which are cyanamide generators) with guanidine.

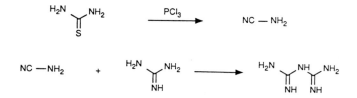

He decided to name this new compound "biguanide", considering that it could be likened to the result of a condensation between two molecules of guanidine with ammonia elimination. Subsequently, Rathke noticed that the resulting new product reacted with a copper sulfate solution to give pink crystals of the copper-biguanide complex, which was then converted into the sulfate (Herth 1880). The synthesis was rapidly improved by using a condensation reaction of cyanoguanidine with an ammoniacal solution of cupric sulfate at 110°C in a sealed tube. The resulting complex was subsequently treated to give the sulfate.

A few years later, it was shown that biguanides can be prepared by treating cyanoguanidine with ammonium chloride in boiling ethanol (Smolka and Friedreich 1888). Finally, it was found that biguanides can be obtained by direct fusion of ammonium chloride with cyanoguanidine at 195°C for a few minutes (Bamberger and Dieckmann 1892). It is noteworthy that, in all cases, the purification was effected by using the intermediate copper complex and the sulfate. A century later, this complexation reaction is still used in the synthesis of substituted biguanides to demonstrate the presence of biguanide (Patereau 1965, unpublished results). This is a very suitable and easy to use method for detecting the presence of biguanide derivatives.

In 1951, Oxley and Short improved the reaction by the use of ammonium benzene sulfonate instead of ammonium chloride. Nevertheless, the yield never exceeded 30%. Concurrently, the same procedures have been followed for the synthesis of N-substituted biguanides, but relatively few compounds have been synthesized. In 1946 *Chemical Abstracts* mentioned only about 100 molecules which still had not found any application.

II. The Golden Age

By contrast, the decade which followed World War II saw a very great deal of productive research, with the discoveries of: paludrine, an effective antimalarial drug (1947), antidiabetic biguanides (1958–1959) and chlorhexidine, a topical antibacterial and disinfectant (1956).

In addition, the anticarcinogenic effect of some biguanides was established. Other compounds revealed an antitubercular or antiviral activity. On the other hand, biguanides are known to be starting materials for a variety of triazines and heterocyclic compounds of medical value. As Kurzer (1955) said, "Although investigations on biguanides have been diverse, much remains to be done."

III. Nomenclature

Since 1972, the nomenclature usually employed by *Chemical Abstracts* has been: "imidodicarbonimidic diamide". Nevertheless, for practical reasons and in conformity with the practice of *Chemical Abstracts* until 1971, the name "biguanide" will be used in the present review. Since the first synthesis of biguanides, several rules for numbering the different atoms have been successively applied. The latter will be adopted in this review.

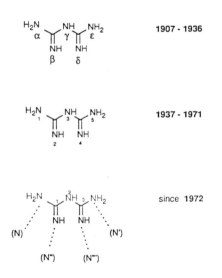

Table 1. History of oral treatment of diabetes by biguanides – from a fortuitous observation to a new class of drugs

Guanidines	Biguanides	Others
Middle ages – Use of galega officinalis (goat's rue) rich in guanidine substance		
	1879 – RATHKE: first synthesis of biguanide	
1914 – UNDERHILL and BLATHERWICK: The blood sugar is lowered after parathyroidectomy 1916 – BURNS, SHARPE: Guanidine content of the blood is increased after parathyroidectomy 1918 – WATANABE: Guanidine may cause hypoglycemia		
		1921 – BANTING and MACLEOD: Discovery of insulin
	1922 – WERNER and BELL: First synthesis of dimethylbiguanide	
1926 – FRANCK et al.: Agmatine previously discovered in herring sperm decreases the blood sugar level and is less toxic than guanidine 1926–1928 – Synthalin: use and prohibition (hepatotoxic activity and renal complications) 1927 – SIMONNET and TANRET: Clinical application of galegin		Availability of insulin
	1929 – SLOTTA and TSCHESCHE: Hypoglycemic activity of dimethylbiguanide 1929 – HESSE and TAUBMAN: Description of pharmacological and toxicological properties of dimethylbiguanide toxicity/activity ratio	
		1942 – JANBON and LOUBATIERES: Discovery of the hypoglycemic activity of sulfamides

Table 1. *Continued*

Guanidines	Biguanides	Others
		1955 – Launching of carbutamide 1956 – Launching of tolbutamide

The availability of insulin and the clinical use of the sulfamides led to a decrease in interest in biguanides.

1946 – Clinical studies demonstrated the therapeutic efficacy of paludrine (chlorguanide) in malaria treatment.

1947 – CHEN and ANDERSON reported a slight hypoglycemic activity of paludrine. This observation gave a new impetus to the interest in biguanides.

1957 – STERNE and DUVAL: Further investigations were performed on the *N,N*-dimethylbiguanide.

1958 – SHAPIRO and UNGAR: Synthesis of phenethylbiguanide and butylbiguanide.

1959 – Three members of a new class of antidiabetic drugs were launched onto the market: metformin (dimethylbiguanide), phenformin (phenethylbiguanide), and buformin (butylbiguanide).

Table 1 presents an historical overview of the oral treatment of diabetes by biguanides. Table 2 provides a classification of the antidiabetic biguanides.

IV. New Antihyperglycemic Biguanides

BECKMANN's review, published in 1971, covers the literature up to that date. Since then, several structures have been claimed to possess hypoglycemic properties. However, these structures have not been subject to further investigation:

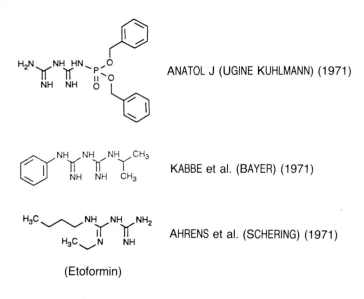

ANATOL J (UGINE KUHLMANN) (1971)

KABBE et al. (BAYER) (1971)

AHRENS et al. (SCHERING) (1971)

(Etoformin)

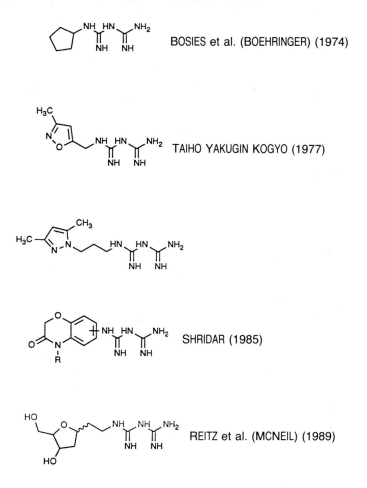

BOSIES et al. (BOEHRINGER) (1974)

TAIHO YAKUGIN KOGYO (1977)

SHRIDAR (1985)

REITZ et al. (MCNEIL) (1989)

Various salts of biguanides were also investigated: (1) isoxazole and pyrazole carboxylates (TAIHO YAKUGIN KOGYO 1977), (2) trimethoxybenz-oates (SHRAMOVA et al. 1981), (3) salts of carnitine and metformin (OTSUKA 1969) and (4) sulfamoyl phenoxyacetate of metformin (ONISCU et al. 1981).

B. Chemistry

I. Synthesis

Since 1878, a large number of biguanides have been described, the synthetic approach to which is standard and is briefly described in Sect. A. In this chapter, we will merely describe the new methods published since the previous review in 1971.

Table 2. Antidiabetic biguanides

	Metformin	Phenformin	Buformin
Formula	H₃C–N(CH₃)–C(=NH)–NH–C(=NH)–NH₂	C₆H₅–CH₂–CH₂–NH–C(=NH)–NH–C(=NH)–NH₂	H₃C–CH₂–CH₂–CH₂–NH–C(=NH)–NH–C(=NH)–NH₂
	$C_4H_{11}N_5 = 129.7$	$C_{10}H_{15}N_5 = 205.27$	$C_6H_{15}N_5 = 157.22$
	N,N-Dimethyl imidodicarbonimidic diamide	N-(2-Phenylethyl) imidodicarbonimidic diamide	N-Butyl imidodicarbonimidic diamide
CA No.	657-24-9	114-86-3	692-13-7
Official synonym	Metformin	Phenformin	Buformin
Unofficial synonym	Dimethylbiguanide	Phenethylbiguanide	Butylbiguanide
Derivatives	4-Chlorophenoxyacetate	Hydrochloride	Hydrochloride
	Glucinan (France)	Adiabetin (Austria)	Adebit (Hungary)
	Embonate	Cronoformin (Italy)	Biforon (Japan)
	Stagid (France)	DB retard (Germany)	Bufomamin (Japan)
	Hydrochloride	Debei (Brazil)	Bulbonin (Japan)
	Diaberit (Italy)	Debeone (USA)	Dibetos (Japan)
	Diabex SR (Fisons)	Diabetal (Brazil)	Gliporal (Mexico)
	Diformin (Finland)	Diabis (Spain)	Glybutid (Russia)
	Diguanil (UK, Italy)	Dibein (Sweden)	Insulamin (Japan)
	Gliformin (Russia)	Dibotin (Bayer, Winthrop)	Panformin (Japan)
	Glucadal (Italy)	Glucopostin (Germany)	Silubin (Grünenthal, Spain)
	Glucadal (Italy)	Insoral (USA)	Sindiatil (Bayer)
	Glucophage (UK, France, Canada, Germany)	Kataglicina (Italy)	Ziavetine (Japan)
	Glufagos (Spain)	Meltrol (USA)	
	Glukofag (UK)	Prontoformin (Italy)	
	Glukoliz (Turkey)		
	Islotin (Argentina)		
	Melbin (Japan)		
	Mellitin (Italy)		
	Metforal (Italy)		
	Metiguanide (UK)		

1. Unsubstituted Biguanides

The standard procedure of preparation by condensation of cyanoguanidine with ammonia or ammonium salts gave only poor and erratic yields. Two new methods have been used:

a)

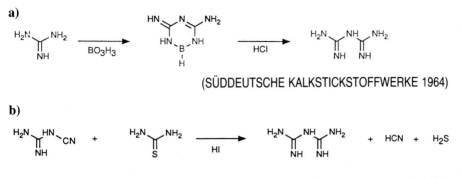

(SÜDDEUTSCHE KALKSTICKSTOFFWERKE 1964)

b)

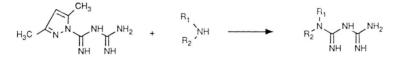

The resulting biguanide is isolated as the diparatoluenesulfonate in 45% yield (JOSHUA and RAJAN 1974).

2. N-Monosubstituted and N,N-Disubstituted Biguanides

This reaction of amines with cyanoguanidine is the most widely used synthetic route to the biguanides. The usual fusion technique in numerous cases yielded several degration products: substituted guanidine and many other by-products including cyclic derivatives such as melamines. A variation of this procedure involves the use of boiling aqueous hydrochloric acid or solvents with a boiling range of 110°–160°C such as: hydrocarbons, toluene, xylene; alcohols, butanol, hexanol; and Cellosolve. Thus, optimum experimental conditions are needed in order to obtain the best yield with the lowest content of impurities. In another synthesis, N-(amino iminomethy)-1-pyrazole carboximidamide and a selected amine are allowed to react and give the target biguanide [SCHENKER and HASSPACHER (SANDOZ) 1966].

A recent modification involves the addition of $FeCl_3$ or $ZnCl_2$ to the reaction mixture (SUYAMA et al. 1989). Thus, the reaction proceeds under milder conditions. The initially formed biguanide complexes are readily hydrolysed to give hydrochlorides of biguanides in good yields.

3. N,N'-Substituted Biguanides

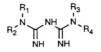

In 1971 Kabbe et al. (Bayer) described a series of antihyperglycemic biguanides with the following structure:

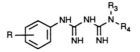

The synthesis was realized by reaction of arylcyanoguanidines with hydrochlorides of amines according to the classical method. The arylcyanoguanidines can be obtained either from sodium dicyanamide:

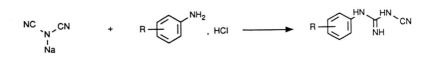

or from diazonium salts:

N,N'-disubstituted biguanides have not been subject to a great deal of development in the domain of antihyperglycemic substances. Nevertheless, it was in the context of investigations on paludrine that the potential antihyperglycemic properties of biguanides attracted renewed interest.

4. N,N''-Substituted Biguanides

Several new N,N''-substituted biguanides were patented in 1971 by Ahrens et al. (Schering) for their antihyperglycemic activity. The different substituent groups were: R_1 = alkyl or aralkyl, R_2 = H or alkyl, and R_3 = alkyl. These biguanides were synthesized by reacting thiourea derivatives, isothiourea or carbodiimide with guanidine or guanidine salt:

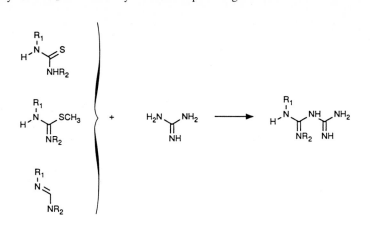

On the other hand, N,N''-substituted biguanides can be obtained by reaction of guanylthiourea or guanyl isothiourea with an amine:

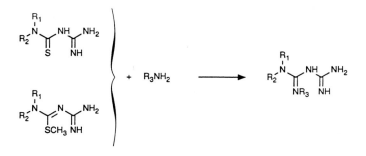

5. N,N',N'',N'''-Substituted Biguanides

CANTELLO (BEECHAM) (1980) described antihyperglycemic compounds such as:

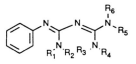

These compounds were obtained by condensation of isocyanide dichlorides successively with an amine and a guanidine:

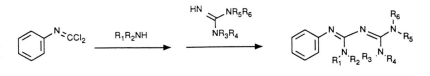

II. Stability Degradation

1. Thermal Decomposition

As noted by Bell et al. (1977), it is remarkable that little work has been done to investigate the thermal behavior of the biguanides. Sugino (1939) studied the thermal decomposition of phenylbiguanide in the presence of ammonium chloride or aniline hydrochloride and obtained guanidine, N-phenyl and N,N'-diphenylguanidine. A large proportion of water-insoluble compound was identified as diphenylmelamine. By studying the thermal decomposition of methylbiguanide, Sugino and Yamashita (1944) obtained analogous results by isolating guanidine, methyl guanidine and melamine derivatives. Patereau (1965, unpublished results), by studying the synthesis of substituted biguanides in the fusion of cyanoguanidines with hydrochlorides of various amines, obtained mixtures containing high proportions of water-insoluble melamines.

During his experiments on the thermal decomposition of biguanide and some of its salts, such as the sulfate, the hydrochloride and the carbonate, Bell et al. (1977) observed a nearly quantitative loss of ammonia at and above 130°C. The resulting solid was identified as melamine. He suggested that the decomposition reaction was achieved according to the equation:

Fusion of biguanide is endothermic and is followed by an exothermic reaction over a temperature range of 134°–170°C with formation of ammonia and melamine. The decomposition of biguanide sulfate and hydrochloride, which are more stable, is also more complex. The biguanide carbonate, which is less stable, decomposes between 80° and 135°C, with a resulting loss of water. At and above 140°C there is a loss of ammonia and water up to 210°C, the resulting residue again being identified as melamine by infrared (IR) and nuclear magnetic resonance (NMR) spectra.

2. Action of Mineral Acids

Biguanides such as phenformin and metformin are rather resistant in aqueous hydrochloric acid at ambient temperature. It is necessary to heat the reaction mixture to obtain an appreciable degradation of the biguanide present. After 1 h in boiling 3 N hydrochloric acid, unchanged phenformin was recovered in 65% yield (Shapiro et al. 1959a). No further information was given about the nature of the degradation products (35%). On the other hand, a solution of metformin in 1 N hydrochloric acid was refluxed for 48 h. The resulting products were analysed by thin-layer chromatography (TLC), gas chromatography (GC) and high-performance liquid chromatography

(HPLC), (POIREE 1991). Unchanged metformin was recovered in 34% yield and two impurities were identified as dimethylurea and amidinourea. Two other by-products were assumed to be dimethylguanylurea and biuret. The stability of metformin is similar to that of phenformin in acidic medium. Arylbiguanides are readily converted into N-aryl-N'-amidinoureas. Thus, treatment of paludrine with $2N$ hydrochloric acid gives N-(4-chlorophenyl)-N'-isopropylamidinourea (CURD et al. 1949).

3. Action of Alkalis

HPLC monitoring showed the complete disappearance of metformin after treatment in boiling $1N$ aqueous sodium hydroxide for 30 min (POIREE 1991). The first resulting degradation products were identified as biuret, dimethylurea and guanylurea. These compounds themselves underwent further degradations, resulting in smaller molecules which could not be detected by HPLC. In the last step of the degradation pathway, dimethylamine was detected by GC. Similar treatment of phenformin resulted in degradation products identified as phenethylguanidine, phenethylurea and phenethylamine (SHAPIRO et al. 1959a).

4. Action of Reducing Agents

The biguanide moiety itself is resistant to various reduction operations specifically acting on the substituent groups. Mention may be made of the following: (1) reduction of an aromatic nitro compound by iron in acetic acid (ROSE 1943) and (2) debenzylation of biguanides in the presence of palladium on carbon (SHAPIRO et al. 1959a). In each case, the biguanide moiety remains unaffected. On the other hand, VICENTE PEDROS and TRIJUEQUE MONGE (1984) studied and discussed the polarographic reduction of various biguanides, and specifically phenformin.

5. Action of Oxidizing Agents

Under mild conditions, the biguanide chain is resistant to oxidation. The unsubstituted biguanide does not react with iodine (KURZER 1955). Under drastic conditions involving reagents such as hydrogen peroxide, lead tetraacetate, potassium permanganate or potassium ferricyanide, paludrine gives non-identifiable products (BIRTWELL 1952). After treatment of metformin by refluxing hydrogen peroxide, degradation products were identified as biuret, dimethylurea, guanylurea and melamine (POIREE 1991). Guanylurea and 1,4-benzoquinone were isolated after treatment of N-phenylbiguanide with Cr^{VI} and Ce^{IV} in sulfuric medium (BANERJEE et al. 1988). Analogous derivatives which are substituted on the phenyl moiety show various behaviors, the nature of the substituents being of importance. No appreciable amount of oxidizing agent was consumed when allowed to react with 4-chlorophenylbiguanide. Under similar conditions, the attempted oxidation of N_1-methylbiguanide was unsuccessful (BANERJEE et al. 1988).

III. Cyclization Reactions

Biguanides are subject to cyclization reactions with derivatives of carboxylic acids, carbonyl compounds, β-difunctional compounds and benzil compounds.

1. Cyclization with Derivatives of Carboxylic Acids

Such cyclizations are the most numerous and the oldest known. By drying on potassium hydroxide a chloroform solution containing a piperidine derivative of biguanide, BAMBERGER and SIEBERGER (1892) observed the formation of a new compound which they identified as 2-amino-4-piperidino-1,3,5-triazine. This compound was also obtained by heating the formiate of the same piperidine derivative of biguanide. Reactions between biguanides and carboxylic acids can be generalized according to the equation:

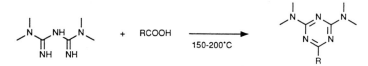

RACKMANN (1910) showed that esters of carboxylic acids and acid chlorides reacted with biguanides under milder conditions. Subsequently the cyclization reaction was extended to other derivatives of carboxylic acids such as lactones, imides, amides, ortho esters and anhydrides. Furthermore, yields were improved by the addition of a base equivalent. A great variety of diaminotriazines were prepared mainly by the American Cyanamid Company, mainly as intermediates in the preparation of resinous compounds. The main pharmaceutical activity was found in the domain of diuretics with chlorazanil [2-amino-4-(4-chloroanilino)-1,3,5-triazine], but its toxicity precludes its use in humans. No hypoglycemic property has been claimed for the 2,4-diaminotriazines.

From an analytical point of view, cyclic derivatives of biguanides with chlorodifluoroacetic anhydride and nitrobenzoyl chlorides have found an important application in:

1. GC because of their great thermal stability and their easy detectability by electron capture detectors.

2. HPLC because of their high UV absorption properties.

2. Cyclization with Carbonyl Compounds

According to BIRTWELL and CURD (1948) and MODEST (1956), biguanides react with carbonyl compounds to give dihydrotriazines:

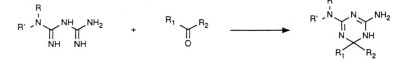

In 1973, YAMANOUCHI claimed hypoglycemic activity for the $R_1 = R_2 = CH_3$ and $NRR' = NH\text{-}CH_2\text{-}CH_2\text{-}C_6H_5$ compound, which can be regarded as a cyclic derivative of phenformin. In 1979, TAIHO YAKUGIN KOGYO claimed the same activity for the following derivatives: (1) $NRR' = N(CH_3)_2$, which is a cyclic derivative of metformin, and (2) $NRR' = NH\text{-}C_4H_9$, which is a cyclic derivative of buformin. R_1 is an isoxazolyl or pyrazolyl group.

3. Cyclization with β-Difunctional Compounds

Most often, these β-difunctional compounds are β-dicarbonyl compounds such as diketones or ketoesters. Cyclization also occurs with ketonitriles or α, β-unsaturated esters. In any case, the biguanide reacts like an amidine to give pyrimidinylguanidines. Few such compounds have been described:

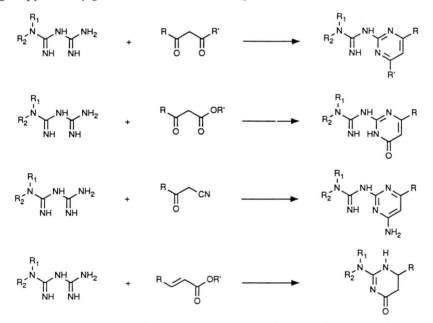

Mention may be made of the paludrine derivatives (CLIFFE et al. 1948), insecticidal and antifungal agents (SHUTO et al. 1974, 1979), diuretics with hypotensive properties (SKULNICK et al. 1985) and antimicrobial agents (EISA et al. 1990). An exception to this type of reaction was reported by FURUKAWA et al. (1972), who found arylbiguanides to react with benzoylacetone on both N and N' nitrogens to give 1,3,5-triazocinediamines:

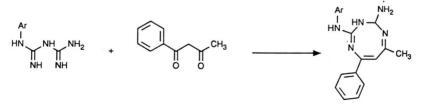

4. Cyclization with Benzil or Benzoin

a) Benzil

In 1972 FURUKAWA et al. obtained 2-guanylidene-4-oxo-5,5-diphenyl imidazolines by reacting benzil with substituted biguanides:

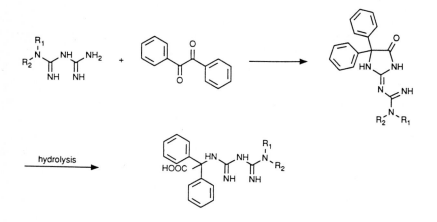

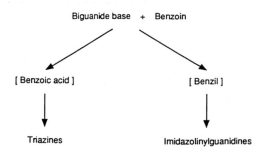

b) Benzoin

According to SCHRAMM et al. (1991) and SCHRAMM and BELAJ (1992), the reaction of free bases of biguanides with benzoin is more complex. Depending on the biguanide used, variations of pH are observed and benzoin is converted either into benzoic acid or into benzil. Benzoic acid then leads to triazine derivatives while benzil leads to imidazolines:

Biguanide base + Benzoin

[Benzoic acid]   [Benzil]

Triazines   Imidazolinylguanidines

5. Miscellaneous Cyclizations

FURUKAWA et al. (1974a) described various cyclizations from arylbiguanides:

Acylisothiocyanates   → triazinoguanidines
Isatin   → dihydrotriazines

IV. Metal–Biguanide Complexes

1. Synthesis and Structure

Biguanides are characterized by their remarkable ability to form chelates with transition elements. This capacity was often used by RATHKE (1878) and HERTH (1880) during their earliest experiments. Thus: biguanides which had been synthesized in the presence of an ammoniacal cupric solution were chelated either during the reaction itself or at the separation stage. The dissociation of chelates by the action of hydrogen sulfide produced either the free base or the sulfate as a result of treatment with sulfuric acid. This method is still efficiently used in the case of biguanides which are particularly difficult to prepare (PATEREAU 1965, unpublished results).

After 1935 RAY synthesized derivatives with chromium, cobalt and nickel, in most cases with unsubstituted biguanides. Later on the synthesis of biguanide complexes was extended to silicon and di- and trivalent metals, such as beryllium, magnesium, and aluminum. Lastly, lanthanide complexes of paludrine have been prepared by Rumanian researchers (see Table 3; see exhaustive review by RAY 1961). All these metal-complex biguanides have deep colors, which are typical of both the metal ion and the biguanide (RAY 1961). See Table 4.

RAY's review in 1961 was updated by SYAMAL in 1987, who proposed that biguanide complex structures should be classified into the following types:

1. Uncharged metal biguanides, which can be represented by [M (Big)$_n$].
2. Uncharged hydrated metal biguanides, which can be represented by [M(Big)$_n$]$_n$H$_2$O. Consequently the chelates such as [M(Big H)$_n$] Xn (where X = Cl$^-$, Br$^-$, 1/2 SO$_4^2$, etc.) should be considered as:
3. Charged (cationic) metal–biguanide complexes.

On the basis of UV data, LCAO (linear combination of atomic orbitals) calculations, and measurements from X-ray data, SEN (1969) suggested the following structures:

4. Uncharged metal–biguanide complex

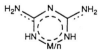

5. Charged metal–biguanide complex

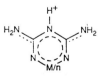

Table 3. Elements which have been complexed with biguanides (represented in clear boxes)

IA	IIA	IIIB	IVB	VB	VIB	VIIB	VIII			IB	IIB	IIIA	IVA	VA	VIA	VIIA	INSERT GASES
1 H 1.0079																	2 He 4.0026
3 Li 6.941	4 Be 9.01218											5 B 10.81	6 C 120.1115	7 N 14.0067	8 O 15.9994	9 F 16.9954	10 Ne 20.183
11 Na 22.9996	12 Mg 24.3050											13 Al 26.9815	14 Si 28.066	15 P 30.9736	16 S 32.086	17 Cl 35.453	18 Ar 39.948
19 K 39.096	20 Ca 40.05	21 SC 44.956	22 Ti 47.96	23 V 50.942	24 Cr 51.996	25 Mn 54.9380	26 Fe 55.847	27 Co 58.9332	28 Ni 58.69	29 Cu 63.546	30 Zn 65.38	31 Ga 69.72	32 Ge 72.61	33 As 71.9216	34 Se 78.96	35 Br 79.904	36 Kr 83.80
37 Rb 35.47	38 Sr 87.52	39 Y 38.805	40 Zr 91.22	41 Nb 22.906	42 Mo 95.94	43 Tc (96)	44 Ru 101.07	45 Rh 102.905	46 Pb 106.42	47 Ag 107.870	48 Cd 112.40	49 In 114.82	50 Sn 118.710	51 Sb 121.75	52 Te 127.60	53 I 126.9044	54 Xe 131.29
55 Cs 132.9054	56 Ba 137.327	57 *La 135.91	72 Hf 178.49	73 Ta 180.948	74 W 183.85	75 Re 186.2	76 Os 190.2	77 Ir 192.22	78 Pt 195.08	79 Au 196.967	80 Hg 200.59	81 Tl 204.38	82 Pb 207.2	83 Bi 206.9854	84 Po (209)	85 At (210)	86 Rn (222)
87 Fr (223)	88 Ra 226.0254	89 **AC 227.0278															

*LANTHANUM SERIES

58 Ce 140.12	59 Pr 140.907	60 Nd 144.24	61 Pm (145)	62 Sm 150.36	63 Eu 151.965	64 Gd 157.25	65 Tb 158.924	66 Dy 162.50	67 Ho 164.930	68 Er 157.28	69 Tm 166.934	70 Yb 173.04	71 Lu 174.97

**ACTINIUM SERIES

90 Th 232.038	91 Pa 231.0359	92 U 238.03	93 Np 237.0482	94 Pu (241)	95 Am (243)	96 Cm (247)	97 Bk (247)	98 Cf (351)	99 Es (254)	100 Fm (257)	101 Md (257)	102 No (250)	103 Lr (260)

Table 4. Colors of metal–comlex biguanides

Metal		Colors of biguanide complexes
V (IV)		Bluish green to light green
Cr (II)	tris	Crimson (base), yellow (salt)
	bis	Violet red to rose red
Mn (III)		Chocolate red (base), yellow (salt)
Mn (IV)		Dark red to chocolate red
Co (II)		Yellow and red
Co (II)	tris	Dark red (base), yellow (salt)
	bis	Red to red violet, yellow
Ni (II)		Orange yellow to yellow
Cu (II)	uni	Blue
	bis	Rose red, violet red
Zn		White
Pd (II)		Light yellow
Ag (III)		Orange red to purple red
Rh (V)		Rose violet (base), brownish yellow (salts)
Os (VI)		Yellow

A better representation with a completely delocalized positive charge was given by CREITZ et al. (1969):

2. Pharmacological Activity of Complexes

a) Antidiabetic Activity

PICCININI et al. (1960) compared the hypoglycemic activity of alkylbiguanides, aromatic biguanides and their copper complexes in the guinea pig. Thus, alkylbiguanides exhibit a higher activity than that of the corresponding complexes, while aromatic biguanides are less potent in relation to the corresponding chelates. By studying phenformin derivatives, FOYE et al. (1961) established a correlation between the complex-forming power and the hypoglycemic activity, but no other report confirming this has appeared in the literature.

A silver biguanide complex was described as antidiabetic by IQBAL KAZMI (1983):

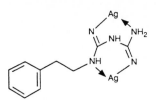

b) Antibacterial Activity

Complexes of Co(II), Ni(II), Cu(II) and Zn(II) with metformin were synthesized and studied for their antibacterial activity by Abu-el Wafa (1987). He compared the biological activity of these complexes against eight microorganisms with that of metformin at different concentrations.

3. Use of Biguanide Complexes in Analytical Procedures

The complex-forming capacity of biguanides can be efficiently used for titration of either the biguanide moiety or the metal ion. Some examples are listed in Table 5.

C. Structure-Activity Relationships

Structure-activity relationships were extensively reviewed by Beckmann in 1971. The results reviewed were from assays which were performed by different authors using a number of variables: (1) animal models: mouse, rat, rabbit and guinea pig; (2) route of administration: oral, subcutaneous and intraperitoneal; (3) animal state of health: normal and pathological; and (4) tested dose. It is somewhat difficult to compare between data publications. The only results which could be utilized were those from series with a significant number of compounds.

Table 5. The complex–forming capacity of biguadines

Compound	Method	Author
Metformin	Atomic absorption of the copper complex	Aly and El Rayes (1983)
Metformin	Conductimetric titration of copper	Maetinez-Calatayud
Moroxydine	complex	et al. (1985)
Buformin	Potentiometric determination with a	Baiulescu et al. (1976)
	copper-sensitive ion electrode	
Phenformin	Complexometric titration of the copper	Alessandro et al. (1974)
Metformin	complex	
V	Atomic absorption (with biguanide)	Chakraborty and Das (1989)
Pd	Gravimetric determination	Ramis Ramos ad Ibanez
	(with biguanide sulfate)	Tomas (1986)
Ni	Gravimetric determination	Spacu and Albescu
	(with paludrine)	(1960)

1. HESSE-TAUBMAN (1929): About 20 N-alkyl-substituted compounds were evaluated in rabbits by oral administration.
2. SHAPIRO (1959): More than 150 N- or N,N-substituted compounds including phenformin were tested in guinea pigs by subcutaneous and oral administration.
3. PROSKE (1962): About 20 compounds including buformin and its analogues were tested in mice, rats, guinea pigs, rabbits and alloxan-pretreated rats.
4. PAUL (1963): Twenty-five derivative compounds of phenformin were tested in guinea pigs.

The products which were studied by these different authors did not show a sufficient structural variety to be a statistically representative sample of the whole class of biguanides. Almost all of them were mono- or di-substituted on the N position by alkyl or aralkyl substiuents. Very few compounds were: (1) tri- or tetra-substituted, (2) N,N''-substituted, (3) substituted by aromatic groups or (4) substituted by functional groups.

Keeping in mind this important structural limitation and the fact that he never used a pathological animal model but only guinea pigs, the conclusions given by SHAPIRO (1959b) were still representative of the knowledge of biguanides in 1971. SHAPIRO's conclusions were as follows:

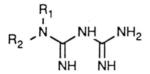

When R_1 is an alkyl substituent, the highest activity is obtained for R_1 = n-pentyl and a decrease in activity is then noted through R_1 = n-octyl. The activity disappears for R_1 = n-decyl. A decrease in activity has also been shown when the R_1 substituent is branched or cyclic. For R_2, a hydrogen atom gives the best response, though biguanides bearing a methyl group as R_2 are also potent. Furthermore when R_1 is an aralkyl substituent, the benzyl derivative is effective and the highest activity is achieved for R_1 = phenethyl. Finally when the chain is lengthened or branched, the activity decreases or disappears.

Two interesting families of compounds have been described since BECKMANN's review:

1. In 1971, BAYER claimed for N-aryl-N'-alkyls or N',N'-dialkyl derivatives a stronger activity than that of buformin in rats (same activity for a lower dose).
2. In 1972, SCHERING claimed for N,N''-di-substituted derivatives an activity "at least as strong" as those of N_1-substituted biguanides, but no data were given. Their goal was to show a lower toxicity than those of phenformin and buformin.

Since these two series of compounds have not been examined further, it is difficult to compare them with previous ones, and SHAPIRO's conclusions are still relevant today.

A recent approach by NOEL and BARBENTON (1992, unpublished results) compares *metformin* and *phenformin*. Molecular lipophilicity potential determinations and a global lipophilic structural visualization have been made for both structures. This work was based on published X-ray diffraction data and modelling methods for molecular volumes and surface determinations. The authors emphasized that packing modes, C=-N double bonds, protonation sites, major hydrophilic and lipophilic characters have clearly different locations. They concluded that metformin and phenformin show very important structural differences which could account for the nonidentical activities of these two molecules with regard to biological interactions.

Acknowledgements. We would like to express our thanks to G. Botton for his expert technical assistance and contribution and to J. Quentin for her secretarial assistance.

References

Abu-El Wafa SM (1987) Formation of metformin complexes with some transition metal ions: their biological activity. inorgan Chim Acta 136:127–131

Ahrens H, Rufer C, Biere H, Schroeder E, Losert W, Loge O, Schillinger E (Schering 1971) 1,2-Disubstituted biguanides. Ger Offenlegungsschriften 2.117.015

Alessandro A, Pieri M, Liguori A (1974) Complexometric determination of biguanides in pharmaceutical preparations. G Med Mil 124(2–3):279–284

Aly FA, El Rayes M (1983) Method for the determination of metformin hydrochloride by atomic absorption of its copper complex. Egypt Pharmacol Sci 24(1–4):169–175

Anatol J, Vidalenc HM, Loiseau GPM (Ugine Kuhlmann 1971) Hypoglycemic sodium salts of phosphoryl biguanides. Ger Offenlegungsschriften 2:130–303

Baiulescu GE, Cosofret VV, Cocu FG (1976) Potentiometric determination of n butyl biguanide with a liquid state copper sensitive electrode. Talanta 23(4):329–331

Bamberger E, Dieckmann W (1892) Zur Kenntniss des Biguanids. Bamberger Ber Dtsch Chem Gesell 25:543

Bamberger E, Sieberger L (1892) Ring Synthesen. Ber 25:525

Banerjee R, Bhattacharya A, Das PK, Chakrabutty AK (1988) Mechanism of reactions of some N[1] substituted biguanides with chromium (VI) in aqueous sulfuric media. J Chem Soc Dalton Trans (6):1557–1560

Beckmann R (1971) Biguanide. In: Maske H (ed) Oral wirksame Antidiabetika. Springer, Berlin Heidelberg New York, pp 439–596 (Handbuch der experimentellen Pharmakologie vol 29)

Bell NA, Hutley BG, Shelton J, Turner JB (1977) Biguanides. Part I. The thermal decomposition and mass spectral behaviour of biguanide and some of its salts. Thermochim Acta 21:255–262

Birtwell S (1952) Attemptsell S to prepare a possible metabolite of "paludrine" and related 1,3,5 triazines. J Chem Soc 1279–1286

Birtwell S, Curd FHS (1948) Synthetic antimalarials. Part XXX. Some N[1] aryl N[4] N[5] dialkyldiguanides and observations on the conversion of guanylthioureas into diguanides. J Chem Soc 1645–1657

Bosies E, Stach K, Schmidt FH, Heerdt R, Weber H (Boehringer) (1974) Biguanides. Ger Offenlegungsschriften 2,426,683

Burns D, Sharpe JS (1917) Guanidine and methylguanidine in the blood and urine in tetania parathyreopriva and in the urine of idiopathic tetany. Q J Exp Physiol 10:345–354

Cantello BCC (Beecham) (1980) Carboxamidine derivatives. European Patent Application EP 34.002

Chakraborty D, Das AK (1989) Indirect determination of vanadium by atomic absorption spectrometry. Anal Chim Acta 218(2):341–344

Chen KK, Anderson RC (1947) The toxicity and general pharmacology of N^1 p-chlorophenyl N^5 isopropyl biguanide. J Pharmacol Exp Ther 91:157–160

Cliffe WH, Curd FHS, Rose FL, Scott M (1948) Synthetic antimalarials. Part XXIII. 2 Aryl guanidino 4-aminoalkylamino pyrimidines. Further variations. J Chem Soc 574–591

Creitz TC, Gsell R, Wampler DL (1969) Crystal structural studies of bis (biguanide) nickel (II) chloride. J Chem Soc (D) 23:1371–1372

Curd FHS, Davey DG, Richardson DN (1949) Synthetic antimalarials. Part XLII. Preparation of guanyl/ureas and biurets corresponding to "paludrine" and related diguanides. J Chem Soc 1732–1738

Dohi T, Yu T, Nakagawa T, Hiraoka K (Otsuka 1972) Antidiabetic carnitine salts of N,N-dimethyl-biguanide hydrohalides. US 3.651.132

Duval D (1960) Contribution l'etude de à l'éction hypoglycémiante des biguanides. Dssertation, University of Paris

Eias HM, Tayel MA, Yousif MY, El-Kerdawy MM (1990) Synthesis of certain N-aryl N^1 (2 pyrimidinyl) guanidine derivatives as potential antimicrobial agents. Chung Hua Yao Hsueh Tsa Chih 42(5):385–389

Foye WO, O'Langhlin RL, Duvall RN (1961) Chelation of β-phenethylbiguanide and other biguanides with copper ion. J Pharm Sci 50:641–644

Frank F, Nothmann M, Wagner A (1926) Über synthetisch dargestellte Körper mit insulinartiger Wirkung auf den normalen und diabetischen Organismus. Klin Wochenschr 5:2100–2107

Furukawa M, Fujino Y, Kojima Y, Hayashi S (1972a) Reaction of biguanides and related compounds. II. Reaction of biguanides with benzil. Chem Pharm Bull (Tokyo) 20(3):521–525

Furukawa M, Kojima Y, Hayashi S (1972b) Reaction of biguanides and related compounds. IV. Reaction of arylbiguanide with benzoylacetone in the presence of a small amount of the aryl-biguanide hydrochloride. Chem Pharm Bull (Tokyo) 20(5):927–930

Furukawa M, Kojima Y, Hayashi S (1973) Reaction of biguanides and related compounds. VII. Condensation of aryl-biguanide with β-unsaturated carboxylic ester in dimethylformamide. Chem Pharm Bull (Tokyo) 21(5):1126–1131

Furukawa M, Goto M, Hayashi S (1974a) Reaction of biguanides and related compounds. X. Cyclisation of biguanides with acylisothiocyanates. Bull Chem Soc Lap 47:1977–1980

Furukawa M, Yoshida T, Hayashi S (1974b) Reacion of biguanides and related compounds. XII. Condensation of aryl biguanides and amidino rsoureas with isatin. Chem Pharm Bull 22:2875–2882

Herth R (1880) Synthèse des "biguanids". Mh Chem 1:88–98

Hesse E, Taubman G (1929) Die Wirkung des Biguanids und seiner Derivate auf den Zuckerstoffwechsel. Arch Exptl Patho Pharmakol 142:290–308

Iqbal Kazmi SA (1983) Synthesis and characterisation of a new antidiabetic complex of α-phenethyl biguanide with silver. J Sci Res [Suppl] (1):49–51

Joshua CP, Rajan VP (1974) Synthesis of biguanide. Chem Ind 12:497–498

Kabbe HJ, Petersen S, Horstmann H, Pluempe H, Puls W (Bafer) (1971) Antihyperglycemic biguanides. German Offenlegungssriften 2009737–2009738 and 2009743

Kurzer F (1955) The oxidation of amidinothiourea. J Chem Soc 1–6

Kurzer F, Pitchfork ED (1968) The chemistry of biguanides. Fortsch Chem Forsch 10:375–472

Martinez-Calatayud J, Campins Falco P, Pascual Marti MC (1985). Metformine and moroxydine determination with Cu (II). Aual Lett ••

Modest EJ (1956) Chemical and biological studies of 1,2 dihydro-s-triazines. J Org Chem 21:1–20

Oniscu C, Bibian-Cilianu S, Braha S, Simionovici M, Boesteanu N, Cristescu Y (1983) Salts of (sulfamoyl phenoxy) acetic acids with N,N-dimethyl biguanide. Institutue de Cercetari Chimico-Farmcentice Rom RO82,052

Oxley P, Short WF (1951) Amidines. Part XV. Preparation of diguanides from cyanoguanidines and ammonium sulphonates. J Chem Soc 1252–1256

Paul SP, Bose AN, Basu UP (1963) Synthesis of biguanides as potential hypoglycaemic agents. Part IV. Structure-activity relationship. Indian J Chem 1:218–220

Piccinini F, Marazzi Uberti E, Lucatelli I (1960) Sull'azione ipoglicemizzante di alcune biguanidi e dei corrispondenti complessi di rame. Il Farmaco Ed Sci 15:521–529

Poirée MA (1991) Stabilité et voies de dégradation du chlorhydrate de metformine. Divertation, Ecole Pratique des Hautes Etudes, France

Proske G, Osterloh G, Beckmann R, Lagler F, Michael G, Mückter M (1962) Tierexperimentelle Untersuchungen mit blutzuckerwirksamen Biguaniden. Arzneimittel forschung 12:314–318

Rackmann K (1910) Untersuchungen über Diguanid und einige daraus hergestellte Verbindungen. Liebigs Ann Chem 376:163–183

Ramis Ramos G, Ibanez Tomas R (1986) Gravimetric deetermination of palladium with biguanide sulfate. Microchem J 33:379–383

Rathke B (1878) Über geschwefeltes Dicyandiamin. Ber Dtsch Chem Gesell 11:962

Rathke B (1879) Über Biguanid. Ber Dtsch Chem Gesell 12:776–784

Ray P (1955) Chemistry of dicyandiamidines and biguanides and their metalic complexes. J Indian Chem Soc 32:142–156

Ray P (1961) Complex compounds of biguanides and guanylureas with metallic elements. Chem Rev 61:313–359

Reitz AB, Tuman RW, Marchione CS, Jordan AD, Bowden CR, Maryanodf BE. (1989) Carbohydrate biguanides as potential hypoglycemic agents. J Med Chem 32(9):2110–2116

Rose FL (1943) British Patent 550–538

Schenker E, Hasspacher K (Sandoz 1966) Hypoglycemic heterocyclic biguanides. Fr 1.51.398

Schramm HW, Schubert-Zsilavecz M, Saracoglu A, Kratky C (1991) Über Reaktionen von Alkylbiguaniden mit Benzoin beim pH der Biguanidbasen. Mh Chem 122:1063–1073

Schramm HW, Belaj F (1992) Über Reaktionen von Arylbiguaniden mit Benzoin beim pH der Biguanidbasen. Mh Chem 123:237–245

Sen D (1969) Ultraviolet spectral studies on metal biguanide complexes. J Chem Soc (A):2900–2903

Shapiro SL, Parrino VA, Freedman A (1959a) Hypoglycemic agents. I. Chemical properties of β-phenethylbiguanide. New hypoglycemic agent. J Am Chem Soc 81:2220–2225

Shapiro SL, Parrino VA, Freedman L (1959b) Hypoglycemic agents. III. and IV. N^1 and $N^1 N^5$ Alkyl and Aralkylbiguanides. J Am Chem Soc 81:3728–3736 and 4635–4646

Shramova ZI, Voronin VG, Aleshina VA, Pleshakow MG, Zuev AP, Zaks K, Kotegov VP, Gasanov SG (1983a) Derivatives of biguanide compounds with hypoglycemic action. USSR. SU 992, 512

Shramova ZI, Voronin VG, Aleshina Trubnikov VL, Pleshakow MG, Zuev AP, Zaks AS, Kotegov VP, Gasanov SG (1983b) Antipyrylbiguanides having hypoglycemic activity. USSR. SU 992.514. 16564

Shridar DR (1985) Synthesis and hypoglycemic activity of (3-oxo-3,4-dihydro-2H 1,4-benzoxazin-6/7-yl) biguanide hydrochlorides. Indian J Chem B, 24B:1293–1294

Shuto Y, Tanigushi E, Maekawa K (1974) Synthesis of guanidino pyrimidine derivatives and their biological activity. J Fac Agric Kyushu Univ 18(4):221–237

Shuto Y, Tamigushi E, Maekawa K (1979) Studies on biologically active guanidino-pyrimidines. Part II. Effect of guanidinopyrimidines on phytopathogens. J Fac Agric Kyushu Univ 23(3–4):125–132

Skulnick HI, Ludens JH, Wendling MG, Glenn EM, Rohlodf NA, Smith RJ, Wierenga W (1986) N substituted 6-phenyl pyrimidinones and pyrimidinediones with diuretic/hypotensive and anti-inflammatory activity. J Med Chem 29(8): 1499–1504

Slotta KH, Tschesche R (1929a) Über Biguanide. I. Zur Konstitution der Schwermetallkomplexverbindungen des Biguanids. Ber Dtsch Chem Gesell 62B: 1390–1398

Slotta KH, Tschesche R (1992b) Über Biguanide. II. Die Blutzucker-senkende Wirkung der Biguanide. Ber Otsch Chen Gesell 62B:1398–1305

Smolka A, Friedreich A (1888) Über eine neue Darstellungweise der Biguanide und Über einige Derivate des Phenylbiguanids. Mh Chem 9:227–241

Spacu P, Albescu I (1960) Determination of nickel. Acad Rep Pop Rom, Fil Goj Stud Cercet Chim 8:85–90

Sterne J, Duval D (1959) Effects hypoglycémiants de la N-N-dimethyl diguanide. In: Koberdisse K, Jahnke K (eds) Diabetes mellitus. III. Kongress der International Diabetes Federation. Dusseldorf 1958. Thieme, Stuttgart, pp 443–452

Süddeutsche Kalkstickstodfwerde AG (1964) Verfahren zur Herstellung von Bignanidsalzen aus Guanidinen. Trostberg

Sugino K. (1939) Some derivatives of calcium cyanamide. VII. Mechanism of the formation/of guanidine and its derivatives from dicyanodiamide 2. Mechanism of the formation of guanidine salts by fusion of dicyanodiamide with ammonium salts. J Chem Soc Japan 60:351–365

Sugino K, Yamashita M (1994) Mechanism of the formation of methylguanidine and guanidine salts by fusion of dicyanodiamide with methylamine hydrochloride. J Chem Soc Japan 65:271–280

Suyama T, Soga T, Miyauchi K (1989) A method for the preparation of substituted biguanides. Nippon Kagaku Kaishi 5:884–887

Symal A (1978) Recent chemistry of metal biguanide complexes. J Sci Ind Res 37:661–685

Taiho Yakuhin Kogyo (1979) 4-Heterocyclyl-2-6-diamino-1,4-dihydro-s-trazines. Jpn Kokai Tokkyo Koho 79 14,986

Taiho Yakuhin Kogyo (1979) Biguanide derivatives. Jpn Kokai Tokkyo Koho 79 12,371

Tanret G, Simonnet H (1927) Hypoglycemic properties of galegine sulfate. Compt Rend 184:1600–1602

Underhill FP, Blatherwick NR (1914) Studies in carbohydrate metabolism. VI. The influence of thyreoparathyroidectomy upon the sugar content of the blood and the glycogen content of the liver. J Biol Chem 18:87–90

Vicente Pedros F, Trijueque Monge J (1983) Reduccion polarografica di la fenformina diprotonada. An Quim 80:498

Watanabe CK (1918) Studies in the metabolic changes induced by administration od guanidine bases. I. Influence of injected guanidine hydrochloride upon blood subgar content. J Biol Chem 33:253–265

Werner EA, Bell J (1922) Preparation of methyl guanidine, and of $\beta\beta$-dimethylguanidine by the interaction of dicyanodiamide and methylammonium and dimethylammonium chlorides respectively. J Chem Soc 121:1790–1795

CHAPTER 11

Physicochemical Properties and Analytical Methods of Determination of Biguanides

E. Prugnard and M. Noel

A. Physical Properties

I. General Properties

As described by Shapiro et al. (1959), biguanides are diacid bases. They have a particularly strong primary dissociation constant and a considerably weaker second dissociation constant. The first salts of biguanides which were prepared were the sulfate (obtained by treatment of the biguanide-copper complex), the hydrochloride and the nitrate because of its insolubility. Other known salts are the toluenesulfonate, the tartrate and the pamoate. In order to obtain a better pharmacological activity some other salts have been prepared by treating biguanides with carboxylic acids. These salts were the adamantate, clofibrate, orotate, nicotinate, isoxazole-carboxylate, pyrazole-carboxylate, trimethoxybenzoate and 4-chlorophenoxyacetate.

Antidiabetic biguanides are commonly used as their hydrochloride, 4-chlorophenoxyacetate or pamoate salts. Biguanide salts are generally monosalts. This can be explained by the fact that, as reported by Shapiro et al. (1959) for the dihydrochloride form, the second basic site is protonated by formation of an intramolecular hydrogen bond which prevents biguanide forming a stable salt with a second molecule of acid.

Generally, biguanides are known to have low melting points and to be sparingly soluble in cold water. The configuration of the side chain is of importance; hence biguanide derivatives bearing a long-chain substituent are more soluble in ethanol or in ethanol-water mixtures.

Biguanides are rapidly carbonated under ambient conditions and are best characterized as their crystalline salts. In addition, their stability is pH dependent. Many biguanides are remarkable for their ability to form chelates with transition metal ions.

II. Dissociation Constants

The first pK values of biguanides were obtained either by potentiometric titration (these compounds being relatively soluble in aqueous medium) or were calculated from measurements of the UV absorption spectra, which are pH dependent as shown in Fig. 1.

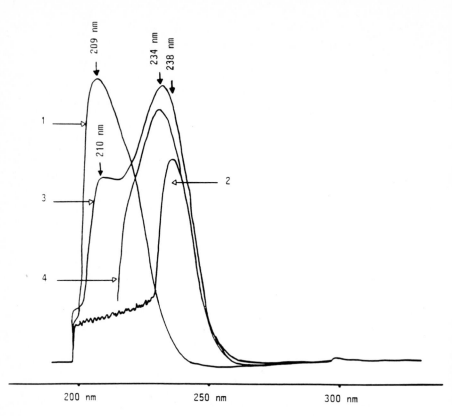

Fig. 1. Ultraviolet absorption spectra of metformin solutions at various pHs (J. BROHON 1989, unpublished results). One hundred milliliters solution contains 1 mg metformin. *1*, HCl 0.1 *N* at pH 1.1; *2*, citrate HCl buffer solution, merck, at pH 4; *3*, phosphate buffer solution, merck, at ph 7; *4*, NaOH 0.1 *N* at pH 13

GAGE (1949) drew the following conclusions:

1. The pK values calculated from these measurements shown that biguanides are diacid bases and exist almost entirely as monocations at physiological pH. Biguanide itself is a very strong base with a pK_1 value nearly equal to 12, and is subject to variations from one measurement method to another. Its pK_2 value (about 3) is lower than might be expected for a singly charged cation considering that the pK_2 value of malondiamidine, which is structurally comparable with biguanide, is equal to 9 (FANSHAWE et al. 1964).

2. In the biguanide structure the four amino groups are almost equivalent and the positive charge is distributed amongst them, which consequently gives a stabilized ion.

The values obtained are listed in Table 1, which shows that alkylbiguanides are stronger bases than arylbiguanides. Thus, the pK values of alkyl-

Table 1. pK_1 and pK_2 values of various biguanides

Compound	pK_1	pK_2	Author and method
Biguanide	12.8	3.1	GAGE (1949) Spectrometry
	13.8	3.2	GAGE (1949) Spectrometry
	11.51	2.94	DE (1950)
	11.52	2.93	DAS SARMA (1952) Complexometry
	11.49	2.95	BANDYAPADHAYA (1952)
Methylbiguanide	11.4	3.00	DAS SARMA (1952) Complexometry
Ethylbiguanide	11.47	3.08	DAS SARMA (1952) Complexometry
Dimethylbiguanide (metformin)	11.52	2.77	DAS SARMA (1952) Complexometry
	12.31	3.24	DORNBOOS (1967) Potentiometry
Diethylbiguanide	11.68	2.53	DAS SARMA (1952) Complexometry
n-Propylbiguanide	11.35	3.10	DUTTA (1961) Complexometry
Isopropylbiguanide	11.35	3.10	DUTTA (1961) Complexometry
Butylbiguanide (buformin)	11.28	2.92	DUTTA (1961) Complexometry
	13.1		BRES (1976) Partition coefficient
	12.8		DUTTA (1961) Complexometry
Hexylbiguanide	11.44	3.30	DUTTA (1961) Complexometry
Cyclohexylbiguanide	11.39	3.30	DUTTA (1961) Complexometry
Benzylbiguanide	11.25	2.70	DUTTA (1961) Complexometry
Phenethylbiguanide (phenformin)	12.7		BRES (1976) Partition coefficient
	13.0		BRES (1976) Partition coefficient
Phenylbiguanide	10.72	2.14	DE (1950)
	10.72	2.16	DAS SARMA (1952)
	10.70	2.16	BANDYAPADHAYA (1952)
N-(4-Tolyl)biguanide	10.84	2.60	DUTTA (1962) Complexometry
N-(4-Chlorophenyl)biguanide	10.4	2.2	GAGE (1949) UV spectrometry
N-(4-Chlorophenyl) N'-Isopropylbiguanide(paludrine)	10.4	2.3	GAGE (1949) UV spectrometry
N-(4-Chlorophenyl) N'''-Methyl N'-isopropylbiguanide	12.2	2.0	GAGE (1949) UV spectrometry

biguanides range between pK values of $11-12$ while those of arylbiguanides remain below 11.

The spectroscopic determination of pK values for paludrine analogues bearing N-(4-chlorophenyl) and N'-isopropyl substituents was performed by GAGE (1949) in order to display a correlation between the antimalarial activity and the UV absorption spectra. This objective was not achieved, but some important conclusions were nevertheless reported by GAGE (1949): (1) the introduction of a 4-chlorophenyl group induces a decrease in the pK value (more than 2 pK units), (2) the supplementary addition of an Isporopyl group does not increase the pK and (3) the introduction of a second alkyl group induces an increase in pK to 12.2.

Table 2. pK values of phenformin and buformin

	NaOH/CHCl₃ Isoamylic alcohol	NaOH/CH₂Cl₂
Phenformin	12.7	13
Buformin	13.1	12.8

Another method can be used: pK values of compounds can be calculated from measurements of partition coefficients. This method can be used with ionizable molecules which are soluble in at least one organic solvent, the latter being immiscible with water. Bres et al. (1976) studied the behavior of metformin, buformin and phenformin in sodium hydroxide – dichloromethane and sodium hydroxide – chloroform – isoamylic alcohol mixtures. This method has been unsuccessfully used to determine the pK value of metformin because of its overly low partition coefficient. Nevertheless, pK values of phenformin and buformin were determined (Table 2).

III. Spectroscopic Data

1. Ultraviolet Spectra

It is well known that the ultraviolet (UV) spectra of biguanides are pH dependent. In 1958 Hirt and Schmitt noted that a single protonation of biguanide has no effect on the maximum absorption but only induces a decrease intensity. A second protonation makes the molecule unable to form a conjugated resonating system and consequently induces the disappearance of the absorption band into the vacuum UV region. On the basis of the above observations, the following neutral and ionized structures of biguanides were suggested by Hirt and Schmitt (1958):

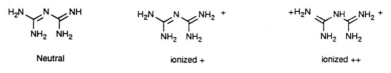

Shapiro (1959) summarized the behavior of phenethylbiguanide at various hydrogen ion concentrations in the following equations:

$$
\begin{array}{lcl}
BH^+\,Cl^-\;(H_2O) & \rightarrow & BH^+ + Cl^- \\
BH^+Cl^-\;(HCl) & \rightarrow & BH_2^{2+} + 2Cl^- \\
BH_2^{++}\,Cl_2^{2-}\;(H_2O) & \rightarrow & BH^+ + 2Cl^- + H^+ \\
B(H_2O) & \rightarrow & BH^+ + OH^- \\
BH^+\;(NaOH) + OH^- & \rightarrow & B + H_2O \\
BH^+\,Cl^-\;(CH_3OH) & \rightarrow & B(CH_3OH)H^+ + Cl^- \\
B(CH_3OH) & \rightarrow & BH^+ + CH_3O^-
\end{array}
$$

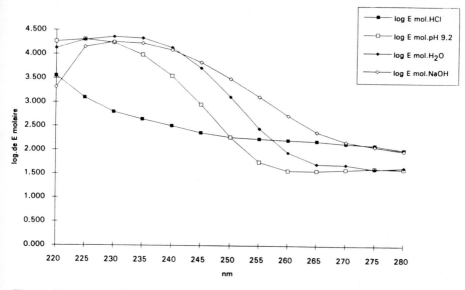

Fig. 2. Biguanide sulfate UV spectra

NANDI (1972) studied seven monohydrochlorides of biguanides which showed and absorption maximum at 230 nm due to the $\pi\pi^*$ transition. UV spectra of metformin were recorded by BROHON (1989, unpublished results) at pH 1, pH 4, pH 7 and pH 13 and produced similar results.

Measurements by UV spectroscopy have been applied in the quantitative determination of N,N-dimethylbiguanide hydrochloride (metformin) and N-phenethylbiguanide hydrochloride (phenformin) in tablets (GOIZMAN et al. 1985). In 1984, OVSEPYAN et al. detemined the metformin contents in galenic forms and biological fluids. This procedure may be used for pharmacokinetics and forensic biological investigations. Furthermore, UV spectra have been used for pK determination.

2. Infrared Spectra

The IR spectra of metformin, phenformin and buformin hydrochlorides exhibit nearly identical strong absorption bands for which the following assignments are given:

3100–3400/ cm ν NH
1480–1660/ cm ν C = N and δC = $\overset{+}{N}$H due to the guanidinium ion (KOJI Nakanishi 1962)

In each spectrum a sharp absorption peak of medium intensity is present in the 1150–1170/cm region.

3. Nuclear Magnetic Resonance Spectra

Nuclear magnetic resonance (NMR) spectra can be helpful for structure determinations. In this way, using the measurements of a proton NMR spectrum they recorded, Wellman et al. (1067) showed that the monocation structure of buformin was symmetrical:

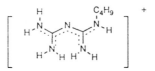

Furthermore, quantitative determination of the metformin content in pellets was achieved by El Khateeb et al. (1988) by using measurements from NMR spectra. ^{1}H-NMR spectra of metformin, phenformin and buformin hydrochlorides exhibit signals in the 6.8- to 7.5-ppm region due to the NH_2 protons of the biguanide moiety (Patereau 1993, unpublished results).

IV. Crystal Structures and Protonation Sites

Crystallographic studies have been previously performed on paludrine (Brown 1967) and moroxydine (Handa and Saha 1973). Subsequently, crystal data and mass spectrometric measurements clearly established the dimeric nature of biguanide dihydrochloride, due to the hydrogen bonding. In addition, another crystallographic investigation showed the crystals to be monoclinic (Syamal 1975). Since 1977, several authors have studied the crystal structures of the biguanides. They have confirmed the structures deduced by Hirt and Schmitt (1958) from UV spectroscopy. The main results are summarized in Table 3. On the basis of work carried out by Herrnstadt et al. (1979) and Hariharan et al. (1989), Noel and Barbenton (1992, unpublished results) have emphasized the differences between phenformin and metformin (see Fig. 3).

1. Phenformin

a) *Protonation site:* The protonation occurs at the terminal imino group.
b) *Packing mode*: Lipophilic elements such as the phenethyl groups and hydrophilic elements such as the protonated terminal imino groups and the chloride anions are stacked one above the other in alternating parallel layers.

2. Metformin

a) *Protonation site*: The protonation occurs at the central amino group.
b) *Packing mode*: The chlorine atoms are interspersed between layers of molecules.

Table 3. Crystal structure and protonation site

Biguanide	Configuration	Positive charge	Structure	Author
Biguanide	Roughly planar			Ernst (1977)
Biguanide HCl	Two individually planar portions	Uniformly delocalized over the four amino groups		Ernst (1977)
Phenformin HCl	Does not achieve an overall planar conformation	At the terminal imino group	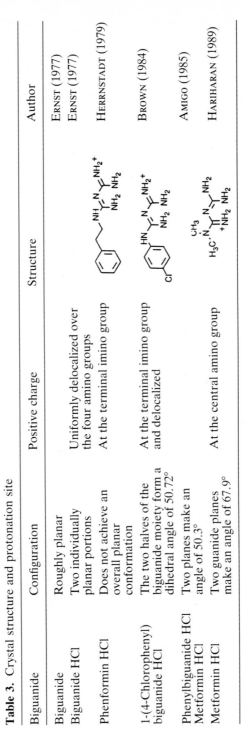	Herrnstadt (1979)
1-(4-Chlorophenyl) biguanide HCl	The two halves of the biguanide moiety form a dihedral angle of 50.72°	At the terminal imino group and delocalized		Brown (1984)
Phenylbiguanide HCl Metformin HCl	Two planes make an angle of 50.3°			Amigo (1985)
Metformin HCl	Two guanide planes make an angle of 67.9°	At the central amino group		Hariharan (1989)

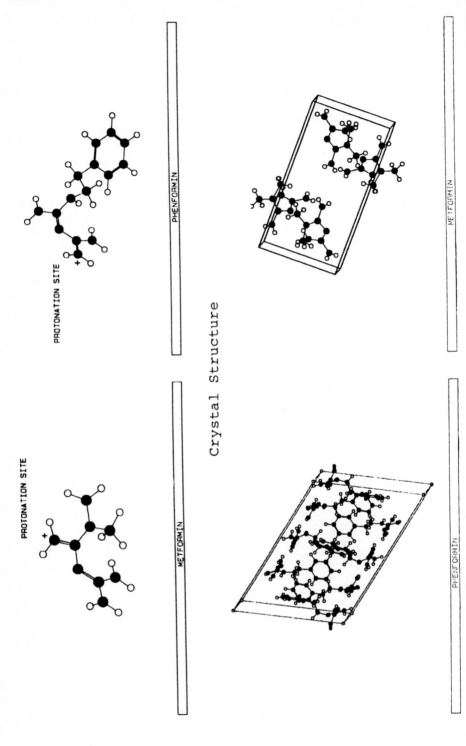

Crystal Structure

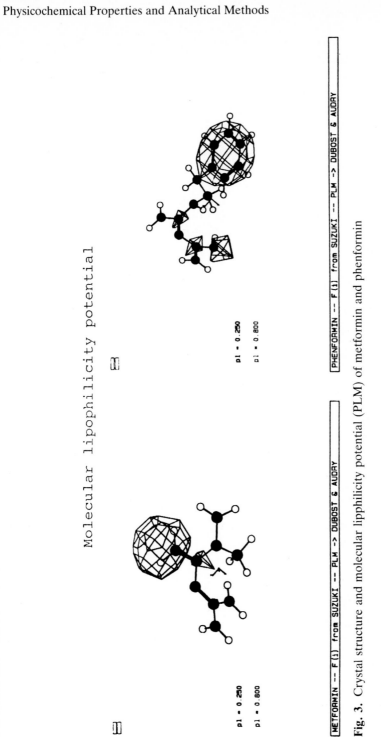

Fig. 3. Crystal structure and molecular lipphilicity potential (PLM) of metformin and phenformin

According to Noel and Barbenton (1992, unpublished results), this different localization of the hydrophilic character, together with the variation of the protonation site, could induce nonidentical enzymatic interactions.

B. Quantitative Determination in Biological Medium

Three kinds of problems need to be considered with respect to the titration methods: (1) sensitivity, (2) specificity and (3) feasibility (whether the method is easy to use).

Two reviews, by Kurzer and Pitchfork in 1968 and by Beckmann in 1971 and Boyer's thesis published in 1984, are the three main sources which allow us to appreciate the evolution of both the techniques and the results obtained.

I. Detection of Antidiabetic Biguanides by Spectrophotometry

Boyer's thesis, which focused on the pharmacokinetics of metformin, contains a rather large introduction dedicated to the analytical problems. This study is particularly interesting because figures are given for the sensitivities obtained and, furthermore, information is given about the specificity and the feasibility of several methods. Following Boyer's thesis, the main steps of the detection methods are listed in Table 4.

As shown in Table 4, the main contributions after the Kurzer-Pitchfork's and Beckmann's reviews are those of Garret and Tsau (Garret and Tsau 1972; Garret et al. 1972), Predescu et al. (1989) and Sastry et al. (1991). None of the techniques described in these studies was sensitive enough and compounds present in concentrations of less than about 1 $\mu g/ml$ could not be determined.

II. From Micrograms to Nanograms: The Chromatographic Revolution

1. Gas Chromatography

a) Direct Determination

In 1972 Wickramasinghe and Shaw explored the gas chromatographic (GC) behavior of buformin and phenformin under different temperature conditions. Unfortunately, this method was suitable for pure compounds only and could not be applied to biological samples of buformin.

b) After Derivatization

The first successful GC measurements of biological samples were performed by Matin et al. (1975), who converted the thermally unstable biguanides

Table 4. Detection of antidiabetic biguanides by spectrophotometry

Titrated biguanide	Titrated medium	Detection method	Sensitivity (μg/ml)	Author
Metformin	Urine	Spectrophotometry	Traces	JUNG (1961)
Metformin	Plasma, serum, red blood cells, urine, saliva, tissue extracts	Spectrophotometry, 400 nm		PIGNARD (1962)
Metformin	Urine	Spectrophotometry, 550 nm	10	SIEST (1963)
Metformin	Water	Spectrophotometry of ion pair, 630 nm	0.13	GARRET and
			1.3	TSAU (1972)
Metformin	Urine	Spectrophotometry of metformin, 232 nm	0.16	GARRET (1972)
Metformin	Blood, bile, plasma	Spectrophotometry of metformin free of its ion pair, 232 nm		
Metformin	Plasma	Spectrophotometry, 565 nm	3	HEUCLIN (1975)
Buformin tosylate	Pharmaceuticals	Spectrophotometry, 527 nm		PREDESCU (1989)
Phenformin	Pharmaceuticals	Spectrophotometry, 440 nm	2–20	SASTRY (1991)

into 1,3,5-triazine derivatives by treatment with chlorodifluoroacetic anhydride.

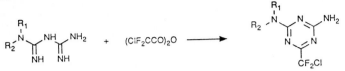

For a few specific types of molecules, the electron capture detector is known to have an exceedingly high sensitivity. In the cases of metformin, buformin and phenformin present in biological fluids, MATIN et al. ächieved a detection limit of less than 1 ng/ml. A giant step forward! In addition, the introduction of a chlorodifluoromethyl group is useful in mass spectrometry. Subsequently, several authors proposed different modifications concerning the nature of derivatization reagents and the type of detectors. The development of HPLC methods without derivatization which are characterized by a treatment of biological medium reduced to a minimum has led to a decrease of interest in GC methods (see Table 5).

2. High-Performance Liquid Chromatography

High-performance liquid chromatography (HPLC) can be performed either with or without chemical derivatization. In a manner similar to that used for GC determinations, Ross (1977) converted metformin into its triazine derivative by cyclization with paranitrobenzoyl chloride. The paranitrobenzoyl derivative was selected because of its good UV absorption capacity:

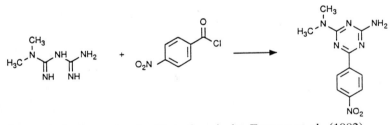

The same method was used with buformin by FUERST et al. (1982).

The spectral properties of 16 nitro and amino derivatives of metformin and buformin were studied by BELLENGER et al. (1983) in order to determine the most suitable derivatives for detection methods such as UV spectrophotometry or fluorescence detection. Thus, the ortho, meta, para nitro and 3,5-dinitro derivatives were synthesized. In addition, the corresponding amines were obtained by catalytic reduction:

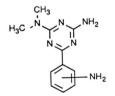

Table 5. Determination of antidiabetic biguanides by gas chromatography

Titrated biguanide	Titrated medium	Derivatization agent	Internal standard	Detection method	Sensitivity (μg/ml)	Author
Buformin Phenformin			Phenylbiguanide	Flame ionization detector		WICKRAMASINGHE (1972)
Metformin Buformin Phenformin	Urine, plasma	Chlorodifluoroacetic anhydride	Propylbiguanide for metformin and buformin Benzylbiguanide for phenformin	Electron capture detector	<0.001	MATIN (1974)
Phenformin and metabolite	Serum, urine	Trifluoroacetic anhydride	Naphthylamine	Flame ionization detector	0.2	MOTTALE (1975)
Metformin Buformin Phenformin		Chlorodifluoroacetic anhydride		Flame ionization detector, electron capture detector		ALKALAY (1976)
Metformin	Urine, plasma	Heptafluorobutyric anhydride	d_6-Metformin	Mass fragmentography	~0.001	SIRTORI (1978)
Metformin	Plasma, whole blood, liver, bile saliva, urine, feces	Chlorodifluoroacetic anhydride	Buformin	Electron capture detector	0.01	LENNARD (1978)
Metformin	Plasma, urine	4-Nitrobenzoylchloride	Propylbiguanide	Nitrogen detector	0.025	BROHON (1978)

The spectral properties of all these compounds were determined and compared. In conclusion, Bellenger recommended the use of the meta nitro derivatives for UV spectrophotometry, whereas the para amino derivatives were preferred for use in fluorescence detection. The relative complexity of the derivatization techniques led to the development of direct methods for determination of phenformin (Hill and Chamserlain 1978), of phenformin and its metabolite (Oates et al. 1980) and of metformin (Charles et al. 1981). Because of its high polarity (octanol/water = 0.01), metformin is particularly difficult to extract. In order to overcome this difficulty, Keal and Somogyi (1986) adapted the ion pair extraction technique which had been developed by Garret (Garret and Tschau 1972; Garret et al. 1972). New methods of derivatization based on fluorometric detection were used by Tanabe et al. (1987), who converted biguanides into their phenanthro [9,10-d] imidazole derivatives by reaction with 9,10-phenanthraquinone:

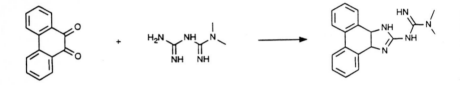

An analogous technique was described by Kobayashi et al. (1988). This procedure, which involved a post-column derivatization with 1,2-naphthoquinone-4-sulfonate, was characterized by a decrease in the duration of analysis. In addition, the sensitivity was improved (see Table 6). Revelle et al. identified the impurities in variously stressed chlorhexidine digluconate solutions. Impurities were isolated at a 2.5 gram scale by quantitative HPLC method and identified by an HPLC-MS method (11 impurities were identified by a comparison with standards).

3. Miscellaneous

a) Radioimmunology

Radioimmunological determinations of phenformin and its metabolites in human fluids such as plasma and serum were perfomed by Alkalay et al. (1978). Determinations of buformin in human serum were performed by Oetting in 1975 and 1982 with a lower limit of detection of 1.5 ng.

b) Turbidimetry

This method, which was used by Martinez-Calatayud and Sampedro (1989), was based on the formation of an ion pair between phenformin and sodium tungstate. The determination was effected by flow-injection analysis (FIA).

Table 6. Determination of antidiabetic Biguanides by HPLC

Titrated biguanide	Titrated medium	Derivatization agent	Internal standard	Detection method	Sensitivity (µg/ml)	Author
Metformin	Urine	4-Nitrobenzoylchloride		UV, 280 nm	0.2	Ross (1976)
Phenformin	Plasma, urine			UV, 233 nm	0.25	Hill (1978)
Phenformin and metabolite	Urine			UV, 230 nm	1	Oates (1980)
Metformin	Plasma, urine		Propylbiguanide	UV, 230 nm	50–100	Charles (1981)
Buformin	Blood	4-Nitrobenzoylchloride				Fuerst (1982)
Metformin Buformin		3-Nitrobenzoylchloride 4-Nitrobenzoylchloride		Spectrophotometry Spectrofluorimetry		Bellenger (1983)
Metformin Phenformin	Plasma, urine			UV, 254 nm	50 12.5	Benzi (1986)
Metformin	Plasma, urine		Propylbiguanide	UV, 234 nm	0.01	Keal (1986)
Metformin Buformin Phenformin	Serum	Phenanthrenequinone		Fluorescence	20 40 40	Tanabe (1987)
Buformin Phenformin	Serum	1,2-Naphthoquinone 4-Sulfonate postcolumn		Fluorescence	0.02 0.02	Kobayashi (1988)
Metformin	Plasma, red blood cells			UV, 232 nm	0.02 0.03	Lacroix (1991)
Metformin	Plasma		Phenformin	UV, 236 nm	0.05	Huuponen (1992)

Acknowledgement. We would like to express our thanks to G. Botton for his expert technical assistance and contribution and to J. Quentin for her secretarial assistance.

References

Alkalay D, Volk J, Bartlett MF (1976) Conversion of biguanides into substituted triazines assayable by GC or mass fragmentography. J Pharm Sci 65:525–529

Alkalay D, Khermani L, Bartlett MF (1978) Radioimmunoassay determination of p-hydroxyphenformin and of apparent phenformin in human plasma and serum. Anal Lett Bll (9):741–751

Amigo JM, Martinez-Calatayud J (1985) Molecular structure of the phenylbiguanide hydrochloride. Bull Soc Chim Belg 94:119–121

Audry E, Dubost JP, Colleter JC, Dallet P (1986) Eur J Med Chem Chim Ther 21:71–72

Bandyapadhaya DN, Gosh NN, Ray P. (1952) Stability of chromium (III) biguanide and chromium (III) phenyl biguanide complexes. J Indian Chem Soc 29:157–168

Beckamnn R (1971) Biguanide. In: Maske H (ed) Ornl Wirksame Antidiabetika Springer, Berlin Heidelberg New York, pp 439–596 (Handbook der experimen tellen Pharmakologie, vol 29)

Bellenger P, Hamon M, Mahuzier G (1983) Synthèse et propriétés spectrales de guanamines substitues utilisables pour la chromatographie des biguanides. Ann Pharm Fr 41(4):327–337

Benzi L, Marchetti P Cecchetti P, Navalesi R (1986) Determination of metformin and phenformin in human plasma and urine by reversed-phase high performance liuid chromatography. J Chromatogr 375:184–189

Boyer F. (1984) Pharmacocinétique de la Metformine chez l'homme. Doctoral Dissertation, Montpellier University

Bres J, Bressolle F, Huguet MT (1976) Importance de la dissociatin ionique des médicaments en pharmacocinétique. Méthode de détermination de leur pKa. Trav Soc Pharm Montpellier 36(4):331–364

Brohon J, Noël M (1978) Determinatio of metformin in plasma at therapeutic levels by gas liquid chromatography using a nitrogen detector. J Chromatogr 146(I):148–151

Brown CJ (1967) The crystal structure of 1-(p-chlorophenyl-5-isopropylbiguanide (Paludine). J Chem Soc A:60–65

Brown CJ, Sengier L (1984) 1-(p-Chlorophenyl) biguanide hydrochloride. Acta Crystallogr C 40:1294–1295

Charles BG, Jacobsen NW, Ravenscroft PJ. (1981) Rapid liquid chromatographic determination of metformin in plasma and urine. Clin Chem 27(3):434–436

Das Sarma B (1952) Acid dissociation constants and basicity of biguanides and dibiguanides. J Indian Chem Soc 29:217–224

De AK, Ghosh NN, Ray P (1950) Stability of cobaltic biguanide complexes. J Indian Chem Soc 27:493–508

Dornboos DA (1967) The determination of the acid dissociation constants of L-cysteine, D-penicillamine, N-acetyl D-penicillamine and some biguanides by an accurate method for pH measurement. Pharm Weekbl 102:269–287

Dutta RL, Sengupta R (1961) Acid dissociation constants of the N' substituted biguanides and dibiguanides. J Indian Chem Soc 38:741–746

El Khateeb SZ, Assaad HN, El Bardicy MG, Ahmad AS (1988) Determination of metformin hydrochloride in tablets by nuclear magnetic resonance spectrometry. Anal Chim Acta 208:321–324

Ernst SR (1977) Biguanide hydrochloride. Acta Cryst allogr B 33:237–240

Ernst SR, Cagle Jr FW (1977) Biguanide. Acta Cryst allogr B 33:235–237

Fanshawe WJ, Bauer VJ, Ullman EF, Safir SR (1964) Synthesis of unsymmetrically substituted malnamidimes. J Org Chem 29(2):308–311

Fuerst W, Schmidt A, Stuetz B (1982) Problems of high-pressure liquid chromatography after derivatization. Wiss Z Ernst-Moritz-Arndt-Univ Greifsw Math-Naturwiss Reihe 31(2):49–50

Gage JC (1949) Synthetic antimalarials. Part XXXIV. Physicochemical studies on the diguanides. J Chem Soc:221–226

Garrett ER, Tsau J (1972) Application of ion-pair methods to drug extraction from biological fluids. I. Quantitative determination of biguanides in urine. J Pharm Sci 61(9):1404–1410

Garrett ER, Tsau J, Hinderlign PH (1972) Application of ion-pair methods to drug extraction from biological fluids. II. Quantitative determinatilon of bgiuanides in biological fluids and comparison of proteins binding estimates. J Pharm Sci 61(9):1411–1418

Goizman MS, Sarkisyan SO, Sarkysian AA, Persianova IV (1985) Differential spectrophotometric determination of biguanide derivatives. Pham Chem J 19:503–508

Handa R, Saha N (1973) Crystal and molecular structures of morpholine biguanide hydrobromide. Acta Crystallog:544–568

Hariharan M, Rajan SS, Srinivasan R (1989) Structure of metformn hydrochloride. Acta Crystallog C 45(6):911–913

Herrnstadt C, Mootz D, Wunderlich H (1979) Protonation sites of orgaic bases with several nitrogen functions: crystal structures of salts of chlrodiazepoxide dihydralazine and phenformin. J Chem Soc Perkin Trans II:735–740

Heuclin C, Pene F, Savouret JF, Assan R (1975) Characterization of phenformin and metabolites in plasma. Diabète et Métabolismè 1:235–240

Hill HM, Chamberlain J (1978) Determination of oral anitdiabetic agents in human body fluids using high-performance liquid chromatography. J Chromatogr 149:349–358

Hirt RC, Schmitt RG (1958) Ultraviolet absorption spectra of derivatives of striazine. II. Oxotriazines and their acyclin analogs. Spectrochim Acta 12:127–138

Huuponen R, Ojala-Karlsson P, Rouru J, Koulu M (1992) Determination of metformin in plasma by high performance liquid chromatography. J Chromatogr Biomed Appl 583:270–273

Jung L, Wermuth CG, Morand P (1961) Identification and determination of synthetic hypoglycemic agents in urine. Travaux Soc Pharm Montpellier 21:170–175

Keal J, Somogyi A (1986) Rapid and sensitive high-performance liquid chromatographic assay for metformin in plasma and urine using ion-pair extraction techniques. J Chromatogr 378:503–508

Kobayashi Y, Kubo H, Kinoshita T (1988) Fluorimetric determination of biguanides in serum by high-performance liquid chromatography with reagent containing mobile phase. J Chromatogr 430:65–71

Koji Nakanishi (1962) Infrared absorption spectroscopy. Holden-Day, San Francisco

Kurzer F, Pitchfork ED (1968) The chemistry of biguanides. Fortschr Chem Forsch 10:375–472

Lacroix C, Danger P, Wojciechowski F (1991) Microdosage de la metformine plasmatique et intra-erythrocytaire par chromatographie en phase liquide. Ann Biol Clin (Paris) 49:98–101

Lennard MS, Casey C, Trucker GT, Woods HF (1978) Determination of metformin in biological samples. Br J Clin Pharmacol 6(2):183–184

Matin SB, Karam JH, Forsham PF (1975) Simple electron capture gas chromatographic method for the determination of oral hypoglycemic biguanides in biological fluids. Anal Chem 47(3):545–548

Martinez-Calatayud JM, Sampedro AS (1989) Turbimetric determination of phenformin by flow-injection analysis. Analysis 17:413–416

Mottale M, Stewart CJ (1975) Gas chromatographic determinatilon of β-phenethylbiguanide and its metabolite p-hydroxy β-phenethylbiguanide in serum and urine. J Chromatogr 106:263–270

Nandi SD (1972) Spectrophotometric (UV) investigation on biguanide and substituted biguanides. Tetrahedron 28(3):845–853

Oates NS, Shah RR, Idle JR, Smith RL (1980) On the urinary dispositin of phenformin and 4-hydroxy-phenformin and their rapid simultaneous measurement. J Pharm Pharmacol 32(10):731–732

Oetting F (1975) Radioimmuological determination of n-butylbiguanide in human serum. Arznei Forsch 25(4):524–526

Oetting F (1982) Radioimmunoassay of n-butylbiguanide. Methods Enzymol 84: 577–585

Ovsepyan AM, Fialkova MA, Mikhailova NV, Kobyskov VV, Kochkina SN, Belai VE, Panov VP (1984) Spectrophotometric determination of N,N-dimethylbiguanide hydrochloride. Pharm Chem J 18:366–368

Pignard P (1962) Dosage spectrophotométrique du NN dimethylbiguanide dans le sang et l'urine. Ann Biol Clin (Paris) 20:325–333

Predescu I, Moisescu S, Cenuse M (1989) Adaptation of a color reaction to the quantitative determinatio of buformin tosylate Farmacia (Bucharest) 37(1):45–52

Ross MSF (1977) Determination of metformin in biological fluids by derivatization followed by high-performance liquid chromatography. J Chromatogr 133: 408–411

Sastry CSP, Rao T, Sailaya A (1991) Spectrophotometric determination of phenformin hydrochloride in dosage forms. Indian Drugs 28(8):378–379

Shapiro SL, Parrino VA, Freedman AL (1959) Hypoglycemic agents. I. Chemical properties of β-phenethylbiguanide. New hypoglycemc agent. J Am Chem Soc 81:2220–2225

Siest G, Roos F, Gabou JJ (1963) Dosage du NN dimethylbiguanide par le diacetyle en milieu alcalin. Bull Soc Pharm Nancy 58:29–38

Sirtori CS. Franceschini G, Galli-Kienle M, Cighetti G, Galli G, Bondioli A (1978) Disposition of metformin (N,N-dimethylbiguanide) in man. Clin Pharmacol Ther 24(6):683–93

Suzuki T, Kudo Y (1990) J Comput Aided Mol Design 4:155–158

Sayamal A (1975) Crystal data of biguanide dihydrochloride. Indian J Phys 49(9): 707–808

Tanabe S, Kobayashi T, Kawanabe K (1987) Determination of oral hypoglycemic biguanides by high performance liquid chromatography with fluorescence detection. Anal Sci 3(1):69–73

Wellman KM, Harris DL, Murphy PJ (1967) Structure of mono, di, and tri protonated biguanides. Chem Commun:568–569

Wickramasinghe JAF, Shaw SR (1972) Gas chromatographic behaviour of buformin hydrochloride, phenformin hydrochloride and phenylbiguanide. J Chromatogr 71:265–273

CHAPTER 12
Preclinical Pharmacology of Biguanides

N.F. WIERNSPERGER

A. Introduction

Probably few classes of chemicals have generated as many publications as have the biguanides. In the previous review on the preclinical pharmacology of the biguanides in 1971, BECKMANN discussed in great detail the pharmacological properties of mainly three biguanide derivatives: phenformin, buformin and metformin. Since that time, important changes have occurred in the form of an almost complete withdrawal of PHEN and BUF from the international markets. On the other hand, a large number of publications have appeared during the last 25 years, most of which have been oriented towards the elucidation of the mode of action of biguanides in diabetes. There are therefore new aspects which must now be considered, for example, prediabetic or non-diabetic insulin resistance, insulin receptor/cell-signaling effects and other properties of potential new applications such as vascular effects.

Due to the limited size of this article, a necessary – but nevertheless extensive – selection of data had to be made. Thus, exhaustive description of individual data can be found in the previous review by BECKMANN (1971). Since the clinical use of biguanides is practically limited today to Metformin (Glucophage), the present review is mainly devoted to this drug, but references to PHEN and BUF are included when appropriate. Similarly, clinical pharmacological data are included when considered important for illustration and comprehension. The author suggests that, for further information and an understanding of the experimental pharmacology of the biguanides, the reader refers to the following pertinent reviews: BECKMANN (1971), LOSERT et al. (1972), MUNTONI (1974) and BAILEY (1992).

During the course of reading this review, the reader who is not yet familiar with the biguanides may be surprised by the high number of apparent discrepancies present among the data. A closer examination shows that this is not simply a consequence of varying experimental procedures over about 40 years of pharmacological research devoted to these compounds. Biguanides are pleiotropic drugs, which renders a simplistic analysis of their mode(s) of action illusive. As will be seen, a series of factors can be cited which explain this situation: species, physiological state of animals (or organs), type of tissue, duration and time of treatment application, type of biguanide, experimental conditions (in vitro vs. in vivo) (normo- vs. hyper-glycemia), etc. However, the most important factor may be the dosage used, since this parameter appears to be of paramount importance for the action of these drugs. Despite the data grouping and analysis, the reader must be aware that data must still be considered under the individual experimental conditions used.

B. Absorption and Distribution

I. Absorption

BECKMANN (1971) showed BUF to be rapidly absorbed, since measurements with labeled drug showed a plasma peak of $0.4\,\mu g$/ml within 1–2 h following an oral administration of 50 mg/kg to rats. The absorption of PHEN was obviously less than that of BUF, appearing to occur passively, with a more rapid effect in the ileum than in the jejunum (KOJIMA et al. 1976). There was no major difference between normal and diabetic animals. A non-negligible enterohepatic cycling has been described (HALL et al. 1968; BAILEY 1993).

Plasma levels of MET were $6\,\mu g$/ml 2 h after an oral administration of 150 mg/kg to rats. In portal blood, MET concentrations peaked at $50\,\mu M$ 30 min after oral intake of 50 mg/kg in rats (WILCOCK and BAILEY 1994). The percentage absorbed decreased with increasing drug concentrations (NOEL 1979). In recent in vitro investigations, we found that MET was absorbed through a paracellular route in isolated perfused rat ileojejunal segments (CUBER et al. 1994). In man, it was shown that there was no preferential site for MET absorption and that obviously the entire gastrointestinal tract was necessary (VIDON et al. 1988).

II. Tissue Distribution

Many studies have shown that the highest biguanide concentrations are found in the small intestine, where values reached $10^{-2}\,M$, i.e., 1000 times above plasma levels (BECKMANN 1971; BAILEY 1992). This intestinal accumulation occurred even after intravenous or subcutaneous administration (YOH 1967; BECKMANN 1969, 1971; WILCOCK and BAILEY 1994). Buformin was reported to accumulate more specifically in intestinal fat (BECKMANN 1965). Our recent investigations in vitro failed to show any presence of MET in the enterocytes, but in stained histological sections the radiolabeled drug was found in villeous lacteals and in submucosal capillaries (CUBER et al. 1994). The reason for the intestinal tropism of biguanides is unknown. Kinetic studies suggest that MET does not stay in the gut wall during its direct absorption, but rather accumulates upon its recirculation through the intestine (WILCOCK and BAILEY 1994). A similar process has been suggested for PHEN (NICHOLLS and LEESE 1984).

Hepatic levels of biguanides are usually higher than the plasma concentration but only for a short duration. Thus in rat liver 16% of ingested PHEN was found after 1 h, whereas at 12 h only 1% remained (BECKMANN 1971; WICK et al. 1960). Additional investigations showed that most of this compound was represented by its metabolite, since PHEN is hydroxylated by the liver. Likewise, BUF is metabolized up to 50%; its tissue levels are higher in the liver, kidney, pancreas and gut than in plasma, but only for several hours (BECKMANN 1971). The tissue distribution of MET is essentially the same as BUF, but MET is not metabolized. After oral ingestion, MET accumulates for short periods in liver, kidney, salivary gland and gastrointestinal tract (BECKMANN 1969). In the isolated perfused liver, an uptake rate of about 15% was found (LENNARD et al. 1978). Whole body autoradiographic pictures showed that chronic oral MET treatment in mice did not result in any significant accumulation (COHEN and HIRSCH 1968; GIANNATTASIO et al. 1968).

In skeletal muscle, the PHEN present was essentially constituted by its metabolite (HALL et al. 1968), the efficacy of which has been questioned. Buformin levels were only 40% of hepatic concentrations (YOH 1967). After several hours, BUF or MET levels were usually lower in skeletal muscle than in plasma. No drug was found in the brain. There is no accumulation of biguanides in any tissue 24 h after oral administration. It should also be mentioned that the tissue distribution in diabetic rats is not significantly different from that in normal animals.

The elimination of biguanides occurs mainly by the urine (80% or more) and the feces. Accordingly, the half-life of biguanides is quite short (1.5–2.5 h) (BECKMANN 1971). In blood, biguanides are distributed almost equally between plasma and erythrocytes (BECKMANN 1965; RAPIN, unpublished data). Fasting levels in chronically treated animals are in the range of 0.5–5 µg/ml, depending on the dosage used, which is not noticeably different from human data.

An important question is the binding of biguanides to plasma proteins. Differing results have been reported: whereas it was claimed that BUF or MET did not bind to plasma proteins (BECKMANN 1971; SIRTORI et al. 1978), other studies showed that MET might bind to proteins. In vitro the addition of MET to rat plasma showed 8% binding, whereas this value increased to 21% after in vivo administration. Using human blood, we found that 27% was bound to globulins, 18% to β-globulins and 24% to albumin (BROHON, unpublished results). Using gradient centrifugation, we found that in plasma of orally MET-treated rats (100 mg/kg per hour, 12 h fasted), 28% of the drug was free whereas the remainder was almost entirely bound to very low density lipoprotein (VLDL) and LDL-rich plasma fractions (WIERNSPERGER and RAPIN, unpublished results). In a study of the binding of MET in rabbit lymph, we found that the main part of the drug was free or weakly bound. Discrepancies between these reports may be due to differences in the techniques used. When testing the bound MET in a dialysis system, we found that most could be dialysed, suggesting that the binding of MET as a consequence of its cationic properties is mainly noncovalent.

C. Efficacy

Most studies on biguanide efficacy were reviewed by Beckmann (1971); more recent publications mostly deal with investigations on their mechanisms of action. It is thus sufficient to recall here that biguanides are ineffective in normal animals, unless very high doses are used. Hypoglycemia never occurs in normal organisms, because at high drug concentrations glycogenolysis is supposed to compensate for glycemia-lowering actions in peripheral tissues. Only when animals are fasted and the liver cannot increase its glucose production may hypoglycemia occur.

In diabetic animals, glycemia-lowering action is observed more easily than in normal animals, though the biguanide efficacy strongly depends on the model used. Thus, in severe insulinopenic diabetes induced in rodents by alloxan or streptozotocin, a high concentration must be used. In mild diabetes, in cortisone-induced hyperglycemia, in genetic strains such as the KK mouse or in models of insulin resistance such as fructose feeding, lower doses (sometimes very close to the human dosage) are efficacious against hyperglycemia or hyperinsulinemia. The reader is referred to the remarkable articles published by Losert et al. (1972), who showed the various effects induced by biguanides in different tissues according to their dosage.

Interestingly, at low dosages biguanides were shown to be active on the various mechanisms involved in the etiology of insulin resistance and hyperglycemia. Differences in cell type sensitivity to biguanides could partly explain such observations: for example, at low concentrations BUF was very active on adipocytes but not on skeletal muscle glucose transport, whereas higher doses induced the opposite reaction (Daweke and Bach 1963). Thus, low biguanide concentrations requiring insulin could be active on defects underlying insulin resistance, while higher doses, which are at least partly insulin-independent, may rather have direct glycemia-lowering effects. In recent investigations, several marked effects of low MET concentrations were observed on insulin-induced *Xenopus* oocyte maturation (Stith, personal communication), protein synthesis (Zaibi et al. 1994) or peripheral insulin sensitivity (Halimi et al. 1994).

Finally and most importantly, it should be stressed here again that the antihyperglycemic action of biguanides cannot be explained by mild toxic effects at the level of mitochondrial respiration. Many arguments, already stated in Beckmann's review (1971) and in more recent articles (Bailey 1992, 1993), have proved that there is no link between the capacity of drugs to interfere with oxidative phosphorylation and their antihyperglycemic potency. The moderate increase seen in blood lactate level is very likely due to a stimulation of intestinal glucose transformation into lactate, and it should be recalled that, except in situations of concomitant pathologies such as renal failure, the plasma lactate levels remain within the physiological range. No signs of toxicity are seen histologically at the biguanide concentrations normally used, even in tissues subject to very high drug levels such as the gut wall. In certain pathological conditions of elevated anaerobic metabolism, MET has remarkable protective properties (see Sect. D.V.2.b).

Recent studies on cellular respiration have shown that high concentrations of MET interfere with the redox state of whole cells but that ATP production and oxygen consumption are only mildly reduced (ARGAUD et al. 1993; FISCHER et al. 1995). However, with the exception of the small intestine and the liver for a short duration after oral drug intake, MET concentrations which interfere with mitochondrial respiration (i.e., at least $5 \times 10^{-3} M$) are not encountered.

Not only the physiological state but also the animal species is an important parameter in biguanide efficacy: thus, guinea pigs appear to be much more sensitive to biguanides than other rodents. This is especially true for PHEN and BUF, as these compounds are not metabolized in this species. Generally speaking, the order of potency among these three biguanides is PHEN > BUF > MET when studied at similar concentrations.

Biguanides are efficacious when administered by the oral, intraperitoneal or subcutaneous route. Whether they are active intravenously is an open question: in man no effect was seen when plasma concentrations were matched with those attained after oral ingestion (SUM et al. 1992).

D. General Pharmacology

I. Blood Pressure

Here again most information can be found in the review by BECKMANN (1971). In normotensive animals, biguanides do not significantly influence blood pressure, except for a slight fall when administered intravenously. The chronic oral treatment also has no consequences on systemic blood pressure, unless individuals are hypertensive. In this case, but not always, MET is able to lower blood pressure (see also Sect. D.V.3).

II. Varia

Antispastic effects have been shown with high drug concentrations in isolated intestinal or uterine segments contracted with $BaCl_2$ or acetylcholine (BECKMANN 1971). However, others were not able to reproduce this finding (STERNE 1964). No variation in coronary flow was found with MET in the Langendorff heart preparation (STERNE 1969). Emetic effects have been seen in cats and dogs. Respiration is unchanged except for a depression after i.v. administration. There is also no influence on body temperature. Finally, biguanides are neither analgesic nor anesthetic.

III. Hormones

Adrenal activation has been demonstrated after i.v. administration of BUF (LOSERT et al. 1972), but this effect does not occur if other routes of drug administration are used (STERNE 1969). In chronically MET-treated rats, we found no changes in blood levels of epinephrine, vasopressin, aldosterone or corticosterone (FREMINET and WIERNSPERGER, unpublished results). Glucagon levels are unchanged by biguanides; albeit an increase of entero-glucagon has been reported after oral glucose load. In humans, MET did not affect the levels of counterregulatory hormones, cortisol, catecholamines and growth hormone (LANDIN et al. 1994). The histological appearance of the pituitary gland was not changed by MET (STERNE 1969). A reduction in thyroid function was reported in MET-treated rats (HORN and PALKOVITS 1964).

IV. Antitumoral Properties

PHEN, BUF and MET have been shown to limit the growth of HeLa or KB cells (RIKIMARU et al. 1968; NEUMANN and TYTELL 1962). In Ehrlichascites tumors, biguanides decreased the number of cells as well as the liquid volume (CORBELLINI and TORTI 1967). This effect was explained by a very high sensitivity of tumor cells to the respiration inhibitory effects of biguanides compared with normal cells (LUGARO and GIANNATTASIO 1968).

V. Vasculoprotective Effects

1. Macrocirculation

a) Ischemia/Atherosclerosis

Remarkable improvements have been achieved in the leg blood flow of hyperlipidemic, non-diabetic patients suffering arteritis (SIRTORI et al. 1984). This effect was partly due to improved postocclusive blood flow, an effect which was also observed in experimental ischemia/reperfusion of the hamster cheek pouch (BOUSKELA and WIERNSPERGER 1993). In numerous studies it was demonstrated that chronic treatment of hypercholesterolemic rabbits with MET strongly inhibited the development of atherosclerosis in aorta and other large vessels (AGID and MARQUIE 1973; MARQUIE and LAFONTAN 1974). This effect could be explained by the drug effects on disturbed lipid metabolism as well as by its inhibitory effects on vascular smooth muscle cell proliferation (KOSCHINSKY et al. 1988; SIRTORI et al. 1991). In cardiac ischemia in rats, MET strongly reduced the infarct size (CHARLON et al. 1988). A reduction in the myocardial reinfarction was noted in MET-treated patients followed up for several years (SGAMBATO et al. 1980)

b) Platelet Aggregation

In vitro, MET has relatively weak effects on ADP-induced platelet aggregation (WILSON et al. 1986; DE CATERINA et al. 1989). In vivo MET diminished platelet aggregation induced by collagen in hypercholesterolemic rabbits; this effect was not mediated through actions on the cyclooxygenase pathway (TREMOLI et al. 1982). In diabetic patients, MET significantly reduced the disease-related hyperaggregability of platelets (PARISI et al. 1979). Studies performed on models of in vivo-induced thrombus formation showed remarkable antithrombotic effects with MET which were equivalent or superior to those observed with standard drugs such as aspirin (Fig. 1) (MASSAD et al. 1988; WEICHERT and BREDDIN 1988).

c) Fibrinolysis

Early in the history of biguanide development it was seen that these drugs improved fibrinolysis (BECKMANN 1971). In monkeys and dogs, PHEN enhanced the fibrinolytic activity reduced by insulin administration (BACK et al. 1968). PHEN increased the plasminogen activator levels in monkeys (REGOECZI and WALTON 1967). In patients suffering vascular occlusive diseases, PHEN and MET reduced the blood fibrinolytic time and the

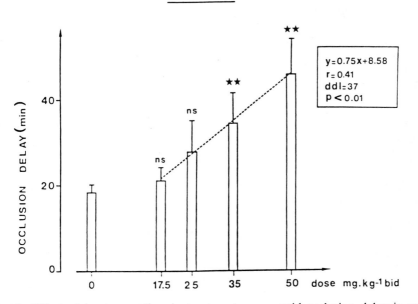

Fig. 1. Effect of in vivo metformin treatments on carotid occlusion delay in rats induced in situ by electrical parietal stimulation. (From MASSAD et al. 1988)

euglobulin lysis time; however, MET, in contrast to PHEN, had no effect on plasma fibrinogen (Hocking et al. 1967).

Over recent years, this problem has been reassessed, in particular when it was found that MET caused a concomitant reduction in insulin levels and in plasminogen activator capacity in obese, non-diabetic subjects (Vague et al. 1987) and in non-insulin dependent diabetes mellitus (NIDDM) patients (Nagi and Yudkin 1993). It was discovered that these effects of MET were explained by an inhibition of the production of plasminogen activator inhibitor-1 (PAI-1) by hepatocytes and endothelial cells (Anfosso et al. 1993). Accordingly, basal and postvenous occlusion levels of PAI-1 were reduced in MET-treated NIDDM patients (Grant et al. 1991). Recently, in non-obese, non-diabetic humans, MET was shown to lower t-PA antigen both during the postabsorptive state and during a hyperinsulinemic clamp (Landin et al. 1994). In an in vitro examination of thrombin-clotted plasma, MET accelerated the lysis time of fibrin networks (Nair et al. 1991).

2. Microcirculation

There are practically no drugs available which have direct effects on microvessels, so the discovery of such properties with the biguanide MET has opened exciting challenges for this drug.

a) Arteriolar Vasomotion

Besides its antihyperglycemic action, MET has remarkable protective effects in vascular pathology. A unique property of this drug is its selective action on small vessels, mainly arterioles. It stimulates a key regulatory mechanism of microflow distribution, namely the precapillary vasomotion. This physiological phenomenon consists of periodic phases of contraction and dilatation of precapillary arterioles, which allows the alternating perfusion of so-called microvascular units (Intaglietta 1988). Vasomotion disappears in diabetes and can be restored by chronic treatment of hamsters or bats with MET (Bertuglia et al. 1988; Bouskela 1988). The same effect was observed in hamsters submitted to hemorrhagic shock, in which vasomotion was restored and accompanied by a 50% reduction in leukocytes sticking to small vessel walls as well as by an almost complete reperfusion of the capillary bed (Bertuglia et al. 1989). The reestablishment of such processes regulating the supply of nutrients through the capillaries may play a cardinal role in the response of small vessels to an increased supply of glucose and insulin to peripheral tissues in diabetes (Zemel et al. 1990; Baron and Brechtel 1993; Feldman and Bierbrier 1993; Wiernsperger 1994).

b) Permeability

In diabetes, MET strongly reduced the exaggerated permeability of small vessels (Colantuoni et al. 1988; Valensi et al. 1994). A similar result was obtained clinically in women suffering cyclic edema (VALENSI, personal communication). Reduction in vascular permeability and edema has also been seen in ischemic situations such as hemorrhagic shock (with a remarkable survival rate of awake animals) (BERTUGLIA et al. 1989; BOUSKELA and WIERNSPERGER 1993) as well as in peripheral and brain ischemia in rats (RAPIN et al. 1988a).

c) Hemorheology

Metformin exerts beneficial effects on hemorheological parameters. In particular, at low concentrations it improved the erythrocyte deformability impaired by diabetes (KIESEWETTER et al. 1987; BARNES et al. 1988; RAPIN et al. 1988b). Plasma viscosity was unchanged.

d) Diabetic Microangiopathy

Chronic MET treatment of sand rats rendered diabetic by laboratory chow feeding inhibited the development of microangiopathies such as basal membrane thickening in skin and kidney (BOUGHERRA 1985). A protection against the incidence of glomerulosclerosis was also described in chronically treated KK mice (REDDI and JYOTHIRMAYI 1993).

e) Angiogenesis

Angiogenesis, the formation of new vessels, is the pathological process underlying proliferative retinopathy. In vitro, low-dose MET inhibited the proliferation of human vascular endothelial cells cultured in a hypoxic environment (FRANKE et al. 1988). In vivo, MET strongly inhibited the neovascularization induced in rabbit cornea by the implantation of growth factor-loaded material, as well as in neonate kittens submitted to a hyperoxia/hypoxia challenge (KISSUN 1988). The most impressive effect was observed as an almost total inhibition of preretinal neovascularization in mini-pig eyes submitted to branch vein retinal occlusion by laser (POURNARAS et al. 1988).

3. Hypertension

Diabetes is frequently associated with hypertension. In NIDDM patients MET, in contrast to glibenclamide, reduced systolic and diastolic blood pressure. This antihypertensive effect was accompanied by weight reduction and an improved lipoprotein profile (CHAN et al. 1993). In non-diabetic women suffering polycystic ovary syndrome, MET reduced insulin levels and systemic blood pressure (VELASQUEZ et al. 1993). More interestingly, MET lowered blood pressure in hypertensive, non-diabetic and non-obese patients (LANDIN-WILHELMSEN 1992). Investigations in rats of the SHR strain showed that MET decreased insulin levels and blood pressure, and that hypertension could be reestablished by artificially raising insulin plasma levels (MORGAN et al. 1992; VERMA et al. 1994). In in vitro studies on rat tail isolated arterial rings, MET reduced the vasoconstriction induced by nore-pinephrine (ZAMMAN et al. 1994).

E. Organ Pharmacology

I. Gastrointestinal Tract

1. In Vitro

a) Intestinal Glucose Absorption

This has been one of the most extensively studied mechanisms of biguanide pharmacology (see review by CASPARY 1977). Experiments on everted sacs, using high biguanide concentrations, originally revealed inhibiting effects on glucose absorption. This observation, together with the known local accumulation of biguanides in the gut (see Sect. B.II), promptly led to the hypothesis that interference with intestinal glucose absorption was one key mechanism whereby these drugs reduced glycemia. This observation was not uniform, however (LOVE 1969; COUPAR and McCOLL 1974; CASPARY 1977). In such experiments, MET was clearly less active than PHEN or BUF, the minimal efficacious dose of MET being around $5 \cdot 10^{-3} M$ (LORCH 1971; CASPARY and CREUTZFELDT 1971a). These authors claimed that biguanides inhibited the intestinal absorption of actively transported sugars, but not of fructose, for example. So they hypothesized that due to local high concentrations of biguanides, the Na^+-dependent, energy-requiring component of mucosal intestinal sugar transport would be impaired (CASPARY and CREUTZFELDT 1971).

However, in agreement with previous findings (CASPARY 1977; KESSLER et al. 1975), recent investigations using isolated rat enterocytes failed to show any effect of even high MET doses on the Na^+/K^+ ATPase (Table 1) (CUBER et al. 1994).

Table 1. Effect of various concentrations of metformin on lactate, pyruvate and CO_2 production from U-^{14}C-glucose by freshly isolated rat enterocytes. (From CUBER et al. 1994)

Metformin concentration (M)	Lactate (μM/g dry wt.)	Pyruvate (μM/g dry wt.)	CO_2 (% of control)	Na^+/K^+ ATPase (% of control)
0	288.5 ± 19.2	13.5 ± 0.7	100	112 ± 13
10^{-6}	251.8 ± 42.8	12.5 ± 1.3	99.6 ± 11.8	ND
10^{-5}	266.2 ± 49.6	13.8 ± 0.3	101.2 ± 8.4	98 ± 3
10^{-4}	263.5 ± 1.0	12.8 ± 0.6	110.6 ± 13.8	101 ± 4
10^{-3}	274.1 ± 4.0	16.0 ± 0.9	98.7 ± 0.4	91 ± 11

It must be recalled that many data originated from experiments using everted intestinal sacs, a model in which the oxygen supply across the whole thickness of the gut wall is a critical issue. Furthermore, extremely high drug concentrations were usually used, based on findings showing biguanide accumulation in this tissue.

The effect of MET on the mucosal side was found to be very weak (GHAREEB et al. 1969; KESSLER et al. 1975; WILCOCK and Bailey 1991a), whereas on the serosal side it reduced glucose transfer by only 12% (WILCOCK and BAILEY 1991).

Reconsidering the problem in vitro, measurements of glucose transfer across the isolated perfused rat ileojejunal segment failed to demonstrate any effect of MET ($10^{-2} M$ in the intestinal lumen) on the appearance of intact sugar in the effluent blood (Fig. 2) (CUBER et al. 1994). Examination of tissue distribution of radiolabeled MET showed that, in contrast to expectations, the drug was totally absent from the enterocytes, but was found in villous lacteals and in submucosal blood vessels. These data indicate that, upon intraintestinal addition of MET, the drug does not penetrate the gut wall through the enterocytes but most likely by the paracellular route (CUBER et al. 1994). In contrast, when therapeutic doses ($1 \mu g/ml$) of MET were added to the blood perfusion medium, a slight inhibition of glucose absorption was found together with an increase in effluent lactate levels. The latter was dose independent.

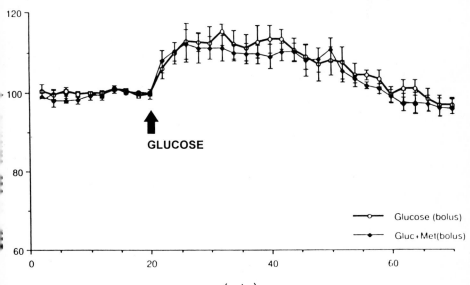

Fig. 2. Effect of intraluminal addition of metformin on glucose absorption in isolated, perfused rat intestine. (From CUBER et al. 1994)

b) Intestinal Glucose Metabolism

These increases in lactate production by biguanides were attributed to "toxic" effects of the local high drug levels. Again, however, this finding was not uniform (Beckmann 1971) and our own experiments in freshly isolated enterocytes failed to show any increase in lactate production by MET concentrations up to $10^{-3} M$ (Table 1) (Cuber et al. 1994). Two hours after intrajejunal administration of MET to rats, rings of the small intestine showed an average 22% reduction in glucose oxidation (Bailey et al. 1992). It must be recalled here that anaerobic handling of glucose is a normal physiological function of the intestine and that any stimulation of this process will automatically result in an elevated lactate production.

2. In Vivo

a) Intestinal Glucose Absorption

In addition to the in vitro findings, one major argument for explaining the antihyperglycemic effect of biguanides by their action on intestinal glucose absorption was that in several studies they were active against oral – but not intravenous – glucose tolerance tests (Czyzyk et al. 1968). However, this was not a uniform finding (Beckmann 1971). The appearance of lower glucose levels in portal blood was considered to reflect lower absorption by the gut (Czyzyk et al. 1968; Berger and Kunzli 1970). In vivo, controversial findings were obtained with the models and drug used (reviewed in Beckmann 1971). Inhibition was found in the small intestine of guinea pigs (Ghareeb et al. 1969) or dogs (Czyzyk et al. 1968), although in the latter the correlation between this effect and the glycemic reduction was poor. On the other hand, reports also existed which failed to show any significant effect of BUF (Creutzfeldt et al. 1962; Förster et al. 1965) or MET (Coupar and Mccoll 1974; Lorch 1971; Mainguet et al. 1972). Very small effects were found in rats which received 200 mg/kg MET (Kakemi et al. 1983; Cuber et al. 1994), whereas 20 mg/kg (approximately the human dose) was without effect on glucose absorption. It must be emphasized that also for this parameter MET appeared much weaker than BUF or PHEN (Caspary and Creutzfeldt 1971; Creutzfeldt 1972).

The possibility that kinetic changes could explain the apparent reduction in postload glycemia was unfortunately underestimated. Thus, rather than comparing one-sample levels, it is preferable to consider the area under the curve for the total glucose absorbed over a number of hours. When this parameter is integrated, it appears that biguanides may simply delay the glucose absorption in rats (Förster et al. 1965; Creutzfeldt et al. 1992) or in man (Fossati et al. 1985). Such a delay was attributed to restrictions in gastric emptying (Förster et al. 1965; Ghareeb et al. 1969; Yoh 1967). However, neither transit times in mice (Bailey et al. 1991a) nor gastric emptying in man (Leatherdale and Bailey 1986; Eisner and Berger 1971) were found to be slowed down with chronic MET treatment. In any case, the delayed appearance of glucose in the circulation did not modify the total amount absorbed over a number of hours (Fossati et al. 1985; Jackson et al. 1987). Thus many reports on the reduced amounts of absorbed glucose simply failed to take into account the fact that initial glycemic values were lower: when correcting for this parameter it was usually found that the incremental increase in plasma glucose after an oral load was not or only marginally decreased by MET. Finally, it should be recalled that biguanides have no glycemia-lowering activity in animals or patients totally deprived of insulin, which confers little significance to any possible interference with intestinal glucose absorption for their antidiabetic efficacy.

b) Intestinal Glucose Metabolism

Lower glucose and higher lactate levels in portal blood are common findings with biguanides (Berger and Kunzli 1970; Bailey et al. 1992; Cuber et al. 1994), suggesting that what was believed to be a reduction in glucose absorption was in fact the consequence of enhanced glucose consumption by the gut wall. Kinetic studies showed that PHEN and MET accumulated in the wall after peak concentrations were seen in the portal blood (Nicholls and Leese 1984) (Wilcock and Bailey 1994). The in vitro observation of intravascular MET enhancing lactate levels in the effluent blood of the isolated rat intestine also supports the concept that biguanides, once absorbed and having entered the general circulation, penetrate into and accumulate within the small intestine. In one layer – which may be the mucosa (Bailey et al. 1994) – they boost the anaerobic handling of glucose by the gut, which is known to be deficient in diabetes (Lovejoy and Digirolamo 1990). Metformin was found to promote lactate formation without reducing fructose absorption, supporting the concept that lactate production was not the consequence of direct sugar metabolization during its absorption process (Caspary and Creutzfeldt 1971). A positive consequence of this mechanism could consist of a reduction in the glucose load of the liver. It has been proposed that also here MET might act by a potentiation of insulin (Bailey and Mynett 1994), since the transformation of glucose into lactate in the intestinal tissue is partly controlled by this hormone (Kellett et al. 1984).

c) Absorption of Other Nutrients

Biguanides were claimed to inhibit the absorption of many substances such as lipids, amino acids, vitamins or metals (Caspary 1971). Thus, in rabbits MET was shown to decrease the absorption of cholesterol and glycerides (Agid and Marquie 1972). In rats, MET decreased the absorption of fatty acid and triglyceride glycerol moieties (Curtis-Prior 1982). One main aspect of this topic is the malabsorption of vitamin B_{12}, in particular in the case of MET (Beckmann 1971) (Shaw et al. 1993). The extent of this effect was a matter of controversy (Berger et al. 1972) and, although severe vitamin B_{12} deficiency may not occur, a regular estimation is indicated (Tomkin 1973).

d) Gastrointestinal Hormones

In a study in humans, no effect was found on gastrin, gastric inhibitory polypeptide (GIP) or secretin levels in patients treated with MET. VIP levels were slightly increased (Molloy et al. 1980). However, GLI (glucagon-like immunoreactive) levels were strongly increased, pointing to an effect on glucagon-like material of intestinal rather than pancreatic origin. Similar effects on "enteroglucagon" had been described in rats (Kakemi et al. 1983). We were unable to confirm the latter finding in vitro (unpublished results). In the absence of clear-cut knowledge about the physiological meaning of intestinal glucagon-like peptides, it is difficult to consider the functional relevance of these findings; they could, however, see renewed interest according to recent developments on glucagon-like peptide 1 (GLP-1 or GLIP) as a potential compound in diabetes therapy.

e) Conclusion

There is no solid evidence to suggest that reasonable biguanide concentrations significantly interfere with intestinal glucose absorption. In any case, quali-tative and quantitative aspects rule out this process as being an important component of the antihyperglycemic activity of these drugs. In contrast, upon recirculation biguanides accumulate in the intestinal wall, where they stimulate a normal physiological function of the gut, i.e., the anaerobic transformation of glucose into lactate. In terms of glucose utilization, the intestine therefore represents a major target tissue for biguanides (see also review by Bailey, submitted).

II. Pancreatic Hormone Secretion

1. In Vitro

Perfusion of the denervated pancreas with biguanide did not increase insulin secretion in dogs or rats (Mehnert et al. 1962; Schillinger et al. 1970). These compounds also did not influence pancreatic glucagon secretion. They were also inactive on the production of insulin by cultured B cells (Moore and Cooper 1991), either in the basal or in the glucose-stimulated situation (Schatz et al. 1972). On the other hand, the perfusion of isolated pancreas with MET under conditions of hyperglycemia or in the presence of arginine reportedly enhanced insulin secretion in rats (Bobbioni et al. 1978; Gregorio et al. 1991; Son 1992). In such experiments, PHEN kinetics, as were also the underlying mechanism, were different from those of MET (Loubatieres et al. 1971; Gregorio et al. 1991).

2. In Vivo

Although in vivo some increase in portal insulinemia was recorded in rats (Bobbioni et al. 1978) or pig (Kühl et al. 1979), long clinical experience with biguanides showed no significant increase in either plasma insulin or C peptide levels (Pedersen et al. 1989; Perriello et al. 1994; Johnson et al. 1993). Biguanides are still active on hyperglycemia in pancreatectomized animals (Beckmann 1971). There is thus a discrepancy between the in vitro and in vivo data, whereby the in vitro results appear irrelevant since, if portal insulin levels were increased in vivo, then at least C peptide levels would increase in peripheral blood. The absence of an elevation in circulating insulin is a major distinctive feature between the biguanides and the sulfonylureas.

III. Liver

1. In Vitro

a) Gluconeogenesis

Biguanides are claimed to reduce postabsorptive hyperglycemia mainly by interfering with gluconeogenesis. Many studies have demonstrated their capacity to reduce the synthesis of glucose from various precursors: in isolated or perifused hepatocytes (Lloyd et al. 1975; Hotta et al. 1991; Owen and Halestrap 1992; Argaud et al. 1993), in liver slices (Losert et al. 1972) or in the isolated perfused liver (Haeckel and Haeckel 1972; Ho and Kelly 1980; Zhang et al. 1994), MET, BUF and PHEN were all shown to interfere with this process. However, no direct effect was found in concentrations below $10^{-3} M$ MET or even PHEN (Cook 1978; Alengrin et al. 1987; Wollen and Bailey 1988; Argaud et al. 1993). In this case, interference of high drug concentrations with cellular respiration has been proposed to explain their antigluconeogenic potential (Altschuld and Kruger 1966): mild mitochondrial respiratory chain inhibition, leading to decreases in the ATP/ADP ratio, would accordingly stimulate channeling through pyruvate kinase (Argaud et al. 1993) and/or inhibit pyruvate carboxylase (Owen and Halestrap 1992) to divert the gluconeogenic precursors from the glucose synthetic pathways (Beckmann 1971). The antigluconeogenic capacity of MET was, however, much lower than for other biguanides (Medina et al. 1971; Schäfer 1979; Hotta et al. 1991). These data raise the question of the relevance of such findings for in vivo situations, since the liver does not accumulate biguanides and the hepatic drug concentration is low in the postabsorptive state. Hepatocytes are subject to biguanide levels in the range of $10^{-4}-10^{-3} M$ only for short periods after their oral intake. On the other hand, the antigluconeogenic action of biguanides appears to be better expressed in diabetic livers in the presence of insulin (WOLLEN and BAILEY 1988), and studies performed in isolated perfused livers suggested that their efficacy would be much greater in fasted than fed states (HAECKEL and HAECKEL 1972). In addition it was found that human hepatocytes were more sensitive than rodent liver cells to the action of biguanides (BECKMANN 1971). An indirect effect of biguanides on gluconeogenesis could also be induced by an inhibition of the uptake of gluconeogenic precursors by the liver, as shown for lactate (COHEN and ILES 1977; ZHANG et al. 1994) or alanine (HOTTA et al. 1991).

b) Hepatic Glycogen Metabolism

Alternatiely or in addition to reducing gluconeogenesis, biguanides could also restrain hepatic glucose production by inhibiting glycogenolysis. Data supporting this mechanism are limited but suggest that, in contrast to the normal liver, these drugs may block the glycogenolysis in diabetic livers (PAVEL et al. 1964). A reduction in G6Pase has been described for PHEN (WILLIAMS et al. 1957), BUF (HILDMANN and LIPPMANN 1963) and MET (PEARS et al. 1990). An increase in hepatocyte glycogen synthesis was found with MET in hepatoma cells (PURRELLO et al. 1988). Another study reported a glycogenic effect of MET in the presence of insulin but not under basal conditions (MELIN et al. 1990). Data concerning hepatic glycogen levels are extremely variable, likely due to differences in drug concentrations and in the initial physiological state of the livers. Thus it was also reported that biguanides may easily behave in a glycogenolytic manner (PROSKE et al. 1962; BECKMANN 1971; ALENGRIN et al. 1987). These discrepant data must therefore be interpreted with caution, since experimental conditions were very different in terms of medium glucose content, initial glycogen stores and duration of drug exposure.

c) Hepatic Glucose Oxidation

Here again data look variable: whereas increased lactate production by biguanide-treated liver cells was described (JALLING and OLSEN 1984; ALENGRIN et al. 1987), other reports failed to show reductions in glucose oxidation (MEYER 1960; PAVEL et al. 1964; BAILEY et al. 1992). No drug effect was found on hepatic glucokinase by BUF (HILDMANN and LIPPMANN 1963) or MET (KANEKO 1965).

d) Antiglucagon Effects

Since glucagon is a potent counterregulatory hormone, biguanides may reduce hepatic glucose output by blocking its hyperglycemic action. Remarkable antiglucagon effects have indeed been shown in studies on MET: reduced glucagon-stimulated gluconeogenesis and amino acid transport in vitro (ALENGRIN et al. 1987; KÜHNLE et al. 1990; ARGAUD et al. 1993;

KOMORI et al. 1993; UBL et al. 1994; YU et al. 1994). When studying the stimulation of adenylate cyclase, it was found that MET reduced the maximal glucagon effect by 25% and restored insulin's ability to suppress this enzyme in diabetic animals (GAWLER et al. 1988).

2. In Vivo

a) Hepatic Glucose Production

In most – but not all (LOSERT et al. 1972; PENICAUD et al. 1989) – studies, biguanides were reported to decrease hepatic glucose production, which largely explains their effect on fasting hyperglycemia (ROSSETTI et al. 1990; JOHNSON et al. 1993; PERRIELLO et al. 1994; CUSI and CONSOLI 1994; BARZILAI and SIMONSON 1988).

b) Hepatic Blood Flow

Under basal conditions, no effect was found in rats (BAILEY and MYNETT 1994; WIERNSPERGER and RAPIN,unpublished data; KÜHL et al. 1979) or humans (SIGNORE and POZZILI, submitted). In conditions of glucose loading, some findings, however, pointed to an effect of MET on elevations in portal blood flow (OHNHAUS et al. 1978; WIERNSPERGER and RAPIN, unpublished data). Increased hepatic blood flow has already been described for PHEN (TRANQUADA et al. 1960), but recent investigations with MET in diabetic patients failed to demonstrate such an effect (SIGNORE and POZZILI, submitted). Such drug action could conceivably occur as a consequence of increased portal lactate levels.

In addition to flow, biguanides might modify the composition of portal blood supplying the liver, which would then be expected to be largely influenced by changes in blood flow as well as hormone or gluconeogenic precursor levels. For example, the ratio of insulin to glucose can be increased, which would potentially benefit the suppressive effect of insulin on hepatic glucose production (BERGER and KUNZLI 1970).

c) Gluconeogenesis

In vivo data on gluconeogenesis are partly contradictory: no effect of BUF (BECKMANN 1971) and MET (FRAYN, 1976; BAILEY 1992; CUSI and CONSOLI 1994) or reductions in glucose production from alanine (HERTZ et al. 1989) or under cortisone stimulation (MEYER et al. 1967) have been found.

Drugs can affect gluconeogenesis directly or by interfering with the levels or cellular uptake of the gluconeogenic precursors. Thus the lack of glycemia-lowering effect of MET in portal-structured rats and the concomitant increase in circulating precursors (SCHLIENGER et al. 1979; MARCHETTI et al. 1989) support such an effect. However, due to the multiple possibilities of metabolic channeling of various precursors, these are not necessarily elevated in plasma (TESSARI et al. 1990). Whereas early experiments failed to find a reduction in substrate uptake by biguanides (LESCURE and VOLFIN 1974), more recent studies showed inhibitory effects on alanine and lactate uptake by the liver (KOMORI et al. 1993; ZHANG et al. 1994). Therapeutic biguanide concentrations may reduce gluconeogenesis only in situations where this process is activated by either hormones or excessive substrate delivery, and limited to the diabetic state (MUNTONI 1974). According to another hypothesis, biguanides reduce gluconeogenesis by inhibiting the oxidation of fatty acids: the subsequent generation of acetyl coenzyme A of lipid origin would then be reduced as well as the activity of pyruvate carboxylase (MUNTONI 1974).

d) Hepatic Glycogen Metabolism

As in vitro, data concerning hepatic glycogen are very variable due to experimental conditions of treatment duration, fasting or fed state, etc. Early in the development of biguanides, stimulation of hepatic glycogenolysis was reported and partly considered to be a counterregulatory mechanism compensating for hypoglycemic mechanisms of extrasplanchnic origin, thereby explaining the absence of absolute hypoglycemia with biguanides, in contrast to sulfonylureas (STERNE 1969; BECKMANN 1971; SÖLING et al. 1963). Glycogenogenesis was particularly reduced in the presence of biguanides when stimulants such as cortisone (CATHELINE 1974), stress (FRAYN 1976) or a high-fat diet (MEYER 1960) were used. More recently, no net change was found in MET-treated diabetic rodents (ROSSETTI et al. 1990; REDDI and JYOTHIRMAYI 1992; HUUPPONEN et al. 1993) but in the two first studies an increase in the hepatic glycogenic index was reported. On the other hand, MET decreased glucose-6-phosphatase in diabetic rats (SALEH 1974).

e) Hepatic Glucose Oxidation

In contrast to in vitro studies, in vivo measurements have shown increased hepatic ATP levels in diabetic animals (KANEKO 1965; KÜHNLE et al. 1984; DUCH 1992). A stimulation of glucokinase was also reported (KANEKO 1965). In normal rats, usually no effect was found (ALTSCHULD and KRUGER 1966; KANEKO 1965; BECKMANN 1971).

f) Antiglucagon Effects

Similarly to in vitro findings, biguanides appeared to decrease the hepatic glucose production in rats when stimulated by glucagon (MAGGI 1968), as well as to reduce insulin and C peptide levels in patients receiving glucagon (FERLITO et al. 1983). The production of "enteroglucagon" was reported in MET-treated rats (KAKEMI et al. 1983) and humans (MOLLOY et al. 1980). In the absence of precise knowledge of the role these intestinal glucagon-related peptides play in the regulation of glucose homeostasis, a clear-cut conclusion appears difficult; albeit the recent developments on GLP-1 could motivate reconsiderations of such biguanide effects.

IV. Adipose Tissue

Although not considered to be an important tissue in quantitative terms when considering the whole body glucose consumption (except in very obese individuals), fat may, however, play a cardinal role in qualitative terms. Indeed this tissue is the source of many lipidic modifications which might interfere in a determinant manner with glucose metabolism in other tissues.

1. In Vitro

a) Glucose Uptake

In the absence of insulin, biguanides usually have no effect on glucose uptake by the adipose tissue (BUCKMANN 1971; CIGOLINI et al. 1984; JACOBS et al. 1986; MATTHAEI et al. 1989; PEDERSEN et al. 1989). In guinea pig adipocytes, or with high biguanide concentrations, increased glucose uptake was reported (JANGAARD et al. 1968; BECKMANN 1971). In contrast, a stimulation of glucose uptake was obtained when adding insulin (JACOBS et al. 1986; MATTHAEI et al. 1989).

b) Glucose Oxidation

This function seems to be closely linked to the concentration of biguanide used; low doses had either no effect (CIGOLINI et al. 1984; WILCOCK and BAILEY 1990) or stimulated glucose oxidation (SCHÄFER and MEHNERT 1962; DAWEKE and BACH 1963; FANTUS and BROSSEAU 1986; L'AGE et al. 1963), whereas high doses decreased basal and insulin-stimulated glucose oxidation (BECKMANN 1971; SCHÄFER and MEHNERT 1962). Addition of insulin was not found to have additional effects to those of MET in one study (FANTUS and BROSSEAU 1986), while MET was found to increase glucose oxidation in the presence of insulin in adipocytes from normal – but not from diabetic – animals (WILCOCK and BAILEY 1990). The sensitivity of adipocytes toward biguanides was claimed to be higher than that of skeletal muscle cells (DAWEKE and BACH 1963).

2. In Vivo

a) Glucose Uptake

In fat cells from in vivo-treated Zucker rats, no increase in glucose uptake was found in one study (PENICAUD et al. 1989), whereas another study showed improved glucose transport in the adipose after MET administration (HANDBERG et al. 1993).

b) Glucose Oxidation

There was no effect of biguanides on glucose oxidation in fat cells from biguanide-treated animals (BECKMANN 1971; PENICAUD et al. 1989) or humans (PEDERSEN et al. 1989).

V. Skeletal Muscle/Heart

1. In Vitro

a) Glucose Uptake

Direct in vitro effects of biguanides on skeletal muscle glucose uptake obviously require the presence of stimulants of the sugar transport. Indeed no effect was seen under normal basal conditions with BUF (STROHFELDT et al. 1972, 1975), PHEN (WILLIAMSON et al. 1963; BOLINGER et al. 1960) or MET (FRAYN et al. 1973; GALUSKA et al. 1991). In human skeletal muscle, addition of insulin did not evoke an action of MET in normal muscle (FRAYN et al. 1973) but the combination of MET + insulin improved the obesity/-diabetes-induced deficient glucose uptake (Fig. 3) (GALUSKA et al. 1991).

In cultured muscle cells, MET was active in the presence of high glucose (SARABIA et al. 1992) as well as in L6 myotubes but not in myoblasts (HUNDAL et al. 1992). In rat diaphragm, MET increased glucose uptake in muscle from diabetic but not from normal rats (FRAYN and ADNITT 1972). An important point was raised by comparing low and relatively high drug concentrations, where it was found that low doses required the presence of insulin whereas high doses increased glucose uptake in a hormone-independent manner (BOLINGER et al. 1960; FISCHER et al. 1995). This point is discussed in more detail in Sect. G.III.

b) Glucose Metabolism

Glucose oxidation was not affected by biguanides in vitro (PAVEL et al. 1964; CORSINI et al. 1974; WILCOCK and BAILEY 1990). Glycogen synthesis was found to be increased in rat diaphragm independently of the presence of insulin (FRAYN and ADNITT 1972), whereas in skeletal muscle insulin was required to obtain an increased glycogen synthesis with biguanides (KEMMER et al. 1977). A study using BUF resulted in decreased glycogen levels in skeletal muscle, but these data were considered to be artefactual (STROHFELDT et al. 1975).

2. In Vivo

a) Glucose Uptake

Many data are available, especially because of the recent development and clinical utilization of the clamp technique. As far as assumptions are

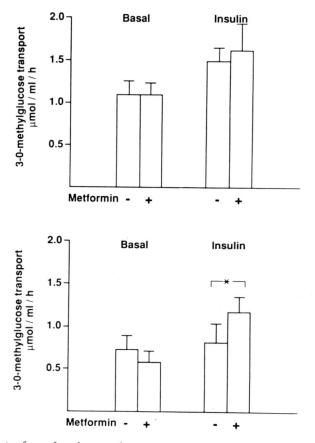

Fig. 3. Effect of metformin on glucose transport in isolated muscle strips from healthy (*upper panel*) and insulin-resistant subjects (*lower panel*). (From GALUSKA et al. 1991)

valid, the information obtained in such protocols is mainly derived from the skeletal muscle mass (see Chap. 14, this volume). Most clamp studies revealed an increase in glucose uptake after biguanide treatment. In animals, biguanides also elevated skeletal muscle glucose uptake, in particular in the presence of hyperglycemia, a finding in agreement with clinical observations using hyperglycemic clamps (LOSERT et al. 1972; BAILEY and PUAH 1986; BAILEY and MYNETT 1994). Using the radioactive 2-deoxyglucose technique, we found that chronic treatment with MET had no effect on basal glucose uptake in skeletal muscles of normal rats but normalized the deficient glucose uptake in mildly diabetic rats (WIERNSPERGER and RAPIN, unpublished results).

b) Glucose Metabolism

Most studies found no change in skeletal muscle glucose oxidation of animals or man (DAWEKE and BACH 1963; KEMMER et al. 1977; BAILEY and PUAH 1986; PENICAUD et al. 1989; JOHNSON et al. 1993). No increase in lactate production could be seen (LOSERT et al. 1972; WILCOCK and BAILEY 1990). Only one study reported slight increases in glucose oxidation with low – but not with higher – biguanide concentrations and hexokinase was not modified (BECKMANN 1971) (LOSERT et al. 1972). Most clinical studies failed to demonstrate significant effects of MET on muscle glucose oxidation.

In contrast to glucose oxidation, glycogen synthesis can clearly be affected by biguanides. Whereas no or only mild changes occurred when given to normal animals (KRONEBERG and STOEPEL 1958; PROSKE et al. 1962; PAVEL et al. 1964; SCHILLINGER et al. 1970), studies on diabetic animals clearly showed that chronic biguanide treatment improved the deficient glycogen synthetic capacity of skeletal muscle (SCHILLINGER et al. 1970; STROHFELDT et al. 1972; PAVEL et al. 1964; LORD et al. 1985; BAILEY and PUAH 1986; ROSSETTI et al. 1990; REDDI et al. 1992). Similarly to the effect in the liver, the glycogenic index of skeletal muscle was stimulated by MET in nSTZ rats (Fig. 4) (ROSSETTI et al. 1990); in diabetic KK mice, however, a net glycogen synthesis was measured, due to the activation of glycogen synthase but not of phosphorylase (JYOTHIRMAYI et al. 1992). Therefore, biguanides seem to correct one key mechanism which underlies the diabetes pathophysiology, namely the capacity of skeletal muscle to store glucose as glycogen. Similar data have been reported in humans, as illustrated by an increase in the nonoxidative glucose disposal in diabetic or insulin-resistant, non-diabetic patients explored by the clamp techniques (HOTHER-NIELSEN et al. 1989; WIDEN et al. 1992).

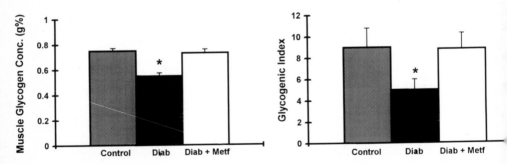

Fig. 4. Muscle glycogen concentration and muscle glycogenic rate following insulin infusion in control, diabetic and metformin-treated diabetic rats. (From ROSSETTI et al. 1990)

VI. Other Tissues

1. Blood Cells

The erythrocyte mass potentially plays an important role because it represents a significant part of the whole body glucose uptake and could also be a major source of circulating lactate levels. Red blood cells (RBCs) may be involved in the postprandial glucose homeostasis since they are able to store glucose in a transient manner (FERRANNINI and BJORKMAN 1986). In diabetic animals and humans, RBC glucose uptake is defective (KATO et al. 1990; CIVELEK et al. 1991; CONGET et al. 1991; RAPIN et al. 1991; YOA et al. 1993; COMI and HAMILTON 1994). Also steroids, which are increased in the late night period, decrease glucose uptake in erythrocytes (MAY and DANZO 1988). MET (BAILEY 1992) and PHEN (LLOYD et al. 1975) do not stimulate the lactate production by these cells. Studies with MET recently showed that this drug was able to restore the deficient glucose uptake in RBC from type 1 (RAPIN et al. 1991) as well as from glucose-intolerant or type 2 diabetic patients (KANIGÜR-SULTUYBEK et al. 1993; YOA et al. 1993). Similar data was obtained with MET in IM-9 human lymphocytes, an effect which was due to an increase in V_{max} and not in K_m (PURRELLO et al. 1987). As shown in Fig. 5, MET shifted the glucose metabolism from glycolysis to glycogen formation under hyperglycemic conditions (YOA et al. 1993).

2. Varia

Metformin did not affect glucose oxidation in renal cells (MEYER 1960), in isolated enterocytes (CUBER et al. 1994) or in a series of other tissues such as liver, skin, brain or renal medulla (WILCOCK and BAILEY 1990).

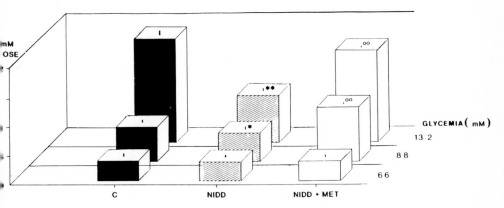

Fig. 5. Glycemia-dependent effect of metformin ($5 \mu g/ml$) on glucose uptake by erythrocytes from normal or type 2 (NIDD) diabetic individuals. $*P < 0.05$, $**P < 0.01$ vs. control; $^{\circ\circ}P < +$ vs. untreated NIDD (From YOA et al. 1993 with permission)

F. Lipid Metabolism

Throughout the clinical use of biguanides, a major observation has been the changes in lipidic profiles of patients. This particularly concerned triglyceride (TG) levels, which are usually diminished to a quite large extent by such drugs. Such modifications can be considered to be simply positive side effects linked with protective actions on vascular complications, but they may also play an integrated role in their antihyperglycemic efficacy. Thus interruption of free fatty acid (FFA) oxidation would directly improve glucose oxidation in muscle tissue, according to the so-called Randle cycle. Such an effect has even been considered by MUNTONI (1974) as the main – if not unique – mode of action of the biguanides.

I. Lipogenesis

Basal lipid synthesis in hepatocytes was not affected by MET (ANFOSSO et al. 1993; MELIN et al. 1990). Recently, however, alanine-induced lipid synthesis was found to be decreased by both MET and BUF (HOTTA et al. 1994). In the isolated perfused liver, BUF was ineffective in concentrations up to $10^{-4} M$ (BECKMANN 1971); in the presence of insulin, MET increased lipid synthesis in hepatocytes (MELIN et al. 1990). In vivo, glucose incorporation into lipids was either unchanged (WILLIAMS et al. 1958; KETEKOU et al. 1969) or stimulated (JYOTHIRMAYI et al. 1992). In fat cells, low doses ($<10^{-5} M$) were usually without significant effect on lipogenesis (BECKMANN 1971). Insulin-stimulated lipogenesis was reduced by higher doses of BUF (DITSCHUNEIT et al. 1968). In vivo, MET was shown to increase lipid synthesis in adipocytes from chronically treated diabetic KK mice (JYOTHIRMAYI et al. 1992).

II. Lipolysis

Low biguanide concentrations did not interfere with basal lipolysis (BECKMANN 1971; CIGOLINI et al. 1984) or reduced it (DUCH 1992). whereas higher doses were inhibitory (STONE and BROWN 1968). Stimulated lipolysis was reported to be decreased by PHEN (STONE and BROWN 1968) but not by BUF or MET (BECKMANN 1971).

III. Fatty Acids

At doses which did not interfere with mitochondrial oxidative phosphorylation, PHEN and MET decreased palmitic or butyrate oxidation in rat diaphragm (MUNTONI et al. 1973; CORSINI et al. 1974; FRAYN et al. 1973) without directly affecting glucose oxidation. Fatty acid oxidation was also reduced by biguanides in vivo (LOSERT et al. 1972; SCHÖNBORN et al. 1975; RICCIO et al. 1991; PERRIELLO et al. 1994). An action on hepatic carnitine transferase was described for MET, which might explain the action on long-chain fatty acids (SANDOR et al. 1979). At therapeutic concentrations BUF increased the uptake of unesterified fatty acids in the presence of glucose, an effect similar to that of insulin (L'AGE et al. 1963). Higher doses decreased their utilization, however. In man MET reduced the turnover of FFAs (SCHÖNBORN et al. 1975; MUNTONI et al. 1978). Similar data were obtained with PHEN (TAGLIAMONTE et al. 1973; STOUT et al. 1974). In view of the recent demonstration of the Randle cycle operating in skeletal muscle, such drug effects could be of cardinal importance in the mechanism leading to a reduction in hyperglycemia through the interplay between lipid and glucose metabolism. Therefore the hypotheses elaborated by MUNTONI (1974), based on pharmacological drug profiles opposite to those of fatty acid oxidation, would be worth reconsideration.

IV. Triglycerides

In hypertriglyceridemic patients or in diabetic patients with increased triglyceride levels, MET reduced fasting TG levels (SIRTORI et al. 1977) by as much as 45%. Possibly of even greater interest is the newly reported decrease in postprandial TG and TG-rich lipoproteins of intestinal origin in patients under MET treatment (JEPPESEN et al. 1994; SHEPHERD and KUSHAHA 1994). VLDL-TG removal was increased by MET in a high-fat diet (ZAVARONI et al. 1984). Reduced triglyceride synthesis by biguanides could be the result of decreased hepatic lipogenesis, of decreased fatty acid supply to the liver as a consequence of diminished lipolysis in fat cells or of a direct effect on fatty acid oxidation.

V. Lipoproteins

The production and clearance of VLDLs was reduced by PHEN (STOUT et al. 1974). A decrease in VLDL secretion by MET was also observed in high-fat-fed rabbits (LACOMBE and NIBBELINK 1981) as well as in hyper-insulinemic rats (ZAVARONI et al. 1984). In cholesterol-fed rabbits, MET modified the VLDL composition and increased their clearance from blood (SIRTORI et al. 1977). The uptake of VLDL by the aorta was strongly inhibited, an effect explained by an action of MET on the complexing factor bewteen VLDL and aortic cells. In human fibroblasts, MET slightly increased LDL binding, uptake and internalization (MAZIERE et al. 1988).

VI. Cholesterol

Clinical studies often showed a slight increase in HDL with a concomitant decrease in LDL. In human fibroblasts as well as in macrophages, cholesterol esterification was diminished by MET, whether LDLs were present or not (MAZIERE et al. 1988). In leukocytes, no effect on cholesterol synthesis was induced by MET (MOORE et al. 1990). Metformin was shown to decrease human menopausal gonadotropin (HMG)-CoA reductase in diabetic rat enterocytes (MOORE et al. 1988; MAZIERE et al. 1988). The activity of the intestinal ACAT was reduced by MET (MAZIERE et al. 1988; SCOTT and TOMKIN 1983). In contrast, the activity of these enzymes was not modified in the liver.

G. Cellular Effects

I. Need for Insulin?

Many clinical studies, using the clamp technique, showed that MET was able to decrease insulin resistance in diabetic patients. Positive effects of MET on insulin resistance have even been obtained in nondiabetic, insulin-resistant animals (ZAVARONI et al. 1981; HALIMI et al. 1994) and in humans such as relatives of diabetic parents (ERIKSSON et al. 1990) or women with polycystic ovary syndrome (VELAZQUEZ et al. 1993). The results were usually better in hyperglycemic than in euglycemic clamps, suggesting that MET could also promote the glucose-mediated glucose uptake (DE FRONZO et al. 1991).

Several experiments have shown that, for lowering hyperglycemia, biguanides required the presence of insulin (STERNE 1969). In insulin-dependent diabetes, biguanides were found to be inactive per se, but very significantly reduced the amount of exogenous insulin needed to normalize hyperglycemia (LOSERT et al. 1972; GIN et al. 1985). On the other hand, some authors observed that "trace" amounts of insulin were sufficient to obtain the efficacy of MET (STERNE 1964; BAILEY and MYNETT 1994; FISCHER et al. 1995). These latter findings suggested that the relationship between MET and insulin might not be simply in the direction of a potentiation of the hormone by the drug, but that also reverse interactions could occur. Thus MET may exert pleiotropic effects, of which some (but not necessarily all) are insulin dependent whereas others may be glucose dependent or directly drug induced (WIERNSPERGER and RAPIN 1995).

II. Potentiation of Insulin Actions

In the liver, insulin does not increase the transport of glucose but blocks gluconeogenesis; this reaction has been shown to be potentiated by biguanides in diabetes (see Sect. E.III, 1.a, 2c).

In extrasplanchnic tissues, insulin promotes the utilization and the storage of glucose by skeletal muscle and fat. At least two main defects underlie the impaired glucose consumption by these tissues in NIDDM: one at the level of transmembrane glucose transport and another at the level of glycogen synthesis. Very early, investigators showed that biguanides "increased the permeability of cells to glucose," an effect which is now explained by an action on the recently identified glucose transporters. The GLUT-4 subtype is specifically activated by insulin and appears to be deficient in NIDDM, whereby glucose transport becomes a major limiting factor for the clearance of excessive plasma glucose. In vitro and ex vivo, MET was shown to increase glucose uptake by human skeletal muscle (GALUSKA et al. 1991) or adipocytes from Zucker rats (Fig. 6) (MATTHAEI et al. 1993) in the presence of insulin. This effect was shown to be due to an increased translocation of GLUT-4 from intracellular low-density microsomes to the cell plasma membrane (Fig. 7) (MATTHAEI et al. 1993; KOZKA and HOLMAN 1993).

The fact that MET did not reduce – but in fact potentiated – the insulin-stimulated transport of glucose speaks against any "toxic" effect (GOULD 1984). In Zucker rats treated with MET, a positive effect on GLUT-4 was observed only in fat cells but not in skeletal muscle (HANDBERG et al. 1993). This result may be explained by a possible absence of defective GLUT-4 translocation in skeletal muscles fo this rat strain (GALANTE et al. 1993; ZARJEVSKI et al. 1992). In contrast, GLUT-1 expression was decreased by MET in Zucker rats in which it was previously overexpressed by the underlying insulin resistance (HANDBERG et al. 1994).

In addition to improving the GLUT-4 related glucose transport when given curatively, MET also prevented the downregulation of GLUT-4 in

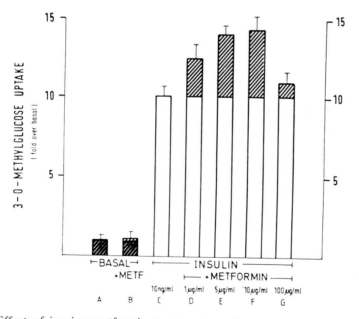

Fig. 6. Effect of in vivo metformin treatment on glucose transport in adipocytes from lean and obese Zucker rats. *Hatched bars* illustrate the additional effect of MET compared with INS alone. (From MATTHAEI et al. 1993)

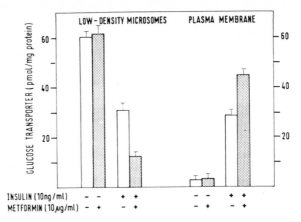

Fig. 7. Effect of in vivo metformin treatment on GLUT-4 in Zucker rat adipocyte membranes. (From MATTHAEI et al. 1993)

adipocytes submitted to chronic in vitro insulin treatment (KOZKA and HOLMAN 1993). Although the data by MATTHAEI and GRETEN (1991) had been obtained with MET concentrations within the therapeutic range, other findings on increased GLUT-4 translocation by MET obtained at higher doses and after longer-lasting incubation may be linked with another process: the recruitment of another pool of GLUT-4, which is responsive to muscle contraction or hypoxia (CARTEE et al. 1991).

Besides lipid synthesis and glucose transport, insulin also stimulates protein synthesis. Investigations on albumin production by freshly cultured rat hepatocytes revealed that, in vitro as well as ex vivo, MET potentiated the action of insulin (ZAIBI et al. 1994). In contrast, amino acid uptake was not increased by concentrations of MET which increase glucose uptake (PURRELLO et al. 1988; HUNDAL et al. 1992).

Altogether these results show that MET is able to increase the rates of lipid synthesis, glucose transport or protein synthesis above the maximum reached with insulin alone. How this effect might be mechanistically explained is discussed in the next sections.

III. Mechanisms of Insulin Potentiation

This aspect is very new, since the understanding of how insulin acts is still in complete, in spite of tremendous developments over the last decade. For this reason data on biguanides are logically limited to MET. Metformin could interfere in multiple steps with insulin's action, from the hormone production in the pancreas down to its ultimate intracellular degradation. Only a summary will be given here since more detail is available in another review (WIERNSPERGER and RAPIN 1995).

1. Plasma Insulin

As discussed in Sect. A.II, MET does not increase the absolute amounts of circulating insulin. It may, however, have beneficial effects on the levels of ultimately active insulin available for the target cells, either by correcting the disturbed proinsulin/insulin ratio (HAUSMANN and SCHUBOTZ 1975; MANLEY et al. 1994) or by increasing the rate of free to bound insulin in the plasma (STERNE 1964; WIERNSPERGER and RAPIN 1995). If such mechanisms were operating, biguanides could allow more intact insulin to reach cells without apparent modifications in plasmatic values.

2. Insulin Receptor Binding/Phosphorylation

Data about biguanide influence on insulin receptor binding are numerous but discordant: no effect was found with MET in cultured fibroblasts (FRORATH et al. 1985; MOUNTJOY et al. 1987) or human adipocytes (PEDERSEN et al. 1988). However, another study performed on various cell types reported a 40%–100% increase in insulin binding in vitro when using PHEN or MET (GOLDFINE et al. 1984). Increased low-affinity receptor binding was found in red blood cells (HOLLE et al. 1981). Frequently, however, no clear relationship between binding modifications and antihyperglycemic activity could be demonstrated (FANTUS and BROSSEAU 1986), and most studies showed that responsiveness rather than sensitivity was changed by MET (WIERNSPERGER and RAPIN 1995).

In vitro studies showed no changes in phosphorylation of either purified insulin receptors (Fig. 7) or adipocytes (JACOBS et al. 1986; MATTHAEI et al. 1991) (Fig. 8). Some studies performed after previous in vivo treatment of animals with MET showed, however, that phosphorylation might have been

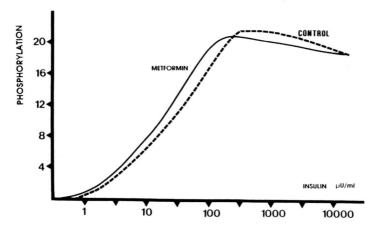

Fig. 8. Autophosphorylation on tyrosine of purified hepatocyte insulin receptors and effect of metformin in vitro. (GRIGORESCU, personal communication)

increased (GRIGORESCU et al. 1991; ROSSETTI et al. 1990). Since changes at the receptor levels were not conclusive, the potentiating action of biguanides on insulin was attributed to postreceptor actions taking place along the signaling cascade of insulin.

3. Postreceptor Mechanisms

This area of research is only in its infancy, as unraveling of the mechanisms whereby insulin stimulates various biological functions is a recent event. The most logical step to investigate is the expression or activity of the protein insulin receptor substrate 1 (IRS-1). Recent data do not, however, support this hypothesis for MET (TURNBOW et al. 1995). In view of the demonstrations that MET can potentiate most – if not all – biological functions controlled by insulin, one can hypothesize that the site(s) of MET action is likely located in the close vicinity of IRS-1, possibly at the level of docking of proteins with IRS-1, before the branching of various signaling pathways related to specific distal reactions.

Other hypotheses would concern plasma membrane-borne second messengers of the phosphatidyl-inositol type. In studies on *Xenopus* oocytes, MET potentiated insulin-induced maturation; preliminary data showed that the drug stimulated insulin action through an increase in inositol triphosphate (IP3) and intracellular calcium (STITH, personal communication). Another possibility would be an action of MET on phosphorylation levels of enzymes involved in the insulin-signaling cascade, for example at the level of phosphatases, some of which might be stimulated by MET (WORM et al. 1994).

Finally, comparing the action of MET in the presence of various agonists in different experimental approaches revealed that this drug exerted specific effects towards insulin or at least hormones using tyrosine kinase pathways (Fig. 9) (ZAÏBI et al., submitted; LAURENT et al., submitted; GRIGORESCU et al. 1994).

4. Mechanisms Not Linked with Potentiation

A set of very exciting data has shown that biguanides can still be efficacious in the presence of very low, trace amounts of insulin (STERNE 1964; BAILEY and MYNETT 1994). These studies revealed that MET increased glucose transport in the presence of insulin concentrations which were totally inactive alone (Fig. 10) (FISCHER et al. 1995). Using erythrocytes, a cell selected for its absence of reaction to insulin for glucose transport, we recently found that a short preincubation with normal levels of insulin ($10\,\mu$U/ml) significantly potentiated the effect of MET (WIERNSPERGER and RAPIN 1995). It thus appears that insulin may have a permissive effect on the action of biguanides, as was already suggested many years ago by observations with PHEN (FAJANS 1960; GOULD and CHAUDRY 1970).

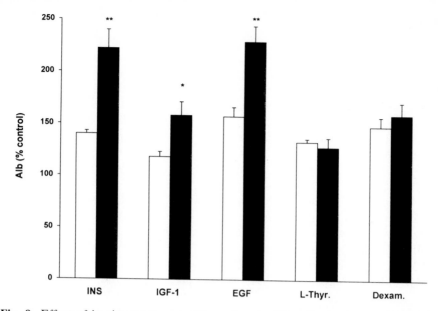

Fig. 9. Effect of in vivo treatment with metformin (50 mg/kg per day os) of mildly diabetic rats on primary cultured rat hepatocyte albumin production stimulated in vitro by insulin, IGF-1, EFG, thyroxine or dexamethasone. *Black columns*, with metformin; *white columns*, control; *P < 0.05; **P < 0.01. (From ZAÏBI et al. 1995 with author's permission)

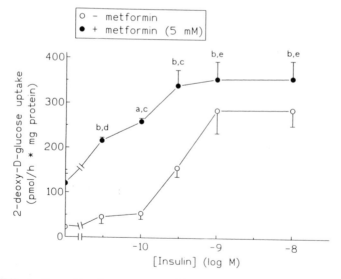

Fig. 10. Effect of combined treatments with metformin and insulin on in vitro 2-deoxyglucose transport in cardiomyocytes from normal rats. (From FISCHER et al. 1995)

5. Insulin-Independent Effects

Several studies performed on various cell types have shown that MET does not affect basal glucose transport in normoglycemic conditions. However, hyperglycemic conditions applied to cells which are mainly or only equipped with GLUT-1 transporters (muscle cells in culture, erythrocytes or vascular cells) strongly enhanced the efficacy of MET. Thus in L6 yotubes as well as in human cultured muscle cells, the effect or MET on transmembrane glucose transport was higher in the presence of 15 mM than 5 mM glucose in the medium (KLIP et al. 1992; SARABIA et al. 1992). Similar data have been obtained with vascular smooth muscle cells and endothelial cells (SASSON et al. 1994).

While these studies deal with high MET concentrations, experiments performed with doses corresponding to the therapeutic plasma levels $(1-5 \mu g/ml)$ showed that MET restored the deficient glucose transport in erythrocytes from diabetic rats or patients (RAPIN et al. 1991; YOA et al. 1993; KANIGÜR-SULTUYBEK et al. 1993).

In cultured human fibroblasts, gene expression of GLUT-1 was elevated by a subchronic exposure to MET (HAMANN et al. 1993). A glucose transporter subtype which is also not regulated by insulin, GLUT-3, was also activated by an incubation of large duration with MET (ESTRADA et al. 1994). These data suggest that MET increases the intrinsic activity of membrane-trapped glucose transporters, since in red blood cells there is neither translocation nor new protein synthesis.

Thus MET may exert insulin-independent effects under hyperglycemic conditions which confer to this drug synergistic actions together with the hormone. In fact, additive effects have been noticed. Thus synergism can be envisaged, for example, for glucose transport in insulin-sensitive cells, in which insulin could promote the translocation of GLUT-4 transporters, whereas MET may increase their intrinsic activity once inserted into the plasma membrane. At the level of the liver, hyperglycemia could favor the insulin-independent component of the antigluconeogenic activity of MET (YU et al. 1994).

That hyperglycemia is a favoring factor for the efficacy of biguanides is a well-known observation (STERNE 1969; HERMAN and MELANDER 1992; BERGER and KUNZLI 1970; YU et al. 1994). Bearing in mind that about 75% of the whole body glucose uptake is an insulin-independent process, MET might exert a permanent, direct effect on hyperglycemia through its interaction with insulin-independent regulatory mechanisms.

IV. Biguanides: Membrane-Active Drugs

The precise cellular sites where biguanides exert their effects is still a matter of extreme complexity. Many arguments support the notion that these drugs are mainly (if not exclusively) membrane-active compounds (SCHÄFER 1983).

On the other hand, from experiments on drug distribution and uptake there are many data which have shown that drastic differences sometimes exist between the in the vitro and in vivo situation. If the latter assumptions are true, then one must admit that there are basic differences between the physicochemical behavior of biguanides in vitro and in vivo, which also casts doubt on the relevance of a non-negligible part of the long-lasting collection of data on these drugs.

1. Supporting Arguments

Biguanides have no direct effect on soluble enzymes, whereas they affect enzymatic activities in whole cells. Similarly, MET seems to affect the phosphorylation of insulin receptors from in-vivo-treated animals, while having no effect on purified receptors. However, the latter effects could have been mediated indirectly, through the improvement of the diabetic state of the animals.

In intact leukocytes, biguanides, at high concentrations, reduce mitochondrial phosphorylation whereas at least MET does not affect isolated mitochondria of the same cells (MEYER 1960; LEVERVE, personal communication). In *Xenopus* oocytes, incubation with MET increased the levels of IP3, which is generated by membrane-active phospholipases.

Another series of argument favoring membrane effects consists of various protective actions of biguanides on cellular integrity observed in different pathological situations. Thus, the myocardial infarct size of rats with a permanent coronary artery ligation was markedly reduced by chronic MET treatment (CHARLON et al. 1988). Phenformin and MET were shown to inhibit proteolysis in rat myocardium (THORNE and LOCKWOOD 1990). In studies on primary cultured rat hepatocytes, the yield of viable hepatocytes from collagenase-perfused rat livers and their subsequent attachment to the culture substratum was improved by a previous in vivo treatment of rats with MET (DUCH 1992; ZAIBI et al. 1995). Finally, the fragility of erythrocyte membranes to hypo- or hyperosmolar shock was remarkably well preserved by MET (DUCH 1992).

2. Membrane Binding/Cellular Uptake

a) Biguanides in General

The investigations performed by SCHÄFER in the 1970s have provided much information about the relationships between physicochemical properties of biguanides and their biological activity. Globally the conclusion was that, because they affect cellular and mitochondrial membranes, biguanides interfere with transfer processes and inhibit mitochondrial respiration. It was shown that biguanides bound to membranes in a nonspecific manner and that there was practically no contribution of electrostatic forces to this binding (SCHÄFER 1979). Because biguanides are weak bases with a pK_a of

11-12, they exist in a cationic form at physiological pH and are expected to bind to negatively charged membrane components, the phospholipids in this case being likely binding sites for these drugs (SCHÄFER 1976, 1979). After binding to the plasma membrane, biguanides remain membrane bound and less than 10% reach the submembraneous region.

However, these experiments were: (1) performed mainly directly on isolated mitochondria, (2) with very high drug concentrations and (3) with biguanides of varying size and side-chain length. Unfortunately, only few data were obtained with MET, the smallest biguanide of series and most findings were thus extrapolated from larger (eventually toxic) biguanides to MET, assuming that there was a common behavior of all derivatives of this class. However, the experimental investigations and the reassessment of the pharmacology of MET clearly showed that most assumptions originating from SCHÄFER's investigations did not apply to MET.

b) Metformin

α) Membrane Binding. Recent investigations confirmed that MET binds to cell membranes and remains largely bound even after repetitive washing. At therapeutically relevant concentrations this binding was, however, limited to membrane proteins: in red cells, MET interacted with proteins (FREISLEBEN et al. 1992) and in various in vitro systems using cells or artificial membranes, only binding to protein was found, unless very high MET concentrations were used, at which it bound also to phospholipids (HONEGGER and WIERNSPERGER, unpublished results; TEISSIE, personal communication). Thus, in contrast to larger biguanides, MET preferentially binds to membrane proteins, where it may modify their conformation as recently suggested by studies on erythrocyte membrane protein glycation (FREISLEBEN et al., submitted). This would conceivably provide an explanation for the mechanism whereby MET modifies the biochemistry of cells and possibly reestablishes active pathways which are governed by membrane-bound receptors and related phosphorylation reactions or docking of proteins in the subplasmalemmal region.

β) Cellular Uptake. In view of the finding with other biguanides and the risk of interference with mitochondrial oxidative phosphorylation, a major question is the relation between intracellular drug concentrations and subcellular organelles.

Very interestingly, in vitro biguanides bind to the membrane but practically no MET can be found in the cell interior. Thus, there was no uptake of radiolabeled MET in human fibroblasts or macrophages or in rat astrocytes (HONEGGER and WIERNSPERGER, unpublished results) or hepatocytes (WILCOCK et al. 1991b). Less than 1/1000 was found in the cytoplasm when

Xenopus oocytes were incubated in $10^{-5} M$ MET (KHAN et al. 1994), whose pK_a of 11.5 is expected to severely limit its ability to cross membranes, as is expected from an ionized molecule at physiological pH.

The situation appears completely different in vivo, where the drug might lose its cationic charge to a variable extent when it binds to various plasma components. This may permit an easy access of MET to abluminal cells in spite of the multiple membrane barriers the drug has to cross (intestine, endothelium, cell plasma membrane). In autoradiographic sections of rat liver and skeletal muscle, we found that the radiolabeled drug was present around and within both cell types (WIERNSPERGER and RAPIN, unpublished results). In studies on subcellular fractionation, it was found that, 30 min after oral intake of 50 mg/kg MET, the rat liver contained drug levels which were three times higher than plasma concentrations. In these hepatocytes, 78% of the total radioactivity was in the cytosol, 3% in the nucleus, 9% in the mitochondrial/lysosome fraction and the remainder in mixed membranes. In erythrocytes incubated with $10^{-5} M$ MET, a maximum of 100 pM for 10^8 cells was found, of which 82% was in the cytosol (WILCOCK et al. 1991).

Therefore there is a clear discrepancy between the cellular uptake of MET in vitro and in vivo. The extent to which this difference affects the pharmacological data is unknow. In studies on *Xenopus* oocytes in which MET potentiates the insulin-activated maturation, a direct injection of 130 nM/oocyte (corresponding to the level normally present in the cell interior when incubated in $10^{-5} M$ MET) also potentiated insulin-induced maturation, although the rate was less and the duration was prolonged (KHAN et al. 1994)

The difference in cellular uptake in vitro and in vivo could be explained either by differences in the physicochemical properties (cationic charge) or by the existence of a membrane transporter which, whatever the reason, would only operate in vivo. There is no evidence available for a "MET receptor or transporter," although technical difficulties limit the strength of this conclusion. The similarity between MET and arginine led us to investigate whether there was competition between these compounds for their transport in 3T3 cells. The transport of MET appeared strongly Na^+- dependent and amiloride inhibited the transport by 75%. Therefore, sodium might be the source of the energy required for MET transport: additional data showing an inhibition of MET transport by the ionophore gramicidin further support the conclusion that the Na^+ electrochemical gradient is required. Addition of arginine or lysine, but not asparagine or polylysine, stimulated the transport of MET into these cells (Fig. 11) (KHAN et al. 1992). Moreover, MET stimulated the uptake of arginine. Thus MET appears to be taken up via the y^+ transport system for cationic amino acids. In addition, MET inhibited the transport of spermidine, while the latter failed to affect the transport of the former (KHAN et al. 1992). Clearly many more investigations are needed to clarify the question of MET cellular transport.

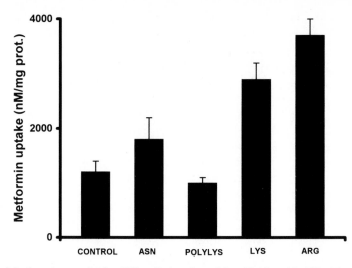

Fig. 11. Metformin uptake by 3T3 cells incubated for 60 min with 200 μM metformin and various amino acids. (Modified from KHAN et al. 1992)

γ) Membrane Effects. In contrast to the predictions from the work by SCHÄFER (1976) on larger biguanides, MET does not reduce the fluidity of membranes. Fluorescence-bleaching techniques as well as electron paramagnetic resonance spectroscopic studies on human red blood cells and other cells show that MET increases cell membrane fluidity (FREISLEBEN et al. 1992; CANDILOROS et al. 1994; TEISSIE et al., unpublished results). When erythrocytes were incubated in high glucose or high insulin, their membrane fluidity decreased and the addition of MET at therapeutic concentrations (0.5–5 μM) normalized this parameter (Fig. 12) (FREISLEBEN et al. 1992). Thus MET was capable of completely inhibiting the increase in membrane viscosity induced by 20 mM glucose. The use of various labels showed that the fluidizing effects of MET occured mainly in the polar interface of the membrane. To some extent, MET must penetrate into the cell plasma membrane to potentiate insulin (KHAN et al. 1994).

More recently it was demonstrated that 20 mM glucose reduced the exposure of thiol groups in the erythrocyte membrane and that this defect could also be corrected to normal values by the combination of insulin and MET (FREISLEBEN et al., submitted). Although it is difficult to directly link these major membrane physical changes to the action of the drug in diabetes, it should be recalled that decreases in membrane fluidity are linked with reduced insulin-stimulated glucose transport (CZECH 1980). Conversely, increased fluidity enhances the low-affinity insulin binding (GINSBERG et al. 1981), an effect which was reported for MET (HOLLE et al. 1981), as well as the transport of glucose (READ and MCELHANEY 1976; PILCH et al. 1980).

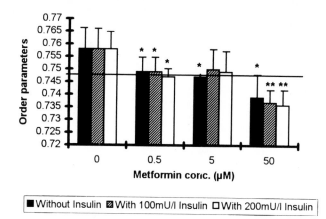

Fig. 12. Influence of various metformin concentrations within the therapeutic range on the order parameter of erythrocyte membranes coincubated with 20 mM glucose, with or without insulin (100 and 200 μU/ml). *$P < 0.05$; **$P < 0.01$ vs. zero value. *Horizontal line* represents the control value = 0.749. (From FREISLEBEN et al. 1992)

V. Intracellular Effects

As mentioned above, at least at high doses, biguanides are considered to interfere with mitochondrial respiration, and, despite much contradictory evidence, claims can still be found that this "pseudotoxic" process would explain their antihyperglycemic efficacy.

However, work by SCHÄFER and BOJANOWSKI (1972) and MEYER (1960) has clearly shown that the binding of MET to mitochondrial membranes was very weak when compared with larger biguanides (Fig. 13). In vitro, MET does not significantly change the mitochondrial membrane potential (UBL et al. 1993), although high MET concentrations can also interfere with the mitochondrial respiration. Data on intrahepatocyte drug concentrations show that the amount of MET bound to liver cell mitochondria at the time of the highest plasma levels was weak. Therefore, and in contrast to previous expectations, the interference which might occur at very high concentrations (FISCHER et al. 1995) is not likely to be due to direct drug effects at the level of the mitochondrial membrane. This conclusion is further corroborated by the absence of an effect in isolated hepatocyte mitochondria (LEVERVE and WIERNSPERGER, unpublished results). Thus, at normal concentrations, the antidiabetic effect of MET is not likely to be explained by an effect on mitochondrial respiration, since: (1) there is only a weak binding and concentration of MET at the level of the mitochondrial membrane and (2) there is no direct effect on isolated mitochondria. This mechanism has been overestimated as an explanation for the antidiabetic efficacy of biguanides since, among the many arguments against this theory (see also BECKMANN 1971), we should at least mention that biguanides usually in-

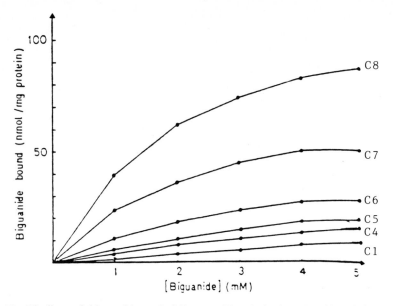

Fig. 13. Binding of biguanides of different side chain lengths (C1–C8) to non-energized rat liver mitochondria (metformin = C1). (From SCHÄFER and BOJANOWSKI 1972)

crease both lactate and pyruvate levels and also that no link has been found between this effect and the antihyperglycemic potency of biguanides (DAVIDOFF 1970).

In contradiction with the pure membrane effect is the repetitive observation of long lag periods for the action of MET in vitro (SARABIA et al. 1992; PURRELLO et al. 1987; KOZKA and HOLMAN, 1993; FISCHER et al. 1995). Several hours of incubation are frequently necessary to observe a change in function (SARABIA et al. 1992). Interestingly, this effect is not modified by the addition of a protein synthesis inhibitor such as cycloheximide. On the other hand, this observation is not universal, since the effect is almost immediate in erythrocytes (YOA et al. 1993) and present after 2 h incubation in the case of adipocytes from Zucker rats (MATTHAEI et al. 1991). This observation could mean that, at least to some degree, MET needs to enter the cell in order to act on some of the biological processes.

VI. Conclusions

Evidence favors the cell membrane as the main site of action of biguanides and more particularly MET. Due to its low lipophilic characteristics, MET may behave very differently from the larger biguanides, and in particular has much weaker effects on mitochondrial metabolism. Even in this case, the effects of MET do not take place directly at the mitochondrial membrance since the intact cell is required to observe an effect. There are important differences in cellular MET uptake according to whether the drug is applied in vitro or in vivo. This cellular uptake is almost absent in vitro, whereas it may be important in vivo. Possible explanations for these discrepant data are either differences in the physicochemical states or the existence and operation of a transporter in vivo such as system y^+. Metformin binds to membrane proteins, and only at high concentrations $(10^{-3} M)$ does it bind to phospholipids. The drug may modify the conformation of proteins in the sense of restoring or directly stimulating the activity of enzymatic systems. Metformin fluidizes membranes, which may favor such structural modifications as well as lipid-protein interactions. These data, which open a new concept on how MET and insulin act in concert to reduce hyperglycemia, do not exclude intracellular effects in addition to those occurring at the cell membrane. These, however, remain to be demonstrated.

H. Conclusions

This review tries to give, in the limited space available, an analytical description of the essential present knowledge of the preclinical pharmacology of biguanides. It is fascinating to note that since the exhaustive review by BECKMAN in 1971 only few questions have been answered whereas so many have been raised. This is due to the inherent properties of the biguanides, which are chaotropic and thus pleiotropic drugs, and also to the impressive progress made in the fundamental knowledge on the pathophysiology of diabetes.

The analysis of so many publications appearing since 1971 confirms that MET, the only biguanide in clinical use today, behaves differently from PHEN or BUF. We can assume that both main hypotheses which were claimed to explain the glycemia-lowering effects of biguanides are not valid – or at least not sufficient in the particular case of MET. Thus, absorption of glucose by the intestine is only marginally affected by MET, to an extent which cannot explain its antihyperglycemic efficacy. However, biguanides stimulate a physiological function of the intestine, which is the catabolism of glucose to lactate. Several arguments also allow the exclusion of any toxic cellular effects as a source of lactate as long as no concomitant renal or hepatic damage is present. In any case, the slight rise in blood lactate remains wihtin the physiological range.

From the intensive research performed during the last 15 years, we have learned that MET, at its therapeutic dosage, reduces diabetic hyperglycemia by acting in the splanchnic bed (gut, liver) as well as in peripheral tissues such as skeletal, muscle, fat or blood cells. Low concentrations, such as those found in the postabsorptive state, have clear-cut effects on the mechanisms responsible for insulin resistance. At higher concentrations, an inhibition of hepatic gluconeogenesis occurs, which is likely to represent a key mechanism underlying the antihyperglycemic efficacy of these drugs.

The mechanism(s) of action of biguanides is in fact an addition of various individual effects, whose sequence and respective importance varies according to the physiological state during the day and the local tissue concentrations of the drug. The complexity of the biguanide mode of action is continuously illustrated by the variability of data obtained according to parameters such as time, dosage and tissue drug levels, species, model and severity of diabetes. This is due to a series of effects on insulin postreceptor mechanisms and glucose transporters which are actually highlighted. Accordingly, some effects of MET are insulin-dependent, whereas others are not. Data support an action of MET on both basal and insulin-stimulated glucose transport. Recent observations have shown that, for reasons which remain to be elucidated, hyperglycemia clearly favors the action of these compounds. Generally it looks like improvement of insulin resistance could be obtained with low concentrations, while higher doses may be needed to reduce hyperglycemia. In addition, effects on lipid oxidation in particular may play an important role in the overall antidiabetic efficacy of biguanides, and this aspect needs deeper investigation.

Accompanying the rapid progress of fundamental research in diabetes, mechanisms have been unraveled which should soon allow an understanding of how MET potentiates the action of insulin at the cellular level. Recent evidence suggests that the interaction between insulin and MET may be bidirectional. The main site of action of biguanides appears to be the cell membrane, from which secondary modifications may govern the biochemical machinery in the subplasmalemmal region and in the cell interior.

In addition to effects on glucose and lipid metabolism, MET exerts effects on the blood vessels and hemostatic regulation. Of even greater interest are the unique and potent properties of MET on microcirculation, which could find interesting applications in syndrome X as well as in the diabetic complications (microangiopathy), two situations for which preclinical and clinical evidence for a benefit of MET is accumulating.

In conclusion, biguanides appear to be complex drugs whose action is extremely variable and multifactorial, depending on many physiological parameters. Their antihyperglycemic efficacy can thus not be explained by any single mechanism. From the growing knowledge accumulated over several decades, it appears that this class of drugs has fascinating properties, many of which remain to be identified and further elucidated and which may lead to a complete reconsideration of both their use in diabetes and therapeutic applications.

References

Agid R, Marquie G (1972) Effets inhibiteurs de certains biguanides antidiabétiques sur l'absorption intestinale des lipides. C R Acad Sci [III] 275:1787–1790
Agid R, Marquie G (1973) The effect of metformin on lipid-induced atherosclerosis. Adv Metab Disord 2 [Suppl 2]:575–586

Alengrin F, Grossi G, Canivet B, Dolais-Kitabgi J (1987) Inhibitory effects of metformin on insulin and glucagon action in rat hepatocytes involve post-receptor alterations. Diabete Metab 13:591–598

Altschuld R, Kruger FA (1966) The mechanism of the hypoglycemic action of phenethylbiguanide (DBI). Clin Res 14:439

Anfosso F, Chomiki N, Alessi MC, Vague P, Juhan-Vague I (1993) Plasminogen activator inhibitor-1 synthesis in the human hepatona cell line HepG2. J Clin Invest 91:2185–2193

Argaud D, Roth H, Wiersperger N, Leverve XM (1993) Metformin decreases gluconeogenesis by enhancing the pyruvate kinase flux in isolated rat hepatocytes. Eur J Biochem 213:1341–1348

Back N, Wilkens H, Barlow B, Czarnecki J (1968) Fibrinolytic studies with biguanide derivatives. Ann NY Acad Sci 148:691–713

Bailey CJ (1992) Biguanides and NIDDM. Diabetes Care 15:755–772

Bailey CJ (1993) Metformin – an update. Gen Pharmacol 24:1299–1309

Bailey CJ, Mynett KJ (1994) Insulin requirement for the antihyperglycemic effect of metformin. Br J Pharmacol 111:793–796

Bailey CJ, Puah J (1986) Effect of metformin on glucose metabolism in mouse soleus muscle. Diabete Metab 12:212–218

Bailey CJ, Wilcock C, Day C, Turner S (1989) Metformin effects in transhepatic glucose and lactate concentrations. Diabetic Med 6 [Suppl 1]:A14

Bailey CJ, Flatt PR, Wilcock C, Day C (1991a) Antihyperglycemic mechanism of action of metformin. In: Shafrir E (ed) Lessons from animal diabetes, vol 3. Smith-Gordon, London, pp 277–282

Bailey CJ, Wilcock C, Wyer ND, Turner SL (1991b) Subcellular localization of metformin. Diabetologia 34 [Suppl 2]:115

Bailey CJ, Wilcock C, Day C (1992) Effect of metformin on glucose metabolism in the splanchnic bed. Br J Pharmacol 105:1009–1013

Bailey CJ, Mynett KJ, Page T (1994) Importance of the intestine as a site of metformin-stimulated glucose utilization. Br J Pharmacol 112:671–75

Barnes AJ, Willars EJ, Clark PA, Hunt WB, Rampling M (1988) Effects of metformin on haemorheological indices in diabetes. Diabete Metab 14:608–609

Baron A, Brechtel G (1993) Insulin differently regulates systemic and skeletal muscle vascular resistance. Am J Physiol 265:E61–E67

Barzilai N, Simonson D (1988) Mechanism of metformin action in NIDDM, Diabetes 37 [Suppl 1]:244

Beckmann R (1965) Resorption Verteilung im Gewebe und Ausscheidung von 1-butyl-biguanid-[^{14}C]-hydrochlorid. Arzneimittelforschung 15:761–764

Beckmann R (1969) Resorption Verteilung im Organismus und Ausscheidung von Metformin. Diabetologia 5:318–324

Beckmann R (1971) Biguanide (Experimenteller Teil). In: Maske H (ed) Oral wirksame Antidiabetika. Springler, Berlin Heidelberg New York, pp 439–596 (Handbook of Experimental Pharmacology, vol. 29)

Berger W, Kunzli H (1970) Effect of dimethylbiguanide on insulin glucose and lactive acid contents observed in portal vein blood and peripheral venous blood in the course of intraduodenal glucose tolerance tests. Diabetologia 6:37

Berger W, Lauffenburger T, Denes A (1972) The effect of metformin on the absorption of vitamin B12. Horm Metab Res 4:311–312

Bertuglia S, Coppini G, Colantuoni A (1988) Effects of metformin on arteriolar vasomotion in normal and diabetic Syrian hamsters. Diabete Metab 14:554–559

Bertuglia S, Colantuoni A, Donato L (1989) Effects of metformin on microcirculation during hemorrhagic shock. Excerpta Med Int Congr Ser 868:1189–1194

Bobbioni E, Coscelli C, Zavaroni I, Alpi O, Capretti L (1978) The effect of metformin on the insulin response in vivo and in vitro to arginine and glucose in the normal rat. Excerpta Med Int Congr Ser 454:353–358

Bolinger RE, Mckee W, Davis JW (1960) Comparative effects of DBI and insulin on glucose uptake of rat diaphragm. Metabolism 9:30–35

Bouguerra S (1985) Effets d'un biguanide antidiabétique (NN, dimethylbiguanide) sur l'évolution du syndrome diabétique chez le rat des sables (Psammomys obesus) et les complications vasculaires. Thesis, University of Algiers

Bouskela E (1988) Effects of metformin on the wing circulation of normal and diabetic bats. Diabete Metab 14:560–565

Bouskela E, Wiernsperger N (1993) Effects of metformin on hemorrhagic shock blood volume and ischemia/reperfusion in nondiabetic hamsters. J Vasc Med Biol 4:41–46

Candiloros H, Denet S, Ziegler O, Muller S, Donner M, Drouin R (1994) Effet de la metformin in vitro sur la fluidité membranaire des érythrocytes. Diabete Metab 20:18A

Cartee GD, Douen AG, Ramlal T, Klip A, Holloszy JO (1991) Stimulation of glucose transport in skeletal muscle by hypoxia. J Appl Physiol 70:1593–1600

Caspary WF (1971) Effect of biguanides on intestinal transport of sugars, amino acids and calcium. Naunyn Schniedebergs Arch Pharmakol 269:421–422

Caspary WF (1977) Biguanides and intestinal absorptive function. Acta Hepatogas-troenterol stuttg 24:473–480

Caspary WF, Creutzfeldt W (1971) Analysis of the inhibitory effects of biguanides on glucose absorption: inhibition of active sugar transport. Diabetologia 7:379–385

Catheline M (1974) Contribution biochimique à l'étude du N1N1-dimethylbiguanide. Thesis, University of Rennes

Chan JCN, Tomlinson B, Critchley AJH, Cockram CS, Walden RJ (1993) Metabolic and haemodynamic effects of metformin and glibenclamide in normotensive NIDDN patients. Diabetes Care 16:1035–1038

Charlon V, Boucher F, Mouhieddine S, De Leiris J (1988) Reduction of myocardial infarct size by metformin in rats submitted to permanent left coronary artery ligation. Diabete Metab 14:591–595

Cigolini M, Bosello O, Zancanaro G, Oralandi PG, Fezzi O, Smith U (1984) Influence of Metformin on metabolic effect of insulin in human adipose tissue in vitro. Diabete Metab 10:311–315

Civelek V, Yilmaz T, Satman I, Onen S, Gumustas K, Arioglu E, Demiroglu Y, Deveim S (1991) Impaired glucose utilization as measured by 3H$_2$O production in erythrocytes of patients with type 1 diabetes mellitus. Diabetes 40 [Suppl 1]:160

Cohen RD, Iles RA (1977) Lactic acidosis: some physiological and clinical considerations. Clin Sci Mol Med 53:405–410

Cohen Y, Hirsch C (1968) Etude autoradiographique chez la souris d'un antidiabétique oral marqué au ^{14}C, le N,N-dimethylbiguanide, après administrations tépétées. Therapie 23:1185–1191

Colantuoni A, Bertuglia S, Donato L (1988) Effects of metformin on microvascular permeability in diabetic Syrian hamsters. Diabete Metab 14:549–553

Comi RJ, Hamil Ton H (1994) Reduction of red cell glucose transporter intrinsic activity in diabetes running. Horm Metab Res 26:26–32

Conget JI, Sarri Y, Gonzalez-Clemente JM, Gomis R, Malaisse WJ (1991) Impaired utilization of glucose by insulin-insensitive cells: study in erythrocytes from normal and diabetic subjects. Diabetologia 34:P471

Cook DE (1978) The effects of phenformin in normal vs diabetic isolated perfused rat liver. Res Commun Chem Pathol Pharmacol 22:119–134

Corbellini A, Torti G (1967) Research on the antitumoral activity of biguanides. Arch Ital Patol 10:197–210

Corsini GU, Sirigu F, Tagliamonte P, Muntoni S (1974) Effects of biguanides on fatty acid and glucose oxidation in muscle. Pharmacol Res Commun 6:253–261

Coupar IM, McColl I (1974) Glucoe absorption from the rat jejunum during acute exposure to metformin and phenformin. J Pharm Pharmacol 26:997–998

Creutzfeldt W (1972) Effects of biguanides on intestinal absorption in vivo and in vitro. Isr J Med Sci 8:691–696

Creutzfeldt W, Söling HD, Moench A, Rauh E, Bol M (1962) Die Wirkung von N1, n-butylbiguanide (W37) und N1, β-phenyläthylbiguanid (W32) auf den Alloxan- und Phlorizin-Diabetes und die intestinale Glucoseabsorption von Ratten. Naunyn Schmiedebergs Arch Exp Pathol Pharmakol 244:31–47

Cuber JC, Bosshard A, Vidal H, Vega F, Wiernsperger N, Rapin JR (1994) Metabolic and drug distribution studies do not support direct inhibitory effects of met-formin on intestinal glucose absorption. Diabete Metab 20:1–8

Curtis-Prior PB (1982) Reduction of the absorption of the fatty acid and glycerol moieties of ingested triglycerides by biguanides: a possible contribution to their antiobesity, antihypertriglyceridemic and anti-diabetes properties. Int J Obes 6:229–306

Cusi K, Consoli A (1994) Effect of metformin on glucose and lactate metabolism in NIDDM. Diabetes 43 [Suppl 1]:817

Czech M (1980) Insulin action and the regulation of hexose transport. Diabetes 29:399–409

Czyzyk A, Tawecki J, Sadowski J, Ponikowska I, Szcepanik Z (1968) Effect of biguanides on intestinal absorption of glucose. Diabetes 17:492–498

Davidoff F (1970) Parameters of biguanide action in vitro which correlate with hypoglycemic activity. Diabetes 19 [Suppl 1]:368

Daweke H, Bach I (1963) Experimental studies on the mode of action of biguanides. Metabolism 12:319–332

De Caterina R, Marchetti P, Bernini W, Giannarelli R, Giannessi D, Navalesi R (1989) The direct effects of metformin on platelet function in vitro. Eur J Pharmacol 37:211–213

De Fronzo RA, Barzilai N, Simonson DC (1991) Mechanism of metformin action in obese and lean noninsulin-dependent diabetic subjects. J Clin Endocrinol Metab 73:1294–1301

Ditschuneit H, Rott WH, Faulhaber JD (1968) Effekt von Biguaniden auf den Stoffwechsel isolierter Fettzellen. In: Oberdisse K, Daweke H, Michael G (eds) 2 Internationales Biguanid Symposium Düsseldorf 1967. Thieme, Stuttgard, pp 62–73

Duch A (1992) Etude du métabolisme intermédiaire chez le rat. Influence de l'entrainement à l'exercice musculaire de l'exposition intermittent au froid et au froid couplé à l'hypoxie, et du diabète induit par la streptozotocine associé à un traitement à la metformine. Thesis, University of Lyon

Eisner M, Berger W (1971) Biguanides and gastric emptying in man. Digestion 4:309–313

Eriksson J, Widen E, Saloranta C (1990) Metformin improves insulin sensitivity in insulin resistant normoglycemic relatives of patients with NIDDM. Diabetes 39 [Suppl 1]:434

Estrada DE, Elliott E, Zinman B, Poon I, Liu Z, Klip A, Daneman D (1994) Regulation of glucose transport and expression of GLUT 3 transporters in human circulating mononuclear cells: studies in cells from insulin-dependent diabetic and nondiabetic individuals. Metabolism 43:591–598

Fajans SS (1960) Discussion on "a new hypoglycemic agent, phenformin (DBI)." Diabetes 9:216

Fantus IG, Brosseau R (1986) Mechanism of action of metformin: insulin receptor and postreceptor effects in vitro and in vivo. J Clin Endocrinol Metab 63:898–905

Feldman RD, Bierbrier GS (1993) Insulin-mediated vasodilation: impairment with increased blood pressure and body mass. Lancet 342:707–709

Ferlito S, Del Campo F, De Vincenzo S, Damante G, Coco R, Branca S, Fichera C (1983) Effect of metformin on blood glucose, insulin and C-peptide responses to glucagon in non-insulin dependent diabetics. Farmaco 38:248–254

Ferrannini E, Bjorkman O (1986) Role of red blood cells in the regulation of blood glucose levels in man. Diabetes 35 [Suppl 1]:39

Fischer Y, Thomas J, Rösen P, Kammermeier H (1995) Action of metformin on glucose transport and glucose transporter GLUT 1 and GLUT 4 in heart muscle cells from healthy and diabetic rats. Endocrinology (in press)

Förster H, Hager E, Mehnert H (1965) Der Einfluß von Butylbiguanid im Tierversuch auf die Resorption von Glucose und Fructose. Arzneimittelforschung 15: 1340–1344

Fossati P, Fontaine P, Beuscart R, Romon M, Bourdelle-Hego MF, Lepoutre-Vaast D (1985) Les diabètes non insulino-dépendants échappant au contrôle des antidiabétiques oraux. Rev Fr Endocrinol Clin 26:105–116

Franke RP, Fuhrmann R, Schnittler HJ, Petrow W, Simons G (1988) Inhibition of human endothelial cell proliferation by metformin during states of hypoxia. Diabete Metab 14:571–574

Frayn KN (1967) Effects of metformin on insulin resistance after injury in the rat. Diabetologia 12:53–60

Frayn KN, Adnitt PI (1972) Effects of metformin on glucose uptake by isolated diaphragm from normal and diabetic rats. Biochem Pharmacol 21:3153–3162

Frayn KN, Adnitt P, Turner P (1973) The use of human skeletal muscle in vitro for biochemical and pharmacological studies of glucose uptake. Clin Sci 44:55–62

Freisleben HJ, Ruckert S, Wiernsperger N, Zimmer G (1992) The effects of glucose, insulin and metformin on the order parameters of isolated red cell membranes. An electron paramagnetic resonance spectroscopic study. Biochem Pharmacol 43:1185–1194

Frorath B, Dreyer M, Rüdiger HW (1985) Three different classes of oral antidiabetic drugs do not increase insulin binding and insulin-induced RNA synthesis in human fibroblast cultures. Res Exp Med 185:45–49

Galante P, Maerker E, Scholz R, Rett K, Herberg L, Mosthaf L, Häring HU (1993) Insulin-induced translocation of GLUT-4 in skeletal muscle of insulin-resistant Zucker rats. Exp Clin Endocrinol 101 [Suppl 2]:186–189

Galuska D, Zierath J, Thorne A, Sonnenfeld T, Wallberg-Henriksson H (1991) Metformin increases insulin-stimulated glucose transport in insulin-resistant human skeletal muscle. Diabete Metab 17 (1 bis):159–163

Gawler D, Milligan G, Houslay MD (1988) Treatment of streptozotocin diabetic rats with metformin restores the ability of insulin to inhibit adenylate cyclase activity and demonstrates that insulin does not exert this action through the inhibitory guanine nucleotide regulatory protein Gi. Biochem J 249:537–542

Ghareeb A, Btros M, Saba JA, El-Asmar F, El-Shawarby K, Wahba N (1969) Mechamism of action of biguanides on glucose metabolism. Effect of biguanides on intestinal absorption of glucose: an in vivo and in vitro study. Ain Shams Med J 20:313–322

Giannattasio G, Torti G, Ferrara G, Lugaro G (1968) Tissue distribution and excretion of N,N-dimethylbiguanides-^{14}C in mouse. Arch Ital Patol Clin Tumori 11:331–345

Gin H, Messerzchmitt C, Brottier E, Aubertin J (1985) Metformin iproves insulin resistance in type I Insulin-dependent diabetic patients. Metabolism 34:923–925

Ginsberg BH, Brown TJ, Simon I, Spector AA (1981) Effect of the membrane lipid environment on the properties of insulin receptors. Diabetes 30:773–780

Goldfine ID, Iwamoto Y, Pezzino V, Trischitta V, Purrello F, Vigneri R (1984) Effects of biguanides and sulfonylureas on insuln receptors in cultured cells. Diabetes Care 7 [Suppl 1]:54–58

Gould MK (1984) Multiple roles of ATP in the regulation of sugar transport in muscle and adipose tissue. TIBS December 1984:524–527

Gould MK Chaudry IH (1970) The action of insulin on glucose uptake by isolated rat soleus muscle. II. Dissociation of a priming effect of insulin from its stimulatory effect. Biochim Biophys Acta 215:258–263

Grant PJ, Stickland MH, Booth NA, Prentice CR (1991) Metformin causes a reduction in basal and post-venous occlusion plasminogen activator inhibitor-1 in type 2 diabetic patients. Diabetic Med 8:361–365

Gregorio F, Ambrosi F, Cristallini S, Marchetti P, Navalesi R, Brunetti P, Filipponi P (1991) Do metformin and phenformin potentiate differently β-cell response to high glucose? An in vitro study on isolated rat pancreas. Diabete Metab 17:19–28

Grigorescu F, Laurent A, Chavanieu A, Capony JP (1991) Cellular mechanism of metformin action. Diabete Metab 17:146–149

Grigorescu F, Bacara MT, Rouard M, Renard E (1994) Insuln and IGF-1 signaling in oocyte maturation. Horm Res 42:55–61

Haeckel R, Haeckel H (1972) Inhibition of gluconeogenesis from lactate by phenylethylbiguanide in the perfused guinea pig liver. Diabetologia 8:117–124

Halimi S, Rossini E, Benhamou PY, Faure P, Andre P (1994) Insulin resistance induced by post-weaning high fructose diet in Wistar rats: reversal by metformin and pentobarbital. IDF Congress, Kobe

Hall H, Ramachander G, Glassman JM (1968) Tissue distribution and excretion of phenformin in normal and diabetic animals. Ann NY Acad Sci 148:601–611

Hamann A, Benecke H, Greten H, Matthaei S (1993) Metformin increases glucose transporter protein and gene expression in human fibroblasts. Biochem Biophys Res Commun 196:382–387

Handberg A, Kayser L, Hoyer PE, Voldstedlund M, Hansen HP, Vinten J (1993) Metformin ameliorates diabetes but does not normalize the decreased GLUT 4 content in skeletal muscle of obese (fa/fa) Zucker rats. Diabetologia 36:481–486

Handberg A, Kayser L, Hoyer PE, Micheelsen J, Vinten J (1994) Elevated GLUT 1 level in crude muscle membranes from diabetic Zucker rats despite a normal GLUT 1 level in perineurial sheaths. Diabetologia 37:443–448

Hausmann L, Schubotz R (1975) Proinsulin und Insulinsekretion bei übergewichtigen Frauen vor und nach Gabe von Metformin. Arzneimittelforschung Res 25: 668–675

Herman LS, Melander A (1992) Biguanides: basic aspects and clinical uses. In: Alberti KGMM, De Fronzo RA, Keen H, Zimmet P (eds) International textbook on diabetes mellitus. Wiley, London, pp 774–795

Hertz Y, Epstein N, Abraham M, Madar Z, Hepber B, Gertler A (1989) Effects of metformin on plasma insulin glucose metabolism and protein synthesis in the common carp Cyprinus carpio L. Aquaculture 80:175–187

Hildmann W, Lippmann HG (1963) Aktivitäten von Glukose-6-phosphatase und hexokinase unter der Einwirkung von N1, n-butylbiguanid. Acta Biol Med Ger 14:345–352

Ho RS, Kelly LA (1980) Effects of two glucose absorption inhibitors: phenformin and 43–522 on hepatic gluconeogenesis. J Pharm Pharmacol 32:554–557

Hocking ED, Chakrabarti R, Evans J, Fearnley GR (1967) Effect of biguanides and Atromid on fibrinolysis. J Atheroscler Res 7:121–130

Holle A, Mangels W, Dreyer M, Kühnau J, Rüdiger H (1981) Biguanide treatment increases the number of insulin receptor sites on human erythrocytes. N Engl J Med 305:563–566

Horn Z, Palkovits M (1964) L'effet du NN-diméthylbiguanide sur la thyroïde. Therapie 19:619–623

Hother-Nielsen O, Schmitz O, Andersen PH, Beck-Nielsen H, Pedersen O (1989) Metformin improves peripheral but not hepatic insulin action in obese patients with type II diabetes. Acta Endocrinol (copenh) 120:257–265

Hotta N, Komori T, Kobayashi M, Sakakibara F, Koh N, Sakamoto N (1991) A new possible mechanism of hypoglycemic effect of biguanides. Diabetologia 34 [Suppl 2]:116

Hotta N, Mori Y, Nakamura J, Koh N, Sakakibara F, Hamada Y (1994) Effect of biguanides on alanine-induced lipid synthsis in hepatocytes of WKY-fatty rats. Diabetologia 37 [Suppl 1]:238

Hundal JS, Ramlal T, Reyes R, Leiter LA, Klip A (1992) Cellular mechanism of metformin action involves glucose tranpsorter translocation from an intracellular pool to the plasma membrane in L6 muscle cells. Endocrinology 131:1165–1173

Huupponen R, Pyykkö K, Koulu M, Rouru J (1993) Metformin and liver glycogen synthase activity in obese Zucker rats. Res Commun Chem Pathol Pharmacol 79:219–227

Intaglietta M (1988) Arteriolar vasomotion: normal physiological activity or defence mechanism. Diabete Metab 14:489–494

Jackson RA, Hawa MI, Jaspan JB, Sim BM, Disilvio L, Featherde D, Kurtz AB (1987) Mechanism of metformin action in non-insulin-dependent diabetes. Diabete 36:632–640

Jacobs D, Hayes G, Truglia J, Lockwood D (1986) Effects of metformin on insulin receptor tyrosine kinase activity in rat adipocytes. Diabetologia 29:798–801

Jalling O, Olasen C (1984) The effect of metformin compared to the effects of phenformin on the lactate production and the metabolism of isolated parenchymal rat liver cells. Acta Pharmacol Toxicol 54:327–332

Jangaard N, Pereira JN, Pinson R (1968) Metabolic effects of the biguanides and possible mechanism of action. Diabetes 17:96–104

Jeppesen J, Zhou MY, Chen YD, Reaven GM (1994) Effect of metformin on postprandial lipemia in patients with fairly to poorly controlled NIDDM. Diabetes Care 17:1093–1099

Johnson AB, Webster JM, Sum CF, Heseltine L, Argyraki M, Cooper BG, Taylor R (1993) The impact of metformin therapy on hepatic glucose production and skeletal muscle glycogen synthase activity in overweight type II diabetic patients. Metabolism 42:1217–1222

Jyothirmayi GN, Jayasundaramma B, Reddi A (1992) In vivo glycogen and lipid synthesis by various tissues from normal and metformin-treated KK mice. Res Commun Chem Pathol Pharmacol 78:113–116

Kakemi M, Sasaki H, Saeki K, Endoh M, Katayama K, Koizumi T (1983) Pharmacological effects of metformin in relation to its disposition in alloxandiabetic rats. J Pharm acobiodyn 6:71–87

Kaneko T (1965) Studies on the mode of action of hypoglycemic biguanides. j Jpn Soc Int Med 52:78–95

Kanigür-Sultuybek G, Hatemi H, Güven M, Ulutin T, Tezcan V, Ulutin ON (1993) The effect of metformin and glicazide on platelets and red blood cell glucose transport mechanisms in impaired glucose tolerance and type II diabetic patients with vasculopathy. Thromb Haemorrh Dis 7:17–21

Kato S, Kawabe T, Nakai A, Miyamoyo T, Masunaga R, Ito M, Mukuno T, Sawai T, Tanabe E, Kataoka K, Nakagawa H, Aono T, Nagasaka A (1990) Glucose uptake by erythrocytes in NIDDM. Diabetes 39 [Suppl 1]:1117

Kellet GL, Jamal A, Robertson JP, Wollen N (1984) The acute regulation of glucose absorption transport and metabolism in rat small intestine by insulin in vivo. Biochem J 219:1027–1035

Kemmer F, Berger M, Herberg L, Gries F (1977) Effects of metformin on glucose metabolism of isolated perfused rat skeletal muscle. Arzneimittelforschung 27:1573–1576

Kessler M, Meier W, Storelli C, Semenza G (1975) The biguanide inhibition of D-glucose transport in membrane vesicles from small intestinal brush borders. Biochim Biophys Acta 413:444–452

Ketekou F, Rous S, Favarger P (1969) The effect of chlorpropamide and phenethyl-biguanide on lipogenesis. Med Exp 19:1–9

Khan NA, Wiernsperger N, Quemener V, Havouis R, Moulinoux JP (1992) Characterization of metformin transport system in NIH 3T3 cells. J Cell Physiol 152:310–316

Khan NA, Wiernsperger N, Quemener V, Moulinoux JP (1994) Internalization of metformin is necessary for its action on potentiating the insulin-induced Xenopus laevis oocyte maturation. J Endocrinol 142:245–250

Kiesewetter H, Jung F, Gerhards M, Roggenkamp HG, (1987) Rheological effect of metformin on the blood of patients with dietetically controlled type-II b diabetes. Clin Hemorheol 7:781–791

Kissun R (1988) Inhibition of induced neovascularization in the rabbit cornea: a preliminary study. Diabete Metab 14:575–579

Klip A, Guma A, Ramlal T, Bilan PJ, Lam L, Leiter LA (1992) Stimulation of hexose transport by metformin in L6 muscle cells in culture. Endocrinology 130:2535–2544

Kojima S, Tanaka R, Hamada C (1976) Intestinal absorption characteristics of buformin and phenformin in rats. Chem Pharm Bull Tokyo 24:1555–1560

Komori T, Hotta N, Kobayashi M, Sakakibara F, Koh N, Sakamoto N (1993) Biguanides may produce hypoglycemic action in isolated rat hepatocytes through their effects on L-alanine transport. Diabetes Res Clin Pract 22:11–17

Koschinsky T, Bünting CE, Rütter R, Gries FA (1988) Influence of metformin on vascular cell proliferation. Diabete Metab 14:566–570

Kozka IJ, Holman GD (1993) Metformin blocks downregulation of cell surface GLUT 4 caused by chronic insulin treatment of rat adipocytes. Diabetes 42: 1159–1165

Kroneberg G, Stoepel K (1958) Untersuchungen über die Guanid-Hyperglykämie und die Beeinflussung der Adrenalinwirkung durch β-Phenyläthylbiguanid und andere Guanidin-Verbindungen. Arzneimittelforschung 8:470–475

Kühl C, Jensen L, Vagn Nielsen O, Pedersen J (1979) The effect of metformin on the arginine-induced insulin and glucagon release in pigs. Acta Pharmacol Toxicol 44:235–237

Kühnle HF, Schmidt FH, Deaciuc IV (1984) In vivo and in vitro effects of a new hypoglycemic agent, 2–3 methylcinnamylhydrazano propionate, BM 42–304, on glucose metabolism in guinea pigs. Biochem Pharmacol 33:1437–1444

Kühnle HF, Wolff HP, Schmidt FH, Reiter R (1990) Blood glucose lowering activity of 2-(3-phenylpropoxyimido-)-butyrate (BM 13677). Biochem Pharmacol 40: 1821–1825

L'age M, Stehr J, Wahl P (1963) Der Einfluß von N1-n-butylbiguanide auf das Verhalten de unveresterten Fettsäuren (UFS) bei Normalpersonen und bei Diabetikern und am epididymalen Fettgewebe der Ratte. Klin Wocharschr 41:659–662

Lacombe C, Nibbelink M (1981) Changes in the lipoproteins in rabbits on a high fat cholesterol-free diet: preventive action of metformin. Experientia 37:854–855

Landin K, Tengborn L, Smith U (1994) Effects of metformin and metoprolol CR on hormones and fibrinolytic variables during a hyperinsulinemic euglycemic clamp in man. Thromb Haemost 71:783–787

Landin-Wilhelmsen K (1992) Metformin and blood pressure. J Clin Pharmacol Ther 17:75–79

Leatherdale BA, Bailey CJ (1986) Acute antihyperglycemic effect of metformin without alteration of gastric emptying. ICRS Med Sci 14:1085–1086

Lennard MS, Casey C, Tucker GT, Woods HP (1978) Determination of metformin in biological samples. Br J Clin Pharmacol 6:183–184

Lescure B, Volfin P (1971) Importance of Mn in the regulation of gluconeogenesis in cellular and acellular kidney cortex systems: new data on mechanism of action of biguanides. Biochimie 53:391–397

Lloyd MH, Iles RA, Walton B, Hamilton CA, Cohen RD (1975) Effect of phenformin on gluconeogenesis from lactate and intracellular pH in the isolated perfused guinea-pig liver. Diabetes 24:618–624

Lorch E (1971) Inhibition of intestinal absorption and improvement of oral glucose tolerance by biguanides in the normal and in the streptozotocin-diabetic rat. Diabetologia 7:195–203

Lord JM, Puah JA, Atkins TW, Bailey CJ (1985) Postreceptor effect of metformin on insulin action in mice. J Pharm Pharmacol 37:821–823

Losert W, Schillinger E, Kraaz W, Loge O, Jahn P (1972) Tierexperimentelle Untersuchungen zur Wirkungsweise der Biguanide. Arzneimittelforschung 22: 1157–1169, 1540–1552

Loubatieres AL, Mariani MM, Jallet F (1971) The role of the pancreas and extrap-ancreatic tissues in the hypoglycemic action provoked by the biguanides. J Pharmacol 2:201–202

Love AHG (1969) The effects of biguanides on intestinal absorption. Diabetologia 5:422

Lovejoy J, Digirolamo M (1990) Acute lactate production and insulin sensitivity during intravenous glucose and insulin administration in lean and obese subjects. Diabetes 39 [Suppl 1]:1108

Lugaro G, Giannattasio G (1968) Effect of biguanides on the respiration of tumour cells. Experientia 24:794–795

Maggi G (1968) Sul mecanismo d'azione della fenetilbiguanide (PEBG). XI. Effetti sulla glicolisi e sulla glicogenosintesi nell'animale normale. Boll Soc Ital Biol Sper 44:155–159

Mainguet P, Lavaux JP, Franckson J (1972) Study of intestinal glucose absorption in diabetic patients after acute administration of dimethylbiguanide. Diabete 20:39–42

Manley SE, Mussett S, Sutton PJ, Morris ER, Trinick TR, Cull CA, Holman RR, Turner RC (1994) Total proinsulin and specific insulin in type 2 diabetic patients randomized to diet, sulphonylurea, insulin and metformin therapy Diabetic Med 11 [Suppl 2]:P34

Marchetti P, Mastello P, Benzi L, Cecchetti P, Fierabracci V, Giannarelli R, Gregorio F, Brunetti P, Navalesi R (1989) Effects of metformin therapy on plasma amino acid pattern in patients with maturity onset diabetes. Drugs Exp Clin Res 15:565–570

Marquie G, Lafontan M (1974) Inhibition par le NN-dimethylbiguanide de la biosyn-thèse "in vivo" des lipides dans l'aorte de lapin normal à partir de l'acétate 14C. J Physiol (paris) 69:271A–272A

Massad L, Plotkine M, Alix M, Boulu RG (1988) Antithrombotic drugs in a carotid occlusion model: beneficial effects of the antidiabetic agent metformin. Diabete Metab 14:544–548

Matthaei S, Greten H (1991) Evidence that metformin ameliorates cellular insulin resistance by potentiating insulin-induced translocation of glucose transporters to the plasma membrane. Diabete Metab 17 (1 bis):150–158

Matthaei S, Reibold JP, Hamann A, Klein HH, Greten H (1989) Effect of in vivo metformin-treatment on insulin resistance in the obese (fa/fa) Zucker rat. Diabetes 38 [Suppl 2]:855

Matthaei S, Hamann A, Klein HH, Benecke H, Kreymann G, Flier JS, Greten H (1991) Association of metformin's effect to increase insulin-stimulated glucose transport with potentiation of insulin-induced translocation of glucose transpor-ters from intracellular pool to plasma membrane in rat adipocytes. Diabetes 40:850–875

Matthaei S, Reibold JP, Hamann A, Benecke H, Häring HU, Greten H, Klein HH (1993) In vivo metformin treatment ameliorates insulin resistance: evidence for potentiation of insulin-induced translocation and increased functional activity of glucose transporters in obese (fa/fa) rat adipocytes. Endocrinology 133:304–311

May JM, Danzo BJ (1988) Photolabeling of the human erythrocyte glucose carrier with androgenic steroids. Biochim Biophys Acta 943:199–210

Maziere JC, Maziere C, Gardette J, Salmon S, Auclair M Polonowski J (1988) The antidiabetic drug metformin decreases cholesterol metabolism in cultured human fibroblasts. Atherosclerosis 71:27–33

Medina JM, Sanchez-Medina Mayor F (1971) Effect of phenformin on gluconeo-genesis in perfused rat liver. Rev Esp Fisiol 27:253–256

Mehnert H, Schäfer G, Kaliampetsos G, Stuhlfauth K, Engenhardt W (1962) Die Insulinsekretion des Pankreas bei extracorporaler Perfusion. II. Durchströmun-gen der Bauchspeicheldrüse mit Pertison, Glucose, Carbutamid und Biguaniden. Klin Wochen schr 40:1146–1151

Melin B, Cherqui G, Blivet MJ, Caron M, Lascols O, Capeau J, Picard J (1990) Dual effect of metformin in cultured rat hepatocytes: potentiation of insulin action and prevention of insulin-induced resistance. Metabolism 39:1089–1095

Meyer F (1960) Etude sur le mode d'action des biguanides hypoglycémiants. C R Acad Sci [III] 251:1928–1930

Meyer F, Ipaktchi M, Clauser H (1967) Specific inhibition of gluconeogenesis by biguanides. Nature January 14:203–204

Molloy AM, Ardill J, Tomkin GH (1980) The effect of metformin treatment on gastric acid secretion and gastrointestinal hormone levels in normal subjects. Diabetologia 19:93–96

Moore CX, Cooper GJ (1991) Co-secretion of amylin and insulin from cultured islet beta-cells: modulation by nutrient secreta-gogues, islet hormones and hypoglycemic agents. Biochem Biophys Res Commun 179:1–9

Moore UM, Tighe OP, Collins PB, Johnson AH, Tomkin GH (1988) The in vitro effect of insulin and metformin on cholesterol biosynthesis in cultured enterocytes from diabetic rats. Diabetologia 31:524A

Moore UM, Lyons D, Tighe OP, Johnson AH, Tomkin GH, Collins PB (1990) Metformin-induced modulation of serum lipids is not reflected in leucocyte cholesterogenic rates. Diabetologia 33:A205

Morgan DA, Ray CA, Balon TW, Mark AL (1992) Metformin increases insulin sensitivity and lowers arterial pressure in spontaneously hypertensive rats. Clin Res 40:740A

Mountjoy KG, Finlay GJ, Holdaway IM (1987) Effects of metformin and glibenclamide on insulin receptors in fibroblasts and tumor cells in vitro. J Endocrinol Invest 10:553–557

Muntoni S (1974) Inhibition of fatty acid oxidation by biguanides: implications for metabolic physiopathology. Adv Lipid Res 12:311–377

Muntoni S, Tagliamonte P, Pintus F (1978) Metformin and plasma free fatty acid turnover in man. In: Carlson LA (ed) International Conference on Atherosclerosis. Raven, New York, pp 333–338

Muntoni S Tagliamonte P, Sirigu F, Corsini GU (1973) Demonstration of the mechanism of action of biguanides. Acta Diabetol Lat 10:1300–1307

Nagi DK, Yudkin JS (1993) Effects of metformin on insulin resistance risk factors for cardiovascular disease, and plasminogen activator inhibitor in NIDDM subjects. Diabetes Care 16:621–629

Nair CH, Azhar A, Wilson JD, Dhall DP (1991) Studies on fibrin network structure in human plasma. Part II, Clinical application: diabetes and antidiabetic drugs. Thromb Res 64:477–485

Neumann RE, Tytell AA (1962) Stimulated glycolysis of KB cell cultures by guanidine derivatives and other compounds affecting respiration. Proc Soc Exp Biol 110:622–626

Nicholls TJ, Leese HJ (1984) The effects of phenformin on the transport and metabolism of sugars by the rat small intestine. Biochem Pharmacol 33:771–777

Noel M (1979) Kinetic study of normal and sustained release dosage forms of metformin in normal subjects. Res Clin Forum 1:35–43

Ohnhaus EE, Berger W, Nars PW (1978) The effect of different doses of dimethyl-lniguanide on liver blood blow, blood glucose and plasma immunoreactive insulin in anesthetized rats. Biochem Pharmacol 27:789–793

Owen MR, Halestrap AP (1992) The inhibition of gluconeogenesis by mild respiratory chain inhibitors suggests a possible mode of action for the biguanide hypoglycemic agents. Diabetic Med 9 [Suppl 1]:A29

Parisi R, Cavaliere R, Innocenti M, Porta M (1979) Modifications de l'agrégation plaquettaire induites par la metformine chez les diabétiques. Gaz Med Fr 86:169–172

Pavel I, Sdrobici D, Chisiu N, Mihalache N, Tanasescu N, Bonaparte H (1964) Recherches concernant le mécanisme d'action du dimethylbiguanide sur le métabolisme du glucose chez le rat diabétique alloxanisé. Rev Roum Med 1:361–367

Pears JS, Jung RT, Burchell A (1990) Contrasting effects of metformin on glucose-6-phosphatase from fed and diabetic livers. Diabetic Med 7 [Suppl 2]:12A

Pedersen O, Hother-Nielsen O, Bak J, Richelsen B, Beck-nielsen H, Schwartz-Soerensen N (1988) The effects of metformin on adipocyte insulin action and metabolic control in obese subjects with type 2 diabetes. Diabetic Med 6:249–256

Penicaud L, Hittier Y, Ferre P, Girard J (1989) Hypoglycemic effect of metformin in genetically obese (fa/fa) rats results from an increased utilization of blood glucose by intestine. Biochem J 262:881–885

Perriello G, Misericordia P, Volpi E, Santucci A, Santucci C, Ferrannini E, Ventura M, Santeusanio F, Brunetti P, Bolli GB (1994) Acute antihyperglycemic mechanisms of metformin in NIDDM. Evidence for suppression of lipid oxidation and hepatic glucose production. Diabetes 43:920–928

Pilch PF, Thompson PA, Czech MP (1980) Coordinate modulation of D-glucose transport activity and bilayer fluidity in plasma membranes derived from control and insulin-treated adipocytes. Proc Natl Acad Sci USA 77:915–918

Pournaras C, Strommer K, Tsacopoulos M, Gilodi N (1988) Experimental branch vein occlusion in miniature pigs: effects of metformin on the evolution of the ischemic microangiopathy. Diabete Metab 14:580–586

Proske G, Osterlob G, Beckmann R, Lagler F, Micheal G, Mückler H (1962) Tierexperimentelle Untersuchungen mit blutzuckerwirksamen Biguaniden. Arzneimittelforschung 12:314–318

Purrello F, Gullo D, Brunetti A, Buscema M, Italia S, Goldfien I, Vigneri R (1987) Direct effects of biguanides on glucose utilization in vitro. Metabolism 36:774–776

Purrello F, Gullo D, Buscema N, Pezzino V, Vigneri R, Goldfine ID, (1988) Metformin enhances certain insulin actions in cultured rat hepatoma cells. Diabetologia 31:385–389

Rapin JR, Lamproglou I, Jacques W, Leponcin M (1988a) Effects of metformin on metabolic indices of cerebral and peripheral ischemia. Diabete Metab 14:587–590

Rapin JR, Lespinasse P, Yoa RG, Raymon L, Vaillant G, Brun JM (1988b) Effects of metformin on erythrocyte deformability in the presence of insulin. In vitro study on erythrocytes from diabetic patients. Diabete Metab 14:610–612

Rapin JR, Lespinasse C, Yoa R, Wiernsperger N (1991) Erythrocyte glucose consumption in insulin-dependent diabetes: effect of Metformin in vitro. Diabete Metab 17:164–167

Read BD, Mcelhaney RN (1976) Influence of membrane lipid fluidity on glucose and uridine facilitated diffusion in human erythrocytes. Biochim Biophys Acta 419:331–341

Reddi AS, Jyothirmayi GN (1992) Effect of chronic metformin treatment on hepatic and muscle glycogen metabolism in KK mice. Biochem Med Metab Biol 47:124–132

Reddi AS, Jyothirmayi GN (1993) Effect of metformin treatment on glucose tolerance and glomerulosclerosis in KK mice. Diabete Metab 19:44–51

Regoeczi E, Walton PI (1967) Metabolism of ^{125}I-fibrinogen in normal monkeys and in those with pharmacologically induced plasminogen activator release. Clin Sci 33:559–568

Riccio A, Del Prato S, Vigili DE, Kreutzenberg S, Tiengo A (1991) Glucose and lipid metabolism in non-insulin dependent diabetes. Effect of metformin. Diabete Metab 17 (1bis):180–184

Rikimaru M, Nishikawa T, Shimizu Y, Ishida N (1965) Relationship between tissue culture cytotoxicity and acute toxicity in mice of biguanide derivatives. J Antibiot 18:196–199

Rossetti L, De Fronzo R, Gherzi R, Stein P, Andraghetti G, Falzetti G, Shulman G, Klein-Robbenhaar E, Cordera R (1990) Effect of metformin treatment on insulin action in diabetic rats: in vivo and in vitro correlations. Metabolism 39:425–435

Saleh S (1974) Studies on the effect of oral hypoglycemic agents on hepatic glycogenolysis. Pharmacol Res Commun 6:539–550

Sandor A, Kerner J, Alkonyi I (1979) Role of carnitine in promoting the effect of antidiabetic biguanides on hepatic ketogenesis. Biochem Pharmacol 28:969–974

Sarabia V, Lam L, Burdett E, Leiter LA, Klip A (1992) Glucose transport in human skeletal muscle cells in culture. J Clin Invest 90:1386–1395

Sasson S, Gorowitz N, Boukobza-Vardi N, Cerasi E, King GL, Kaiser N (1994) Regulation of the hexose transport system in vascular cells by glucose and metformin Diabetologia 37 [Suppl 1]:237

Schäfer G (1976) On the mechanism of action of hypoglycemia-producing biguanides. A reevaluation and a molecular theory. Biochem Pharmacol 25:2005–2014

Schäfer G (1979) Biguanides: molecular mode of action. Res Clin Forum 1:21–32

Schäfer G (1983) Biguanides: a review of history, pharmacodynamics and therapy. Diabete Metab 9:148–163

Schäfer G, Bojanowski D (1972) Interaction of biguanides with mitochondrial and synthetic membranes. Eur J Biochem 27:364–375

Schäfer G, Mehnert H (1962) Vergleichende Untersuchungen zur Wirkung von Biguaniden auf die Glukoseoxydation am epididymalen Fettanhang der Ratte und am subcutanen Fettgewebe des Menschen. Klin Wochenschr 12:654–655

Schatz H, Katsilambros N, Nirele C, Pfeiffer EE (1972) The effect of biguanides on secretion and biosynthesis of insulin in isolated islets of rats. Diabetologia 8:402–407

Schillinger E, Kraaz W, Loge O, Jahn P, Losert W (1970) Verstärkung der hypoglykämischen Wirkung von Insulin durch Buformin bei Ratten. Naunyn Schmiedebergs Arch Exp Pathol Pharmakol 266:437–438

Schlienger JL, Frick A, Marbach J, Fruend H, Imler M (1979) Effects of biguanides on the intermediate metabolism of glucose in normal and portal-structured rats. Diabete Metab 5:5–9

Schönborn J, Heim K, Rabast U, Kasper M (1975) Oxidation rate of plasma FFA in maturity onset diabetes. Effects of metformin. Diabetologia 11:246

Scott LM, Tomkin GH (1983) Changes in hepatic and intestinal cholesterol regulatory enzymes. The influence of metformin. Biochem Pharmacol 32:827–830

Sgambato S, Varricchio M, Tesaura P, Passariello N, Carbone L (1980) L'uso della metformina nella cardiopatia ischemica. Clin Ter 94:77–85

Shaw S, Jayatilleke E, Bauman W, Herbert V (1993) Mechanism of B12 malabsorption and depletion due to metformin discovered by using serial serum holotranscobalamin II (HolotCII) (B12 on TCII) as a surrogate for serial Schilling tests. Blood 82 [Suppl 1]:423A

Shepherd M, Kushwaha R (1994) Effect of metformin on basal and postprandial lipid and carbohydrate metabolism in NIDDM subjects. Diabetes 43 [Suppl 1]:245

Sirtori CR, Catapano A, Ghiselli GC, Innocenti AL, Rodriguez J (1977) Metformin: an antiatherosclerotic agent modifying very low density lipoproteins in rabbits. Atherosclerosis 26:79–89

Sirtori CR, Franceschini G, Galli-Kienle M, Clghetti G, Galli G, Bondioli A, Conti F (1978) Disposition of metformin in man. Clin Pharmacol Ther 24:686–693

Sirtori CR, Franceschini G, Gianfranceschi G (1984) Metformin improves peripheral vascular flow in nonhyperlipidemic patients with arterial disease. J Cardiovasc Pharmacol 6:914–923

Sirtori CR, Manzoni C, Lovati MR (1991) Mechanisms of lipid-lowering agents. Cardiology 78:226–235

Söling HD, Werchau H, Creutzfeldt W (1963) Untersuchungen zur Stoffwechselwirkung von blutzuckersenkenden Biguaniden bei verschiedenen Tierspezies. Naunyn Schmiedebergs Arch Exp Pathol Pharmakol 244:290–310

Son HS (1992) Effects of metformin on glucose-stimulated insulin secretion in the perfused rat pancreas. J Cathol Med Coll 45:133–140

Sterne J (1964) The present state of knowledge on the mode of action of the antidiabetic biguanides. Metabolism 13:791–798

Sterne J (1969) Pharmacology and mode of action of the hypoglycemic guanidine derivatives. In: Campbell GD (ed) Oral hypoglycemic agents. Academic, New York, pp 193–245

Stone D, Brown JD (1968) In vitro effects of phenformin hydrochloride: observations using isolated fat cells. Ann NY Acad Sci 148:623–630

Stout RW, Brunzell JD, Porte D, Bierman EL (1974) Effect of phenformin on lipid transport in hypertriglyceridemia. Metabolism 23:815–828

Strohfeldt P, Ehrhardt M, Kettl H, Weinges KF (1972) Experimental studies on the glycogen content and the incorporation of radioglucose into the glycogen of the diaphragm of rats after initial and short term administration of buformin. Diabetologia 8:37–40

Strohfeldt P, Kettl H, Obermaier U, Weinges KF (1975) Immediate effects of buformin on muscle metabolism. Horm Metab Res 7:355

Strohfeldt P, Strubel-Obermaier U, Kettl H (1977) Effect of buformin on the regulation of glycogen metabolism in the skeletal muscle of normal rats. Arzneimittelforschung 27:1034–1036

Sum CF, Webster JM, Johnson AB, Catalano C, Cooper BG, Taylor R (1992) The effect of intravenous metformin on glucose metabolism during hyperglycemia in type 2 diabetes. Diabetic Med 9:61–65

Tagliamonte P, Sirigu F, Corsini GU, Muntoni S (1973) Influence of phenformin on plasma FFA turnover in man. Riv Farmacol Ter 4:151–157

Tessari P, Biolo G, Bruttomesso D, Inchiostro S, Panebianco G, Fongher C, Sabadin L, Vettore M, Carlini M, Tiengo A (1990) Metformin treatment does not affect amino acid metabolism in type II diabetes. Diabetologia 33:A126

Thorne DP, Lockwood D (1990) Effects of insulin biguanide antihyperglycemic agents and β-adrenergic agonists on pathways of myocardial proteolysis. Biochem J 266:713–718

Tomkin GH (1973) Malabsorption of vitamin B12 in diabetic patients treated with Phenformin: a comparison with metformin. Br Med J 3:673–675

Tranquada R, Kleeman C, Brown J (1959) The acute effect of DBI on human hepatic intermediary metabolism. Clin Res 7:110–111

Tranquada RE, Kleeman C, Brown J (1960) Some effects of phenethylbiguanide on human hepatic metabolism as measured by hepatic vein catheterization. Diabetes 9:207–214

Tremoli E, Ghiselli G, Maderna P, Colli S, Sirtori CR (1982) Metformin reduces platelet hypersensitivity in hypercholesterolemic rabbits. Atherosclerosis 41:53–60

Turnbow MA, Smith LK, Garner CW (1995) The oxazolidinedione CP-92, 768-2 partially protects IRS-1 from dexamethasone down-regulation in 3T3-L1 adipocytes. Endocrinology 136 (in press)

Ubl JJ, Chen S, Stucki JW (1993) Inhibition of hormone-induced cytosolic (Ca^{++}) oscillations by biguanides in single rat hepatocytes. Experientia 49:260

Ubl JJ, Chen S, Stucki JW (1994) Antidiabetic biguanides inhibit hormone-induced intracellular Ca^{++} concentration oscillations in rat hepatocytes. Biochem J 304:561–567

Vague P, Juhan-Vague I, Alessi MC, Badier C, Valadier J (1987) Metformin decreases the high phasminogen activator inhibition capacity, plasma insulin and triglyceride levels in non-diabetic obese subjects. Thromb Haemost 57:326–328

Valensi P, Attalah M, Behar A, Attali JR (1994) Capillary permeability in diabetes. Sang Thromb Vaisseaux 6:473–481

Velazquez EM, Mendoza S, Glueck CJ, Hamer T, Sosa F (1993) Metformin in polycystic ovary syndrome reduces hyperinsulinemia, hyperandrogenemia and systolic blood pressure, allowing normal menses and pregnancy. Circulation 88 [Suppl]:1928

Verma S, Bhanot S, McNeill JH (1994) Metformin decreases plasma insulin levels and systolic blood pressure in spontaneously hypertensive rats. Am J Physiol 267:H1250–H1253

Vidon N, Chaussade S, Noel M, Franchisseur C, Huchet B, Bernier JJ (1988) Metformin in the digestive tract. Diabetes Res Clin Pract 4:223–229

Weichert W, Breddin HK (1988) Antithrombotic effects of metformin in laser injured arteries. Diabete Metab 14:540–543

Wick AN, Stewart CJ, Serif GS (1960) Tissue distribution of ^{14}C-labeled beta-phenethylbiguanide. Diabetes 9:163–166

Widen EIM, Eriksson JG, Groop LC (1992) Metformin normalizes nonoxidative glucose metabolism in insulin-resistant normoglycemic first-degree relatives of patients with NIDDM. Diabetes 41:354–358

Wiernsperger N (1994) Vascular defects in the aetiology of peripheral insulin resistance in diabetes. A critical review of hypotheses and facts. Diabetes Metab Rev 10:287–307

Wiernsperger N, Rapin JR (1995) Metformin-insulin interactions: from organ to cerll. Diabetes Metab Rev 11 (Suppl 1):S3–S12

Wilcock C, Bailey CJ (1990) Sites of Metformin-stimulated glucose metabolism. Biochem Pharmacol 39:1831–1834

Wilcock C, Bailey CJ (1991) Reconsideration of inhibitory effect of metformin on intestinal glucose absorption. J Pharm Pharmacol 43:120–121

Wilcock C, Bailey CJ (1994) Accumulation of metformin by tissues of the normal and diabetic mouse. Xenobiotica 24:49–57

Wilcock C, Wyre ND, Bailey CJ (1991) Subcellular distribution of metformin in rat liver. J Pharm Pharmacol 43:442–444

Williams RH, Tyberghein J, Hyde PM, Nielsen RL (1957) Studies related to the hypoglycemic action of phenethylbiguanide. Metabolism 6:311–319

Williams RH, Tanner DC, Odell WD (1958) Hypoglycemic actions of phenethyl-, amyl-, and isoamyldiguanide. Diabetes 7:87–92

Williamson JR, Walker RS, Renold AE (1963) Metabolic effects of phenethylbiguanide (DBI) on the isolated perfused rat heart. Metabolism 12:1141–1152

Wilson AP, Nathan M, Betteridge DJ (1986) Effect of metformin on platelet aggregation and prostanoid generation in vitro. Diabetologia 29 [Suppl]:607A

Wollen N, Bailey CJ (1988) Inhibition of hepatic gluconeogenesis by metformin. Biochem Pharmacol 37:4343–4358

Worm D, Handberg A, Vinten J, Beck-Nielsen H (1994) A method for measurement of muscle phosphotyrosine phosphatase activity towards the human insulin receptor applied on insulin resistant obese rats. 15th IDF Congress Kobe

Yoa RG, Rapin JR, Wiernsperger N, Martinand A, Belleville I (1993) Demonstration of defective glucose uptake and storage in erythrocytes from non-insulin dependent diabetic patients and effects of metformin. Clin Exp Pharmacol Physiol 20:563–567

Yoh YA (1967) Distribution of n-butylbiguanide-C hydrochloride in mouse tissues. Jpn J Pharmacol 17:439–449

Yu B, Pugazhenti S, Khandelwal RL (1994) Effects of metformin on glucose and glucagon regulated gluconeogenesis in cultured normal and diabetic hepatocytes. Biochem Pharmacol 48:949–954

Zaibi MS, Rapin JR, Wiernsperger N, Padieu P (1994) Metformin increases albumin production of primary cultured rat hepatocytes in a hormone-selective manner. Diabetologia 37 [Suppl 1]:631

Zamman HY, Soltis EE, Sowers JR, Peuler JD (1994) Effects of insulin-sensitizing drugs on arterial contractile responses. Hypertension 24:P97

Zarjevski N, Doyle P, Jeanrenaud B (1992) Muscle insulin resistance may not be a primary etiological factor in the genetically obese fa/fa rat. Endocrinology 130:1564–1570

Zavaroni I, Dall'aglio E, Coscelli C (1981) Effect of metformin on dietary induced hypertriglyceridemia in the rat. Diabetologia 21:345

Zavaroni I, Dall'aglio E, Bruschi F, Alpi O, Coscelli C, Butturini U (1984) Inhibition of carbohydrate induced hypertriglyceridemia by Metformin. Horm Metab Res 16:85–87

Zemel MB, Reddy S, Shehin S, Lockette W, Sowers JR (1990) Vascular reactivity in Zucker obese rats: role of insulin resistance. J Vasc Med Biol 2:81–85

Zhang Z, Wiernsperger N, Radziuk J (1994) Metformin reduces both lactate uptake and gluconeogenesis by the perfused rat liver. Diabetologia 37 [Suppl 1]:239

CHAPTER 13

Toxicology of Biguanides

F. Schmidt, F. Hartig, W. Rebel, and P. Ochlich

A. Introduction

The use of biguanides as oral antidiabetic agents has elicited considerable problems due to their lack of sufficient tolerability. The first derivative, phenylethylbiguanide (phenformin), was given marketing approval in the United States in 1957. Due to the serious side effects, however, the compound was withdrawn from the market in the United States and the major European countries about 20 years later. Similar problems occurred with N-butylbiguanide (buformin), which was given market approval in Europe in 1960. N-Dimethylbiguanide (metformin) is a clearly less active agent from this class, but it is the only derivative which is approved in a number of countries including the United States for therapeutic application. Diabetes research into guanidine derivatives focused from 1956 nearly exclusively on biguanides. Analogues of special interest are as follows:

1. N_1,n-Amylbiguanide (DBB, AB4)
2. N_1-Iso-amylbiguanide
3. N_1-Phenethylbiguanide (phenformin, DBI)
4. $N_1 n$-Butylbiguanide (buformin)
5. N_1,N_1-Dimethylbiguanide (metformin, LA6023)

While the first two substances had no therapeutic impact as antidiabetic agents, phenformin, metformin and buformin have been investigated in more detail.

B. General Pharmacology and Toxicology

Biguanides exert markedly different hypoglycemic activities in various animal species, as shown in Table 1. The reason for these differences may be, on the one hand, a different species-dependent pattern of hormones and enzymes which are involved in the maintenance of blood glucose homoeostasis. On the other hand, there are indications of distinct differences between various animal species in the pharmacokinetics and also the route of administration of biguanides. Very high doses of biguanides have to be administered orally, in contrast to parenteral injection, to achieve blood glucose lowering effects, e.g., the intestinal absorption rate of phenformin is

Table 1. Blood glucose effects in different animal species (*minimal* blood glucose decrease of 20%) after parenteral administration (modified from Beckmann 1971)

Species	Administration	Phenformin (mg/kg)	Buformin (mg/kg)	Metformin (mg/kg)
Mouse	s.c.	200	60	100
	i.v.	20	100	200
Rat	s.c.	75	30–75	350
	i.v.			
Guinea pig	i.p.	15	25	100
Dog	s.c.	25	25	
	i.v.		25	
Monkey	s.c.	6		

Table 2. Minimal active doses of biguanides in fasted guinea pigs after s.c. dosage and therapeutic doses in type 2 diabetics during treatment (daily dose for effective treatment)

Compound	Guinea pigs (s.c.) (mg/kg)	Humans (type 2 diabetics) (mg/patient)
Phenformin	7.5–10	50–100
Buformin	20	100–200
Metformin	75	500–2000

rather low and its metabolism to parahydroxy-phenformin occurs fairly rapidly. Thus it is small wonder that no hypoglycemic effect was seen in healthy rats but only in partly nephrectomized rats (Hrstka et al. 1977b). The most favorable animal model for testing blood glucose lowering activity of biguanides seems to be the guinea pig. The minimal effective dose of three biguanides in guinea pigs correlates well with the mean therapeutic doses in diabetic patients (Table 2). The guinea pig may also be a suitable model for discriminating between lethal hypoglycemic and toxic effects of biguanides.

C. Acute and Chronic Toxicity

I. Phenethylbiguanide (Phenformin)

1. Biochemical Deviations in Blood Serum and Liver Mitochondria

Since the time course of blood glucose and lactate concentrations of guinea pigs were inversely correlated after administration of phenformin (Fig. 1), we investigated metabolic alterations in phenformin-dosed animals after

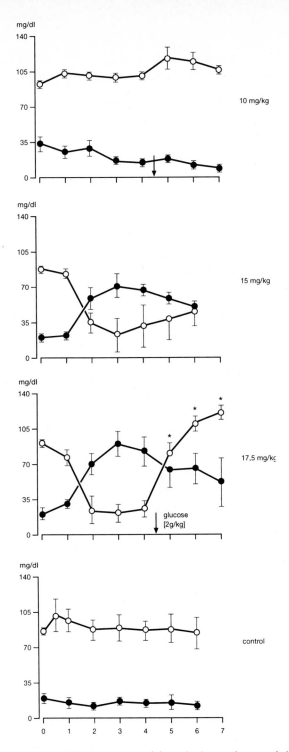

Fig. 1. Concentrations of blood glucose (o) and plasma lactate (●) after different doses of phenformin (i.p.) in male guinea pigs (mg/100 ml)

establishing a very sensitive microtechnique for lactate determinations (Schmidt et al. 1984). The following results were obtained: Intraperitoneal injection of 20 mg/kg phenformin caused a typical lactic acidotic state in guinea pigs. The isolated liver mitochondria of these animals showed severe functional damages. Table 3 shows blood pH, pO_2 and lactate in controls and pretreated animals. It demonstrated a fall in pH to 7.1, a decrease in pO_2 by 50% and an increase in lactate to sixfold higher levels. The oxidation capacity of the mitochondria dramatically decreased in comparison to that of the control. The inhibition of respiration was more pronounced with pyridine nucleotide-linked substrates than with succinate, which is oxidized in a FAD-dependent way. This shows that NADH-linked reoxidation is suppressed and the equilibration NADH/NAD is shifted to H' accumulation. Increasing the substrate concentrations in the reaction mixture had no effect on the oxidation rate, indicating that the substrate carrier systems are probably not involved in the inhibition of respiration. Albumin in the incubating media – as the binding structure for free fatty acids – was shown to increase the ADP-stimulated respiration, suggesting an additional role of free fatty acids liberated from the mitochondrial structure in the inhibited respiration. In contrast to earlier publications, no significant increase in the resting respiration due to phenformin was detected. Under in vitro conditions, phenformin caused a dose- and temperature-dependent progressive inhibition of the ADP-stimulated respiration without affecting the resting oxygen uptake (Table 4). It is shown that phenformin at nearly $0.5\,mM$, which may be required as an effective concentration, induced a reduction of respiration (10%–20%).

The results obtained with different artificial electron donors and acceptors make it probable that more than one step of the electron transport chain is inhibited by phenformin. Neither activation of the Mg-ATPase nor inhibition of DNP-stimulated ATPase of mitochondria by phenformin was detected (Somogyi et al. 1979). No significant structural damage of mitochondria from lactic acidotic animals was seen by electron microscopic examinations. It may be concluded that phenformin exerts multiple actions on mitochondria (see also Schäfer 1974).

Table 3. Blood parameters of control and phenformin in pretreated guinea pigs (average value of four animals)

		Control	Pretreated
Blood	pH	7.45	7.1
	pO_2	66.0	33.0
	pCO_2	41.0	57.0
	HCO_3	27.0	15.0
	Lactate (mmol/l)	2.5	14.1

Table 4. Oxidation capacity of isolated liver mitochondria from control and phenformin-pretreated guinea pigs

Substrate	N atoms oxygen/mg protein per minute		
	Control	Without albumin	With 10 mg/ml albumin
Malate 1 mM	8.5 ± 2.5	5.4 ± 1.8	7.8 ± 2.2
Glutamate 5 mM			
ADP 0.5 mM	90.9 ± 5.0	15.2 ± 3.4	2.4 ± 4.1
Respective control ratio	10.7	2.9	3.4
Succinate 5 mM	25.0 ± 3.5	21.9 ± 3.2	22.1 ± 2.8
ADP 0.5 mM	162.0 ± 7.5	103.1 ± 5.5	137.8 ± 5.7
Respective control ratio	6.5	4.7	6.2
Succinate 5 mM	21.5 ± 4.2	20.2 ± 3.9	18.6 ± 2.9
Rotenon 2 μM			
ADP 0.5 mM	140.4 ± 8.3	96.0 ± 6.2	107.5 ± 5.9
Respective control ratio	6.6	4.8	5.8
Malate 1 mM	10.4 ± 3.6	6.4 ± 1.5	7.6 ± 1.9
Pyruvate 1 mM			
ADP 0.5 mM	84.5 ± 6.2	13.4 ± 3.1	22.1 ± 3.3
Respective control ratio	8.1	2.1	2.9

Respiratory inhibition by phenformin may depend on several factors:

1. Rate of penetration of phenformin into the mitochondria
2. Dramatic decrease of the intracellular pH
3. Different sensitivity of the oxidation of various substrates
4. Loss of free fatty acids from mitochondrial structures

The direct inhibition of the electron transport chain is primarily involved in the inhibitory effects on the respiratory capacity and ATP production from ADP and inorganic phosphate.

2. Serum Levels and Organ Distribution (Kidneys, Duodenum, Liver and Organelles)

Guinea pigs were dosed with 20 mg/kg phenformin i.p. and were put to death 2 h later. Phenformin levels were determined by means of a specific radioimmunoassay (HRSTKA et al. 1977a) in serum and liver tissue as well as in fractions of liver tissue homogenate after ultracentrifuging, so that different subcellular compartments including cytosol, mitochondria and microsomes could be determined separately.

Phenformin serum concentrations were about 2000 ng/ml, which represents a total quantity of 8–10 μg phenformin in this compartment, accounting for about 0.1% of the total administered phenformin dose. With regard to the extracellular compartment of about 20% total body weight, this serum concentration would correspond to about 200 μg phenformin, which represents only about 4% of the total administered dose. With these

Animal 1

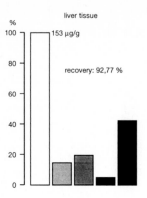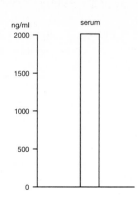

Animal 2

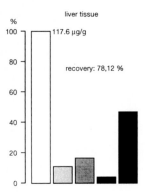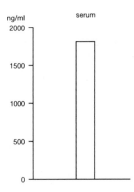

Animal 3

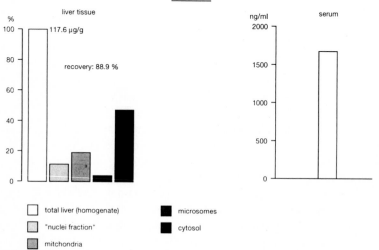

Fig. 2. Phenformin concentration in serum and hepatic tissue (total liver, "nucleic fraction," mitochondria, microsomes and cytosol) of guinea pigs after a single dose of phenformin (20 mg/kg i.p.) in three animals

considerations in mind, a major proportion of the phenformin dose should be located in different organs. Phenformin levels in liver and subcellular fractions reached levels of around $130 \mu g/g$, which is clearly above the serum levels. These results suggest that about 20%–25% of the dosed quantity is localized in the liver.

About 20% of the hepatic proportion can be demonstrated in the mitochondrial fraction, which contained phenformin at a concentration of around $10^{-4} M$ local concentration in mitochondria ex vivo (Fig. 2). Similar concentrations of the drug were found in organs and subcellular fractions (liver) of patients who died from lactic acidosis. Molar concentrations of around 10^{-4}–$10^{-3} M$ in vitro were demonstrated by STEINER and WILLIAMS (1958) to induce a considerable stimulation of anaerobic glycolysis. These findings may confirm the authors' observation that the Pasteur effect was abolished. It means that in the presence of O_2 the glycolysis is not inhibited, resulting in proton accumulation and excess lactate formation. The inhibition of hepatic gluconeogenesis by phenformin (Fig. 3) may also account for lactate accumulation.

3. Acute Toxicity

The acute toxicity of phenformin in various animal species was summarized by BECKMANN (1971) (Table 5). A number of authors did not report whether

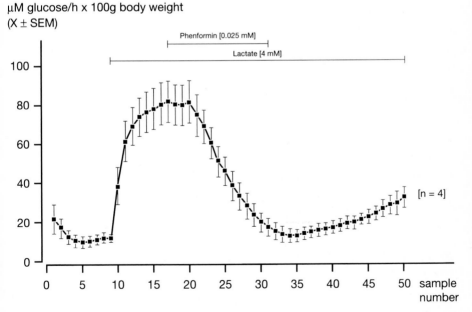

Fig. 3. Effect of phenformin $(0.02 \, \text{m}M)$ on glucose production $(\mu M/h \times 100 \, \text{g}$ body weight) from lactate in isolated perfused guinea pig liver (fasting state) (mean $\pm$ SEM)

Table 5. LD_{50} (mg/kg) of phenformin in different species after oral and parenteral administration (Beckmann 1971)

Species	Administration	LD_{50} (mg/kg)
Mouse	p.o.	450, 800, 410
	s.c.	230, 235
	i.p.	200, 150, 202
	i.v.	19, 16
Rat	p.o.	1050, 800, 938, 150, 650
	s.c.	100, 200, 88, 190
	i.p.	172
Guinea pig	p.o.	47, 37.5, 38
	s.c.	19, 16, 26
	i.v.	12
Rabbit	p.o.	100
	s.c.	About 150
Dog	p.o.	75, <50
	s.c.	48, about 50
	i.v.	About 15
Cat	p.o.	50, 63
	i.v.	14
Monkey	p.o.	15
	s.c.	12
Hamster	p.o.	1620
	s.c.	450
	i.p.	180

the LD_{50} was estimated from nonfasted or fasted animals. As shown by Scholz and Schütz (1969), this causes a marked difference: The oral LD_{50} for phenformin in rats was 2340 (2107–2588) mg/kg body weight in non-fasted animals vs. 1151 (980–1338) mg/kg body weight in fasted animals.

4. Chronic Toxicity

In a carefully performed study, Scholz and Schütz (1969) fed young rats with a formula containing phenformin, granulate and albumin VM 1025 (Internal Report, Hoechst AG, Frankfurt). In a period of 6 months, the animals (20 female and 20 male per dose) received phenformin in concentrations of 2100 ppm, 6720 ppm and 21 000 ppm as a food additive. The calculated daily intake was 42, 150 and 275 mg/kg body weight. There were no toxic alterations in organs and no reduction of body weight with the lowest and middle dosage. With the high dose (275 mg/kg), food consumption and body weight gain were reduced.

In dogs, 5, 16 and 32 mg/kg over a period of 6 months led to a body weight reduction at the highest dosage and to monocystosis. No other findings were observed, nor was there a reduction of blood glucose or toxic organ alterations (Scholz and Schütz 1969).

Table 6. LD_{50} (mg/kg) of buformin in different species after oral and parenteral administration (BECKMANN 1971)

Species	Administration	LD_{50} (mg/kg)
Mouse	p.o.	380
	s.c.	227
	i.p.	223, 213
	i.v.	105
Rat	p.o.	320, 300
Guinea pig	p.o.	58
	s.c.	23, 18
Rabbit	p.o.	>500
	i.v.	75–100
Dog	p.o.	50–100
	i.v.	About 25

a) Reproduction Toxicology, Mutagenicity

Studies of the teratogenic and embryotoxic effects and mutagenicity of phenformin are not available.

b) Carcinogenicity

Tests for carcinogenicity were performed in rats and mice at dose levels of 1500–3000 and 1200–2500 ppm in the animal diet over 78 weeks. There was no evidence of carcinogenic effects at these dose levels, which were shown to induce some toxicity in terms of decreased body weight from 26 weeks of treatment onwards (GUEST et al. 1980).

II. Butylbiguanide (Buformin)

1. Acute Toxicity

The LD_{50} of this biguanide in various animal species was summarized by BECKMANN (1971) (Table 6). No additional data have been reported.

2. Chronic Toxicity

Chronic toxicity studies have been evaluated by BECKMANN (1971). Withdrawal of buformin from the market may be the reason that no additional data have been reported.

III. Dimethylbiguanide (Metformin)

1. Acute Toxicity

The LD_{50} values of metformin in different animals species from the literature are presented in Table 7.

Table 7. LD_{50} (mg/kg) of metformin in different species
after oral and parenteral administration (Beckmann 1971)

Species	Administration	LD_{50} (mg/kg)
Mouse	p.o.	2150, 3500
	s.c.	225–500, 750
	i.p.	620
	i.v.	200
Rat	p.o.	1000
	s.c.	300
Guinea pig	p.o.	500
	s.c.	150, 146
	i.m.	200
	i.p.	200
Rabbit	p.o.	350
	s.c.	150
Chicken	s.c.	150

2. Subchronic and Chronic Toxicity

(Data were made available by Lipha Arzneimittel, Essen.) Metformin was tested in a 6-month oral study in CR-CD *rats*, which were administered doses of 120, 300, 600 and 900 mg/kg daily via the feed. It was only at the high dose in the females that a slightly more frequent vacuolization in renal tubuli could be diagnosed histopathologically.

Beagle dogs were administered during a 6-month study daily doses of 50, 100 and 150 mg/kg in two equal portions in capsules. The 100-mg/kg dose level and the higher dose induced nausea, diarrhea and effects on the central nervous system, especially convulsions. Histopathological alterations were demonstrated in the brain, myocardium, kidneys, stomach and duodenum, which was the case especially in animals from the high-dose group.

Chronic studies covering 52 weeks or longer periods were conducted in mice, rats, dogs and monkeys. *Mice* received metformin at dose levels of 150, 450 and 1500 mg/kg daily in the diet during a 52-week study. There were no compound-related effects at the low- and mid-dose level, but the high dose of 1500 mg/kg reduced body weight gain and induced enlargement of renal tubuli.

Rats (Sprague-Dawley) were administered doses of 0, 150, 300, 600 and 900 mg/kg daily in the diet during a 52-week toxicity study. Although both genders showed reduced body weight gain at the 600- and 900-mg/kg dose levels, no pathological alterations were demonstrated. In a 78-week toxicity study, rats were dosed metformin at 120, 300, 600 and 900 mg/kg in the diet. The no-effect level was determined to be smaller than 120 mg/kg daily. The main effect at higher dose levels was the reduction in body weight gain and an increased number of benign tumors of the uterus.

In a 78-week study of *dogs*, metformin was administered at dose levels of 50 and 100 mg/kg daily in gelatin capsules. The clinical signs observed were similar to those which occurred during a 6-month study; however, microscopy revealed no alterations that could be ascribed to metformin administration.

Rhesus monkeys were administered 60, 180 and 360 mg metformin/kg body weight as an aqueous solution by gavage during a 2-year study. The mid and high doses induced clinical signs of intolerance with emesis and diarrhea. There were no adverse effects at 60 mg/kg daily. It should be pointed out that six out of eight animals from the high-dose group died for unknown reasons before completion of the study.

3. Mutagenicity, Fertility and Teratogenicity

The Ames test with various strains in a number of *Salmonella typhimurium* mutants revealed no evidence of a mutagenic effect of metformin, in either the absence or presence of liver enzyme extract for metabolic activation. Metformin dose levels of 120, 300 and 600 mg/kg per day administered over 9 weeks to 12 male and 24 female rats induced no detectable effets on fertility or organ differentiation of the embryos, whereby the females were treated from day 6 to day 15 of pregnancy. Metformin was devoid of teratogenic effects in rabbits at doses of 50, 100 and 140 mg/kg daily (New Zealand rabbits, $n = 20$). The compound was administered between days 6 and 18 of pregnancy. A slight increase in the fetus resorption rate was reported by TUCHMANN-DUPLESSIS and MERCIER-PAROT (1961) for high doses of metformin.

4. Carcinogenicity Studies

In a 91-week toxicity study in mice, metformin was administered at dose levels of 150, 450 and 1500 mg/kg diet. There was a slight increment in mortality rates in male animals from the mid- and high-dose groups. There were non-neoplatic alterations of the kidneys that were ascribed to metformin administration.

The most frequent occurrence was an aggregation of cystic nephropathy in all male dose groups. These were mainly characterized by polycysts and they were often seen in combination with a shortening of the papilla and with hydronephrosis. Kidney adenomas were observed in two males from the high-dose group.

In a 2-year toxicity study in Sprague-Dawley rats, metformin doses of 150, 300, 600 and 900 mg/kg were tested by administering the compound in the diet. Elevated mortality rates occurred in males at dose levels of 300 mg/kg and upwards and in females at the high-dose level of 900 mg/kg. While necropsy revealed no differences from control, histopathological examinations revealed slight alterations of endocrine tissues. Leydig cell hyperplasia was observed at 300 mg/kg and higher dose levels. The number of Leydig cell tumors was elevated. In the females, there was a tendency toward lower occurrences of atrophic ovaries. Taken altogether, there were no occurrences of tumors that would have exceeded the normal age-related spectrum. Metformin was devoid of a direct carcinogenic effect in rats and mice and did not exert effects on the level of sexual hormones.

D. Comparative Evaluations and Critical Remarks

Great caution must be applied in the endeavor to extrapolate data from acute and chronic tolerance/toxicity studies in animals to the situation in man. An assessment of the tolerability or toxicity of biguanides in man can hardly be obtained by considering the data obtained after oral administration in acute or chronic studies with rats and dogs. The exceptional tolerance of rats with regard to biguanides may be caused in part by the pharmacokinetic behavior of these agents in this animal species. In dogs, however, biguanides are not well tolerated and did not exert blood glucose lowering correlation in acute and subacute studies with antihyperglycemic effects in acute and chronic experiments.

The only species which shows good pharmacodynamic correlation in acute and subacute studies with antihyperglycemic effects in man is the guinea pig. This also holds true in a broader sense with regard to tolerability. Since marginal overdosage of biguanides can induce lactacidosis, as was seen in tolerance experiments, the therapeutic dose is an individual dose and accordingly has to be explored carefully.

Metformin represents the least hydrophobic biguanide from the three therapeutically used agents and is consequently the compound with the lowest affinity to membrane structures. The less frequent occurrence of lactacidosis in man is considered to be due to the shorter half-life than that in the guinea pig. The prevailing pattern of side effects of metformin relates to gastrointestinal disturbances. Metformin is the only biguanide derivative which has been approved in a large number of countries for the treatment of type 2 diabetes.

References

Beckmann R (1971) Biguanide. In: Maske H (ed) Oral wirksame Antidiabetika. Springer, Berlin Heidelberg New York, pp 439–596 (Handbuch der experimentellen Pharmakologie, vol 29)
Guest D, King LJ, Margetts G, Parke DV (1980) The dose-dependent toxic effects of phenformin in the rat. Biochem Pharmacol 29:2291–2295
Hrstka VE, Pick K, Reiter J, Kühnle HF, Schmidt FH (1977a) Untersuchungen zur Pharmakokinetik von Phenformin. 12th annual meeting of the German Diabetes-Society, Homburg/Saar, 19–21 May 1977 (abstract)
Hrstka VE, Pick K, Reiter J, Kühnle HF, Schmidt FH (1977b) Investigations on the pharmacokinetics of phenformin. Diabetologia 13:403
Schäfer G (1974) Interaction of biguanides with mitochondrial and synthetic membranes. Eur J Biochem 45:57–66

Schmidt FH (1981) Labormethoden. In: Robbers H, Sauer H, Willms B (eds) Praktische Diabetologie, 2nd edn. Banaschewski, Munich, pp 44–54

Schmidt FH, Kühnle HF, von Dahl K (1984) Internal report. Medical Research, Metabolic Unit of Boehringer, Mannheim

Scholz J, Schütz E (1969) Internal report. Hoechst AG, Frankfurt

Somogyi J, Varro A, Schmidt FH, Hrstka VE, Kuehnle HF, Willig F (1979) Is mitochondrial damage the cause or the consequence of lactate acidosis induced by phenformin in guinea-pigs? In: Waldhäusl W, Alberti KGMM (eds) Congress of the International Diabetes Federation (10). Excerpta Medica, Amsterdam, pp 220–221

Steiner DF, Williams RH (1958) Respiratory inhibition and hypoglycemia by biguanides and decamethylenediguanide. Biochem Biophys Acta 30:325–340

Tuchmann-Duplessis H, Mercier-Parot L (1961) Repercussions sur la gestation et le development foetal du rat d'un hypoglycémiant le chlorhydrate de N,N-diméthylbiguanide. C R Acad Sci (Paris) 253:321–323

CHAPTER 14

Clinical Pharmacology of Biguanides

L.S. HERMANN

A. Introduction

The biguanide drugs have been used to treat hyperglycaemia in non-insulin-dependent diabetes mellitus (NIDDM) for almost 40 years (STERNE 1969; HERMANN 1979; SCHÄFER 1983; BAILEY 1992). Repeated reports of phenformin-associated lactic acidosis led to the withdrawal of this biguanide in the United States in 1977 and in most other countries. Buformin shared the same fate, but metformin escaped withdrawal, even if the use became more restricted. At the same time the University Group Diabetes Program (UGDP) investigation created a discussion about potential cardiovascular morbidity and mortality of oral antidiabetic agents. Although the findings of the UGDP have been challenged and the conclusions criticized, the trial had a considerable impact on the prescription. After the withdrawal of phenformin and buformin, clinical pharmacological research on biguanides has largely been confined to metformin (HERMANN and MELANDER 1992, 1995). The intended pharmacodynamic action of metformin is to reduce elevated blood glucose levels. The therapeutic effect is dependent on the metabolic state of the subject and counterregulatory mechanisms. Metformin and other biguanides do not reduce blood glucose concentrations in non-diabetic subjects and do not normally lower the blood glucose below the normal levels in diabetics. Hence, metformin has been labelled an *antihyperglycaemic* drug (BAILEY 1992) rather than a hypoglycaemic agent. An essential feature is the lack of any direct insulinotropic effect. The precise mechanism behind the antihyperglycaemic action of metformin is still unsettled, but several factors seem to interact in a multifactorial process. Clinical pharmacological evidence indicates amelioration of insulin resistance as an important basis for the therapeutic action, not only to reduce elevated blood glucose concentrations and improve glucose tolerance, but also to improve the dyslipidaemia of NIDDM and other abnormalities associated with insulin resistance (CAMPBELL 1990; HERMANN 1994; WIDÉN and GROOP 1994). The potential long-term clinical benefits from these pharmacodynamic effects, including a reduction of hyperinsulinaemia because of improved insulin action, remain to be settled. Metformin is included in the ongoing United Kingdom Prospective Diabetes Study (UKPDS), which compares long-term effects of different therapies in patients with newly diagnosed NIDDM (UKPDS 1995).

The final results are due to appear in 1998. The potential benefits of metformin in non-diabetic subjects with insulin resistance are still unexplored,
but some preliminary studies have been performed in subjects with impaired
glucose tolerance (IGT). Metformin has recently undergone extensive investigations in the United States and received marketing approval from the
Food and Drug Administration (FDA) at the end of 1994. No serious safety
problems were identified in the two pivotal United States studies (DeFronzo
et al. 1995). The place of metformin in therapy has recently been reviewed
(Dunn and Peters 1995).

B. Pharmacodynamics

I. Antihyperglycaemic Effect

Biguanides reduce elevated blood glucose concentrations without stimulating
insulin secretion. Compared with placebo, metformin has shown reductions
amounting to 15%–40% of pretreatment fasting glucose values (Table 1)
and 1.4–3.9 mmol/l in absolute terms (Hermann 1994). Comparative studies
with metformin and sulphonylurea are summarized in Table 2, showing the
antihyperglycaemic efficacy of metformin and sulphonylurea to be of the
same magnitude. In a comparative study metformin gave better long-term
fasting plasma glucose control than glipizide (Campbell et al. 1994). A
decrease in glycosylated haemoglobin during metformin treatment is observed alongside the reduction in fasting glucose concentrations. In a meta-
analysis of comparative studies with metformin and sulphonylurea
(Campbell and Howlett 1995), the mean reduction of glycosylated
haemoglobin was 1.2% (12.5% fall from baseline) with both therapies. For
fasting plasma glucose it was 1.3 mmol/l (14% fall from baseline) after
metformin and 1.8 mmol/l (19% fall) after sulphonylurea (NS). A minimum
of 6 weeks treatment was required in this analysis. Discontinuation of
biguanide therapy leads to deterioratoin of glucose control (Hermann et al.
1991a). Metformin reduces fasting hyperglycaemic in both obese and non-
obese NIDDM patients (Tables 1, 2). The antihyperglycaemic effect is
independent of the initial body weight and body mass index (Rains et al.
1988; Hermann et al. 1994a; Campbell et al. 1994). No correlation between
the antihyperglycaemic response to metformin and weight changes has been
found (DeFronzo et al. 1991; Nagi and Yudkin 1993; Hermann et al.
1994a), except in part of the follow-up period in the study by Campbell et
al. (1994). The decrease in fasting glucose concentrations after metformin is
correlated to initial hyperglycaemia (Campbell et al. 1994; Hermann et al.
1994a), i.e. the greater the baseline level, the greater the reduction. In our
study, the probability of achieving optimum glucose control was lower at
high baseline fasting glucose but was not affected by age, sex, duration of
diabetes, lipid status or blood pressure (Hermann et al. 1994a). High initial

Table 1. Placebo-controlled studies of metformin as monotherapy and added to sulphonylurea in NIDDM. Percentage reduction of fasting glucose versus placebo and effects on body weight, insulin, C peptide, lipids and blood pressure. See also Table 4

First author	Year	n	ob	SU	Time (months)	Design	BW	Percentage reduction fasting glucose vs. placebo	INS	CP	TG	CH	HDL	LDL	BP
HIGGINBOTHAM	79	17	+/0	GB	2	RdBX	0	20	0	–	–	–	–	–	–
PRAGER	83	10	+	0	0.5	dBX	0	37	0	0	–	–	–	–	–
RIZKALLA	86	6	+	GB	0.5	RdBX	0	15	0	–	–	–	–	–	–
JACKSON	87	10	0	0	5	RX	0	40	–	–	–	–	–	–	–
LALOR	90	19	+	0	3	RdBX	0	25	0	0	(↓)	0	0	–	0
DORNAN	91	30	+	0	8	RdBP	0	33	0	0	0	(↓)	0	↓	0
NAGY	93	27	+	0	3	RdBX	0	26	0	0	→ 0	→	→	→	–
DEFRONZO	95	143	+	0	7	RdBP	0	22	–	–	0	→	0	→	–

n, number of patients treated with metformin; ob, obese; SU, sulphonylurea; BW, body wt.; INS, insulin; CP, C peptide; TG, triglyceride; CH, total cholesterol; HDL, high-density lipoprotein cholesterol; LDL, low-density lipoprotein cholesterol; BP, blood pressure; GB, glibenclamide; R, randomized; dB, double-blind; X, crossover; P, parallel groups.

Table 2. Comparative studies of metformin versus sulphonylurea in NIDDM. Percentage reduction of fasting glucose or comparative efficacy (~). Effects on body weight, insulin, C peptide, lipids and blood pressure (Boyd et al. 1992: see Table 4)

First author	Year	n	ob	SU	Time (months)	Design	BW		Percentage reduction fasting glucose		INS		CP		TG		CH		HDL		LDL		BP	
							M	SU	M	SU	M	SU	M	SU	M	SU	M	SU	M	SU	M	SU	M	SU
Clarke	68	77	+	C	12	X	→	←	44	43 (nf)	–	–	–	–										
Clarke	77	58	0	C	12	X	→	←	49	51 (nf)	–	–	–	–										
Taylor	82	23	+	GB	12	P	→	0	39–50	45–52	–	–	–	–									0	0
Rains	88	34	+/0	GB	3	RX	0	←	23	23	–	–	–	–	→	0	(↓)→	0	0	0	→	0	–	–
McAlpine	88	21	+	GZ	3	X	→	←	M	GZ	M <GZ		–	–	0	–	→	–	0	–		0		
Collier	89	12	+	GZ	6	RP	0	←	36	48	0	0	–	–	→	0	→	0	→	0			0	0
Josephkutty	90	20	+/0	T	3	RdBX	→	←	~	T	→	(↑)	–	–	0	0	0	0	0	0			→	0
Noury	91	30	+/0	GZ	3	RP	→	←	14	18	0	0	0	0	0	0	→	0	←	0	0	0	0	0
Hermann	91	22	0	GB	6	RX	→	0	~	M	21 GB	0	←	0	0	0	0	→	0	0	0	0	0	–	–
Chan	93	12	0	GB	1	RX	0	←	22	23	0	–	0	–	0	0	→	0	0	0	0	0		
Hermann	94	19	+/0	GB	6	RdBP	→	0	36	24	0	←	0	0	0	0	→	0	0	0	(↓)→	0	(↓)→	0
Campbell	94	24	+	GP	12	RP	→	←	36	23	→	←	–	–	0	0	0	0	0	0	0	0	0	0
UKPDS	95	262	+	C/GB	36	RP	←	←	7	9/2 *	→	←	–	–	–	–	–	–	–	–			–	–

n, number of patients treated with metformin; ob, obese; SU, sulphonylurea; BW, body wt.; M, metformin; INS, insulin; CP, C peptide; TG, triglyceride; CH, total cholesterol; HDL, high-density lipoprotein cholesterol; LDL, low-density lipoprotein cholesterol; BP, blood pressure; nf, non-fasting; C, chlorpropamide; T, tolbutamide; GB, glibenclamide; GZ, gliclazide; GP, glipizide; R, randomized; dB, double-blind; X, crossover; P, parallel groups.
* $P < 0.05$.

C peptide values predicted a good antihyperglycaemic response with both metformin and glibenclamide in this study. A reduction of postprandial hyperglycaemia by metformin has been observed in studies recording diurnal plasma glucose profiles (HOTHER-NIELSEN et al. 1989; WU et al. 1990; McINTYRE et al. 1991; JEPPESEN et al. 1994). Metformin seems to exert a greater effect on postprandial than on fasting glucose concentrations (WU et al. 1990) and the effect is persistent over 24 h (McINTYRE et al. 1991). In the meta-analysis mentioned above, the mean reduction of postprandial plasma glucose was 7.3 mmol/l (44.6% fall from baseline) for both therapies. Metformin has been shown repeatedly to improve oral glucose tolerance in diabetics (CARPENTIER et al. 1975; CAPORICCI et al. 1979; PRAGER and SCHERNTHANER 1983; FANTUS and BROSSEAU 1986; JACKSON et al. 1987). In our study (HERMANN et al. 1994b) the integrated area under the curve during the 3 h of a meal stimulation test improved by 19%. Both the total and incremental area was reduced by metformin and there was no difference between metformin and the sulphonylurea. The intravenous glucose tolerance can also be improved by the drug (CAPORICCI et al. 1979). As mentioned earlier, biguanides do not lower normal blood glucose levels. In non-diabetic subjects no effects on glucose concentrations can be demonstrated unless these are artificially raised. The antihyperglycaemic effect seems to be dose related. In a dose-response study in diabetic patients, MARCHETTI et al. (1990) showed higher metformin concentrations and lower diurnal glucose concentrations at the high metformin dose (1.7 g daily) compared with a lower dose (1.0 g daily). In an earlier study these authors found no correlation between metformin levels and fasting plasma glucose concentrations and between dose and drug concentration (MARCHETTI et al. 1987). A dose-effect relationship was observed in another study (McINTYRE et al. 1991) both for fasting plasma glucose and for total 24-h area under the curve (AUC), and to some extent for the incremental AUC. Glucose utilization was also dose dependent in this study, in both the basal and insulin-stimulated condition. Our double-blind controlled trial included a dose-response study (HERMANN et al. 1994a). In a subgroup of patients with more severe diabetes, who needed higher dose or combined treatment, a dose-effect relation was seen for fasting blood glucose over the whole dose range of metformin (1–3 g daily). whereas the micronized formulation of glibenclamide used in this study had no additional effect above 3.5 mg daily.

II. Weight-Stabilizing Effect

In contrast to sulphonylureas, metformin has never been associated with an increase in body weight (Tables 1, 2). Early clinical experience showed that diabetic patients often lost weight during metformin treatment (HERMANN 1979), but weight loss is not a consistent finding in clinical trials. In the meta-analysis by CAMPBELL and HOWLETT (1994), the mean weight changes were −1.2 kg with metformin and +2.8 kg with sulphonylureas. A mean difference of $-2.6\,\text{kg}$ (95% confidence interval -4.2 to $-1.1\,\text{kg}$) between metformin and glibenclamide was found in a 12-month crossover study (HERMANN et al. 1991b). In the UKPDS, body weight remained unchanged over 3 years in the obese patients treated with metformin or diet alone, whereas a significant weight increase was seen in the sulphonylurea and insulin groups (UKPDS 1995). Metformin may induce weight reduction in non-diabetic subjects (PAPOZ et al. 1978; FENDRI et al. 1993), but not consistently (GUSTAFSON et al. 1971; GIUGLIANO et al. 1993a; LANDIN et al. 1994a). When metformin induced weight loss there was no obvious association with the extent of reduced hyperglycaemia or hyperinsulinaemia, except for a recent finding (CAMPBELL et al. 1994), as mentioned in Sect. B.I. The mechanism of the potential weight loss associated with metformin and other biguanides remains obscure. Thermogenic effects and increased futile substrate recycling have been suggested (BAILEY 1992). In a study of combined treatment with metformin and sulphonylurea (GROOP et al. 1989), lean body mass decreased slightly during this therapy as opposed to a weight gain in the insulin-treated group. The patients were non-obese with NIDDM of long duration and secondary sulphonylurea failure. The clinical significance of this finding is uncertain and conclusions cannot be generalized to the diabetic population. Energy expenditure decreased in this study after the oral combination, but was unaffected in recent studies with metformin (PERRIELLO et al. 1994; STUMVOLL et al. 1995), and in a comparative study of metformin and sulphonylurea (LESLIE et al. 1987). A recent study of women with insulin resistance and the polycystic ovary syndrome (VELAZQUEZ et al. 1994) showed a decrease in waist-hip ratio after metformin, indicating an effect on fat distribution (see also Sect. B.IV.3). However, this ratio was unchanged in diabetic patients treated with metformin (NAGI and YUDKIN 1993) and when metformin was administered to non-diabetic hypertensive subjects (LANDIN et al. 1991, 1994a). Preliminary results of the BIGPRO investigation (Biguanides and the Prevention of the Risk of Obesity) showed no change in the elevated waist-hip ratio after metformin in these non-diabetic subjects exhibiting the android type of fat distribution, but the patients lost weight after 12 months (RUDNICHI et al. 1994).

III. Lipid-Lowering Effect

1. Triglycerides, Cholesterol and Lipoproteins

It has long been known that metformin and other biguanides may reduce triglyceride and cholesterol levels, both in NIDDM and in non-diabetic subjects (HERMANN 1979; BAILEY 1992; HERMANN and MELANDER 1995), especially in patients with hyperlipidaemia (GUSTAFSON et al. 1971; CAPORICCI et al. 1979; SIRTORI et al. 1985; HOLLENBECK et al. 1991). A triglyceride-lowering effect of metformin in NIDDM patients has been observed in most studies, but could neither be confirmed in our patients, who had rather low initial values (HERMANN et al. 1991c, 1994b), nor in some other studies (Tables 1, 2, 4). Several factors may affect lipid levels in patients with NIDDM, e.g. the glycaemic control, obesity and sex, factors which might modify the response to therapy. The triglyceride-lowering effect of biguanides seems unrelated to the antihyperglycaemic effect, as it is also observed in non-diabetic subjects (SIRTORI et al. 1985). In a placebo-controlled study of non-diabetic obese women without hyperlipidaemia (VAGUE et al. 1987), metformin reduced triglyceride levels by 26%, and this decrease was associated with a reduction of hyperinsulinaemia. Similar changes were seen after metformin in hypertensive, obese women (GIUGLIANO et al. 1993a). The decrease in triglycerides in non-diabetic subjects with hyperlipoproteinaemia may be even higher (SIRTORI et al. 1985). Studies in diabetic subjects have shown that the triglyceride-lowering effect of metformin may appear only in certain subgroups, e.g. patients with elevated cholesterol concentrations (LALOR et al. 1990) or only in the more obese patients (RAINS et al. 1988). Sex differences were seen in the comparative study by TAYLOR et al. (1982). Plasma triglyceride concentrations were lowered by metformin when measured hourly during daytime (WU et al. 1990). The hypotriglyceridaemic effect of metformin is associated with compositional changes in lipoprotein particles (SCHNEIDER 1991), especially very low density lipoproteins (VLDLs) (LALOR et al. 1990; WU et al. 1990; HOLLENBECK et al. 1991). A direct effect of metformin on VLDL catabolism has been suggested (SIRTORI et al. 1985). A triglyceride-lowering effect has also been observed when metformin was combined with glipizide (REAVEN et al. 1992), but was not seen after metformin + glibenclamide in our study (HERMANN et al. 1994b). The effect of metformin on postprandial lipidaemia was highlighted in a recent study (JEPPESEN et al. 1994), which measured retinyl ester concentrations to quantify the concentration of triglyceride-rich lipoproteins of intestinal origin. Metformin also reduces total cholesterol concentrations, but to a lesser degree than the reduction of triglycerides, and not consistently (Tables 1, 2, 4). A few studies have shown decreased LDL cholesterol and others have shown increased HDL cholesterol after metformin (Tables 1, 2, 4). LDL cholesterol decreased slightly after metformin in our study, but this change was significantly different from the slight increase in the glibenclamide group only in patients concluding 6 months maintenance treatment (HERMANN et al. 1991c). The difference between groups was insignificant in the intention-to-treat analysis (HERMANN et al. 1994b). HDL cholesterol increased slightly after metformin compared with glibenclamide in our crossover study (HERMANN et al. 1991b). HDL subfractions were unchanged by metformin in a comparative study with glibenclamide (RAINS et al. 1988) and in another with glipizide (CAMPBELL et al. 1994). Even if HDL cholesterol was unchanged after adding metformin to glipizide in the study by

JEPPESEN et al. (1994), the ratio total cholesterol/HDL cholesterol was reduced. Lipid changes after metformin may be dose dependent (MARCHETTI et al. 1987).

2. Free Fatty Acids and Glycerol

Inhibition of free fatty acid (FFA) oxidation was suggested by MUNTONI (1974) as an explanation for biguanide action. An inhibitory effect of metformin on FFA oxidation was found in NIDDM patients in an early study with [^{14}C]palmitate (SCHÖNBORN et al. 1975). Over the years several investigations included determination of plasma FFA and sometimes glycerol concentration in order to evaluate lipolysis, but the results have been inconclusive (HERMANN and MELANDER 1995). After metformin, FFAs were either decreased (CARPENTIER et al. 1975) or unchanged (CAMPBELL et al. 1987). Compared with glibenclamide, phenformin and metformin were associated with slightly elevated diurnal FFA and glycerol concentrations (NATTRASS et al. 1977). These authors suggested that gluconeogenesis from glycerol was inhibited by the two biguanides. Metformin did not affect glycerol concentrations after glucose ingestion in a study by JACKSON et al. (1987). Recently there has been renewed interest in the effect of metformin on FFA (Table 4). In the study by WU et al. (1990) and in a further study by HOLLENBECK et al. (1991) in patients with mild diabetes and hypertriglyceridaemia, the improvement in glycaemic control by metformin was associated with lower daytime FFA levels, but no improvement in insulin action as judged from the clamp studies. When added to glipizide, metformin also reduced diurnal FFA levels (REAVEN et al. 1992; JEPPESEN et al. 1994) but did at the same time improve insulin-stimulated glucose disposal. A study of obese and non-obese NIDDM patients (DEFRONZO et al. 1991) measuring fasting FFA concentrations showed a decrease by metformin only in the obese group. RICCIO et al. (1991) have confirmed the early observatons of reduced FFA oxidation after metformin in the basal state but found no effect during hyperinsulinaemic clamp. Glucose metabolism was improved concurrently as shown by increased glucose oxidation in the basal state and increased non-oxidative glucose metabolism in the insulin-stimulated state. Moreover, it has recently been shown that acute oral administration of metformin reduced FFA concentrations, but not glycerol, and suppressed lipid oxidation (PERRIELLO et al. 1994). This placebo-controlled study eliminated the influence of long-term non-specific effects of the removal of glucotoxicity, which can be seen during chronic antihyperglycaemic therapy of any kind. The suppression of FFA and lipid oxidation induced by metformin correlated with suppression of hepatic glucose production (HGP). A slight increase in glucose oxidation was observed. It is well known that FFA can inhibit glucose utilization and stimulate gluconeogenesis. The various effects were seen in both obese and non-obese NIDDM patients. In a study of combined metformin + sulphonylurea treatment, GROOP et al. (1989)

found reduced basal lipid oxidation and accentuated FFA suppression during hyperinsulinaemic clamp. These findings were associated with increased glucose disposal, mainly as non-oxidative metabolism. In relatives of NIDDM patients, metformin slightly suppressed basal lipid oxidation, whereas FFA levels were unchanged compared with placebo (WIDÉN et al. 1992). Drug treatment was very short in this study, starting the day before the investigation. In a chronic study (JOHNSON et al. 1993), metformin did not change FFA, glycerol and lipid oxidation. When ketone bodies have been measured after chronic biguanide treatment, the results were inconclusive (NATTRASS et al. 1977, 1979; CAMPBELL et al. 1987; JOHNSON et al. 1993). Acute administration of metformin suppressed plasma β-hydroxybutyrate concentrations both in obese and in non-obese NIDDM patients (PERRIELLO et al. 1994). As FFA plays a role in insulin-resistant conditions, a decrease in FFA levels and FFA oxidation by metformin could have positive metabolic consequences.

IV. Insulin-Sensitizing Effect

1. Insulin Secretion

Animal experiments show that biguanides do not stimulate insulin secretion in isolated pancreatic islets. Clinical pharmacological support for this observation is derived from measurements of insulin and C peptide levels in the blood (Tables 1, 2, 4). Using an intravenous bolus dose of metformin in non-diabetic subjects, BONORA et al. (1984) found no changes in C peptide, insulin, glucagon, growth hormone or fasting blood glucose up to 30 min after the injection. In a double-blind, crossover study in NIDDM patients, the acute metabolic response to an intravenous metformin infusion was compared with the response to saline (SUM et al. 1992). Basal insulin levels and serum insulin concentrations during a hyperglycaemic clamp were the same in the two experiments, and there were no acute effects of the drug on HGP and glucose disposal by this route of administration. In the study by PERRIELLO et al. (1994), acute oral administration of metformin induced suppression of HGP, but this could not be related to increased insulin secretion as plasma C peptide and estimated portal plasma insulin concentrations were the same as with placebo. In an experiment with arginine infusion (CARPENTIER et al. 1975), metformin had no effect on plasma insulin and glucagon in diabetic patients. In the study by JACKSON et al. (1978), the B-cell response to glucose, expressed as the incremental area of insulin divided by the corresponding area of the glucose curve (after glucose ingestion), was not affected by metformin. Glucagon levels were also unchanged. Even if there is no direct drug effect on insulin secretion, it is to be expected that improvement of insulin action and reduction of hyperglycaemia by any means will lead to an improved insulin secretion by breaking the vicious circle of glucose toxicity. This was demonstrated in a study where metformin was compared with tolbutamide and diet therapy (FERNER et al. 1988). The B-cell function improved after reduction of the hyperglycaemia, irrespective of the mode of treatment. This was shown by the increase in insulin and C peptide in response to an intravenous glucose stimulus; the increase was correlated to the reduction of fasting glucose and was not associated with weight changes. The improvement of B-cell function was related to the second phase of insulin secretion. Both B-cell function and insulin action improved after 12 weeks of metformin treatment in a placebo-controlled study in a random sample of NIDDM patients from two ethnic groups (NAGI and YUDKIN 1993). The results were based on homeostatic model assessment using fasting C peptide and glucose concentrations. Fasting C peptide in itself was unchanged by treatment. The change in fasting plasma glucose concentration was correlated to the improved insulin action, and it was believed that B-cell function improved secondarily. All other studies in diabetic patients have also shown unchanged C peptide levels after metformin (Tables 1, 2, 4). C peptide and insulin in the fasting state and during a meal-stimulation test were unaffected by metformin in our study (HERMANN et al. 1994b) in contrast to the increase observed after glibenclamide at 6 months. In another comparative study (BOYD et al. 1992), the response after 6 weeks was unrelated to the mode of treatment (insulin, glibenclamide, metformin), and insulin and C peptide during an oral glucose tolerance test were unchanged. C peptide decreased after metformin in non-diabetic, hypertensive subjects (LANDIN et al. 1991, 1994a), which was believed to be secondary to a metformin-induced reduction in insulin resistance.

2. Insulin Levels

Some results have already been mentioned in the foregoing section. Table 4 shows the findings in clinical pharmacological studies using the clamp technique. After metformin, insulin levels are unchanged or reduced. In contrast to sulphonylurea, metformin never increases plasma insulin concentrations (Table 2). It appears that the insulin-lowering effect is exerted in individuals with elevated insulin levels. Only one study (NAGI and YUDKIN 1993) measured true insulin as well as immunoreactive insulin, and both were unaffected by metformin. These authors later reported reduced levels of proinsulin and split products in NIDDM patients treated with metformin (NAGI et al. 1994). Daytime insulin levels were decreased after 3 months treatment with metformin compared with gliclazide (MCALPINE et al. 1988) and in two further studies (WU et al. 1990; HOLLENBECK et al. 1991); the decrease was observed especially in the afternoon and the reduction could be as high as 30% of pretreatment values. The consistent findings of reduced hyperglycaemia without any increase, but often a decrease, in plasma insulin concentrations after metformin indicate that the drug acts on insulin resistance. Although it seems that insulin is required for the action of biguanides, no clinical studies have revealed a relationship between glucose disposal and circulating insulin concentrations. However, any such relationship could be obscured in diabetic patients due to the influence of insulin resistance (BAILEY 1993). Even if biguanides cannot substitute for insulin therapy in IDDM, it has been shown in a balance study by hepatic vein catheterization that phenformin had an effect in patients with total pancreatic insufficiency (TRANQUADA et al. 1960). Non-insulin-mediated actions of metformin have been discussed in recent years (KLIP and LEITER 1990; BAILEY 1993). In human muscle cells the effect of metformin seems to be additive to that of insulin (SARABIA et al. 1992). Two clamp studies (WU et al. 1990; HOLLENBECK et al. 1991) showed improved glycaemia in the absence of changes in insulin action, and in a further clamp study (MCINTYRE et al. 1991) it appeared that increased non-insulin-mediated glucose uptake could have contributed to the increased total glucose utilization observed. The demonstration of an effect of the drug on glucose uptake and glycogen formation in erythrocytes from diabetic patients (YOA et al. 1993) supplied further evidence for non-insulin-mediated mechanisms of metformin, as erythrocytes are insulin insensitive. However, the total body of evidence is in favour of a more significant pharmacodynamic effect on processes involved in the action of insulin.

3. Insulin Action and Insulin Resistance

By its effect on hyperglycaemia, obesity and dyslipidaemia, metformin counters the impact of insulin resistance in diabetes. The support for an amelioration of insulin resistance by metformin is summarized in Table 3. BUTTERFIELD and WICHELOW (1968) showed increased peripheral glucose uptake by phenformin in the forearm preparation. Indirect evidence for an effect of metformin on insulin resistance came from repeated observations of reduced insulin requirements in IDDM when metformin was added to insulin (GIN et al. 1985; SLAMA 1991). This was also found in NIDDM by GIUGLIANO et al. (1993b) in a placebo-controlled study (Sect. F.II). The first attempt to estimate insulin sensitivity more directly in relation to biguanide action was made by LISCH et al. (1980), who used an insulin tolerance test. In this study, a short treatment with metformin, phenformin or buformin

Table 3. Support for amelioration of insulin resistance with metformin

- Increase of insulin-stimulated peripheral glucose disposal in most clamp studies
- Reduction of hepatic glucose production in some studies
- No increase of plasma insulin levels, reduction of hyperinsulinaemia in some clinical studies
- Effect on insulin receptors and postreceptor sites in isolated tissues
- Reduction of insulin requirements in IDDM and reduced insulin dose in NIDDM
- Effect on different components of the metabolic syndrome (body weight, lipids, blood pressure, fibrinolysis)
- No direct effect on insulin secretion

increased insulin sensitivity, but only in those patients who had an adequate blood glucose response to long-term treatment. The improved insulin sensitivity was correlated to this previous response. Later, a whole range of studies on insulin receptor binding focused the interest on the effect of biguanides on insulin-mediated processes. The results showed that an effect on postreceptor sites was more important for the action of biguanides than an effect on receptors. In accordance with this, studies using the euglycaemic, hyperinsulinaemic clamp technique and infusion of tritiated glucose have confirmed that metformin is able to increase insulin-stimulated peripheral glucose disposal and decrease HGP in NIDDM patients (Table 4). However, two studies (Wu et al. 1990; HOLLENBECK et al. 1991) have failed to show an effect on peripheral glucose uptake as well as HGP. Furthermore, in the study by JACKSON et al. (1987) in non-obese diabetic men, metformin added to an ongoing sulphonylurea therapy did not improve peripheral insulin sensitivity (expressed as the initial increment of forearm glucose uptake after glucose loading in relation to insulin). This study also allowed calculation of glucose kinetics and it was shown that the disappearance of radio-labelled glucose, like the forearm glucose uptake, was not enhanced by the drug, compared with placebo. The study and a recent one (STUMVOLL et al. 1995) implicated the liver as the site of metformin action in NIDDM but did not preclude an effect on peripheral insulin resistance. Such an effect has been demonstrated in several clamp studies (Table 4) and in the study by NAGI and YUDKIN (1993), who used glucose infusion for the determination of steady-state plasma glucose and metabolic clearance rate of glucose. Both glucose utilization and insulin sensitivity were increased after 3 months by metformin compared with placebo. The improved insulin action in this study correlated with the decrease in fasting plasma glucose. Steady-state concentrations of glucose and insulin decreased after metformin in the study by JEPPESEN et al. (1994), where metformin was added to sulphonylurea. However, this improvement in insulin action was modest. Body weight did not change in these two studies. The increase in glucose utilization observed in clamp studies (Table 4) mostly amounts to 20%–30% at submaximal and maximal insulin concentrations but could be even higher under basal conditions (MCINTYRE et al. 1991). In this study it was dose related and in another (PRAGER et al. 1986) it was correlated to the decrease in fasting glucose. Concurrent with an increased glucose utilization, a reduction in HGP by metformin was seen in some studies (Table 4). In the study by BOYD et al. (1992), HGP decreased only at high initial fasting plasma glucose concentrations (>12 mmol/l) irrespective of the treatment modality, i.e. metformin, glibenclamide or insulin. No peripheral effect of these therapies was seen in this study. In the study by DEFRONZO et al. (1991), glucose utilization with euglycaemic clamp increased only in the obese group, and was associated with weight loss. A hyperglycaemic clamp was also performed in the study, showing an increased glucose disposal in both non-obese and obese patients by 25% on average, indicating that the insulin-independent mass action of

Table 4. Clamp studies of metformin as monotherapy and added to sulphonylurea in NIDDM. Percentage increase of glucose disposal (at low and high insulin concentration) by metformin, and percentage reduction of hepatic glucose output and change in glucose oxidation. Effects on body weight, insulin, C peptide and lipids

First author	Year	n	ob	SU	Time (weeks)	Design	BW	INS	CP	TG	CH	HDL	LDL	FFA	Glucose disposal	Glucose output	Glucose oxidation
PRAGER	86	12	+	GB	4	Open	0	—	—	—	—	—	—	—	←	—	—
NOSADINI	87	7	+	0	4	Open	0	→0	0	—	—	—	—	—	← 43/22	↓9	—
HOTHER-N	89	10	++	0	4	dBXPl	0	0	0	—	—	—	—	—	← 16/18	0	0
GROOP	89	12	0	GB	24	OpenRP	0	→	0	—	—	—	—	—	← 32	0	—
WU	90	12	+	0	12	Open	0	→0	0	→	→	0	0	→	0	0	0
HOLLENBECK	91	9	+/0	0	12	Open	0	→→	—	→→	→→	←	0	→/0	0	0	—
DEFRONZO	91	8/6	+	0	12	Open	↓/0	→→	—	→→	→→	←	0	—	← ob 21	↓27	—
MCINTYRE	91	9	—	0	6	Open	0	→0	0	—	—	—	—	—	← 23–29	0	—
RICCIO	91	6	0	0	4	Open	↓0	→0	0	—	←	←	—	—	0	0	—
BOYD	92	8	+	0	6	OpenRP	0	→0	0	→	→	←	0	→	← 33/31	(↓)	(↑)
REAVEN	92	13	++	GP	12	Open	0	→0	0	→→	→→	←←	0	→→	← 24	16	0
JOHNSON	93	8	+	0	12	RdBXPl	0	0	0	0	←	←	—	→	← 23	↓18	0
MARENA	94	10	+	GB	6	RdBXPl	0	0	—	→	0	←	—	—	← 29	62	—

n, number of patients treated with metformin; ob, obese; SU, sulphonylurea; BW, body weight; INS, insulin; CP, C peptide; TG, triglyceride; CH, total cholesterol; HDL, high-density lipoprotein cholesterol; LDL, low-density lipoprotein cholesterol; FFA, free fatty acids; GB, glibenclamide; GP, glipizide; R, randomized; dB, double-blind; X, crossover; P, parallel groups; Pl, placebo.

glucose to promote its own uptake was enhanced by metformin. Most clamp studies have only included obese patients, and it is noteworthy that the body weight did not change in these studies, which excludes the possibility of improved insulin action solely by weight reduction. A confounding factor is the influence of sulphonylureas in some studies, where metformin was added to an insufficient sulphonylurea therapy in some or all of the patients (Table 4). A further confounding factor is the influence of removing glucose toxicity, which was addressed by PERRIELLO et al. (1994) in the acute investigation mentioned in Sect. B.III.2. This hyperglycaemic clamp study indicated the liver as the site of action. The decrease in HGP was about 30% compared with placebo, and glucose uptake did not increase. However, an effect on glucose utilization was not ruled out as glucose oxidation increased slightly. Finally, study design is also important to consider. The euglycaemic clamp is highly reproducible but prone to observer bias. There is a wide variability in insulin sensitivity between subjects. For these reasons, a double-blind, placebo-controlled, crossover design is preferable (Table 4).

Whether the insulin-mediated effect is more important in the periphery than in the liver is not settled. The relation between peripheral and hepatic effects seems to vary in different individuals and during different experimental conditions. Reduced HGP without an increase in glucose utilization by clamp has been observed only in non-obese patients (DEFRONZO et al. 1991; BOYD et al. 1992). Using another technique, JACKSON et al. (1987) showed the same pattern of an exclusive increase in hepatic insulin sensitivity in their patients, who were also non-obese. Metformin was given in the fasting state 90 min before HGP measurements were taken, i.e. at the time of peak drug concentrations rather than at steady state. In the postabsorptive phase a peripheral effect might be more important, but metformin also improves basal glucose disposal (McINTYRE et al. 1991). The peripheral effect was exclusive in the obese patients of this study and in the very obese patients included in the study by HOTHER-NIELSEN et al. (1989). Increased peripheral glucose disposal after metformin has also been observed in a clamp study in IDDM patients (GIN et al. 1985). Hyperglycaemia is not a prerequisite for the effect of metformin on insulin resistance but could have some significance for the results seen in some investigations with the clamp technique (DEFRONZO et al. 1991; BOYD et al. 1992; PERRIELLO et al. 1994). Hyperglycaemia can compensate for the impaired glucose utilization in NIDDM by a mass action, which appears to be enforced by metformin (DEFRONZO et al. 1991). Obesity is not necessary for the action of the drug on insulin resistance but seems to have some impact, as there is a preferential hepatic effect of metformin in non-obese NIDDM patients. It is not known to what extent coexisting dyslipidaemia affects the outcome. One study (HOLLENBECK et al. 1991) included NIDDM patients with hypertriglyceridaemia and mild hyperglycaemia, but the results were not much different from a similar study in patients with lower lipid levels and higher glucose concentrations (WU et al. 1990). Glucose utilization was improved without an effect on the liver in a short-term study in non-obese, insulin-resistant, normoglycaemic, first-degree relatives of NIDDM patients (WIDÉN et al. 1992). Metformin or placebo was given twice the day before the clamp and 1h before the investigation; therefore the results of this study reflect acute effects of the drug, which are at variance with the findings in NIDDM patients by PERRIELLO et al. (1994). In a study of non-diabetic, obese women (FENDRI et al. 1993), metformin improved insulin-mediated glucose disposal, but this change was significant only in subjects who lost weight. A 2-year controlled study in men with IGT(PAPOZ et al. 1978) showed no effect of metformin and glibenclamide on glucose-stimulated insulin levels.

Insulin resistance in hypertension is well documented and hypertension is part of the metabolic insulin resistance syndrome. In an uncontrolled clinical trial, LANDIN et al. (1991) treated non-diabetic, non-obese, hypertensive men for 6 weeks with metformin in order to study insulin resistance in hypertension. There was amelioration of this condition as shown by clamp, and moreover metformin reduced the blood pressure, whereas body weight remained unchanged. These findings were not reproduced in a further controlled trial (LANDIN et al. 1994a), where metformin was compared with placebo and metoprolol. The patients in this study were not insulin resistant. However, GIUGLIANO et al. (1993a) recently confirmed an improvement of glucose utilization after metformin treatment for 3 months in non-diabetic, hypertensive, obese women. This clamp study was placebo controlled and had a double-blind, randomized crossover design. Glucose oxidation increased and lipids and blood pressure improved (Sect. B.V.1). Changes in insulin action were uncorrelated to the blood pressure reduction and lipid changes, but correlated with a decrease in fasting insulin; body weight did not change. Recently, it was found that metformin countered the insulin resistance in women with the polycystic ovary syndrome and reduced hyperandrogenaemia and blood pressure, facilitating normal menses and pregnancy (VELAZQUEZ et al. 1994). Insulin sensitivity was estimated from insulin and glucose concentrations during an oral glucose tolerance test and not by clamp. Because of this and the fact that the study was uncontrolled, the results should be interpreted with caution. In special insulin-resistant conditions, metformin may have dramatic effects (DIPAOLO et al. 1992). In conclusion, the evidence presented in this section demonstrates that metformin can ameliorate insulin resistance both in NIDDM and in non-diabetic subjects. This action is apparently not the sole explanation for the antihyperglycaemic efficacy of metformin in NIDDM, which also includes non-insulin-mediated mechanisms (Sect. B.III.2, B.IV.2). Metformin should be regarded as an "insulin sensitizer" with a direct effect on insulin action and no effect on insulin secretion. The metabolic pathways involved in this mechanism are described in Sect. C.I.

V. Vascular Effects

1. Blood Pressure and Renal Effects

A blood pressure reduction in diabetic patients treated with metformin has
been observed occasionally in uncontrolled clinical trials. HAUPT et al.
(1991) examined a great number of patients on combined therapy with metformin
+ sulphonylurea. These patients also lost weight along with the decrease in
blood pressure, triglycerides and cholesterol. Blood pressure results from
controlled studies in NIDDM patients are shown in Table 1 and 2. In a
crossover trial of metformin and glibenclamide in normotensive diabetic
patients (CHAN et al. 1993), blood pressure decreased slightly after both
therapies, but the diastolic pressure significantly more after metformin.
There was also a difference between treatments regarding systemic vascular
resistance, which increased slightly after the sulphonylurea but remained
constant after metformin. GIUGLIANO et al. (1993b) observed a blood pressure reduction after adding metformin to insulin. NAGI and YUDKIN (1993)
found no changes in blood pressure and ankle/brachial ratio after metformin,
and CAMPBELL et al. (1987) showed unchanged blood pressure and heart rate
after 6 months treatment. Reduced urinary albumin excretion (UAE) rate
was observed after metformin in the two last-mentioned studies. Whereas
this was unrelated to blood glucose changes and blood pressure in the study
by NAGI and YUDKIN (1993) of mostly normoalbuminuric patients, CAMPBELL
et al. (1994) found that the reduced UAE in their patients with elevated
levels was correlated to the improvement in glycaemic control whether by
metformin or glipizide. The unexpected blood pressure reduction after metformin observed by LANDIN et al. (1991) in non-diabetic men (Sect. B.IV.3)
was unconfirmed in the subsequent study, also in men (LANDIN et al. 1994a).
Catecholamines and other hormones were unchanged, as reported separately
(LANDIN et al. 1994b). In contrast, blood pressure and catecholamines
decreased after metformin compared with placebo in obese, hypertensive
women (GIUGLIANO et al. 1993a). Left ventricular mass index was decreased
by metformin in this study. Although blood pressure was unaffected in
another controlled study in hypertension (SEMPLICINI et al. 1993), metformin
corrected insulin resistance and increased renal sodium excretion and
glomerular filtration rate compared with placebo; fasting insulin was reduced
and a renal vasodilatation was suggested. It is not excluded that metformin
may have an antihypertensive effect in some patients with insulin resistance
with or without diabetes, but further studies are needed in order to assess
the relation between metformin and blood pressure.

2. Fibrinolysis, Haemorrheology and Thrombosis

Disturbances in blood flow properties in NIDDM are often corrected when
metabolic control is improved by any means. Early observations of an
increased fibrinolysis after metformin were confirmed by VAGUE et al. (1987)

in a placebo-controlled study of obese, non-diabetic women and in further studies in NIDDM patients (GRANT 1991; NAGI and YUDKIN 1993), all showing decreased plasminogen activator inhibitor (PAI-1) after metformin. Improved fibrinolytic activity (t-PA) was found in hypertensive patients receiving metformin (LANDIN et al. 1991, 1994a), but PAI-1 was unchanged in both these studies. The effect of the drug on PAI-1 was independent of glycaemic control and changes in insulin sensitivity in the study by NAGI and YUDKIN (1993), but it was correlated to changes in body mass index (BMI) and total cholesterol. Insulin levels were unchanged in this study. The reduction of PAI-1 in non-diabetic subjects might be related to decreased fasting insulin (VAGUE et al. 1987). Reduction of PAI-1, insulin levels and insulin dose has been observed after adding metformin to insulin in NIDDM patients (GIUGLIANO et al. 1993b). As elevated PAI-1 may be part of the mechanism whereby insulin resistance is associated with coronary heart disease, an effect of metformin on fibrinolysis could have long-term beneficial effects. Metformin may also have positive rheological effects. Red cell flow was improved by metformin in vitro, in a study with cells obtained form insulin-treated diabetic patients (BARNES et al. 1988). Except for a decrease in whole blood clogging rate, there were, however, no changes in rheological parameters (including viscosity) after metformin in a study in IDDM patients (JANSSEN et al. 1991). It has been shown that arterial blood flow improved after the drug in men with peripheral arterial disease (SIRTORI et al. 1985; MONTANARI et al. 1992). Fibrinogen is of significance for the blood flow, and it could be noted that low plasma fibrinogen concentrations, not different from those in matched controls, have been found in biguanide-treated patients, compared with other treatment groups (DESILVA et al. 1979). However, this has not been confirmed in prospective studies in NIDDM (COLLIER et al. 1989; NAGI and YUDKIN 1993). Fibrinogen decreased by 24% after metformin in non-diabetic, hypertensive women (GIUGLIANO et al. 1993a), but was unchanged by the drug in hypertensive men (LANDIN et al. 1991; 1994a). Decreased sensitivity to platelet-aggregating agents after metformin has been demonstrated in NIDDM (COLLIER et al. 1989) and IDDM (GIN et al. 1989a; JANSSEN et al. 1991). Effects on platelets and coagulation factors are difficult to evaluate as changes may be related to blood glucose control. Several platelet variables returned towards normal in the study by COLLIER et al. (1989), when glycaemic control and lipids improved by either metformin or gliclazide (Table 2). However, NAGI and YUDKIN (1993) observed no effects of metformin, either independently or secondarily to changes in glycaemic control, on spontaneous or ADP-induced platelet aggregation in whole blood or platelet-rich plasma. Moreover, there were no effects of the drug on markers of in vivo platelet aggregation (β-thromboglobulin and platelet factor 4). There is a whole range of animal experiments showing potentially useful vascular effects of the drug (BAILEY 1992).

VI. Gastrointestinal Effects

Inhibition of intestinal glucose absorption by phenformin was demonstrated in early clinical pharmacological studies (CZYZYK et al. 1968). Metformin accumulates in the tissues of the small intestine and inhibits glucose absorption in animal experiments, but such inhibition has never been confirmed in human studies. In the study by JACKSON et al. (1987) using radiolabelled glucose, the systemic appearance of ingested glucose was unaffected by metformin in NIDDM patients. Thus, metformin had no effect on the total amount of glucose absorbed. By looking at the data, it seems, however, that the rate of glucose absorption after the oral challenge was slightly decreased. A possible delay of glucose absorption may help to reduce the glycaemic impact of a meal, but seems insignificant for the antihyperglycaemic effect of metformin in NIDDM. Metformin increases intestinal glucose utilization in animal experiments (BAILEY 1992). This has been proposed as a source of lactate formation supplying gluconeogenic substrate to the liver. In this way the inherent effect of the drug to reduce hepatic gluconeogenesis may be modified and clinical hypoglycaemia prevented. As mentioned in Sect. C.I, phenformin increases gluconeogenesis and glucose-lactate interconversions in non-diabetic subjects. In the study by JACKSON et al. (1987), an increased splanchnic lactate output was found after metformin. The lactate did not appear to result form increased lactate production in the periphery, where the uptake of lactate was increased. The results are in accordance with an increased intestinal (splanchnic) glucose utilization by biguanides. Biguanides have been associated with inhibited absorption of a variety of different substances, such as hexoses, amino acids, calcium and bile salts. There are no human studies supporting a clinical significance of such findings. Intestinal biopsies in healthy subjects showed decreased activity of brush border disaccharidases during metformin treatment (BERCHTOLD et al. 1971). Malabsorption of vitamin B_{12} may occur during metformin treatment of long duration (TOMKIN et al. 1971; ADAMS et al. 1983). The mechanism is unknown. Folic acid absorption may also be affected by the drug (BERCHTOLD et al. 1971). There are only few clinical pharmacological studies on other aspects of the gastrointestinal function in relation to biguanide treatment, and a pathophysiological explanation for the gastrointestinal side effects frequently observed has not been offered. No consistent changes in gastric emptying were observed either in NIDDM patients (LEATHERDALE and BAILEY 1986) or in healthy volunteers (VIDON et al. 1988), although duodenogastric reflex was seen. Gastric acid secretion and gastrointestinal hormone levels showed some changes after a short metformin treatment in normal subjects (MOLLOY et al. 1980). Maximal and peak acid output increased, indicating a weak histamine (H_2) agonist action of the drug. Salivary and gastric secretions were unaffectd in the study by VIDON et al. (1988).

C. Effects on Metabolic Pathways in Human Diabetes

I. Glucose Metabolism

The main effect of metformin is exerted on glucose metabolism, but there are also important effects on lipid metabolism. A proposed integrated pharmacodynamic and metabolic effect of metformin based on human studies is summarized in Fig. 1. Early animal studies associated the action of biguanides with an increased anaerobic glucose utilization by glycolysis, resulting in an increased lactate production. This was also shown after phenformin in isotope experiments with ^{14}C-glucose in non-diabetic subjects (KREISBERG et al. 1970). This and other studies demonstrated futile recycling of substrates, i.e. increased glucose-lactate interconversions in the Cori cycle. The increased lactate formation indicates inhibition of glucose oxidation by phenformin. Whereas this cannot explain the antihyperglycaemic effect of the drug, an enhanced anaerobic glycolysis seems to be part of the potential toxic effect in cases of phenformin-associated lactic acidosis

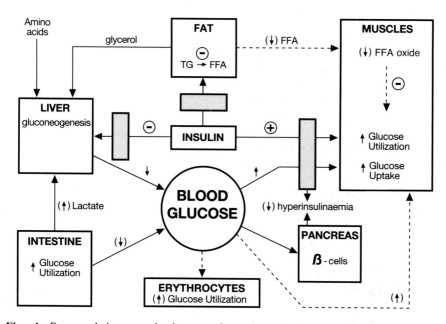

Fig. 1. Proposed integrated pharmacodynamic action of metformin in glucose homeostasis adapted form results of human studies. Effects on various metabolic processes are *marked with arrows, if less significant or consistent in parenthesis. Bars* symbolize impairment of insulin action in target cells (insulin resistance) and +/− indicate the normal actions of insulin. Glucose utilization in muscles is inhibited by FFA oxidation. Part of the muscular glucose uptake is non-insulin mediated (*broken line*), as glucose utilization in erythrocytes. The main effects of metformin are increased glucose disposal and decreased hepatic glucose output

(HERMANN 1973). A study using ^{13}C-glucose (LEFEBVRE et al. 1978) showed that buformin increased glucose oxidation and improved glucose tolerance in obese subjects with glucose intolerance. Metformin does not inhibit oxidative processes in general (STERNE 1969; FANTUS and BROSSEAU 1986; KLIP and LEITER 1990; BAILEY 1992), but an increased intestinal glucose utilization after the drug is associated with increased lactate production in the wall of the gut (Sect. B.VI). Glucose oxidation was unchanged by metformin in the clamp study by JOHNSON et al. (1993), but increased in the acute study by PERRIELLO et al. (1994) and in the clamp study in non-diabetic, hypertensive, obese women by GIUGLIANO et al. (1993a). A slight increase was observed by RICCIO et al. (1991) under basal conditions (Table 4). An increased glucose oxidation by metformin has also been demonstrated in adipose tissue from non-diabetic subjects (CIGOLINI et al. 1984), but this was not confirmed in adipose tissue obtained from NIDDM patients (PEDERSEN et al. 1989). On the other hand, an increased non-oxidative glucose metabolism after metformin has been confirmed, both in NIDDM patients (GROOP et al. 1989; JOHNSON et al. 1993) and in their relatives (WIDÉN et al. 1992), even if it was not seen as an acute effect of metformin (PERRIELLO et al. 1994). An effect on non-oxidative glucose metabolism indicates an effect on glycogenesis, which has also been shown in animal experiments (BAILEY 1992). In the study by JOHNSON et al. (1993), there was no effect of metformin on glycogen synthase activity in muscle biopsies from NIDDM patients and control subjects. An effect of metformin on glycogen synthesis in erythrocytes has been observed as mentioned in Sect. B.IV.2.

An increased muscular glucose uptake by metformin was found in an early in vitro study using muscle biopsies from non-diabetic patients undergoing surgery (FRAYN et al. 1973). However, this effect was seen only when insulin and sodium butyrate was added to metformin in the incubation medium. Similarly, GALUSKA et al. (1991) demonstrated that high metformin concentrations increased glucose uptake in vitro in human muscle strip preparations obtained from insulin-resistant subjects and examined under euglycaemic conditions. In a further study in which insulin-resistant skeletal muscle was exposed to a hyperglycaemic environment (GALUSKA et al. 1994), the effect on insulin-stimulated 3-0-methylglucose transport was not verified at therapeutic metformin concentrations in the incubation medium, but was still seen at higher levels. These two studies suggested an effect of the drug eigher due to accumulation in the extracellular space of muscle tissue, or an effect distal to the glucose transport step. Metformin also increases insulin-independent glucose uptake in cultured human muscle cells (KLIP and LEITER 1990; SARABIA et al. 1992). An increased cellular glucose uptake may be mediated via translocation of glucose transporters (KLIP and LEITER 1990; BAILEY 1993). A study using monocytes from NIDDM patients (BENZI et al. 1990) points to an improvement of insulin internalization and processing (degradation) after metformin. An effect on muscle insulin receptor tyrosine kinase activity by metformin has been found in diabetic rats but has not been confirmed in human tissues (BAILEY 1993). Other potential mechanisms involve the insulin-signaling process and the transendothelial insulin transfer (see Chap. 12, this volume).

Suppression of hepatic gluconeogenesis by metformin and other biguanides has been observed repeatedly in animal experiments. Small elevations of gluconeogenic precursors have been noticed during biguanide therapy, especially in phenformin-treated patients (NATTRASS et al. 1977, 1979; MARCHETTI et al. 1989). However, this has not been demonstrated consistently (CAMPBELL et al. 1987; JACKSON et al. 1987; JOHNSON et al. 1993). The acute effect of metformin on the liver as demonstrated by PERRIELLO et al. (1994) was associated with a slight elevation of alanine and lactate at high insulin level, but no change of glycerol and pyruvate. In various tolerance tests, hyperlactataemia was induced by exercise (HERMANN et al. 1973; BJÖRNTORP et al. 1978), glucose loading (JACKSON et al. 1987) or other means. These studies have shown some impairment of lactate elimination by phenformin, but only minor changes with metformin. Lactate is mainly eliminated by gluconeogenesis, and therefore these results support an inhibitory action of biguanides on this process. In non-diabetic subjects, gluconeogenesis is not inhibited by biguanides unless the process is accelerated in advance (DIETZE et al. 1978). Increased gluconeogenesis from lactate in non-diabetic subjects induced by phenformin was seen in isotope experiments as mentioned above. An unpublished study using radiolabelled lactate indicates inhibition of glycogenolysis by metformin (DEFRONZO 1994). A recent study (STUMVOLL et al. 1995) confirms that the drug inhibits gluconeogenesis from lactate in NIDDM.

II. Lactate Metabolism

There are no specific disturbances in lactate metabolism in NIDDM but it seems that diabetes predisposes to hyperlactataemia. The effect of biguanides on lactate metabolism involves lactate production and lactate elimination (HERMANN 1973). These processes are integrated with glucose metabolism as described in Sect. C.I. High concentrations of phenformin promote lactate production by enhancement of anaerobic glycolysis, whereas metformin does not inhibit oxidative processes in general. Lactate oxidation increased after metformin in the study by STUMVOLL et al. (1995). Phenformin augmented lactate production in hypoxic conditions in a study of diabetic patients (HERMANN et al. 1973), and the elimination of lactate accumulated during hypoxia and exercise was decreased. Comparative studies show higher lactate levels after phenformin than after metformin (NATTRASS et al. 1977; BJÖRNTORP et al. 1978; CAVALLO-PERIN et al. 1989). The increased lactate production by metformin from increased intestinal glucose utilization has been mentioned earlier (Sects. B.VI, C.I). Peripheral lactate production by muscle seems to be unaffected after metformin (BAILEY 1992; STUMVOLL et al. 1995). Most commonly, blood lactate concentrations are normal during metformin treatment (Nattrass et al. 1977; Hermann 1979; Campbell et al. 1987; DeFronzo et al. 1991; Johnson et al. 1993), but may be slightly elevated after glucose loading (Jackson et al. 1987) and meals (Pedersen et al. 1989). The average diurnal lactate concentration was slightly raised by metformin in a dose-response study (Marchetti et al. 1990) and was correlated with the dose and plasma concentration of the drug. Compared with sulphonylurea, lactate levels on metformin therapy may be slightly higher (Nattrass et al. 1977; McAlpine et al. 1988) or the same (Josephkutty and Potter 1990; Hermann et al. 1994b). Combined with glibenclamide, metformin was associated with slightly higher lactate levels than during sulphonylurea mono-therapy in the study by Nattrass et al. (1979), but this has not been confirmed by others (Higginbotham and Martin 1979; Hermann et al. 1994b).

III. Lipid Metabolism

The effect of metformin on lipid metabolism is discussed in Sect B.III. The dyslipidaemia associated with NIDDM includes hypertriglyceridaemia and decreased HDL cholesterol. Increased levels of FFA and increased FFA turnover and oxidation may have a pathophysiological significance as described in Sect. B.III.2. FFAs also promote hepatic VLDL-triglyceride secretion, leading to hypertriglyceridaemia. Apart from the studies already discussed in Sect. B.III, there are virtually no human data on the action of biguanides on specific biochemical processes in lipid metabolism. The effect of metformin on adipose tissue is unclear (Bailey 1992). In a study using adipose tissue biopsies from non-diabetic subjects, Cigolini et al. (1984) found that incubation with metformin stimulated glucose conversion to both triglycerides and CO_2 without an effect on receptors, i.e. increasing lipogenesis and glucose oxidation. This was not confirmed in a similar study on adipocytes from NIDDM patients (Pedersen et al. 1989). There was no effect on lipolysis in these experiments. A suppression of lipid oxidation has been observed by both acute (Widén et al. 1992; Perriello et al. 1994) and chronic (Groop et al. 1989) metformin treatment, but not consistently (Johnson et al. 1993). It has been suggested that metformin modifies lipoprotein metabolism, resulting in the formation of less atherogenic lipoprotein particles (Sirtori et al. 1985; Schneider 1991). There are no consistent changes in apoprotein levels after metformin. A decrease in postprandial concentrations of triglyceride-rich lipoproteins of intestinal origin was seen in a recent investigation (Jeppesen et al. 1994). It should be noted that normalization of glycaemia affects the diabetic dyslipidaemia but it seems that metformin has some specific effects on lipid metabolism independently of the antihyperglycaemic effect.

D. Indications

Metformin is the only biguanide recommended for clinical use. It is indicated for the treatment of NIDDM patients with diet failure and as an adjunct to sulphonylurea in patients with secondary sulphonylurea failure. Primary combination therapy has also been proposed (Sect. F.I). Metformin can be used in both obese and non-obese NIDDM patients, and it is often preferred to initiate oral treatment with metformin in the obese subject. IDDM is not an approved indication for metformin, but the drug has been used for its insulin-sparing effect. Such use should be very cautions because of the potential risk of lactic acidosis superimposed on the risk of ketosis, and because of the prevalence of renal impairment in IDDM. The potential use of metformin in IGT and other non-diabetic, insulin-resistant conditions is being explored at present. Metformin has been used successfully as a lipid-lowering agent in non-diabetic subjects with hyperlipidaemia.

E. Contraindications and Precautions

Contraindications are summarized in Table 5. Most of these are justified by an increased resk of lactic acidosis. Impaired renal function is of particular importance because of drug accumulation (Sect. G). If an acute complication arises, metformin should be temporarily withheld and insulin given until the condition is stable. This is also the case in patients undergoing radiological examinations involving parenteral administration of iodinated contrast materials, as such investigations in rare cases have precipitated acute alterations of renal function. Pregnancy is normally regarded as a contraindication but metformin has been administered to pregnant NIDDM patients during the second and third trimester without any particular problems (COETZEE and JACKSON 1986). Metformin treatment has not been linked with adverse embryonic effects in pregnant diabetic women (HELLMUTH et al. 1994). Alcohol abuse is a contraindication because of the possibility of coexisting liver disease (affecting lactate elimination), and because ethanol itself

Table 5. Contraindications for metformin

- Ketosis-prone diabetes
- Acute complications (severe infections, major operations and trauma)
- Pregnancy
- Impaired renal function
- Iodine contrast X-ray
- Liver damage
- Alcoholism
- Severe cardiovascular disease
- Severe respiratory disease
- B_{12}, folic acid and iron deficiency
- Poor general condition (malnutrition, dehydration)
- Old age (decreased renal function)

decreases lactate oxidation and gluconeogenesis from lactate. The inhibition of gluconeogenesis by alcohol also promotes hypoglycaemia. Old age is a contraindication if glomerular function is reduced. Serum creatinine, liver function tests and serum B_{12} should be monitored regularly. Estimation of creatinine clearance by nomogram is useful (HERMANN et al. 1981).

F. Combination Therapy

I. Metformin Plus Sulphonylurea

This therapy is well established in patients with secondary sulphonylurea failure. An increased clinical efficacy of the combination compared with previous sulphonylurea monotherapy has been confirmed in several studies (HERMANN 1990); see also Tables 1 and 4. The oral combination could be as good as insulin (PEACOCK and TATTERSALL 1984; GROOP et al. 1989), or even better, as far as postprandial plasma glucose is concerned (TRISCHITTA et al. 1992). In this study the superiority of oral combination was seen at high baseline postglucagon C peptide. The combination was inferior to insulin in the study by HOLMAN et al. (1987). The large United States pivotal study on combination therapy with metformin + glibenclamide confirmed the superiority of this combination over monotherapy with either drug (DEFRONZO et al. 1995). The pharmacodynamic action of the combination reflects the action of the two components. However, a further increase in insulin secretion cannot be obtained when metformin is added to sulphonylurea in patients with secondary sulphonylurea failure (GROOP et al. 1989). This study showed increased peripheral insulin action by the combination. Two other clamp studies (REAVEN et al. 1992; MARENA et al. 1994) showed both increased peripheral glucose disposal and decreased hepatic glucose output. By combining metformin with glibenclamide. weight increase and hyperinsulinaemia can be avoided (HERMANN et al. 1994b). The two drugs may have synergistic effects on blood glucose (HERMANN et al. 1994a). Insulin levels were reduced in the study of metformin + glipizide by REAVEN et al. (1992), but were unchanged in other studies combining metformin and sulphonylurea (Tables 1, 4). Lipid levels are unchanged (GROOP et al. 1989; TRISCHITTA et al. 1992; HERMANN et al. 1994b) or improved (REAVEN et al. 1992). Two studies showed decreased FFAs after combination (GROOP et al. 1989; REAVEN et al. 1992). Primary or early secondary combination therapy is rational from a pathophysiological viewpoint and may have therapeutic advantages (HERMANN 1994). It might also be expected that side effects could be reduced by lowering the dose each drug. However, this was not verified in our study (HERMANN et al. 1994b).

II. Metformin Plus Insulin

NIDDM is essentially a non-insulin-dependent condition, but eventually many patients with secondary sulphonylurea failure need insulin. These patients are often very insulin resistant and may need increasing insulin doses, which is associated with weight gain. Sometimes insulin has been combined with a sulphonylurea, but another option is to combine insulin with metformin, which seems more rational, especially in obese, insulin-resistant patients receiving high insulin doses. There are several reports of the insulin-sparing effect of metformin in IDDM (GIN et al. 1985; SLAMA 1991), but until recently there were only sporadic reports of the use of this combination in NIDDM. In a randomized, double-blind, placebo-controlled trial (GIUGLIANO et al. 1993b), the combination of metformin and insulin was shown to improve glycaemic control and risk factor profile considerably. The patients were obese and had been treated with insulin for at least 3 months after secondary sulphonylurea failure. The mean daily glucose concentration decreased by 34% after 6 months, and the reduction of HbA_{1c} was 1.8% in absolute terms. Fasting insulin decreased but C peptide was unchanged. Lipids and blood pressure improved and the insulin dose could be reduced. These changes were most marked in patients with good glycaemic response. Lactate levels were unchanged, body weight (BMI) remained stable and the treatment was well tolerated. These promising results need to be confirmed.

G. Pharmacokinetics

A pharmacokinetic profile of metformin is shown in Table 6 based on data from pharmacokinetic studies in healthy volunteers and NIDDM patients (SIRTORI et al. 1978; NOEL 1979; PENTIKÄINEN et al. 1979; TUCKER et al. 1981). Metformin is incompletely absorbed. Estimated from plasma data and urinary recoveries, the bioavailability of a single oral dose of 0.5–1.5 g is 50%–60%. The amount recovered in the faeces is 20%–30%. The difference between absorbed and available drug is believed to reflect presystemic clearance or binding to the intestinal wall. In the study by NOEL (1979), the

Table 6. Pharmacokinetic profile of metformin in man

- Bioavailability 50%–60%
- Peak plasma concentration (C_{max}) 1.5–2.0 μg/ml at 2–3 h (t_{max})
- Absorption slower than elimination, completed within 6 h
- Volume of distribution (V_d) 63–276 l
- No binding to plasma proteins
- Elimination half-life ($t_{1/2}$) 1.5–4.5 h, from deep compartment 9–19 h
- High renal clearance rate 335–544 ml/min, by glomerular filtration and tubular secretion; elimination almost complete after 8 h
- Not measurably metabolized

fraction of the dose recovered in the urine fell from 0.86 to 0.42 as the dose
was raised from 0.25 g to 2.0 g. A decrease in absorption with increasing
dosage was also suggested from urinary and faecal recovery data in the study
by TUCKER et al. (1981). That higher doses are proportionally less available
indicates a saturable absorption process. Absorption is slower than elimination and is rate limiting for the disposition. The absorption is reduced at
decreased renal function (TUCKER et al. 1981) and by concomitant intake of
guar gum (GIN et al. 1989b). There might be a slight delay and decrease
after food intake (BROOKES et al. 1991). Absorption is completed within 6 h.
Although there is absorption over the whole range of the intestine (VIDON et
al. 1988), the main part of the drug appears to be absorbed at a confined
area in the upper part of the intestine (TUCKER et al. 1981). The volume of
distribution is high and there is no binding to plasma proteins. Peak plasma
concentrations of $1.5 - 3.0\,\mu g/ml$ are reached 2–3 h after a single dose of 0.5
– 1.0 g. The distribution is rapid, but there is a slow association of the drug
with blood cells (TUCKER et al. 1981). Metformin accumulates in the tissues
of the small intestine and elsewhere (see Chap. 12, this volume).

Metformin is eliminated rapidly by renal excretion, including both
glomerular filtration and tubular secretion. The clearance of metformin is
correlated to creatinine clearance (SIRTORI et al. 1978). The mean plasma
elinimation half life is 1.5–4.5 h, with 90% being cleared in 12 h. Slower
elimination from a minor deep compartment may have some significance in
the case of drug accumulation. This compartment might be erythrocytes and
/or intestinal tissue. Metformin does not appear to be metabolized in
humans (PENTIKÄINEN et al. 1979), but this question is not completely resolved. Data from the study by SIRTORI et al. (1978) showed incomplete
recovery in the urine after intravenous administration, and 20% of the dose
was unaccounted for in the study by TUCKER et al. (1981). Some metabolic
transformation may therefore occur, but neither conjugates nor other
metabolites have been identified (TUCKER et al. 1981). In contrast, phenformin is metabolized by hydroxylation in all species including man. Phenformin accumulates in subjects with low hydroxylation capacity (BOSISIO et
al. 1981). Biguanide metabolites have no biological action. No metformin
seems to be exhaled by the lungs (PENTIKÄINEN et al. 1979). There are no
published data on breast milk levels of metformin, but it is not excluded that
the drug enters breast milk in small amounts. Metformin does not appear to
cross the placental barrier (see also Sect. E).

H. Safety and Tolerance

I. Lactic Acidosis

Lactic acidosis is the safety problem of most concern with biguanides. Most
cases of this life-threatening condition have been associated with phenformin

(LUFT et al. 1978). Buformin-associated cases were fewer, and metformin-associated lactic acidosis (MALA) was very rare at that time (HERMANN 1979). In our 1992 review (HERMANN and MELANDER 1992), we quoted a total of 90 MALA cases reported in published papers. When updating this review (HERMANN and MELANDER 1995), the number amounted to 119, including 18 cases from the period 1977–1991 described in a recent Swedish survey (WIHOLM and MYRHED 1993). There are additional cases not published in the literature but reported to various regulatory agencies. The true frequency of MALA is not known, but recently the United States Food and Drug Administration accepted a figure of 0.03/1000 patient years (i.e. 1:33 000). Not all cases of MALA are in fact causesd or precipitated by metformin (LAMBERT et al. 1987; LALAU et al. 1994; SIRTORI and PASIK 1994). In some cases there is no metformin accumulation, but hypoxia from other causes. Most MALA cases are seen in patients with renal failure, but even in this condition there is a certain clearance of the drug (HERMANN et al. 1981). MALA is a very serious complication, with a mortality rate of about 50%. CAMPBELL (1984) found the mortality risks for MALA and glibenclamide-induced hypoglycaemia identical (0.02 and 0.03/1000 patient years). Lactic acidosis can develop acutely, with the emergence of drowsiness and coma within a few hours. The treatment is complicated and haemodialysis is often necessary. MALA can probably be prevented by proper prescription of the drug, including strict observance of contraindications. Except in a very few cases, contributing factors other than the drug have been present in MALA. The pathophysiological basis for biguanide-induced lactic acidosis is increased lactate production and decreased lactate elimination (Sect. C.II).

II. Gastrointestinal and Other Adverse Effects

Gastrointestinal side effects are rather frequent during metformin therapy, especially diarrhoea (DANDONA et al. 1983). In our double-blind sutdy (HERMANN et al. 1994a,b), digestive tract symptoms were more frequent with metformin than with glibenclamide, but the frequency of all treatment-emergent events was the same in the two treatment groups and not different from that in combination therapy. The symptoms are generally transient and can be reduced by taking the drug with food. There is no clear dose dependence for side effects (HERMANN et al. 1994a). The gastrointestinal side effects are probably related to accumulation of the drug in intestinal tissue, inducing disturbances in motility. A proposed H_2-agonist action of metformin is mentioned in Sect. B.VI. Malabsorption of vitamin B_{12} and folic acid has also been discussed. Three cases of megaloblastic anaemia during metformin treatment have been published (HERMANN 1994). Very rarely, metformin has given rise to rashes and other hypersensitivity reactions. Clinical hypoglycaemia is virtually unknown with metformin monotherapy, but is seen wheen the drug is combined with sulphonylurea. However, hypoglycaemia with metformin monotherapy was observed in our double-blind study (HERMANN et al. 1994b), probably provoked by rapid blood glucose reduction. It was also seen in a few cases in the United States pivotal studies and in the UKPDS. It is conceivable that patients can experience hypoglycaemia during metformin treatment, if there are additional factors promoting hypoglycaemia, for example, fasting and alcohol. Hypoglycaemia could also be of the reactive type. Otherwise, there is a "safeguard" against hypoglycaemia by the increased intestinal glucose utilization supplying lactate to the liver as a gluconeogenic substrate (Sect. B.VI, C.I,II).

I. Interactions

A possible interaction with guar gum is mentioned in Sect. C. There are no known pharmacokinetic interactions with sulphonylurea, but a preliminary report described reduced bioavailability of metformin when acarbose was coadministered in healthy volunteers (SCHEEN et al. 1994). An increased elimination of the anticoagulant phenprocoumon (Marcoumar) has been reported during metformin treatment in diabetic patients (OHNHAUS et al. 1993), associated with increased liver blood flow. Cimetidine increased the availability of metformin and reduced its clearance in a study in healthy subjects (SOMOGYI et al. 1987). The lactate-pyruvate ratio was elevated when both agents were given. The results indicate competitive inhibition of renal tubular secretion. Alcohol potentiates the glucose-lowering and hyperlactataemic effect of biguanides and should be limited (Sect. E).

References

Adams JF, Clark JS, Ireland JT, Kesson CM, Watson WS (1983) Malabsorption of Vitamin B_{12} and intrinsic factor secretion during biguanide therapy. Diabetologia 24:16–18
Bailey CJ (1992) Biguanides and NIDDM. Diabetes Care 15:755–772
Bailey CJ (1993) Metformin – an update. Gen Pharmacol 24:1299–1309
Barnes AJ, Willars EJ, Clark PA, Hunt WB, Rampling M (1988) Effects of metformin on haemorrheological indices in diabetes. Diabete Metab 14:608–609
Benzi L, Trischitta V, Ciccarone A, Cecchetti P, Brunetti A, Squatritio S, Marchetti P, Vigneri R, Navalesi R (1990) Improvement with metformin in insulin internalization and processing in monocytes from NIDDM patients. Diabetes 39:844–849
Berchtold P, Dahlqvist A, Gustafson A, Asp NG (1971) Effects of a biguanide (metformin) on vitamin B_{12} and folic acid absorption and intestinal enzyme activities. Scand J Gastroenterol 6:751–754
Björntorp P, Carlström S, Fagerberg SE, Hermann LS, Holm AGL, Scherstén B, Östman J (1978) Influence of phenformin and metformin on exercise induced lactataemia in patients with diabetes mellitus. Diabetologia 15:95–98
Bonora E, Cigolini M, Bosello O, Zancanaro C, Capretti L, Zavaroni I, Coscelli C, Butturini U (1984) Lack of effect of intravenous metformin on plasma concentrations of glucose, insulin, C-peptide, glucagon and growth hormone in nondiabetic subjects. Curr Med Res Opin 9:47–51

Bosisio E, Galli Kienle M, Galli G, Ciconali M, Negri A, Sessa A, Morosati S, Sirtori CR (1981) Defective hydroxylation of phenformin as a determinant of drug toxicity Diabetes 30:644–649

Boyd K, Rogers C, Boreham C, Andrews WJ, Hadden DR (1992) Insulin, glibenclamide or metformin treatment for non insulin dependent diabetes: heterogeneous responses of standard measures of insulin action and insulin secretion before and after differing hypoglycaemic therapy. Diabetes Res 19:69–76

Brookes LG, Sambol NC, Lin ET, Gee W, Benet LZ (1991) Effect of dosage form, dose and food on the pharmacokinetics of metformin (abstract). Pharm Res Oct 8 [Suppl]:32;

Butterfield WJH, Whichelow MJ (1968) Effect of diet, sulphonylureas and phenformin on peripheral glucose uptake in diabetes and obesity. Lancet II:785–788

Campbell IW (1984) Metformin and glibenclamide: comparative risks. BMJ 289:289

Campbell IW (1990) Sulphonylureas and metformin: efficacy and inadequacy. In: Bailey CJ, Flatt PR (eds) New antidiabetic drugs. Smith-Gordon Nishimura, London, pp 33–51

Campbell IW, Howlett HCS (1995) Worldwide experience of metformin as an effective glucose lowering agent. A meta-analysis. Diabetes Metab Rev 11:S57–S62

Campbell IW, Duncan C, Patton NW (1987) The effect of metformin on glycaemic control, intermediary metabolism and blood pressure in non-insulin-dependent diabetes mellitus. Diabet Med 4:337–341

Campbell IW, Menzies DG, Chalmers J, McBain AM, Brown IRF (1994) One year comparative trial of metformin and glipizide in type 2 diabetes mellitus. Diabete Metab 21:394–400

Caporicci D, Mori A, Pepi R, Lapi E (1979) Effetti della dimetilbiguanide (metformina) sulla clearance periferica dell'insulina e sulla biosintesi lipidica in pazienti obesi dislipidemici con e senza malatta diabetica. Clin Ter 88:371–386

Carpentier J-L, Luyckx AS, Lefebvre PJ (1975) Influence of metformin on arginine-induced glucagon secretion in human diabetes. Diabete Metab 1:23–28

Cavallo-Perin P, Aluffi E, Estivi P, Bruno A, Carta Q, Pagano G, Lenti G (1989) The hyperlactatemic effect of biguanides: a comparison between phenformin and metformin during a 6-month treatment. Eur Rev Med Pharmacol Sci 11:45–49

Chan JCN, Tomlinson B, Critchley JAJH, Cockram CS, Walden RJ (1993) Metabolic and hemodynamic effects of metformin and glibenclamide in normotensive NIDDM patients. Diabetes Care 16:1035–1038

Cigolini M, Bosello O, Zancanaro C, Orlandi PG, Fezzi O, Smith U (1984) Influence of metformin on metabolic effect of insulin in human adipose tissue in vitro. Diabete Metab 10:311–315

Clarke BF, Campbell IW (1977) Comparison of metformin and chlorpropamide in non-obese, maturity-onset diabetics uncontrolled by diet. BMJ 2:1576–1578

Clarke BF, Duncan LJP (1968) Comparison of chlorpropamide and metformin treatment on weight and blood-glucose response of uncontrolled obese diabetics. Lancet I:123–126

Coetzee EJ, Jackson WPU (1986) The management of non-insulin dependent diabetes during pregnancy. Diabetes Res Clin Pract 1:281–287

Collier A, Watson HHK, Patrick AW (1989) Effect of glycaemic control, metformin and gliclazide on platelet density and aggregability in recently diagnosed type 2 (non-insulin-dependent) diabetic patients. Diabete Metab 15:420–425

Czyzyk A, Tawecki J, Sadowski J (1968) Effect of biguanides on intestinal absorption of glucose. Diabetes 17:492–498

Dandona P, Fonesca V, Mier A, Beckett AG (1983) Diarrhea and metformin in a diabetic clinic. Diabetes Care 6:472–474

DeFronzo RA (1994) Mechanism of metformin action: clinical studies in NIDDM. In: Groupe Lipha (eds) Glucophage International Symposium Heidelberg, NIDDM: prevention and treatment (abstracts), p 15

DeFronzo RA, Barzilai N, Simonson DC (1991) Mechanism of metformin action in obese and lean noninsulin-dependent diabetic subjects. J Clin Endocrinol Metab 73:1294–1301

DeFronzo RA, Goodman AM, and the Multicenter Metformin Study Group (1995) Efficacy of metformin in patients with non-insulin-dependent diabetes mellitus. N Engl J Med 333:541–549

De Silva SR, Shawe JEH, Patel H, Cudworth AG (1979) Plasma fibrinogen in diabetes mellitus. Diabete Metab 5:201–206

Di Paolo S (1992) Metformin ameliorates extreme insulin resistance in a patient with anti-insulin receptor antibodies: description of insulin receptor and postreceptor effects in vivo and in vitro. Acta Endocrinol (Copenh) 126:117–123

Dietze G, Wicklmayr M, Mehnert H, Czempiel H, Henftling HG (1978) Effect of phenformin on hepatic balances of gluconeogenetic substrates in man. Diabetologia 14:243–248

Dornan TL, Heller SR, Peck GM, Tattersall RB (1991) Double-blind evaluation of efficacy and tolerability of metformin in NIDDM. Diabetes Care 14:342–344

Dunn CJ, Peters DH (1995) Metformin: a review of its pharmacological properties and therapeutic use in diabetes mellitus. Drugs 49:721–749

Fantus I, Brosseau R (1986) Mechanism of action of metformin: insulin receptor and postreceptor effects in vitro and in vivo. J Clin Endocrinol Metab 63:898–905

Fendri S, Debussche X, Puy H, Vincent O, Marcelli JM, Dubreuil A, Lalau JD (1993) Metformin effects on peripheral sensitivity to insulin in non diabetic obese sujects. Diabete Metab 19:245–249

Ferner RE, Rawlins MD, Alberti KGMM (1988) Impaired β-cell responses improve when fasting blood glucose concentration is reduced in non-insulin-dependent diabetes. Q J Med New Ser 66:137–146

Frayn KN, Adnitt PI, Turner P (1973) The use of human skeletal muscle in-vitro for biochemical and pharmacological studies of glucose uptake. Clin Sci 44:55–62

Galuska D, Zierath J, Thörne A, Sonnenfeld T, Wallberg-Henriksson H (1991) Metformin increases insulin-stimulated glucose transport in insulin-resistant human skeletal muscle. Diabete Metab 17:159–163

Galuska D, Nolte LA, Zierath JR, Wallberg-Henriksson H (1994) Effect of metformin on insulin-stimulated glucose transport in isolated skeletal muscle obtained from patients with NIDDM. Diabetologia 37:826–832

Gin H, Messerchmitt C, Brottier E, Aubertin J (1985) Metformin improved insulin resistance in type I, insulin-dependent diabetic patients. Metabolism 34:923–925

Gin H, Freyburger C, Boisseau M, Aubertin J (1989a) Study of the effect of metformin on platelet aggregation in insulin-dependent diabetics. Diabetes Res Clin Pract 6:61–67

Gin H, Orgerie MB, Aubertin J (1989b) The influence of guar gum on absorption of metformin from the gut in healthy volunteers. Horm Metab Res 21:81–83

Giugliano D, DeRosa N, DiMario G, Marfella R, Acapora R, Buoninconti R, D'Onofrio F (1993a) Metformin improves glucose, lipid metabolism, and reduces blood pressure in hypertensive, obese women. Diabetes Care 16:1387–1390

Giugliano D, Quatraro A, Consoli G, Minei A, Ceriello A, De Rosa N, D'Onofrio F (1993b) Metformin for obese, insulin-treated diabetic patients: improvement in glycaemic control and reduction of metabolic risk factors. Eur J Clin Pharmacol 44:107–112

Grant PJ (1991) The effects of metformin on the fibrinolytic system in diabetic and non-diabetic subjects. Diabete Metab 17:168–173

Groop L, Widén E, Franssila-Kallunki A, Ekstrand A, Saloranta C, Schalin C, Eriksson J (1989) Different effects of insulin and oral antidiabetic agents on glucose and energy metabolism in type 2 (non-insulin-dependent) diabetes mellitus. Diabetologia 32:599–605

Gustafson A, Björntorp P, Fahlén M (1971) Metformin administration in hyperlipidemic states. Acta Med Scand 190:491–494

Haupt E, Knick B, Koschinsky T, Liebermeister H, Schneider J, Hirche H (1991) Oral antidiabetic combination therapy with sulphonylureas and metformin. Diabete Metab 17:224–231

Hellmuth E, Damm P, Mølsted-Pedersen L (1994) Congenital malformations in offspring of diabetic women treated with oral hypoglycaemic agents during embryogenesis. Diabet Med 11:471–474

Hermann LS (1973) Biguanides and lactate metabolism. A review. Dan Med Bull 20:65–79

Hermann LS (1979) Metformin: a review of its pharmacological properties and therapeutic use. Diabete Metab 5:233–245

Hermann LS (1990) Biguanides and sulfonylureas as combination therapy in NIDDM. Diabetes Care 13 [Suppl 3]:37–41

Hermann LS (1994) Biguanides (metformin)/øral combination therapy with sulphonylurea + metformin. In: Hermann LS (ed) Metformin as monotherapy and combined with glibenclamide in patients with non-insulin dependent diabetes mellitus (thesis). Studentlitteratur, Lund, pp 38–62

Hermann LS, Melander A (1992) Biguanides: basic aspects and clinical uses. In: Alberti KGMM, DeFronzo RA, Keen H, Zimmet P (eds) International textbook of diabetes mellitus. Wiley, Chichester, pp 773–795

Hermann LS, Melander A (1995) Biguanides: basic aspects and clinical uses. In: Alberti KGMM, DeFronzo RA, Keen H, Zimmet P (eds) International textbook of diabetes mellitus, 2nd edn. Wiley, Chichester (in press)

Hermann LS, Nathan E, Ebbesen I (1973) The influence of phenformin on lactate metabolism in diabetic patients in relation to hypoxia and exercise. Acta Med Scand 194:111–116

Hermann LS, Magnusson S, Möller B, Casey C, Tucker T, Woods HF (1981) Lactic acidosis during metformin treatment in an elderly diabetic patient with impaired renal function, Acta Med Scand 209:519–520

Hermann LS, Bitzén P-O, Kjellström T, Lindgärde F, Scherstén B (1991a) Comparative efficacy of metformin and glibenclamide in patients with non-insulin-dependent diabetes mellitus. Diabete Metab 17:201–208

Hermann LS, Karlsson J-E, Sjöstrand Å (1991b) Prospective comparative study in NIDDM patients of metformin and glibenclamide with special reference to lipid profiles. Eur J Clin Pharmacol 41:263–265

Hermann LS, Kjellström T, Nilsson-Ehle P (1991c) Effects of metformin and glibenclamide alone and in combination on serum lipids and lipoproteins in patients with non-insulin-dependent diabetes mellitus. Diabete Metab 17:174–179

Hermann LS, Scherstén, Melander A (1994a) Antihyperglycaemic efficacy, response prediction and dose-response relations of treatment with metformin and sulphonylurea, alone and in primary combination. Diabet Med 11:953–960

Hermann LS, Scherstén B, Bitzén P-O, Kjellström T, Lindgärde F, Melander A (1994b) Therapeutic comparison of metformin and sulphonylurea, alone and in various combinations: a double-blind controlled study. Diabetes Care 17:1100–1109

Higginbotham L, Martin FIR (1979) Double-blind trial of metformin in the therapy of non-ketotic diabetics. Med J Aust 2:154–156

Hollenbeck CB, Johnston P, Varasteh BB, Ida Chen Y-D, Reaven GM (1991) Effects of metformin on glucose, insulin and lipid Metabolism in patients with mild hypertriglyceridaemia and non-insulin dependent diabetes by glucose tolerance test criteria. Diabete Metab 17:483–489

Holman RR, Steemson J, Turner RC (1987) Sulphonylurea failure in type 2 diabetes: treatment with a basal insulin supplement. Diabet Med 4:457–462

Hother-Nielsen O, Schmitz O, Andersen PH, Beck-Nielsen H, Pedersen O (1989) Metformin improves peripheral but not hepatic insulin action in obese patients with type II diabetes. Acta Endocrinol (Copenh) 120:257–265

Jackson RA, Hawa MI, Jaspan JB, Sim BM, DiSilvio L, Featherbe D, Kurtz AB (1987) Mechanism of metformin action in non-insulin-dependent diabetes. Diabetes 36:632–640

Janssen M, Rillaerts E, De Leeuw I (1991) Effects of metformin on haemorheology, lipid parameters and insulin resistance in insulin-dependent diabetic patients (IDDM). Biomed Pharmacother 45:363–367

Jeppesen J, Zhou M-Y, Ida Chen Y-D, Reaven GM (1994) Effect of metformin on postprandial lipemia in patients with fairly to poorly controlled NIDDM. Diabetes Care 17:1093–1099

Johnson AB, Webster JM, Sum C-F, Heseltine L, Argyraki M, Cooper BG, Taylor R (1993) The impact of metformin therapy on hepatic glucose production and skeletal muscle glycogen synthase activity in overweight type II diabetic patients. Metabolism 42:1217–1222

Josephkutty S, Potter JM (1990) Comparison of tolbutamide and metformin in elderly diabetic patients. Diabet Med 7:510–514

Klip A, Leiter LA (1990) Cellular mechanism of action of metformin. Diabetes Care 13:696–704

Kreisberg RA, Pennington LF, Boshell BR (1970) Lactate turnover and gluconeogenesis in obesity. Effect of phenformin. Diabetes 19:64–69

Lalau JD, Lacroix C, De Cagny B, Fournier A (1994) Metformin-associated lactic acidosis in diabetic patients with acute renal failure. A critical analysis of its pathogenesis and prognosis. Nephrol Dial Transplant 9 [Suppl 4] 126–129

Lalor BC, Bhatnagar D, Winocour PH, Ishola M, Arrol S, Brading M, Durrington PN (1990) Placebo-controlled trial of the effects of guar gum and metformin on fasting blood glucose and serum lipids in obese, type 2 diabetic patients. Diabet Med 7:242–245

Lambert H, Isnard F, Delorme N, Claude D, Bollaert PE, Straczek J, Larcan A (1987) Approche physiopathologique des hyperlactatémies pathologiques chez le diabétique. Intérêt de la metforminémie. Ann Fr Anesth Reanim 6:88–94

Landin K, Tengborn L, Smith U (1991) Treating insulin resistance in hypertension with metformin reduces both blood pressure and metabolic risk factors. J Intern Med 229:181–187

Landin K, Tengborn L, Smith U (1994a) Metformin and metoprolol CR treatment in non-obese men. J Intern Med 235:335–341

Landin K, Tengborn L, Smith U (1994b) Effects of metformin and metoprolol CR on hormones and fibrinolytic variables during a hyperinsulinemic, euglycemic clamp in man. Thromb Haemost 71:783–787

Leatherdale BA, Bailey CJ (1986) Acute antihyperglycaemic effect of metformin without alteration of gastric emptying. IRCS Med Sci 14:1085–1086

Lefebvre P, Luyckx A, Mosora F, Lacroix M, Pirnay F (1978) Oxidation of an exogenous glucose load using naturally labelled C-13-glucose. Diabetologia 14:39–45

Leslie P, Jung RT, Isles TE, Baty J (1987) Energy expenditure in non-insulin dependent diabetic subjects on metformin or sulphonylurea therapy. Clin Sci 73:41–45

Lisch H-J von, Sailer S, Braunsteiner H (1980) Die Wirkung von Biguaniden auf die Insulinempfindlichkeit von Altersdiabetikern. Wien Klin Wochenschr 92:266–269

Luft D, Schmülling RM, Eggstein M (1978) Lactic acidosis in biguanide-treated diabetics. A review of 330 cases. Diabetologia 14:75–87

Marchetti P, Benzi L, Cecchetti P, Giannarelli R, Boni C, Ciociaro D, Ciccarone AM, Di Cianni G, Zappella A, Navalesi R (1987) Plasma biguanide levels are correlated with metabolic effects in diabetic patients. Clin Pharmacol Ther 41:450–454

Marchetti P, Masiello P, Benzi L, Cecchetti P, Fierabracci V, Giannarelli R, Gregorio F, Brunetti P, Navalesi R (1989) Effects of metformin therapy on plasma amino acid pattern in patients with maturity-onset diabetes. Drugs Exp Clin Res 15:565–570

Marchetti P, Gregorio F, Benzi L, Giannarelli R, Cecchetti P, Villani G, Di Cianni G, Di Carlo A, Brunetti P, Navalesi R (1990) Diurnal pattern of plasma metformin concentrations and its relation to metabolic effects in type 2 (non-insulin-dependent) diabetic patients. Diabete Metab 16:473–478

Marena S, Tagliaferro V, Montegrosso G, Pagano A, Scaglione L, Pagano G (1994) Metabolic effects of metformin addition to chronic glibenclamide treatment in type 2 diabetes. Diabete Metab 20:15–19

McAlpine LG, McAlpine CH, Waclawski ER, Storer AM, Kay JW, Frier BM (1988) A comparison of treatment with metformin and gliclazide in patients with non-insulin-dependent diabetes. Eur J Clin Pharmacol 34:129–132

McIntyre HD, Paterson CA, Ma A, Ravenscroft PJ, Bird DM, Cameron DP (1991) Metformin increases insulin sensitivity and basal glucose clearance in type 2 (non-insulin dependent) diabetes mellitus. Aust NZ J Med 21:714–719

Molloy AM, Ardill J, Tomkin GH (1980) The effect of metformin treatment on gastric acid secretion and gastrointestinal hormone levels in normal subjects. Diabetologia 19:93–96

Montanari G, Bondioli A, Rizzato G, Puttini M, Tremoli E, Mussoni L, Mannucci L, Pazzucconi F, Sirtori CR (1992) Treatment with low dose metformin in patients with peripheral vascular disease. Pharmacol Res 25:63–73

Muntoni S (1974) Inhibition of fatty acid oxidation by biguanides: implications for metabolic physiopathology. Adv Lipid Res 12:311–377

Nagi DK, Yudkin JS (1993) Effects of metformin on insulin resistance, risk factors for cardiovascular disease, and plasminogen activator inhibitor in NIDDM subjects. A study of two ethnic groups. Diabetes Care 16:621–629

Nagi DK, Mohamed AV, Yudkin JS (1994) Effects of metformin on intact and des 31,32 proinsulin in subjects with non-insulin-dependent diabetes (abstract). Diabet Med Apr 11 [Suppl 1]:25–26

Nattrass M, Todd PG, Hinks L, Lloyd B, Alberti KGMM (1977) Comparative effects of phenformin, metformin and glibenclamide on metabolic rhythms in maturity-onset diabetics. Diabetologia 13:145–152

Nattrass M, Hinks L, Smythe P, Todd PG, Alberti KGMM (1979) Metabolic effects of combined sulphonylurea and metformin therapy in maturity-onset diabetics. Horm Metab Res 11:332–337

Noel M (1979) Kinetic study of normal and sustained release dosage forms of metformin in normal subjects. Res Clin Forums 1:35–44

Nosadini R, Avogaro A, Trevisan R, Valerio A, Tessari P, Duner E, Tiengo A, Velussi M, Del Prato S, De Kreutzenberg S, Muggeo M, Crepaldi G (1987) Effect of metformin on insulin-stimulated glucose turnover and insulin binding to receptors in type II diabetes. Diabetes Care 10:62–67

Noury J, Nandeuil A (1991) Comparative three-month study of the efficacies of metformin and gliclazide in the treatment of NIDD. Diabete Metab 17:209–212

Ohnhaus EE, Berger W, Duckert F, Oesch F (1983) The influence of dimethyl-biguanide on phenprocoumon elimination and its mode of action. Klin Wochenschr 61:851–858

Papoz L, Job D, Eschwege E, Aboulker JP, Cubeau J, Pequignot G, Rathery M, Rosselin G (1978) Effect of oral hypoglycaemic drugs on glucose tolerance and insulin secretion in borderline diabetic patients. Diabetologia 15:373–380

Peacock I, Tattersall RB (1984) The difficult choice of treatment for poorly controlled maturity onset diabetes: tablets or insulin? BMJ 288:1956–1959

Pedersen O, Hother Nielsen O, Bak J, Richelsen B, Beck-Nielsen H, Schwartz Sørensen N (1989) The effects of metformin on adipocyte insulin action and metabolic control in obese subjects with type 2 diabetes. Diabet Med 6:249–256

Pentikäinen PJ, Neuvonen PJ, Penttilä A (1979) Pharmacokinetics of metformin after intravenous and oral administration to man. Eur J Clin Pharmacol 16:195–202

Perriello G, Misericordia P, Volpi E, Santucci A, Santucci C, Ferrannini E, Ventura MM, Santeusanio F, Brunetti P, Bolli GB (1994) Acute antihyperglycemic mechanisms of metformin in NIDDM. Evidence for suppression of lipid oxidation and hepatic glucose production. Diabetes 43:920–928

Prager R, Schernthaner G (1983) Insulin receptor binding to monocytes, insulin secretion, and glucose tolerance following metformin treatment. Diabetes 32:1083–1086

Prager R, Schernthaner G, Graf H (1986) Effect of metformin on peripheral insulin sensitivity in non insulin dependent diabetes mellitus. Diabete Metab 12:346–350

Rains SGH, Wilson GA, Richmond W, Elkeles RS (1988) The effect of glibenclamide and metformin on serum lipoproteins in type 2 diabetes. Diabetic Med 5:653–658

Reaven GM, Johnston P, Hollenbeck CB, Skowronski R, Zhang J-C, Goldfine ID, Ida Chen Y-D (1992) Combined metformin-sulfonylurea treatment of patients with noninsulin-dependent diabetes in fair to poor glycemic control. J Clin Endocrinol Metab 74:1020–1026

Riccio A, Del Prato S, Vigili de Kreutzenberg S, Tiengo A (1991) Glucose and lipid metabolism in non-insulin-dependent diabetes. Effect of metformin. Diabete Metab 17:180–184

Rizkalla SW, Elgrably F, Tchobroutsky G, Slama G (1986) Effects of metformin treatment on erythrocyte insulin binding in normal weight subjects, in obese non diabetic subjects, in type 1 and type 2 diabetic patients. Diabete Metab 12:219–224

Rudnichi A, Fontbonne A, Safar M, Bard J-M, Vague P, Juhan-Vague I, Eschwege E, and the BIGPRO Study Group (1994) The effect of metformin on the metabolic anomalies associated with android type body fat distribution. Results of the BIGPRO trial. Diabetes 43 [Suppl 1] 150A

Sarabia V, Lam L, Burdett E, Leiter LA, Klip A (1992) Glucose transport in human skeletal muscle cells in culture. Stimulation by insulin and metformin. J Clin Invest 90:1386–1395

Scheen AJ, Ferreira Alves de Malaghaes AC, Salvatore T, Lefebvre PJ (1994) Reduction of the acute bioavailability of metformin by the α-glucosidase inhibitor acarbose in normal man. Eur J Clin Invest 24 [Suppl 3]:50–54

Schneider J (1991) Effects of metformin on dyslipoproteinemia in non-insulin-dependent diabetes mellitus. Diabete Metab 17:185–190

Schäfer G (1983) Biguanides: a review of history, pharmacodynamics and therapy. Diabete Metab 9:148–163

Schönborn J, Heim K, Rabast U, Kasper H (1975) Oxidation rate of plasma free fatty acids in maturity-onset diabetics. Effect of metformin. Diabetologia 11:375

Semplicini A, Del Prato S, Giusto M, Campagnolo M, Palatini P, Rossi GP, Valle R, Dorella M, Albertin G, Pessina AC (1993) Short-term effects of metformin on insulin sensitivity and sodium homeostasis in essential hypertensives. J Hypertens 11 [Suppl 5]:276–277

Sirtori CR, Pasik C (1994) Re-evaluation of a biguanide, metformin: mechanism of action and tolerability. Pharmacol Res 30:187–228

Sirtori CR, Franceschini G, Galli-Kienle M, Cighetti G, Galli G, Bondioli A, Conti F (1978) Disposition of metformin (N,N-dimethylbiguanide) in man. Clin Pharmacol Ther 24:683–693

Sirtori CR, Lovati MR, Franceschini G (1985) Management of lipid disorders and prevention of atherosclerosis with metformin. In: Krans HMJ (ed) Diabetes and metformin. A research and clinical update. International Congress and Symposium Series number 79, Royal Society of Medicine, London, pp 33–44

Slama G (1991) The insulin sparing effect of metformin in insulin-treated diabetic patients. Diabete Metab 17:241–243

Somogyi A, Stockley C, Keal J, Rolan P, Bochner F (1987) Reduction of metformin renal tubular secretion by cimetidine in man. Br J Clin Pharmacol 23:545–551

Sterne J (1969) Pharmacology and mode of action of the hypoglycaemic guanidine derivatives. In: Campbell GD (ed) Oral hypoglycaemic agents. Academic, New York, pp 193–245

Stumvoll M, Nurjhan N, Perriello G, Dailey G, Gerich JE (1995) Metabolic effects of metformin in non-insulin-dependent diabetes mellitus. N Engl J Med 333: 550–554

Sum C-F, Webster JM, Johnson AB, Catalano C, Cooper BG, Taylor R (1992) The effect of intravenous metformin on glucose metabolism during hyperglycaemia in type 2 diabetes. Diabet Med 9:61–65

Taylor KG, John WG, Matthews KA, Wright AD (1982) A prospective study of the effect of 12 months treatment on serum lipids and apolipoproteins A-I and B in type 2 (non-insulin-dependent) diabetes. Diabetologia 23:507–510

Tomkin GH, Hadden DR, Weaver JA, Montgomery DAD (1971) Vitamin-B_{12} status of patients on long-term metformin therapy. BMJ 2:685–687

Tranquada RE, Kleeman C, Brown J (1960) Some effects of phenethylbiguanide on human hepatic metabolism as measured by hepatic vein catheterization. Diabetes 9:207–214

Trischitta V, Italia S, Mazzarino S, Buscema M, Rabuazzo AM, Sangiorgio L, Squatritio S, Vigneri R (1992) Comparison of combined therapies in treatment of secondary failure to glyburide. Diabetes Care 15:539–542

Tucker GT, Casey C, Phillips PJ, Connor H, Ward JD, Woods HF (1981) Metformin kinetics in healthy subjects and in patients with diabetes mellitus. Br J Clin Pharmacol 12:235–246

United Kingdom Prospective Diabetes Study Group (1995) UKPDS 13: relative efficacy of randomly allocated diet, sulphonylurea, insulin, or metformin in patients with newly diagnosed non-insulin dependent diabetes followed for three years. BMJ 310:83–88

Vague P, Juhan-Vague I, Alessi MC, Badier C, Valadier J (1987) Metformin decreases the high plasminogen activator inhibition capacity, plasma insulin and triglyceride levels in non-diabetic obese subjects. Thromb Haemost 57:326–328

Velazquez EM, Mendoza S, Hamer T, Sosa F, Glueck CJ (1994) Metformin therapy in polycystic ovary syndrome reduces hyperinsulinemia, insulin resistance, hyperandrogenemia, and systolic blood pressure, while facilitating normal menses and pregnancy. Metabolism 43:647–654

Vidon N, Chaussade S, Noel M, Franchisseur C, Huchet B, Bernier JJ (1988) Metformin in the digestive tract. Diabetes Res Clin Pract 4:223–229

Widén EIM, Eriksson JG, Groop LC (1992) Metformin normalizes nonoxidative glucose metabolism in insulin-resistant normoglycaemic first-degree relatives of patients with NIDDM. Diabetes 41:354–358

Widén E, Groop L (1994) Biguanides: metabolic effects and potential use in the treatment of the insulin resistance syndrome. In: Marshall SM, Home PD (eds) The Diabetes Annual/8, Elsevier, Amsterdam, pp 227–241

Wiholm B-E, Myrhed M (1993) Metformin-associated lactic acidosis in Sweden 1977–1991. Eur J Clin Pharmacol 44:589–591

Wu M-S, Johnston P, Sheu WHH, Hollenbeck CB, Jeng C-Y, Goldfine ID, Ida Chen Y-D, Reaven GM (1990) Effect of metformin on carbohydrate and lipoprotein metabolism in NIDDM patients. Diabetes Care 13:1–8

Yoa RG, Rapin JR, Wiernsperger NF, Martinand A, Belleville I (1993) Demonstration of defective glucose uptake and storage in erythrocytes from non-insulin dependent diabetic patients and effects of metformin. Clin Exp Pharmacol Physiol 20:563–567

Section III
Glucosidase Inhibitors

CHAPTER 15
Chemistry and Structure-Activity Relationships of Glucosidase Inhibitors

B. Junge, M. Matzke, and J. Stoltefuss

A. Introduction

When intestinal sucrase and pancreatic α-amylase were identified at Bayer as new targets for improved diabetes therapy (Puls et al. 1973), the problem remained of the best way to find a potent and selective inhibitor of these enzymes. The medicinal chemist today has two alternatives in the search for a first lead compound, firstly the screening of thousands of compounds with maximum structural diversity in a random screening approach, or secondly the rational design of a lead compound, when the biochemical mechanism and the structure of the enzyme are well known. In the late 1960s, when we started to look for such potent and selective inhibitors, we had to choose the random screening approach, relying mainly on testing of extracts of the culture broths of microorganisms.

Today the mechanism of enzymatic splitting of sucrose by intestinal sucrase is fairly well understood. First the glucosidic oxygen atom of sucrose is protonated via a carboxyl group of the enzyme. The splitting of the glucosidic C-O bond is further facilitated by the glucosyl cation being stabilized by a second carboxylate group of the enzyme (Cogoli and Semenza 1975). Finally, the glucosyl cation reacts with water to give the products D-glucose and D-fructose. The protonated sucrose molecule I and the glucosyl cation II are two high-energy intermediates of the enzymatic reaction. Today we know that mimics of such high-energy intermediates are often potent inhibitors of the enzyme.

Accordingly, compounds similar in structure to D-glucose but with an easily protonated basic N-atom either in the position of the anomeric oxygen atom (inhibitor type I) or in the position of the ring oxygen atom (inhibitor type II) should be potent inhibitors of sucrase (Fig. 1). Interestingly, both types of inhibitors represented by either acarbose or 1-deoxynojirimycin were found in our random screening. In the following sections these two types of inhibitors are not discussed in a historical order but under structural criteria. Not included in this discussion are the α-amylase inhibitors of protein nature because there is no evidence that they will find any therapeutic application.

Acarbose type inhibitors

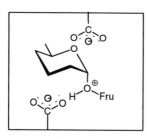

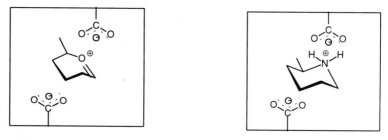

High energy intermediate I   Inhibitor type I

1-Deoxynojirimycin type inhibitors

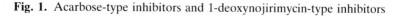

High energy intermediate II   Inhibitor type II

Fig. 1. Acarbose-type inhibitors and 1-deoxynojirimycin-type inhibitors

B. Pseudoglucosylamines

I. Validamine, Valienamine, and Valiolamine

The enzymatic cleavage of α-glucosides by sucrase/isomaltase takes place via
a high-energy intermediate, an oxonium ion *1* formed by protonation of the
glycosidic oxygen atom (Cogoli and Semenza 1975). Compounds with structural similarities to such high-energy intermediates or to the substrate in the
transition state of the enzymatic reaction have been described as potent
inhibitors of the enzyme (Brodbeck 1980). For instance, glucosylamines are
effective inhibitors of a series of glucosidases (Legler 1990) since the
ammonium ion *2* is a suitable mimic for the oxonium ion *1*. However,
because of the higher basicity of the N atom, *2* is more stable than *1* and
thus the enzyme-inhibitor complex has lower energy than the enzyme-
substrate complex.

Because of the low chemical stability and high reactivity of the glucosylamines, these compounds are not suitable for therapeutic use. In contrast,
the amines *3–5*, derived from pseudosugars, are chemically stable analogues

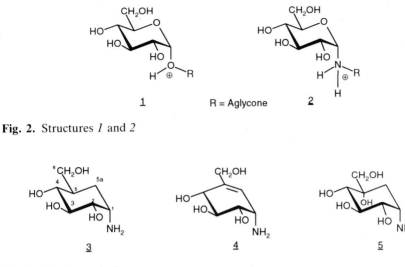

Fig. 2. Structures *1* and *2*

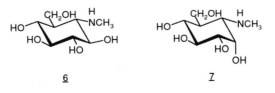

Fig. 3. Structures *3–5*

of the oxonium ion *1*. These compounds were isolated from cultures of *Streptomyces hygroscopicus* (KAMEDA et al. 1984) or were obtained by degradation of validamycin (KAMEDA and HORII 1972; KAMEDA et al. 1980). The 5a-carba-α-D-glucopyranosylamine *3*, validamine (HORII et al. 1971a), is a potent inhibitor of sucrase from the porcine small intestine (IC$_{50}$ = 7.5 × 10^{-6} *M*) and other intestinal disaccharidases (maltase, isomaltase). Microbial α- and β-glucosidases are hardly inhibited. Valienamine (*4*), with a double bond in the cyclitol ring, is somewhat less effective (IC$_{50}$ = 5.3 × 10^{-5} *M*), whereas the 5-hydroxyl derivative valiolamine (*5*) is more than two orders of magnitude more effective (IC$_{50}$ = 4.9 × 10^{-8} *M*) than *3*. The reason for this increase in activity upon introduction of an axial hydroxyl group in the 5-position is not known, but one could speculate that the hydroxyl stabilizes the enzyme's carboxyl group, responsible for protonation of the glycosidic O-atom of the substrate, by forming a hydrogen bond.

The exact position of the amine group is important for the good efficacy of these inhibitors. Therefore the 5a-carba-5a-amino-α,β-D-glucopyranose derivatives *6* and *7* have no effect on sucrase from the rat small intestine (PEET et al. 1991).

Fig. 4. Structures *6* and *7*

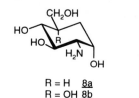 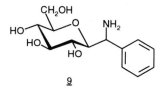

R = H 8a
R = OH 8b

9

Fig. 5. Structures 8 and 9

The activities of the isomers of validamine *8a* and valiolamine *8b* with the amine group occupying position 2 (OGAWA and TSUNODA 1992) and the C-glycoside *9* with an amine group in the α-position of the aglycone function (SCHMIDT and DIETRICH 1991) on intestinal α-glucosidases are not known, but the changed position of the amine group hardly suggests high activity.

What is the significance of the four hydroxyl groups of validamine for good inhibitory activity? The deoxy derivatives *10* ($IC_{50} = 2.8 \times 10^{-4} M$) (KAMEDA et al. 1985) and *11* ($IC_{50} = 2.4 \times 10^{-5} M$) (HORII and FUKASE 1985) as well as the epivaliolamine (*12*) ($IC_{50} = 4.7 \times 10^{-4} M$) (KAMEDA et al. 1985) have a markedly lower effect than the parent substances. These efficacy comparisons make it clear that the hydroxyl groups in the 3- and 6-positions have considerable significance for good efficacy of these inhibitors. In analogy to the better-examined structure-activity relationships of the derivatives of 1-deoxynojirimycin, one can, however, assume that the hydroxyl groups in positions 2 and 4 also contribute significantly to the binding to the enzyme.

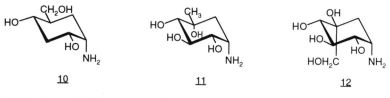

10   11   12

Fig. 6. Structures *10–12*

The pseudoglucosylamines discussed here were first prepared by fermentation or microbiological degradation of natural products. The first synthesis of optically active valienamine (PAULSEN and HEIKER 1981) began with L-quebrachite as the starting material and led to the target molecule in 22 steps. Larger quantities of valienamine, e.g., for systematic derivatization to optimize the efficacy, could not be prepared efficiently by this route.

In his validamine synthesis, OGAWA et al. (1985a) started from a readily available carboxylic acid obtained from furan and acrylic acid by a Diels-Alder reaction. The (−)-enantiomer (*13*) was first isolated by resolution of the racemate. This is converted into the triacetate (*14*) by hydroxylation of

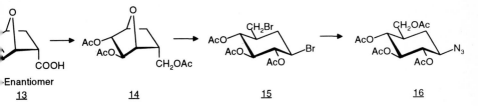

Fig. 7. Synthesis of validamine. (OGAWA et al. 1985a)

the double bond and reduction of the carboxyl group. Opening the bicycle with HBr gives the dibromide (*15*), which after successive treatment with acetate and azide yields compound *16*. The *N,O*-pentaacetate of validamine is obtained by hydrogenation with Raney nickel and deacetylation. Ogawa also prepared valienamine from the dibromide *15* (OGAWA et al. 1985b).

It seems obvious to use D-glucose as the starting material and make use of the homologous stereochemistry in the synthesis strategy. This concept was adopted by several groups and put into practice in different ways in detail. D-Glucose can be converted, via the intermediate *17*, into the carbocyclic ketone *18* by a procedure developed by FERRIER (1979) and optimized by SEMERIA et al. (1983). C-C coupling takes place between D-glucose carbons 1 and 6, carbon 6 becoming carbon 5a of the pseudosugar. Schmidt (KÖHN and SCHMIDT 1987) and Kuzuhara (HAYASHIDA et al. 1988) chose *18* as the starting material for their syntheses of valienamine and valiolamine. The missing carbon 6 of the pseudosugar is introduced into *18* via the carbonyl group (cf. KNAPP et al. 1992).

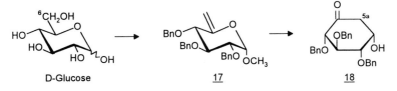

Fig. 8. Ferrier rearrangement of D-glucose

In another synthesis strategy the cyclitol ring was built up by coupling D-glucose carbons 1 and 5 using a reactive C-1 synthon. We mention here, as an example, S. Horii's very well elaborated valiolamine synthesis (FUKASE and HORII 1992). Starting from the perbenzylated D-glucono-1,5-lactone (*19*), readily accessible from D-glucose, reaction with dichloromethane and lithium diisopropylamide gives the adduct *20*. This is first reduced and then oxidized to the diketone *21*, which cyclizes spontaneously to dichloroinosose (*22*).

Reductive dehalogenation and treatment with hydroxylamine gives the oxime *23*. Catalytic hydrogenation of the oxime followed by removal of the

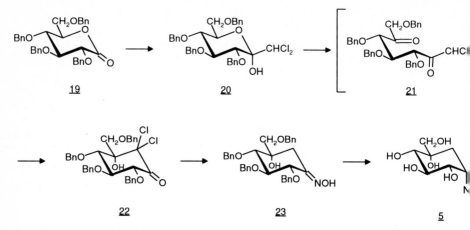

Fig. 9. Synthesis of valiolamine. (FUKASE and HORII 1992)

protective benzyl groups gives valiolamine (*5*). Similar synthesis routes to pseudosugars were followed by PAULSEN and DEYN (1987) and by ALTENBACH (1993), whereas Kitagawa, in his syntheses of valiolamine and valienamine, first coupled the C-1 synthon nitromethane with the 5-position of D-glucose and then completed the ring closure by an additional coupling to position 1 (YOSHIKAWA et al. 1988).

Valienamine (*4*), readily available by microbiological degradation of validamycin, can be converted in a few steps into the more effective valiolamine (*5*) by way of the cyclic carbamate *24* (HORII and FUKASE 1985).

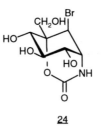

24

Fig. 10. Structure *24*

The pseudoglucosylamines discussed here are only weakly effective against microbial α-glucosidases. However, valiolamine in particular is a relatively potent inhibitor of the enzymes glucosidase I and II, involved in glycoprotein trimming, and of lysosomal α-glucosidase from rat liver (TAKEUCHI 1990).

II. N-Substituted Valiolamine Derivatives

Valiolamine and valienamine, as mimics of the oxonium ion *1*, still have the defect of the missing aglycone function R (a fructose or glucose residue). The question of which part of the aglycone residue mainly contributes to binding to the enzyme has been answered very systematically by HORII et al. (1986) by synthesis of numerous N-substituted valiolamine and valienamine derivatives. It was found that, as a first approximation, only the 3'-hydroxyl group of the aglycone of maltose *25* or sucrose *26* contributes significantly to this binding.

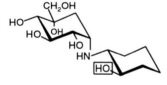

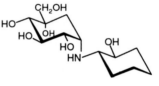

25   26

Fig. 11. Structures *25* and *26*

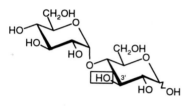

27   28

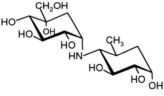

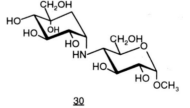

29   30

	IC$_{50}$ (M)		IC$_{50}$ (M)
Sucrase:	3.6×10^{-8}	Sucrase:	8×10^{-8}
Maltase:	6.8×10^{-8}	Maltase:	7.2×10^{-8}

Fig. 12. Structures *27–30*

Thus, of the two isomeric N-(*trans*-2-hydroxycyclohexyl)-valiolamines *27* and *28*, compound *27* is more effective than valiolamine: tenfold more active towards sucrase and 1000-fold towards maltase. *28* has a similar efficacy to valiolamine. Compounds such as *29* and *30*, which have structures even more similar to the enzyme substrate (maltose), are not more effective than *27*.

Whereas the introduction of an aglycone-like residue into the valiolamine molecule only leads to a modest increase in effectiveness towards sucrase, this procedure leads to a clear increase in efficacy in the case of valienamine (see Table 1).

If one assumes that the 3'-hydroxyl group of the disaccharide-like derivatives also interacts with the enzyme carboxyl group which protonates the glycosidic O-atom, this 3'-hydroxyl group would compete with the 5-hydroxyl group in the case of the valiolamine derivatives. This would explain the smaller improvement in efficacy. As the last example in the table shows, even very simple valiolamine derivatives are extremely potent sucrase and maltase inhibitors. *34*, with the generic name voglibose (AO-128), is an α-

Table 1. Inhibition of sucrase and maltase by valiolamine and valienamine derivatives

| | | | IC_{50} (M) | | IC_{50} (M) | |
			Sucrase	Maltase	Sucrase	Maltase
R = H	5:		4.9×10^{-8}	2.2×10^{-6}	4: 5.3×10^{-5}	3.4×10^{-4}
R = (structure)	30:		8×10^{-8}	7.2×10^{-8}	31: 3.2×10^{-7}	7.2×10^{-6}
R = (structure)	32:		1×10^{-8}	4.9×10^{-9}	33: 1.6×10^{-7}	3.2×10^{-6}
R = —CH (with CH$_2$OH, CH$_2$OH)	34:	4.6×10^{-9}	1.5×10^{-8}	35: 1.8×10^{-7}	5.9×10^{-6}

glucosidase inhibitor from Takeda, which is now on the marked in Japan for the treatment of diabetes.

Voglibose is obtained by reductive alkylation of valiolamine with dihydroxyacetone (HORII et al. 1986) or, alternatively, by reductive amination of the synthetically accessible inosose *36* with 2-aminopropane-1,3-diol and subsequent removal of the protecting groups (FUKASE and HORII 1992).

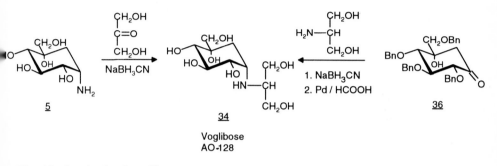

Fig. 13. Synthesis of voglibose

III. Acarviosin Derivatives

By degradation of acarbose (see Sect B.IV) with 2N-methanolic HCl at 60°C, one can obtain a mixture of anomeric methylglucosides of acarviosin *33* and *37*, from which the pure anomers have been obtained (JUNGE et al. 1984). The α-anomer *33* is a somewhat more potent sucrase inhibitor than acarbose and 2 orders of magnitude more effective than valienamine, whereas the β-anomer *37* is very weakly effective.

In addition to *33*, HORII et al. (1986) have synthesized the 6′-hydroxyl derivative *31*. Despite the greater structural similarity of this compound to the substrate of maltase, *31* is not a stronger inhibitor than *33* (see Table 1). The isomers of *31* and *33*, with inverted stereochemistry at C-4′ of the amino sugar unit (galacto configuration), were also formed in these syntheses. They are clearly less effective and only achieve the efficacy of valienamine.

OGAWA (1992) synthesized a series of derivatives of α-methylacarviosin modified in the sugar residue, e.g., by treatment of a protected valienamine

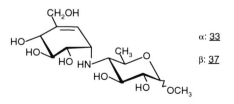

Fig. 14. Structures *33* and *37*

derivative with a suitable 3,4-anhydro sugar derivative. Unfortunately, the compounds obtained were only tested for their activity against α-glucosidase from yeast.

However, as far as can be ascertained, the structure-activity relationships agree with those found for N-substituted valiolamine derivatives towards sucrase. Thus, methylation of the OH group in the 3′-position (38) leads to a strong loss of efficacy, whereas methylation in the 2′-position (39) does not change the efficacy to any substantial extent.

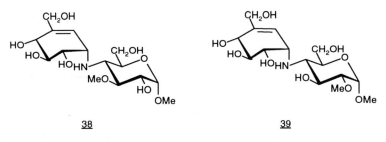

Fig. 15. Structures 38 and 39

An acarviosin derivative synthesized by Heiker (1982), in which an ethylene residue bridges the OH-group in position 1′ and 2′, has a good efficacy on intestinal sucrase.

Interestingly, in his work Ogawa (1990) discovered some potent pseudodisaccharide inhibitors of yeast α-glucosidase (e.g., 40 and 41), which at first glance have structures very different from acarviosin. 40 and 41 inhibit the α-glucosidase with IC_{50} values of 5.6×10^{-7} and $4.5 \times 10^{-6} M$, respectively. As an explanation, one can postulate that in these compounds, with a $^1C_4$ conformation in the sugar part of the molecule, the OH group in the 2′-position takes over the role of the 3′-OH group in acarviosin derivatives.

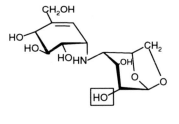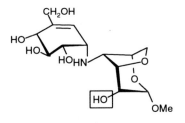

Fig. 16. Structures 40 and 41

The concept of using disaccharide-like pseudoglucosylamines resembling the natural substrates to inhibit glucoside-cleaving enzymes is not restricted to sucrase. For example, the high efficacy of validoxylamine A (*42*) towards various trehalases (K_i values from 10^{-7} to $10^{-10}\,M$; KAMEDA et al. 1987) is comprehensible if one compares its structure with that of trehalose (*43*).

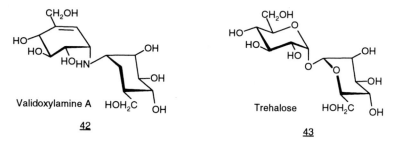

Validoxylamine A   *42*

Trehalose   *43*

Fig. 17. Structures *42* and *43*

IV. Acarbose

Towards the end of the 1960s, Bayer searched for inhibitors of pancreatic α-amylase and intestinal disaccharidases, in particular sucrase. High-moleculear-weight inhibitors of α-amylase were at first found in culture broths of microorganisms. Later it was possible to control the fermentation so that mainly low-moleculear-weight inhibitors were formed. These inhibitors were highly active on intestinal sucrase. The *Actinoplanes* strain SE 50 finally yielded a potent sucrase inhibitor, acarbose (*44*) [$IC_{50} = 0.5 \times 10^{-6}\,M$

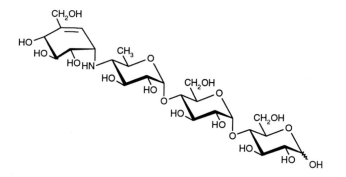

O - {4,6 - Dideoxy - 4 - [1(*S*) - (1, 4, 6/5) - 4,5,6 - trihydroxy - 3 -
-hydroxymethyl - 2 - cyclohexene - 1 - yl] amino - α - D - glucopyranosyl}
- (1 → 4) - *O* - α - D - glucopyranosyl - (1→ 4) - D - glucose

44

Fig. 18. Structure *44*

(pig)], which was selected for more detailed examination and finally for clinical development for the treatment of diabetes mellitus (SCHMIDT et al. 1977). After several years of intensive clinical development, in 1990 acarbose became the first α-glucosidase inhibitor to be launched in Germany. Acarbose provides the physician with the first new therapeutic principle for the treatment of diabetes in nearly 40 years.

Determination of the structure of the pseudotetrasaccharide acarbose (*44*) is based partly on spectroscopic studies and also on a series of degradation reactions which break acarbose down into several small fragments with known structures (JUNGE et al. 1984a).

Acid hydrolysis under vigorous reaction conditions splits off the two glucose units of acarbose. The remaining part of the molecule, which we called acarviosin (*45*), cannot be isolated. It is not stable under the reaction conditions but condenses to a tricyclic compound *46*. The oxazolidine ring in this molecule is easily opened by reduction with NaBH₄, yielding a bicyclic compound *47*, which is further cleaved with Pt/H₂.

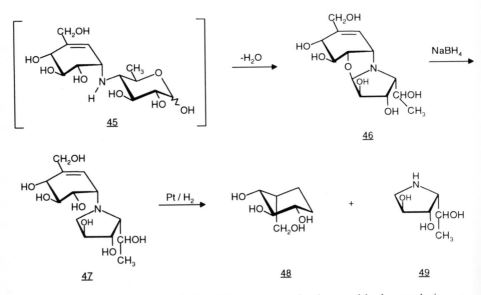

Fig. 19. Degradation of acarviosin: dehydration, reduction, and hydrogenolysis

The products of this cleavage are two small molecules, one of which, validatol (*48*), is known from the literature (HORII et al. 1971b). The isolation of *48* gave important information on the stereochemistry of the cyclitol part of acarbose. The other compound was identified as the pyrrolidine derivative *49*. Acarviosin, the core structure of acarbose, can, however, be isolated in the form of its methylglycosides if one degrades acarbose with methanolic HCl under carefully controlled conditions. A mixture of the anomeric acarviosinides *33* and *37* is obtained, which can be cleaved further into the cyclitol and amino sugar fractions *48* and *50* with Pt/H$_2$. Validatol (*48*) is easily separated and, after acetylation, so is *51*, which is identical with the known methyl-tri-*O*,*N*-acetyl-α-D-viosaminide (STEVENS et al. 1966).

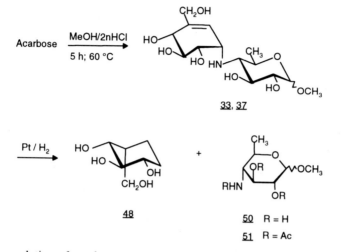

Fig. 20. Degradation of acarbose: methanolysis and hydrogenolysis

Hydrogenation of acarbose itself with Pt/H$_2$ yields, among other products, validatol and a trisaccharide *52*, which is further degraded with acetic anhydride/sulfuric acid to give another known derivative of the amino sugar viosamine (*53*), besides penta-*O*-acetyl-α-D-glucospyranose *54* (JUNGE

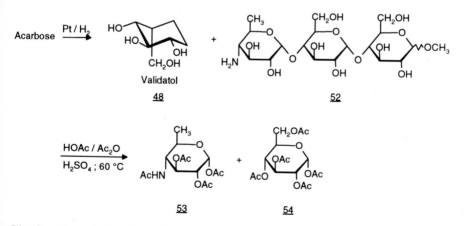

Fig. 21. Degradation of acarbose: hydrogenolysis and acetolysis

et al. 1984a; see STEVENS et al. 1966). These simple degradation reactions, yielding small molecules known from the literature, unambiguously proved the structure of acarbose.

The first total synthesis of acarbose was achieved by Ogawa, who treated a valienamine derivative 55, accessible in six steps from 15, with an epoxide (56), obtained from maltotriose (SHIBATA and OGAWA 1989).

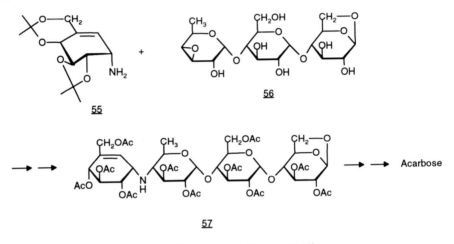

Fig. 22. Synthesis of acarbose. (SHIBATA and OGAWA 1989)

Reaction of 55 and 56, after removal of the protecting isopropylidene groups, acetylation, and chromatography, gives the 1,6-anhydro compound 57 in 19% yield. After acetolysis and deacetylation, one finally obtains acarbose. The acarbose derivative with a hydroxyl group in the 6-position of the amino sugar was prepared in a similar way.

Using a similar synthesis strategy, Kuzuhara attempted to prepare acarbose by reductive alkylation of a valienamine derivative with a trisaccharide ketone. He obtained only a very low yield of an acarbose isomer with the 4-epi configuration in the amino sugar unit (HAYASHIDA et al. 1989). The inverse synthesis strategy was used more successfully for the production of dihydroacarbose. Treatment of the ketone 58 with the amino sugar 59 gave the dihydroacarbose derivative 60 in 30% yield. 60 can be converted into dihydroacarbose 61 in a few steps (HAYASHIDA et al. 1986, 1989).

Coupling of the cyclitol and amino sugar fractions should in principle also be possible by the reaction of an allyl bromide with an amino sugar. However, treatment of the bromide 62, accessible from Ferrier's ketone, with the amino sugar derivative 63, accessible from maltose, gives the target molecule 64 in only 5% yield after a difficult separation (SAKAIRI and KUZUHARA 1982).

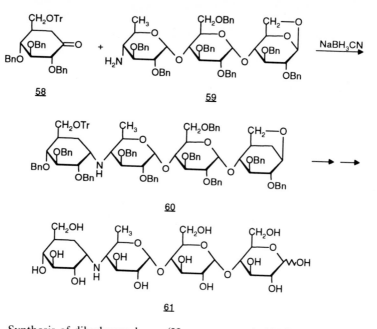

Fig. 23. Synthesis of dihydroacarbose. (HAYASHIDA et al. 1989)

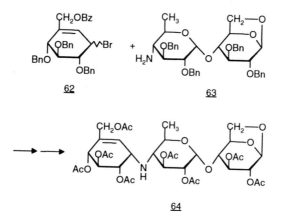

Fig. 24. Synthesis of intermediate *64*. (SAKAIRI and KUZUHARA 1982)

The two D-glucose units of acarbose are of subordinated importance for the in vitro activity on sucrase (JUNGE et al. 1984a). This is shown by the fact that α-methylacarviosin (*33*) is somewhat more effective than acarbose. On the other hand, the pseudotrisaccharide obtained by cleavage of one glucose unit from acarbose by mild acid (component 2, amylostatin XG) is a somewhat weaker sucrase inhibitor than acarbose in vitro (MÜLLER 1989).

Numerous β-glucosides of acarbose have been prepared, which are hardly more effective than acarbose itself (Junge et al. 1980). Cleavage of the cyclitol fraction of acarbose gives a basic trisaccharide 52, which has no effect on sucrase.

The other structure-activity relationships of acarbose derivatives fit seamlessly into the findings already discussed for valienamine and acarviosin derivatives. Dihydroacarbose (Junge et al. 1984a; Bock and Meldal 1991) is just as effective as acarbose, whereas the derivatives with saturated cyclitol fractions, having the L-ido configuration and inverse $^1C_4$ conformation, obtained by hydrogenation of acarbose, are ineffective on sucrase (Junge et al. 1984a). The 6-deoxy derivative of dihydroacarbose, similarly formed by hydrogenation, is also ineffective. Adiposine-2, the pseudotetra-saccharide with a 6-hydroxyl group in the amino sugar, has a similar activity to acarbose (Namiki et al. 1982b; Kangouri et al. 1982).

V. Higher Pseudo-oligosaccharides

Compounds with more than two glucose units coupled to the acarviosin core (65) are formed as by-products of the fermentation of acarbose (Truscheit et al. 1981). Inhibitors of this type have also been described as amylostatins (Murao and Ohyama 1975, 1979). Whereas acarbose is the most potent inhibitor of intestinal α-glucosidases, and in particular of sucrase, the pancreatic α-amylase inhibition of the compounds increases with ascending glucose units (Müller et al. 1980). Acarviosin is also the central unit of the inhibitor AI-5662 (Vertesy et al. 1986, Hoechst) and the trestatines A, B, and C (Yokose et al. 1983, Hoffmann-La Roche). In the trestatines a trehalose molecule is coupled 1,4-α-glucosidically with the reducing end of acarbose.

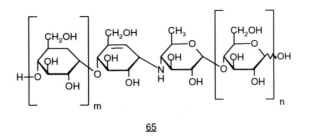

Fig. 25. Structure 65

Various microbial inhibitors with structures derived from acarviosin-like oligostatins (Itoh et al. 1981; Omoto et al. 1981), epoxy derivatives (Takeda et al. 1983), adiposins (Otani et al. 1979; Namiki et al. 1982a), and derivatives of AI-5662 (Vertesy et al. 1988) have been isolated. However, none of these high-molecular inhibitors have reached clinical use or market.

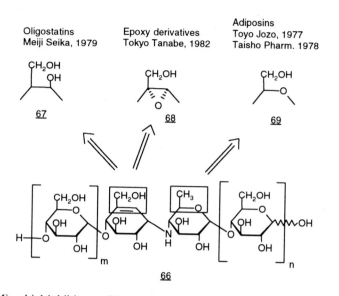

Fig. 26. Microbial inhibitors with structures derived from acarviosin

C. Polyhydroxypiperidines and Polyhydroxypyrrolidines

I. Nojirimycin

In 1965 a new antibiotic R-468 was discovered in fermentation broths of *Streptomyces roseochromogenes* R-468 (NISHIKAWA and ISHIDA 1965). Shortly thereafter the two antibiotics SF-425 and SF-426 were isolated as fermentation products of the species *S. lavendulae* SF-425 and *S. nojiriensis* n. sp. SF-426 (from a soil near Lake Nojiri) (TSURUOKA et al. 1966; SHOMURA et al. 1966; ISHIDA et al. 1967a; ISHIDA et al. 1967b; ISHIDA et al. 1968). The antibiotics R-468, SF-425, and SF-426 proved to be identical and were named nojirimycin (INOUYE et al. 1966; ISHIDA et al. 1967a). Nojirimycin has also been obtained by fermentation of *S. ficellus* NRRL 8067 (ARGOUDELIS et al. 1976) and in a screening program for sucrase inhibitors in culture broths of bacilli of various provenances (SCHMIDT et al. 1979).

The structure of nojirimycin [(+)-5-amino-5-deoxy-D-glucopyranose] (*70*) is closely related to D-glucose (*71*), with a nitrogen atom in place of the ring oxygen. Nojirimycin exists in aqueous solution in a cyclic form as a mixture of α/β-anomers in a ratio of 60:40 (INOUYE et al. 1966; PINTO and WOLFE 1982). *70* is thus the first "heterose" (PAULSEN 1966) to be found in nature.

The slightly basic semicrystalline compound (specific rotation $[\alpha]_D^{21}$ = 49, c = 1, water) melts with decomposition at about 115°C, is relatively stable in basic media, but readily dehydrates in neutral or acidic solution to form a pyridine derivative. Nojirimycin can be stabilized as a crystalline

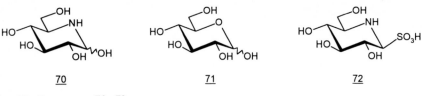

Fig. 27. Structures 70–72

bisulfite adduct 72 with a cyclic structure both in the solid state and in aqueous solution.

The sulfonate group almost exclusively adopts an equatorial orientation (LEGLER and BECHER 1982; KODAMA et al. 1985). The bisulfite adduct 72 was also the target of the first chemical synthesis of nojirimycin (INOUYE et al. 1968; INOUYE 1970). The furanose derivative 73 was prepared in four steps starting from glucose (71). After oxidation of the 5-hydroxyl group in 73 to a ketone function (74), treatment with hydroxylamine produced an oxime which, by a largely stereoselective reduction, produced the amino compound 75 predominantly in the desired conformation. After removal of the protecting groups, the action of sulfurous acid produced the bisulfite adduct 72.

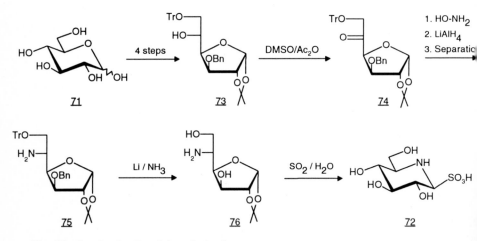

Fig. 28. Synthesis of nojirimycin by Inouye

The derivative 76, synthesized via various intermediates (WHISTLER and GRAMERA 1964; KLEMER et al. 1979), forms a key compound in numerous synthetic procedures for nojirimycin which were published in the years which followed (SAEKI and OHKI 1968a; SAEKI and OHKI 1968b; VASELLA and VOEFFRAY 1982; TSUDA et al. 1988; TSUDA et al. 1989; STASIK et al. 1990; ANZEVENO and CREEMER 1990). Other syntheses of nojirimycin start from tartaric acid esters (IIDA et al. 1987) or myo-inositol derivatives (CHIDA et al. 1989). The latter procedure also enabled (−)-nojirimycin 77, the aza-derivative of L-glucose, to be produced for the first time. *77* was later also obtained starting from serine (DONDONI et al. 1993).

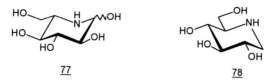

<u>77</u>   <u>78</u>

Fig. 29. Structures 77 and 78

In 1970 an inhibitory effect of (+)-nojirimycin (*70*) on nonmammalian α-glucosidases (takadiastase and glucoamylase from *Rhizopus niveus* with phenyl-α-D-glucoside as substrate) and β-glucosidases (from apricot emulsin and *Trichoderma viride*; substrate, *p*-nitrophenyl-β-D-glucoside) was found (NIWA et al. 1970). Further studies showed that the nojirimycin bisulfite adduct *72* inhibits α- and β-glucosidases of plant and microbiological origin (REESE et al. 1971; see also, e.g., MÜLLER 1985; LEGLER 1990).

Only several years later was it reported that nojirimycin is an effective inhibitor of α-glucosidases from the mucosa of the pig small intestine (SCHMIDT et al. 1979).

(−)-Nojirimycin (*77*) is only known to have a weak activity against yeast α-D-glucosidase (CHIDA et al. 1989).

II. 1-Deoxynojirimycin

The preparation of 1-deoxynojirimycin (1,5-dideoxy-1,5-imino-D-glucitol) *78* was first referred to by Paulsen in a review article in 1966 (PAULSEN 1966). A detailed description of the synthesis of *78* starting from L-sorbose (*79*) followed 1 year later (PAULSEN et al. 1967) (Fig. 30). Meanwhile *78* had also been synthesized by reduction of *70* using either catalytic hydrogenation over platinum oxide or sodium borohydride (INOUYE et al. 1966). 1-Deoxynojirimycin is stable towards bases, acids, and weak oxidants. It melts between 195°C and 206°C and has a specific rotation $[\alpha]_D^{21}$ of +47 (*c* = 1.045, water) (INOUYE et al. 1966; PAULSEN et al. 1967; SCHMIDT et al. 1979). During subsequent years preferentially variations of the routes of Inouye (Fig. 28; cf. the literature in Sect. C.I. on the preparation of *70* and BROXTERMAN et al. 1987; SCHMIDT RR et al. 1989; DAX et al. 1990a) and Paulsen (Fig. 30) (KÖBERNICK and FURTWÄNGLER 1981; BEAUPERE et al. 1989; BEHLING et al. 1991) were used to synthesize 1-deoxynojirimycin.

In addition, new approaches to the preparation of *78* were developed via different intermediates derived from carbohydrates (VASELLA and VOEFFRAY 1982; BERNOTAS and GANEM 1984; BERNOTAS and GANEM 1985; SETOI et al. 1986; FLEET et al. 1987; FLEET et al. 1990a; ERMERT and VASELLA 1991; OVERKLEEFT et al. 1993; SMID et al. 1993; POITOUT et al. 1994; KIGUCHI

430   B. JUNGE et al.

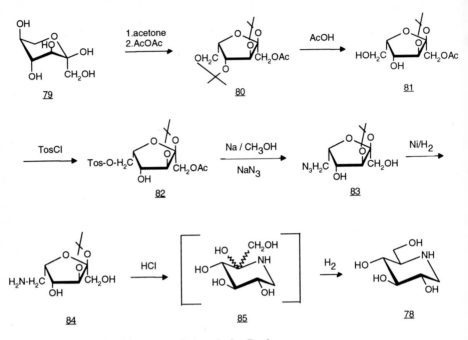

Fig. 30. Synthesis of 1-deoxynojirimycin by Paulsen

et al. 1995). Alternatively, the necessary chirality was transferred from
tartaric acid (IIDA et al. 1987), pyroglutamic acid (IKOTA 1989), serine
(DONDONI et al. 1993), phenylalanine (RUDGE et al. 1994), or myo-inositol
(CHIDA et al. 1992). Also obtained from myo-inositol was (−)-1-deoxyno-
jirimycin, the enantiomer of the naturally occurring (+)-1-deoxynojirimycin
(cf. LIU et al. 1991).

A very economical synthesis of 1-deoxynojirimycin was introduced by
the development of a method involving a combination of chemical synthesis
and biotransformation (KINAST and SCHEDEL 1980; KINAST and SCHEDEL
1981) (Fig. 31). D-Glucose is converted into 1-deoxy-1-aminosorbitol by
catalytic reductive amination. After protection of the amino group, compound 87 is selectively oxidized at the hydroxyl group in position 5 by
fermentation with *Gluconobacter oxydans*. The protecting group is then
removed by catalytic hydrogenation. By simultaneous stereoselective ring
closure, 1-deoxynojirimycin (78) is obtained in high yield. Further optimization of this procedure was achieved with acid-, base- and penicillin-acylase-
labile protecting groups (KINAST et al. 1982; SCHRÖDER and STUBBE 1987;
SCHUTT 1991).

A similar principle of stereoselective ring formation was used for the
double reductive amination of 5-keto-D-glucose. The preparation of this
keto sugar, however, is not without problems (REITZ and BAXTER 1990;
BAXTER and REITZ 1994).

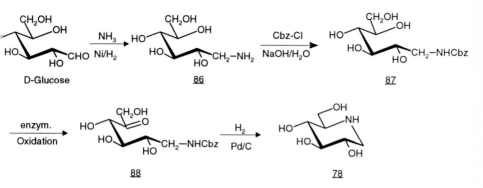

Fig. 31. Combined chemical and microbiological process

A combined chemical-enzymatic procedure can also be carried out with an achiral starting material (ZIEGLER et al. 1988; STRAUB et al. 1990; PEDERSON et al. 1993; PEDERSON and WONG 1989; VON DER OSTEN et al. 1989; LOOK et al. 1993; EFFENBERGER et al. 1993; WONG et al. 1995). The chirality is introduced by stereoselective aldol addition of dihydroxyacetone phosphate (*89*) to 3-azido-2-hydroxypropanal (*90*) under catalysis by rabbit muscle aldolase. 1-Deoxynojirimycin is obtained after cleaving the phosphate group in *91* with phosphatase and subsequent catalytic hydrogenation (Fig. 32) (for a detailed description of the syntheses of 1-deoxynojirimycin see the recently published review article of HUGHES and RUDGE 1994).

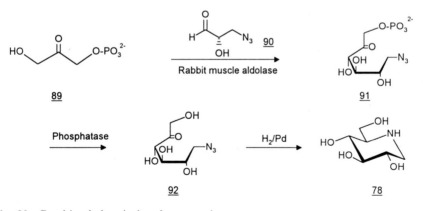

Fig. 32. Combined chemical and enzymatic process

1-Deoxynojirimycin (*78*) was first found in nature 10 years after its initial chemical synthesis. It is a constituent of the roots and bark of the mulberry tree and was designated moranolin (YAGI et al. 1976; MURAI et al. 1977; DAIGO et al. 1986; cf. YAMADA et al. 1993; ASANO et al. 1994a,b). 1-

Deoxynojirimycin was later found by extraction of the plants *Jacobinia suberecta* and *J. tinctoria* (MATSUMURA et al. 1980; SHIBATANI et al. 1980).

In a screening program for inhibitors of mammalian intestinal α-glucosidases, 1-deoxynojirimycin was found to be produced by numerous species of the genus *Bacillus* (FROMMER et al. 1978a, 1979a; SCHMIDT et al. 1979; FROMMER and SCHMIDT 1980; STEIN et al. 1984; HARDICK and HUTCHINSON 1993). It was also discovered in the culture broths of various species of *Streptomyces lavendulae* (MATSUMURA et al. 1979c; MITSUO and SAWAO 1980; MURAO and MIYATA 1980; EZURE et al. 1985) and *S. subrutilus* (HARDICK et al. 1991, 1992). Compared with nojirimycin, 1-deoxynojirimycin is a weaker inhibitor of nonmammalian and mammalian β-glucosidases (NIWA et al. 1970; see also: TRUSCHEIT et al. 1981; MÜLLER 1985; SCOFIELD et al. 1986; TRUSCHEIT 1987; ELBEIN 1987; TRUSCHEIT et al. 1988; YOSHIKUNI et al. 1988; LEGLER 1990; NISHIMURA 1991; YOSHIKUNI 1991; ROBINSON et al. 1992; HUGHES and RUDGE 1994). However, the discovery of the inhibitory effect on mammalian α-glucosidases opened the possibility of a therapeutic application for 1-deoxynojirimycin. Nippon Shinyaku (OHATA et al. 1977; MURAI et al. 1977) and Bayer AG (FROMMER et al. 1978b, 1979b; SCHMIDT et al. 1979) independently discovered that both 1-deoxynojirimycin and nojirimycin are strong inhibitors of α-glucosidases in the digestive tract and thus suited to the treatment of diabetes and obesity.

Table 2 shows that the levels of inhibition achieved in vitro by 1-deoxynojirimycin (*78*) and nojirimycin (*70*) towards various disaccharidases solubilized from porcine brush border mucosa are virtually identical (SCHMIDT et al. 1979). More detailed studies have shown that the IC_{50} values of 1-deoxynojirimycin for intestinal α-glucosidases of various origins (mouse, rat, dog, monkey) are in the range of $6.6-16 \times 10^{-8} M$ for sucrase and $7.4-63 \times 10^{-8} M$ for maltase (MURAO and MIYATA 1980; YOSHIKUNI 1988, 1991).

III. N-Substituted Derivatives of 1-Deoxynojirimycin

1-Deoxynojirimycin (*78*) is an excellent starting material for chemical derivatization and can be selectively substituted at the nitrogen atom either by treatment with reactive alkyl halides (path A) or by reductive alkylation

Table 2. Molar Concentrations of nojirimycin and 1-deoxynojirimycin required for a 50% inhibition of intestinal α-glucosidases. A mixture of the enzymes (Pig small intestinal mucosal homogenates) (SCHMIDT et al. 1979) and sucrose, maltose, isomaltose, and starch as substrates were used

Inhibitor	Sucrase (M)	Maltase (M)	Isomaltase (M)	Glucoamylase (M)
70	5.6×10^{-7}	1.7×10^{-6}	2.5×10^{-7}	7.6×10^{-7}
78	2.2×10^{-7}	1.3×10^{-7}	1.3×10^{-7}	9.6×10^{-8}

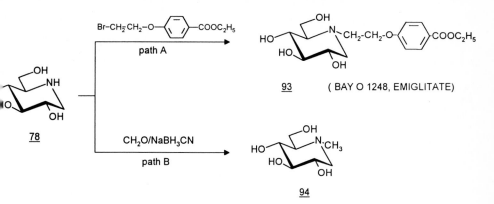

Fig. 33. Alkylation methods of 1-deoxynojirimycin

with carbonyl compounds (path B), without using protecting groups (JUNGE et al. 1979; MATSUMURA et al. 1979a,b; MURAI et al. 1979; JUNGE et al. 1981; VAN DEN BROEK et al. 1994).

Further preparative methods were developed for the *N*-hydroxyethyl derivative *97* (BAY m 1099, miglitol) (KINAST et al. 1982; KÖBERNICK 1981; HINSKEN 1990). The route via the bicyclic *N,O*-hemiacetal *95*, obtained by treating 1-deoxynojirimycin with glyoxal, is interesting from a chemical standpoint. Catalytic hydrogenation followed by reduction with sodium borohydride converts *95* into miglitol (*97*) (HINSKEN 1991) (Fig. 34).

N-Acyl derivatives of 1-deoxynojirimycin are obtained by selective acylation of *78* without using protecting groups (JUNGE et al. 1979) (Fig. 35).

A general route to *N*-aryl derivatives consists of the treatment of compound *82* with anilines at 100°–150°C to form arylaminosorbose derivatives

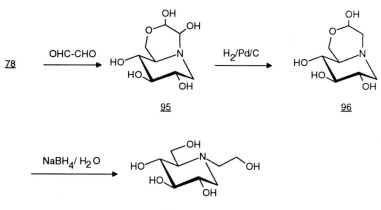

Fig. 34. Synthesis of miglitol

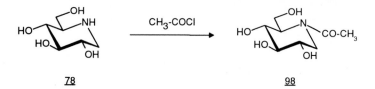

Fig. 35. Preparation of *N*-acetyl-1-deoxynojirimycin

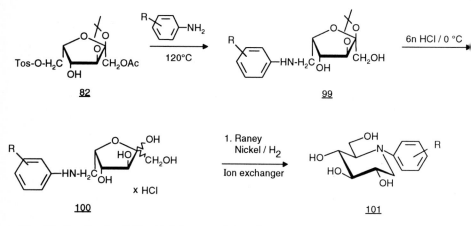

Fig. 36. Synthesis of *N*-aryl-1-deoxynojirimycins

of formula *99*. This reaction generally gives yields of around 20%. By cleaving off the isopropylidene-protecting group with 6*N* hydrochloric acid at 0°C, one obtains the derivative *100*. After hydrogenation and neutralization, the free base *101* is obtained (JUNGE et al. 1989) (Fig. 36).

Only in rare cases can *N*-aryl-1-deoxynojirimycins be produced directly by the reaction of 1-deoxynojirimycin with activated aryl halides. In the example given below the *p*-nitrophenyl substituent forces the piperidine ring to adopt the inverted $^1C_4$-conformation (cf. Fig. 39) (JUNGE et al. 1989) (Fig. 37). The *N*-amino-1-deoxynojirimycin (*104*) can be obtained with surprising ease by treating *78* with nitrous acid and subsequent reduction of the intermediate nitroso compound (KINAST et al. 1980; JUNGE et al. 1989) (Fig. 38).

Despite the excellent α-glucosidase inhibitory activity of 1-deoxynojirimycin *78* in vitro, its efficacy in vivo is only moderate. We therefore prepared a large number of 1-deoxynojirimycin derivatives in the hope of increasing the in vivo activity. Acarbose was used as an internal standard in our investigations of the inhibitory effect of 1-deoxynojirimycin derivatives towards the raw sucrase complex. No K_i values were determined, the figures cited being relative values for the inhibitory effect in comparison

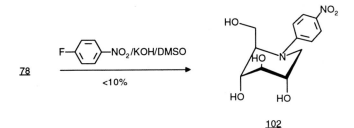

Fig. 37. Direct arylation of 1-deoxynojirimycin

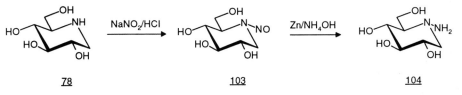

Fig. 38. Preparation of *N*-amino-1-deoxynojirimycin

to acarbose. These data differ in a few cases from some results in the literature, which are also not always consistent (LEMBCKE et al. 1985; SCOFIELD et al. 1986; SAMULITIS et al. 1987; YOSHIKUNI et al. 1988; ROBINSON et al. 1991, 1992). This is partly due to the fact that the disaccharidases used are not pure enzymes and originated from the small intestines of various mammalian sources (pig, rabbit, rat, mouse). Since the inhibition values also depend on substrate concentration and incubation time, a two- to fourfold difference can easily result and should not be overinterpreted. In those cases where the data show large differences, this will be specifically indicated. Tables 3–7 below give an insight into the in vitro structure-activity relationships of a large number of N-substituted deoxynojirimycin derivatives.

Whereas, according to data in the literature, the inhibition of intestinal sucrase from rabbits by 1-deoxynojirimycin (*78*) is very marked and only approached by a few derivatives, we found using raw porcine intestinal mucosa disaccharidase complex that there are derivatives giving clearly better inhibition than *78* (Table 3). Thus, the *N*-methyl derivative *94*, in contrast to the data in the literature, is about seven times as effective as *78* and 44 times more effective than acarbose. The efficacy falls steeply with increasing alkyl group chain length up to the propyl compound *106*, and then increases again with longer alkyl residues. As the alkyl chain length is further increased, the inhibitory effect is rapidly lost. Branching of the alkyl residue in the immediate neighborhood of the nitrogen atom leads to a complete loss of activity (*114, 115*). In contrast, a noteworthy increase in activity compared with saturated alkyl chains occurs with alkenyl residues

Table 3. In vitro activity of N-substituted 1-deoxynojirimycin derivatives

Inhibitor	R	Relative sucrase inhibition[a] (Acarbose = 1)	IC_{50} (μM)[b]	
			Sucrase	Maltase
78	—H	6.0	0.06 (m)	0.37 (m)
			0.12, 0.2 (r)	0.12 (r)
			0.41	1.0
			0.22 (p)	0.13 (p)
94	—CH$_3$	44.0	0.38	2.6
105	—CH$_2$—CH$_3$	5.3	0.79	2.2
106	—CH$_2$—CH$_2$—CH$_3$	0.5	1.4	10.9[c]
107	—CH$_2$—(CH$_2$)$_2$—CH$_3$	1.0		
108	—CH$_2$—(CH$_2$)$_3$—CH$_3$	2.4		
109	—CH$_2$—(CH$_2$)$_6$—CH$_3$	2.9	0.51[c]	3.6[c]
110	—CH$_2$—(CH$_2$)$_7$—CH$_3$	4.3	0.76[c]	7.4[c]
111	—CH$_2$—(CH$_2$)$_8$—CH$_3$	7.5	0.65[c]	6.7[c]
112	—CH$_2$—(CH$_2$)$_9$—CH$_3$	12.7	0.82[c]	9.0[c]
113	—CH$_2$—(CH$_2$)$_{10}$—CH$_3$	10.8	1.6[c]	3.6[c]
114	—CH(CH$_3$)$_2$	<0.1		
115	—cyclohexyl	<0.1		
116	—CH$_2$—CH=CH$_2$	13.2		
117	—CH$_2$—C≡CH	2.2		
118	—CH$_2$—CH=CH—CH$_3$	26.1		
119	—CH$_2$—(CH=CH)$_2$—CH$_3$	40.0		

Enzyme Preparation. Crude disaccharidase homogenates from swine small intestine mucosa by tryptic digestion (substrate, sucrose).
Assay. (1) Preincubation of 100 μl enzyme solution (0.15 U/ml) with 10 μl inhibitor solution at pH 6.25, 37°C, for 10 min, (2) reaction with 100 μl 0.4 M sucrose solution, (3) reaction stopped with 0.5 M TRIS buffer, D-glucose determined with D-glucose dehydrogenase reagent.
Internal Standard. Acarbose with 77.7 SIU/mg.
IC_{50}, rabbit or mouse(m), rat(r), pig(p) intestine mucosal homogenates with corresponding substrate.
[a] Relative sucrase inhibitory activity: SIU/mg test compound/SIU/mg acarbose (MÜLLER et al. 1980).
[b] Data from SCHMIDT et al. (1979), SCOFIELD et al. (1986), YOSHIKUNI (1988), YOSHIKUNI et al. (1988), ROBINSON et al. (1991, 1992 and references therein).
[c] Tosylate.
[d] Dipicrate.

containing a double bond in the allyl position (*116, 118, 119*). Derivatives with a triple bond, however, have a weaker relative effect (*117*).

Very variable effects were achieved using substituents in the alkyl residues (Table 4). Polar substituents (*120–122*) give mainly almost inactive compounds. The hydroxyethyl residue is, however, an exception. In this case the effect is retained (*97*, miglitol). The phenoxy group, in contrast to the methoxy group, gives a further increase in the inhibitory effect (*123, 124*), which declines again with longer alkyl chain length (*125–127*). Here again the effect is clearly increased by the introduction of an allyl double bond (*128*). As expected, polar substituents on the phenyl ring (*93*) again

Table 4. In vitro activity of N-substituted 1-deoxynojirimycin derivatives

Inhibitor	R	Relative sucrase inhibition[a] (Acarbose = 1)	IC_{50} (μM)[b] Sucrase	Maltase
93	$-(CH_2)_2-O-C_6H_4-COOC_2H_5$	8.1	0.41 (r)	1.7 (r)
97	$-CH_2-CH_2-OH$	6.0	2.99 0.62 (r)	7.29 2.9 (r)
120	$-CH_2-CH_2-CH_2-NH_2$	<0.1		
121	$-CH_2-CH_2-COOH$	0.35		
122	$-CH_2-CH_2-CN$	<0.1		
123	$-CH_2-CH_2-O-CH_3$	3.3		
124	$-(CH_2)_2-O-C_6H_5$	17.0		
125	$-(CH_2)_3-O-C_6H_5$	0.6		
126	$-(CH_2)_4-O-C_6H_5$	1.7		
127	$-(CH_2)_5-O-C_6H_5$	3.7		
128	$-CH_2-CH=CH-CH_2-O-C_6H_5$	20.0		
129	MDL 73945 (structure)		0.2 (r)	1.0 (r)

See Table 3 for explanation of footnotes.

lower the inhibitory effect. However, compound *93* (emiglitate) is characterized by a longer lasting effect in vivo than for *124* (see Table 9). The *N*-glucosyl derivative *129* (MDL 73945) also shows a marked inhibitory effect on sucrase and maltase.

Aralkyl derivatives of 1-deoxynojirimycin (*130–133*) are very strong inhibitors of porcine intestinal sucrase (Table 5). It is noticeable that in each case the compound with a three-carbon chain, both in the pure alkyl series and also among the substituted derivatives, gives the least inhibition (*106*, *125*, *132*).

Whereas polar groups on the phenyl ring clearly reduce the activity again (*134*), introduction of a double bond achieves the increase in activity already known from other derivatives (*135*, *136*). The polar N-amino compound *104* (Table 6) still is clearly active. Derivatives of *104* are less effective (*137*). *N*-Acyl (*98*, *138*) and *N*-aryl (*102*, *139*) derivatives as well as compounds with quarternary nitrogen atom (*104*, *141*) are completely inactive.

Table 5. In vitro activity of N-substituted 1-deoxynojirimycin derivatives

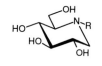

Inhibitor	R	Relative sucrase inhibition[a] (Acarbose = 1)	IC_{50} (μM)[b] Sucrase	Maltase
130	—CH₂—⟨phenyl⟩	6.1	1.90[c]	22.1[c]
131	—(CH₂)₂—⟨phenyl⟩	11.8	4.55[c]	34.1[c]
132	—(CH₂)₃—⟨phenyl⟩	2.4	1.06[c]	10.6[c]
133	—(CH₂)₄—⟨phenyl⟩	9.7	0.32[c]	5.99[c]
134	—CH₂—⟨phenyl⟩—COOH	0.25		
135	—CH₂—CH=CH—⟨phenyl⟩	23.0–40.0	0.79	10.0
136	—CH₂—CH=CH—⟨phenyl⟩ (+)(CH₂)₃N—(CH₂)₂O		2.86[d]	5.13[d]

See Table 3 for explanation of footnotes.

Table 6. In vitro activity of N-substituted 1-deoxynojirimycin derivatives

Inhibitor	R	Relative sucrase inhibition[a] (Acarbose = 1)
98	—COCH$_3$	<0.1
102		<0.1
104	—NH$_2$	3.9
137	—NH—CO—CH$_3$	0.3
138		<0.1
139		<0.1
140		0.4
141		~0.05

See Table 3 for explanation of footnote.

This is due to the reduced basicity and especially to the conformation of these derivatives. X-ray analysis indicated that 1-deoxynojirimycin preferentially adopts the $^4C_1$ *gauche-trans* conformation. However, *N*-alkyl derivatives such as miglitol (*97*) adopt the *gauche-gauche* conformation, in which the C$_6$-hydroxyl group occupies a position perpendicular to the pyranose ring (Fig. 39) (A. Göhrt, Bayer AG, private communication; cf. Linden et al. 1994). These observatons are in accordance with results from NMR studies. A preference of the *gauche-trans* conformation in 1-deoxynojirimycin (*78*) and the *gauche-gauche* conformation in N-substituted deoxynojirimycin derivatives was found not only in aqueous solution (Glaser and Perlin 1988) but also in organic solvents (Tan et al. 1991; Van den Broek et al. 1993). Additionally, these findings were corroborated by molecular modeling studies of *78* and *94* (Kajimoto et al. 1991a). Probably a *gauche-gauche* conformation achieves an additional stabilizing effect of the enzyme/inhibitor

Table 7. In vitro activity of bridged deoxynojirimycin derivatives

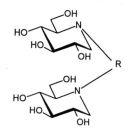

Inhibitor	R	Relative sucrase inhibition[a] (Acarbose = 1)	IC$_{50}$ (μM)[b] Sucrase	Maltase
142	—CH$_2$—CH$_2$—		42.6	162
143	—(CH$_2$)$_5$—	1.9		
144	—(CH$_2$)$_8$—	0.9		
145	—CH$_2$—CH=CH—CH$_2$—	1.1		
146	—CH$_2$—(CH=CH)$_2$—CH$_2$—	9.4		
147	—CH$_2$—CH=CH— / —CH$_2$—CH=CH— (benzene)	14.1	1.46	6.03
148	—CH$_2$—CH$_2$—O—⟨benzene⟩—O—CH$_2$—CH$_2$—	6.1		
149	—CH$_2$—CH$_2$—O—⟨benzene⟩—CH$_2$—	4.2		
150	(CH$_2$)$_3$⟨ O—⟨benzene⟩—CH=CH—CH$_2$— / O—⟨benzene⟩—CH=CH—CH$_2$— ⟩		0.58	4.73
151	—CH$_2$—⟨benzene⟩—CH$_2$—	1.8		

See Table 3 for explanation of footnotes.

complex (cf. HEMPEL et al. 1993). This assumption is supported by a recent structure determination of a complex of deoxynojirimycin (78) with glucoamylase from *Aspergillus awamori var. X100*, because 78 fits best with the hydroxyl group at C-6 orientated axially to the piperidine ring (HARRIS et al. 1993). This might be the reason that most of the N-alkyl deoxynojirimycin derivatives with preformed axial orientation of the C-6 hydroxyl group are somewhat more active than 1-deoxynojirimycin itself.

In contrast acylated piperidinoses do not adopt the normal $^4C_1$ conformation but the inverted $^1C_4$ conformation (KISO et al. 1992; see also PAULSEN et al. 1968) and thus present a totally new structure to the active center of the receptor.

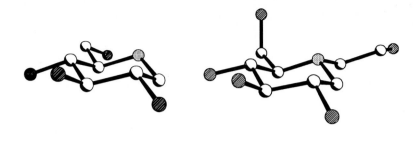

1-Deoxynojirimycin   Miglitol

Fig. 39. X-ray structure of 1-deoxynojirimycin and miglitol

Interestingly, one also finds this inversion for the 4-nitrophenyl compound *102*, as can be seen by comparing the [¹H]NMR spectra of the acetyl
compounds of *139* and *102* (Fig. 40). Whereas the phenyl derivative, like all
the alkyl derivatives, retains the normal chair form and has large vicinal α,α-
coupling constants, the ring nitrogen in compound *102* is apparently forced
into a planar sp² configuration by the high electronic demands of the *p*-nitro
group. This leads to a severe steric interaction between the nitrophenyl
substituent and the neighboring hydroxymethyl group. The molecule therefore switches to the inverted chair form, as can be seen from the absence of
the large vicinal coupling constants (JUNGE et al. 1989).

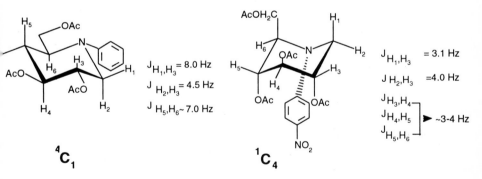

Fig. 40. Different conformations of N-substituted 1-deoxynojirimycins

To increase the residence time in the intestine, compounds which could
be expected to be less readily absorbed were synthesized by linking two or
more 1-deoxynojirimycin molecules with bridges (BÖSHAGEN et al. 1980;
MATSUMURA et al. 1982) (Table 7). Some of the compounds obtained were
several times more effective than 1-deoxynojirimycin in our test system, the
most active ones being *146*, *147*, *148*, and *150*.

This concept was valid up to a point. However, in the case of the trimer *152*, mutual steric hindrance occurs to such an extent that only marginal inhibition is achieved either in vitro or in vivo.

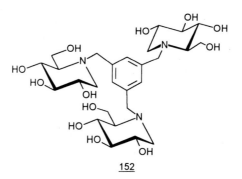

152

Fig. 41. Structure *152*

Nojirimycin (*70*) and 1-deoxynojirimycin (*78*) are fully competitive inhibitors of sucrase. Their K_i values are $1.3 \times 10^{-7}\,mol/l$ and $3.2 \times 10^{-8}\,mol/l$, respectively, at pH 6.8, the optimum of sucrase activity. According to further kinetic studies the enzyme/inhibitor association is slow: the steady state is reached within minutes (HANOZET et al. 1981).

The rate constant for enzyme-inhibitor dissociation K_{off} of, e.g., 1-deoxynojirimycin is 100–1000 times larger than that of the quasi-irreversible inhibitor castanospermine (see Sect. C.VI.1). Since castanospermine is a tertiary amine but 1-deoxynojirimycin, like nojirimycin and acarbose, is a secondary amine, it was thought that tertiary amines might bind more strongly to the enzyme than secondary amines (DANZIN and EHRHARD 1987; ROBINSON et al. 1992).

$$\text{Enzyme-inhibitor complex} \overset{K_{off}}{\underset{K_{on}}{\rightleftharpoons}} \text{enzyme} + \text{inhibitor}$$

In a comparative study on sucrase from rat small intestine it was indeed found that the values of K_{off} for the simplest 1-deoxynojirimycin derivative with a tertiary nitrogen atom, N-methyl-1-deoxynojirimycin (*94*), were 16 times smaller than those for 1-deoxynojirimycin. In this test the pseudo-isomaltase derivative *129* (MDL 73945), also a tertiary amine, even had a dissociation rate from the enzyme of comparable slowness to that of castanospermine and can thus also be viewed as a quasi-irreversible inhibitor of sucrase (Table 8).

However, this sort of kinetic data and the in vitro results shown in Tables 3–7 do not always reflect the conditions in the animal model. Pharmacokinetic parameters, especially absorption from the gastrointestinal tract, enterohepatic circulation, rate of enzyme synthesis, etc., all play a

Table 8. Kinetic constants for tight-binding inhibitors of sucrase from rat small intestine (ROBINSON et al. 1992 and references therein)

Inhibitor	k_{on} $(M^{-1}s^{-1})$	k_{off} (s^{-1})	$t_{1/2}$ for activity regain (h)	K_i (μM)
Acarbose	1.8×10^3	6×10^{-4}	0.3	0.34
78	8.8×10^4	2.1×10^{-3}	0.1	0.024
94	6.0×10^4	1.3×10^{-4}	1.5	0.002[a]
MDL 73945 (129)	1.0×10^2	$<9 \times 10^{-7}$	>200	<0.009
Castanospermine	6.5×10^3	3.6×10^{-6}	54	0.0005

[a] Corrected value.

major role in the animal model. Studies with pre-incubated enzymes do indeed in many cases allow a correlation with the in vivo effect, but generally applicable predictions are not always possible.

Thus, the inhibitor MDL 73945 (129) should be an excellent long-term inhibitor on the basis of its extremely low K_{off} value. In fact MDL 73945 has a stronger effect in studies in carbohydrate-loaded rats if it is administered *1h before* the sucrose instead of *simultaneously*, but is less effective if it is administered *4h before* the sucrose. MDL 73945 is ineffective in starch-loaded rats even at doses of 3 and 10 mg/kg if it is administered 4h before the starch (ROBINSON et al. 1991). In contrast the 1-deoxynojirimycin derivative 93 (emiglitate) is still active if it is administered 17h before the sucrose loading, possibly because it is recycled to the intestinal enzymes via the enterohepatic circulation.

Table 9 shows the reduction in blood glucose caused by some deoxynojirimycin derivatives in carbohydrate-loaded rats. It is clear from the data obtained that 1-deoxynojirimycin (78), which is highly effective in vitro, is substantially less effective than many of the N-substituted 1-deoxynojirimycin derivatives shown in Tables 3–7 in carbohydrate-loaded rats. The compounds 93, 94, 97, and 105 and the bridged bis-deoxynojirimycin derivatives 146 and 147 were particularly effective in our test system. Literature data also show that the substituted deoxynojirimycin derivatives are more effective than 78, although the absolute numbers are generally a factor of 10–30 above our values (possibly as a result of different experimental arrangements and measurement periods). Some of the compounds shown in Table 9 were still highly effective when they were administered 4 or 17h before the sucrose load. This is particularly true for the bridged bis-deoxynojirimycin derivatives 146 and 147, but also for compound 93, which has an ED_{50} of 1.5 mg/kg. Since, for use as a glucosidase inhibitor, a steady inhibition of starch and sucrose digestion plays a major role alongside the effectiveness, we have selected 97 (miglitol) and 93 (emiglitate for long-term inhibition) for development (PULS et al. 1984; JUNGE et al. 1984b, 1986). The tight-binding inhibitor MLD 73945 (129) is also undergoing preclinical studies (Merrell Dow). The compound is very effective in vivo in sucrose-loaded

Table 9. In vivo activity in carbohydrate-loaded rats

Compound No.	ED_{50} (mg/kg)			
	Sucrose[a]	Cooked starch[a]	Sucrose[b]	Starch[b]
78	0.8	≫8	21 (10^d)	16
93	0.2 (1.5^c)	0.3		
94	0.1	0.9	5.8	11
97	0.24	0.5	7.6 (10^d)	10
105	~0.3	≤1	5	
116	~0.1	~1		
129			1.0^d	~9.8^d
132	~0.3	~1.0		
133			7	
135			1.0	
136			0.4	
146	1.5^c			
147	<1		0.23	0.42
150			0.62	

[a] Fasting male rats received sucrose (500 mg–2 g/rat) or cooked starch (300 mg–2 g/rat) in the absence of (controls) or with simultaneous administration of one of the inhibitors shown in Tables 2–6 by gavage. The ED_{50} was determined from the reduction in the postprandial blood glucose increase during the first 90 min (area under the curve).
[b] Inhibition of postprandial hyperglycemia expressed as an increment in the area under the blood glucose curve up to 180 min when 2 g/kg sucrose or soluble starch was given to rats (YOSHIKUNI 1988; YOSHIKUNI et al. 1988).
[c] The inhibitor was administered 17 h before sucrose loading.
[d] (ROBINSON et al. 1991, 1992).

rats (ED_{50} = 1–1.9 mg/kg), but is five to ten times less effective in starch-loaded rats (ED_{50} = 9.8 mg/kg) (ROBINSON et al. 1991, 1992).

IV. Branched and Chain-Extended Deoxynojirimycin Derivatives

1. Derivatives Branched at C-1

The starting material for the synthesis of C-1 branched deoxynojirimycin derivatives is in most cases the nojirimycin-bisulfite adduct *72* obtainable by Inouye's procedure. Treatment with barium hydroxide and hydrocyanic acid gives the α-cyano derivative *153*, which can be hydrogenated to the aminomethyl derivative. *154* is then available for further derivatization (JUNGE et al. 1989) (Fig. 42).

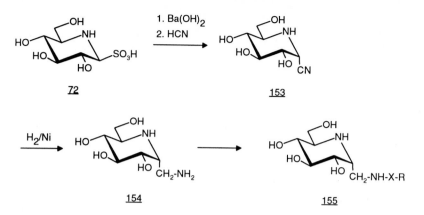

Fig. 42. Synthesis of 1-aminomethyl derivatives of 1-deoxynojirimycin

The cyano compound *153* can also be converted into 1-hydroxymethyl-1-deoxynojirimycin, giving either the α-homonojirimycin (*156*) or the β-isomer (*158*), depending on the reaction procedure (JUNGE et al. 1989; cf. HOLT et al. 1994) (Fig. 43).

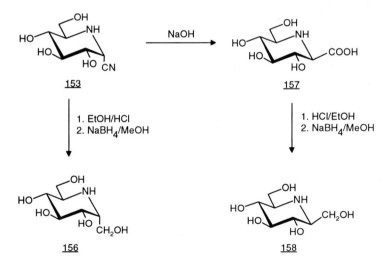

Fig. 43. Preparation of α- and β-homonojirimycin

Other synthetic routes to α-homonojirimycin (*156*) have been described (LIU 1987; ANZEVENO et al. 1989; AOYAGI et al. 1990). *156* is also contained in the foliage of *Omphalea diandra* (KITE et al. 1988). After silylating the cyano compound *153*, the C-1 alkyl or aryl derivatives of 1-deoxynojirimycin *160* can be synthesized by treatment with organometallic reagents with elimination of hydrogen cyanide (BÖSHAGEN et al. 1981). Mixtures of α and β anomers are formed but they can be separated by column chromatography (Fig. 44). C-1 alkyl derivatives with an oxirane substituent have also been synthesized (LIOTTA et al. 1991).

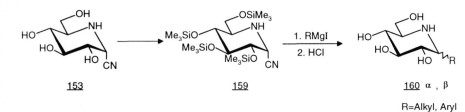

Fig. 44. Access to 1-alkyl- and aryl-1-deoxynojirimycins

Table 10 shows the effectiveness of C-1 branched derivatives in our sucrase inhibition test. Alkyl and aryl derivatives (*162–164*) have only a small effect, with β-derivatives being, if anything, weaker. In contrast the paramount importance of stereochemistry is clear in compounds *156* and *158*. Whereas the β-hydroxymethyl derivative *158* has no effect, the α-compound *156* (homonojirimycin) was three times as effective as 1-deoxynojirimycin in our assay, and is thus the most effective of any of the C-1 branched deoxynojirimycins. KITE et al. (1988) found that *156* has an efficacy comparable with that of 1-deoxynojirimycin.

Glycosylation of the aglycon *156* at the 1-hydroxymethyl group leads to MDL 25637 (*161*) (LIU 1987; ANZEVENO et al. 1989) with a clearly lower efficacy in vitro, but a surprisingly good efficacy in vivo ($ED_{50} = 4\,mg/kg$; sucrose load) (RHINEHART et al. 1987a). This may be due to the disaccharide-like structure of *161*, which might lead to reduced absorption from the intestine.

The α-carboxylic acid derivatives (*153*, *157*) and the aminomethyl compound *154* are only weakly active. If, however, one acylates compound *154*, there is a clear increase in efficacy. Amides with longer chains (*166*) and, in particular, sulfonamides (*168* and *171*) exceed the effect of 1-deoxynojirimycin. Ureas (*167*) and aromatic amides (*165*) reach the activity of acarbose. It is interesting that a methyl group at the ring nitrogen atom (*169*) is still tolerated, but, in contrast to the 1-deoxynojirimycin derivatives, substitution with longer alkyl chains (*170* and *172*) leads to complete loss of efficacy (JUNGE et al. 1989).

2. Derivatives Branched at C-5

The intermediate *85*, formed during Paulsen's synthesis of 1-deoxynojirimycin (see Sect. C.II), serves as the starting material for C-5 branched deoxynojirimycin derivatives. Reaction with hydrogen cyanide gives the cyano deri-

Table 10. In vitro activity of 1-deoxynojirimycin derivatives branched at C-1

Inhibitor	R	R^1	α, β	Relative sucrase inhibition[a] (Acarbose = 1)
153	—CN	H	α	0.25
154	—CH$_2$—NH$_2$	H	α	0.2
156	—CH$_2$OH	H	α	25 (IC$_{50}$ = 0.08 μM)[b] (m)
157	—COOH	H	α	0.05
158	—CH$_2$OH	H	β	<0.1
161	MDL 25637	H	α	IC$_{50}$ = 3.5 μM[c] (r)
162	—CH$_3$	H	α	0.35
163	—CH$_3$	H	β	0.1
164	Phenyl	H	α	0.1
165	—CH$_2$—NH—CO—〈 〉	H	α	0.5
166	—CH$_2$—NH—CO—(CH$_2$)$_6$—CH$_3$	H	α	9.1
167	—CH$_2$—NH—CO—NH—(CH$_2$)$_3$—CH$_3$	H	α	2.4
168	—CH$_2$—NH—SO$_2$—CH$_3$	H	α	13.0
169	—CH$_2$—NH—SO$_2$—CH$_3$	CH$_3$	α	6.7 (15)
170	—CH$_2$—NH—SO$_2$—CH$_3$	CH$_2$—CH$_2$—OH	α	<0.1
171	—CH$_2$—NH—SO$_2$—〈 〉	H	α	9.6
172	—CH$_2$—NH—CO—〈 〉—O—CH$_3$	(CH$_2$)$_5$—CH$_3$	α	<0.1

[a] Relative sucrase inhibitory activity: SIU/mg test compound/SIU/mg acarbose (MÜLLER et al. 1980).
[b] KITE et al. (1988)
[c] RHINEHART et al. (1987a); ROBINSON et al. (1992).

vative *173*. In analogy to the 1-cyano derivative (see Sect. C.IV.1), *173* can be further derivatized either via the aminomethyl compound *176* or via the carboxylic acid ester *174*. From *174*, 5-hydroxymethyl-1-deoxynojirimycin (*175*) is easily obtained (STOLTEFUSS et al. 1980) (Fig. 45).

Table 11 shows some representative C-5 branched deoxynojirimycins. Aliphatic (*177*) and aromatic (*178*) amides are only weakly effective. So are the polar cyano (*173*), aminomethyl (*176*), and carboxylic acid (*184*) compounds. In contrast to the C-1 branched derivatives, the sulfonamides (*179*) only have a very weak effect. Ureas (*180*, *181*), particularly with aromatic

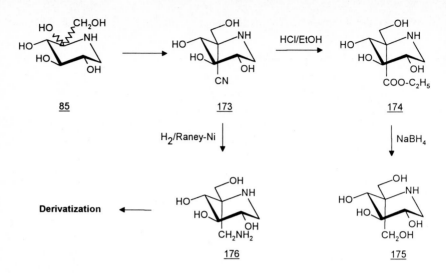

Fig. 45. Reaction of 5-hydroxy-1-deoxynojirimycin

Table 11. In vitro activity of 1-deoxynojirimycin derivatives branched at C-5

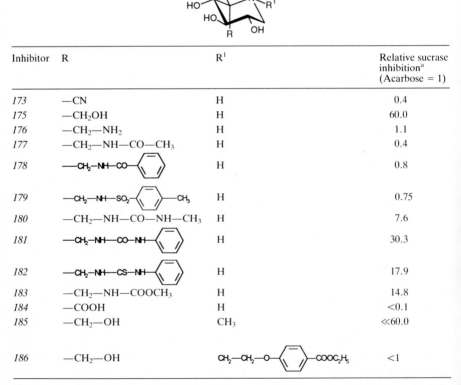

Inhibitor	R	R^1	Relative sucrase inhibition[a] (Acarbose = 1)
173	—CN	H	0.4
175	—CH$_2$OH	H	60.0
176	—CH$_2$—NH$_2$	H	1.1
177	—CH$_2$—NH—CO—CH$_3$	H	0.4
178	—CH$_2$—NH—CO—⟨⟩	H	0.8
179	—CH$_2$—NH—SO$_2$—⟨⟩—CH$_3$	H	0.75
180	—CH$_2$—NH—CO—NH—CH$_3$	H	7.6
181	—CH$_2$—NH—CO—NH—⟨⟩	H	30.3
182	—CH$_2$—NH—CS—NH—⟨⟩	H	17.9
183	—CH$_2$—NH—COOCH$_3$	H	14.8
184	—COOH	H	<0.1
185	—CH$_2$—OH	CH$_3$	≪60.0
186	—CH$_2$—OH	CH$_2$—CH$_2$—O—⟨⟩—COOC$_2$H$_5$	<1

[a] Relative sucrase inhibitory activity: SIU/mg test compound/SIU/mg acarbose (MÜLLER et al. 1980).

substituents (*181*), thioureas (*182*), and urethanes (*183*), are, however, strong inhibitors. Surprisingly the hydroxymethyl compound *175*, with 60 times the inhibitory activity of acarbose, was the strongest sucrase inhibitor of all 1-deoxynojirimycin derivatives under our assay conditions.

In the case of C-5 branched deoxynojirimycins, as for the C-1 branched derivatives, additional N-alkylation leads to virtually inactive compounds (*185*, *186*) (JUNGE et al. 1989).

3. Derivatives Chain-Extended at C-6

C-6 chain-extended derivatives are prepared from 1-deoxynojirimycin by the targeted use of protecting groups. The treatment of *78* with ethyl chloroformate yields a cyclic carbamate, which is fully protected by benzylation to give *187*. Access to the free 6-hydroxy position is made available by cleavage of the carbamate and subsequent reductive benzylation. After oxidation to the aldehyde *189*, the alkyl group is introduced by reaction with Grignard reagents. One finally obtains the deoxynojirimycin derivatives *191* as a mixture of diastereoisomers by cleaving off the protecting groups (Fig. 46).

A second method of preparation starts from intermediate *75* of the nojirimycin synthesis (see Sect. C.I). Acetylation of the free amino group followed by cleaving off the trityl group in position 6 gives the free alcohol *192*, which is converted into the aldehyde *193* by Swern oxidation. In this

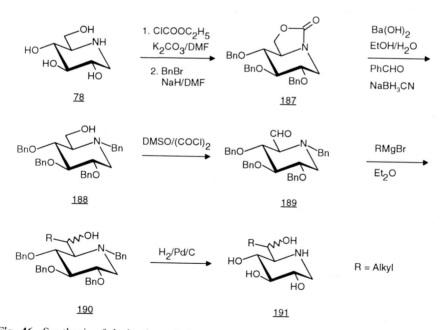

Fig. 46. Synthesis of derivatives chain-extended at position 6

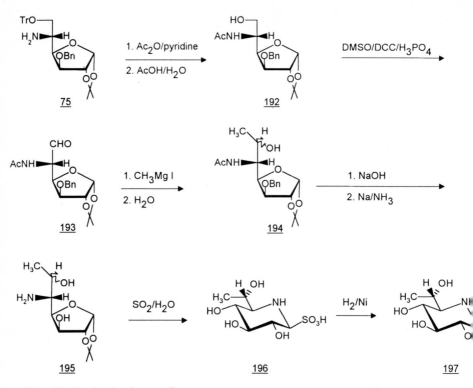

Fig. 47. Synthesis of pure diastereomers

case the chain is also lengthened by addition of a Grignard reagent at the carbonyl double bond (Fig. 47). This synthetic route results in only one diastereomer which is active in the sucrase inhibition test (KINAST et al. 1981; JUNGE et al. 1989). By comparison with the recently synthesized (6S)-6-C-ethyl-1-deoxynojirimycin (BERGER et al. 1992) and data in Meiji Seika's patent application (ISHII et al. 1993), the new chiral center at the C-6 position in *197* was determined to have S-configuration.

As can be seen from Table 12, these derivatives are also very potent inhibitors of sucrase. The methyl compound *197* is about 50 times more effective than acarbose and almost 10 times more effective than 1-deoxynojirimycin. The diastereoisomeric mixture *198* is clearly less effective. The activity decreases with ascending alkyl chains (*202*, *203*). Also in the case of derivatives chain-extended at C-6, an additional N-alkylation leads to substantially less effective compounds (*199–201*).

V. Deoxy, Amino, and Halogen Derivatives

Deoxy, amino, and halogen derivatives of 1-deoxynojirimycin are, if at all, very weak inhibitors of α-glucosidases. Therefore these compounds will be

Table 12. In vitro activity of 1-deoxynojirimycin derivatives chain-extended at C-6

Inhibitor	R	R¹	Relative sucrase inhibition[a] (Acarbose = 1)
197	—CH₃[b]	H	50.0
198	—CH₃[c]	H	23.2
199	—CH₃[b]	—CH₃	7.3
200	—CH₃[b]	—CH₂—CH₃	1.3
201	—CH₃[b]	—CH₂—(CH=CH)₂—CH₃	0.3
202	—CH₂—CH₃[c]	H	20.9
203	—(CH₂)₃—CH₃[c]	H	9.1

[a] Relative sucrase inhibitory activity: SIU/mg test compound/SIU/mg acarbose (MÜLLER et al. 1980).
[b] Pure diastereomer.
[c] 1:1 mixture of diastereomers.

only briefly reviewed. Starting from 1-deoxynojirimycin, the central building blocks for derivatizations are compounds with a protected nitrogen atom and hydroxyl functions depending on the requirements of the subsequent syntheses. Thus, the cyclic carbonate *204*, produced by treating 1-deoxynojirimycin with ethyl chloroformate under basic conditions, is an important key intermediate in providing access to positions 2, 3, 4, and 6 (JUNGE et al. 1989) (Fig. 48).

4,6-Acetals *215a–c* open access to various 1-deoxynojirimycin derivatives such as 1,2,3-trideoxynojirimycin (*216*) (JUNGE et al. 1989), 4-amino- (*218*) and 2-acetylamino-deoxynojirimycins (*217*) (JUNGE et al. 1989; KISO et al. 1991), and fluoro derivatives such as 3-fluoro-1-deoxynojirimycin (*219*) (GETMAN and DE CRESCENZO 1991, 1992a,b; cf. ARNONE et al. 1993) (Fig. 49).

Another strategy for the preparation of substituted deoxynojirimycin derivatives follows the classical protecting group chemistry starting from monosaccharides. Fluoro (DAX et al. 1990b; LEE et al. 1993; DI et al. 1992), acylamino (FLEET and SMITH 1986; KAPPES and LEGLER 1989), and deoxy derivatives of 1-deoxynojirimycin (FLEET and SMITH 1985; FLEET et al. 1987; JUNGE et al. 1989; FLEET and WITTY 1990) can be obtained by such a procedure. Some more recent chemoenzymatic syntheses consist, in principle, of a combination of an enzyme-catalyzed aldol-condensation and a palladium-catalyzed reductive alkylation and lead to deoxynojirimycin derivatives in relatively few steps (ZIEGLER et al. 1988; PEDERSON et al. 1988;

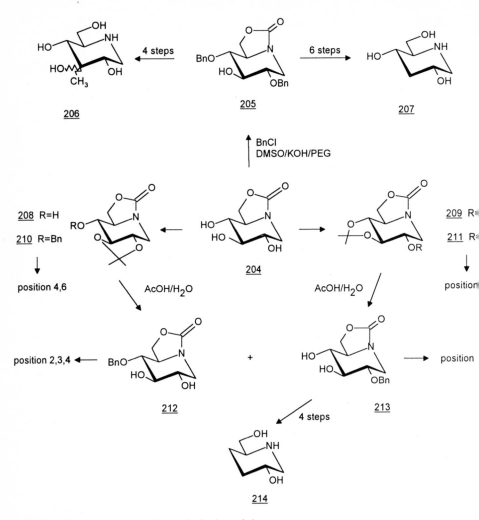

Fig. 48. Access to positions 2, 3, 4, and 6

PEDERSON and WONG 1989; VON DER OSTEN et al. 1989; STRAUB et al. 1990; KAJIMOTO et al. 1991a,b; EFFENBERGER and NULL 1992; LOOK et al. 1993; EFFENBERGER et al. 1993) (cf. Sect. C.I).

A special group of substituted deoxynojirimycins are the 1-amino compounds *220a–c* and their derivatives (TONG et al. 1990; GANEM and PAPANDREOU 1991; PAPANDREOU et al. 1993).

Unlike the chair conformation adopted by most sugar derivatives, compounds *220a–c* adopt a flattened half-chair conformation. The substituents on the ring no longer exist in true axial-equatorial positions, rather in positions somewhere in between. This structural ambiguity may explain

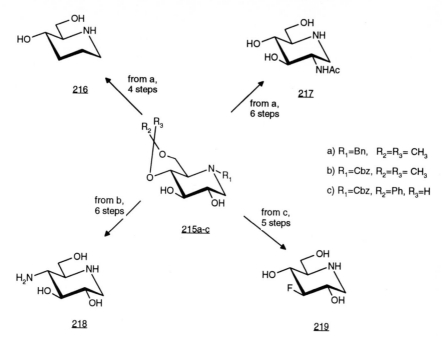

Fig. 49. Derivatization of 1-deoxynojirimycin starting from 4,6-acetals

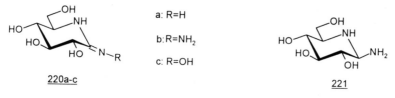

Fig. 50. Structures *220a–c* and *221*

why these compounds inhibit not only α- and β-glucosidases but also mannosidases and galactosidases.

The mannose derivatives corresponding to compound *220* also show this lack of selectivity and are inhibitors with a broad overall spectrum against various glycosidases (PAN et al. 1992). In contrast, 1-β-amino-1-deoxynojirimycin (*221*) is a *selective* inhibitor of β-glucosidases because it occurs in the chair form with β-orientation at the anomeric carbon (YOON et al. 1991).

VI. Polyhydroxypiperidines with Altered Configuration

After the discovery of the glucosidase-inhibiting properties of deoxynojirimycin, polyhydroxypiperidines with configurations other than D-gluco were synthesized. PAULSEN et al. (1980) prepared the D-galacto configured piperidine 1,5-dideoxy-1,5-imino-D-galactitol (also called deoxygalactonojirimycin) and found that it had β-galactosidase but no glucosidase-inhibiting activity. This finding stimulated intensive activity in the synthesis and biological testing of polyhydroxyheterocycles as aza analogues of monosaccharides (see, e.g., FLEET 1989; WINCHESTER and FLEET 1992). In general it turned out that a polyhydroxypiperidine (e.g., 1,5-dideoxy-1,5-imino-D-galactitol) has the strongest affinity to that hydrolase (e.g., D-galactosidase) which cleaves the glycoside with the configuration (e.g., D-galacto), corresponding to that of the inhibitor. Because none of these polyhydroxypiperidines with altered configuration possesses a strong glucosidase-inhibiting activity, they are not discussed further in our context.

VII. Bicyclic Derivatives of Deoxynojirimycin

1. Castanospermine

In addition to the monocyclic polyhydroxypiperidines, some naturally occurring bicyclic polyhydroxyheterocycles with indolizidine and pyrrolizidine structures also exhibit glycosidase-inhibiting activity. The former group includes castanospermine (222) and its analogues. (+)-Castanospermine (222) was isolated in 1981 from the seeds of the Australian tree *Castanospermum australe* (HOHENSCHUTZ et al. 1981), and later from the dried pods of *Alexa leiopetala* (NASH et al. 1988a). This 1,6,7,8-tetrahydroxyindolizidine may be regarded as a bicyclic derivative of deoxynojirimycin, with an ethylene bridge between the hydroxymethyl group and the ring nitrogen.

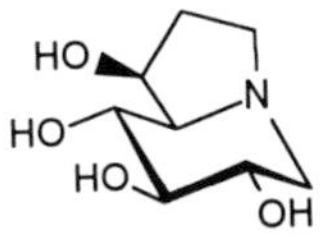

222

Fig. 51. Structure *222*

X-ray crystallography showed that the chiral centers of the six-membered ring of castanospermine correspond to the gluco configuration (HOHENSCHUTZ et al. 1981). The absolute configuration of *222* was proved after its synthesis from D-glucose (BERNOTAS and GANEM 1984, see below). Like deoxynojirimycin, castanospermine is a powerful inhibitor of various glucosidases. The synthesis and the structure-activity relationships of castanospermine and its analogues are described below.

As in the synthesis of polyhydroxypiperidines, carbohydrates are used as starting materials in the majority of cases. The various syntheses of cas-

Table 13. Syntheses of castanospermine

Starting material	Reference
D-Glucose	BERNOTAS and GANEM (1984)
D-Mannose	SETOI et al. (1985)
D-Glucose	HAMANA et al. (1987)
(1S, 4S)-7-Oxabicyclo-[2,2,1]hept-5-en-2-one	REYMOND and VOGEL (1989)[a], REYMOND et al. (1991)
D-Glucofuranurono-6,3-lactone	ANZEVENO et al. (1990)
3-Acetoxyproline, 3-ketoproline	BHIDE et al. (1990)
D-Glucono-1,5-lactone	MILLER and CHAMBERLIN (1990)
D-Glucono-1,5-lactone	GERSPACHER and RAPOPORT (1991)
chiral allylic alcohol from dimethyl L-tartrate	INA and KIBAYASHI (1991, 1993)
D-Xylose	MULZER et al. (1992)[b]
D-Glucofuranurono-6,3-lactone	GRAßBERGER et al. (1993)
(3S)-2,3-Dihydroxy-tetrahydrofuran	KIM et al. (1993)

[a] Synthesis of racemate.
[b] Synthesis of (−)-enantiomer.

tanospermine (*222*) are summarized in Table 13. The first synthesis, in which D-glucose was used as the starting material, was published in 1984 (BERNOTAS and GANEM 1984). The blocked D-glucose derivative *223* was converted in several steps, with inversion of the configuration at C-5 into the open-chain amino epoxide *224*, which spontaneously cyclized into the deoxynojirimycin derivative *225* and an isomeric azepane after removal of the trifluoroacetyl group. The 1:1 mixture of the diastereoisomers *227* and *228* was then obtained from oxidation of the primary alcohol function in *225* followed by condensation of *226* with *tert*-butyl lithioacetate. Saponification of the ester and hydrogenolysis of the benzyl ethers led to cyclization with formation of the lactams *229* and *230*, from which castanospermine (*222*) and its 1-epimer *231* were obtained by reduction.

The group of GANEM later improved the synthesis of *222* (HAMANA et al. 1987). Starting from the known deoxynojirimycin derivative *226*, the key step is the chain extension on C-6 via a highly diastereoselective chelate-controlled Sakurai allylation (*232*). Ring closure to form the indolizidine heterocycle *236* took place after ozonolytic chain shortening by one carbon atom (*233*) followed by reduction of the terminal aldehyde (*234*) and mesylation of the resulting alcohol (*235*). The desired compound was then liberated by hydrogenolytic removal of the benzyl ether protective groups. 6-Epi- and 8-epi-castanospermine (D-manno and D-galacto configuration) have been prepared analogously.

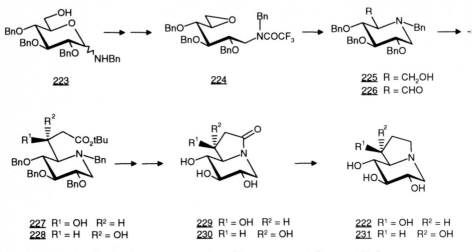

227 R¹ = OH R² = H
228 R¹ = H   R² = OH

229 R¹ = OH R² = H
230 R¹ = H   R² = OH

222 R¹ = OH R² = H
231 R¹ = H   R² = OH

Fig. 52. Synthesis of castanospermine. (BERNOTAS and GANEM 1984)

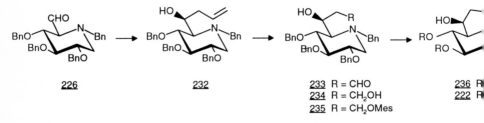

226

232

233 R = CHO
234 R = CH₂OH
235 R = CH₂OMes

236 R
222 R

Fig. 53. Synthesis of castanospermine. (HAMANA et al. 1987)

A comprehensive survey of the syntheses of castanospermine and its epimers and analogues has been published by BURGESS and HENDERSON (1992). Castanospermine (*222*) is a strong inhibitor of various intestinal glucosidases (SCOFIELD et al. 1986; DANZIN and EHRHARD 1987; RHINEHART et al. 1987b). In addition, *222* inhibits a wide range of α- and β-glucosidases derived from plant as well from animal and human sources (SAUL et al. 1983; PAN et al. 1983; HORI et al. 1984; SAUL et al. 1984; SASAK et al. 1985; GROSS et al. 1986; CHAMBERS and ELBEIN 1986; TRUGNAN et al. 1986; ELLMERS et al. 1987; SUNKARA et al. 1989).

Unlike acarbose and deoxynojirimycin, castanospermine behaves towards lysosomal α-1,4-glucosidase (ELLMERS et al. 1987), sucrase (DANZIN and EHRHARD 1987; TRUGNAN et al. 1986), and isomaltase (DANZIN and EHRHARD 1987) as a quasi-irreversible tight-binding inhibitor. The rate constant of the enzyme-inhibitor dissociation k_{off} is smaller than that of acarbose or deoxynojirimycin by a factor of 100–1000 (cf. Sect. C.II.3, Table 8). A half-life of 54 h has been found for the reactivation of sucrase, while no appreciable reactivation was detectable after 24 h in the case of isomaltase (DANZIN and EHRHARD 1987).

2. Castanospermine Derivatives

a) Epimers, Deoxy Derivatives, and Homologues

Detailed structure-activity relationships have been demonstrated for the polyhydroxypiperidines. In the case of deoxynojirimycin, the D-gluco configuration in the piperidine ring was found to be essential for the inhibition of glucosidases. The influence of the configuration of the hydroxyl groups on the inhibitory spectrum has also been investigated for polyhydroxyindolizidines. However, the relationships discovered here are not as clear cut as for the polyhydroxypiperidines. 6-Epicastanospermine (*237*), which was isolated from *Castanospermum australe* (MOLYNEUX et al. 1986) and later synthesized (FLEET et al. 1990b), has the D-manno configuration in the piperidine ring, but, unlike the corresponding deoxymannojirimycin, it does not inhibit mannosidases. As expected, the Trimming glucosidases I and II from mung beans are not inhibited by *237* (MOLYNEUX et al. 1986). Nothing is known at present concerning the inhibitory action on intestinal α-glucosidases.

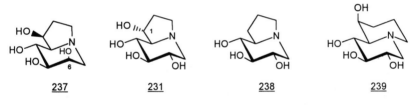

Fig. 54. Structures *231, 237, 238, 239*

The 1-epimer of castanospermine (*231*) was synthesized in the course of various castanospermine syntheses (BERNOTAS and GANEM 1984; SETOI et al. 1985; ANZEVENO et al. 1990; INA and KIBAYASHI 1993). In a comparison of the inhibition of sucrase by various castanospermine derivatives, *231* falls between castanospermine (*222*) and the weak inhibitor 1-deoxy-castanospermine (*238*) on the activity scale (HENDRY et al. 1987, 1988; WINCHESTER et al. 1990; ROBINSON et al. 1992) (Table 14). Like the hydroxyl group on C-6 of deoxynojirimycin, therefore, the hydroxyl group with the (S)-configuration on C-1 of castanospermine is essential for optimal glucosidase activity (see also HEMPEL et al. 1993).

The inhibitory effects of various castanospermine derivatives on glucosidases, mannosidases, and fucosidase have been investigated by WINCHESTER et al. (1990). Briefly, the results for activity towards glucosidases show that alteration of the configuration in the piperidine ring of castano-

Table 14. Inhibition of sucrase (rat) by castanospermine
and derivatives (ROBINSON et al. 1992)

Inhibitor	R^1	R^2	IC_{50} (μM)
222	OH	H	0.02
231	H	OH	0.9
238	H	H	5.0

spermine or removal of hydroxyl groups leads to distinctly weaker α- and β-glucosidase inhibitors.

For the 1,2,3,9-tetrahydroxyquinolizidine 239, which may be regarded as a homologue of castanospermine, an inhibitory effect on mammalian α-glucosidases is unknown. α-Glucosidase from yeast is not inhibited (GRADNIG et al. 1991).

b) Acylated and Glucosylated Derivatives of Castanospermine

Regioselective acylation of castanospermine (222) has been achieved by selective use of protecting groups (LIU et al. 1990) or via stannylidene stannylidene ketals (ANDERSON et al. 1990). Mono and diacyl derivatives have also been obtained by the aid of enzymatic catalysis (DELINCK and MARGOLIN 1990; MARGOLIN et al. 1990). Acylation of castanospermine on OH-6 and OH-7 leads in some cases to compounds whose inhibitory effects on glucosidase I from mouse SC-1 cells are comparable to or even distinctly stronger than that of castanospermine itself. The most active compound in this study was 6-O-butyrylcastanospermine (240), the IC_{50} of which was lower than that of castanospermine by a factor of about 15 (SUNKARA et al. 1989).

The inhibition of glucosidase I should correlate with the in vitro inhibition of plaque formation in cells infected with Moloney leukemia virus (MOLV) and of syncytium formation due to HIV-1. It is assumed that the inhibition of glucosidase I leads to modification of the glycosidation pattern of the viral envelope glycoprotein gp 120, with the result that the infec-

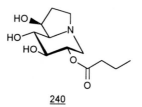

240

Fig. 55. Structure 240

tiousness of the virus is reduced or destroyed. *240* is in fact the most active compound in these assays. However, identical antiviral activities and toxicological profiles (RUPRECHT et al. 1991) found for *240* and castanospermine (*222*) in vivo (mouse) leads one to assume that *240* is a prodrug of *222*. No information is available at present concerning the activity of the acylated compounds towards intestinal disaccharidases.

Because of its toxicity, which is probably due to the nonspecific inhibition of various glucosidases, castanospermine is unsuitable for therapeutic use in diabetes mellitus. Attempts have therefore been made to increase the specificity of the inhibition of glucosidases by glucosidation of castanospermine.

O-Glucosylated derivatives of castanospermine have been synthesized in moderate yields and selectivities (LIU and KING 1992). It was found that 7-(α-glucopyranosyl)-castanospermine (*241*) and 8-(α-glucopyranosyl)-castanospermine (*242*) have inhibitory effects on sucrase comparable to that of castanospermine itself (Table 15). On the other hand, the activities of the 8-β-glucoside (*243*) and the 1-α-glucoside (*244*) were much weaker. Like castanospermine, glucosylated derivatives exhibit quasi-irreversible binding to sucrase.

The strong in vitro activity correlates with a long-lasting in vivo activity. In equimolar concentrations administered 4 h before sucrose challenge to rats, *241* and *242* reduced the blood glucose concentration to a greater degree than castanospermine itself, whereas castanospermine was more effective than *242* after an isomaltose challenge, in agreement with the relationships found in vitro. The authors concluded that *241* and *242* are not prodrugs of castanospermine. However, it is not entirely impossible that the cleavage of castanospermine glucosides takes place in the active site of the

Table 15. Inhibitor of glucosidases by *O*-glucosides of castanospermine

Inhibitor	n^a	R^{nb}	Sucrase[c] IC_{50} $(\mu M)^{d,e}$	Isomaltase[c] IC_{50} $(\mu M)^e$
222	1, 2, 3	OH	0.02	0.3
241	1	α-D-glc	0.04	3.9
242	2	α-D-glc	0.03	3.0
243	2	β-D-glc	0.4	9.9
244	3	α-D-glc	8.0	60

[a] Number of residue.
[b] All other residues represent hydroxyl.
[c] Rat small intestine mucosal homogenates with corresponding substrate.
[d] ROBINSON et al. (1989).
[e] RHINEHART et al. (1990).

enzyme, and that castanospermine then remains quasi-irreversibly bound to the enzyme. Despite their potency and the specificity achieved towards intestinal disaccharidases, none of these glucosylated derivatives appears to be undergoing clinical development at present, so that the idea of modified castanospermines for the treatment of diabetes has not been realized.

VIII. Polyhydroxypyrrolidines

1. Monocyclic Pyrrolidine Derivatives

The polyhydroxypyrrolidines can be divided into three subgroups. Figure 56 compares the monosaccharides with the corresponding aza analogues. The synthesis and the inhibitory activity of the hexose analogues, which have the D-gluco and D-manno configurations, and of the pentose analogues, which have the arabino configuration, will be discussed below.

a) 2,5-Dideoxy-2,5-iminohexitols

Like other glucosidase inhibitors, 2(R),5(R)-dihydroxymethyl-3(R),4(R)-dihydroxypyrrolidine (246, often abbreviated to DMDP in the literature;

Furanose: Polyhydroxypyrrolidine:

β-D-Fructofuranose (245)   2,5-Dideoxy-2,5-imino-
                                    - D-mannitol (246)

D-Glucofuranose (247)   1,4-Dideoxy-1,4-imino-
                                   -D-glucitol (248)

D-Arabinofuranose (249)   1,4-Dideoxy-1,4-imino-
                                   -D-arabinitol (250)

Fig. 56. Subtypes of polyhydroxypyrrolidines and their corresponding D-gluco-configured furanoses

2,5-dideoxy-2,5-imino-D-mannitol) was first isolated form natural sources. DMDP has been found in the leaves of the plants *Derris elliptica* (WELTER et al. 1976), in the seeds of *Lonchocarpus sericeus* (EVANS et al. 1985), in the neotropic liana *Omphalea diandra* L. (KITE et al. 1988), in *Omphalea queenslandiae*, and in *Endospermum medullosum* (KITE et al. 1991).

The first total synthesis of DMDP (*246*) started from L-sorbose (CARD and HITZ 1985). After introduction of appropriate protective groups (*251*), the 5-hydroxyl group was converted to an azide function, with inversion. Removal of the protective groups followed by hydrogenation of the azido derivative *253* led to the C_2-symmetrical DMDP. DMDP has also been prepared from D-glucose (FLEET and SMITH 1985, 1987), from D-mannitol (DUREAULT et al. 1991), and by an enzyme-catalyzed reaction of 2-azido-3-hydroxypropanal with dihydroxyacetone phosphate (DHAP), in which rabbit muscle aldolase (RAMA) was used as the catalytic enzyme (HUNG et al. 1991).

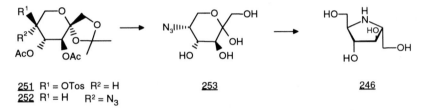

251 R^1 = OTos R^2 = H   **253**   **246**
252 R^1 = H R^2 = N_3

Fig. 57. Synthesis of DMDP (*246*). (CARD and HITZ 1985)

DMDP is a distinctly weaker inhibitor of intestinal disaccharidases from mice (SCOFIELD et al. 1986), rabbits (YOSHIKUNI 1988), and chickens (KITE et al. 1990) than deoxynojirimycin or castanospermine (Table 16). This can undoubtedly be explained by the weaker structural similarity of DMDP to the transition state of the cleavage of α-glucosides.

On the other hand, DMDP inhibits α- and β-glucosidases from plant sources (EVANS et al. 1985; FLEET et al. 1985), both α- and β-glucosidases from human liver (CENCI DI BELLO et al. 1985), glucosidase I from MDCK cells (mouse) (ELBEIN et al. 1984), and, unlike deoxynojirimycin and deoxymannojirimycin, *246* is an inhibitor of invertase (β-fructosidase) (CARD and HITZ 1985; EVANS et al. 1985).

b) 1,4-Dideoxy-1,4-iminohexitols

1,4-Dideoxy-1,4-imino-D-glucitol (*248*) may be considered as an aza analogue of 1-deoxy-D-glucofuranose. *248* was synthesized from open-chain D-glucose derivatives by PAULSEN et al. (1969) and by KUSZMANN and KISS (1986). FLEET and SON (1988), Bernotas (1990), and BUCHANAN et al. (1990) used

Table 16. Inhibition of intestinal disaccharidase activity by deoxynojirimycin (DNJ), castanospermine (*222*), DMDP (*246*), 1,4-dideoxy-1,4-imino-D-arabinitol (*250*), and 1,4-dideoxy-1,4-imino-L-arabinitol (*259*)

Inhibitor	Sucrase[a]	Maltase[a]	Isomaltase[a]
DNJ	0.06[b] 0.41[c]	0.37 1.04[c]	0.12
222	0.042	0.83	3.1
246	42 190[c] 54[d]	200 325[c] >330[d,e]	23
250	23	35	4.0
259	2.2	2.5	0.066

[a] Mouse small intestine mucosal homogenates with corresponding substrate (Scofield et al. 1986).
[b] IC_{50} (μM).
[c] Rabbit intestine mucosal homogenates (Yoshikuni 1988).
[d] Chicken small intestine mucosal homogenates (Kite et al. 1990).
[e] Less than 50% inhibition at $330\,\mu M$.

galactose derivatives as starting materials for their syntheses of *248*. The elegant synthesis by Bernotas (1990) will be described here as an example. Key steps are the selective bromination in position 6 to *255* followed by a reductive elimination/reductive amination procedure. Subsequent cyclization to form the vinylpyrrolidine *257* took place under Mitsunobu conditions. The double bond of *257* was then dihydroxylated with a high degree of diastereoselectivity, with formation of *258*, which has the D-gluco configuraton. After removal of the protective groups, *248* was obtained.

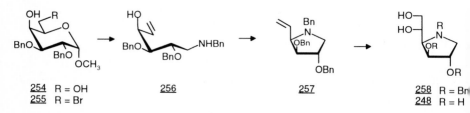

Fig. 58. Synthesis of *248*. (Bernotas 1990)

A comparison of the positions of the hydroxyl groups in *248* and deoxynojirimycin in relation to the ring nitrogen shows that *248* lacks the critical hydroxyl group corresponding to the 2-OH group of deoxynojirimycin. The importance of the 2-hydroxyl group of deoxynojirimycin to the glucosidase-inhibiting activity has already been mentioned. This structural difference and the additional difference in the side chain in comparison with deoxynojirimycin may well explain the weak effects of *248* on α- and β-glucosidases from microbial and plant sources (KUSZMANN and KISS 1986; BUCHANAN et al. 1990).

c) 1,4-Dideoxy-1,4-iminopentitols

Formal removal of one hydroxymethyl group from *246* leads to 1,4-dideoxy-1,4-imino-D-arabinitol (*250*), which has been found in *Arachniodes standishii* (FURUKAWA et al. 1985) and *Angylocalyx botiqueanus* (NASH et al. 1985). The D-enantiomer *250* was prepared for the first time by FLEET et al. from D-xylose (FLEET et al. 1985; FLEET and SMITH 1986). An azido group was first introduced, with inversion, on C-2 of the D-xylose derivative *261*. Removal of the isopropylidene group from the D-lyxo derivative *262*, selective introduction of a leaving group in position 5, and reduction of the azido group resulted in cyclization to form a bicyclic amine, which was then converted into the benzyloxycarbonyl-protected derivative *263*. FLEET and WITTY (1990) later synthesized the bicyclic amines *264* and *265* by an inverse strategy. The cyclization was carried out from *266*, which has an azido group on C-5 and a leaving group on C-2, with inversion at C-2. *250* was liberated from the bicyclic products by glycoside cleavage, reduction of the lactol, and removal of the protective groups. *250* has also been synthesized from (S)-pyroglutamic acid by IKOTA and HANAKI (1987), and also by aldolase-catalyzed C-C linkage (Ziegler et al. 1988; VON DER OSTEN et al. 1989; PEDERSON and WONG 1989; KAJIMOTO et al. 1991b).

Fig. 59. Structures *259* and *260*

FLEET et al. (1985)
FLEET and SMITH (1986)

FLEET and WITTY (1990)

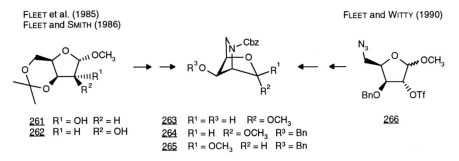

261 R¹ = OH R² = H
262 R¹ = H   R² = OH

263 R¹ = R³ = H   R² = OCH₃
264 R¹ = H R² = OCH₃ R³ = Bn
265 R¹ = OCH₃ R² = H R³ = Bn

266

Fig. 60. Synthetic intermediates for 1,4-dideoxy-1,4-imino-D-arabinitol (*250*)

Jones et al. (1985) synthesized the L-enantiomer *259* from the L-arabino-pyranoside *267* by azide formation at C-4 with double inversion (*269*), removal of protective groups and glycoside cleavage, and reduction. Further syntheses of *259* starting from different xylose derivatives have been published (Fleet et al. 1985; Fleet and Smith 1986; Axamawaty et al. 1990; Naleway et al. 1988; Van der Klein et al. 1992). Surprisingly, intestinal α-glucosidases (mouse) were inhibited more strongy by the L-arabino isomer *259* than by its D-enantiomer *250* (Scofield et al. 1986). The IC_{50} of *259* for isomaltase was even lower than that of deoxynojirimycin (Table 16). However, in comparison *250* was still more active than DMDP (*246*).

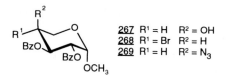

267 R^1 = H   R^2 = OH
268 R^1 = Br R^2 = H
269 R^1 = H   R^2 = N$_3$

Fig. 61. Structures *267–269*

One could speculate about the reasons for the greater activity of *259* towards intestinal α-glucosidases. The conformation of *250* is probably similar to that of *246*. Moreover, *250* has no function corresponding in its position to the 2-OH group of deoxynojirimycin, which is important to the activity of the latter (see Sect. C.II). On the other hand, the steric structures of the L-arabino derivative *259* and deoxynojirimycin can be fitted together in such a manner that the hydroxyl groups on C-5, C-3, and C-2 of *259* coincide with the hydroxyl groups on C-2, C-4, and C-6 of deoxynojirimycin when the two ring nitrogens also coincide (Robinson et al. 1992). Thus, in *259* only the hydroxyl group corresponding to OH-3 of deoxynojirimycin is missing. Therefore, for good activity towards α-glucosidases, the orientation of the hydroxyl groups relative to the ring nitrogen appears to play a dominant role. Less important is whether or not the ring nitrogen and the hydroxyl groups are part of a piperidine or a pyrrolidine structure.

In investigations on the anti-HIV activity of glycosidase inhibitors, moreover, 1,4-dideoxy-1,4-imino-L-arabinitol (*259*) was found to inhibit syncytium formation more strongly than other polyhydroxypyrrolidines (Fleet et al. 1988a; Karpas et al. 1988). On the other hand, in vitro studies on microbial glucosidases indicated that the D-enantiomer *250* is the stronger inhibitor of α-D-glucosidase (yeast) ($IC_{50} = 0.18 \mu M$). The L-enantiomer *259* was weaker by a factor of 55 ($IC_{50} = 10 \mu M$) (Fleet et al. 1985). It is once again evident here that activity towards intestinal glucosidases cannot be predicted on the basis of investigations on microbial or plant glucosidases.

The polyhydroxypyrroline nectrisine (*260*), isolated as a metabolite from the fungus *Nectria lucida* (Shibata et al. 1988), is known to be a powerful inhibitor of α-D-glucosidase from yeast (IC_{50} = 80 nM) (KAYAKIRI et al. 1988).

2. Bicyclic Pyrrolidine Derivatives

There are two groups of bicyclic polyhydroxyheterocycles, with two five-membered rings or one five-membered ring and one six-membered ring fused together. The best-known compound of the first group is swainsonine (*270*), which is a powerful inhibitor of α-mannosidases, and will not therefore be discussed here. The structure-activity relationships of swainsonine and its analogues for the inhibition of α-mannosidases have been described by CENCI DI BELLO et al. (1989). Glucosidases are not inhibited by swainsonine (SCOFIELD et al. 1986). For this work the bicyclic pyrrolidine derivatives belonging to the polyhydroxypyrrolizidine group are more interesting.

The first compound of this group to be isolated from natural sources was (+)-alexine   [(1R,2R,3R,7S,7aS)-3-hydroxymethyl-1,2,7-trihydroxypyrrolizidine)] (*271*). This substance was isolated from *Alexa leiopetala* and identified by X-ray crystallography (NASH et al. 1988b). At about the same time, a similar compound was isolated by MOLYNEUX et al. (1988) from *Castanospermum australe*. This compound, which was given the name australine (*272*), was found to be the 7a-epimer of alexine.

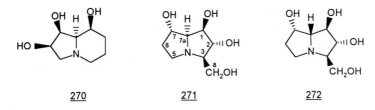

Fig. 62. Structures *270–272*

Australine may be regarded as a derivative of DMDP (*246*) with an ethylene bridge between the hydroxymethyl group and the ring nitrogen. The following epimers were later isolated from *Castanospermum australe*: 3,7a-diepialexine (3-epiaustraline) (*273*) (NASH et al. 1988c), 1,7a-diepialexine (1-epiaustraline) (*274*) (HARRIS et al. 1989), and 7,7a-diepialexine (7-epiaustraline) (*275*) together with *274* (NASH et al. 1990).

7a-Epialexaflorine (*276*) was isolated from *Alexa grandiflora*. This is a derivative of 7a-epialexine (australine) with a carboxyl group in place of the hydroxymethyl group on C-3 (PEREIRA et al. 1991).

Alexine, together with its 3- and 7-epimers, was synthesized for the first time by FLEET et al. (1988b) from D-glucose. 1,7a-Diepialexine (*274*) has been synthesized by CHOI et al. (1991) and by IKOTA (1992). Additionally, 7,7a-diepialexine (*275*) and 7-epialexine have been prepared by PEARSON and HINES (1991).

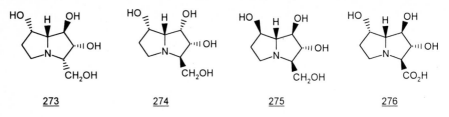

Fig. 63. Structures *273–276*

The naturally occurring polyhydroxypyrrolizidines *271–276* have been investigated for glucosidase-inhibiting activity. All of the pyrrolizidines investigated were weaker inhibitors of sucrase (mouse) than castanospermine (Table 17). 1,7a-Diepialexine (*274*) and 7,7a-diepialexine (*275*) are at best only mediocre inhibitors of intestinal α-glucosidase activity (Nash et al. 1990).

Australine (*272*), 1,7a-diepialexine (*274*), 3,7a-diepialexine (*273*), and castanospermine exhibit comparably good inhibition of amyloglucosidase (1,4-α-glucosidase), and the IC$_{50}$ of 7-epiaustraline (*275*) is even lower, by a factor of 10 (Table 17). No sucrase inhibition value is known for *275*.

A comparative study on the inhibition of α-glucosidase I (pig kidney) by polyhydroxypyrrolizidines has been carried out by Taylor et al. (1992). The

Table 17. Inhibition of glucosidases by polyhydroxypyrrolizidines, DMDP, and castanospermine (IC$_{50}$ [μM])

Inhibitor	Sucrase[a]	α-Gluco-sidase I[b]	1,4-α-Gluco-sidase[c,d]
271	>330[e,f]	18	11
272	≈28[g]	720 (≈20)[h]	1.5 (6)[i]
273	210[e]	70	2.1
274		16	1.5 (26)[i]
275		4.5	0.13
276	≈330[l]		(140)[k]
246	42[l]		
222	0.042[l]	0.1 (≈1)[h]	1.5 (8)[i]

[a] Mouse intestinal sucrase.
[b] Pig kidney α-glucosidase I (Taylor et al. 1992).
[c] 1,4-α-Glucosidase from *Aspergillus niger*.
[d] Nash et al. (1990).
[e] Nash et al. (1988c).
[f] Less than 50% inhibition at 330 μM.
[g] Rat intestinal sucrase (Tropea et al. 1989).
[h] From mung bean seedlings (Tropea et al. 1989).
[i] Molyneux et al. (1991).
[k] Pereira et al. (1991).
[l] Scofield et al. (1986).

anti-HIV activity was measured at the same time, and was found to correlate with the glucosidase inhibition. According to this study, 7,7a-diepialexine (*275*) is again the strongest inhibitor of this enzyme, but its activity is still weaker by a factor of 45 than that of castanospermine. Australine is the weakest compound of this series.

No simple structure-activity relationships found for polyhydroxypiperidines have been identified here. The activity of the polyhydroxypyrrolizidines towards intestinal disaccharidases is distinctly weaker than that of acarbose, deoxynojirimycin, or castanospermine.

References

Altenbach HJ (1993) Flexible, stereocontrolled routes to sugar mimics via convenient intermediates. In: Krohn K, Kirst H, Maas H (eds) Antibiotics and antiviral compounds. VCH, Weinheim, pp 361–363

Anderson WK, Coburn RA, Gopalsamy A, Howe TJ (1990) A facile selective acylaton of castanospermine. Tetrahedron Lett 31:169–170

Anzeveno PB, Creemer LJ (1990) Efficient synthesis of (+)-nojirimycin and (+)-1-deoxynojirimycin. Tetrahedron Lett 31:2085–2088

Anzeveno PB, Creemer LJ, Daniel JK, King CHR, Liu PS (1989) A facile, practical synthesis of 2,6-dideoxy-2,6-imino-7-O-beta-D-glucopyranosyl-D-glycero-L-gulo-heptitol (MDL 25637). J Org Chem 54:2539–2542

Anzeveno PB, Angell PT, Creemer LJ, Whalon MR (1990) An efficient, highly stereoselective synthesis of (+)-castanospermine. Tetrahedron Lett 31:4321–4324

Aoyagi S, Fujimaki S, Kibayashi C (1990) Total synthesis of (+)-alpha-homonojirimycin. J Chem Soc Chem Commun 20:1457– 1459

Argoudelis AD, Reusser F, Mizsak SA, Baczynskyj L (1976) Antibiotics produced by Streptomyces ficellus. II. Feldamycin and nojirimycin. J Antibiot 29: 1007–1014

Arnone A, Bravo P, Donadelli A, Resnati G (1993) Asymmetric synthesis of fluorinated analogues of 1-deoxynojirimycin. J Chem Soc Chem Commun: 984–986

Asano N, Tomioka E, Kizu H, Matsui K (1994a) Sugars with nitrogen in the ring isolated from the leaves of Morus bombycis. Carbohydr Res 253:235–245

Asano N, Oseki K, Tomioka E, Kizu H, Matsui K (1994b) N-containing sugars from Morus alba and their glycosidase inhibitory activities. Carbohydr Res 259: 243–255

Axamawaty MTH, Fleet GWJ, Hannah KA, Namgoong SK, Sinnott ML (1990) Inhibition of the α-L-arabinofuranosidase III of Monilinia fructigena by 1,4-dideoxy-1,4-imino-L-threitol and 1,4-dideoxy-1,4-imino-L-arabinitol. Biochem J 266:245–249

Baxter EW, Reitz AB (1994) Expeditious synthesis of azasugars by the double reductive amination of dicarbonyl sugars. J Org Chem 59:3175–3185

Beaupere D, Stasik B, Uzan R, Demailly G (1989) Selective azidation of L-sorbose. Application to the rapid synthesis of 1-deoxynojirimycin. Carbohydr Res 191: 163–166

Behling J, Farid P, Medich JR, Scaros MG, Prunier M, Weier RM, Khanna, I (1991) A short and practical synthesis of 1-deoxynojirimycin. Synth Commun 21: 1383–1386

Berger A, Dax K, Gradnig G, Grassberger V, Stütz AE, Ungerank M. Legler G, Bause E (1992) Synthesis and biological activity of C-6 modified derivatives of the glucosidase inhibitor 1-deoxynojirimycin. Bioorg Med Chem Lett 2:27– 32

Bernotas RC (1990) A short, versatile approach to polyhydroxylated pyrrolidines utilizing a reductive elimination-reductive amination as a key step. Tetrahedron Lett 31:469–472

Bernotas RC, Ganem B (1984) Total syntheses of (+)-castanospermine and (+)-deoxynojirimycin. Tetrahedron Lett 25:165–168

Bernotas RC, Ganem B (1985) Efficient preparation of enantiomerically pure cyclic aminoalditols, total synthesis of 1-deoxynojirimycin and 1-deoxymannojirimycin. Tetrahedron Lett 26:1123–126

Bhide R, Mortezaei R, Scilimati A, Sih CJ (1990) A chemoenzymatic synthesis of (+)-castanospermine. Tetrahedron Lett 31:4827–4830

Bock K, Meldal M (1991) Controlled reduction of acarbose and the resulting saturated products. Carbohydr Research 221:1–16

Böshagen H, Junge B, Stoltefuß J, Schmidt D, Krause HP, Puls W (1980) Neue Derivate von 3,4,5-Trihydroxypiperidin, Verfahren zu ihrer Herstellung und ihre Verwendung als Arzneimittel und in der Tierernährung. Ger Offen 2922760

Böshagen H, Geiger W, Junge B (1981) Reaction of 1-alpha-cyano-deoxynojirimycin with Grignard compounds. Complete exchange of the CN group. Angew Chem 93:800–801

Böshagen H, Heiker FR, Schüller AM (1987) The chemistry of the 1-deoxynojirimycin system. Synthesis of 2-acetamido-1,2-dideoxynojirimycin from 1-deoxynojirimycin. Carbohydr Res 164:141–148

Brodbeck U (1980) On the design of transition state analog enzyme inhibitors and their future in medicinal and agricultural chemistry. In: Brodbeck U (ed) Enzyme inhibitors, Verlag Chemie, Weinheim, pp 3–17

Broxterman HJG, Van der Marel GA, Neefjes JJ, Ploegh HL, Van Boom JH (1987) Synthesis of the antibiotic 1,5-dideoxy-1,5- imino-D-glucitol; concomitant formation of the D-mannitol analog. Recl Trav Chim Pays-Bas 106:571–576

Buchanan JG, Lumbard KW, Sturgeon RJ, Thompson DK, Wightman RH (1990) Potential glycosidase inhibitors: synthesis of 1,4-dideoxy-1,4-imino derivatives of D-glucitol, D- and L-xylitol, D- and L-allitol, and D-gulitol. J Chem Soc Perkin Trans: 699–706

Burgess K, Henderson I (1992) Synthetic approaches to stereoisomers and analogues of castanospermine. Tetrahedron 48:4045–4066

Card PJ, Hitz WD (1985) 2(R),5(R)-bis(hydroxymethyl)-3(R),4(R)-dihydroxypyrrolidine. A novel glycosidase inhibitor. J Org Chem 50:891–893

Cenci di Bello I, Dorling P, Evans S, Fellows L, Winchester B (1985) Inhibition of human α- and β-glucosidases and α- and β-D-mannosidases by 2,5-dihydroxymethyl-3,4-dihydroxypyrrolidine. Biochem Soc Trans 13:1127–1128

Cenci di Bello I, Fleet G, Namgoong SK, Tadano K-I, Winchester B (1989) Structure-activity relationship of swainsonine. Biochem J 259:855–861

Chambers JP, Elbein AD (1986) Effects of castanospermine on purified lysosomal alpha-1,4-glucosidase. Enzyme 35:53–56

Chida N, Furuno Y, Ogawa S (1989) New synthesis of (+) and (−)-nojirimycin from myo-inositol. J Chem Soc Chem Commun: 1230–1231

Chida N, Furuno Y, Ikemoto H, Ogawa S (1992) Synthesis of (+)- and (−)-nojirimycin and their 1-deoxy derivatives from myo-inositol. Carbohydr Res 237:185–194

Choi S, Bruce I, Fairbanks AJ, Fleet GWJ, Jones AH, Nash RJ, Fellows LE (1991) Alexines from heptonolactones. Tetrahedron Lett 32:5517–5520

Cogoli A, Semenza G (1975) A probable oxocarboniumion in the reaction mechanism of small intestinal sucrase and isomaltase. J Biol Chem 250:7802–7809

Daigo K, Inamori Y, Takemoto T (1986) Studies on the constituents of the water extract of the root of mulberry tree (Morus bombycis Koidz). Chem Pharm Bull 34:2243–2246

Dale MP, Ensley HE, Kern K, Sastry KAR, Byers LD (1985) Reversible inhibitors of β-glucosidase. Biochemistry 24:3530–3539

Danzin C, Ehrhard A (1987) Time-dependent inhibition of sucrase and isomaltase from rat small intestine by castanospermine. Arch Biochem Biophys 257:472–475

Dax K, Gaigg B, Graßberger V, Kölblinger B, Stütz AE (1990a) Simple syntheses of 1,5-dideoxy-1,5-imino-D-glucitol (1-deoxynojirimycin) and 1,6-dideoxy-1,6-imino-D-glucitol from D-glucofuranurono-6,3-lactone. J Carbohydr Chem 9:479–499

Dax K, Graßberger V, Stütz AE (1990b) Einfache Synthese von 1,5,6-Tridesoxy-6-fluor-1,5-imino-D-glucit, dem ersten fluorhaltigen Derivat des Glucosidaseninhibitors 1-Desoxynojirimycin. J Carbohydr Chem 9:903–908

Delinck DL, Margolin AL (1990) Enzyme-catalyzed acylation of castanospermine and 1-deoxynojirimycin. Tetrahedron lett 31:3093–3096

Di J, Rajanikanth B, Szarek WA (1992) Fluorinated 1,5-dideoxy-1,5-iminoalditols – synthesis of 1,5,6-trideoxy-6-fluoro-1,5-imino-D-glucitol (1,6-dideoxy-6-fluoronojirimycin) and 1,4,5-trideoxy-4-fluoro-1,5-imino-D-ribitol (1,2,5-trideoxy-2-fluoro-1,5-imino-L-ribitol). J Chem Soc Perkin Trans I 17:2151–2152

Dondoni A, Merino P, Perrone D (1993) Totally chemical synthesis of azasugars via thiazole intermediates. Stereodivergent routes to (−)-nojirimycin, (−)-mannojirimycin and their 3-deoxy derivatives from serine. Tetrahedron 49:2939–2956

Duréfault A, Portal M, Depezay JC (1991) Enantiospecific syntheses of 2,5-dideoxy-2,5-imino-D-mannitol and L-iditol from D-mannitol. Synlett: 225–226

Effenberger F, Null V (1992) Enzyme-catalyzed reaction. 13. A new, efficient synthesis of fagomine. Liebigs Ann Chem:1211–1212

Effenberger F, Null V, Straub A (1993) Enzymic carbon-carbon bonding. DECHEMA Monogr 129:197–207

Elbein AD (1987) Inhibitors of the biosynthesis and processing of N-linked oligosaccharide chains. Annu Rev Biochem 56:497–534

Elbein AD, Mitchell M, Sanford BS, Fellows LE, Evans SV (1984) The pyrrolidine alkaloid 2,5-dihydroxymethyl-3-4-dihydroxypyrrolidine inhibits glycoprotein processing. J Biol Chem 259:12409–12413

Ellmers BR, Rhinehart BL, Robinson KM (1987) Castanospermine: an apparent tight-binding inhibitor of hepatic lysosomal alpha-glucosidase. Biochem Pharmacol 36:2381–2385

Ermert P, Vasella A (1991) Synthesis of a glucose-derived tetrazole as a new β-glucosidase inhibitor. A new synthesis of 1-deoxynojirimycin. Helv Chim Acta 74:2034–2053

Evans SV, Fellows LE, Shing TKM, Fleet GWJ (1985) Glycosidase inhibition by plant alkaloids which are structural analogues of monosaccharides. Phytochemistry 24:1953–1955

Ezure Y, Maruo S, Miyazaki K, Kawamata M (1985) Moranoline (1-deoxynojirimycin) fermentation and its improvement. Agric Biol Chem 49:1119–1125

Ferrier RJ (1979) Unsaturated carbohydrates. A carbocylic ring closure of a hex-5-enopyranoside derivative. J Chem Soc Perkins Trans I part 21:1455–1458

Fleet GWJ (1989) Homochiral compounds from sugars. Chem Br 25:287–292

Fleet GWJ, Smith PW (1985) Enantiospecific syntheses of deoxymannojirimcin, Fagomine and 2R,5R-dihydroxymethyl-3R,4R-dihydroxypyrrolidine from D-glucose. Tetrahedron Lett 26:1469–1472

Fleet GWJ, Smith PW (1986) The synthesis from D-xylose of the potent and specific enantiomeric glucosidase inhibitors, 1,4-dideoxy-1,4-imino-D-arabinitol and 1,4-dideoxy-1,4-imino-L-arabinitol. Tetrahedron 42:5685–5692

Fleet GWJ, Smith PW (1987) Methyl 2-azido-3-O-benzyl-2-deoxy-α-D-mannofuranoside as a divergent intermediate for the synthesis of polyhydroxylated piperidines and pyrrolidines: synthesis of 2,5-dideoxy-2,5-imino-D-mannitol [2R,5R-dihydroxymethyl-3R,4R-dihydroxypyrrolidine]. Tetrahedron 43:971–978

Fleet GWJ, Son JC (1988) Polyhydroxylated pyrrolidines from sugar lactones: synthesis of 1,4-dideoxy-1,4-imino-D-glucitol from D-galactonolactone and syntheses of 1,4-dideoxy-1,4-imino-D-allitol, 1,4-dideoxy-1,4-imino-D-ribitol, and (2S,3R,4S)-3,4-dihydroxyproline from D-gulonolactone. Tetrahedron 44: 2637–2647

Fleet GWJ, Witty DR (1990) Synthesis of homochiral β-hydroxy-α-amino acids [(2S,3R,4R)-3,4-dihydroxyproline and (2S,3R,4R)-3,4-dihydroxypipecolic acid]. and of 1,4-dideoxy-1,4-imino-D-arabinitol [DAB1] and fagomine [1,5-imino-1,2, 5-trideoxy-D-arabino-hexitol]. Tetrahedron Asymmetry 1:119–136

Fleet GWJ, Nicholas SJ, Smith PW, Evans SV, Fellows LE, Nash RJ (1985) Potent competitive inhibition of α-galactosidase and α-glucosidase activity by 1,4-dideoxy-1,4-iminopentitols: syntheses of 1,4-dideoxy-1,4-imino-D-lyxitol and of both enantiomers of 1,4-dideoxy-1,4-iminoarabinitol. Tetrahedron Lett 26: 3127–3130

Fleet GWJ, Smith PW, Nash RJ, Fellows LE, Parekh RB, Rademacher TW (1986) Synthesis of 2-acetamido-1,5-imino-1,2,5-trideoxy-D-mannitol and of 2-acetamido-1,5-imino-1,2,5-trideoxy-D-glucitol, a potent and specific inhibitor of a number of β-N-acetylglucosaminidases. Chem Lett: 1051–1054

Fleet GWJ, Fellows LE, Smith PW (1987) Synthesis of deoxymannojirimycin, fagomine, and deoxynojirimycin, 2-acetamido-1,5-imino-1,2,5-trideoxy-D-mannitol, 2-acetamido-1,5-imino-1,2,5-trideoxy-D-glucitol, 2S,3R,4R,5R-trihydroxypipecolic acid and 2S,3R,4R,5S-trihydroxypipecolic acid from methyl 3-O-benzyl-2,6-dideoxy-2,6-imino-α-D-mannofuranoside. Tetrahedron 43: 979–990

Fleet GWJ, Karpas A, Dwek RA, Fellows LE, Tyms AS, Petursson S, Namgoong SK, Ramsden NG, Smith PW, Son JC, Wilson F, Witty DR, Jacob GS, Rademacher TW (1988a) Inhibition of HIV replication by amino-sugar derivatives. FEBS Lett 237:128–132

Fleet GWJ, Haraldsson M, Nash RJ, Fellows LE (1988b) Synthesis from D-glucose of alexine [(1R,2R,3R,7S,8S)-3-hydroxymethyl-1,2-7-trihydroxypyrrolizidine], 3-epialexine and 7-epialexine. Tetrahedron Lett 29:5441–5444

Fleet GWJ, Carpenter NM, Petursson S, Ramsden NG (1990a) Synthesis of deoxynojirimycin and of nojirimycin δ-lactam. Tetrahedron Lett 31:409–412

Fleet GWJ, Ramsden NG, Nash RJ, Fellows LE, Jacob GS, Molyneux RJ, Cenci di Bello I, Winchester B (1990b) Synthesis of the enantiomers of 6-epicastanospermine and 1,6-diepicastanospermine from D- and L-gulonolactone. Carbohydr Res 205:269–282

Frommer W, Müller L, Schmidt D, Puls W, Krause HP (1978a) Glucoside hydrolase inhibitors production from bacteria. Belg Pat 826 165 (prior DE 2658563, 23.12.76)

Frommer W, Müller L, Schmidt D, Puls W, Krause HP (1978b) Inhibitoren für α-Glucosidasen. Ger Offen 2658561 (prior 23.12.76)

Frommer W, Müller L, Schmidt D, Puls W, Krause HP (1979a) Inhibitors for glucoside hydrolases from bacilli. Ger Offen 2726899 (prior 15.06.77)

Frommer W, Müller L, Schmidt D, Puls W, Krause HP (1979b) Inhibitoren für α-Glucosidasen. Ger Offen 2726898 (prior 15.06.77)

Frommer W, Schmidt D (1980) 1-Desoxynojinimycin-Herstellung. Eur Pat 15 388 Prior.: Ger Offen 2907190

Fukase H, Horii S (1992) Synthesis of a branched-chain inosose derivative, a versatile synthon of N-substituted valiolamine derivatives from D-glucose. J Org Chem 57:3642–3650

Fukuhara K, Murai H, Murao S (1982) Isolation and structure-activity relationship of some amylostatins (F-1b fraction) produced by Streptomyces diastatiens subsp. amylostatins no. 9410. Agric Biol Chem 46:1941–1945

Furukawa J, Okuda S, Saito K, Hatanaka S-I (1985) 3,4-Dihydroxy-2-hydroxymethylpyrrolidine from Arachnoides standishii. Phytochemistry 24:593–594

Ganem B, Papandreou G (1991) Mimicking the glucosidase transition state: shape/ charge considerations. J Am Chem Soc 113:8984–8985

Gerspacher M, Rapoport H (1991) 2-Amino-2-deoxyhexoses as chiral educts for hydroxylated indolizidines. Synthesis of (+)-castanospermine and (+)-6-epicastanospermine. J Org Chem 56:3700–3706

Getman DP, DeCrescenzo GA (1991) Preparation of dideoxyfluoronojirimycins as glycosidase inhibitors. Eur Pat 410953 (prior 27.08.89)

Getman DP, DeCrescenzo GA (1992a) Intermediates for 1,2-dideoxy-2-fluoronojirimycin. US Pat 5175168 (prior 27.06.89)

Getman DP, DeCrescenzo GA (1992b) Glucosidase inhibiting 1,4-dideoxy-4-fluoronojirimycin. US Pat 5128347 (prior 18.10.90)

Glaser R, Perlin AS (1988) ^{1}H- and ^{13}C-N.M.R.studies on N-methyl-1-deoxynojirimycin, an α-D-glucosidase inhibitor. Carbohydr Res 182:169–177

Gradnig G, Berger A, Grassberger V, Stütz AE, Legler G (1991) First synthesis of (1R,2R,3S,9S,9aR)-1,2,3,9-tetrahydroxyquinolizidine, a novel isosteric homologue of the glucosidase inhibitor castanospermine. Tetrahedron Lett 32: 4889–4892

Graßberger V, Berger A, Dax K, Fechter M, Gradnig G, Stütz AE (1993) Synthese von (+)-Castanospermin und 1-Epicastanospermin aus D-Glucofuranurono-6,3-lacton durch Reformatsky-Reaktion. Liebigs Ann Chem: 379–390

Gross V, Tran-Thi T-A, Schwarz RT, Elbein AD, Decker K, Heinrich PC (1986) Different effects of the glucosidase inhibitors 1-deoxynojirimycin, N-methyl-1-deoxynojirimycin and castanospermine on the glycosylation of rat α_1-proteinase inhibitor and α_1-acid glycoprotein. Biochem J 236:853–860

Hamana H, Ikota N, Ganem B (1987) Chelate selectivity in chelation-controlled allylations. A new synthesis of castanospermine and other bioactive indolizidine alkaloids. J Org Chem 52:5492–5494

Hanozet G, Pircher HP, Vanni P, Oesch B, Semenza G (1981) An example of enzyme hysteresis. The slow and tight interaction of some fully competitive inhibitors with small intestinal sucrase. J Biol Chem 256:3703–3711

Hardick DJ, Hutchinson DW (1993) The biosynthesis of 1-deoxynojirimycin in Bacillus subtilis var niger. Tetrahedron 49:6707–6716

Hardick DJ, Hutchinson DW, Trew SJ, Wellington EMH (1991) The biosynthesis of deoxynojirimycin and deoxymannonojirimycin in Streptomyces subrutilius. J Chem Soc Chem Commun 10:729–730

Hardick DJ, Hutchinson DW, Trew SJ, Wellington EMH (1992) Glucose is a precursor of 1-deoxynojirimycin and 1-deoxymannonojirimycin in Streptomyces subrutilus. Tetrahedron 48:6285–6296

Harris CM, Harris TM, Molyneux RJ, Tropea JE, Elbein AD (1989) 1-Epiaustraline, a new pyrrolizidine alkaloid from Castanospermum australe. Tetrahedron Lett 30:5685–5688

Harris EMS, Aleshin AE, Firsov LM, Honzatko RB (1993) Refined structure for the complex of 1-deoxynojirimycin with glucoamylase from Aspergillus awamori var. ×100 to 2.4-Å resolution. Biochemistry 32:1618–1626

Hayashida M, Sakairi N, Kuzuhara H (1986) Synthesis of dihydroacarbose, a potent α-glucosidase inhibitor Carbohydr Res 158:C5–C8

Hayashida M, Sakairi N, Kuzuhara H (1988) Novel synthesis of penta-N, O-acetylvaliolamine. J Carbohydr Chem 7:83–94

Hayashida M, Sakairi N, Kuzuhara H, Yajima M (1989) Synthesis of dihydroacarbose, an α-D-glucosidase inhibitor having a pseudo-tetrasaccharide structure. Carbohydr Res 194:233–246

Heiker F-R (1982) Synthesis of acarviosin-glycosides. Isolation and modification of the central structural unit of acarbose. Lecture Stockholm, 20–24 June 1982

Heiker F-R, Böshagen H, Junge B, Müller L, Stoltefuß J (1982) Studies designed to localize the essential structural unit of glycoside-hydrolase inhibitors of the acarbose type. In: Creutzfeldt W (ed) First international symposium on acarbose. Excerpta Medica, Amsterdam, pp 137–141

Hempel A, Camerman N, Mastropaolo D, Camerman A (1993) Glucosidase inhibitors: structures of deoxynojirimycin and castanospermine. J Med Chem 36: 4082–4086

Hendry D, Hough L, Richardson AC (1987) Enantiospecific synthesis of 1-deoxycastanospermine, (6S,7R,8R,8aR)-trihydroxyindolizidine, from D-glucose. Tetrahedron Lett 28:4597–4600

Hendry D, Hough L, Richardson AC (1988) Enantiospecific synthesis of polyhydroxylated indolizidines related to castanospermine: 1-deoxy-castanospermine. Tetrahedron 44:6143–6152

Hinsken W (1990) Herstellung von N-(ω-Hydroxyalkyl)-iminopyranosen als α-Glycosidaseinhibitoren. Ger Offen 3906463 (prior 01.03.89)

Hinsken W (1991) Herstellung von Zwischenprodukten für N-(2-Hydroxyethyl)-2-hydroxymethyl-3,4,5-trihydroxypiperidine. Ger Offen 3936295 (prior 01.11.89)

Hohenschutz LD, Bell EA, Jewess PJ, Leworthy DP, Pryce RJ, Arnold E, Clardy J (1981) Castanospermine, a 1,6,7,8-tetrahydroxyoctahydroindolizine alkaloid, from seeds of Castanospermum australe. Phytochemistry 20:811–814

Holt KE, Leeper FJ, Handa S (1994) Synthesis of β-1-homonojirimycin and ß-1-homomannojirimycin using the enzyme Aldolase. J Chem Soc Perkin Trans 1:231–234

Hori H, Pan YT, Molyneux RJ, Elbein AD (1984) Inhibition of processing of plant N-linked oligosaccharides by castanospermine. Arch Biochem Biophys 228: 525–533

Horii S, Fukase H (1985) Stereoselective conversion of valienamine and validamine into valiolamine. Carbohydr Res 140:185–200

Horii S, Fukase H (1992) Synthesis of valiolamine and its N-substituted derivatives AO-128, validoxylamine G, and validamycin G via branched-chain inosose derivatives. J Org Chem 57:3651–3658

Horii S, Iwasa T, Kameda Y (1971a) Studies on validamycins, new antibiotics. V. Degradation studies. J Antibiot 24:57–58

Horii S, Iwasa T, Mizuta E, Kameda Y (1971b) Studies on validamycins, new antibiotics. VI. Validamine, hydroxyvalidamine and validatol, new cyclitols. J Antibiot 24:59–63

Horii S, Fukase H, Matsuo T, Kameda Y, Asano N, Matsui K (1986) Synthesis and α-D-glucosidase inhibitory activity of N-substituted valiolamine derivatives as potential oral antidiabetic agents. J Med Chem 29:1038–1046

Hughes AB, Rudge AJ (1994) Deoxynojirimycin: Synthesis and biological activity. Nat Prod Rep 11:135–162

Hung RR, Straub JA, Whitesides GM (1991) α-Amino aldehyde equivalents as substrates for rabbit muscle aldolase: synthesis of 1,4-dideoxy-D-arabinitol and 2(R),5(R)-bis(hydroxymethyl)-3(R),4(R)-dihydroxypyrrolidine. J Org Chem 56: 3849–3855

Iida H, Yamazaki N, Kibayashi C (1987) Total synthesis of (+)-nojirimycin and (+)-1-deoxynojirimycin. J Org Chem 52:3337–3342

Ikota N (1989) Synthesis of (+)-1-deoxynojirimycin from (S)-pyroglutamic acid. Heterocycles 29:1469–1472

Ikota N (1992) Stereocontrolled synthesis of 1,7a-diepialexine. Tetrahedron Lett 33:2553–2556

Ikota N, Hanaki A (1987) Synthesis of (−)-swainsonine and optically active 3,4-dihydroxy-2-hydroxymethylpyrrolidines. Chem Pharm Bull 35:2140–2143

Ina H, Kibayashi C (1991) A total synthesis of (+)-castanospermine. Tetrahedron Lett 32:4147–4150

Ina H, Kibayashi C (1993) Total syntheses of (+)-castanospermine and (+)-1-epicastanospermine and their 1-O-acyl derivatives from a common chiral building block. J Org Chem 58:52–61

Inouye S (1970) Studies on the synthesis of nojirimycin and 6-deoxynojirimycin. Sci Reports Meiji Seika Kaisha 11:52–74

Inouye S, Tsuruoka T, Niida T (1966) Structure of nojirimycin, sugar antibiotic with nitrogen in the ring. J Antibiot Ser A 19:288–292

Inouye S, Tsuruoka T, Ito T, Niida T (1968) Structure and synthesis of nojirimycin. Tetrahedron 24:2125–2144

Ishida N, Kumagai K, Niida T, Hamamoto K, Shomura T (1967a) Nojirimycin, a new antibiotic. I. Taxonomy and fermentation. J Antibiot Ser A 20:62–65

Ishida N, Kumagai K, Niida T, Tsuruoka T, Yumoto H (1967b) Nojirimycin, a new antibiotic. II. Isolation, characterization, and biological activity. J Antibiot Ser A 20:66–71

Ishida N, Kumagai K, Nishikawa T, Niida T, Tsuruoka T, Ueda M, Watanabe K (1968) Nojirimycin, a new antibiotic. Jpn Pat 43/760 (prior 15.10.65)

Ishii Y, Usui T, Shibahara M, Nagaoka K, Inoe S (1993) Preparation of deoxynojirimycins as glucosidase-inhibiting virucides. Jpn Pat 05043545 (prior 15.08.91)

ltoh J, Omoto S, Shomura T, Ogino H, Iwamatsu K, Inouye S, Hidaka H (1981) Oligostatins, new antibiotics with amylase inhibitory activity I. Production, isolation and characterisation. J Antibiot 34:1424–1428

Jones DWC, Nash RJ, Bell EA, Williams JM (1985) Identification of the 2-hydroxymethyl-3,4-dihydroxypyrrolidine (or 1,4-dideoxy-1,4-iminopentitol) from Angylocalyx boutiqueanus and from Arachnroides standishii as the (2R,3R,4S)-isomer by the synthesis of its enantiomer. Tetrahedron Lett 26: 3125–3126

Junge B, Krause HP, Müller L, Puls W (1979) Neue Derivate von 3,4,5-Trihydroxypiperidin, Verfahren zu ihrer Herstellung und ihre Verwendung. Ger Offen 2758025 (prior 24.12.77)

Junge B, Böshagen H, Stoltefuß J, Müller L (1980) Derivatives of acarbose and their inhibitory effects on α-glucosidases. In: Brodbeck U (ed) Enzyme inhibitors, Verlag Chemie Weinheim, pp 123–137

Junge B, Stoltefuß J, Müller L, Krause HP, Sitt R (1981) Derivate des 3,4,5-Trihydroxypiperidins. Ger Offen 3007078

Junge B, Heiker F-R, Kurz J, Müller L, Schmidt DD, Wünsche C (1984a) Untersuchungen zur Struktur des α-D-Glucosidaseinhibitors Acarbose. Carbohydr Res 128:235–268

Junge B, Böshagen H, Kinast G, Krause HP, Müller L, Puls W, Schedel M, Stoltefuß J (1984b) BAY m 1099 and BAY o 1248, new α-glucosidase inhibitors and potential antidiabetic agents. In: Proceedings of the 8th Intern Symp on Med Chem, Uppsala

Junge B, Aubell R, Bischoff H, Böshagen H, Frommer W, Heiker FR, Hillebrand J, Kinast G, Krause HP, Müller L, Puls W, Schmidt D, Stoltefuß J, Truscheit E (1986) Delaying carbohydrate absorption by means of glucosidase inhibitors. 192nd ACS National Meeting, Anaheim

Junge B, Böshagen H, Heiker FR, Hillebrand J, Kinast G, Müller L, Puls W, Schedel M, Schmidt D, Schüller M, Stoltefuß J (1989) Deoxynojirimycin and derivatives: chemistry and antidiabetic activity. 198th ACS National Meeting, Miami Beach

Kajimoto T, Liu KKC, Pederson RL, Zhong Z, Ichikawa Y, Porco JA Jr, Wong CH (1991a) Enzyme-/catalyzed aldol condensation for asymmetric synthesis of azasugars: synthesis, evaluation, and modeling of glycosidase inhibitors. J Am Chem Soc 113:6187–6196

Kajimoto T, Chen L, Liu KKC, Wong CH (1991b) Palladium-mediated stereocontrolled reductive amination of azido sugars prepared from enzymic aldol condensation: a general approach to the synthesis of deoxy aza sugars. J Am Chem Soc 113:6678–6680

Kameda Y, Horii S (1972) The unsaturated cyclitol part of the new antibiotics, the validamycins. J Chem Soc Chem Comm 1972:746–747

Kameda Y, Asano N, Teranishi M, Matsui K (1980) New cyclitols, degradation of validamycin by Flavobacterium saccharophilum. J Antibiot 33:1573–1574

Kameda Y, Asano N, Yoshikawa M, Takeuchi M, Yamaguchi T, Matsui K, Horii S, Fukase H (1984) Valiolamine, a new α-glucosidase inhibiting aminocyclitol produced by Streptomyces hygroscopicus. J Antibiot 37:1301–1307

Kameda Y, Asano N, Takeuchi M, Yamaguchi T, Matsui K, Horii S, Fukase H (1985) Epivaliolamine and deoxyvalidamine, new pseudo-aminosugars produced by Streptomyces hygroscopicus. J Antibiot 38:1816–1818

Kameda Y, Asano N, Yamaguchi T, Matsui K (1987) Validoxylamines as trehalase inhibitors. J Antibiot 40:563–565

Kangouri K, Namiki S, Nagate T, Hara H, Sugita K, Omura S (1982) Studies on the α-glucoside hydrolase inhibitor adiposin. III. α-Glucoside hydrolase inhibitory activity and antibacterial activity in vitro. J Antibiot 35:1160–1166

Kappes E, Legler G (1989) Synthesis and inhibitory properties of 2-acetamido-2-deoxynojirimycin (2-acetamido-5-amino-2,5-dideoxy-D-glucopyranose) and 2-acetamido-1,2-dideoxynojirimycin(2-acetamido-1,5-imino-1,2,5-trideoxy-D-glucitol). J Carbohydr Chem 8:371–388

Karpas A, Fleet GWJ, Dwek RA, Petursson S, Namgoong SK, Ramsden NG, Jacob GS, Rademacher TW (1988) Aminosugar derivatives as potential anti-human immunodeficiency virus agents. Proc Natl Acad Sci USA 85:9229–9233

Kayakiri H, Takase S, Setoi H, Uchida I, Terano H, Hashimoto M (1988) Structure of FR 900483, a new immunomodulator isolated from a fungus. Tetrahedron Lett 29:1725–1728

Kayakiri H, Nakamura K, Takase S, Setoi H, Uchida I, Terano H, Hashimoto M, Tada T, Koda S (1991) Structure and synthesis of nectrisine, a new immunomodulator isolated from a fungus. Chem Pharm Bull 39:2807–2812

Kiguchi T, Tajiri K, Ninomiya I, Naito T, Hiramatsu H (1995) A novel and concise synthesis of aminocyciopentitols and 1-deoxynojirimycin via radical cyclization of oxime ethers. Tetrahedron Lett 36:253–256

Kim N-S, Choi J-R, Cha JK (1993) A concise, enantioselective synthesis of castanospermine. J Org Chem 58:7096–7099

Kinast G, Schedel M (1980) N-Substituted derivatives of 1-deoxynojirimycin. Ger Offen 2853573 (prior 12.12.78)

Kinast G, Schedel M (1981) A four-stage synthesis of 1-deoxy-nojirimycin with a biotransformation as the central reaction step. Angew Chem 93:799–800

Kinast G, Müller L, Puls W, Sitt R (1980) N-Amino-3,4,5-trihydroxypiperidine. Ger Offen 2835069

Kinast G, Müller L, Sitt R, Puls W (1981) 2-Hydroxyalkyl-3,4,5-trihydroxypiperidine. Eur Pat 027908 prior.: Ger Offen 2942365 (1979)

Kinast G, Schedel M, Köbernick W (1982) N-Substituted derivatives of 1-desoxynojirimycin. Eur Pat 49858 (prior DE 3038901, 15.10.80)

Kiso M, Kitagawa M, Ishida H, Hasegawa A (1991) Studies on glycan processing inhibitors: synthesis of N-acetylhexosamine analogs and cyclic carbamate derivatives of 1-deoxynojirimycin. J Carbohydr Chem 10:25–45

Kiso M, Katagiri H, Furui H, Hasegawa A (1992) Studies on 1-deoxynojirimycin-containing glycans – synthesis of novel disaccharides related to lactose, lactosamine, and chitobiose. J Carbohydr Chem 11:627–644

Kite GC, Fellows LE, Fleet GWJ, Liu PS, Scofield AM, Smith NG (1988) α-Homonojirimycin [2,6-dideoxy-2,6-imino-D-glycero-L-gulo-heptitol] from Omphalea diandra I.: isolation and glucosidase inhibition. Tetrahedron Lett 29:6483–6486

Kite GC, Horn JM, Romeo JT, Fellows LE, Lees DC, Scofield AM, Smith NG (1990) α-Homonojirimycin and 2,5-dihydroxymethyl-3,4-dihydroxypyrrolidine: alkaloidal glycosidase inhibitors in the moth Urania fulgens. Phytochemistry 29:103–105

Kite GC, Fellows LE, Lees DC, Kitchen D, Monteith GB (1991) Alkaloidal glycosidase inhibitors in nocturnal and diurnal uraiine moths and their respective food plant genera, Endospermum and Omphalea. Biochem Syst Ecol 19:441–445

Klemer A, Hofmeister U, Lemmes R (1979) Eine neue Synthese von 5-Amino-5-desoxy-1,2,-O-isopropyliden-α-Glucofuranose. Carbohydr Res 68:391–395

Knapp S, Naughton ABJ, Murali Dhar TG (1992) Intramolecular amino delivery reactions for the synthesis of valienamine and analogues. Tetrahedron Lett 33:1025–1028

Kodama Y, Tsuruoka T, Niwa T, Inouye S (1985) Molecular structure and glycosidase – inhibitory activity of nojirimycin bisulfite adduct. J Antibiot 38: 116–118

Köbernick W (1981) 1,5-Dideoxy-1,5-imino-D-glucitol and its N-derivatives. Eur Pat 55431 (prior 30.12.80)

Köbernick W, Furtwängler HR (1981) Verfahren zur Herstellung bekannter und neuer 6-Amino-6-desoxy-2,3-O-isopropyliden-α-L-sorbofuranose-Derivate sowie neue Zwischenprodukte des Verfahrens. Eur Pat 25140 (prior DE 2936240, 07.09.79)

Köhn A, Schmidt RR (1987) α-Glucosidase inhibitors 5: Investigations towards a synthesis of C_1-branched cyclitols from D-glucose. Liebigs Ann Chem: 1045–1054

Kuszmann J, Kiss L (1986) Synthesis of 1,4-dideoxy-1,4-imino-D-glucitol, a glucosidase inhibitor. Carbohydr Res. 153:45–53

Laszlo E, Hollo J, Hoschke A, Sarosi G (1978) A study by means of lactone inhibition of the role of a "half-chair" glycosyl conformation at the active centre of amylolytic enzymes. Carbohydr Res 61:387–394

Lee CK, Jiang H, Koh LL, Xu Y (1993) Synthesis of 1,3-dideoxy-3-fluoronojirimycin. Carbohydr Res 239:309–315

Legler G (1990) Glycoside hydrolases: mechanistic information from studies with reversible and irreversible inhibitors. Adv Carbohydr Chem Biochem 48:319–384

Legler G, Becher W (1982) Dissociation constants and dissociation rate of the nojirimycin-hydrogen sulfite adduct and related compounds. Carbohydr Res 101:326–329

Lembcke B, Foelsch UR, Creutzfeldt W (1985) Effect of 1-desoxynojirimycin derivatives on small intestinal disaccharidase activities and on active transport in vitro. Digestion 31:120–127

Linden A, Hoos R, Vasella A (1994) 1-Deoxynojirimycin hydrochloride. Acta Cryst C50:746–749

Liotta LJ, Lee J, Ganem B (1991) Effect of 1-epoxyalkyl-1-deoxynojirimycins on exoglucosidases. Tetrahedron 47:2433–2447

Liu PS (1987) Total synthesis of 2,6-dideoxy-2,6-imino-7-O-(B-D-glucopyranosyl)-D-glycero-L-gulo-heptitol hydrochloride. A potent inhibitor of α-glucosidases. J Org Chem 52:4717–4721

Liu PS, King C-HR (1992) Synthesis of castanospermine glucosides. Syn Comm 22:2111–2116

Liu PS, Hoekstra WJ, King C-HR (1990) Synthesis of potent anti-HIV agents: esters of castanospermine. Tetrahedron Lett 31:2829–2832

Liu KK-C, Kajimoto T, Chen L, Zhong Z, Ichikawa Y, Wong C-H (1991) Use of dihydroxyacetone phosphate dependent aldolases in the synthesis of deoxysugars. J Org Chem 56:6280–6289

Look GC, Fotsch CH, Wong CH (1993) Enzyme-catalyzed organic synthesis: practical routes to aza sugars and their analogs for use as glycoprocessing inhibitors. Acc Chem Res 26:182–190

Margolin AL, Delinck DL, Whalon MR (1990) Enzyme-catalyzed regioselective acylation of castanospermine. J Am Chem Soc 112:2849–2854

Matsumura S, Enomoto H, Aoyagi Y, Ezure Y, Yoshikuni Y, Yagi M (1979a) Verfahren Zur Gewinnung von Moranolin und N-Methylmoranolin. Ger Offen (prior.: Jap Pat P 52-135506)

Matsumura S, Enomoto H, Aoyagi Y, Yoshikuni Y, Kura K, Yagi M, Shirahase I (1979b) Neue N-substituierte Moranolinderivate. Ger Offen 2915 037 (prior.: Jap Pat 53-77, 03.06.1978)

Matsumura S, Enomoto H, Aoyagi Y, Yoshikuni Y, Yagi M (1979c) Moranolin. Ger Offen 2850467 (prior Jap Pat 77-140126, 21.11.77)

Matsumura S, Enomoto H, Aoyagi Y, Yoshikuni Y, Yagi M (1980) Moranolin preparation from Jacobinia plant. Jap Pat 55027136 (prior 14.08.78)

Matsumura S, Enomoto H, Aoyagi Y, Yoshikuni Y, Yagi M, Kura K, Shirahase I (1982) Bis-moranolinderivate. Ger Offen DE 3102769 (prior 28.01.80)

Miller SA, Chamberlin AR (1990) Enantiomerically pure polyhydroxylated acyliminium ions. Synthesis of the glycosidase inhibitors (−)-swainsonine and (+)-castanospermine. J Am Chem Soc 112:8100–8112

Mitsuo S, Sawao M (1980) Microbial production of 1-deoxynojirimycin. Jap Pat 55120792 (prior 09.03.79)

Molyneux RJ, Roitman JN, Dunnheim G, Szumilo T, Elbein AD (1986) 6-epicastanospermine, a novel indolizidine alkaloid that inhibits α-glucosidase. Arch Biochem Biophys 251:450–457

Molyneux RJ, Benson M, Wong RY, Tropea JE, Elbein AD (1988) Australine, a novel pyrrolizidine alkaloid glucosidase inhibitor from Castanospermum australe. J Nat Prod 51:1198–1206

Molyneux RJ, Pan YT, Tropea JE, Benson M, Kaushal GP, Elbein AD (1991) 6,7-Diepicastanospermine, a tetrahydroxyindolizidine alkaloid inhibitor of amyloglucosidase. Biochemistry 30:9981–9987

Mulzer J, Dehmlow H, Buschmann J, Luger P (1992) Stereocontrolled total synthesis of the unnatural enantiomers of castanospermine and 1-epi-castanospermine. J Org Chem 57:3194–3202

Murai H, Ohata K, Enomoto H, Yoshikuni Y, Kono T, Yagi M (1977) 2-Hydroxymethyl-3,4,5-trihydroxypiperidine and extraction process for its manufacture. Ger Offen 2656602 (prior Jap Pat 75-157423, 29.12.75)

Murai H, Enomoto H, Aoyagi Y, Yoshikuni Y, Yagi M, Shirahase I (1979) N-Alkylpiperidin Derivate. Ger Offen 2824761 (prior.: Jap Pat 77-75936, 25.06.1977)

Murao S, Miyata S (1980) Isolation and characterization of a new trehalase inhibitor, S-GI. Agric Biol Chem 44:219–221

Murao S, Ohyama K (1975) New amylase inhibitor (S-AI) from Streptomyces diastatious var amylostaticus No. 2476. Agric Biol Chem 39:2271–2273

Murao S, Ohyama K (1979) Chemical structure of an amylase inhibitor, S-AI. Agric Biol Chem 43:679–681

Müller L (1985) Microbial glycosidase inhibitors. In: Rehm H-J, Reed G (eds) Biotechnology, vol 4, VCH, Weinheim, pp 1–37

Müller L (1989) Chemistry, biochemistry and therapeutic potential of microbial α-glucosidase inhibitors. In: Demain AL, Somkuti GA, Hunter-Cevera JC, Rosmoore HW (eds) Novel microbial products for medicine and agriculture. Elsevier, Amsterdam, pp 109–116

Müller L, Junge B, Frommer W, Schmidt DD, Truscheit E (1980) Acarbose (BAY G 5421) and homologous α-glucosidase inhibitors from Actinoplanaceae. In: Brodbeck U (ed) Enzyme inhibitors. Verlag Chemie, Weinheim, pp 109–122

Naleway JJ, Raetz CRH, Anderson L (1988) A convenient synthesis of 4-amino-4-deoxy-L-arabinose and its reduction product, 1,4-dideoxy-1,4-imino-L-arabinitol. Carbohydr Res 179:199–209

Namiki S, Kangouri K, Nagate T, Hara H, Sugita K, Omura S (1982a) Studies on the α-glucoside hydrolase inhibitor adiposin. II. Taxonomic studies on the producing microorganism. J Antibiot. 35:1156–1159

Namiki S, Kangouri K, Nagate T, Hara H, Sugita K, Omura S (1982b) Studies on the α-glucoside hydrolase inhibitor adiposin. I. J Antibiot 35:1234–1236

Nash RJ, Bell EA, Williams JM (1985) 2-Hydroxymethyl-3,4-dihydroxypyrrolidine in fruits of Angylocalyx boutiqueanus. Phytochemistry 24:1620–1622

Nash RJ, Fellows LE, Dring JV, Stirton CH, Carter D, Hegarty MP, Bell EA (1988a) Castanospermine in Alexa Species. Phytochemistry 27:1403–1404

Nash RJ, Fellows LE, Dring JV, Fleet GWJ, Derome AE, Hamor TA, Scofield AM, Watkin DJ (1988b) Isolation from Alexa leiopetala and X-ray crystal structure of alexine, (1R,2R,3R,7S,8S)-3-hydroxymethyl-1,2,7-trihydroxypyrrolizidine, [(2R,3R,4R,5S,6S)-2-hydroxymethyl-1-azabicyclo[3.3.0]octan-3,4,6-triol], a unique pyrrolizidine alkaloid. Tetrahedron Lett 29:2487–2490

Nash RJ, Fellows LE, Plant AC, Fleet GWJ, Derome AE, Baird PD, Hegarty MP, Scofield AM (1988c) Isolation from Castanospermum australe and X-ray crystal structure of 3,8-diepialexine, (1R,2R,3S,7S,8R)-3-hydroxymethyl-1,2,7-trihydroxypyrrolizidine [(2S-3R,4R,5S,6R)-2-hydroxymethyl-1-azabicyclo[3.3.0]octan-3,4,6-triol]. Tetrahedron 44:5959–5964

Nash RJ, Fellows LE, Dring JV, Fleet GWJ, Girdhar A, Ramsden NG, Peach JM, Hegarty MP, Scofield AM (1990) Two alexines [3-hydroxymethyl-1,2,7-trihydroxypyrrolizidines] from Castanospermum australe. Phytochemistry 29:111–114

Nishikawa T, Ishida N (1965) A new antibiotic R-468 active against drug-resistant Shigella. J Antibiot Ser A 18:132–133

Nishimura Y (1991) The synthesis and biological activity of glycosidase inhibitors. J Synth Org Chem Jpn 49:846–857

Niwa T, Inouye S, Tsuruoka T, Koaze Y, Niida T (1970) "Nojirimycin" as a potent inhibitor of glucosidase. Agr Biol Chem 34:966–968

Ogawa S, Tsunoda H (1992) New synthesis of 2-amino-5a-carba-2-deoxy-α-DL-glucopyranose and its transformation into valienamin and valiolamine analogues. Liebigs Ann Chem: 637–641

Ogawa S, Iwasawa Y, Nose T, Suami T (1985a) Total Synthesis of (+)-(1,2,3/4,5)-2,3,4,5-Tetrahydroxycyclohexane-1-methanol and (+)-(1,3/2,4,5)-5-Amino-2,3,4-trihydroxycyclohexane-1-methanol [(+)-Validamine]. X-Ray Crystal Structure of (3S)-(+)-2-exo-Bromo-4,8-dioxatricyclo [4.2.1.0^{3,7}]nonan-5-one[1]

Ogawa S, Iwasawa Y, Nose T, Suami T (1985a) Total synthesis of (+)-(1,2,3/4,5)-2,3,4,5-tetrahydroxy-cyclohexane-1-methanol and (+)-(1,3/2,4,5)-5-amino-2,3,4-trihydroxycyclohexane-1-methanol [(+)-validamine] X-ray, crystal structure of (3S)-(+)-2-exo-bromo-4,8-dioxatricyclo[4.2.1.0$^{3,×}$]nonan-5-one. J Chem Soc Perkin Trans I 1985:903–906

Ogawa S, Shibata Y, Nose T, Suami T (1985b) Synthetic studies on the validamycins. IXX. Synthesis of optically active valienamine and validatol. Bull Chem Soc Jpn 58:3387–3388

Ogawa S, Shibata Y, Kosuge Y, Yasuda K, Mizukoshi T, Uchida C (1990) Synthesis of potent α-glucosidase inhibitors: methyl acarviosin analogue composed of 1,6-anhydro-β-D-glucopyranose residue. J Chem Soc Chem Commun: 1387–1388

Ohata M, Enomoto H, Yoshikuni Y, Kono T, Yagi M (1977) A piperidine derivative as a hypoglycemic drug. Jap Pat 52–83951 (prior 01.01.76)

Omoto S, Itoh J, Ogino H, Iwamatsu K, Nishizawa N, Inouye S (1981) Oligostatins, new antibiotics with amylase inhibitory activity. II. Structures of oligostatins C, D and E. J Antibiot 34:1429–1433

Otani M, Saito T, Satoi S, Mizoguchi J, Muto N (1979) Jpn Kokai JP 54-92909

Overkleft HS, van Wiltenburg J, Pandit UK (1993) An expedient stereoselective synthesis of gluconolactam. Tetrahedron Lett 34:2527–2528

Pan YT, Hori H, Saul R, Sanford BA, Molyneux RJ, Elbein AD (1983) Castanospermine inhibits the processing of the oligosaccharide portion of the influenza viral hemagglutinin. Biochemistry 22:3975–3984

Pan YT, Kaushal GP, Papandreou G, Ganem B, Elbein AD (1992) D- Mannonolactam amidrazone. A new mannosidase inhibitor that also inhibits the endoplasmic reticulum or cytoplasmic α-mannosidase. J Biol Chem 267:8313–8318

Pan YT, Ghidoni J, Elbein AD (1993) The effects of castanospermine and swainsonine on the activity and synthesis of intestinal sucrase. Arch Biochem Biophys 303:134–144

Papandreou G, Tong MK, Ganem B (1993) Amidine, amidrazone, and amidoxime derivatives of monosaccharide aldonolactams: synthesis and evaluation as glycosidase inhibitors. J Am Chem Soc 115:11682–11690

Paulsen H (1966) Kohlenhydrate mit Stickstoff oder Schwefel im "Halbacetal"-Ring. Angew Chem 78:501–556

Paulsen H, Heiker F-R (1981) Synthese von enantiomerenreinem Valienamin aus Quebrachit. Liebigs Ann Chem: 2180–2203

Paulsen H, von Deyn W (1987) Synthese von Pseudozuckern aus D-Glucose durch intramolekulare Horner-Emmons-Olefinierung. Liebigs Ann Chem: 125–131

Paulsen H, Sangster I, Heyns K (1967) Synthese und Reaktionen von Keto-piperidinosen. Chem Ber 100:802–815

Paulsen H, Todt K, Ripperger H (1968) Konformation und anomerer Effekt von N-substituierten 2-Alkyl-piperidin-Derivaten. Chem Ber 101:3365–3376

Paulsen H, Propp K, Heyns K (1969) Monosaccharide mit stickstoffhaltigem Ring XXI 4-amino-4-desoxy-D-glucose und 4-amino-4-desoxy-D-galactose. Tetrahedron Lett:683–686

Paulsen H, Hayauchi Y, Sinnwell V (1980) Synthese von 1,5-Didesoxy-1,5-imino-D-galactit. Chem Ber 113:2601–2608

Pearson WH, Hines JV (1991) A synthesis of (+)-7-epiaustraline and (−)-7-epialexine. Tetrahedron Lett 32:5513–5516

Pederson RL, Wong C-H (1989) Enzymatic aldol condensation as a route to heterocycles: synthesis of 1,4-dideoxy-1,4-imino-D-arabinitol, fagomine, 1-deoxynojirimycin and 1-deoxymannojirimycin. Heterocycles 28:477–480

Pederson RL, Kim MJ, Wong CH (1988) A combined chemical and enzymic procedure for the synthesis of 1-deoxynojirimycin and 1-deoxymannojirimycin. Tetrahedron Lett 29:4645–4648

Peet NP, Huber EW, Farr RA (1991) Diastereoselectivity in the intramolecular nitrone, oxime, and nitrile oxide cycloaddition reactions. Synthesis of amino inositol derivatives as α-glucosidase inhibitors. Tetrahedron 47:7537–7550

Pereira AC de S, Kaplan MAC, Maia JGS, Gottlieb OR, Nash RJ, Fleet G, Pearce L, Watkin DJ, Scofield AM (1991) Isolation of 7a-epialexaflorine from leaves of Alexa grandiflora, a unique pyrrolizidine amino acid with a carboxylic acid substituent at C-3. Tetrahedron 47:5637–5640

Pinto BM, Wolfe S (1982) Orbital energy and interactions: the anomeric effect in nojirimycin. Tetrahedron Lett 23:3687–3690

Poitout L, Le Merrer Y, Depezay J-C (1994) Polyhydroxylated piperidines and azepanes from D-mannitol. Synthesis of 1-deoxynojirimycin and analogues. Tetrahedron Lett 35:3293–3296

Puls W, Keup U (1973) Influence of an α-amylase inhibitor (BAY d 7791) on blood glucose, serum insulin, and NEFA (non esterified fatty acids) in starch loading tests in rats, dogs, and man. Diabetologia 9:97–101

Puls W, Keup U, Krause HP, Müller L, Schmidt DD, Thomas G, Truscheit E (1980) Pharmacology of a glucosidase inhibitor. Front Horm Res 7:235–247

Puls W, Krause HP, Müller L, Schutt H, Sitt R, Thomas G (1984) Inhibitors of the rate of carbohydrate and lipid absorption by the intestine. Int J Obes 8:181–190

Reese ET, Parrish FW, Ettinger M (1971) Nojirimycin and D-glucono-1,5-lactone as inhibitors of carbohydrases. Carbohydr Res 18:381–388

Reitz AB, Baxter EW (1990) Pyrrolidine and piperidine amino sugars from dicarbonyl sugars in one step. Concise synthesis of 1-deoxynojirimycin. Tetrahedron Lett 31:6777–6780

Reymond J-L, Vogel P (1989) A highly stereoselective total synthesis of ($\pm$)-castanospermine. Tetrahedron Lett 30:705–706

Reymond J-L, Pinkerton AA, Vogel P (1991) Total, asymmetric synthesis of (+)-castanospermine, (+)-6-deoxycastanospermine, and (+)-6-deoxy-6-fluorocastanospermine. J Org Chem 56:2128–2135

Rhinehart BL, Robinson KM, Liu PS, Payne AJ, Wheatley ME, Wagner SR (1987a) Inhibition of intestinal disaccharidases and suppression of blood glucose by a new α-glucohydrolase inhibitor – MDL 25637. J Pharmacol Exp Ther 241: 915–920

Rhinehart BL, Robinson KM, Payne AJ, Wheatley ME, Fisher JL, Liu PS, Cheng W (1987b) Castanospermine blocks the hyperglycemic response to carbohydrates in vivo: a result of intestinal disaccharidase inhibition. Life Sci 41: 2325–2331

Rhinehart BL, Robinson KM, King C-HR, Liu PS (1990) Castanospermine-glucosides as selective disaccharidase inhibitors. Biochem Pharmacol 39:1537–1543

Robinson KM, Rhinehart BL, Begovic ME, King C-HR, Liu PS (1989) Castanospermine-glucosides are potent, selective, long acting sucrase inhibitors. Pharmacol Exp Ther 251:224–229

Robinson KM, Begovic ME, Rhinehart BL, Heineke EW, Ducep JB, Kastner PR, Marshall FN, Danzin C (1991) New potent α-glucohydrolase inhibitor MDL 73945 with long duration of action in rats. Diabetes 40:825–830

Robinson KM, Rhinehart BL, Ducep J-B, Danzin C (1992) Intestinal disaccharidase inhibitors. Drugs Future 17:705–720

Romero PA, Friedlander P, Fellows L, Evans SV, Herscovics A (1985) Effects of manno-1-deoxynojirimycin and 2,5-dihydroxymethyl-3,4-dihydroxypyrrolidine on N-linked oligosaccharide processing in intestinal epithelial cells. FEBS Lett 184:197–201

Rudge AJ, Collins I, Holmes AB, Baker R (1994) An enantioselective synthesis of deoxynojirimycin. Angew Chem Int Ed 33:2320–2322

Ruprecht RM, Bernard LD, Bronson R, Gama Sosa MA, Mullaney S (1991) Castanospermine vs. its 6-O-butanoyl analog: a comparison of toxicity and antiviral activity in vitro and in vivo. J Acquir Immune Defic Syndr 4:48–55

Saeki H, Ohki E (1968a) Synthesis of nojirimycin, 5-amino-5-deoxy-D-glucopyranose. Chem Pharm Bull 16:962–964

Saeki H, Ohki E (1968b) 5,6-Epimino-D-glucofuranose and synthesis of nojirimycin (5-amino-5-deoxy-D-glucose). Chem Pharm Bull 16:2477–2481

Sakairi N, Kuzuhara H (1982) Synthesis of amylostatin (XG), α-glucosidase inhibitor with basic pseudotrisaccharide structure. Tetrahedron Lett 23:5327–5330

Samulitis BK, Goda T, Lee SM, Koldovsdy O (1987) Inhibitory mechanism of acarbose and 1-deoxynojirimycin derivatives on carbohydrases in rat small intestine. Drugs Exp Clin Res XIII:517–524

Sasak VW, Ordovas JM, Elbein AD, Berninger RW (1985) Castanospermine inhibits glucosidase I and glycoprotein secretion in human hepatoma cells. Biochem J 232:759–766

Saul R, Chambers JP, Molyneux RJ, Elbein AD (1983) Castanospermine, a tetrahydroxylated alkaloid that inhibits β-glucosidase and β-glucocerebrosidase. Arch Biochem Biophys 221:593–597

Saul R, Molyneux RJ, Elbein AD (1984) Studies on the mechanism of castanospermine inhibition of α- and β-glucosidases. Arch Biochem Biophys 230:668–675

Schmidt RR, Dietrich H (1991) Amino substituted β-benzyl-C'-glycosides, novel β-glycosidase inhibitors. Angew Chem 103:1348–1349

Schmidt DD, Frommer W, Müller L, Junge B, Wingender W, Truscheit E (1977) α-Glucosidase inhibitors, new complex oligosaccharides of microbial origin. Naturwissenschaften 64:535–536

Schmidt DD, Frommer W, Müller L, Truscheit E (1979) Glucosidase inhibitors from bacilli. Naturwissenschaften 66:584–585

Schmidt RR, Michel J, Rücker E (1989) Synthese von 1,6-Anhydro-D-glucose- und -D-galactose-Derivaten. Herstellung des 1-Desoxynojirimycins. Liebigs Ann Chem: 423–428

Schröder T, Stubbe M (1987) Verfahren zur Herstellung von 1-Desoxynojirimycin und dessen N-Derivaten. Ger Offen 3611841 (prior 09.04.86)

Schutt H (1991) Enzymatic deacylation of acyl-amino-sorboses. US Pat 5177004 (prior DE 4030040, 22.09.90)

Scofield AM, Fellows LE, Nash RJ, Fleet GWJ (1986) Inhibition of mammalian digestive disaccharidases by polyhydroxy alkaloids. Life Sci 39:645–650

Semeria D, Philippe M, Delaumeny JM, Sepulchre AM, Gero SD (1983) A general synthesis of cyclitols and aminocyclitols from carbohydrates. Synthesis: 710–713

Setoi H, Takeno H, Hashimoto M (1985) Total synthesis of (+)-castanospermine from D-mannose. Tetrahedron Lett 26:4617–4620

Setoi H, Takeno H, Hashimoto M (1986) Synthesis of 1-deoxynojirimycin and 1-deoxymannojirimycin. Chem Pharm Bull 34:2642–2645

Shibata Y, Ogawa S (1989) Total synthesis of acarbose and adiposin-2. Carbohydr Res 189:309–322

Shibata Y, Nakayama O, Tsurumi Y, Okuhara M, Terano H, Kohsaka M (1988) A new immunomodulator, FR-900483. J Antibiot 41:296–301

Shibata Y, Kosuge Y, Mizukoshi T, Ogawa S (1992) Chemical modification of the sugar part of methyl acarviosin: synthesis and inhibitory activities of nine analogues. Carbohydr Res 228:377–398

Shibatani H, Yokoi T, Okago Y (1980) Moranolin preparation from Jacobinia plant. Jpn Pat 55027147 (prior 16.08.78)

Shomura T, Hamamoto K, Yoshida J, Moriyama C, Niida T (1966) Anti-Xanthomonas oryzae substance produced by Streptomyces. IV. Some biological characteristics of SF-425 (nojirimycin). Meiji Shika Kenkyu Nempo Yakuhin Bumon (CA 67:52327q)

Smid P, Schipper FJM, Broxterman HJG, Boons GJPH, van der Marel GA, van Boom JH (1993) Use of (chloromethyl)dimethylphenylsilane in sugar chemistry Stereo-controlled approach to destomic acid and 1-deoxy-nojirimycin. Recl Trav Chim Pays-Bas 112:451–456

Stasik B, Beaupere D, Uzan R, Demailly G, Morin C (1990) A new approach to the synthesis of nojirimycin. C R Acad Sci Ser 2, 311:521–523

Stein DC, Kopec LK, Yasbin RE, Young FE (1984) Characterization of Bacillus subtilis DSM 704 and its production of 1-deoxynojirimycin. Appl Environ Microbiol 48:280–284

Stevens CL, Blumbergs P, Daniker FA, Otterbach DH, Taylor KG (1966) Synthesis and chemistry of 4-amino-4,6-dideoxy sugars. II. Glucose. J Org Chem 31:2822–2829

Stoltefuß J, Müller L, Puls W (1980) 3,4,5-Trihydroxy-piperidin-Derivate. Ger Offen 2838309 (prior 09.09.78)

Straub A, Effenberger F, Fischer P (1990) Aldolase-catalyzed carbon-carbon bond formation for stereoselective synthesis of nitrogen containing carbohydrates. J Org Chem 55:3926–3932

Sunkara PS, Taylor DL, Kang MS, Bowlin TL, Liu PS, Tyms AS, Sjoerdsma A (1989) Anti-HIV activity of castanospermine analogues. Lancet I:1206

Takeda H, Nakagawa Y, Kiuchi A (1983) Amino-oligosaccharide derivatives with saccharase inhibiting, and lipid lowering activity. Jpn Kokai JP 58-172400

Takeuchi M, Kamata K, Yoshida M, Kameda Y, Matsui K (1990) Inhibitory effect of pseudoaminosugars on oligosaccharide glucosidases I and II and on lysosomal α-glucosidase from rat liver. J Biochem (Tokyo) 108:42–46

Tan A, Van den Broek L, Van Boeckel S, Ploegh H, Bolscher J (1991) Chemical modification of the glucosidase inhibitor 1-deoxynojirimycin. Structure-activity relationships. J Biol Chem 266:14504–14510

Taylor DL, Nash R, Fellows LE, Kang MS, Tyms AS (1992) Naturally occurring pyrrolizidines: inhibition of α-glucosidase 1 and anti-HIV activity of one stereoisomer. Antiviral Chem Chemother 3:273–277

Tong MK, Papandreou G, Ganem B (1990) Potent broad-spectrum inhibition of glycosidases by an amidine derivative of D-glucose. J Am Chem Soc 112:6137–6139

Tropea JE, Molyneux RJ, Kaushal GP, Pan YT, Mitchell M, Elbein AD (1989) Australine, a pyrrolizidine alkaloid that inhibits amyloglucosidase and glycoprotein processing. Biochemistry 28:2027–2034

Trost BM (1993) Antibiotics, A challenge for new methodology. In: Krohn K, Kirst H, Maas H (eds) Antibiotics and antiviral compounds. VCH, Weinheim, pp 3–4

Trugnan G, Rousset M, Zweibaum A (1986) Castanospermine: a potent inhibitor of sucrase from the human enterocyte-like cell line caco-2. FEBS Lett 195: 28–32

Truscheit E (1987) Mikrobielle α-Glucosidasen-Inhibitoren. In: Kleemann A, Lindner E, Engel J (eds) Arzneimittel, Fertschritte 1972–1985. VCH, Weinheim

Truscheit E, Frommer W, Junge B, Müller L, Schmidt DD, Wingender W (1981) Chemistry and biochemistry of microbial α-glucosidase inhibitors. Angew Chem Int Ed 20:744–761

Truscheit E, Hillebrand I, Junge B, Müller L, Puls W, Schmidt D (1988) Microbial α-glucosidase inhibitors: chemistry, biochemistry, and therapeutic potential. Progress in clinical biochemistry and medicine, vol 7, Springer, Berlin Heidelberg New York

Tsuda Y, Okuno Y, Kanemitsu K (1988) Utilization of sugars in organic synthesis. XX. Practical synthesis of nojirimycin. Heterocycles 27:63–66

Tsuda Y, Okuno Y, Iwaki M, Kanemitsu K (1989) Regio- and stereo-selective transformation of glycosides to amino-glycosides: practical synthesis of amino-sugars, 4-amino-4-deoxy-D-galactose, 4-amino-4-deoxy-L-arabinose, 3-amino-3-deoxy-D-allose, 3-amino-3-deoxy-D-glucose, 3-amino-3-deoxy-D-ribose, 3-amino-3-deoxy-D-xylose, 2-amino-2-deoxy-D-mannose, and 5-amino-5-deoxy-D-glucose (nojirimycin). Chem Pharm Bull 37:2673–2678

Tsuruoka T, Yumoto H, Niida T (1966) Anti-Xanthomonas oryzae substance produced by Streptomyces. III. Isolation and characterization of substance SF-425. Meiji Shika Kinkyu Nempo Yakuhin Bumon 8:1–6 (CA 67:73801e)

Van den Broek LAGM, Vermaas DJ, Heskamp BM, Van Boeckel CAA, Tan MCAA, Bolscher JGM, Ploegh HL, van Kemenade FJ, de Goede REY, Miedema F (1993) Chemical modification of azasugars, inhibitors on N-glycoprotein-processing glycosidases and of HIV-I infection. Review and structure-activity relationships. Recl Trav Chim Pays-Bas 112:82–94

Van den Broek LAGM, Vermaas DJ, van Kemenade FJ, Tan MCCA, Rotteveel FTM, Zandberg P, Butters TD, Miedema F, Ploegh HL, Van Boeckel CAA (1994) Synthesis of oxygen-substituted N-alkyl 1-deoxynojirimycin derivatives: aza sugar α-glucosidase inhibitors showing antiviral (HIV-1) and immunosuppressive activity. Recl Trav Chim Pays-Bas 113:507–516

Van der Klein PAM, Filemon W, Broxterman HJG, Van der Marel GA, Van Boom JH (1992) A cyclic sulfate approach to the synthesis of 1,4-dideoxy-1,4-imino derivatives of L-xylitol, L-arabinitol and D-xylitol. Synth Comm 22:1763–1771

Vasella A, Voeffray R (1982) Total synthesis of nojirimycin. Helv Chim Acta 65: 1134–1144

Vertesy L, Bender R, Fehlhaber H-W (1986) New α-glucosidase inhibitor. Eur Pat Appl 0173950, Hoechst AG

Vertesy L, Betz J, Fehlhaber H-W, Geisen K (1988) Oxirane pseudooligosaccharides, method for their synthesis, and pharmaceutical use as α-glucosidase inhibitors. Eur Pat Appl 0257418 A2, Hoechst AG

Von der Osten CH, Sinskey AJ, Barbas III CF, Pederson RL, Wang Y-F, Wong C-H (1989) Use of a recombinant bacterial fructose-1,6-diphosphate aldolase in aldol reactions: preparative syntheses of 1-deoxynojirimycin, 1-deoxymannojirimycin, 1,4-dideoxy-1,4-imino-D-arabinitol, and fagomine. J Am Chem Soc 111:3924–3927

Welter A, Jadot J, Dardenne G, Marlier M, Casimir J (1976) 2,5-Dihydroxymethyl 3,4-dihydroxypyrrolidine dans les feuilles de Derris elliptica. Phytochemistry 15:747–749

Whistler RL, Gramera RE (1964) Substitution of the exocyclic secondary hydroxyl group by an amino group in a D-glucofuranose structure. J Org Chem 29: 2609–2610

Winchester B, Fleet GWJ (1992) Amino-sugar glycosidase inhibitors: versatile tools for glycobiologists. Glycobiology 2:199–210

Winchester B, Cenci di Bello I, Richardson AC, Nash RJ, Fellows LE, Ramsden NG, Fleet G (1990) The structural basis of the inhibition of human glycosidases by castanospermine analogues. Biochem J 269:227–231

Wong C-H, Halcomb RL, Ichikawa Y, Kajimoto T (1995) Enzyme in der organischen Synthese: das Problem der molekularen Erkennung von Kohlenhydraten (Teil 1). Angew Chem 107:453–474

Yagi M, Kouno T, Aoyagi Y, Murai H (1976) The structure of moranoline, a piperidine alkaloid from Morus species. Nippon Nogei Kagaku Kaishi 50: 571–572 (CA 86:167851r)

Yamada H, Oyal I, Nagai T, Matsumoto T, Kiyohara H, Omura S (1993) Screening of alpha-glucosidase II inhibitor from Chinese herbs and its application on the quality control of mulberry bark. Shoyakugaku Zasshi 47:47–55 (CA 119:146433)

Yokose K, Ogawa K, Suzuki Y, Umeda I, Suhara Y (1983) New α-amylase inhibitor, trestatin-structure determinations of trestatins A, B and C. J Antibiot 36:1166–1175

Yoon H, King SB, Ganem B (1991) Synthesis of 1-β-amino-1-deoxynojirimycins: a new family of glucosidase inhibitors. Tetrahedron Lett 32:7199–7202

Yoshikawa M, Cha BC, Okaichi Y, Takinami Y, Yokokawa Y, Kitagawa I (1988) Synthesis of validamine, epi-validamine, and valienamine three opticaly active pseudo-amino-sugars, from D-glucose. Chem Pharm Bull 36:4236–4239

Yoshikuni Y (1988) Inhibition of intestinal α-glucosidase activity and postprandial hyperglycemia by moranoline and its N-alkyl derivatives. Agric Biol Chem 52:121–128

Yoshikuni Y (1991) α-Glucosidase inhibitor, 1-deoxynojirimycin and its N-substituted derivatives as antidiabetic agents or anti-human immunodeficiency virus agents. TIGG 3:184–192

Yoshikuni Y, Ezure Y, Aoyagi Y, Enomoto H (1988) Inhibition of intestinal α-glucosidase and postprandial hyperglycemia by N-substituted moranoline derivatives. J Pharmacobiodyn 11:356–362

Ziegler T, Straub A, Effenberger F (1988) Enzym-katalysierte Synthese von 1-Desoxymannojirimycin, 1-Desoxynojirimycin und 1,4-Didesoxy-1,4-imino-D-arabinitol. Angew Chem 100:737–738

CHAPTER 16
Analytical Methods of Determination of Glucosidase Inhibitors

H.J. Ploschke, H. Schlecker, S. Seip, and C. Wünsche

This short review is limited to enzyme inhibitors with a carbohydrate or pseudosaccharide structure. It describes the physical methods used to isolate the substances and determine their physicochemical properties: column chromatography for isolation, liquid and gas chromatography for separation and quantification, special preparative methods such as Craig distribution, and spectroscopic techniques for structure determination. Individual data such as melting points, boiling points, and specific rotations have not been compiled, since they are not very informative for such diverse chemical species. Summaries of all the physicochemical properties of the individual glucosidase inhibitors can be found in the original literature, cited in Chap.15, this volume, e.g., for acarbose in Takahashi et al. (1989). Nuclear magnetic resonance (NMR) and mass spectroscopic data for MDL 25637 (Merrell-Dow) have been described by Anzeveno et al. (1989) and Liu (1987). The physicochemical properties including NMR data to AO 128 (Voglibose) were presented by Fukase and Horii (1992)

    A survey of analytical methods for sugars appears in the reports of the meeting of the International Commission for Uniform Methods of Sugar Analysis (ICUMSA), which appear as periodicals [Publication Department (eds) Sugar analysis, British Sugar PLC Research Laboratories, Colney, Norwich, UK].

A. Chromatographic Techniques

I. Analytical Methods

Chromatographic methods predominate in the determination of pseudoglycosylamines and polyhydroxypiperidines from batch fermentation. Combined procedures, in which the culture filtrate is successively separated using acidic and basic ion-exchange resins, have proved particularly successful (Müller 1985).

    Intermediates and final products are also purified using column chromatography during the synthesis and derivatization of pseudoglycosylamines. The stationary phases in this case, unlike in the isolation of natural products, are usually silica gel phases. The most generally applicable method of checking the analytical purity of the fractions is thin-layer chromatography (TLC). Silica gel stationary phases containing fluorescence indicators with alcohol/acetic acid/water mixtures as the solvent systems are very suitable. By using this method it is possible to separate the aminocyclitols validamine, valienamine, and valiolamine (KAMEDA et al. 1984) and also a large number of N-substituted valiolamines (HORII et al. 1986). The parent compounds have been visualized on the TLC plates by spraying with Ninhydrin (triketohydrindene hydrate) as a color reagent. Sugars which contain several rings, e.g., acarbose and its derivatives and degradation products, can also be separated by TLC on silica gel. Either n-butanol/ethanol/water (9:7:4) or ethylacetate/methanol/water/dimethylsulfoxide (10:6:4:1) has been used as the solvent system. The usual color agent is thymol/sulfuric acid (TAKAHASHI et al. 1989). Paper chromatography or two-dimensional paper electrophoresis can also be used as less common alternatives to TLC for the separation of basic sugars. The suitability of these techniques has been described repeatedly using the example of nojirimycin (NISHIKAMA and ISHIDA 1965; INOUYE et al. 1966; ISHIDA et al. 1967; DAIGO et al. 1986).

High-performance techniques such as high-pressure liquid chromatography (HPLC) or gas chromatography (GC) are at present rarely used for the separation and quantification of basic sugars because of the difficulties of detection, stability, and derivatization. HPLC methods, combined, e.g., with ultraviolet (UV) detection, are used in those cases in which the sugars contain chromophores which give sufficiently intense signals at short wavelengths. Acarbose and a large number of the by-products formed in its fermentation have been detected in this way because of the double bond present in the molecule (Fig. 1).

A substantially higher detection sensitivity can be achieved using electrochemical detection. A method of determining 1-deoxynojirimycin, 1-deoxymannonojirimycin, and the bicyclic indolizine alkaloids castanospermine and swainsonine in plant extracts or in material of microbiological origin was developed by DONALDSON et al. (1990). The procedure uses a cation-exchange column as the stationary phase under isocratic conditions, and post-column derivatization with 300 mM NaOH and pulsed amperometric detection. Capillary electrophoresis, which in recent years has gained in importance, has also been used for the separation and quantification of sugars. Since the commercial capillary electrophoresis instruments are usually equipped with UV detectors, the method is limited to sugars which contain chromophores (Fig. 2).

II. Preparative Methods

The preparative separation of unchanged glucosidase inhibitors by chromatographic procedures is most commonly carried out using liquid-chromatographic methods of analysis (MEYER 1986; ENGELHARDT 1975; VERHAAR and KUSTER 1981). The properties of the sample components and

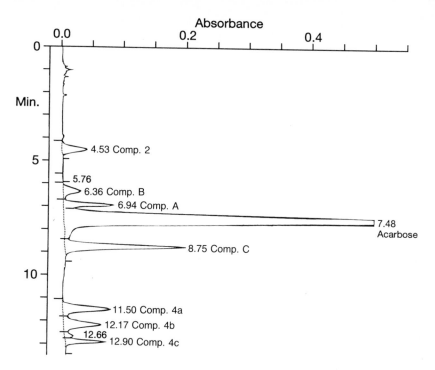

Fig. 1. HPLC chromatogram of acarbose and its by-products. *Column*, Shandon Hypersil APS-1 125 × 4.6 mm; *eluent*, phosphate buffer pH 6.6, acetonitrile; *flow*, 1.5 ml/min; *temperature*, 38°C; *detection*, UV 210 nm (HPLC chromatogram with permission of H. Scheer, Verfahrensentwicklung Biochemic, Bayer AG)

the specific requirements generally determine which method is most suitable. Every preparative separation should in principle be preceded by an analytical test. The following possible types of stationary phase interaction are available for the present class of substances:

1. Reverse-Phase Chromatography

Reverse-phase chromatography on the so-called amino phase, in which aminopropyl groups are bound chemically to the silica. This phase can be used under anhydrous conditions for less polar substances or as a reverse phase for polar compounds, and is especially suitable for carbohydrates and as a weakly basic anion exchanger in acidic aqueous solutions. In hydrogen-bonding interactions, the amino function can act both as a proton acceptor and as a proton donor. The main eluents used for the present substances are acetonitrile/water or acetonitrile/buffer. The strength of the eluent decreases with increasing proportion of acetonitrile. It is advantageous to maintain the column at 35°–55°C to improve the selectivity.

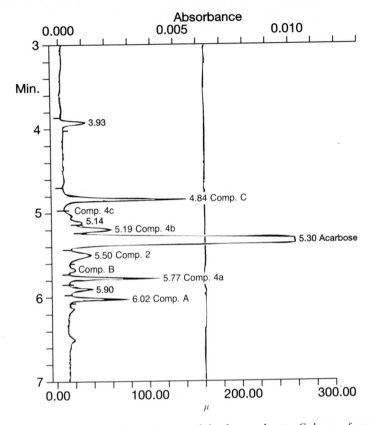

Fig. 2. Electropherogram of acarbose and its by-products. *Column*, fused silica; *length*, *I*, 57 cm; detection window at 50 cm; *buffer*, 0.1 *M* borate, pH 9.4; *temperature*, 50°C; *detection*, UV 214 nm

Since the amino groups are easily oxidized, peroxides (e.g., in diethyl ether, dioxane, or tetrahydrofuran) need to be avoided. Aldehydes and ketones must also be avoided, since they react to form Schiff bases. The amino phase is available commercially for semipreparative high-performance separations in 5-, 7-, and 10-μm particle sizes in stainless steel columns up to 250 mm long and of 10, 20, and 25 mm internal diameter.

For preparative separations the amino phase is used in 25- to 40-μm particle size ranges and 40- to 63-μm stainless steel columns up to 600 mm long and 200 mm internal diameter. The eluent flow-rate can be calculated from the characteristics of an analytical column containing the same sorbent:

$$\frac{X_A}{r^2_A} = \frac{X_P}{r^2_P}$$

where X_A = flow rate in the analytical system, X_P = flow rate in the preparative system, r_A = radius of the analytical column, and r_P = radius of the preparative column.

The capacity is between 10^{-3} and 10^{-2} g sample/g sorbent. The amino phase is very well suited, e.g., to the difficult separation of acarbose by-products (Fig. 1).

2. Ion-Exchange Chromatography

Ion-exchange chromatography generally uses cation exchangers based on dextran, cellulose, or artificial resins with SO_3H (strongly acid) or OCH_2COOH (weakly acid) groups (DORFNER 1970; JANDERA and CHURACEK 1974; HAVLICEK and SAMUELSON 1975; VERHAAR and KUSTER 1981). The choice is based on the basicity of the sample and the objectives (BRENDEL et al. 1967; TIKHOMIROW et al. 1978; GECKELER and ECKSTEIN 1987; FALLON et al. 1987).

Substances with different pK values can be separated from each other, as can basic compounds from salts. Isomers generally cannot be separated by ion-exchange chromatography. High-pressure separations are possible only with ion exchangers based on silica gel.

The capacity is in the region of 2–5 meq/g. The flow rate is between 40 and 100 ml/cm^2 per hour. Before use, the ion exchangers (except those based on silica gel) must be allowed to swell in the mobile phase for between 3 and 24 h. Ethanol/water, water, or buffers can be used as eluents. When carrying out preparative separations, it is sensible to use volatile buffers, such as ammonium formate, ammonium acetate, or pyridine acetate, as otherwise the salts must be removed after the separation. An example, with a direct comparison with the amino phase, is shown in Fig. 3. The type of exchanger, particle size, buffer, acidity (pH value), temperature, ionic strength of the buffer, and flow rate all have a direct effect on the separation, so that optimization can be very time-consuming.

3. Size-Exclusion Chromatography

Size exclusion chromatography (SEC) (GECKELER and ECKSTEIN 1987; DETERMANN 1967) uses a different separation principle than the methods so far described, which all separate substances on the basis of their differing physicochemical properties. In SEC the sample components are eluted in order of decreasing molecular weight. The choice of a column with a stationary phase of suitable pore size is crucial for the separation.

One column can cover approximately two orders of magnitude in molecular weight. According to the type of material, ligand-exchange and distribution processes may play a role alongside the size exclusion. An absolute precondition for work with most of these stationary phases is thorough swelling before use. This process takes between 3 h and 3 days at room temperature (RT), and is accelerated by heating. Here again smaller grain sizes give sharper separation. The effect of temperature on the separation is often considerable.

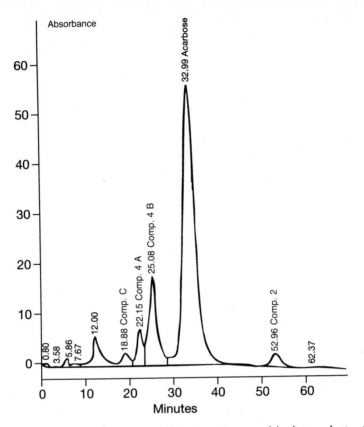

Fig. 3. Cation-exchange chromatography of acarbose and its by-products. *Column*, Shodex SH 10011, 300 × 3 mm; *flow rate*, 1 ml/min; *mobile phase*, 62.5 m*M* phosphate buffer (pH 5); *column temperature*, 60°C; *detector wavelength*, UV 200 nm

4. Adsorption Chromatography

Adsorption chromatography was often used for separation despite the polarity of the present substances. Solvent systems which give *Rf* values (retention factors) of about 0.3 in TLC (Kaizuka and Takahashi 1983; Schleich and Engelhardt 1989) can often be transferred directly to an analytical and corresponding preparative separation (Table 1).

5. Craig Distribution

Craig distribution (liquid-liquid distribution) is of limited use because of the polarity of this class of substances, but, depending on the objective, it can offer advantages over the above methods (Hecker 1955; Ploschke 1991). In contrast to chromatographic techniques, the method can easily be transferred from the analytical to the preparative scale. For example, the sepa-

Table 1. Possible components in mobile solvent systems for HPLC

Solvent system	Chloroform	1-Butanol	Methanol	Ethanol	Water	25% ammonia	Iso-propanol	Acetic acid	Ethyl acetate	Acetone	3% boric acid	Ethylmethyl ketone
Solvent A	×		×			×						
Solvent B	×	×	×			×						
Solvent C					×		×	×				
Solvent D				×		×						
Solvent E						×				×		
Solvent F									×	×		
Solvent G						×			×	×	×	
Solvent H		×		×	×							×

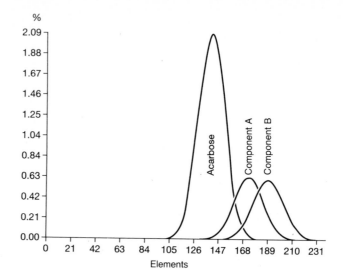

Fig. 4. Theoretical distribution curves of acarbose and its by-products. Components A and B in *n*-butanol/ethanol/water (5.2/1.1/3.7) after *n* = 2000 transfers. $V_{upper}/V_{lower} = 1.3/1$

ration shown in Fig. 4 could easily be carried out in a commercially available apparatus with a 1000 ml phase volume using 1 kg starting material. The choice of phases is limited by the polarity of the substances. Suitable systems are butanol/water (1:1 v/v), butanol/ethanol/water (5.2:1.2:3.7 v/v), and butanol/25% ammonia (1:1 v/v).

If none of these methods leads to a satisfactory solution to the problem of separation, success may be achieved by carrying out separation after derivatization.

6. Detection

For detection in preparative chromatographic separations, one can use either UV detectors with preparative cells (detection wavelength 190–210 nm), refractive index (RI) detectors, or light-scattering detectors.

B. Spectroscopic Techniques

I. Nuclear Magnetic Resonance

NMR spectroscopy provides an important contribution to the structural characterization of glucosidase inhibitors (JUNGE et al. 1984; SCHMIDT et al. 1979; ARAI et al. 1986). In addition to the fact that it is a nondestructive method, i.e., the measurements are done on the intact molecule, the range of structurally relevant data which can be determined by NMR is considerable. The characterization of oligosugars or of structurally related compounds is not always easy because of a massive overlap of signals.

To resolve this problem it is often necessary to resort to derivatization or degradation of the sugar chain. Structural elucidation of acarbose (JUNGE et al. 1984) will be used as an example. Whereas most structural characterizations are now carried out by multi-dimensional NMR methods (DABROWSKI 1986), the structure of acarbose was clarified using one-dimensional techniques alone. The numerous overlapped signals could, however, only be dealt with by a laborious analysis and comparison with degradation products.

The [^{1}H] NMR spectrum of acarbose in D$_2$O shows the typical features of an oligosugar (JUNGE et al. 1984; Fig. 5): most of the sugar ring protons resonate at 3.5–4.5 ppm. The resonances of the anomeric protons one observes at a lower field, between 4.5 and 5.7 ppm. They can easily be recognized because of their position and because they do not overlap with other protons. One also finds in this region a signal due to the olefinic proton of the cyclitol ring. The labile OH protons are not visible in D$_2$O.

The 1–4 linking of the sugar residues has been derived from difference nuclear-Overhauser-effect (NOE) measurements (NEUHAUS and WILLIAMSON 1989). After first irradiating the anomeric proton H(1), one observes an increase in the intensity of nearby protons in the 1-D spectrum, including an increase in the intensity of signal H(4) of the sequentially neighboring monomer. Other modes of linking can easily be demonstrated by analogy.

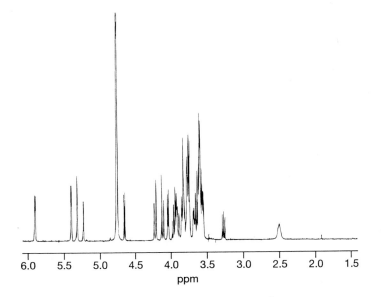

Fig. 5. [^{1}H]NMR spectrum of acarbose in D$_2$O (500 MHz)

Acarbose exists as a (D-glucose-1)-α/β (2:3) anomeric mixture. For this reason one finds two signals for H(1), the anomeric proton of the terminal glucose residue, one for the α-form and one for the β-form. They were unambiguously assigned to the α- and β-anomers by a determination of the H(1):H(2) coupling constant. For the β-anomer one finds and H(1):H(2) coupling constant of 8.0 Hz, whereas H(1) of the α-anomer [gauche with respect to H(2)] only shows a splitting of 3.5 Hz. The same argument has been used to demonstrate that all the monomeric units in acarbose are α-coupled.

The constitution of monomers and their stereostructure can also be determined by analyzing the coupling patterns of individual resonances. In the case of acarbose it was possible to determine the entire configuration in this way, partly by comparison with known sugars (JUNGE et al. 1984). NMR-based studies of biosynthesis and metabolism have also been reported (FLOSS et al. 1986; DEGWERT et al. 1987). Important evidence concerning the biosynthesis pathway to acarbose was obtained using ^{13}C-labeled glycerol (BEALE et al. 1987).

The characteristic connectivities of the carbon skeleton were demonstrated using heteronuclear techniques, giving information about the incorporation of fragments during biosynthesis. It was also possible to identify the essential metabolites with the aid of NMR spectroscopy (BOBERG et al. 1990).

Other papers have concerned conformation analysis (GOLDSMITH et al. 1987; GLASER and PERLIN 1988) and studies of the mode of action of glucosidase inhibitors by methods of NMR. Thus, the conformation of acarbose was investigated using HSEA (hardspheres exo-anomeric effect) calculations (BOCK and PEDERSON 1984). The reaction kinetics of the interaction between amylglycosidase and modified inhibitors were followed by NMR (BOCK and PEDERSEN 1986).

The stereochemistry of the sugar-enzyme binding was clarified using ^{19}F-labeled sugar derivatives (WITHERS and STREET 1988). These authors found evidence that the binding of glucosidases to substrates is covalent. Structures of glucosidase-bound inhibitors have not yet been clarified by NMR.

II. Mass Spectrometry

The structural diversity of the glucosidase inhibitors summarized in the preceding section demands the use of a variety of mass-spectrometric techniques for their characterization. Electron-impact mass spectrometry (EI-MS) rapidly reaches its limit because of the relatively high polarity and low volatility of the polyhydroxyl compounds. It is only possible to produce molecular ions of sufficient intensity for the simplest bicyclic representatives, e.g., the castanospermine derivatives, by using EI-MS (MOLYNEUX et al. 1986). Suitable derivatives have therefore usually been sought. Hakamori methylation and acetylation (with acetic anhydride in pyridine) was successful for pseudoglycosylamines. Thus, mol-ions were detected at 1191 amu by EI-MS for peracetylated acarbose (JUNGE et al. 1984). Silylated derivatives suitable for gas chromatography/mass spectrometry (GC-MS) were prepared for nojirimycin-type monocyclic compounds such as miglitol (INOUYE et al. 1972; STRAUB et al. 1990).

The introduction of soft ionization techniques represented a major improvement for the mass-spectrometric analysis of underivatized glucosidase inhibitors. Desorption chemical ionization (DCI), field desorption (FD), fast atom bombardment (FAB), laser desorption (LD), and secondary ion mass spectrometry (SIMS) have been tested on numerous representatives of this group. FUKUSHIMA (1981) recorded FD spectra of bioactive substances including nojirimycin.

A comparison of the various methods of ionization was carried out on derivatives of nojirimycin (WÜNSCHE et al. 1984). Molecular weight determinations could usually be achieved unproblematically by DCI for compounds with relatively low molecular weights. Ammonia proved to be a suitable reactant gas. The monocyclic validamine- and valienamine-type compounds could also be analyzed by this method. FAB has a wider applicability as a method of ionization. For smaller molecules 3-nitrobenzylalcohol seems to be the most suitable liquid matrix. With the pseudoglycosylamines, however, the cation yield decreases with increasing number of sugar residues. A changeover to glycerol as the matrix and spiking with Li^+, Na^+ or K^+ is necessary to achieve sufficiently intense quasi-molecular ions. In the positive ion FAB spectrum of acarbose, for example, the base peak was the M + Na ion at 668 amu, and the M + H ion at 646 amu had 20%–90% relative intensity, depending on the Na^+ concentration. Apart from matrix ions (glycerol), one finds a typical fragment at 304 amu, corresponding to the tricyclic degradation product formed by acid hydrolysis of acarbose (JUNGE et al. 1984).

The FAB mass-spectrum of acarbose was published by TAKAHASHI et al. in 1989 together with its NMR, UV, and IR spectra and other physicochemical data.

Very good results were obtained for acarbose and its higher molecular weight derivatives by the SIMS technique (BENNINGHOVEN et al. 1990). On preparation of an approximately $10^{-3} M$ solution on a silver target, positive secondary ion MS showed almost exclusively M + Na and M + Ag ions.

The different capabilities of the individual soft ionization techniques were demonstrated impressively in studies on the in vitro and in vivo metabolization of acarbose. On anaerobic degradation by intestinal bacteria, smaller fragments can be detected by DCI and, in part, by EI (PFEFFER and SIEBERT 1986; BOBERG et al. 1990). Higher homologues and isomers require FAB as a method of detection or DCI measurements after acetylation.

In in vivo degradation of acarbose (rat, dog, and man) conjugated metabolites (sulfates, glucuronides) also occur (AHR et al. 1989; BOBERG et al. 1990). In this case it is necesary to use FAB in the anion mode. Even better results are obtained by the SIMS technique. The trisalt of an aromatic trisulfate could be detected as an M – K anion among the reference substances, which was not possible by any other MS technique.

Despite the good results from mass spectrometry in the determination of molecular weights and elemental compositions (high-resolution FAB), there is inadequate information obtainable about positional and stereochemical isomers in this class of substances. In soft ionization spectra, which give little fragmentation, collision-induced secondary fragmentation in tandem MS (MS/MS) is usually needed to distinguish between isomers, as has been shown with the example of synthetic oligosaccharides with amino-sugar residues (LAINE et al. 1991 and literature therein). The necessity, inter alia, to carry out extensive comparisons against substances with authentic structures makes it more effective to distinguish and assign (stereo-)isomers of compounds on the basis of NMR.

References

Ahr HJ, Boberg M, Krause HP, Maul W, Müller FO, Ploschke HJ, Weber H, Wünsche C (1989) Pharmacokinetics of acarbose part 1: absorption, concentration in plasma, metabolism and excretion after single administration of [^{14}C] acarbose to rats, dog and man. Arzneimittelforschung 39(II):1254–1260

Anzeveno PB, Creemer LJ, Daniel KJ, King RC, Liu PS (1989) A facile, practical synthesis of 2,6-dideoxy-2,6-imino-7-0-β-D-glucopyranosyl-D-glycero-L-gulohep-titol (MDL 25637). J Organic Chem 54:2539–2542

Arai M, Sumida M, Fukuhara K, Kainosho M, Murao S (1986) Isolation and characterization of amylase inhibitors. Agricultural Biol Chem 50:639–644

Beale JM, Cottrel CE, Keller PJ, Floss HG (1987) Development of triple-quantum "INDAEQUATE" for biosynthetic studies. J Magn Reson 72:574–578

Benninghoven A, Musche H, Wünsche C (1990) SIMS (secondary ion mass spectrometry) in pharmaceutical research. In: Benninghoven A, Evans CA, McKeegan KD, Stroms HA, Werner HW (eds) Secondary ion mass spectrometry SIMS VII. Wiley, New York, p 293

Boberg M, Kurz J, Ploschke HJ, Schmitt P, Scholl H, Schüller M, Wünsche C (1990) Isolation and structural elucidation of biotransformation products from acarbose. Arzneimittelforschung 40(I):555–563

Bock K, Pedersen H (1984) The solution conformation of acarbose. Carbohydr Res 132:142–149

Bock K, Pedersen H (1986) Protein-carbohydrate interactions: the substrate specificity of amyloglucosidase. FEMS Symp (Protein-Carbohydr Interact Biol Syst) 31:173–182

Brendel K, Roszel NO, Wheat RW, Davidson EA (1967) Ion-exchange separation and automated assay of some hexosamines. Anal Biochem 18:147–160

Dabrowski J (1986) 2D NMR analysis of oligosaccharides. In: Croasmun WR, Carlson RMK (eds) Methods in stereochemical analysis: two-dimensional NMR spectroscopy. Verlag Chemie, Weinheim, p 349

Daigo K, Inamori Y, Takemoto T (1986) Studies on the constituents of the water extract of the root of mulberry tree (Morus bombycis Koidz). Chem Pharm Bull 34:2243–2246

Degwert U, Van Huelst R, Pape H, Richard E, Beale JM, Keller PJ, Lee J, Floss HG (1987) Studies on the biosynthesis of the alpha-glucosidase inhibitor acarbose, a m-C7N unit not derived from the shikimate pathway. J Antibiot 40:855–861

Determann H (1967) Gelchromatographie, Springer, Berlin Heidelberg New York

Donaldson MJ, Broby H, Adlard MW, Bucke C (1990) High pressure liquid chromatography and pulsed amperometric detection of castanospermine and related alkaloids. Phytochem Anal 1:18–21

Dorfner K (1970) Ionenaustauscher, 3rd edn. De Gruyter, Berlin

Engelhardt H (1975) Hochdruckflüssigkeitschromatographie. Springer, Berlin Heidelberg New York

Fallon A, Booth RFG, Bell LD (1987) Carbohydrates. In: Burdan RH, van Knippenberg PH (eds) Laboratory techniques in biochemistry and molecular biology, vol 17. Elsevier, Amsterdam, p 213

Floss HG, Keller PJ, Beale JM (1986) Studies on the biosynthesis of antibiotics. J Nat Prod 49(6):957–970

Fukase H, Horii S (1992) Synthesis of valiolamine and its N-substituted derivatives AO-128, validoxylamine G, and validamycin G via branched-chain inosose derivatives. J Organic Chem 57: 3651–3658

Fukushima K (1981) Field desorption mass spectrometry of bioactive substances. Nippon Kagaku Kaishi 5:874–882

Geckeler KE, Eckstein H (1987) Analytische und präparative Labormethoden. Vieweg, Braunschweig

Glaser R, Perlin AS (1988) Proton and carbon studies on N-methyl-1-deoxynojirimycin, an alpha-D-glucosidase inhibitor. Carbohydr Res 182:169–177

Goldsmith EJ, Fletterick RJ, Withers S-G (1987) The three-dimensional structure of acarbose bound to glycogen phosphorylase. J Biol Chem 262:1449–1455

Havlicek J, Samuelson O (1975) Separation of oligosaccharides by partition chromatography on ion exchange resins. Anal Chem 47:1854–1857

Hecker E (1955) Verteilungsverfahren im Laboratorium. Verlag Chemie, Weinheim

Horii S, Fukase H, Matsuo T, Kameda Y, Asano U, Matsui K (1986) Synthesis and α-D-glucosidase inhibitory activity of N-substituted valiolamine derivatives as potential oral antidiabetic agents. J Med Chem 29:1038–1046

Inouye S, Tsuruoka T, Niida T (1966) The structure of nojirimicin, a piperidinose sugar antibiotic. J Antibiotics Ser A XIX:288–292

Inouye S, Omoto S, Tsuruoka T, Niida T (1972) Analysis of nojirimicin by gas liquid chromatography and mass spectrometry. Meiji Seiko Kenkyu Nempo 12:96–103

Ishida N, Kumagai K, Niida T, Tsuruoka T, Yumoto H (1967) Nojirimicin, a new antibiotic II. J Antibiotics Ser A XX:66–72

Jandera P, Churacek J (1974) Ion-exchange chromatography of aldehydes, ketones, ethers, alcohols, polyols and saccharides. J Chromatogr 98:55–104

Junge B, Heiker FR, Kurz J, Müller L, Schmidt DD, Wünsche C (1984) Structure of the alpha D glucosidase inhibitor acarbose. Carbohydr Res 128:235–268

Kaizuka H, Takahashi K (1983) High-performance liquid chromatographic system for a wide range of naturally occuring glycosides. J Chromatogr 258:135–146

Kameda Y, Asano N, Yoshikawa M, Takeuchi M, Yamaguchi T, Matsui K (1984) Valiolamine, a new α-glycosidase – inhibiting aminocyclitol produced by Streptomyces hygroscopicus. J Antibiotics (Tokyo) XXXVII:1301–1307

Liu PS (1987) Total synthesis of 2,6-dideoxy-2,6-imino-7-O-β-D-glucopyranosyl-D-glycero-L-gulo-heptitol hydrochloride: a potent inhibitor of α-glucosidases. J Org Chem 52:4717–4721

Laine AR, Yoon E, Mahier TJ, Abbas S, DeLappe B, Jain R, Matta K (1991) Nonreducing terminal linkage position determination in intact and permethylated synthetic oligosaccharides having a penultimate aminosugar: fast atom bombardment ionization, collisional inducecd dissociation and tandem mass spectrometry. Biol Mass Spectrom 20:505–514

Meyer V (1986) Praxis der Hochdruckflüssigkeitschromatographie, 4th edn. Diesterweg Salle Sauerländer, Frankfurt

Molyneux RJ, Roitman JN, Dunnheim G, Szumilo T, Elbein AD (1986) 6-Epicasta-nospermin, a novel indolizidine alkaloid that inhibits alpha glucosidase. Arch Biochem Biophys 251:450–457

Müller L (1985) Microbial glycosidase inhibitors. In: Rehm HJ, Reed G (eds) Biotechnology, vol 4. Verlag Chemie, Weinheim, p 531

Neuhaus D, Williamson M (1989) The nuclear Overhauser effect in structural and conformational analysis. VCH, Weinheim

Nishikawa T, Ishida N (1965) A new antibiotic R-468 active against drug-resistant Shigella. J Antibiotics Ser A XVIII:132–133

Pfeffer M, Siebert G (1986) Prefeeding-dependent anaerobic metabolization of xenobiotics by intestinal bacteria – methods for acarbose metabolites in an artificial colon. Z Ernaehrungswiss 25:189–195

Ploschke J (1991) Gegenstromverteilung. In: Nürnberg E, Surmann P (eds). Hagers Handbuch der pharmazeutischen Praxis, 5th edn, vol 2, Methoden. Springer, Berlin Heidelberg New York, p 411

Schleich W, Engelhardt H (1989) Möglichkeiten der HPLC in der Zuckeranalytik. GIT Fachz Lab 33:624–630

Schmidt DD, Frommer W, Mueller L, Truscheit E (1979) Glucosidase inhibitors from bacilli. Naturwissenschaften 66(11):584–585

Straub A, Effenberger F, Fischer P (1990) Enzyme-catalyzed reactions. IV. Aldo-lase-catalyzed carbon-carbon bond formation for stereoselective synthesis of nitrogen containing carbohydrates. J Organic Chem 55:3926–3932

Takahashi Y, Sakaguchi F, Morimoto K, Hashimoto K, Funaba T, Hayauchi Y (1989) Physicochemical properties and stability of acarbose (in Japanese). Iyakuhin Kenkyu 20:769–783

Tikhomirov MM, Khorlin A, Voelter W, Bauer H (1978) High-performance liquid chromatographic investigation of the amino acid, amino sugar and neutral sugar content in glycoproteins. J Chromatogr 167:197–203

Verhaar LAT, Kuster BFM (1981) Liquid chromatography of sugars on silica-based stationary phases. J Chromatogr 220:313–328

Withers SG, Street IP (1988) Identification of a covalent alpha-D-glucopyranosyl enzyme intermediate formed on a beta-glucosidase. J Am Chem Soc 110: 8551–8553

Wünsche C, Benninghoven A, Eicke A, Heinen HJ, Ritter HP, Taylor LCE, Veith J (1984) Comparison of soft ionization techniques with electron impact mass spectrometry for desoxinojirimycin and folic acid derivatives. Organic Mass Spectrom 19:176–182

CHAPTER 17

Pharmacology of Glucosidase Inhibitors

W. Puls

A. Introduction

Since the discovery of insulin (Banting and Best 1922), diabetes mellitus has been associated with deficiency or complete absence of this hormone because countless diabetics have been safeguarded against diabetic coma and premature death by parenteral injection of insulin (Marble 1974). Further advances in the drug treatment of diabetes were made in the late 1950s when the sulphonylureas carbutamide and tolbutamide, which increase insulin secretion and are β-cytotropic, were introduced with great success. These were followed by the biguanides phenformin, buformin and metformin (see Chaps. 4 and 10). These drugs not only provided more user-friendly treatment for type 2 diabetics because of their oral efficacy, but they were also used to investigate the pathogenesis and pathophysiology of the disease, which was then known as "adult-onset diabetes." At the same time, basic diabetes research gained considerably from the development of a radio-immunoassay which allowed the blood levels of insulin to be determined.

Yalow and Berson (1960) and Grodsky et al. (1963) found no insulin deficiency in the blood of adults with mild diabetes. Bagdade et al. (1967) tested insulin concentrations in adults of normal weight in comparison with obese adults with and without diabetes before and after the oral glucose tolerance test. Non-obese diabetics had insulin deficiency postprandially whereas obese diabetics had pronounced hyperinsulinaemia at baseline and postprandial levels. As insulin not only lowers blood glucose but also promotes the synthesis and storage of lipids, doubts arose as to whether treating obese diabetics with sulphonylureas and thus worsening their hyperinsulinaemia was justifiable (Perley and Kipnis 1966). This raised the question of whether hyperglycaemia could be treated effectively by a different method that would simultaneously reduce hyperinsulinaemia. One of the hypothetical approaches considered and tested for pharmacological relevance was the inhibition of intestinal α-glucosidases.

B. Intestinal Digestion of Dietary Di-, Oligo- and Polysaccharides by α-Glucosidases

Dietary carbohydrates should account for at least 50% of the daily supply of calories in adult diabetic patients (Alberti and Gries 1988). Monosaccharides play only a minor role as dietary carbohydrates since they consist mainly of starch (~60%) and sucrose (~30%) (REISER 1976). The latter carbohydrates must be hydrolysed by α-glucosidases to monosaccharides before they can be transported through the mucosa of the small intestine (see ELSENHANS and CASPARY 1987). Starch is partially hydrolysed to maltose, maltotriose and so-called α-limit dextrins by pancreatic α-amylase in the duodenum and upper part of the jejunum, that is, in the lumen as well as at the brush-border membrane. The oligosaccharides originating from starch cleavage are degraded to glucose by maltase (EC 3.2.1.20), glucoamylase (EC 3.2.1.3) and isomaltase (EC 3.2.1.10). In addition to these brush-border located α-oligosaccharidases, a dextrinase may be involved in the cleavage of α-limit dextrins (GRAY 1983). The disaccharide sucrose is only hydrolysed into glucose and fructose by a single mucosal α-glucosidase, namely sucrase (EC 3.2.1.48).

The average maximum mucosal glucosidase activity was found distal from the ligament of Treitz in the upper third of the jejunum of healthy subjects, but was still present in considerable amounts in the lower jejunum and also in the entire ileum (NEWCOMER and MCGILL 1966). The digestion of oligo- and polysaccharides and the absorption of monosaccharides mainly takes place in the upper half of the small intestine (BOOTH 1968). Due to a great individual variation of the α-glucosidase activity in the small intestine (NEWCOMER and MCGILL 1966), it may be that the digestion rate of di-, oligo or polysaccharides differs accordingly. An impaired glucosidase activity in the upper part causes a corresponding involvement of the distal part of the small intestine (CASPARY 1987). Since the transport of undigested carbohydrates from the duodenum to the ileum takes time, their degradation rate is reduced, the absorption of monosaccharides is delayed and consequently the steep increase in blood glucose concentration following the ingestion of a carbohydrate containing meal is flattened.

C. Exploratory Investigations on the Feasibility of α-Glucosidase Inhibition

SCHMIDT was the first to prepare an extract from wheat flour which inhibited pancreatic α-amylase (SCHMIDT and PULS 1971; TRUSCHEIT et al. 1981). Oral administration of this inhibitor preparation (BAY d 7791) dose dependently reduced the postprandial increase in blood glucose (Fig. 1) and insulin in starch loading tests on rats, dogs and healthy volunteers (PULS and KEUP 1973) (Fig. 1). The inhibition of a brush-border-located α-glucosidase was studied by using TRIS. This commonly used buffer inhibits the sucrase in vitro (DAHLQUIST 1964) and proved to be active in sucrose loading tests on rats and man, as was indicated by a dose dependently reduced postprandial increase in blood glucose and insulin (PULS and KEUP 1975a).

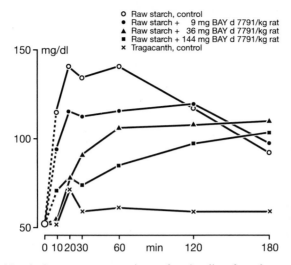

Fig. 1. Mean blood glucose concentrations after loading fasted rats with 2.5 g raw ·starch ± graded doses of BAY d 7791

Both inhibitors had no effect on blood glucose concentrations in fasting animals or in glucose tolerance tests. It was concluded that the antihyperglycaemic effect in starch or sucrose loading tests was caused by inhibition of their corresponding α-glucosidases. Since BAY d 7791 had only a weak effect in diabetic patients (FRERICHS et al. 1973), the search for more potent inhibitors continued. This aim was achieved when FROMMER et al. (1972) found a strong α-glucosidase inhibitory activity in preparations (e.g., BAY e 4609) originating from culture filtrates of *Actinoplanes*.

The inhibitors mentioned above were markedly less or not at all effective in loading tests with carbohydrates containing starch as well as sucrose (PULS et al. 1980). However, a Western diet usually consists of di-, oligo- and polysaccharides. In order to meet the demands of remedial diabetes treatment, the concept of glucosidase inhibition (Fig. 2) was established (PULS 1980; PULS et al. 1980), which took account of the inhibition of those enzymes involved both in starch degradation and in sucrose digestion (Fig. 2).

An appropriate active agent of microbial origin (BAY g 5421, acarbose) inhibits pancreatic α-amylase, other glucosidases involved in the degradation of oligosaccharides, and sucrase in vitro (SCHMIDT et al. 1977), and is also effective in carbohydrate loading tests on rats and healthy volunteers, as was indicated by a reduction of postprandial blood glucose and an insulin increase (PULS et al. 1977). This inhibitor was introduced onto the market (Glucobay) in 1990. In the meantime a number of α-glucosidase inhibitors have been reported (see Chap. 15).

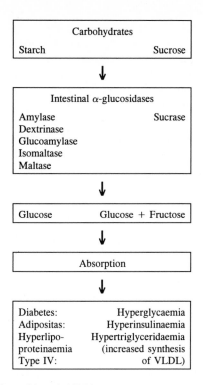

Fig. 2. Concept of α-glucosidase inhibition to reduce the intestinal degradation rate of di-, oligo- and polysaccharides to monosaccharides and their absorption in diabetic, obese or hypertriglyceridaemic patients

D. Primary Effects of Glucosidase Inhibitors

This section will focus on inhibitors which have been reported to effectively reduce postprandial elevations of blood glucose and plasma insulin in animals in loading tests with starch and sucrose: acarbose (BAY g 5421, Glucobay), miglitol (BAY m 1099), emiglitate (BAY o 1248), voglibose (AO-128, Basen), MDL 25637 and MDL 73945

In situ perfusion of the rat small intestine with sucrose or maltose revealed that the disappearance rate of the sucrose was reduced by 50% by 3.2 μg acarbose/ml perfusion medium in comparison to control animals, whereas a 10- to 20-fold higher concentration of acarbose was required in perfusion tests with maltose to achieve a similar effect (KRAUSE et al. 1982b). A recovery of sucrase activity in the rat jejunum, which was perfused for ½ h with sucrose and acarbose (30 mg/l), was seen approximately 75 min after withdrawal of the inhibitor from the perfusion medium. The recovery of the sucrase activity was associated with a similar recovery of water and sodium absorption (TAYLOR and BARKER 1982). The recovery of maltase activity was achieved in this experimental model within 1 h but was limited over 2 h when sucrose was present in the medium (TAYLOR et al. 1984). Similar results were obtained with lower concentrations of miglitol or emiglitate in perfusion studies with starch, maltose or sucrose (TAYLOR et al. 1985). The luminal extraction of lactose, glucose or fructose was not affected.

The carbohydrate content in the stomach, small and large intestine of rats was determined 2 h after administration of 2 g cooked starch or 2 g sucrose/kg body weight (PULS 1980). In untreated rats no undigested carbohydrate remained in the gastrointestinal tract, indicating complete digestion of the carbohydrates. Addition of 1, 2 or 4 mg acarbose/kg body weight brought about a dose-dependent increase in undigested and partially digested starch in the small intestine, but not in the large intestine (Fig. 3a). Very similar effects were seen 2 h after administration of sucrose with and without equal doses of acarbose (Fig. 3b). No disaccharide was detectable in the large bowel after administration of 1 or 2 mg acarbose/kg body weight, but significant amounts of sucrose were found when 4 mg acarbose/kg had been given. In order to determine the concentration of sucrose and its hydrolysis products in the gut segments, twice the ED_{50} of acarbose (2.25 mg/kg body weight) was administered with ^{14}C-labelled sucrose and 2.5 g unlabelled sucrose to fasted meal-fed rats. The radioactivity was determined 2 h postloading in the gastrointestinal tract. No differences in radioactivity remaining in the stomach were found between control and inhibitor-treated animals. In the small and large bowel, however, the percentage of radioactivity was significantly higher (two to five times) in acarbose-treated animals (PULS 1980). These results differ from the above-mentioned findings. However, the figures would not be inconsistent if one assumed that the disaccharide-related radioactivity was accounted for by labelled products resulting from degradation of sucrose by intestinal enzymes or microorganisms.

Corresponding results were obtained in loading tests with unlabelled sucrose with and without voglibose (0.03 and 0.1 mg/kg body weight) (ODAKA et al. 1991). Intraduodenal infusion of sucrose with graded doses of acarbose (0, 0.7–5.6 mg/kg body weight) over a 150-min period in rats resulted in a significant, dose-dependent inhibition of the glucose elevation in the portal venous blood (MADARIAGA et al. 1988).

These experiments clearly show that the primary effect of glucosidase inhibitors is located in the small intestine and, furthermore, that overdosage of inhibitors can be accompanied by a transport of undigested carbohydrates into the large bowel.

E. Secondary Effects of Glucosidase Inhibitors

I. Investigations to Evaluate the Potency of Glucosidase Inhibitors

The inhibitory spectrum of glucosidase inhibitors and their kinetic constants can readily be estimated in vitro assays using pancreatic amylase and preparations of intestinal mucosa (see Chap. 15). In vitro active agents are

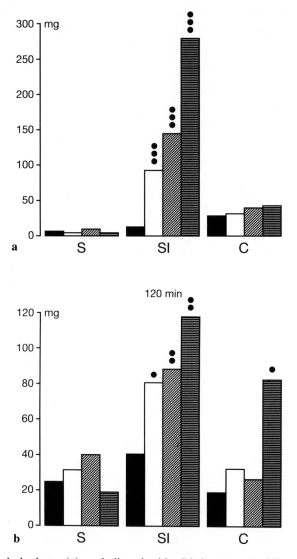

Fig. 3a,b. Carbohydrate (**a**) and disaccharide (**b**) in stomach (*S*), small intestine (*SI*) and large bowel (*C*) of rats, 2 h after oral administration of 2 g boiled starch (**a**) or 3.5 g sucrose (**b**) (■) with 1 mg (□), 2 mg (▨) and (▤) 4 mg acarbose/kg body weight (P-values vs. control = ● < 0.05, ●● < 0.01, ●●● < 0.001)

expected to exert corresponding effects in vivo. However, the intestinal digestion of carbohydrates in animals and man is more complex than the degradation of specific substrates in vitro. The in vivo effects of glucosidase inhibitors depend not only on their inhibitory activity, their binding constants (e.g. K_i, K_{on}, K_{off}), but also on their pharmacokinetics, the amount and species of orally administered carbohydrates and the heterogeneous intestinal glucosidase activity. Thus it is not surprising that distinct differences have been reported with regard to the in vitro versus in vivo potency in rats (PULS et al. 1980; YOSHIKURI et al. 1988; ROBINSON et al. 1992).

In order to select the most favourable inhibitor out of a large number of in vitro active agents, the in vivo potency was studied in carbohydrate loading test with fasted rats (PULS et al. 1977; PULS et al. 1980). The integrated postprandial blood glucose (plasma insulin) increment was evaluated in a cumulative manner as the area under the curve (AUC), which is formed by the difference in the blood glucose (plasma insulin) concentration between rats loaded with carbohydrates with or without inhibitor and loaded with saline. The inhibitor dose (mg/kg body weight) which reduces the postprandial increase in blood glucose (plasma insulin) by 50% (ED_{50}) in comparison to carbohydrate-loaded control animals (= 100) was calculated by interpolation of the percentage AUC of at least three or more different single inhibitor doses (low, intermediate, high).

It is likely that the postprandial blood glucose concentration in nondiabetic animals is affected by the postprandial insulin concentration and does not exactly reflect the rate of intestinal carbohydrate digestion and monosaccharide absorption. But there is every reason to believe that the rate of intestinal glucose absorption dominates the rate of glucose disappearance from peripheral blood as long as the postprandial blood glucose concentrations increase and vice versa. The inhibitory potency of glucosidase inhibitors should therefore be calculated without the AUC which is formed by declining blood glucose levels. Since the glucose concentration in venous blood is more balanced by insulin than in arterial blood, the latter should be preferred for blood glucose measurements in studies like these.

The inhibitory active agents which are equipotent (ED_{50}) in loading tests with starch in comparison to loading tests with sucrose appear to be more favourable antidiabetic inhibitors with regard to their therapeutic application than those with markedly discrepant potencies in these tests.

II. Blood Glucose and Plasma Insulin

1. Single-Dose Studies

The effect of a glucosidase inhibitor on blood glucose elevations is exemplified in the upper part of Table 1 (PULS 1980). The reduction in blood glucose increments is accompanied by correspondingly lower postprandial insulin concentrations (Table 1, lower half).

The most potent inhibitors in sucrose loading tests on rats are voglibose (ED_{50} ~0.03–0.1 mg/kg body weight) (HORII et al. 1986), emiglitate and miglitol (ED_{50} = 0.16 mg and 0.24 mg/kg body weight, respectively) (BISCHOFF et al. 1985). MDL 25637 and MDL 73945 appear to be more effective on postprandial glucose and insulin response when administered 30–60 min be-

Table 1. Mean blood glucose values (mg/dl ± SD) and mean plasma insulin values (μU/ml ± SD) of fasting male rats ($n = 6$) at intervals after oral administration of 2.5 g sucrose ± 1.5 mg acarbose/kg body weight

	10 min	20 min	30 min	60 min
Blood glucose				
Control, saline	67 ± 1*	72 ± 4.6*	68 ± 5.3*	66 ± 5.0*
Sucrose	134 ± 15	140 ± 16	126 ± 12	117 ± 20
Sucrose + acarbose	96 ± 6.2*	93 ± 6.2*	92 ± 5.6*	92 ± 3.4*
Plasma insulin				
Control, saline	8.8 ± 1.8*	9.8 ± 1.4*	10.8 ± 2.3*	10.7 ± 4.6
Sucrose	22.2 ± 11	35.0 ± 14	19.3 ± 5.0	16.7 ± 5.3
Sucrose + acarbose	9.9 ± 4.3	11.1 ± 3.5*	7.6 ± 1.7*	14.8 ± 3.8

* Difference from sucrose load, $P < 0.05$.

Table 2. ED_{50} of acarbose, miglitol and emiglitate for postprandial blood glucose increments in rats following administration of various carbohydrates with and without dietary fat

Oral load	Approx. ED_{50} mg/kg body weight		
	Acarbose	Miglitol	Emiglitate
Glucose	>60	>100	>30
Fructose + glucose	>40	NR	NR
Maltose	~12	NR	~7
Raw starch	0.2	0.48	NR
Cooked starch	1.5	0.50	0.3
Cooked starch + olive oil	1.5	NR	NR
Sucrose	1.2	0.24	0.16
Sucrose + olive oil	1.5	NR	NR
Starch + sucrose	2.5	0.36	NR

NR, not reported.

fore a sucrose load than when given simultaneously with sucrose (RHINEHART et al. 1987; ROBINSON et al. 1991).

The ED_{50} of acarbose (PULS et al. 1983), miglitol and emiglitate (BISCHOFF et al. 1985) on postprandial blood glucose has been calculated and compared from loading tests using various dietary carbohydrates with and without alimentary fat (Table 2).

Each of these inhibitors appears to be somewhat more potent in loading tests with cooked starch than with sucrose, but the difference is very small. Voglibose has been reported to be fivefold more potent when administered with sucrose than with starch (HORII et al. 1986) and a similar difference in inhibitory activity has been reported for MDL 73945 (ROBINSON et al. 1991). Loading tests with graded amounts of sucrose require increased doses of acarbose to achieve equal reductions of blood glucose elevations (PULS et al. 1980). This finding supports the view that acarbose competitively inhibits α-glucosidases not only in vitro but also in vivo.

Whilst acarbose and miglitol had no effect on the intestinal degradation of starch and sucrose in rats, when administered 1 h before carbohydrate loading, emiglitate was very effective when administered 4 or more h before administration of carbohydrates (PULS et al. 1984). The longer lasting effect of emiglitate may be due to its pharmacokinetics (see Chap. 19). The long-lasting effect of MDL 73945 is caused by quasi-irreversible binding to α-glucosidases (ROBINSON et al. 1991).

A dose-related or marked inhibitory effect of glucosidase inhibitors is not limited to rats but was also seen in loading tests in mice (PULS and KRAUSE 1979a; RHINEHART et al. 1987), rabbits (HEITLINGER et al. 1992) and monkeys (HO and ARANDA 1983; ROBINSON et al. 1991). Single-dose studies on glucosidase inhibitors have been reported using artificially or geneticaly diabetic rats (YAMASHITA et al. 1984; TULP et al. 1988; ROBINSON et al. 1991) and diabetic mice (ODAKA et al. 1992). The flattening effect of glucosidase inhibitors on postprandial insulin elevations in hyperinsulinaemic diabetic rats appeared to be more pronounced than on postprandial hyperglycaemia. Concomitant administration of acarbose with either insulin or a sulphonylurea (glisoxepide, tolbutamide) had a synergistic effect on postprandial blood glucose reduction in normal and diabetic rats. The dose dependently increased plasma insulin concentrations in normal rats after administration of glisoxepide or tolbutamide in carbohydrate loading tests were markedly reduced by simultaneous administration of acarbose (HAMADA et al. 1989b; PULS 1980).

2. Repeated Administration

In feeding experiments of several days to several months duration, glucosidase inhibitors were administered as a food-drug mixture ensuring the simultaneous ingestion of carbohydrates and active agents. The metabolic status of the animals used in these studies was very heterogeneous: normal, diabetic due to insulin deficit, diabetic combined with hyperinsulinaemia, obese, hyperlipidaemic or a combination of two or more abnormalities (Table 3).

As could be expected from single-dose studies, elevated blood glucose concentrations and/or the urinary volume and glucose excretion and, last but not least, basal or prandial hyperinsulinaemia in diabetic rats and mice were dose dependently and markedly reduced by administration of acarbose (AXEN and SCLAFANI 1990; AXEN 1993; COHEN and KLEPSER 1991; COHEN et al. 1991; COHEN 1993; FOELLMER et al. 1993; FRIEDMAN et al. 1991; GRAY and OLEFSKY 1982; HAMADA et al. 1989a; KATOVICH et al. 1991; MAROT and LEMARCHAND-BRUSTEL 1988; PETERSON et al. 1988; PETERSON 1991; PETERSON et al. 1993; TULP et al. 1988a; TULP et al. 1988b; TULP et al. 1993; VASSELLI et al. 1982; YAMASHITA et al. 1984). Combined treatment of streptozotocin

Table 3. Main animal models used for testing α-glucosidase inhibitors

	Metabolic status	Glucosidase inhibitor[a]
Rats		
Wistar	Intact	Acarbose
		Miglitol
		Emiglitate
		Voglibose
Wistar	STZ-diabetic	Acarbose
Wistar fatty	Diabetic	Acarbose
	Obese	Voglibose
	Hyperinsulinaemic	
	Hyperlipidaemic	
Sprague-Dawley	Intact	Acarbose
Sprague-Dawely	STZ-diabetic	Acarbose
		Miglitol
		Emiglitate
Sprague-Dawley	VMH-lesioned	Acarbose
	Obese	
	Hyperinsulinaemic	
Fa.-Zucker	Intact	Acarbose
Fa.-Zucker	VHM-lesioned-obese	Voglibose
fa,fa-Zucker	Obese	Acarbose
	Hyperinsulinaemic	Miglitol
	Hyperlipidaemic	Voglibose
	iGT	
Zucker (ZDF/Drt-fa)	Obese	Acarbose
	Diabetic	
	Hyperlipidaemic	
LA/N	Intact	Acarbose
LA/N-cp	Obese	Acarbose
	iGT	Miglitol
	Hyperinsulinaemic	
	Hyperlipidaemic	
SHR/N	Intact	Acarbose
SHR/N-cp	Diabetic	Acarbose
	Obese	
	Hyperinsulinaemic	
	Hyperlipidaemic	
JCR:La-cp	Obese	Acarbose
	Hyperinsulinaemic	
	Hyperlipidaemic	
GK	Diabetic	Voglibose
	Hyperinsulinaemic	
BB/Wor	Diabetic (IDDM)	Acarbose
Mice		
C57 BL *db/db*	Diabetic	Acarbose
		Emiglitate
C57 BL/K-J-*db/db*	Diabetic	Acarbose
	Hyperinsulinaemic	Emiglitate
	Obese	
KKAY	Diabetic	Voglibose
	Obese	
	Hyperinsulinaemic	
	Hyperlipidaemic	

STZ, streptozotocin; iGT, impaired glucose tolerance.
[a] Repeated administration in pharmacologically relevant concentrations.

(STZ)-diabetic rats with insulin and acarbose was more effective than treatment with low-dosed insulin alone (HAMADA et al. 1989b; KATOVICH and MELDRUM 1993a), as is indicated by ameliorated hyperglycaemia, polydipsia, urinary glucose excretion and glycated haemoglobin (GHb).

Administration of voglibose improved the glycaemic control of rats and mice with diabetes or impaired glucose tolerance (IKEDA et al. 1991a; IKEDA et al. 1991b; MATSUO et al. 1992; ODAKA et al. 1992; TAKAMI et al. 1991). Similar beneficial effects have been reported on miglitol (DEBOUNO et al. 1989; MADAR 1989) and emiglitate (LEE et al. 1987; MADAR and OLEFSKY 1986).

GHb, a long-term indicator of glycaemic control, was attenuated or normalized after administration of acarbose (COHEN et al. 1991; COHEN and KLEPSER 1991; COHEN 1993; FOELLMER et al. 1993; HAMADA et al. 1989a; KATOVICH and MELDRUM 1993a; MAGGIO et al. 1993; PETERSON 1991; PETERSON et al. 1988, 1993; TULP et al. 1988b; TULP et al. 1993; VASSELLI et al. 1993), miglitol (DEBOUNO et al. 1989, 1993) and voglibose (ODAKA et al. 1992). In insulin-dependent BB/Wor rats the GHb was not significantly reduced by acarbose (KOEVARY 1993). Basal glucose concentrations of hyperinsulinaemic rats, diabetic rats and mice were significantly ameliorated by acarbose (FOELLMER et al. 1993; FRIEDMAN et al. 1991; PETERSON et al. 1993; PULS et al. 1982a). No effect of acarbose on basal blood glucose was reported by other authors (COHEN 1993; FRIED et al. 1993; MAGGIO et al. 1993; RUSSEL et al. 1993a; VASSELLI et al. 1993). In the latter experiments, blood samples were collected from rats after a very short time period of food withdrawal (3 h only) or with the animals under halothane anaesthesia.

Acarbose-treated rats with normal (ZDF/Drt-fa) or deficient insulin secretion (STZ-induced) had higher fasting plasma insulin levels than untreated control animals. The insulin content in the pancreas of STZ-diabetic rats was more than twice the level in control rats (FRIEDMAN et al. 1991; MAGGIO et al. 1993). Furthermore, it was reported that in untreated genetically db/db mice the initial hyperinsulinaemia decreased continuously during 33 weeks of the experiment whereas the plasma insulin in acarbose-treated mice was maintained at a significantly higher level. These important findings have not yet been confirmed but may provoke speculation about a protective effect of glucosidase inhibitors against functional exhaustion of pancreatic B cells.

The intestinal absorption of orally administered glucose is not affected by glucosidase inhibitors; nevertheless a 4-week treatment of genetically obese Zucker rats with acarbose brought about a markedly improved glucose tolerance accompanied by a reduced insulin response (PULS et al. 1982a).

This unexpected effect was confirmed by other authors with acarbose (TULP et al. 1988a,b), with miglitol (DEBOUNO et al. 1989, 1993) and with voglibose (TAKAMI et al. 1991). An improved glucose tolerance was also reported when glucose was injected intraperitoneally (TULP et al. 1988; AXEN 1993) or intravenously (RUSSEL et al. 1993a). An explanation may be that a peripheral insulin insensitivity of these hyperinsulinaemic animals was ameliorated due to an improved metabolic control as a consequence of treatment with glucosidase inhibitors. More profound studies revealed that reduction in chronic hyperglycaemia by acarbose treatment prevents the decrease in skeletal muscle glucose transporter 4 (GLUT 4) expression that occurs in the ZDF/Drt-fa diabetic rat (DOHM et al. 1993). In studies with a dietary-induced insulin insensitivity, the insulin-stimulated glucose uptake of skeletal muscle was increased by 50% in acarbose-treated rats in comparison to control animals (YOUNG et al. 1993).

Injection of STZ can induce an insulin-deficient diabetes. Administration of acarbose as food admixture 1 week before to 12 days after intraperitoneal injection of 70 mg STZ/kg body weight reduced the incidence and attenuated the severity of diabetes in Wistar rats (GODA et al. 1981). An attenuation of STZ-induced diabetes in acarbose-treated rats has been confirmed (MAGGIO et al. 1993). The incidence of diabetes in BB/Wor rats (autoimmune type 1 diabetes) was not affected by acarbose treatment (KOEVARY 1993).

We conclude that the inhibition of intestinal α-glucosidases not only exerts short-term effects, e.g. in carbohydrate loading tests, but improves the glycaemic control in long-term experiments, as is indicated by decreased GHb levels and ameliorated glucose tolerance, particularly in obese, hyperinsulinaemic mice and rats with diabetes or impaired glucose tolerance.

III. Lipid Metabolism

1. Blood Lipids

The intestinal absorption of dietary fat is not affected by a single dose of acarbose (PULS 1980) or voglibose (ODAKA et al. 1991). Alterations of blood lipid concentrations during or after treatment of animals with glucosidase inhibitors are in all probability not direct effects of these agents but rather a consequence of changed metabolic control.

After a 34-week administration of acarbose, the elevated concentrations of plasma triglycerides, cholesterol and free fatty acids (FFAs) in genetically obese Zucker rats were dose dependently reduced (PULS et al. 1977). More profound experiments revealed that a hypertriglyceridaemia induced by feeding a sucrose-containing or fat-free diet was decreased or normalized by acarbose administration due to a reduced formation and secretion of very low density lipoproteins (VLDLs) (KRAUSE et al. 1982a; PULS et al. 1980; ZAVARONI and REAVEN 1981). The reduction of elevated plasma cholesterol concentrations may result from a lowering of the plasma intermediate density or low-density lipoprotein concentrations as a consequence of decreased VLDL degradation. The increased triglyceride content in the liver and elevated lipoprotein lipase activity in adipose tissues were dose dependently reduced in acarbose-treated rats in comparison to control animals.

An improvement of an artificial or genetic hyperlipidaemia in rats and mice of heterogeneous strains with or without diabetes has been confirmed in a number of experiments by administration of acarbose (ABDOLLAHI et al. 1993; BENNO et al. 1988; CARSWELL et al. 1993; TULP et al. 1988a; TULP et al. 1988b; TULP et al. 1991a; VASSELLI et al. 1993), miglitol (BENNO et al. 1988) and voglibose (IKEDA et al. 1991; IKEDA et al. 1991b; MATSUO et al. 1992; ODAKA et al. 1992). A low dose of miglitol did not improve the lipid profile of lean, obese and obese diabetic rats (TULP et al. 1991b). Acarbose failed to affect plasma triglycerides during a 38-week treatment of genetically diabetic mice with non-insulin-dependent diabetes mellitus (NIDDM) (FOELLMER et al. 1993) and JCR:LA-cp rats (RUSSEL et al. 1993a), although the total plasma cholesterol and cholesterylesters were significantly reduced in acarbose-treated rats.

Studies with regard to secondary effects of acarbose on the hepatic synthesis and circulating concentrations of cholesterol have been conducted on obese LA/N-cp rats and obese diabetic SHR/N-cp rats (TULP et al. 1991). It was reported that the elevated total serum cholesterol in both strains of acarbose-treated rat was reduced by approximately 40%–50% in comparison to control rats. The authors suggest that this effect of acarbose occurred somewhat independently of differences in body weight gain or hepatic HMG-CoA-synthase activity but was rather brought about by the acarbose-induced reduction of hyperisulinaemia in these animal models.

2. Tissue Lipids

Ingested dietary calories exceeding energy expenditure will be stored to a minor degree as protein or glycogen in skeletal muscles or liver, whilst the major part will be accumulated as triglyceride in adipocytes. The storage of energy in adipocytes requires the participation of insulin.

In acute loading tests with ^{14}C-sucrose, less radioactivity was detectable in the lipids of perirenal and epididymal adipose tissues in acarbose-treated rats than in control animals (PULS 1980). This effect of acarbose could be due to low availability of sucrose, reduced lipogenesis or increased lipolysis in adipose tissues. However, if rats were given ^{14}C-triolein dissolved in olive oil together with unlabelled sucrose, the radioactivity in the adipose tissue lipids was also significantly lower in acarbose-treated rats than in control rats, but no effect of acarbose was achieved when sucrose was replaced by directly absorbable glucose. The outstanding regulatory role of insulin with regard to lipogenesis was elucidated in acute loading tests in rats which received a mixed meal consisting of ^{14}C-triolein, olive oil, protein and complex carbohydrates. Addition of a β-cytotropic sulphonylurea (glisoxepide) to this mixed meal brought about an approximately twofold higher radioactivity in the lipids of perirenal adipose tissue in comparison to control animals (PULS and BISCHOFF 1988). When, in addition to glisoxepide, acarbose was administered, the ^{14}C-triolein radioactivity in the adipose tissue lipids was reduced to less than half that of the sulphonylurea-treated rats. Thus the main effect of acarbose on reduced storage of dietary fat ingested with carbohydrates is probably due to a decreased postprandial secretion of lipogenic insulin.

Long-term administration of glucosidase inhibitors to animals of heterogeneous strains resulted in alterations of, e.g., body weight, food intake, adipose tissue mass, adipose tissue cellularity, lipogenesis, activity of lipogenic enzymes, activity of lipase and lipoprotein lipase, insulin sensitivity in adipocytes and skeletal mucles and the amount of lipids, protein and water in the carcass. Most of these experiments were conducted by means of acarbose (ABDOLLAHI et al. 1993; AXEN 1993; BENNO et al. 1988; BJÖRNTORP et al. 1983; DODANE et al. 1991; FRIED et al. 1993; GLICK and BRAY 1982; GLICK and BRAY 1983; KOTLER et al. 1982; KOTLER et al. 1984; KRAUSE et al. 1982a; LEMARCHAND-BRUSTEL et al. 1990; MAGGIO et al. 1987; MAGGIO and VASSELLI 1989; MAURY et al. 1993; PULS et al. 1977; PULS et al. 1980; RUSSEL et al. 1993a; TULP et al. 1991a; VASSELLI et al. 1980; VASSELLI et al. 1982; VASSELLI et al. 1983; VASSELLI et al. 1987; VASSELLI et al. 1993; VEDULA et al. 1991; WILLIAM-OLSSON 1986; WILLIAM-OLSSON and SJÖSTRÖM 1986). These studies have generally shown that acarbose has a consistent body weight-lowering action, which was generally dose dependent and did not appear to be associated with any consistent change in food intake, although food intake was frequently increased. However, genetically or STZ-induced diabetic rats or mice exhibited no reduction of body weight or even an increased body weight gain during long-term administration of acarbose in comparison to diabetic control animals (ABDOLLAHI et al. 1993; COHEN 1993; FAILLA and SEIDEL 1993; FOELLMER et al. 1993; KATOVICH and MELDRUM 1993a; KOEVARY 1993; LEE 1982; MAGGIO et al. 1993; PETERSON et al. 1993; TULP et al. 1993; YAMASHITA et al. 1984). The beneficial effect of acarbose on an impaired body weight development of diabetic animals still needs to be explained. We may speculate that an improved metabolic control, e.g., an ameliorated glucose utilization, contributes to this unexpected effect. Miglitol exerted only weak effects on body weight gain and hyperlipidaemia of rats (BENNO et al. 1988; DEBOUNO et al. 1989, 1993; TULP et al. 1991a). Emiglitate dose dependently reduced body weight gain of obese Zucker rats (PULS et al. 1982b) but did not affect food intake or body weight gain of genetically diabetic mice (LEE et al. 1987). Voglibose brought about a marked reduction of weight gain in intact, diabetic or VMH- (ventral middle hypothalamus-) lesioned rodents (FUJIOKA et al. 1991; IKEDA et al. 1991a; IKEDA et al. 1991b; KOBATAKE et al. 1989; MATSUO et al. 1992; ODAKA et al. 1992; TAKAMI et al. 1991).

The hepatic triglyceride content of obese Zucker rats on a fat-free diet was dose dependently lowered to the normal range after a 17-day administration of acarbose (THOMAS et al. 1982). The activities of hepatic acetyl-CoA carboxylase and fatty acid synthase were significantly reduced in intact as well as in STZ-diabetic rats after 3 weeks of acarbose feeding (VASSELLI et al. 1993). Administration of voglibose gave rise to a diminished hepatic triglyceride content in obese-diabetic rats and in genetically diabetic mice (ODAKA et al. 1992; IKEDA et al. 1991a). The cholesterol concentration in the liver of rats was slightly affected, and concentration of phospholipids was increased. Corresponding examinations with miglitol, emiglitate, MDL 25637 and MDL 73945 have not been reported.

IV. Protein Metabolism

The faecal nitrogen excretion of meal-fed rats was dose dependently increased when acarbose had been administered as a food additive (KIRCHGESSNER et al. 1981). The excess nitrogen found in the faeces may be of bacterial origin, since acarbose administration stimulates the growth of microorganisms in the large bowel (KIRCHGESSNER et al. 1981). This explanation is in accordance with experiments conducted with rats fed ad libitum. The protein concentration of these rats was not decreased but rather increased after a 42-day administration of acarbose in spite of a reduced body weight gain (low-dose acarbose) or a loss of body weight (high-dose acarbose) (PULS et al. 1980). Analogous results have been observed in experiments with obese diabetic rats which had received voglibose (IKEDA et al. 1991b).

Investigations on a molecular basis have shown that markedly depressed levels of glucose transporter 2 (GLUT-2) on pancreatic B cells of diabetic rats were somewhat higher in acarbose-treated rats than in control animals (PETERSON et al. 1993). Decreased levels of GLUT-4 protein in red quadriceps and mixed gastrocnemius muscles of genetically diabetic rats were significantly increased and GLUT-4 mRNA in gastrocnemius muscle maintained normal levels after a 19-week administration of acarbose (FRIEDMAN et al. 1991).

Administration of acarbose gave rise to a significant decrease in elevated GHb in diabetic rats and mice (COHEN et al. 1991; COHEN and KLEPSER 1991; FOELLMER et al. 1993; KATOVICH and MELDRUM 1993a; LEE 1982; MAGGIO et al. 1993; PETERSON et al. 1993; TULP et al. 1993; VASSELLI et al. 1993). In BB/Wor rats, acarbose significantly reduced elevated GHb levels only for a limited period during a 5-month experiment (KOEVARY 1993). Significant reductions of GHb were also reported after administration of miglitol (DEBOUNO et al. 1989, 1993) and voglibose (ODAKA et al. 1992).

Levels of glycated protein in glomerulus basement membranes and fluorescent advanced glycated end products (AGEs) in skin and tail tendon collagen of diabetic rats were markedly decreased after acarbose treatment (COHEN 1990; COHEN et al. 1991). Glucosidase inhibitors given as a food additive obviously cause a persistent decrease in diabetic hyperglycaemia in mice and rats and thus inhibit excess non-enzymatic glycation and AGE formation. Up to now there have been no data available with regard to the effect of glucosidase inhibitors on the activity of proteinases which are capable of degrading AGEs.

512   W. Puls

V. Hormones

1. Pancreatic Insulin

In intact rats a 10-day feeding of emiglitate (50 mg/100 g food) did not affect the pancreatic insulin concentration (GÖKE et al. 1986). In diabetic rats the pancreatic insulin concentration was markedly decreased 11 weeks after injection of STZ, as compared to normal rats (MAGGIO et al. 1993). Acarbose-treated diabetic rats displayed twice the pancreatic insulin content of diabetic control rats, i.e., they maintained nearly 80% of the pancreatic insulin of normal rats. An 11-week feeding of voglibose to genetically diabetic GK rats did not affect the pancreatic insulin concentration and total pancreatic insulin content (TAKAMI et al. 1991). Fibrosis and degranulation of the B cells in the islets of GK rats treated with voglibose were reduced in comparison to GK control rats.

Pancreatic islets, isolated from VMH-lesioned hyperinsulinaemic diabetic rats pretreated with acarbose, exhibited lower baseline insulin release than did islets from untreated rats (AXEN 1993). The glucose (11 mM)-stimulated absolute insulin response was not different between the two groups.

Isolated pancreatic islets from normal rats, exposed to glucose or α-ketoisocaproate, showed an increased insulin release which was not affected by concomitant incubation with an extraordinarily high concentration of acarbose (1 mg/ml) (AXEN 1993). A very high concentration of emiglitate (1 mM ≅ 0.36 mg/ml) brought about a slight, but significant reduction of glucose-stimulated insulin release from isolated rat islets. Lower concentrations of emiglitate (1–100 μM) exhibited no effect (GÖKE et al. 1986).

In experiments with isolated perfused pancreata from rats, the addition of emiglitate (10 μM) to the perfusion medium brought about a slight, but significant decrease in the late insulin secretory response to a half-maximal glucose (10 mM) load and was without effect when the glucose concentration was doubled or replaced by arginine (GÖKE et al. 1986). When isolated pancreata from rats, pretreated with emiglitate (50 mg/100 g food × 10 days), were used in this experimental model, the early glucose-stimulated (10 mM) insulin release was significantly reduced (10–20 min), whereas the late response (20–60 min) was not different from that in control rats (GÖKE et al. 1986).

In order to test the hypothesis that islet acid amyloglucosidase is involved in the glucose-stimulated insulin secretory processes, acarbose (LUNDQUIST and PANAGIOTIDIS 1992; SALEHI and LUNDQUIST 1993a), miglitol (SALEHI and LUNDQUIST 1993b) and emiglitate (SALEHI et al. 1993; SALEHI and LUNDQUIST 1993a; SALEHI and LUNDQUIST 1993c) have been used as tools in in vivo and in vitro experiments to inhibit this lysosomal glycogenolytic enzyme. The dosages of the glucosidase inhibitors exceeded the pharmacologically relevant ED$_{50}$ by a factor of 1000. The inhibitors brought about a decrease in acid amyloglucosidase activity and a diminished glucose-stimulated insulin secretion, whereas the enhanced insulin secretion induced by glibenclamide, carbachol or isobutylmethylxanthine remained unaffected. The authors concluded that the stimulation of insulin secretion can be induced by a variety of pathways.

2. Catecholamines and Thyroid Hormones

The hypertrophy of adrenal medullary cells of genetically diabetic KK mice was reported to be reduced after a 16-week administration of acarbose (Hamada et al. 1989). The concentration of norepinephrine and its potassium-stimulated release in the caudal tail artery of STZ-diabetic rats is higher than in non-diabetic rats. An 8-week treatment of STZ-diabetic rats with acarbose brought about a statistically significant reverse of both parameters in male rats (Katovich et al. 1991) and a statistically insignificant reduction in female rats (Katovich and Meldrum 1993a).

Injection of isoproterenol causes an increase in the tail-skin temperature in intact rats but not in STZ-diabetic rats. When isoproterenol was injected into acarbose-treated male and female STZ-diabetic rats, the increase in tail-skin temperature was significantly improved in comparison to control animals, whereas the decreased isoproterenol-induced temperature increase in the colon of STZ-diabetic rats was not affected by acarbose (Katovich et al. 1991; Katovich and Meldrum 1993a).

The serum total thyroxine (T_4) and free T_4 are significantly depressed in rats rendered diabetic by STZ injection (Katovich and Meldrum 1993b). An 8-week treatment with acarbose or insulin or a combination of acarbose and insulin significantly restored total T_4 levels to above those observed in untreated diabetic rats, and free T_4 was similarly improved. The combined therapy was more effective than treatment with acarbose or insulin alone. Serum total T_3 was not significantly altered in any of the experimental groups.

3. Other Hormones

The content of enteroglucagon, cholecystokinin, vasoactive intestinal polypeptide, substance P, neurotensin and bombesin was increased by many times in the small intestine of rats afte a 4-week feeding of acarbose. The plasma concentrations of gastrin and enteroglucagon were markedly increased, the concentration of neurotensin slightly so (Uttenthal et al. 1982). Other authors found no alterations of plasma gastric inhibitory peptide, gastrin and cholecystokinin (Creutzfeldt et al. 1985) or immunoreactive glucagon and gastrin (Rolston et al. 1985), whereas the glucagon-like immunoreactivity was increased. The biological significance of reported enterohormone alterations remains unclear.

VI. Vitamins and Trace Metals

It is a reasonable assumption that intestinal α-glucosidase inhibition interferes with the absorption of dietary vitamins and trace metals. The addition of voglibose to an oral single load of rats with ascorbic acid and sucrose did not affect the plasma concentrations of this water-soluble vitamin. As an example of the absorption of lipid-soluble vitamins, loading tests were done with retinyl palmitate in oil plus sucrose. The initial plasma concentration of this vitamin was decreased, but no difference was found in comparison with control rats after 6–10 h (Odaka et al. 1991).

Investigations with regard to the absorption and retention of trace metals have been conducted on STZ-diabetic rats. Treatment with acarbose markedly increased the absorption of zinc and reduced the loss of endogenous zinc. Increased concentration of bone zinc after a 45-day acarbose feeding indicated long-term improvement of zinc status. Bone calcium was maintained at the non-diabetic control level. Copper, iron and manganese levels in liver, kidneys and bone were generally not different in untreated and acarbose-treated diabetic rats. The authors (Failla and Seidel 1993) concluded that acarbose treatment did not adversely affect the compensatory mechanisms required to maintain adequate tissue levels of essential trace metals.

VII. Exocrine Pancreas

Initial investigations with regard to an indirect effect or feedback of glucosidase inhibition on the composition of pancreatic enzymes were done by feeding rats BAY e 4609. This amylase inhibitory preparation caused reduced pancreatic weight, decreased pancreatic amylase activity and increased pancreatic trypsinogen content and left lipase activity unchanged. These alterations were reversible within 1 week by withdrawal of BAY e 4609 or could be attenuated by the addition of glucose to a standard feed containing BAY e 4609 (Puls and Keup 1975b). Similar results were reported by other researchers (Fölsch et al. 1976) and extended by enzyme secretion experiments. In response to an infusion of cholecystokinin combined with secretin, the secretion of amylase was reduced and the output of trypsin was increased (Fölsch et al. 1981).

Analogous findings could have been expected from the administration of acarbose, which exerts strong inhibitory activity on amylase and brush-border-located glucosidases. However, a 34-week feeding of acarbose (20, 40, 80 mg/100 g fed) did not affect the pancreatic weight of Wistar and genetically obese Zucker rats of either sex but dose dependently reduced the pancreatic amylase content (Puls 1980).

Corresponding results have been observed 20 days after feeding an excessively high dose of acarbose (150 mg/100 g standard feed) to Wistar rats. The pancreatic weight and the basal pancreatic amylase output did not differ from the control animals but the caerulein-stimulated amylase secretion rate was significantly diminished (Otsuki et al. 1983a). The content of protein, DNA, lipase and trypsinogen in the pancreas of rats was not altered afte a 10-day administration of acarbose (150 mg/100 g feed) in a diet containing sucrose, whereas the secretion of insulin in response to glucose or caerulein was reduced and the exocrine release of amylase in response to caerulein was decreased. When sucrose in the diet was replaced by glucose, no alterations were observed, demonstrating that it was not acarbose itself but a block of carbohydrate digestion which caused pancreatic alterations (Otsuki et al. 1983b; Otsuki et al. 1986).

No differences in the pancreatic content of amylase, trypsin and lipase or the cholecystokin/secretin-stimulated output of these enzymes were observed after a 20-day administration of acarbose (80 mg/100 g feed) in a sucrose- or maltose-enriched diet (Fölsch and Creutzfeldt 1985). Again, different results were obtained with voglibose. The pancreatic weight of genetically obese-diabetic rats was significantly increased but unchanged in lean rats (Ikeda et al. 1991b) and was reduced in spontaneously diabetic GK rats (Takami et al. 1991). A 10-day feeding of the long-acting glucosidase inhibitor emiglitate to rats did not affect the pancreatic weight, the protein, DNA, amylase or trypsin contents or the basal secretion of these enzymes, but decreased the cholecystokinin/secretin-induced secretion of amylase and trypsin (Göke et al. 1986). An explanation of the heterogeneous effects of glucosidase inhibition on the exocrine pancreas of rats has not yet been found.

VIII. Small and Large Bowel

The effects of glucosidase inhibition on the weight and length of the gut wall, mucosa enzyme activity, enterohormones, luminal microflora and faeces have been studied in experiments with rats.

1. Weight, Length and Microflora

The wet weight of the small intestine was increased when acarbose was given as a food-drug mixture to normal and genetically diabetic mice for a period of 15 days (Lee et al. 1983) or 10 weeks (Lee and Koldovsky 1993). An increased weight of the upper, middle and lower part of the small intestine of rats was reported after a 3-week (Puls et al. 1983) or 4-week feeding (Uttenthal et al. 1982) of acarbose. The absolute jejunal wet weight and the length of this intestinal segment of genetically obese rats were increased after 3- and 5-month acarbose administration (Dodane et al. 1991). An unexplained finding is a delayed repletion of the gut mass of rats during a 4-day refeeding period with acarbose following a fasting for 3 days (Kotler et al. 1982).

The wet weight and the length of the small intestine, the weight of the intestinal mucosa and its protein concentration were increased after a 4-week feeding of voglibose to lean and fatty Wistar rats (Ikeda et al. 1991b). The protein content in the intestinal wall of diabetic mice which had received the absorbable inhibitor emiglitate for 3, 7 or 84 days was not different from that in control animals (Lee et al. 1987).

Bile acids are involved in emulsifying alimentary fat in the intestine, reduce net absorption of water and electrolytes and cause diarrhoea if they escape ileal absorption. The absorption of ^{14}C-taurocholic acid was examined in acute perfusion studies and after an 8-week feeding of acarbose and miglitol (Walsh and Harnett 1986). The intestinal absorption of taurocholic acid was not affected by the administration of acarbose or miglitol.

Electron microscopic investigations of the intestines (including liver and pancreas) of rats treated with acarbose for 3 weeks revealed no ultrastructural changes as compared to control animals (Voigt et al. 1976). In an additional study with a single dose of acarbose (5 g/kg body weight), no histological changes in the brush border or intracellular elements in the intestinal mucosa of rats were revealed by electron microscopic investigation (Voigt and Puls 1976).

The large intestine, in particular the caecum of rats, is heavily colonized by bacteria. Indigestible carbohydrates (e.g., cellulose) or undigested oligo- and polysaccharides which leave the ileum and enter the large bowel can be degraded by bacterial hydrolysis and fermentation, generating short-chain fatty acids (acetate, lactate, propionate, butyrate) and gas (H_2, CO_2, CH_4). The absorption or passive diffusion of short-chain fatty acids through the gut wall thus prevents or reduces osmotic diarrhoea and loss of energy. If more short-chain fatty acids are produced in the colon than can be absorbed, the consequences will be a fall in the faecal pH and osmotic diarrhoea.

Alterations of microflora and increased concentrations of short-chain fatty acids in the faeces of hyperlipidaemic patients after a 4- to 8-week administration of acarbose have been reported (Shimoyama et al. 1982). Analogous examinations were carried out with rats which received acarbose or miglitol for a period of up to 4 weeks (Benno et al. 1988). In the rats treated with acarbose, the caecum weight was dose dependently increased, the number of bifidobacteria and clostridia was higher, the caecal pH reduced, the concentration of caecal moisture slightly increased and the ammonia decreased by more than 50% in comparison to control animals. No alterations of this kind were found in the group treated with miglitol, which in contrast to acarbose is completely absorbed in the small intestine of rats. The authors distinctly mentioned that no diarrhoea was seen, either in acarbose- or in miglitol-treated animals. A 4-week feeding of voglibose resulted in a dose-dependent increased caecum weight of lean and fatty Wistar rats as compared with pair-fed control animals (Ikeda et al. 1991b).

Measurements of the exhaled hydrogen as an indicator of intestinal fermentation were carried out on fasted rats after a single load with sucrose (by gavage) with and without a very high concentration of acarbose (10 mM). Sucrose load increased the rate of exhaled H_2 by 39%, and the addition of acarbose increased the H_2-exhalation by 95% as compared with baseline values (Ostrander et al. 1982).

Determinations of the pH values in the ileum and caecum of rats on a high-carbohydrate, fat-free diet with and without 100, 200, 400 and 800 ppm acarbose revealed slight, but dose-dependent reductions in the ileum (control group, pH 7.0; 400 ppm, 6.3; 800 ppm, 6.0) and more pronounced reductions in the caecum (control group, pH 7.0; 200 ppm, 5.7; 400 ppm, 5.5; 800 ppm, 5.1). The gas volume in the small and large intestine increased dose dependently from 1.3 ml/rat in the control group to 2.3, 3.7, 6.7 and 7.2 ml, respectively. Simultaneous administration of 400 ppm acarbose with the antacid Talcid and the antibiotic neomycin nearly eliminated the consequences of bacterial carbohydrate fermentation with no change in the effect of acarbose on blood glucose concentration (Puls and Krause 1979b).

Megacolon, extreme defecation and the death of several rats under the influence of acarbose have been reported (Glick and Bray 1983). The rats received a diet high in raw starch and very low in fat. These effects did not occur when the diet contained a standard amount of raw starch and fat or when the raw starch was replaced by cooked starch (Oshiro and Glick 1985). Thus, the deleterious effects mentioned above were obviously not caused by acarbose itself but by its pharmacodynamic actions under these particular experimental conditions.

We may conclude that the influence of glucosidase inhibitors (e.g., acarbose, miglitol, emiglitate, voglibose) on the large bowel depends not only on their binding constants, pharmacokinetics (Lembcke et al. 1987) or the dose of the inhibitor, but equally on the composition of the diet and probably also on the physical condition of the animals (healthy, diabetic, obese, etc.) used in such feeding experiments. These factors may also be relevant for the problems of maldigestion and malabsorption, i.e., a relevant loss of dietary energy with the faeces.

2. Enzyme Activity Alterations in the Small Intestine

Under the influence of acarbose, the distal part of the small intestine becomes involved in the digestion of alimentary di-, oligo- and polysaccharides (Puls et al. 1982a; Krause et al. 1982b). Examination of glucosidase activity in the median and distal small intestine of mice has shown sucrase activity to be significantly increased after 15 days of acarbose administration, but not after 5 or 10 days (Lee et al. 1983). Surprisingly, the activity of lactase, a β-glucosidase which is not inhibited by acarbose, was dose dependently increased in normal and in diabetic mice. In analogy to lactase, the isomaltase activity, which is not affected by acarbose, was increased in the middle and distal segments of rats after a 21-day acarbose feeding (PULS et al. 1983). These findings invite the conclusion that the transport of undigested carbohydrates into the distal intestinal segment results in increased formation of glucosidases; it may be of secondary importance, e.g. due to experimentally induced atrophy of the exocrine rat pancreas (CREUTZFELDT et al. 1985), feeding of difficult-to-digest oatmeal and dried peas (PULS et al. 1983) or hyperphagia of rats rendered diabetic by injection of STZ (CASPARY et al. 1972) or alloxan (YOUNOSZAI and SCHEDL 1972). A more important factor for increases in glucosidase activity analogous to the growth of intestinal mucosa seems to be that the brush border comes into contact with carbohydrates which are degraded by mammalian enzymes. Infusions of maltose, sucrose or lactose into the small intestine of rats on total parenteral nutrition (for 7 days) brought about a mucosa proliferation which was not found after infusions of glucose or lactulose, a non-hydrolysable disaccharide (WESER et al. 1984). Since the growth of the mucosa did not occur when a sucrose solution, containing acarbose in a very high concentration and thus blocking the sucrase, was infused, it can be concluded that the hydrolysing process itself is an important growth-stimulating factor. However, the exact mechanisms underlying the mucosa proliferation or enzyme activity increments remain incompletely understood.

Administration of emiglitate (5 and 10 mg/100 g laboratory chow) to healthy and diabetic mice failed to increase the lactase and reduced the sucrase and maltase activity in the proximal, middle and distal intestinal segments (LEE et al. 1987). The enzyme assays were done after an overnight withdrawal of the emiglitate-containing food. It may be that the depressed enzyme activity results from systemic absorption or enterohepatic circulation in addition to a passage of the long-acting emiglitate or an inhibitory active metabolite from the blood to the gut lumen, which thus inhibits the sucrase and maltase despite washout of the intestine. The sharp contrasts between emiglitate and acarbose may also lie in the different inhibitory spectrum of these inhibitors on rat α-glucosidases (GODA et al. 1981; SAMULITIS et al. 1987). Decreased sucrase and maltase activities have also been found after administration of voglibose in the intestine of obese diabetic mice (ODAKA et al. 1992), obese diabetic rats (IKEDA et al. 1991b) and normal or obese Wistar rats (IKEDA et al. 1991a). The enzyme assays after feeding voglibose for several weeks were carried out in the fed condition of the animals, i.e., in the presence of this inhibitor.

3. Enterohormones in the Intestinal Gut Wall

Acarbose, given as a food mixture in a semisynthetic diet, caused a severalfold increase in the content of enteroglucagon, cholecystokinin, vasoactive intestinal polypeptide, substance P, neurotensin and bombesin in the enlarged wall of the small intestine. The crypt cell production in the terminal ileum was increased threefold. Immunocytochemical examinations revealed an increased enteroglucagon cell proliferation in the hypertrophied mucosa (UTTENTHAL et al. 1982). The impact of these enterohormone increments in the gut wall on systemic regulators remains an open question.

IX. Liver

1. Glycogen

The liver participates in the glucose homoeostasis mainly by gluconeogenesis, glycogenolysis and glycogen storage. The conversion of monosaccharides from alimentary carbohydrates to hepatic glycogen could hypothetically be impaired by administration of glucosidase inhibitors owing to a reduced postprandial insulin response and a decreased monosaccharide concentration in the portal vein blood.

Oral sucrose loading tests with and without acarbose (PULS et al. 1982a) and voglibose (ODAKA et al. 1991) have been carried out with consecutive determination of the liver glycogen content of rats at various times up to 4 and 6 h, respectively. Acarbose (0.5 mg and 1.0 mg/kg body weight) and voglibose (0.03 mg/kg body weight) surprisingly strengthened the hepatic glycogen content in comparison to sucrose-loaded rats without inhibitor despite a reduced rate of intestinal carbohydrate digestion and monosaccharide absorption and a decreased insulin response. The effect of acarbose on the activity of glycogen synthase a (EC 2.4.1.11) was not examined, but emiglitate was reported not to activate this enzyme (BOLLEN et al. 1988). Thus we can suppose that hepatic extraction of monosaccharides from the portal vein blood is more effective, for whatever reason, when the postprandial monosaccharide peak concentration is flattened. The formation of glycogen is of course decreased or abolished if the monosaccharide absorption is blocked by overdosing glucosidase inhibitors, as was demonstrated by the administration of 0.1 mg voglibose/kg body weight (ODAKA et al. 1991). A reduced hepatic glycogen content was also seen in refeeding experiments on previously fasted rats which had reveived 50 mg acarbose/ 100 mg feed (BJÖRNTORP et al. 1983). The authors suggested that this effect was not related to carbohydrate malabsorption but rather to reduced postprandial serum insulin concentrations since the muscle glycogen content was not reduced and the weight of epididymal and perirenal fat pads was significantly decreased.

Feeding a diet supplemented with sucrose and starch (high-carbohydrate diet) for 7 and 21 days led to an elevated hepatic glycogen concentration in normal and STZ-diabetic rats as compared with control animals on standard feed (HAUGARD et al. 1984; HESS et al. 1986). The addition of acarbose to the high-carbohydrate diet caused a significant reduction of liver glycogen in comparison to the corresponding control rats, but without a significant difference compared to normal rats on standard feed. The hepatic glycogen of STZ-diabetic rats was much lower than in normal animals. Administration of acarbose in standard feed did not affect the liver glycogen of STZ-diabetic rats but caused a hepatic glycogen decrease in STZ-diabetic rats on a high-carbohydrate diet.

The most important pathway of glycogenolysis is catalysed by interactions of glycogen phosphorylase (EC 2.4.1.1) and debranching enzyme (EC 2.4.1.25 and EC3.2.1.33). Glycogenolysis is stimulated, e.g., by endogenous glucagon and epinephrine in the postabsorptive status, thus maintaining glucose homoeostasis or counteracting episodic hypoglycaemia.

In vivo and in vitro experiments were done to explore the effects of glucosidase inhibition on glycogenolysis. Oral administration of acarbose did not significantly affect the glucagon-induced glycogenolysis in rats, even at ~600 mg/kg body weight, whereas the administration of 50 mg miglitol and 5 mg emiglitate/kg body weight or higher doses caused a dose-dependent and significant depression of glycogenolysis. This effect was confirmed in in vitro experiments with isolated rat hepatocytes, where no alteration of the phosphorylase activity was found (BOLLEN et al. 1988). The reduced glycogenolysis is most probably caused by inhibition of α-1,6-glucosidase (EC 3.2.1.33) in the presence of high inhibitor concentrations (BOLLEN and STALMANS 1988).

The gluconeogenesis in isolated rat hepatocytes was not affected in the presence of high concentrations of miglitol or emiglitate, as was indicated by unchanged rates of glucose production from lactate (BOLLEN and STALMANS 1988). The gluconeogenesis in the kidney cortex of rats was significantly increased after a 14-day feeding of acarbose in an excessively high concentration (450 mg/100 g feed), but was not affected after a single extremely high dose of acarbose (150 mg/kg body weight), demonstrating that the increased gluconeogenesis after feeding acarbose was not due to the inhibitor itself, but rather to a lack of alimentary carbohydrates (PULS et al. 1983). The feeding of pharmacologically relevant doses of acarbose (5, 10 or 20 mg/100 g feed) did not alter the gluconeogenesis in weaned rats (MAURY et al. 1993).

A number of experiments have been done to simulate lysosomal glycogen storage (Pompe's disease), predominantly by parenteral administration of excessive doses of acarbose, miglitol and emiglitate (CALDER and GEDDES 1989; GEDDES et al. 1983; GEDDES and TAYLOR 1985a; KONISHI et al. 1989; KONISHI et al. 1990; LEMBCKE et al. 1991; LUELLMANN-RAUCH 1981; LUELLMANN-RAUCH 1982; LUELLMANN-RAUCH and WATERMANN 1987). It was shown that inhibition of the lysosomal acid 1,4-glucosidase was followed by an increased lysosomal content of glycogen in various organs under these experimental conditions.

2. Lipids

An increased lipid content in the liver can result from enhanced deposition of alimentary fat, an intensified hepatic lipogenesis and/or a diminished lipid release from the liver into the blood. Investigations of the effects of α-glucosidase inhibitors on the deposition of alimentary fat in the liver of fasted rats were done in acute loading tests with mixed meals containing starch, sucrose, protein, olive oil and triolein with ^{14}C-labelled fatty acids. The addition of acarbose did not affect their passage through the gastrointestinal tract, the blood stream or the liver; this was assessed by radioactivity measurements up to 4 h after giving the mixed meals by gavage (PULS 1980). Corresponding experiments with other glucosidase inhibitors have not been reported.

The effects of glucosidase inhibitors on hepatic lipogenesis have been studied in experiments with female genetically obese Zucker rats. Feeding a fat-free diet for 17 days caused a strong elevation of hepatic triglycerides compared with rats on standard feed (152 vs. 31 mg triglyceride/g liver wet weight). The elevated triglyceride content could be due to an enhanced hepatic lipogenesis or a reduced triglyceride secretion as VLDLs from the liver into the blood. In order to understand the accountable mechanism, the secretion of VLDLs was tested by determinating the serum triglyceride concentrations of obese Zucker rats before and 2 h after the injection of Triton WR-1339, which does not affect the secretion of VLDLs, but prevents their extraction from blood by blocking the lipoprotein lipase activity in peripheral tissues. The increase in triglycerides in the serum of rats on a fat-free diet was approximately twice that of rats on standard feed (2962 vs. 1730 mg triglyceride/dl). From this it appears that the elevated hepatic triglyceride concentration in rats on a fat-free diet is not caused by a reduced secretion of VLDLs, but by an enhanced hepatic lipogenesis. Administration of 10, 20 or 40 mg acarbose/100 g fat-free diet caused a dose-dependent reduction of hepatic triglyceride concentrations to the level of rats on standard feed. The acarbose-related decrease in hepatic lipogenesis may result from reduced concentrations of monosaccharides, insulin and FFAs in the portal vein blood (PULS 1980).

Examinations of the hepatic lipid composition of lean and obese-diabetic rats have shown that the concentrations of triglycerides and cholesterol were increased and the concentration of phospholipids was decreased in genetically obese-diabetic Wistar rats on standard feed as compared with lean rats. A 4-week administration of voglibose (1 or 5 mg/100 g feed) did not significantly affect the concentration of hepatic triglycerides and cholesterol, but brought about a reduction of the lipid components in obese-diabetic rats. The concentration of phospholipids was significantly increased in lean and obese rats treated with voglibose (IKEDA et al. 1991b). The treatment of genetically obese-diabetic KKA mice with voglibose (1 or 5 mg/100 g feed) during 12 weeks resulted in a dose-dependent reduction of the liver weight and the concentration of hepatic triglycerides and cholesterol and also an increase in hepatic phospholipids (ODAKA et al. 1992).

It remains an open question whether the hepatic lipid alterations are caused by a reduced food ingestion under the influence of glucosidase inhibitors or must be accounted for by unknown factors, e.g., hereditary disposition accompanied by a kind of "metabolic syndrome." It may be significant that a comparative analysis of the portal vein blood of lean and obese Zucker rats showed much higher concentrations of glucose, insulin, cholesterol, triglycerides and FFAs in obese than in lean rats (Kobatake et al. 1989).

X. Heart

Only a few studies have been reported concerning the indirect effects of glucosidase inhibitors on the myocardium. Experiments were done with normal and STZ-diabetic rats which received acarbose for up to 3 weeks in standard feed or a carbohydrate-fortified diet. Changes were observed with regard to cardiac glycogen concentrations, basal and isoproterenol-stimulated phosphorylase *a* activity, uridine kinase activity, and basal and isoproter-enol-induced increase in contractile force of the isolated perfused heart (Haugard et al. 1984; Hess et al. 1986). The acarbose-related cardiac changes have not been evaluated by the authors.

Male, but not female JCR:LA-cp rats aged 9 months have been reported to exhibit raised intimal lesions on the aortic arch and a range of myocardial lesions that are apparently of ischaemic origin. Histological examinations of the myocardium were done 8 months after feeding laboratory chow with and without acarbose (Russel et al. 1993a; Russel et al. 1993b). Acarbose treatment markedly reduced the incidence of old scarred myocardial lesions in comparison to untreated rats. The authors suggest that this effect was due to reduced circulating insulin concentrations as a consequence of acarbose administration.

XI. Skeletal Muscle

The basal glucose update of the isolated soleus muscle of mice, rendered obese and hyperinsulinaemic by i.v. injection of gold thioglucose, was signi-ficantly reduced in comparison to lean animals (Lemarchand-Brustel et al. 1990). Feeding of acarbose for a period of 4 months normalized the basal, but did not affect the insulin-stimulated, glucose incorporation into the soleus muscle of obese mice. A decrease in insulin binding, such as was found both in intact soleus muscle and in partially purified insulin receptor preparations, was partially prevented by administration of acarbose. How-ever, the autophosphorylation of insulin receptors and their tyrosine kinase activity were not altered by acarbose treatment. The authors conclude that acarbose is insufficient to reverse or prevent peripheral insulin resistance, since a defective tyrosine kinase activity of the insulin receptor, which was not ameliorated, plays a key role in insulin insensitivity.

A decreased basal glucose uptake of hindquarter skeletal muscles was found 9 weeks after feeding Sprague-Dawley rats a highly palatable and high-energy "cafeteria" diet as compared to control rats which received standard feed. Addition of acarbose to the "cafeteria" diet teneded to improve the basal and insulin-stimulated glucose incorporation into skeletal muscles without alterations in insulin binding or receptor tyrosine kinase activity in pooled soleus, red gastrocnemius and red quadriceps muscles (YOUNG et al. 1993).

More significant results with regard to an indirect effect of glucosidase inhibitors on skeletal muscles were obtained in experiments using genetically obese-diabetic Zucker rats (ZDF). In the ZDF rat, obesity and insulin resistance precede the development of hyperglycaemia associated with a deficit of basal and glucose-stimulated plasma insulin concentration. At the age of 26 weeks, the GLUT-4 transporter protein in red fibres of the quadriceps muscle and in mixed fibres of the gastrocnemius muscle of diabetic rats was reduced by 40%–50% in comparison to non-diabetic rats. The decrease in muscle GLUT-4 expression in this model of NIDDM was prevented by the feeding of acarbose during the final 19 weeks (FRIEDMAN et al. 1991; DOHM et al. 1993). The authors concluded that a reduced hyperglycaemia and increased B-cell responsiveness in the acarbose-treated animals caused a normalization of the GLUT-4 transporter protein.

XII. Dangerous Late Complications

The most important aim and challenge in diabetes management remains the retardation or prevention of diabetic late complications, e.g., nephropathy, neuropathy and retinopathy.

1. Nephropathy

The first experiments for testing an indirect preventive effect of α-glucosidase inhibition on diabetic kidney alterations were done by using genetically diabetic mice (C57BLKsJ, *db/db*), a model which in many respects resembles NIDDM in humans, and which has proved very useful for investigating diabetic nephropathy (LEE 1982). The mice received standard feed with and without various doses of acarbose at the age of 5–6 weeks for the subsequent 10 weeks. Although the fasting blood glucose levels and body weight gain did not differ between treated and untreated mice, acarbose caused a dose-dependent amelioration of renal lesions, as indicated by the presence of significantly less immunoglobulins (IgG, IgM, IgA) throughout the glomerular mesangium and a significantly decreased, but not completely obstructed, mesangial area. The author concluded that the attenuated nephropathic alterations in acarbose-treated mice were due to an improvement of the metabolic control, as was indicated by lower HbA values and markedly diminished urinary glucose excretion in comparison to the control mice. Similar results were seen in experiments using genetically diabetic KK mice, which had revieved acarbose for 16 weeks (HAMADA et al. 1989a) or voglibose for 12 weeks (ODAKA et al. 1992).

The kidney weight and glycation of glomerular basement membranes of rats were significantly increased 8 weeks after injection of STZ compared with non-diabetic rats. These renal alterations were almost completely prevented by the administration of an acarbose-supplemented feed (Cohen et al. 1991).

The "Cohen rat" is characterized by sucrose-inductive diabetes mellitus and proneness ot diffuse glomerulosclerosis, which is observed in about 80% of animals at the age of 6–7 months (Cohen and Rosenmann 1990). Treatment of "Cohen rats" with acarbose reduced the incidence and severity of glomerulosclerosis. The number of surviving rats at 7 months of age was reported to be 3/15 in the control group and 11/15 in the acarbose-treated group.

When BB/WOR rats, which spontaneously develop an acute diabetic syndrome resembling human type 1 diabetes, were treated with protamine zinc insulin with and without acarbose as a food ingredient for a period of about 5 months, the glomerular filtration rate tended to be higher in acarbose-treated rats (Koevary 1993).

2. Neuropathy

Attenuation of this dangerous diabetic complication by administration of intestinal glucosidase inhibitors was the subject of two experiments using quite different strains of genetically diabetic rats: the insulin-dependent BB/Wor rats and the insulin-independent Zucker diabetic fatty rats (ZDF/Drt-fa).

BB/Wor rats were maintained on low doses of protamine zinc insulin with and without acarbose-supplemented feed during a period of 4 months (Sima and Chakrabarti 1992). Autonomic polyneuropathy was completely prevented by acarbose treatment as indicated by R-BAR values (a measure of respiration-related variations of heartbeat rates). The development of somatic polyneuropathy was significantly ameliorated by acarbose, with partial prevention of reduced nerve conduction velocity during the first 3 months, but this had ceased at 4 months. Axonal atrophy and axo-glial dysjunction of the sural nerve were significantly but only partially attenuated in acarbose-treated rats. The authors conclude that a decrease in cumulative hyperglycaemia caused by acarbose retarded the development of diabetic polyneuropathies.

Nerve conduction velocity, nerve Na^+, K^+-ATPase activity and the concentration of glucose, fructose and myo-inositol in the sciatic nerve of ZDF/Drt-fa rats were determined after a 19-week treatment with acarbose. The decreased nerve conduction velocity and Na^+, K^+-ATPase activity in diabetic rats were prevented by administration of acarbose (Peterson et al. 1993). The increased concentrations of glucose and fructose in the sciatic nerve of diabetic rats were dose dependently reduced or normalized, and the reduced concentration of myo-inositol was not statistically significantly elevated in the acarbose-treated rats.

The results of these experiments suggest that several diabetes-related neural alterations can be improved or retarded by treating insulin-dependent and insulin-independent rats with acarbose.

3. Retinal Microangiopathy

A characteristic feature of retinal microangiopathy is the thickening of capillary basement membranes. This change has been seen in insulin-dependent BB/Wor rats within 4 months of overt diabetes. Administration of acarbose as a food admixture in addition to low doses of protamine zinc insulin completely prevented the thickening of basement membranes in superficial and deep retinal capillaries compared with rats which received insulin without acarbose (CHAKRABARTI et al. 1993). The authors concluded that acarbose prevented the development of retinal microangiopathy due to an improved glycaemic control of the insulin-dependent BB/Wor rats.

Acknowledgement. I am very grateful to Gudrun Lion and Ursula Neubert for their excellent secretarial help.

References

Abdollahi A, Tulp OL, Schnitzer-Polokoff R (1993) The effects of acarbose on cholesterogenesis in obese and obese-diabetic rats. In: Vasselli JR, Maggio CA, Scriabine A (eds) α-Glucosidase inhibition: potential use in diabetes. Neva, Branford, pp 133–143

Alberti KGM, Gries FA (1988) Management of non-insulin-dependent diabetes mellitus in Europe. A consensus view. Diabetic Med 5:275–281

Axen KV (1993) Insulin secretory response in an acarbose-treated rat model of NIDDM and direct effects of acarbose in normal islet function in vitro. In: Vasselli JR, Maggio CA, Scriabine A (eds) α-Glucosidase inhibition: potential use in diabetes. Neva, Branford, pp 45–55

Axen KV, Sclafani A (1990) Acarbose decreases insulin resistance in a rat model of NIDDM. Diabetes 29 (Abstract 195)

Bagdade JD, Bierman EL, Porte D (1967) The significance of basal insulin response levels in the evaluation of the insulin response to glucose in diabetic and nondiabetic subjects. J Clin Invest 46:1549–1557

Banting FG, Best CH (1922) The internal secretion of pancreas. J Lab Clin Med 7:251–266

Benno Y, Endo K, Shiragami N, Mitsuoka T (1988) Effects of two α-glucosidase inhibitors, acarbose and BAY m 1099, on intestinal microflora, cecal properties, body-weight gains, serum cholesterol and serum lipids of rats. Bifidobacteria Microflora 7:75–85

Bischoff H, Puls W, Krause HP, Schutt H, Thomas G (1985) Pharmacological properties of the novel glucosidase inhibitors BAY m 1099 (miglitol) and BAY o 1248. XII Congress of IDF Madrid. Diabetes Res Clin Pract Suppl 1 (Abstract 133)

Björntorp P, Yang M, Greenwood MR (1983) Refeeding after fasting in the rat: effects of carbohydrate. Am J Clin Nutr 37:396–402

Bollen M, Stalmans W (1988) The antiglycogenolytic action of 1-deoxynojirmycin results from a specific inhibition of the α-1,6-glucosidase. Eur J Biochem 181:775–780

Bollen M, Vandebroeck A, Stalmans W (1988) 1-Deoxynojirimycin and related compounds inhibit glycogenolysis in the liver without affecting the concentration of phosphorylase a. Biochem Pharmacol 37:905–909

Booth CC (1968) Effect of location along the small intestine on absorption of nutrients. In: Code CF (ed) Handbook of physiology, vol 3, part 6. American Physiological Society, Washington, pp 1513–1527

Calder PC, Geddes R (1989) Acarbose is a competitive inhibitor of mammalian lysosomal acid alpha-D-glucosidases. Carbohydr Res 191:71–78

Carswell N, Michaelis OE, Prather ES (1989) Effect of acarbose (Bay-g-5421) on expression of noninsulin-dependent diabetes mellitus in sucrose-fed SHR/N-corpulent rats. J Nutr 119:388–394

Caspary WF, Rhein AM, Creutzfeldt W (1972) Increase of intestinal brush border hydrolases in mucosa of streptozotocin diabetic rats. Diabetologia 8:412–414

Caspary WF (1987) Pathophysiology and clinical aspects of the malabsorption syndrome. In: Caspary WF (ed) Structure and function of the small intestine. Excerpta Medica, Amsterdam, pp 217–247

Chakrabarti S, Cherian PV, Sima AAF (1993) The effect of acarbose on diabetes- and age-related basement membrane thickening in retinal capillaries of the BB/W-rat. Diabetes Res Clin Pract 20:123–128

Cohen MP (1990) Acarbose inhibits excess nonenzymatic glycation (NEG) and advanced glycation end product (AGEP) formation in diabetes. Diabetologia 33 (Suppl) A149:544 (abstract)

Cohen MP (1993) α-Glucosidase inhibition and renal involvement in experimental diabetes. In: Vasselli JR, Maggio CA, Scriabine A (eds) α-Glucosidase inhibition: potential use in diabetes. Neva, Branford, pp 211–218

Cohen MP, Klepser H (1991) Alpha-glucosidase inhibition prevents increased collagen fluorescence in experimental diabetes. Gen Pharmacol 22:607–610

Cohen MP, Klepser H, Wu VY (1991) Effect of alpha-glucosidase inhibition on the non enzymatic glycation of glomerular basement membrane. Gen Pharmacol 22:515–519

Cohen AM, Rosenmann E (1990) Acarbose treatment and diabetic nephropathy in the Cohen diabetic rat. Horm Metab Res 22:511–515

Creutzfeldt W, Foelsch UR, Elsenhans B, Ballmann M, Conlon JM (1985) Adaptation of the small intestine to induced maldigestion in rats. Experimental pancreatic atrophy and acarbose feeding. Scand J Gastroenterol [Suppl 112]:45–53

Dahlquist A (1964) Method for assay of intestinal disaccharidases. Anal Biochem 7:18–25

DeBouno JF, Michaelis OE, Tulp OL (1989) The effects of the intestinal glucosidase inhibitor BAY m 1099 (miglitol) on glycemic status of obese rats. Nutr Res 9:1041–1058

DeBouno JF, Michaelis OE, Tulp OL (1993) The effects of the internal glucosidase inhibitor BAY m 1099 (miglitol) on glycemic status of obese-diabetic rats. Gen Pharmacol 24:509–515

Dodane V, Chevalier J, Ripoche P (1991) Na^+/D-Glucose cotransport and sucrase activity in intestinal brush border membranes of Zucker rats. Effects of chronic acarbose treatment. Nutr Res 11:783–796

Dohm GL, Friedman JE, Peterson RG (1993) Acarbose treatment of non-insulin-dependent diabetic fatty (ZDF/Drt-fa) rats restores expression of skeletal muscle glucose transporter GLUT 4.In: Vasselli JR, Maggio CA, Scriabine A (eds) α-Glucosidase inhibition: potential use in diabetes. Neva, Branford, pp 173–180

Elsenhans B, Caspary WF (1987) Absorption of carbohydrates. In: Caspary WF (ed) Structure and function of the small intestine. Excerpta Medica, Amsterdam, pp 139–159

Failla ML, Seidel KE (1993) The absorption and retention of dietary zinc by type I diabetic rats are increased by chronic treatment with acarbose. In: Vasselli

JR, Maggio CA, Scriabine A (eds) α-Glucosidase inhibition: potential use in diabetes. Neva, Branford, pp 155–161

Foellmer HG, Kitano S, Roa L, Ardito T, Vasselli JR (1993) Ameliorating effects of acarbose on progression of NIDDM in the db/db mouse model. In: Vasselli JR, Maggio CA Scriabine A (eds) α-Glucosidase inhibition: potential use in diabetes. Neva, Branford, pp 237–248

Fölsch UR, Creutzfeldt W (1985) Adaptation of the pancreas during treatment with enzyme inhibitors in rats and man. Scand J Gastroenterol [Suppl 112]:54–63

Fölsch UR, Grieb N, Caspary WF, Frerichs H, Creutzfeldt W (1976) Insulin-, DNS-, Protein- und Enzymgehalt des Pankreas und Bürstensaumenzymes nach Kurz- und Langzeitfütterung eines α-Amylase-Inhibitors (BAY e 4609). Z Gastroenterol 14:250–251

Fölsch UR, Grieb N, Capsary WF, Creutzfeldt W (1981) Influence of short- and long-term feeding of an α-amylase inhibitor (BAY e 4609) on the exocrine pancreas of the rat. Digestion 21:74–81

Frerichs H, Daweke H, Gries FA, Grüneklee D, Hessing J, Jahnke K, Keup U, Miss H, Puls W, Schmidt D, Zumfelde C (1973) A novel pancreatic amylase inhibitor (BAY d 7791) Experimental studies on rats and clinical observations in normal and obese diabetic and non-diabetic subjects. Diabetologia 9:68

Fried SK, Maggio CA, Vasselli JR (1993) Effects of acarbose on insulin sensitivity in adipocytes from lean and obese Zucker and Zucker diabetic fatty rats. In: Vasselli JR, Maggio CA, Scriabine A (eds) α-Glucosidase inhibition: potential use in diabetes. Neva, Branford, pp 85–93

Friedman JE, DeVenté JE, Peterson RG, Dohm GL (1991) Altered expression of muscle glucose transporter GLUT-4 in diabetic fatty Zucker rats (ZDF/Drt-fa). Am J Phsiol 261:E782–E788

Frommer W, Puls W, Schäfer D, Schmidt DD (1972) German Offenlegungsschrift DE 206 4092

Fujioka S, Matsuzawa Y, Tokunaga K, Keno Y, Kobatake T, Tarni S (1991) Treatment of visceral fat obesity Int J Obes 15:59–65

Geddes R, Taylor JA (1985a) Factors affecting the metabolic control of cytosolic and lysosomal glycogen levels in the liver. Biosci Rep 5:315–320

Geddes R, Taylor JA (1985b) Lysosomal glycogen-storage induced by acarbose, a 1,4-alpha-glucosidase inhibitor. Biochem J 228:319–324

Geddes R, Otter DE, Scott GK, Taylor JA (1983) Disturbance of lysosomal glycogen metabolism by liposomal anti-alpha-glucosidase and some anti-inflammatory drugs. Biochem J 212:99–103

Glick Z, Bray GA (1982) The alpha glucosidase inhibitor acarbose stimulates food intake in rats eating a high carbohydrate diet. Nutr Behav 1:15–20

Glick Z, Bray GA (1983) Effects of acarbose on food intake, body weight and fat depots in lean and obese rats. Pharmacol Biochem Behav 19:71–78

Goda T, Yamada K, Hosoya N, Moriuchi S (1981) Effects of α-glucosidase inhibitor BAY g 5421 on rat intestinal disaccharidases. J Jpn Soc Nutr Food Sci 34:139–143

Göke B, Fehmann H-C, Siegel EG, Fölsch UR, Creutzfeldt W (1986) Influence of an absorbable alpha-glucosidase inhibitor (BAY o 1248) on the endocrine and exocrine pancreas of the rat. Z Gastroenterol 24:758–766

Gray GM (1983) Carbohydrate digestion and absorption. In: Creutzfeldt W, Fölsch UR (eds) Delaying absorption as a therapeutic principle in metabolic diseases. Georg Thieme, Stuttgart, pp 7–11

Gray RS, Olefsky JM (1982) Effect of a glucosidase inhibitor on the metabolic response of diabetic rats to a high carbohydrate diet, consisting of starch and sucrose, or glucose. Metabolism 31:88–92

Grodsky GM, Karam JH, Pavlatos FCH, Forsham PH (1963) Reduction by phen-formin of excessive insulin after glucose loading in obese and diabetic subjects. Metabolism 12:278–286

Hamada H, Iida R, Iskimura K, Yamashita S, Makihira T, Kobayashi N (1989a) Effect of acarbose on diabetes in KK mice. Yakari To Chiryo 17:17–28

Hamada H, Iida R, Mitsuzono T, Yamashita S, Makihira T, Kobayashi N (1989b) Effects of acarbose in combination with tolbutamide or insulin on normal and diabetic mice. Yakari To Chiryo 17:29–44

Haugard N, Hess ME, Locke CL, Torbati A, Wildey G (1984) Metabolic effects of acarbose administration in normal and diabetic rats. Biochem Pharmacol 33:1503–1508

Heitlinger LA, Sloan HR, DeVore DR, Lee PC, Lebenthal E, Duffey ME (1992) Transport of glucose polymer-derived glucose by rabbit jejunum. Gastroenterology 102:443–447

Hess ME, Haugard N, Min W, Torbati A (1986) Metabolic effects of acarbose in normal and diabetic rats: long- and short-term administration. Arch Int Pharmacodyn 283:163–176

Horii S, Fukase H, Matsuo T, Kameda Y, Asano N, Matsui K (1986) Synthesis and α-D-glucosidase inhibitory activity of N-substituted valiolamine derivative as potential oral antidiabetic agents. J Med Chem 29:1038–1046

Ho RS, Aranda CG (1983) Influence of acarbose on hyperglycemia induced by various carbohydrates in rats and oral starch tolerance in monkeys. Arch Inter Pharmacodyn 261:147–156

Ikeda H, Odaka H, Matsuo T (1991a) Effect of a disaccharidase inhibitor, AO-128, on a high sucrose-diet-induced hyperglycaemia in female Wistar fatty rats. Jpn Pharmacol Ther 19:155–160

Ikeda H, Odaka H, Matsuo (1991b) Antiobesity and antidiabetic actions of a disaccharidase inhibitor, AO-128, in genetically obese-diabetic rats, Wistar fatty. Jpn Pharmacol Ther 19:283–295

Katovich MJ, Meldrum MJ (1993a) Positive effects of dietary acarbose alone or in combination with insulin in the streptozotocin-induced diabetic rat. In: Vasselli JR, Maggio CA, Scriabine A (eds) α-Glucosidase inhibition: potential use in diabetes. Neva, Branford, pp 109–121

Katovich MJ, Meldrum MJ (1993b) Effects of insulin and acarbose alone and in combination in the female streptozotocin-induced diabetic rat. J Pharm Sci 82:1209–1213

Katovich MJ, Meldrum MJ, Vasselli JR (1991) Beneficial effects of dietary acarbose in the streptozotocin-induced diabetic rat. Metabolism 40:1275–1282

Kirchgessner M, Roth FX, Spoerl L (1981) Nutritive effects of intestinal α-glucosidase inhibition in rats. Res Exp Med 178:211–217

Kobatake T, Matsuzawa Y, Tokunaga K, Fujioka S, Kawamoto T, Keno Y, Inui Y, Odaka H, Matsuo T, Tarui S (1989) Metabolic improvements associated with a reduction of abdominal visceral fat caused by a new α-glucosidase inhibitor, AO-128, in Zucker fatty rats. Int J Obes 13:147–154

Koevary SB (1993) Effects of acarbose on the development of diabetes and its renal complications in the BB/Wor rat. In: Vasselli JR, Maggio CA, Scriabine A (eds) α-Glucosidase inhibition: potential use in diabetes. Neva, Branford, pp 189–197

Konishi Y, Hata Y, Fujimor K (1989) Formation of glycogenosomes in rat liver induced by injection of acarbose, an alpha-glucosidase inhibitor. Acta Histochem Cytochem 22:227–231

Konishi Y, Okawa Y, Hosokawa S, Fujimori K, Fuwa H (1990) Lysosomal glycogen accumulation in rat liver and its in vivo kinetics after a single intraperitoneal injection of acarbose, an alpha-glucosidase inhibitor. J Biochem (Tokyo) 107: 197–201

Kotler D, Tierney A, Kral J, Björntorp P (1982) Acarbose modified the process of energy repletion after starvation. Scand J Gastroenterol 17 [Suppl 78]:37

Kotler D, Tierney AR, Kral JG, Björntorp P (1984) Modification of weight gain by an α-glucosidase inhibitor during refeeding in rats. Am J Clin Nutr 40:270–276

Krause HP, Puls W (1981) Effects of the alpha-glucosidase inhibitor acarbose (BAY g 5421) on carbohydrate-induced hyperlipoproteinaemia in Wistar and (fa,fa) "Zucker" rats. Arch Pharmacol 316 [Suppl]:R11

Krause HP, Keup U, Thomas G, Puls W (1982a) Reduction of carbohydrate-induced hypertriglyceridemia in (fa,fa) "Zucker" rats by the α-glucosidase inhibitor acarbose (BAY g 5421). Metabolism 31:710–714

Krause HP, Keup U, Puls W (1982b) Inhibition of disaccharide digestion in rat intestine by the alpha-glucosidase inhibitor acarbose (BAY g 5421). Digestion 23:232–238

Lee SM (1982) The effect of chronic α-glucosidase inhibition on diabetic nephropathy in the db/db mouse. Diabetes 31:249–254

Lee SM (1988) Effects of acarbose on experimental diabetic nephropathy, metabolic control, and intestinal glucosidase activity in normal and genetically diabetic mice. In: Creutzfeldt W (ed) Acarbose for the treatment of diabetes mellitus. Springer, Berlin Heidellberg New York, p 63

Lee SM, Koldowsky O (1993) Chronic effects of acarbose on intestinal disaccharidase activity in normal and idabetic mice. In: Vasseli J, Maggio CA, Scriabine A (eds) α-Glucosidase inhibition: potential use in idabetes. Neva, Branford, pp 35–43

Lee SM, Bustamante SA, Koldovsky O (1983) The effect of α-glucosidase inhibition on intestinal disaccharidase activity in normal and diabetic mice. Metabolism 32:793–799

Lee SM, Bustamante S, Flores C, Bezerra J, Goda T, Koldovsky O (1987) Chronic effects of an alpha-glucosidase inhibitor (BAY o 1248) on intestinal disaccharidase activity in normal and diabetic mice. J Pharmacol Exp Ther 240:132–137

LeMarchand-Brustel Y, Rochet N, Grémeaux T, Marot I, Van Obberghen E (1990) Effect of an α-glucosidase inhibitor on experimentally-induced obesity in mice. Diabetologia 33:24–30

Lembcke B, Loeser C, Foelsch UR, Woehler J, Creutzfeldt W (1987) Adaptive responses to pharmacological inhibition of small intestinal α-glucosidase in the rat. Gut 28 [Suppl]:181–187

Lembcke B, Lamberts R, Wöhler J, Creutzfeldt W (1991) Lysosomal storage of glycogen as a sequel of α-glucosidase inhibition by the absorbed deoxynojirimycin derivative emiglitate (BAY o 1248). Res Exp Med (Berl) 191:389–404

Luellmann-Rauch R (1981) Lysosomal glycogen storage mimicking the cytological picture of Pompe's disease as induced in rats by injection of an alpha-glucosidase inhibitor. I. Alterations in liver. Virchows Arch B Cell Pathol 38:89–100

Luellmann-Rauch R (1982) Lysosomal glycogen storage mimicking the cytological picture of Pompe's disease as induced in rats by injection of an alpha-glucosidase inhibitor. II. Alterations in kidney, adrenal gland, spleen and soleus muscle. Virchows Arch B Cell Pathol 39:187–202

Luellmann-Rauch R, Watermann D (1987) Fusion of storage lysosomes in experimental lipidosis and glycogenosis. Exp Mol Pathol 46:136–143

Lundquist I, Panagiotidis G (1992) The relationship of islet amyloglucosidase activity and glucose-induced insulin secretion. Pancreas 7:352–357

Madar Z (1989) Metabolic consequences of the alpha-glucosidase inhibitor BAY m 1099 given to nondiabetic and diabetic rats fed a high carbohydrate diet. Am J Clin Nutr 49:106–111

Madar Z, Olefsky J (1986) Effect of the alpha-glucosidase inhibitor Bay o 1248 on the metabolic response of nondiabetic and diabetic rats to a high-carbohydrate diet. Am J Clin Nutr 44:206–211

Madariaga H, Lee PC, Heitlinger LA, Lebenthal E (1988) Effects of graded α-glucosidase inhibition on sugar absorption in vivo. Dig Dis Sci 33:1020–1024

Maggio CA, Vasselli JR (1989) Satiety in the obese Zucker rat – effects of carbohydrate type and acarbose (BAY g 5421). Physiol Behav 46:557–560

Maggio CA, DeCarr LB, Vasselli JR (1987) Differential effects of sugars and the alpha-glucosidase inhibitor acarbose (BAY g 5421) on satiety in the Zucker obese rat. Int J Obes 11 [Suppl 3]:53–56

Maggio CA, Vasselli JR, Pi-Sunyer FX (1993) Acarbose attenuates development of streptozotocin-induced diabetes in sucrose-consuming rats. In: Vasselli JR, Maggio CA, Scriabine A (eds) α-Glucosidase inhibition: potential use in diabetes. Neva, Branford, pp 181–187

Marble A (1974) The natural history of diabetes. Horm Metab Res, [Suppl 4]: 153–158

Marot I, Le Marchand-Brustel Y (1988) A preclinical study on the effects of acarbose on mice rendered obese with gold thioglucose. In: Creutzfeldt W (ed) Acarbose for the treatment of diabetes mellitus. Springer, Berlin Heidelberg New York, p 59

Matsuo T, Odaka H, Ikeda H (1992) Effect of an intestinal disaccharidase inhibitor (AO-128) on obesity and diabetes. Am J Clin Nutr 55 [Suppl 1]:314S–317S

Maury J, Issad T, Perderau D, Gouhot B, Ferré P, Girard J (1993) Effect of acarbose on glucose homeostasis, lipogenesis and lipogenic enzyme gene expression in adipose tissue of weaned rats. Diabetologia 36:503–509

Newcomer AD, McGill DB (1966) Distribution of disaccharidase activity in the small bowel of normal and lactase-deficient subjects. Gastroenterology 51:481–488

Odaka H, Miki N, Ikeda H, Matsuo T (1991) Effect of intestinal disaccharidase inhibitor AO-128 on carbohydrate, protein, fat and vitamin absorption in rats. Jap Pharmacol Therapeutics 19:143–154

Odaka H, Shino A, Ikeda H, Matsuo T (1992) Antiobesity and antidiabetic actions of a new potent disaccharidase inhibitor in genetically obese-diabetic mice, KKAY. J Nutr Sci Vitaminol (Tokyo) 38:27–37

Oshiro A, Glick Z (1985) Effects of acarbose in rats when mixed in raw and in cooked cornstarch. Fed Proc 44 (Abstract 8462):1860

Ostrander CR, Stevenson DK, Neu J, Kerner JA, Moses AW (1982) A sensitive analytical apparatus for measuring hydrogen production rates. I. Application to studies in small animals. Evidence of the effects of an alpha-glucosidehydrolase inhibitor in the rat. An Biochem 119:378–386

Otsuki M, Sakamoto C, Ohki A, Okabayashi Y, Suehiro I, Baba S (1983a) Effect of acarbose on exocrine and endocrine pancreatic function in the rat. Diabetologia 24:445–448

Otsuki M, Sakamoto C, Ohki A, Okabayashi Y, Yuu H, Maeda M, Baba S (1983b) Exocrine and endocrine pancreatic function in rats treated with α-glucosidase inhibitor (acarbose). Metabolism 32:846–850

Otsuki M, Okabayashi Y, Ohki A, Suehiro I, Baba S (1986) Effect of α-glucosidase inhibitor on exocrine and endocrine pancreatic function in rats fed a high-carbohydrate diet consisting of sucrose or glucose. Diabetes Res clin Pract 5:257–264

Perley M, Kipnis DM (1966) Plasma insulin responses to glucose and tolbutamide of normal weight and obese diabetic and nondiabetic subjects. Diabetes 15:867–874

Peterson RG (1991) The effectiveness of acarbose in treating Zucker diabetic fatty rats (ZDF/Drt-fa). Diabetes 40 [Suppl 1]:Abstract 871

Peterson RG, Mell M-A, Little LA, Kincaid JC, Fineberg NS (1988) In: Creutzfeldt W (ed) Acarbose for the treatment of diabetes mellitus. Springer, Berlin Heidelberg New York, pp 64–65

Peterson RG, Doss DJ, Neel M-A, Little LA, Kincaid JC, Eichberg J (1993) The effectiveness of acarbose in treating Zucker diabetic fatty rats (ZDF/Drt-fa). In: Vasselli JR, Maggio CA, Scriabine A (eds) α-Glucosidase inhibition: potential use in diabetes. Neva, Branford, pp 167–172

Puls W (1980) Zur Inhibition intestinaler Glucosidasen – Ein neues Prinzip zur Prävention und Therapie von kohlenhydratabhängigen Stoffwechselerkrankungen. Dissertation, Medical Faculty, University of Düsseldorf

Puls W, Bischoff H (1988) The pharmacological rationale of diabetes mellitus therapy with acarbose. In: Creutzfeldt W (ed) Acarbose for the treatment of diabetes mellitus. Springer, Berlin Heidelberg New York, pp 29–38

Puls W, Keup U (1973) Influence of an α-amylase inhibitor (BAY d 7791) on blood glucose, serum insulin and NEFA in starch loading tests in rats, dogs and man. Diabetologia 9:97–101

Puls W, Keup U (1975a) Inhibition of sucrase by TRIS in rat and man, demonstrated by oral loading tests with sucrose. Metabolism 24:93–98

Puls W, Keup U (1975b) Metabolic studies with an amylase inhibitor in acute starch loading tests in rats and men and its influence on the amylase content of the pancreas. In: Howard A (ed) Recent advances in obesity research: I. Newman, London, pp 391–393

Puls W, Krause HP (1979a) Indicative investigations with BAY g 5421 on mice. Pharma Report no. 8230, Bayer AG

Puls W, Krause HP (1979b) The effect of Talcid^R and neomycin sulfate on carbohydrate malabsorption caused by BAY g 5421 in rats. Pharma Report no. 8055, Bayer AG

Puls W, Keup U, Krause HP, Thomas G, Hoffmeister F (1977) Glucosidase inhibition. A new approach to the treatment of diabetes, obesity and hyperlipoproteinemia. Naturwissenschaften 64:536

Puls W, Keup U, Krause HP, Müller L, Schmidt DD, Thomas G, Truscheit E (1980) Pharmacology of a glucosidase inhibitor. In: Creutzfeldt W (ed) Front Hormone Research, vol 7, Karger, Basel, pp 235–247

Puls W, Keup U, Krause HP, Thomas G, Hoffmeister F (1982a) The concept of glucosidase inhibition and its pharmacological realization. In: Creutzfeldt W (ed) First international symposium on acarbose. Excerpta Medica, Amsterdam, pp 16–26

Puls W, Krause HP, Bischoff H (1982b) BAY o 1248. Pharmacological characterization of a novel glucosidase inhibitor. Pharma report no. 10937 P, Bayer AG

Puls W, Bischoff H, Schutt H (1983) Pharmacology of amylase- and glucosidase-inhibitors. In: Creutzfeldt W, Fölsch UR (eds) Delaying absorption as a therapeutic principle in metabolic diseases. Thieme, Stuttgart, pp 70–78

Puls W, Krause HP, Müller L, Schutt H, Sitt R, Thomas G (1984) Inhibitors of the rate of carbohydrate and lipid absorption by the intestine. Int J Obes [Suppl 1] 8:181–190

Reiser S (1976) Digestion and absorption of dietary carbohydrates. In: Berdanier CD (ed) Carbohydrate metabolism. Regulation and physiological role. Wiley, New York, pp 45–78

Rhinehart BL, Robinson KM, Liu PS, Payne AJ, Wheatley ME, Wagner SR (1987) Inhibition of intestinal disaccharidases and suppression of blood glucose by a new α-glucohydrolase inhibitor MDL 25-637 J Pharmacol Exp Ther 241:915–920

Robinson KM, Begovic ME, Rhinehart BL, Heinecke EW, Ducep J-B, Kastner PR, Marshall FN, Danzin C (1991) New potent α-glucohydrolase inhibitor MDL 73945 with long duration of action in rats. Diabetes 40:825–830

Robinson KM, Rhinehart BL, Ducep J-B, Danzin C (1992) Intestinal disaccharidase inhibitors. Drugs Future 17:705–720

Rolston R, Ghiglione M, Bacarese-Hamilton AJ, Uttenthal LO, Bloom SR (1985) Pentagastrin-stimulated gastric acid output and plasma enteroglucagon in acarbose-treated rats. Digestion 32:124–127

Russel JC, Koeslag DG, Dolphin PJ, Amy RM (1993a) Beneficial effects of acarbose in the atherosclerosis-prone JCR: LA-corpulent rat. Metabolism 42:218–223

Russel JC, Roger MA, Dolphin PJ (1993b) Reduction in myocardial disease in an animal model by acarbose treatment. In: Vasselli JR. Maggio CA, Scriabine A (eds) α-Glucosidase inhibition: potential use in diabetes. Neva, Branford, pp 145–151

Salehi A, Lundquist I (1993a) Changes in islet glucan-1,4-alpha-glucosidase activity modulate sulfonylurea-induced but not cholinergic insulin secretion. Eur J Pharmacol 243:185–191

Salehi A, Lundquist I (1993b) Ca^{2+} deficiency, selective alpha-glucosidase hydrolase inhibition, and insulin secretion. Am J Physiol 265:E1–E9

Salehi A, Lundquist I (1993c) Islet glucan-1,4-alpha-glucosidase: differential influence on insulin secretion induced by glucose and isobutylmethylxanthine in mice. J Endocrinol 138:391–400

Salehi A, Panagiotidis G, Lundquist I (1993) The alpha-glucosidase dehydrolase inhibitor emiglitate induces differential effects on insulin release elicited by glucose and cholinergic stimulation. Med Sci Res 21:87–90

Samulitis BK, Goda T, Lee SM, Koldovsky O (1987) Inhibitory mechanism of acarbose and 1-deoxynojiirimycin derivatives on carbohydrases in rat small intestine. Drugs Exptl Clin Res 13:517–524

Schmidt DD, Frommer W, Junge B, Müller L, Wingender W, Truscheit E (1977) α-Glucosidase inhibitors. New complex oligosaccharides of microbial origin. Naturwissenschaften 64:535–536

Schmidt DD, Puls W (1971) German Offenlegungsschrift DE 2003934

Shimoyama T, Hori S, Tamura K, Tanida N, Hosomi M, Wada M (1982) Effects of acarbose on faecal microflora of hyperlipidaemic patients. In: Creutzfeldt W (ed) First international symposium on acarbose. Excerpta Medica, Amsterdam, pp 123–136

Sima AA, Chakrabati S (1992) Long-term suppression of postprandial hyperglycaemia with acarbose retards the development of neuropathies in the BB/W rat. Diabetologia 35:325–330

Takami K, Odaka H, Tsukada R, Matsuo T (1991) Antidiabetic actions of a disaccharidase inhibitor, AO-128, in spontaneously diabetic (GK) rats. Pharmacol Ther 19:161–171

Taylor RH, Barker HM (1982) Reversibility of sucrase inhibition by acarbose in vivo perfusion of the rat jejunum. Gut 23:A913

Taylor RH, Barker HM, Canfield JE (1984) Sucrose impairs recovery of jejunal maltase activity after inhibition by acarbose. Clin Sci 67 [Suppl 9]:37P

Taylor R, Barker HM, Bowey EA, Canfield JE (1985) Inhibition of dietary carbohydrate absorption by two new α-glucosidase inhibitors and acarbose. Abstracts XII Congress of the International Diabetes Federation, [Suppl 1]:1446

Thomas G, Keup U, Krause HP, Puls W (1982) Pharmacological studies on acarbose. II: Antihyperlipaemic effects. In: Creutzfeldt W (ed) First international symposium on acarbose. Excerpta Medica, Amsterdam, pp 151–155

Truscheit E, Hillebrand I, Junge B, Müller L, Puls W, Schmidt D (1988) Microbial α-glucosidase inhibitors: chemistry, biochemistry and therapeutical potential. In: Drug concentration monitoring – microbial alpha-glucosidase inhibitors: plasminogen activators. Springer, Berlin Heidelberg New York, pp 19–99 (Progress in clinical biochemistry and medicine, vol 7)

Truscheit E, Schmidt DD, Arens A, Lange H, Wingender W (1981) Further characterization of new α-amylase inhibitors from wheat flour. In: Berchtold P, Cairella M, Jacobelli A, Silano V (eds) Regulators of intestinal absorption in obesity, diabetes and nutrition, vol II. Societá Editrice Universo, Rome, pp 157–179

Tulp OL, DeBolt SP, Pietrangelo L, Schnitzer-Polokoff R, Abdollahi A, Hes ME, Haugard N (1988a) Effects of low-dose acarbose on glycemia, adiposity, and cholesterolemia in obese and obese non-insulin-dependent diabetic corpulent rats. In: Creutzfeldt W (ed) Acarbose for the treatment of diabetes mellitus. Springer, Berlin Heidelberg New York, p 58

Tulp OL, Stevens C, Barbee O, Apostolou ML, Michaelis OE (1988b) Comparative effects of acarbose on glycemia, weight gain, and serum lipids in adult male and female diabetic Wistar fatty rats. In: Creutzfeldt W (ed) Acarbose for the treatment of diabetes mellitus. Springer, Berlin Heidelberg New York, p 57

Tulp OL, Abdollahi A, Stevens C, Schnitzer-Polokoff R (1991a) The effects of the intestinal glucosidase inhibitor acarbose on cholesterogenesis in corpulent rats. Comp Biochem Physiol 100:763-768

Tulp OL, Sczepesi B, Michaelis OE, DeBouno JF (1991b) The effects of low-dose BYA m 1099 (miglitol) on serum lipids and liver enzyme activity of obese and obese diabetic corpulent rats. Comp Biochem Physiol 99:241-246

Tulp OL, Haugard N, Hess ME (1993) Metabolic responses to acarbose administration in the corpulent rat. In: Vasselli JR, Maggio CA, Scriabine A (eds) α-Glucosidase inhibition: potential use in diabetes. Neva, Branford, pp 61-73

Uttenthal LO, Harris A, Al-Mukhtar MYT, Yeats JC, Ghatei MA (1982) Acarbose: model for studying intestinal adaptation and gut hormone changes in disaccharidase intolerance. Gut 23:442-443

Vasselli JR, Haraczkiewicz E, Greenwood MR (1980) The effect of reduced carbohydrate absorption in growing Zucker fatty rats on food intake, body weight and lipid deposition. Nutr Metab 1:373

Vasselli JR, Haraczkiewicz E, Pi-Sunyer FX (1982) Effects of acarbose (BAY g 5421) on body weight, insulin, and oral glucose and sucrose tolerance in sucrose-consuming rats. Nutr Behav 1:21-32

Vasselli JR, Haraczkiewicz E, Maggio CA, Greenwood MR (1983) Effects of a glucosidase inhibitor (acarbose, BAY g 5421) on the development of obesity and food motivated behavior in Zucker (fa,fa) rats. Pharmacol Biochem Behav 19:85-95

Vasselli RJ, Flory T, Fried SK (1987) Insulin binding and glucose transport in adipocytes of acarbose-treated Zucker lean and obese rats. Int J Obes 11 [Suppl 3]:71-75

Vasselli JR, DeCarr LB, Velazques N (1993) Effects of α-glucosidase inhibition on lipid and lipoprotein metabolism in normal and insulin-deficient rats. In: Vasselli JR, Maggio CA, Scriabine A (eds) α-Glucosidase inhibition: potential use in diabetes. Neva, Branford, pp 125-132

Vedula U, Schnitzer-Polokoff R, Tulp OL (1991) The effect of acarbose on the food intake, weight gain, and adiposity of LA/N-cp rats. Comp Biochem Physiol 100:477-482

Voigt WH, Puls W (1976) Ultrastructural studies on small intestines of rats after acute administration of extremely high doses of BAY g 5421. Pharma Report (Bayer AG) No. 6082

Voigt WH, Puls W, Keup U (1976) Electron microscopic studies on liver-, pancreas- and small intestinal tissue of rats after administration of a glucosidase inhibitor, BAY g 5421. Pharma Report (Bayer AG) No. 6077

Walsh CT, Harnett KM (1986) Intestinal sucrease inhibitors and bile acid absorption in the rat. Fed Proc 45 (Abstract 1081):340

Weser E, Babbit J, Hoban M, Vandeventer A (1984) Interstinal adaptation. Different growth responses to glucose compared with disaccharides in rat small bowel. Clin Res 32:866A

William-Olsson T (1986) Alpha-glucosidase inhibition in obesity. Acta Med Scand Suppl 706:1-39

William-Olsson T, Sjöström L (1986) Effects of alpha-glucosidase inhibition in growing and adult ad libitum-fed animals. J Obes Weight Regul 5:222-234

Yalow RS, Berson SA (1960) Plasma insulin concentrations in nondiabetic and early diabetic subjects. Diabetes 9:254-260

Yamashita K, Sugawara S, Sakaira I (1984) Effect of an α-glucosidase inhibitor, acarbose, on blood glucose and serum lipids in streptozotocin-induced diabetic rats. Horm Metab Res 16:179-182

Yoshikuri Y, Ezure Y, Ayoagi Y, Enomoto H (1988) Inhibition of intestinal alpha-glucosidase and postprandial hyperglycemia by N-substituted moranoline derivatives. J Pharmacobiodyn 11:356-362

Young JC, Treadway JL, Ruderman NB (1993) Effect of acarbose on glucose uptake in skeletal muscle. In: Vasselli JR, Maggio CA, Scriabine A (eds) α-Glucosidase inhibition: potential use in diabetes. Neva, Branford, pp 75–83
Younoszai MK, Schedl PH (1972) Effects of diabetes on intestinal disaccharidase activity. J Lab Clin Med 79:579–586
Zavaroni I, Reaven GM (1981) Inhibition of carbohydrate-induced hypertriglyceridemia by a disaccharidase inhibitor. Metabolism 30:417–420

CHAPTER 18

General Pharmacology of Glucosidase Inhibitors

W. Puls

A. Introduction

Various pharmacological studies have been carried out on α-glucosidase inhibitors using abnormally high doses in animal experiments or excessively high concentrations in in vitro assays in order to detect adverse effects. This chapter briefly reviews the results of these studies.

B. Neuropharmacological Studies

Krause and Hoffmeister (1980a) found that administration of acarbose did not affect orientation motility or spontaneous motility of mice. The incidence of the tonic spasm component of electroshock-induced convulsions of mice was slightly reduced and the tonic spasm component of pentylenetetrazol-induced convulsions was slightly antagonized. In studies by Stoepel and Keup (1975), acarbose exerted no analgesic, cataleptic, sedative or muscle relaxant effects, and the fighting behaviour of mice was not affected by the administration of acarbose. Hoffmeister (1982) reported that acarbose caused no alterations in EEG of rabbit hippocampus, amygdala or frontal cortex, with no modification of EEG induced by pentylenetetrazol. Acarbose had no mydriatic activity in mice (Stoepel and Keup 1975).

In another study by Krause and Hoffmeister (1980b), miglitol did not affect the orientation or spontaneous motility of mice and had no cataleptic effect. A slight anticonvulsive effect of miglitol was seen in electroshock-induced convulsions, but not in pentylenetetrazol-induced convulsions, and Hexobarbital anaesthesia was slightly prolonged.

Jacobi and Neuser (1994) found that emiglitate did not adversely affect locomotor function, balancing and traction ability of mice and exerted no cataleptic, antinociceptive or anticonvulsive properties. The mono- and polysynaptic reflexes in cats were also not modified.

Voglibose, at i.v. doses of up to 100 mg/kg, has been found to produce a very slight decrease in body tone of mice (Irwin test), but not to affect spontaneous motor activity or skeletal muscle coordination of mice or normal body temperature of rats (Kito 1991). This study also showed that voglibose had no anticonvulsant, hypnosis-potentiating or analgesic activity in mice, and no effect on the spontaneous EEG or behaviour in conscious cats, spinal reflex in anaesthetized cats or the neuromuscular junction in isolated rat phrenic nerve-diaphragm preparations.

C. Cardiovascular/Respiratory Studies

The effect of acarbose on haemodynamic parameters has been studied in anaesthetsized dogs (Stoepel 1980). The mean arterial blood pressure, heart rate, cardiac output, rate of increase of intraventricular pressure (dp/dt), end-diastolic pressure, arterial pCO_2 and PO_2, calculated stroke volume and peripheral resistance were not found to be significantly altered by acarbose. In other experiments with isolated guinea pig atria (Stoepel and Keup 1975), no significant inotropic or chronotropic effects were found and in studies with isolated perfused guinea pig hearts acarbose did not display any significant coronary vasodilator activity (Stoepel and Keup 1975). The systolic arterial pressure of experimentally hypertensive rats was not affected. The blood flow in isolated perfused cat hind limb preparations was not altered by intra-arterial injection of acarbose.

Miglitol and emiglitate have been examined with regard to effects on haemodynamic parameters in analogy to acarbose in anaesthetized dogs (Garthoff 1981b). A slight but not significant and not dose-dependent reduction of heart rate was seen after administration of miglitol. None of the other parameters was altered. Garthoff (1982) also found administration of emiglitate to be associated with a slight insignificant increase in the mean arterial pressure, rate of rise of intraventricular pressure (dp/dt) and calculated peripheral resistance.

Voglibose, at an i.v. dose of 10 mg/kg, has been found to bring about transient decreases in blood pressure and increases in mesenteric and renal blood flow in anaeshetized dogs, with no change in heart rate (Kito 1991). Voglibose had no effect on respiration, blood pressure, heart rate or ECG in anaesthetized cats. In isolated guinea pig atria, voglibose had no effect on the spontaneous beating rate of the right atria or the developed tension of electrically paced left atria. The autonomic nervous system of anaesthetized cats was not affected by voglibose.

D. Gastrointestinal Function Studies

Samueli et al. (1984, 1985) found that acarbose did not affect basal or histamine-stimulated gastric acid secretion, did not induce gastric mucosa lesions and did not alter the formation of indomethacin-induced ulcers in rats. Intestinal charcoal transit in mice was not affected by acarbose. Kito (1991) reported that voglibose had no effect on the intestinal transport of charcoal meal in mice or on gastric secretion (volume, pH) in pylorus-ligated rats (Kito 1991).

E. Haematological Studies

Studies on acarbose in male rats (SEUTER 1990) have examined thrombe-
lastography, haematocrit, haemoglobin determinations, platelet aggregation
in addition to sedimentation rate, fibrinogen levels, thrombin and thromboplastin time measurements. With the exception of slightly decreased
sedimentation rate, acarbose did not affect any of these parameters. Miglitol
and emiglitate were found to have no anticoagulant, platelet-aggregation
inhibitory or fibrinolytic effects (SEUTER and PERZBORN 1980, 1981).

F. Antigenic, Antiallergic and Pulmonary Activity

Acarbose has been tested for antiallergic or pseudoallergic effects by
measuring the release of histamine from peritoneal mast cells in vitro. The
inhibitor had no effect in these studies (HAMMOND and COOMBER 1984).
Antigenicity studies with acarbose with and without adjuvant have been
performed in guinea pigs. No antibody production to acarbose was seen in
anaphylaxis, the passive haemagglutination test or the Schultz-Dale reaction.
In an additional study with mice no antigenic effect of acarbose was found
(ATAI et al. 1989). In cultured lung tissue explants prepared from rat
embryos, acarbose caused inhibition of lysosomal acid glucosidase, glycogenolysis and reduced formation of certain phospholipids in surfactant,
but not in residual, fractions. DNA concentration was not affected by
acarbose (BOURBON et al. 1987).

G. Antibacterial, Antimycotic and Antiparasitic Activity

Action against gram-positive, gram-negative bacteria, *Trichophyton men-
tagraphytes*, *Candida albicans*, *Aspergillus fumigatus* and a number of
parasites has been investigated in vitro. Acarbose exerted no effects in these
test models. Weak activity against infection by *Klebsiella* 8055 bacteria was
observed after subcutaneous injection of acarbose to mice (THOMAS 1981).

H. Other Studies

Tests with acarbose on isolated guinea pig trachea have shown no effect on
basal trachea tone or histamine- or leukotriene-induced contractions
(GARDNER 1984). Acetylcholine-, histamine-, barium chloride- or serotonin
(5-HT)-induced contractions of isolated guinea pig ileum were not modified
by acarbose (STOEPEL and KEUP 1975; SAMUELI et al. 1984). Voglibose had
no effect on carrageenin-induced paw edema in rats (KITO 1991). In isolated
smooth muscle preparations, voglibose had no effect on the spontaneous
motility of the rabbit ileum and non-pregnant rat uterus, on agonist-induced

contractions of guinea pig ileum or on KCl-induced tension of guinea pig tracheal muscle (Kito 1991).

Urinary volume and Na^+ and K^+ excretion have been measured in male rats after administration of acarbose (Garthoff 1980a), miglitol (Garthoff 1980b) and emiglitate (Garthoff 1981a). The three glucosidase inhibitors did not affect the urine volume and electrolyte excretion with the exception of the highest dose of acarbose (300 mg/kg body weight), which slightly increased Na^+ excretion. No significant effect of voglibose on urinary volume or excretion of sodium and potassium in rats has been observed (Kito 1991).

Acknowledgement. The author gratefully appreciates the help of Mrs. G. Lion for typing the manuscript and Takeda Chemical Industries for making available a report on the general pharmacology of voglibose.

References

Atai H, Tsuda T, Suzuki S (1989) Antigenicity study of BAY g 5421 in guinea pigs and mice. Precl Rep Cont Inst Exp Anim 15:19–28

Bourbon JR, Doucet E, Rieutort M (1987) Role of α-glucosidase in fetal lung maturation. Biochim Biophys Acta 917:203–210

Gardner PJ (1984) BAY g 5421: General/safety respiratory pharmacology: evaluation in the guinea pig isolated trachea. Pharma Report No. 2893, Bayer AG

Garthoff B (1980a) BAY g 5421 – test of the diuretic action in rats. Pharma Report No. 9112, Bayer AG

Garthoff B (1980b) BAY m 1099 – test of the diuretic action in rats. Pharma Report No. 9360, Bayer AG

Garthoff B (1981a) BAY o 1248 – test of the diuretic action in rats. Pharma Report 10085, Bayer AG

Garthoff B (1981b) BAY m 1099 – effect on the haemodynamics and on the cardiac contractility of anaesthetised dogs after oral administration. Pharma Report No. 9849, Bayer AG

Garthoff B (1982) BAY o 1248 – effect on the haemodynamics and on the cardiac contractility of anaethetised dogs after oral administration. Pharma Report No. 10943, Bayer AG

Hammond MD, Coomber JS (1984) Bay g 5421, general respiratory pharmacology: antiallergic and pseudoallergic activity. Pharma Report, Bayer AG, No. R2812

Hoffmeister F (1982) Influence of acarbose with and without glisoxepide on the spontaneous EEG and the electroencephalogram during an infusion of penetrazole in rabbits. Pharma Report No. 10881, Bayer AG

Jacobi H, Neuser V (1994) BAY o 1248: effect on the central nervous system after oral and intraduodenal administration (mouse, rat and cat). Pharma Report No. TR-10713, Bayer AG

Kito G (1991) General pharmacological studies on AO-128. Jpn Pharm Ther 19: 231–245

Krause HP, Hoffmeister F (1980a) BAY g 5421 – test for central nervous actions. Pharma Report No. 9423, Bayer AG

Krause HP, Hoffmeister F (1980b) BAY m 1099 – test for central nervous actions. Pharma Report No. 9437, Bayer AG

Samueli F, Bonabello A, Grassi A (1984) Bay g 5421 – effect on isolated guinea pig ileum spasm, on stimulated gastric secretion and on experimental gastric ulcers in rats. Pharma Report No. 2859, Bayer AG

Samueli F, Bonabello A, Grassi A (1985) BAY g 5421 –effect on intestinal charcoal transit in mouse, on gastric tolerability and on basal secretion in rat. Pharma Report No. 3164, Bayer AG

Seuter F (1980) BAY g 5421 – blood-pharmacological investigations. Pharma Report No. 9224, Bayer AG
Seuter F, Perzborn E (1980) BAY m 1099 – blood-pharmacological investigations. Pharma Report No. 9523, Bayer AG
Seuter F, Perzborn E (1981) BAY o 1248 – blood-pharmacological investigations. Pharma Report No. 10196, Bayer AG
Stoepel K (1980) BAY g 5421 – effect on the haemodynamics and on the cardiac contractibility of anesthesized dogs after oral administration. Pharma Report No. 9120, Bayer AG
Stoepel K, Keup U (1975) BAY g 5421 – esplanin. Pharmacological screening tests. Pharma Report No. 5721, Bayer AG
Thomas H (1981) BAY g 5421 – results of the test for antibacterial, antimycotic and antiparasitic action. Pharma Report No. 10386, Bayer AG

CHAPTER 19

Pharmacokinetics and Metabolism of Glucosidase Inhibitors

H.P. Krause and H.J. Ahr

A. Introduction

α-Glucosidase inhibitors exert their effects topically by inhibition of the carbohydrate-digesting enzymes at the surface of the intestinal mucosa. The systemic availability of the drugs therefore has no impact on the desired therapeutic effect. Investigations of the systemic pharmacokinetics of the α-glucosidase inhibitors in terms of absorption, plasma concentrations, bioavailability and distribution are necessary for safety evolution rather than for description of the pharmacodynamic action of the drug. Balance and distribution studies have been carried out in animals mainly by use of radiolabelled compounds. Together with toxicological data, the results allow the safety evaluation of these inhibitors (see also Chap. 20, this volume). The pharmacokinetic properties of the α-glucosidase inhibitors differ depending on their chemical structure and are summarized in this chapter. In addition, the aspects of the "pharmacokinetic" behaviour of the inhibitors at the site of action, expressed as onset and duration of action as well as the kinetics of enzyme inhibition, also require consideration according to their pharmacodynamic effects (see Chap. 15, this volume).

B. Acarviosin Derivatives

I. Acarbose

1. Introduction

Acarbose $(O$-4,6-dideoxy-4-[[(1S,4R,5S,6S)-4-5,6-trihydroxy-3-(hydroxymethyl)-2-cyclohexen-1-yl]amino]-α-D-glucopyranosyl-(1→4)-O-α-D-glucopyranosyl-(1→4) glucopyranose, BAY g 5421) is a pseudotetrasaccharide of microbial origin with the structure given in Fig. 1.

The absorption, distribution and excretion of acarbose have been studied mainly with carbon-14-labelled drug in rats (2–200 mg/kg), dogs (2 mg/kg) and healthy male volunteers (200 mg/volunteer) (Ahr et al. 1989a,b). The carbon-14-labelled acarbose used was synthesized by fermentation of an *Actinoplanes* mutant with D-[U-^{14}C]glucose. A product with specific activities

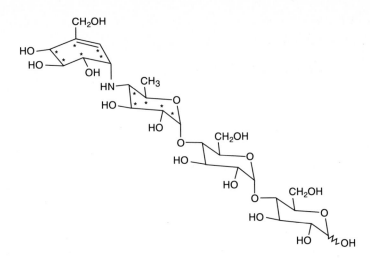

Fig. 1. Chemical structure of acarbose; *position of ^{14}C-label

of 7.77–9.14 MBq/mg and a radiochemical purity of more than 98% was obtained. The labelling position was is the acarviosin part (rings A and B) of the molecule (Maul et al. 1989).

2. Determination Methods

Acarbose is determined by its inhibitory action against sucrase isolated from porcine intestinal mucosa with *p*-nitrophenyl-α-D-maltoheptaoside (NPMH) as substrate (see Chap. 21, this volume). The assay is not completely specific. A metabolite of acarbose (component 2) with a relative inhibitory activity of approximately 14% compared with acarbose may cross-react. The determination of metabolites of acarbose is performed by an ion-pair/high-performance liquid chromatography (HPLC) method on reversed-phase columns with UV detection. Metabolites were detected additionally by off-line scintillation counting following administration of radiolabelled acarbose (Boberg et al. 1990).

3. Absorption

After oral administration, the absorption of unchanged acarbose is low. Less than 2% of the orally administered dose in rats and man and approximately 4% in dogs is absorbed in the active form as concluded from the excretion of inhibitory activity in urine. However, the radioactivity of [^{14}C]acarbose is absorbed to a distinctly higher extent (10%–30% in rats, 8% in dogs, 35% in man). Degradation of [^{14}C]acarbose in the intestinal contents by digestive enzymes and/or microorganisms to inactive but more readily absorbable degradation products is a prerequisite for the absorption of drug-related radioactivity. These reactions take place mainly in the colon and lead to a biphasic absorption profile of drug-related radioactivity (AHR et al. 1989a).

4. Plasma Concentrations

Following intravenous administration, the radioactivity of [^{14}C]acarbose, which represents mainly the unchanged drug, declines very rapidly with at least two distinct phases characterized by half-lives of approximately 0.5 and 24 h in rats and dogs. The plasma clearances, which are identical to the renal clearances following intravenous dosing, are high (0.34 l/h per kilogram in rats and 0.17 l/h per kilogram in dogs) and correspond to the glomerular filtration rate. The volumes of distribution at steady state (V_{ss}) of 0.29 l/kg in rats and 0.22 l/kg in dogs are indicative of a predominantly extracellular distribution (AHR et al. 1989a).

Following oral administration of [^{14}C]acarbose, the plasma concentration versus time profile of radioactivity is strongly influenced by the delayed and biphasic absorption of substance-related radioactivity especially in rats and man. Inhibitory activity (unchanged acarbose and component 2) reaches its peak at 1–2 h. In human volunteers a peak concentration of 49 ng/ml at 2 h is reached following single administration of 200 mg [^{14}C]acarbose. Inhibitory activity is then rapidly eliminated (see Chap. 21, this volume). This lower first maximum is followed by a second maximum of radioactivity, representing inactive metabolites formed in the lower parts of the intestine. Time to reach this maximum is dependent on intestinal transit time, 8 h in rats and 14–24 h in human volunteers. The radioactivity is then eliminated from plasma, with terminal half-lives of 23 h in rats and dogs and 40 h in man. The dose-normalized area under the curve (AUC) values of radioactivity increase from rats (1.7 kg h/l) and dogs (3 kg h/l) to man (9.3 kg h/l). Unchanged acarbose represents only a small fraction of the radioactivity in plasma (1.4% of AUC of radioactivity in humans).

5. Protein Binding

Binding to plasma proteins has been measured with equilibrium dialysis as well as ultrafiltration. At concentrations higher than 1 mg/l, acarbose is virtually not protein-bound in rat and dog plasma. However, there is evidence of a saturable binding to plasma proteins, predominantly albumin, in rats and dogs up to 60%–90% in the lower nanogram per millilitre range. In human plasma there is no binding found at any concentration (AHR et al. 1989a).

6. Distribution

Unchanged acarbose is distributed rapidly and heterogeneously to organs and tissues of rats. The distribution pattern is indicative for a predominantly extracellular distribution. This is underlined by distribution volumes of

0.2–0.3 l/kg in rats and dogs. High concentrations are found in the kidneys and the urine due to the rapid and exclusive renal excretion. Acarbose penetrates the blood-brain barrier only to a negligible extent (AHR et al. 1989b). Following oral administration, high concentrations of radioactivity were found in the kidneys, the liver and the adrenal gland. Follwing repeated oral administration, there was some tendency for accumulation of radioactivity with accumulation factors of 3–9 ($AUC_{0-\tau}$). Terminal elimination half-lives of approximately 10 days were observed. The accumulation is mainly based on incorporation of parts of the radiolabel into the body carbon pool. No specific accumulation or retention of substance-related radioactivity occurs. The radioactivity of [^{14}C]acarbose penetrates the placental barrier slowly and to a small extent, the radioactivity in the fetus being distinctly lower than in the maternal plasma. Acarbose-related radioactivity is secreted into the milk of lactating rats (AHR et al. 1989a).

7. Metabolism

Acarbose administered intravenously is excreted completely unchanged via the renal route and thus is not metabolized systemically. Following oral administration, however, acarbose is cleaved by digestive enzymes and intestinal bacteria. Three major routes of acarbose metabolism in the intestinal tract have been identified (Fig. 2). By the combined effects of intestinal α-glucosidases or the intestinal bacteria, the pseudotetrasaccharide is split (acarbose minus 1 glucose moiety: component 2; acarbose minus 2 glucose: component 1) (PFEFFER and SIEBERT 1986). By these reactions rings C and D are cleaved to produce glucose. The (radiolabelled) ring A is aromatized and reduced to 4-methylpyrogallol by intestinal bacteria. 4-Methylpyrogallol is subsequently absorbed and methylated and/or conjugated with sulphate or glucuronic acid in gut wall or liver (BOBERG et al. 1990). The (radiolabelled) ring B is degraded by the intestinal bacteria to ω-amino carboxylic acids (γ-aminobutyric acid, δ-aminovaleric acid), which are further utilized by the body (incorporation into lipids and poteins).

8. Excretion

Unchanged acarbose is rapidly and completely excreted via urine. There is no relevant metabolization of acarbose in animal or man, as concluded from data following intravenous administration. The renal clearance corresponds to the glomerular filtration rate in rats, dogs and man. After oral administration a non-absorbed fraction of about 60%–80% of the dose was excreted with the faeces. Acarbose and its metabolites formed in the gut and after subsequent absorption were excreted rapidly and almost completely via the renal route. The minor excretion via bile in rats (<5% of the dose) excludes the existence of a relevant enterohepatic circulation. In rats following oral administration of [^{14}C]acarbose, approximately 5% of the dose was exhaled as $^{14}CO_2$ (AHR et al. 1989a).

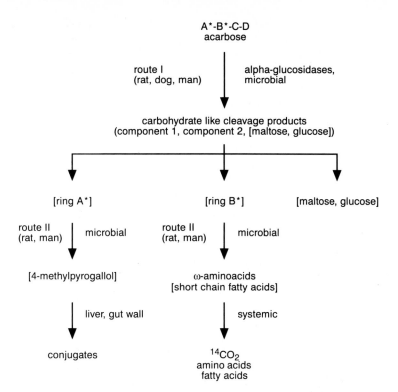

Fig. 2. Acarbose: scheme of metabolic degradation; *carbon 14-labelled; [] not isolated

C. Valiolamine Derivatives

I. AO-128 (Voglibose)

1. Introduction

AO-128 (voglibose, $(+)$-1_L-[1(OH),2,4,5/3]-5-[2-hydroxy-1-(hydroxymethyl) ethyl] amino-1-C-(hydroxymethyl)-1,2,3,4-cyclohexane tetrol) is an α-glucosidase inhibitor from the valiolamine group isolated from *Streptomyces* culture broths. Pharmacokinetic investigations have been carried out using the radiolabelled material (^{14}C-label in the dihydroxypropyl group, specific radioactivity, 2.5–14 MBq/mg, radiochemical purity 97%) (MAESHIBA et al. 1991).

2. Determination Methods

AO-128 was determined in extracts of biological samples after separation of the radiolabelled material by thin-layer chromatography (TLC) on silica gel plates. The zones containing AO-128 were identified on the TLC plates plates using potassium permanganate oxidation and autoradiography on X-ray film or imaging plate. The scraped-off fractions were quantified by liquid scintillation counting. The limit of detection in plasma was 0.5–1 ng/ml; in urine and faeces 0.1% of the dose could be quantified. Total radioactivity was measured by conventional radiometric methods. Whole-body autoradiography was performed using the imaging plate technique.

3. Absorption

The absorption of AO-128 is low. As calculated from the urinary excretion of radioactivity after oral and intravenous administration of $[^{14}C]AO$-128 (1 mg/kg), the fraction absorbed amounts to 5.9% in rats and 2.7% in dogs. From the plasma concentrations of total radioactivity reached after administration of 1, 10 and 100 mg/kg, it can be concluded that the absorption is additionally slightly reduced after the administration of higher doses. Data on the systemic availability of unchanged AO-128 have not been reported.

4. Plasma Concentratons of Total Radioactivity

One hour after oral administration of $[^{14}C]AO$-128 (1 mg/kg) to rats, maximum equivalent concentrations of 17 ng/ml are reached. In dogs, C_{max} values of 80 ng/ml are reachedr after 4 h. The AUC in dogs is about 20 times larger than in rats. The terminal half-life of plasma radioactivity amounts to 10 h in rats and 16 h in dogs.

5. Plasma Concentrations of Unchanged AO-128

The plasma concentrations of the parent compound are only slightly lower than those of total radioactivity: the unchanged drug represents 91% of the plasma radioactivity AUC in rats and 93% in dogs. The terminal half-lives of parent compound and total radioactivity are identical in both species.

6. Distribution

Whole-body autoradiography after intravenous administration of $[^{14}C]AO$-128 to rats reveals a rapid distribution to the tissues, with highest concentrations in kidneys, lung and liver. Due to the low absorption, after oral administration systemic radioactivity could only be detected in the kidneys. Quantitative investigations confirmed this distribution pattern. Maximum concentrations are reached in the tissues within the 1st h after oral administration, with highest values in the intestinal wall and kidneys. The penetration of the blood-brain barrier as well as the erythrocyte membrane is low. Binding of AO-128 to rat and human plasma proteins is low (<15%). In contrast, in dog plasma, 90% protein binding is found at 5 ng/ml, which decreases with increasing concentrations (23% at 500 ng/ml). In pregnant rats, the radioactivity of AO-128 passes the placental barrier only to a limited extent, the fetal concentrations always being below the maternal plasma concentrations. Also, the low secretion of radioactivity into the milk of lactating rats results in concentrations lower than the plasma concentrations.

7. Metabolism

AO-128 is hardly metabolized at all. About 90% of the radioactivity excreted by rats and dogs after oral administration represents unchanged drug. No metabolites have been identified.

8. Excretion

After oral administration of [^{14}C]AO-128, the radioactivity is rapidly and completely excreted via the faeces. The renal excretion (mainly unchanged drug) amounts to less than 5% of the dose in rats and dogs.

D. Deoxynojirimycin Derivatives

I. Miglitol

1. Introduction

Miglitol (1,5-dideoxy-1,5-[2-hydroxyethylimino]-D-glucitol, BAY m 1099) is a deoxynojirimycin derivative with the structure given in Fig. 3.
    The absorption, distribution and excretion have been studies mainly with the tritium – or carbon-14 – labelled drug in rats (1–25 mg/kg), dogs (2–450 mg/kg) and male healthy volunteers (100 mg/volunteer) (AHR et al., in preparation). Tritium labelling was performed in the 1-position of the deoxynojirimycin by hydrogenation of the nojirimycin bisulphite adduct by [^{3}H]borohydride. The specific activity was 2.5 MBq/mg at chemical and radiochemical purities of >99%. Miglitol was labelled with carbon-14 in the hydroxyethyl side chain by condensation of [1,2-^{14}C]ethylene oxide with deoxynojirimycin at specific activities of 1.10–2.36 MBq/mg and radio-chemical purities of >98.2%.

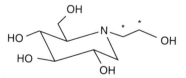

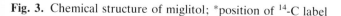

Fig. 3. Chemical structure of miglitol; *position of ^{14}-C label

2. Determination Methods

Miglitol is determined by its inhibitory action against sucrase isolated from porcine intestinal mucosa with p-nitrophenyl α-D-glycopyranoside (NPG) as substrate (see Chap. 21, this volume). The assay is specific for miglitol.

3. Absorption

Following oral administration of miglitol to rats and dogs, the absorption is rapid and complete at doses below 5 mg/kg. At higher doses, the absorption rate and the extent of absorption decreases, indicating a dose/concentration-dependent saturation of the absorption of this hydrophilic compound. In contrast, in humans even at lower doses (<0.7 mg/kg), the extent of absorption is not complete. The absorption mechanism has been investigated in a study with isolated mucosa of guinea pigs. Miglitol is absorbed from the intestine via diffusion through extracellular shunts and even more effectively by transcellular abvsorption via at least one active transport mechanism localized in the jejunum. This transport mechanism is in part identical to the absorption mechanism of glucose (Steinke 1986).

4. Plasma Concentrations

Following intravenous administration, the radioactivity of miglitol, which is representative for the unchanged drug, declines very rapidly, with at least two distinct phases characterized by half-lives of 0.4 (rats) to 1.8 h (man) and 2–4 h, respectively (see Chap. 21, this volume). The total plasma clearances which are identical to the renal clearances are in the range of the glomerular filtration rates in dogs and humans and higher in rats [0.88 l/h per kilogram (rats), 0.18 l/h per kilogram (dogs), 0.10 l/h per kilogram (male volunteers)]. The volumes of distribution at the steady state of 0.8 l/kg (rat), 0.5 l/kg (dog), and 0.3 l/kg (man) indicate a low tissue affinity of miglitol.

Following oral administration of radiolabelled miglitol, peak concentrations of 1.5 mg/l (rats, 2 mg/kg), 2.2 mg/l (dogs, 2 mg/kg) and 1.6 mg/l (humans, 1.4 mg/kg) are reached rapidly in rats (t_{max}, 0.7 h) and more slowly in dogs and man (t_{max}, 2–3 h). After reaching the maximum, the elimination of miglitol is comparable to that after intravenous dosing. The dose-normalized AUC values increase from rat (0.88 kg h/l) to dog (4.32 kg h/l) and man (5.76 kg h/l). At higher doses (≥ 25 mg/kg in rats, ≥ 60 mg/kg in dogs), the dose-normalized peak concentrations and AUC values decrease, indicating a saturable absorption in both species (Ahr et al., in preparation).

5. Protein Binding

Miglitol is freely ultrafiltrable and dialyzable in equilibrium dialysis. Miglitol is not bound to plasma proteins in rat, dog and human plasma.

6. Distribution

Miglitol is distributed rapidly and heterogeneously to organs and tissues of rats. The distribution pattern is indicative of a predominantly extracellular distribution. Very high concentrations in kidney and urine indicate a rapid and exclusively renal excretion of miglitol. Miglitol penetrates the blood-brain barrier only to a negligible extent. The substance-related radioactivity is very rapidly eliminated from organs and tissues. Following oral administration, a high concentration of radioactivity in or at the mucosa of the gastrointestinal tract has been observed in autoradiographic studies for a relatively long period, which corresponds well to the mode of action of miglitol. Otherwise, no specific accumulation or retention of substance-related radioactivity occurs. Following repeated administration of [^{14}C]miglitol to rats, only a slight tendency for accumulation with accumulation factors <4 ($AUC_{o-\tau}$) has been observed. Terminal elimination half-lives of radioactivity of 50–110 h were found. In pregnant rats, [^{14}C]miglitol permeates the placental barrier slowly and to a limited extent. The distribution of radioactivity in the fetuses is, on a distinctly lower concentration level, identical to that in the maternal body. The elimination from the fetuses is similarly rapid to that in the maternal animal. In lactating rats, miglitol is transferred into the milk, reaching concentrations similar to or lower than those in the maternal plasma.

7. Metabolism

Miglitol is excreted quantitatively unchanged via urine both after intravenous and oral administration. No biotransformation of miglitol has been observed in rats, dogs and humans.

8. Excretion

Miglitol is excreted almost exclusively via the renal route. Hence, the amount of miglitol recovered in urine can be considered as a direct measure for the absorbed amount. At low doses in rats and dogs, only minor amounts of miglitol (<1%) have been recovered in faeces. In human volunteers following oral administration of 100 mg miglitol, approximately 30% of the dose is found in faeces, indicating a limited absorption, at this dose. Biliary excretion as tested in rats is insignificant ($\leq 0.16\%$ of the dose).

II. Emiglitate

1. Introduction

Emiglitate   (1,5-dideoxy-1,5-[2-[4-ethoxycarbonylphenoxy]ethylimino)-D-glucitol, BAY o 1248) is a deoxynojirimycin derivative with the structure given in Fig. 4. Its absorption, distribution, metabolism and excretion have been studied with the tritium-labelled drug (1–25 mg/kg) in rats (SUWELACK

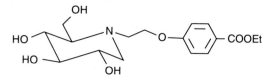

Fig. 4. Chemical structure of emiglitate

and Weber 1983, unpublished data, Bayer). Tritium labelling was performed in the 1-position of the deoxynojirimycin by hydrogenation of the nojirimycin bisulphite adduct by [³H]borohydride. The specific activity was 1.66 MBq/mg at a chemical and radiochemical purity of >99%.

2. Absorption

Following oral administration of emiglitate to rats, the absorption is rapid and complete (≥90% of the dose). No hints on saturation of the absorption of emiglitate have been found at doses between 1 and 25 mg/kg in rats.

3. Plasma Concentrations

Following intravenous administration, the radioactivity of emiglitate declines initially rapidly with a half-life of approximately 1 h. The total plasma clearance of radioactivity is 0.25 l/h per kilogram. Following oral administration of radiolabelled emiglitate, dose-normalized peak concentrations of 0.4 mg/l are reached within 1–2 h. After reaching the maximum concentration, the elimination of [³H]emiglitate occurs similar to that after intravenous administration. The dose-normalized AUC values of radioactivity in rats were approximately 1.7 kg h/l independent of dose between 1 and 25 mg/kg. Data on systemic bioavailability are not available.

4. Distribution

The radioactivity is distributed rapidly and heterogeneously to organs and tissues of rats. Very high concentrations in kidney and urine indicate the rapid and predominantly renal excretion. Biliary excretion is indicated by radioactivity detected in bile ducts and intestinal lumen even after intravenous administration. The distribution pattern indicates a low tissue affinity for the emiglitate-related radioactivity. The blood-brain barrier penetration of [³H]emiglitate is extremely low. The substance-related radioactivity is very rapidly eliminated from the organs and tissues. At 48 h after oral dosing of 20 mg/kg, residual radioactivity is only detected in or at the intestinal mucosa, which corresponds well to the mode of action of emiglitate.

5. Metabolism

The ethylester group of emiglitate is cleaved in the rat to yield the also pharmacologically active free acid, which is predominantly excreted via urine (approximately 80% of the amount excreted renally). Approximately 15% of the radioactivity in urine represents the unchanged drug.

6. Excretion

[^{3}H]emiglitate and its radioactive metabolites are predominantly excreted via the renal route (80%–90% of the dose). Approximately 10% of the dose has been recovered in the faeces after both intravenous and oral administration to rats, indicating a certain contribution of biliary excretion. The amount recovered in bile from bile duct cannulated rat was 11% of the dose.

III. Other Deoxynojirimycin Derivatives

1. 1-Deoxymannojirimycin

Deoxymannojirimycin (dMM; 1,5-dideoxy-1,5-imino-D-mannitol) is a synthetic derivative of deoxynojirimycin, inhibiting mainly mannosidase. Tritiated dMM [1-^{3}H]dMM (specific radioactivity 25 Ci/mmol) has been used for pharmacokinetic investigations. Radioactivity was determined, partly after separation on TLC plates, by liquid scintillation counting and TLC scanning (FABER et al. 1992). After intravenous administration to anaesthetized male Wistar rats, dMM was rapidly eliminated from plasma with a terminal half-life of 50 min. Protein binding (ultrafiltration method) in rat plasma amounted to 5.1%. The steady state volume of distribution (265 ml) was close to total body volume. However, distribution was heterogeneous with high concentrations, e.g., in the kidneys. dMM was described to be metabolized by rat liver only to a minor extent (<5% of the dose). Detailed data were not given. dMM was rapidly excreted, mainly via the renal route. The renal clearance of dMM was similar to the glomerular filtration rate, indicating the absence of active tubular secretion. After 2 h (at the end of the experimental period), 52% of the dose was excreted in the urine. Another 4.9% was recovered in the bile. Preliminary data were reported which demonstrate a slower elimination process, resulting in an excretion of another 19% and 1.4% of the dose via urine and bile, respectively. Blockade of the renal excretion pathway by ligation of renal vessels resulted in nearly complete cessation of the elimination and in markedly increased concentrations of dMM in plasma and bile. The total amount excreted via bile, however, was not increased, most probably due to the reduced bile flow induced by the surgical procedure.

In summary, intravenously administered dMM is rapidly eliminated from rat plasma nearly exclusively via the renal route, without significant metabolization. Renal insufficiency in this model did not result in an increased biliary excretion of dMM.

2. N-Methyl-1-deoxynojirimycin

N-Methyl-deoxynojirinmycin (MedNM, N-methyl-1,5-dideoxy-1,5-imino-D-glucitol) is another glucusidase inhibitor derived from deoxynojirimycin. N-[^{14}C]methyl deoxynojirinmycin ([^{14}C]MedNM, specific radioactivity 15 mCi/mmol, radiochemical purity 98%) has been used for investigation of the pharmacokinetics of MedNM. Radiometric methods partly combined with TLC separation were used for quantification. The more polar compound MedNM after intravenous administration to rats was eliminated from plasma more rapidly than dMM ($t_{1/2}$, 32 min) (FABER et al. 1992). Protein binding in rat plasma was negligible. The volume of distribution in steady state was lower than total body volume. Generally, tissue concentrations of MedNM were low. High concentrations could be demonstrated in the kidneys, indicating the important excretory function of this organ for MedNM. Within 2 h, 80% of the dose was excreted via urine as unchanged MedNM. Biliary excretion was negligible (0.2% of the dose). Even when renal excretion was drastically reduced by ligation of the renal vessels, the biliary excretion remained extremely low (0.5% of the dose). Renal excretion of MedNM was not influenced by increased urine pH. However, in contrast to dMM, the renal clearance of MedNM was at least twice the glomerular filtration rate, indicating that active tubular secretion may be involved.

Summarizing, MedNM was rapidly eliminated from rat plasma via the renal route without systemic metabolization. Tubular secretion may contribute to a renal clearance higher than the glomerular filtration rate.

E. Comparative Discussion

Comprehensive data on preclinical pharmacokinetics are only available for a few α-glucosidase inhibitors either on the market or in late phases of development (acarbose, voglibose, miglitol, emiglitate). No pharmacokinetic information has been published about other glucosidase inhibitors such as castanospermin, MDL25637, MDL73945 or HOE467A. Although similar in their extrasystemic mode of action towards the carbohydrate-digesting enzymes at the surface of the intestinal mucosa, the α-glucosidase inhibitors differ in their chemical structure, their intrinsic activity and also in their pharmacokinetic properties (Table 1).

The most striking difference is observed for the absorption of the different drugs. Neither acarbose nor voglibose are absorbed in their active form to a relevant extent, both drugs thus being confined to their desired site of action. In contrast, the second-generation α-glucosidase inhibitors miglitol and emiglitate are readily and almost completely absorbed at low doses. Miglitol shows a concentration/dose-dependent saturation of absorption based on an active transport process across the intestinal mucosa. It is tempting to speculate that these drugs are mainly active in the upper parts of the small intestine, where most of the carbohydrate digestion occurs. In

Table 1. Summary of pharmacokinetic properties of several glucosidase inhibitors

Drug	Acarbose (BAY g 5421)	Voglibose (AO 128)	Miglitol (BAY m 1099)	Emiglitate (BAY o 1248)
Extent of absorption	Low	Low, dose dependent	High, dose dependent	High
Unchanged drug	<2%	<6%	>96%	>90%
Metabolites	<35%	–	–	–
Bioavailability	<2%	<6%	>96%	–
Clearance	Mainly renal by glomerular filtration	Mainly renal	Mainly renal by glomerular filtration	Mainly metabolic
Protein binding	Low to high Species-dependent Saturable	Low to high Species-dependent Saturable	Low	–
Distribution	Extracellular Low tissue affinity	Low tissue affinity	Extracellular Low tissue affinity	Extracellular Low tissue affinity
Metabolism	Extrasystemic in the intestine	None	None	Ester cleavage (>70% of the dose)
Excretion				
Faecal	>65%	Almost complete	Low	Approx. 10%
Renal	<35%	<5%	>96%	80%–90%
Biliary	<5%	–	<0.2%a	11%

addition, for miglitol and emiglitate a long-lasting presence of substance-related radioactivity at or in the mucosa of the small intestine has been observed in autoradiographic studies, which may be based on the high affinity to the digestive enzymes localized there.

Once absorbed the four drugs exhibit a similar pharmacokinetic behaviour. Their tissue affinity is relatively low, as shown by the low volumes of distribution and a predominantly extracellular distribution pattern in whole-body autoradiography. High concentrations are mainly found in the kidney according to the very predominant renal excretion. The drugs do not penetrate the blood-brain barrier. Protein binding of acarbose, voglibose and miglitol is negligible at high concentrations. In some animal species higher, but saturable binding for acarbose and voglibose was observed at low concentrations. The elimination of these drugs occurs very rapidly. Acarbose, voglibose and miglitol are stable against systemic metabolic attack. They are excreted unchanged via the renal route, probably only via glomerular filtration as shown by the similarity of their clearances to the glomerular filtration rate. For an experimental deoxynojirimycin derivative (N-methyl-deoxynojirimycin), an additional contribution of tubular secretion has been described (Faber et al. 1992). In contrast, for emiglitate one biotransformation step is easily possible based on its ethyl ester structure. The resulting carboxylic acid, however, is also stable against further metabolic attack. The excretion of acarbose and voglibose was mainly faecal according to their low absorbed fraction. Miglitol and emiglitate as well as the absorbed fraction of acarbose and voglibose are excreted rapidly and predominantly via the renal route.

References

Ahr HJ, Boberg M, Krause HP, Maul W, Müller FO, Ploschke HJ, Weber H, Wünsche C (1989a) Pharmacokinetics of acarbose. Part I: absorption, concentration in plasma, metabolism and excretion after single administration of [^{14}C]acarbose to rats, dogs, and man. Arzneimittelforschung 39:1254–1260

Ahr HJ, Krause HP, Siefert HM, Steinke W, Weber H (1989b) Pharmacokinetics of acarbose. Part II: distribution to and elimination from tissues and organs following single repeated administration of [^{14}C]acarbose to rats and dogs. Arzneimittelforschung 39:1261–1267

Ahr HJ, Boberg M, Brendel E, Krause HP, Müller FO, Steinke W (to be published) Pharmacokinetics of miglitol. Absorption, distribution, metabolism, and excretion following administration of tritium of carbon-14 labelled miglitol to rats, dogs and man. Arzneimittelforschung

Boberg M, Kurz J, Ploschke HJ, Schmitt P, Scholl H, Schüller M, Wünsche C (1990) Isolation and structural elucidation of biotransformation products from acarbose. Arzneimittel-forschung 40:555–563

Faber ED, Oosting R, Neefjes JJ, Ploegh HL, Meijer DKF (1992) Distribution and elimination of the glucosidase inhibitors 1-deoxymannojirimycin and N-methyl-1 deoxynojirimycin in the rat in vivo. Pharm Res 9:1442–1450

Maeshiba Y, Kondo T, Kobayashi T, Tanayama H (1991) Metabolic fate of AO-128, a new alpha glucosidase inhibitor, in rats and dogs. Yakuri to Chirio 19:3639–3649

Maul W, Müller L, Pfitzner J, Rauenbusch E, Schutt H (1989) Radiosynthesis of [^{14}C]acarbose. Arzneimittelforschung 39:1251–1253

Pfeffer M, Siebert G (1986) Prefeeding-dependent anaerobic metabolization of xenobiotics by intestinal bacteria – methods for acarbose metabolites in an artificial colon. Z Ernährungswiss 25:189–195

Steinke W (1986) Intestinal absorption mechanism of the glucosidase inhibitor BAY m 1099. Naunyn Schmiedebergs Arch Pharmacol 332:R41

CHAPTER 20

Toxicology of Glucosidase Inhibitors

E.M. Bomhard

A. Acarbose

A large number of toxicity studies have been performed on the general and reproduction toxicology, genotoxicity and carcinogenicity of acarbose. Besides the usual guideline studies required by the international authorities for registration, several special studies have also been initiated to clarify the effects that were observed in the original long-term study of acarbose in rats and which were of questionable relevance for the clinical situation. Some of these results have been published in previous articles (Schlüter 1982a, 1988).

I. General Toxicology

1. Acute Toxicity

Results of the acute toxicity studies performed in mice, rats and dogs via the oral and intravenous route are summarized in Table 1. Based on these figures, acarbose can be characterized as non-toxic after single oral administration. Soft faeces and reduced spontaneous movement were observed a few days or hours after administration in male and female rats, respectively, treated with 20 000 mg/kg body weight. Toxicity was also very low after intravenous administration. While laboured breathing and slight exophthalmus lasting up to 2 min postadministration were the ony signs in mice at a dose of 7680 mg/kg, male and female rats exhibited narrowed eyelids, rough fur, laboured breathing, paleness and cyanosis starting immediately after administration of 3846 mg/kg or more. After 48 h these symptoms disappeared (P. Mürmann 1974, 1979, unpublished data, Bayer; Kamikawa et al. 1989a).

2. Subacute Intravenous Toxicity

To assess the risk of parenteral administration of acarbose to humans, subacute studies with intravenous administration to rats and dogs have been performed as part of an investigation of the pharmacokinetics of this drug.

Table 1. Acute toxicity of acarbose

Species	Strain	Sex	Route	LD_{50} (mg/kg)	Reference
Mouse	CF_1	M	Oral	>15360	P. Mürmann 1974, unpublished data, Bayer
Rat	Wistar	M	Oral	>15360	P. Mürmann 1974, unpublished data, Bayer
Rat	Wistar	M + F	Oral	>20000	Kamikawa et al. 1989a
Dog	Beagle	M + F	Oral	>10000	P. Mürmann 1974, unpublished data, Bayer
Mouse	CF_1	M	i.v.	>7680	P. Mürmann 1974, unpublished data, Bayer
Rat	Wistar	M	i.v.	7362[a]	P. Mürmann 1979 unpublished data, Bayer
	Wistar	F	i.v.	5534[a]	P. Mürmann 1979 unpublished data, Bayer
Rat	Wistar	M + F	i.v.	>6000	Kamikawa et al. 1989a
Dog	Beagle	M + F	i.v.	>3846	Mürmann 1979 unpublished data, Bayer

[a] Confidence intervals for $P \leq 0.05$: 6481–8400 mg/kg for male rats, 4402–6522 mg/kg for female rats.

a) Rat

Groups of ten male and ten female Wistar rats were each administered dose levels of 0, 10, 50 or 250 mg/kg once daily for 14 consecutive days by intravenous injection. The 10-mg/kg dose level was regarded as the no-adverse-effect-level. At higher dose levels erythrocyte counts, haematocrit values and haemoglobin concentration were reduced, and reticulocyte counts and spleen weights were increased (W. Flucke and G. Luckhaus 1979, unpublished data, Bayer).

b) Dog

Groups of two male and two female beagle dogs were administered dose levels of 0, 10, 50 or 250 mg/kg once daily for 18 consecutive days. The 10-mg/kg dose level was tolerated without adverse effects. At higher levels, decreases in body weight and thrombocyte counts in peripheral blood were measured (G. Schlüter and G. Luckhaus 1979, unpublished data, Bayer).

3. Subchronic Oral Toxicity

a) Rat

Groups of 15 male and 15 female Wistar rats each received acarbose at dose levels of 0, 45, 150 or 450 mg/kg body weight once daily for 3 months by gavage. Doses up to and including 450 mg/kg were tolerated with no damage.

A slightly reduced food efficiency in male rats of all treated groups as well as in females at 450 mg/kg was explained by the pharmacodynamic activity (E. LÖSER and G. LUCKHAUS 1975, unpublished data, Bayer).

In a further subchronic toxicity study, 12 male and 12 female Wistar rats per group were administered dose levels of 50, 200 or 800 mg/kg, once daily for 3 months by gavage. In order to study recovery, six extra males and six extra females were allocated to the control, 200 and 800 mg/kg groups. The recovery period lasted 5 weeks. The no-adverse-effect level in this study was 200 mg/kg. At 800 mg/kg, food consumption was increased and serum amylase activity decreased in both males and females, serum alanine amino-transferase increased in females and relative caecum weight increased in males. All changes were reversible (KAMIKAWA et al. 1989b).

b) Dog

Groups of three male and three female beagle dogs were administered daily doses of 0, 50, 150 or 450 mg/kg over a 3-month period. The compound was administered in gelatin capsules 4–6 h before feed supply. This treatment resulted essentially in retarded growth, decreased food efficiency, a slight increase in blood urea nitrogen (within the normal range), decreased α-amylase activities in serum and pancreas, increased absolute and relative pancreas weights and decreased absolute and relative thymus weights at all dose levels. These changes can be interpreted as being the consequences of exaggerated pharmacodynamic effects and do not represent toxic effects (L. MACHEMER et al. 1975, unpublished data, Bayer).

4. Chronic Toxicity

a) Rat

For the study of chronic toxicity in combination with carcinogenicity, ten male and ten female satellite Sprague-Dawley rats were included in a 2-year experiment. They were put to death after 12 months. The animals received 0, 500, 1500 or 4500 ppm acarbose mixed with their diet. A distinctly and dose dependently reduced weight gain was observed in all dose groups, which was more pronounced in males, while feed intake increased simul-taneously. These effects, which were not observed in the subchronic gavage study, occurred because drug administration coincided with the intake of food. In this situation, the glucosidase-inhibiting properties prevent car-bohydrates from being split into monosaccharides and thus from being absorbed, thereby leading to considerable malnutrition with a loss of the isocaloric state. Haematology and histopathology revealed no substance-related changes. Increased plasma activity of alanine aminotransferase (ALT) and aspartate aminotransferase (AST) after 4500 ppm and dose dependently in alkaline phosphatase was also attributed to malnutrition and not to a substance-related hepatotoxicity. Also observed was a dose-dependent trend towards increased plasma urea nitrogen and at 4500 ppm a decrease in

α-amylase activity in the serum, both of which were indirect consequences of the pharmacodynamic activity (W. Neumann and F. Leuschner 1986, unpublished data, Bayer).

b) Dog

Four male and four female beagle dogs were treated with single daily doses of 0, 50, 100, 200 or 400 mg/kg by gavage for a period of 12 months. Again, a pronounced effect on body weight development occurred. Particularly the animals in the higher dose groups lost weight to the point of cachexia during the first 3 months of the experiment. One animal in the 400-mg/kg dose group was completely emaciated and died in the 12th week. All other test parameters, including histopathology, revealed no toxic effects (W. Neumann and F. Leuschner 1982, unpublished data, Bayer).

II. Reproduction Toxicology

1. Fertility and General Reproductive Performance in Rats

Acarbose was administered by gavage to Wistar rats at doses of 0, 60, 180 or 540 mg/kg per day; 24 male and 60 female rats per group were treated for 10 (males) or 3 (females) weeks prior to and during mating through gestation day 7. Whereas the litters of about half of the females were delivered by caesarean section on day 20 of gestation, the remaining dams were allowed to give birth normally and raise their pups. There was no influence on fertility and general reproductive performance, or on intrauterine development, birth process, postnatal development or fertility performances of the F_1 generation at any of the dose levels tested (G. Schlüter 1982, unpublished data, Bayer).

2. Embryotoxicity/Teratogenicity in Rats

Groups of 20–24 pregnant BAY:FB30 (Long Evans derived) rats were administered acarbose from day 6 to day 15 of gestation by gavage at dose levels of 0, 48, 160 or 480 mg/kg. There were no effects on general behaviour, appearance, growth and mortality of the dams as well as no indications of a embryotoxic or teratogenic effect (L. Machemer 1975 unpublished data, Bayer). In a supplementary study, acarbose was administered to 15 inseminated BAY:FB30 rats in doses of 0 or 480 mg/kg body weight on days 6–15 of gestation. The dams reared their pups, which were investigated with respect to their peri- and postnatal development. There were no effects on the dams nor on the peri- and postnatal development of their pups (M. Renhof 1989, unpublished data, Bayer).

3. Peri-/Postnatal Study in Rats

Groups of 50 inseminated female Wistar rats were administered acarbose once daily from day 16 of gestation to day 20 of gestation (Caesarean section group) or to the end of the weaning period (day 22 postpartum) at doses of 0, 60, 180 or 540 mg/kg. Half of the animals underwent caesarean section on day 20; the remainder gave birth normally and raised their pups. In the latter case, a series of function tests including fertility was carried out on the F_1 generation. There were no adverse effects on dams and on the peri- or postnatal development of the pups (G. SCHLÜTER 1982, unpublished data, Bayer).

4. Embryotoxicity/Teratogenicity in Rabbits

Groups of 12 pregnant Himalayan rabbits were administered dose levels of 0, 48, 160 or 480 mg/kg once daily from day 6 to day 18 of gestation by gavage. Caesarean section was on day 29. A dose dependently reduced body weight gain was observed at dose levels of 160 and 480 mg/kg, which was reversible after cessation of treatment. There were no effects on embryos up to and including a dose of 160 mg/kg. A slightly increased rate of resorption and, as a consequence, a reduced average number of fetuses were found at 480 mg/kg and attributed to the maternal effect. There was no increase in the incidence of malformation (L. MACHEMER 1975, unpublished data, Bayer).

III. Genotoxicity/Mutagenicity

An extensive series of tests was performed to cover the three end points of genotoxicity/mutagenicity, namely point mutations, chromosome aberrations and DNA binding/damage. The test systems used, the dose/concentration ranges tested and the results are compiled in Table 2. In summary, in none of these tests were there any indications of a genotoxic/mutagenic potential for acarbose.

IV. Carcinogenicity

1. Rat

a) Two-Year Feeding Study in Sprague-Dawley Rats

Groups of 50 male and 50 female Sprague-Dawley rats were treated with dietary concentrations of 0, 500, 1500 and 4500 ppm acarbose for 2 years. As in the satellite groups (see Sect. I.4.a), there was a marked growth retardation lasting until the end of the experiment. There was a 20%–30% dose-dependent reduction in body weight, and a reduction in body weight gain of up to 50% compared with controls (Fig. 1). The well-known observation that feeding rats restrictively or on low-calorie diets is accompanied by a reduction in certain types of tumours is shown well in the results obtained in this study (Table 3). The total number of pituitary tumours, mammary gland adenomas and carcinomas, liver sarcomas and leukaemias was markedly less

Table 2. Mutagenicity/genotoxicity studies with acarbose

Test system	Dose/concentration	S9 mix	Result	Reference[a]
Point mutation assays				
1. Test compound				
Salmonella typhimurium TA 98	4–2500 µg/plate	±	Negative	B. Herbold 1979
TA 100	4–2500 µg/plate	±	Negative	B. Herbold 1979
TA 1535	4–2500 µg/plate	±	Negative	B. Herbold 1979
TA 1537	4–2500 µg/plate	±	Negative	B. Herbold 1979
Schizosaccharomyces Pombe	100–10 000 µg/plate	±	Negative	V.G. Cocuzza 1985
Adult rat liver cells/HPRT	2–20 000 µg/plate	±	Negative	S.V. Brat 1983
2. Urine from patients	(500–1000 mg/day for 6 years)			
S. typhimurium TA 98	250 + 500 µg/plate	±	Negative	B. Herbold 1983
TA 100	250 + 500 µg/plate	±	Negative	B. Herbold 1983
TA 1535	250 + 500 µg/plate	±	Negative	B. Herbold 1983
TA 1537	250 + 500 µg/plate	±	Negative	B. Herbold 1983
Chromosome aberration assays				
CHO cells in vitro	10–30 mM	±	Negative	C. Loquet 1988
Human lymphocytes (6 years)	500–1000 mg/day	–	Negative	J. Theiss 1983
Micronucleus test (male/female mice, p.o.)	2 × 500 or 1000 mg/kg	–	Negative	B. Herbold 1980
Dominant lethal test (male mice, p.o.)	1 × 2000 mg/kg	–	Negative	B. Herbold 1980
DNA-binding/damage assays				
DNA-binding in SD rat liver and kidneys in vivo	160–200 mg/kg	–	Negative	P. Sagelsdorff et al. 1985
Saccharomyces cerevisiae D4				
ADE2 locus	30–3000 µg/ml	±	Negative	V.G. Cocuzza 1985
TRP5 locus	30–3000 µg/ml	±	Negative	V.G. Cocuzza 1985
Host-mediated assay (*S. cerevisiae* D4), rat p.o.	2 × 3–300 mg/kg	–	Negative.	V.G. Cocuzza 1985
Rat hepatocyte primary culture/ DNA repair	0.1–1000 µg/ml	–	Negative	S.V. Brat 1983

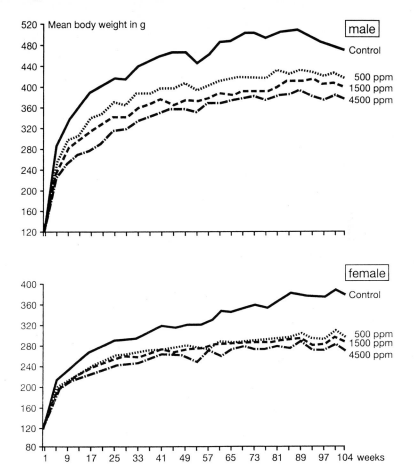

Fig. 1. Body weight curves of male and female Sprague-Dawley rats treated with acarbose in the feed for 24 months.

in all treated groups. This reduction, however, contrasted with an increase in the number of epithelial kidney tumours and a tendency toward an increase in benign Leydig cell tumours. The latter was not statistically significant and roughly in the same range as historical controls and therefore not considered to be treatment related. The incidence of kidney tumours was clearly above the range known for this strain of animals (W. NEUMANN and F. LEUSCHNER 1986, unpublished data, Bayer).

Due to the extreme effects on body weight, which by far exceed the 10% range given by international guidelines as the criterion for a maximum tolerated dose in carcinogenicity experiments, this study could not be extrapolated to the clinical situation. Therefore, a series of long-term studies was initiated to clarify the carcinogenic potential of acarbose.

Table 3. Incidence of selected tumours in a 2-year feeding study with acarbose in Sprague-Dawley rats

Organ/tissue Tumour type	Dose group (ppm)							
	Males				Females			
	0	500	1500	4500	0	500	1500	4500
Liver								
Sarcoma	3	0	0	0	1	0	0	0
Pituitary								
Adenoma	13	5*	5*	10	21	13	11*	12*
Carcinoma	2	1	0	0	1	3	0	0
Mammary gland								
Fibroadenoma	0	0	0	0	4	0	0	1
Carcinoma	0	0	0	0	6	0	0	0
Haemolymphoreticular system								
Leukaemia	2	0	0	0	4	1	1	1
Kidneys								
Adenoma	0	0	3	0	2	3	9*	7
Carcinoma	0	0	0	2	0	1	5*	2
Testes								
Leydig cell tumour (B)	3	7	7	8	–	–	–	–

* Significant at $P \leq 0.05$.
B, benign; M, malignant.

b) Twenty-Six-Month Feeding Study in Sprague-Dawley Rats

This study aimed to examine the reproducibility of the original feeding study results. In addition, a concentration in the feed was included which exerted only moderate pharmacodynamic activity, thereby affecting body weight in the 10% a.m. range. Groups of 50 male and 50 female Sprague-Dawley rats, from the same breeder as in study a) above, were administered dietary concentrations of 0, 150, 500, 1500 or 4500 ppm acarbose for 26 months. Groups of ten male and ten female animals were fed the same concentrations for 14 months. In summary, the results of this study were very similar to those of the original study. Body weights were dose dependently reduced. While at 150 ppm body weight was less than or at the 10% margin compared with controls, with the higher doses body weights were clearly reduced, up to 30% less with the highest dose. Feed consumption was dose dependently increased from 500 ppm, mortality was less in all treated groups with no correlation with dose. There was also a dose-related increase in water consumption. Serum insulin concentration was measured in addition to the standard clinical laboratory parameters. As expected, the data measured at 18 and 24 months of the experiment showed a more or less dose-related decrease (Table 4). In the tumour spectrum, again a shift in various tumour types at various locations is obvious (Table 5). With respect to the reduction

Table 4. Insulin concentrations (μU/ml) in Sprague-Dawley rats fed acarbose

Dose group (ppm)	Males		Females	
	18 months	24 months	18 months	24 months
0	57.4	115.2	48.2	178.8
150	45.7	104.2	44.1	125.2
500	39.0	70.2	27.7**	107.2**
1500	33.0	48.4*	26.7**	95.5*
4500	27.5	48.3*	29.6**	80.2*

Significant differences: * $P \le 0.05$, ** $P \le 0.01$.

Table 5. Incidence of selected tumours in a 26-month feeding study with acarbose in Sprague-Dawley rats

Organ/tissue Tumour type	Dose group (ppm)									
	Males					Females				
	0	150	500	1500	4500	0	150	500	1500	4500
Pancreas										
Islet cell adenoma	7	8	4	2	0	2	2	0	1	0
Pituitary										
Adenoma	23	26	25	20	8	23	24	19	14	18
Adrenal medulla										
Phaeochromocytoma (B)	5	3	5	2	1	2	3	1	1	0
Mammary gland										
Fibroadenoma	0	1	0	0	0	18	2	1	2	3
Kidneys										
Adenoma	1	0	1	4	4	0	0	1	0	2
Carcinoma	1	1	1	2	15	0	1	3	3	5
Testes										
Leydig cell tumour (B)	1	1	12	15	14	–	–	–	–	–

B, benign.

of, e.g., mammary fibroadenomas along with the increase in epithelial kidney tumours and Leydig cell tumours, the effects are more pronounced in this study. The longer duration could be one explanation. It is remarkable that the 150-ppm level did not increase renal and testicular tumours but markedly decreased the incidence of mammary fibroadenomas (E. BOMHARD 1987, unpublished data, Bayer).

c) Twenty-Six-Month Feeding Study in Sprague-Dawley Rats with Glucose Substitution

In this study glucose was administered simultaneously to the animals in the treatment groups to avoid malnutrition due to the pharmacodynamic effect of acarbose. Glucose absorption is not prevented by acarbose. Glucose was administered via the drinking water during the first 4 weeks, afterwards via the feed. Dose levels were 5%–20% up to week 16 and 30% subsequently. Acarbose was administered in concentrations of 0, 150, 500, 1500 and 4500 ppm in the diet. An additional control group received glucose at identical concentrations to the treated animals. Strain and breeder of animals, duration, group size of main and satellite groups and the parameters tested were the same as in study b) above. In males the glucose substitution fully compensated for the effect of acarbose on body weights in concentrations up to 500 ppm. These animals achieved body weights comparable to those of the glucose control group. In males on 1500 ppm body weights were comparable to those of the not substituted control groups, while in the males on 4500 ppm they were up to 10% lower. In females there was a dose-dependent reduction in body weight gain caused by the addition of glucose. At 500 ppm body weights corresponded to the values of the nonsubstituted control group, whereas at higher levels they were slightly less. On the whole, however, development of body weight allows an evaluation of this study with regard to carcinogenicity. The large intestine content was increased in some animals treated with 1500 and 4500 ppm. The incidence of Leydig cell hyperplasia was increased (n, 10/13/23/19 for ascending dose levels compared with 6 in untreated controls and 3 in glucose controls). Acarbose did not increase the incidence of kidney tumours (Table 6). There was also no increase in Leydig cell tumours up to 1500 ppm. There seems to be a slight increase at 4500 pm but this was not statistically significant. There was no reduction in the incidence of certain tumours, which underlines the role calorie reduction or reduced glucose utilization may play in their development (E. Bomhard and O. Vogel 1987, unpublished data, Bayer).

d) Twenty-Six-Month Study with Gavage Administration to Sprague-Dawley Rats

Administration by gavage during the morning largely avoids pharmacodynamic interaction with carbohydrate digestion because food consumption occurs mostly during the night. The doses administered were calculated according to the amounts of acarbose consumed in the study without glucose substitution. Dose groups corresponded to those receiving 150, 1500 and 4500 ppm in the feeding experiment. In addition, there was one control group treated with the vehicle and one untreated control group. The design was otherwise as in study b) above. Body weights were slightly lower in males and females on 1500 and 4500 ppm. Food and water consumption was slightly increased at 4500 ppm. There was a slight increase in the incidence of Leydig cell hyperplasia in the group on 1500 ppm but not in the groups on 150 and 4500 ppm. The incidence of islet cell adenomas of the pancreas, adrenal phaeochromocytomas and mammary tumours was conspicuously less in the animals on 4500 ppm than the controls (Table 7). There was no significant and dose-dependent increase in tumours of the kidney and the testes (E. Bomhard et al. 1987, unpublished data, Bayer).

Table 6. Incidence of selected tumours in a 26-month feeding study with acarbose in Sprague-Dawley rats with glucose substitution

Organ/tissue Tumour type	Dose group (ppm)											
	Males						Females					
	0	0 + Glucose	150[a]	500[a]	1500[a]	4500	0	0 + Glucose	150[a]	500[a]	1500[a]	4500
Pancreas												
Islet cell adenoma	4	11	9	3	1	4	1	1	3	1	1	3
Islet cell carcinoma	0	0	1	0	0	0	0	0	0	0	0	0
Pituitary												
Adenoma	32	34	26	33	16	33	39	36	22	13	22	36
Adrenal medulla												
Phaeochromocytoma (B)	8	9	1	2	1	7	3	2	1	1	2	3
Phaeochromocytoma (M)	1	1	0	2	1	1	0	1	0	1	1	0
Mammary gland												
(Fibro) adenoma	0	0	0	0	1	0	10	10	12	12	11	7
Adenocarcinoma	1	0	1	0	0	0	1	3	6	3	1	4
Kidneys												
Adenoma	0	1	1	2	2	0	0	0	1	1	0	0
Carcinoma	0	0	0	1	0	1	0	0	1	0	0	1
Testes												
Leydig cell tumour (B)	6	2	4	2	5	10	–	–	–	–	–	–

[a] Histopathological examination routinely performed on kidneys, testes and all lesions suspected of being tumours.
B, benign; M, malignant.

Table 7. Incidence of selected tumours in a 26-month study with gavage administration of acarbose to Sprague-Dawley rats

Organ/tissue Tumour type	Dose group (ppm)									
	Males					Females				
	0	0 + H₂O	150ᵃ	1500ᵃ	4500	0	0 + H₂O	150ᵃ	1500ᵃ	4500
Pancreas										
Islet cell adenoma	3	6	2	3	5	1	5	0	1	1
Pituitary										
Adenoma	23	31	22	20	37	15	33	22	24	23
Adrenal medulla										
Phaeochromocytoma (B)	1	7	1	2	3	0	2	1	1	0
Phaeochromocytoma (M)	1	2	0	0	0	0	0	0	0	0
Mammary gland										
Fibroadenoma	0	0	0	1	0	11	15	11	8	9
Adenocarcinoma	1	1	0	1	0	3	2	3	4	2
Kidneys										
Adenoma	1	1	0	0	0	0	0	0	2	1
Carcinoma	0	0	1	1	1	0	2	1	4	1
Testes										
Leydig cell tumour (B)	1	5	3	2	5	–	–	–	–	–
Leydig cell tumour (M)	0	0	0	0	1	–	–	–	–	–

ᵃ Histopathological examination routinely performed on kidneys, testes and all lesions suspected of being tumours.
B, benign; M, malignant.

e) Thirty-Month Feeding Study in Wistar Rats

To investigate a possible strain-specific effect on changes in tumour spectrum, acarbose was also administered to Wistar rats. The design of this study, dose levels, group size of main and satellite groups and parameters investigated were the same as in study b) above. Due to the increased longevity of the Wistar rats compared with the Sprague-Dawley rat, the study had to be extended to 30 months. Food and water consumption were increased from 500 ppm dose dependently. Some males in these groups occasionally had diarrhoea. The number of animals with increased abdominal circumference was increased. In males on 150 ppm there was a slight and transient growth retardation. From 500 ppm upwards growth was retarded markedly and dose dependently. The difference from the control group was often more than 10% in 1500-ppm males and sometimes more than 20% in 4500-ppm males. In females, all dose groups had significantly lower body weights than controls. The differences were essentially less than 10% in animals on 150–1500 ppm and between 10% and 15% in animals on 4500 ppm.

Several of the effects seen in the Sprague-Dawley rat-feeding experiments were reproduced in this experiment (e.g., increased plasma activity of alkaline phosphatase, decreased insulin concentrations in serum both at 500 ppm and above, enlargement of the caecum, which was filled with mushy contents from 500 ppm, increased incidence of Leydig cell hyperplasia and Leydig cell tumours at 1500 and 4500 ppm, decreased incidence of mammary and pituitary tumours; see Table 8). In contrast, there was no increase in the number of epithelial kidney tumours (E. BOMHARD and E. KARBE 1988, unpublished data, Bayer).

f) Thirty-Month Study with Gavage Administration to Wistar Rats

Groups of 50 male and 50 female Wistar rats were administered dose levels corresponding to the drug intake in the animals on 4500 ppm in the Wistar feeding experiment. Groups of ten male and ten female animals were treated accordingly and put to death after 14 months. The same numbers of animals were left untreated and served as controls. The only noteworthy effect was a significantly increased food and water consumption in treated females. The incidence of selected tumours shown in Table 9 shows no treatment-related effect (E. BOMHARD and C. RÜHL-FEHLERT 1988, unpublished data, Bayer).

2. Hamster

a) Eighty-Week Feeding Study in Hamsters

The hamster was chosen as the second species for carcinogenicity testing because mice did not tolerate acarbose administered with the feed. In several experiments (including NMRI strains from two different breeders, Balb/c and CF1), mice developed signs of illness, and mortality was increased

Table 8. Incidence of selected tumours in a 30-month feeding study with acarbose in Wistar rats

Organ/tissue Tumour type	Dose group (ppm)									
	Males					Females				
	0	150	500	1500	4500	0	150	500	1500	4500
Pituitary										
Adenoma	23	19	16	17	14	22	18	10	11	10
Adrenal medulla										
Phaeochromocytoma (B)	9	0	4	5	6	2	0	1	2	3
Phaeochromocytoma (M)	0	2	0	0	1	0	0	0	0	1
Mammary gland										
(Fibro) adenoma	–	–	–	–	–	9	2	2	4	1
Adenocarcinoma	–	–	–	–	–	7	2	1	0	0
Kidneys										
Adenoma	0	0	0	0	1	0	0	0	0	0
Testes										
Leydig cell tumour (B)	11	6	10	16	17	–	–	–	–	–
Leydig cell tumour (M)	1	0	0	0	0	–	–	–	–	–

B, benign; M, malignant.

Table 9. Incidence of selected tumours in a 30-month study with gavage administration of acarbose to Wistar rats

Organ/tissue Tumour type	Dose group (ppm)			
	Males		Females	
	0	4500	0	4500
Pituitary				
Adenoma	16	21	27	21
Adrenal medulla				
Phaeochromocytoma (B)	22	20	5	4
Phaeochromocytoma (M)	3	0	0	0
Mammary gland				
(Fibro) adenoma	0	0	7	8
Adenocarcinoma	0	0	5	4
Kidneys				
Carcinoma	0	0	1	0
Testes				
Leydig cell tumour (B)	9	9	–	–

B, benign; M, malignant.

beginning between weeks 2 and 4. Haemorrhage in the gastric mucosa was the predominant finding at necropsy (E. Bomhard 1981, unpublished data, Bayer; E. Bomhard and G. Kaliner 1982, unpublished data, Bayer; E. Bomhard 1983, unpublished data, Bayer). Postprandial administration via gavage in the morning to NMRI mice, in contrast, revealed no indications of gastro-intestinal disturbances or other signs of toxicity up to 1000 mg/kg even after 13 weeks treatment (E. Bomhard and C. Rühl 1984, unpublished data, Bayer). Feeding of acarbose in concentrations up to 10000 ppm to Syrian golden hamsters for 8 weeks had no effect on mortality and only slight effects on body weight gain (E. Bomhard 1983, unpublished data, Bayer).

The long-term feeding study was therefore conducted with male and female Syrian golden hamsters (*n*, 60/group) in concentrations of 0, 250, 1000 and 4000 ppm in the diet and lasted 80 weeks. At that time the mortality in controls had reached 65% in males and 75% in females. There were no treatment-related effects on body weight, mortality, clinical pathology and incidence of non-neoplastic and neoplastic changes (F. Krötlinger and B. Schilde 1988, unpublished data, Bayer).

b) Eighty-Week Feeding Study in Hamsters with Glucose Substitution

In this study glucose was administered at a concentration of 20% via drinking water throughout the experiment. Acarbose was provided in concentrations of 0, 250, 1000 and 4000 ppm in the diet. An additional control group received 20% glucose dissolved in tap water as drinking fluid. Again there were no adverse and no oncogenic effects at any of the dose levels. Males at 4000 ppm consumed up to 23% more drinking fluid than controls (F. KRÖTLINGER and B. SCHILDE 1988, unpublished data, Bayer).

3. Low-Carbohydrate/Emiglitate Lifetime Carcinogenicity Study in Sprague-Dawley Rats

In order to test the hypothesis that either α-glucosidase inhibition and/or depletion of carbohydrates may be the causative factors for the increased incidence of epithelial kidney tumours after feeding of acarbose to Sprague-Dawley rats, the following study was initiated. Groups of 70 female Sprague-Dawley rats received either 5000 ppm emiglitate or a low-carbohydrate diet or standard diet for up to 137 weeks. Ten animals/group were put to death after week 77 and ten/group after week 104. The low-carbohydrate diet resulted in poor general condition, rough fur, sunken flanks and bloated abdomen. Mortality was appreciably increased from about the 30th week. Only eight animals survived the treatment up to the 105th week of the study. At necropsy the majority of animals showed severe distension of the large intestines with increased contents. The contents were hard in some cases and this led to constipation and subsequent death. The growth was extremely severely retarded with differences in body weights of more than 30% compared with controls.

Feed and water intakes were increased appreciably. Due to the early mortality, it is not possible to say anything about the possible effects on the tumour spectrum. Thus, the available possibilities of diet modification do not permit the establishment of conditions which correspond to simultaneous administration of the normal diet plus α-glucosidase inhibitors such as acarbose. Animals treated with 5000 ppm emiglitate showed an increased excretion of urine and soft faeces. Mortality was slightly less than in standard-diet controls. Growth was much less retarded than after feeding the low-carbohydrate diet. Food consumption was slightly increased part of the time and water consumption markedly increased. At necropsy there was a distension of the caecum, which contained an increased content of soft faeces. The incidence of epithelial kidney tumours was increased. Eleven adenomas and 6 adenocarcinomas were found in the 50 animals scheduled for final death compared with 4 adenomas in standard-diet controls. These results support the hypothesis that the reduced glucose utilization following intestinal α-glucosidase inhibition promotes the occurrence of epithelial kidney tumours in Sprague-Dawley rats (E. BOMHARD and E. KARBE 1994, unpublished data, Bayer).

V. Toxicokinetics

In pharmacokinetic studies it was shown that rats are similar to humans with respect to the pharmacokinetics of acarbose and its metabolites. Unchanged acarbose is only absorbed to a very small extent, but various metabolites are formed in the intestinal lumen by microorganisms and digestive enzymes and subsequently absorbed similarly in both species (Ahr et al. 1989). Radiolabelled acarbose was thus used in supportive toxicokinetic studies lasting 14 days in order to mimic the conditions of the relevant carcinogenicity studies in rats. From these studies an assessment of the systemic exposure to acarbose and its metabolites becomes possible for Sprague-Dawley rats dosed via feed with and without glucose substitution, and dosed by gavage as well as Wistar rats dosed via feed.

The systemic exposure to acarbose and its metabolites increased distinctly with dose (H.J. Ahr et al. 1992, unpublished data, Bayer) (Table 10). At high doses (4500 ppm via feed 360 mg/kg by gavage), the adsorption of acarbose and its extrasystemically formed metabolites tends to be saturated probably due to a limited capacity of the intestinal microorganisms for the degradation of acarbose. No substantial increase in the exposure by further increasing the dose can thus be expected, indicating that the carcinogenicity studies were performed up to the maximum feasible dose. The exposure to acarbose and related metabolites in these rat studies in the high-dose groups was in the range of exposure obtained during therapeutic use in humans.

Under all conditions used the exposure of the rats in the high-dose groups was similar. No influence of glucose substitution, mode of administration or rat strain was observed. Also the metabolic patterns were similar between doses, rat strains and modes of administration (H.J. Ahr et al. 1994, unpublished data, Bayer). Thus the differences in tumour rates under the different study conditions are not related to any differences in the exposure to acarbose and its breakdown products.

Table 10. Acarbose: Area under the curve values from first to last data point (AUD) of radioactivity characterizing the systemic exposure to acarbose and its metabolites in male rats under different conditions. Data were obtained in supportive studies with $[^{14}C]$ acarbose mimicking the conditions of the main carcinogenicity studies. Geometric means and ranges of $N = 5$ are given

Mode of administration	Strain	AUD [mg-eq·h·kg^{-1}] (range)		
		150 ppm (12 mg/kg)	1500 ppm (120 mg/kg)	4500 ppm (360 mg/kg)
Feed	SD	6.1 (4–11)	52.8 (32–93)	62.0 (49–111) 54.5 (29–131)
Feed + glucose	SD	ND	ND	59.6 (18–211)
Gavage	SD	2.9 (1.3–4.4)	8.0 (2.6–21)	37.2 (12–109)
Feed	W	5.4 (3.2–9.9)	36.4 (26–44)	63.7 (19–258)

SD, Spague-Dawley; W, Wistar.

VI. Special Studies

1. Glycogen Storage in Rats

Young adult female Wistar rats received single or multiple (up to 36) intraperitoneal injections at dose levels ranging from 100 to 400 mg/kg. A slowly reversible intralysosomal glycogen storage was observed in hepatocytes and Kupffer cells, in the kidney (collecting duct, renal pelvic transitional epithelium, distal convoluted tubule), in the adrenal cortex (zona reticularis) and in the spleen (trabecular smooth muscle cells). Slight storage was also seen in the adrenal medulla and in the soleus muscle, but was hardly apparent in the cardiac ventricular muscle. Lysosomal glycogen storage was absent from the renal proximal convoluted tubules and from the neurons examined (dorsal root ganglia, supraoptic nucleus). At the single cell level, the experimentally induced alterations closely resembled those occurring in inherited glycogenosis type II (Pompe's disease in man) (LÜLLMANN-RAUCH 1981, 1982).

Treatment of male Wistar rats with a single intraperitoneal dose of 400 mg/kg disturbed liver lysosome metabolism, causing distinct and persistent inhibition of the enzymes and acute disturbances of lysosomal glycogen metabolism. While total liver glycogen was decreased even 5 days after injection, lysosomal glycogen was increased (GEDDES and TAYLOR 1985).

After intraperitoneal injection of 400 mg/kg into male Wistar rats, there was a shift in the light:heavy lysosome fraction ratio from 70%:30% to 40%:60%. Among the heavy lysosome fractions was a new one with a density of >1.14 g/ml, which accounted for 40% of total lysosomes, while the heavy lysosome fractions from control rats had average densities of 1.09 g/ml and 1.11 g/ml. The results after subcutaneous injection of glucagon led to the conclusion that the formation of acarbose-induced glycogenosomes requires glucagon-mediated autophagy (KONISHI et al. 1989).

Within the same experiment, acarbose did not change β-N-acetylglucosaminidase and acid α-glucosidase activity in liver the 7 days following injection but urine collected over 0–24 h from the treated rats showed a strong inhibition of acid α-glucosidase (KONISHI et al. 1990).

Female Wistar rats were administered 1000 mg/kg for 7 days by gavage. Acarbose led to a significant increase in hepatic and soleus muscle lysosomal (but not cytoplasmic) glycogen concentrations in the fasted rat, and total liver glycogen was decreased in rat fed until necropsy (LEMBCKE et al. 1991).

Groups of 15 female Sprague-Dawley were administered single daily doses of 400 mg/kg by gavage or 200 mg/kg by intraperitoneal injection or were treated continuously with a concentration of 4500 ppm in their feed. An untreated group of 15 animals served as control. The intraperitoneal administration led to a massive lysosomal glycogen storage in the liver. This effect was qualitatively similar but quantitatively much less after treatment via feed (Fig. 2) or gavage (E. BOMHARD and A. POPP 1992, unpublished data, Bayer).

Histochemical examination on kidney slices from five animals/group of this experiment demonstrated an enhanced renal accumulation of glycogen preferentially in the collecting ducts with traces of glycogen also in other renal tubular segments and in glomeruli after i.p. injection. Administration by gavage or in feed did not result in renal glycogen storage except for moderate deposits in the collecting ducts (H. ENZMANN and E. BOMHARD 1993, unpublished data, Bayer).

2. Cell Proliferation in Kidney Cortex

In a similarly designed study on female Sprague-Dawley rats using the same acarbose treatment regimen as above but with the addition of a positive control group treated with cisplatin (1 × 6 mg/kg i.v.), cell proliferation was determined in the kidney cortex. After injection of tritiated thymidine 2 h before necropsy, there was no increased incidence of proliferating cells in acarbose-treated groups. In cisplatin-treated rats there was a fivefold increase in cell proliferation (H. ENZMANN and E. BOMHARD 1993, unpublished data, Bayer).

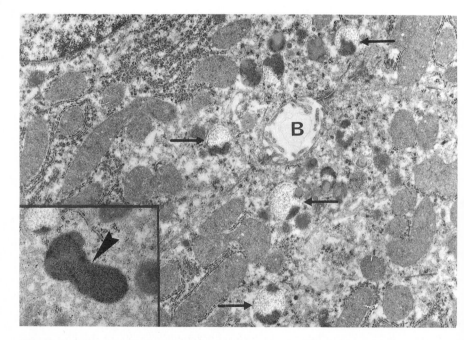

Fig. 2. Intralysosomal glycogen storage. Lysosomes with glycogen and remnants of lysosomal matrix (*arrows*) besides a bile capillary (*B*). *Inset*, normal lysosome (*arrowhead*), ×9000 (*inset*, ×15 000)

3. Electron Microscopic Investigations

Groups of 20 male Sprague-Dawley rats (age 7–8 weeks at start) were treated with concentrations of 0, 500, 1500 and 4500 ppm in their feed for 90 days. No ultrastructural changes were found in the renal cortex or testes. There was a lower content of cytoplasmic glycogen and lipid in centriacinar hepatocytes from the livers of most 4500-ppm animals (A.K. Sykes 1987, unpublished data, Bayer).

4. Investigations of Effects on Endocrinium

Groups of 20 male and female Sprague-Dawley rats with an age of ca. 18 months at the start were treated with concentrations of 0, 500, 1500 or 4500 ppm in their diet for 13 weeks. Treated animals at all dose levels lost body weight while food and water consumption was increased. The measurement of testosterone, luteinizing hormone (LH), follicle-stimulating hormone (FSH), prolactin, adrenocorticotrophic hormone (ACTH) and corticosterone, in serum as well as corticosterone/testosterone and dehydroepiandrosterone in urine revealed no treatment-related effects. Serum insulin concentration was significantly lower in males from 1500 ppm upwards and females of all treated groups, sometimes without a clear-cut dose correlation. Electron microscopic investigations of kidney cortex/testes, ovaries and anterior pituitary (pars distalis) showed no treatment-related changes. Glycogen content within the cytoplasm of centriacinar hepatocytes was found to be slightly lower in 4500-ppm females than in concurrent controls. The number of males with Leydig cell hyperplasia was slightly higher in treated groups (n = 4/5/5 in ascending dosages) compared with the control group (n = 1). The incidence of Leydig cell tumours was not affected (Bomhard and Schilde 1992).

5. Interaction with Other Oral Antidiabetics

Groups of ten male and ten female Wistar rats received Bay g 5421 at dietary concentrations of 0, 150 or 1500 ppm with or without daily cotreatment of 2 × 125 mg metformin/kg, 3 mg glybenclamide/kg or 60 mg chlorpropamide/kg body weight by gavage for 4 weeks. Groups of ten male and ten female rats each receiving twice daily 125 mg metformin/kg, 3 mg glybenclamide/kg or 60 mg chlorpropamide/kg by gavage without cotreatment served as additional controls. Besides a slight growth retardation observed in 1500-ppm male rats in combination with metformin and glybenclamide, there were no effects attributable to the combination of treatments. It is concluded that acarbose in combination with other prescribed oral antidiabetics does not represent an unreasonable risk from the toxicological point of view (E. Bomhard and B. Rühl-Fehlert 1995, unpublished data, Bayer).

VII. Assessment of the Carcinogenic Potential of Acarbose

There are two target organs showing an increase in tumour incidence under certain conditions of treatment of rats with acarbose. They are obviously not correlated to each other and the underlying mechanisms seem to be different. While the increased occurrence of kidney tumours is confined to the Sprague-Dawley rat under conditions of severe malnutrition, effects on Leydig cells (hyperplasia and neoplasia) occurred in Sprague-Dawley and Wistar rats also in feeding studies but irrespective of glucose substitution. Both types of tumours are not increased after postprandial gavage administration. The relevance of these two findings is therefore discussed separately.

1. Significance of Leydig Cell Tumours

Leydig cell tumour as well as hyperplasia incidence is rather variable in untreated or vehicle-treated control groups of Sprague-Dawley and Wistar rats. Taking this into account, a clear-cut effect of acarbose on these cells was to be seen essentially in the 26-month feeding study with Sprague-Dawley rats. The effect was much less pronounced in the 24-month feeding study with Sprague-Dawley rats and the 30-month feeding study with Wistar rats. While in the 26-month study on Sprague-Dawley rats with glucose substitution the slight increase in Leydig cell tumours at the highest dose level might be a questionable treatment-related effect, the increase in Leydig cell hyperplasia at 1500 and 4500 ppm is clearly treatment related. One unusual feature of these effects on Leydig cells is the lack of a dose-reponse correlation between 1500 and 4500 ppm, indicating that with a certain degree of α-glucosidase inhibition (reached at 1500 ppm) the consecutive events leading to Leydig cell effects are at their maximum. Gavage administration did not lead to an increase in Leydig cell tumours either in Sprague-Dawley or in Wistar rats. The conspicuously higher incidence of Leydig cell hyperplasia in Sprague-Dawley rats treated with dose levels corresponding to 1500 ppm in the feeding experiment but not in 4500-ppm animals is therefore, in view of the considerable spontaneous variation, not considered to be treatment related. Neither Leydig cell hyperplasia nor neoplasia were increased in the two carcinogenicity studies with hamsters and 1-year study in dogs. The increase seen in the glucose substitution study argues against a mechanism secondary to severe malnutrition, but rather points to a correlation with the amount of undigested carbohydrates reaching the large intestine. This condition is much more pronounced when acarbose is provided during food consumption and is independent of the presence or absence of glucose. On necropsy the animals generally show a distension especially of the caecum.

If one looks at the literature, there are at least two compounds where there is a very similar constellation of effects on intestinal carbohydrate digestion as well as on Leydig cells, namely lactitol and lactose (Sinkeldam et al. 1992). These compounds are not genotoxic. As with acarbose, feeding of these compounds leads to remarkable caecal enlargement caused by undigested carbohydrate accumulation, increased incidence of Leydig cell hyperplasia and neoplasia with no measurable effects on blood hormone levels, i.e., FSH, LH, testosterone (BÄR 1992). Increased calcium absorption from the gut as speculated for lactose and lactitol as one of the likely contributing factors (BÄR 1992) seems to play no role in acarbose studies. An increased incidence of calcareous deposits in renal medulla, papilla and pelvis found in lactose- and lactitol-treated animals was not seen in any of the long-term rat studies with acarbose. On the other hand, changes in the kidneys indicative of an increased calcium deposition after chronic emiglitate treatment of rats (see below) were not accompanied by a significant increase in Leydig cell hyperplasia or neoplasia. Thus, there is not yet a convincing mechanistic explanation available. Although these mechanisms have not yet been elucidated, there are several lines of evidence that these effects seen in male rats are not relevant to humans. They are summarized by BÄR (1992) for lactose and lactitol but are likewise valid for acarbose as follows:

1. Lack of genotoxicity
2. Rat specificity
3. Long history of the consumption of lactose in humans
4. Insensitivity of the human Leydig cell to agents and conditions that are known to cause neoplastic growth in Leydig cells in rats
5. The generally very low spontaneous incidence of Leydig cell tumours in the human population
6. The absence of any epidemiological evidence establishing a link between nutritional factors and the occurrence of Leydig cell tumours in humans

2. Significance of Epithelial Kidney Tumours in Sprague-Dawley Rats

Male and female Sprague-Dawley rats developed a clearly increased incidence of epithelial kidney tumours (both adenomas and carcinomas) starting with a dose level of 1500 ppm in the diet. With 500 ppm there seemed also to be a slight increase but this was borderline in view of the spontaneous variability. At these dose levels the body weight gain was markedly reduced, e.g., by 32% (first study) and 38% (second study) in males, and by 42% (first study) and 39% (second study) in females, respectively, at 4500 ppm. Thus, the animals exhibited signs of massive malnutrition and loss of isocaloric state. The generally recommended ca. 10% body weight decrement compared with control animals was by far exceeded at these concentrations. Feeding of acarbose at concentrations where body weight decrement is less than 10% (i.e., 150 ppm) did not increase the kidney tumour incidence but still led to a significant reduction in mammary tumours. While under conditions of reduced calorie feeding by means of low-carbohydrate diets or restricted feeding the observation of reduced tumour incidences is well documented (GROSS 1988; KEENAN et al. 1992; PICKERING and PICKERING

1984; Saxton et al. 1948; Tucker 1979), an increase in the incidence of certain tumours has not yet been reported.

The numerous studies performed so far to explain this increase and to clarify the mechanism and its relevance to man can be summarized as follows: In the two studies with Wistar rats and hamsters, the incidence of kidney tumours was not increased. This indicates that this effect is strain specific. The Sprague-Dawley rats used in these experiments differ significantly in at least two aspects from the Wistar rats:

1. The Sprague-Dawley rat has a much higher growth rate, leading to final body weights up to about 50%–100% higher than the Wistar rat.
2. The Sprague-Dawley rat is much more prone to the development of epithelial kidney tumours. Taking all the control groups in the three experiments with acarbose plus the female control group in the emiglitate/low-carbohydrate diet experiment together, then the combined spontaneous incidence of adenomas and carcinomas was 5/300 males and 8/350 females. In contrast, the historical control incidence in Wistar rats from about 70 2-year and 30-month studies was 4/4138 in males 3/4145 in females (Bomhard et al. 1986; E. Bomhard 1992, unpublished data, Bayer; Bomhard and Rinke 1993, unpublished data, Bayer, 1994). Thus, the spontaneous incidence in Sprague-Dawley rats is more than 15 times higher in males and more than 30 times higher in females.

The long-term studies with Sprague-Dawley rats under conditions preventing severe malnutrition (i.e., postprandial gavage administration or glucose substitution) point to a close linkage between treatment-related effects on carbohydrate digestion and tumour development. This assumption is supported by the results of the low-carbohydrate/emiglitate study. These data demonstrate that a chemically and pharmacokinetically different compound but with the same pharmacodynamic activity also induces kidney tumours in Sprague-Dawley rats under conditions of severe malnutrition. Pharmacokinetic studies on the systemic exposure to acarbose and its radioactive metabolites when dosed via food with and without glucose supplementation, or with gavage in addition, revealed that there are no qualitative and quantitative differences in acarbose kinetics and metabolite patterns under the treatment conditions of the different long-term studies which explain the differences in tumour occurrence (H.J. Ahr et al. 1992, 1994, both unpublished data, Bayer). Furthermore, none of the numerous studies on genotoxicity/mutagenicity indicated a potential for interaction with DNA. Therefore, it is justified to conclude that neither acarbose nor its metabolites are carcinogenic. It is the extreme condition of malnutrition (reduced glucose utilization, loss of isocaloric state) which triggers kidney tumour formation.

The relevant question now is whether such a condition can occur with the therapeutic use of acarbose (or other α-glucosidase inhibitors such as emiglitate). Significant body weight reduction or other indicators of severe malnutrition have never been seen in the large number of clinical trials with healthy and diabetic subjects (BALFOUR and McTAVISH 1993).

The overall conclusion is that there is no discernible carcinogenic risk from the therapeutic use of acarbose.

B. Emiglitate

I. General Toxicology

1. Acute Toxicity

Table 11 summarizes the results of the acute toxicity experiments with oral (G. SCHLÜTER 1982, unpublished data, Bayer) or intravenous (M. RENHOF 1983, unpublished data, Bayer) administration to various species. With oral administration, neither symptoms nor lethality was observed up to the maximum of applicable doses. Beginning with i.v. dose levels of 250 (mice), 160 (rats), 200 (rabbits) and 500 (dogs) mg/kg, signs of intoxication were observed. These consisted of tonicoclonic convulsions (all species), gasping respiration and reduced motility (mice, rats, rabbits) and accelerated breathing and narrowing of palpebral fissures (rats and mice). Symptoms and deaths occurred shortly after injection. The only postmortem findings were lungs with dark-red spots in rats that had died.

Table 11. Acute toxicity of emiglitate

Species	Strain	Sex	Route	LD_{50} (mg/kg)	95% Confidence limits	Reference[a]
Mouse	CF$_1$	M + F	Oral	>10000	–	G. SCHLÜTER 1982
Rat	Wistar	M + F	Oral	>10000	–	G. SCHLÜTER 1982
Rabbit	Chinchilla	F	Oral	>5000	–	G. SCHLÜTER 1982
Dog	Beagle	M + F	Oral	>5000	–	G. SCHLÜTER 1982
Mouse	CF$_1$	M	i.v.	436	395–480	M. RENHOF 1983
Mouse	CF$_1$	F	i.v.	503	436–581	M. RENHOF 1983
Rat	Wistar	M	i.v.	328	287–376	M. RENHOF 1983
	Wistar	F	i.v.	322	288–363	M. RENHOF 1983
Rabbit	Chinchilla	F	i.v.	200–250	–	M. RENHOF 1983
Dog	Beagle	M + F	i.v.	ca. 500	–	M. RENHOF 1983

[a] All data unpublished, Bayer AG.

2. Subacute Toxicity

a) Rat

Wistar rats in groups of ten animals per sex were injected intraperitoneally at dose levels of 0, 3, 10 or 30 mg/kg once daily for 4 weeks. After 30 mg/kg, growth was retarded, food and water consumption was reduced in males and there was a higher incidence of inflammatory reactions in the peritoneal cavity (E. Bomhard and P. Gröning 1984, unpublished data, Bayer).

b) Dog

Groups of two male and two female beagle dogs received doses of 0, 5, 16 and 50 mg/kg once daily over a 3-week period by intravenous injection. As a pharmacodynamic effect, diarrhoea was observed throughout the study in all treated animals, with resultant slightly lower body weights than controls and dose dependently reduced serum iron concentrations and elevated free iron-binding capacity. Treatment was otherwise well tolerated both locally and systemically (K. Detzer and O. Vogel 1984, unpublished data, Bayer).

3. Subchronic Toxicity

a) Rat

Groups of 15 male and 15 female Wistar rats were each given dose levels of 0, 50, 200 and 800 mg/kg once daily for 3 months by gavage. Abdominal distension, increased water consumption, soft faeces and body weight retardation were observed after 800 mg/kg, and were more pronounced in males. Plasma activity of aspartate aminotransferase (AST) and alanine aminotransferase (ALT) were increased, and total protein concentrations decreased after 800 mg/kg in males. Increased relative liver weight and a fine honeycomb or vesicular cytoplasm structure of the hepatocytes were seen after administration of 200 and 800 mg/kg (F. Krötlinger and G. Nash 1982, unpublished data, Bayer).

b) Dog

Emiglitate was administered by gavage to three beagle dogs/sex and group at doses of 0, 15, 45 and 135 mg/kg. The animals were provided with standard feed. Diarrhoea occurred in all the animals treated with the test substance soon after the start of the study. As a result, weight development was dose dependently retarded. Serum AST activity was dose dependently increased, pathological levels being reached in the animals at 135 mg/kg (K. Detzer and O. Vogel 1982, unpublished data, Bayer).

4. Chronic Toxicity

a) Rat

Wistar rats (30/sex/dose level) were treated with 0, 250, 1000 or 4000 ppm in their diet for 12 months. Diarrhoea, markedly increased water consumption, poor general condition and rough fur were seen at 4000 ppm. There was a dose-dependent retardation in body weight gain. The difference in body weights to controls was very slight ($\leq 8\%$) at 250 ppm, approximately 10% after 1000-ppm and very marked (up to 43% in males and 25% in females) at 4000 ppm. Clinical pathology revealed increased concentrations of acetoacetic acid and ß-hydroxybutyric acid (in all treated groups) as well as decreased cholesterol, total protein (both from 1000 ppm) and glucose values in plasma. Serum insulin concentrations were lower at 1000 and 4000 ppm. The thromboplastin time was slightly prolonged at 4000 ppm. Alkaline phosphatase activity was increased in 1000-ppm males and 4000-ppm males and females. Gross pathology findings consisted of dilatations of the large intestines, especially the caecum. Histopathology revealed increased lymphohistiocytic infiltration of the intestinal mucous membranes and changes in the mesenteric lymph nodes, which were characterized by dilatation and increased numbers of cells in the sinus at mid- and high-dose level. In the liver, there was a dose-dependent decrease in fat storage, which was expecially pronounced at 4000 ppm. Mineralized deposits in kidneys were seen in female animals in a dose-dependent fashion in all treated groups, whereas they were predominantly present in 4000-ppm male animals. These changes are most likely due to an ionic imbalance (calcium-phosphorous, acid-base equilibrium). As practically all the other effects described, they can be interpreted as being secondary to exaggerated pharmacodynamic effects and not necessarily as primary toxic effects (R. EIBEN 1993, unpublished data, Bayer).

b) Dog

Dose levels of 0, 15, 45 and 135 mg/kg were administered to groups of four male and four female beagle dogs for 52 weeks by gavage. These animals were fed a low-carbohydrate, protein-rich diet. Control groups consisting of two males and two females were provided with a standard diet. The intercurrent death of two animals was due to cachexia as a consequence of pharmacodynamic activity. Serum AST activity was slightly increased in the high-dose groups. Most animals at the mid- and high-dose level had changes in the cytoplasm of hepatocytes. The dose level of 15 mg/kg was considered to be the no-adverse-effect level (K. DETZER and G. KALINER 1994, unpublished data, Bayer).

II. Reproduction Toxicology

1. Fertility and General Reproductive Performance in Rats

Wistar rats (24 male and 60 female/group) were given daily oral doses of 0, 15, 50 or 150 mg/kg by gavage. Males were treated for 10 weeks prior to mating and throughout a 3-week mating period. Females were dosed for 3 weeks prior to mating, during mating and until day 7 of gestation. The only effect seen was a dose-dependent retardation of body weight gain of males at all dose levels (M. Renhof 1986, unpublished data, Bayer).

2. Embryotoxicity/Teratogenicity in Rats

Groups of 25 inseminated Wistar rats were given single oral daily doses of 0, 15, 45 or 135 mg/kg from days 6 to 15 of gestation. In addition, groups of 15 animals were administered doses of 0 and 135 mg/kg. These latter animals were allowed to rear their young. There was no effect on dams, fetuses and postnatal development of the pups (M. Renhof 1986, unpublished data, Bayer).

3. Embryotoxicity/Teratogenicity in Rabbits

Groups of 15 inseminated Himalayan rabbits were treated from the 6th to the 18th day of gestation with daily oral doses of 0, 10, 30 or 100 mg/kg. Body weight gain was significantly less in the dams on 30 and 100 mg/kg during the administration period, probably due to the pharmacodynamic activity. No evidence of any embryotoxicity or teratogenicity was found (M. Renhof 1984, unpublished data, Bayer).

III. Genotoxicity/Mutagenicity

The test systems used, the dose/concentration ranges tested and the results are compiled in Table 12. There were no indications of a genotoxic/mutagenic potential of emiglitate.

IV. Carcinogenicity

1. Rat

Male and female Wistar rats ($n = 50$) were treated with concentrations of 0, 120, 360 and 1000 ppm in their diet for 24 months. Ten animals/sex/dose group were put to death after 1 year for an interim investigation. Body weight gain was retarded in males from 360 ppm and in females at 1000 ppm. Mortality was significantly increased in treated males of all groups. At necropsy the caecum was filled with soft faeces. In 360 and 1000 ppm males honeycomb-like cytoplasmatic rarefaction of hepatocytes and a reduced

Table 12. Mutagenicity/genotoxicity studies with emiglitate

Test system	Dose/concentration	S9 mix	Result	Reference[a]
Point mutation assays				
Salmonella typhimurium TA 98	20–12 500 µg/plate	±	Negative	B. HERBOLD 1981
TA 100	20–12 500 µg/plate	±	Negative	B. HERBOLD 1981
TA 1535	20–12 500 µg/plate	±	Negative	B. HERBOLD 1981
TA 1537	20–12 500 µg/plate	±	Negative	B. HERBOLD 1981
Schizosaccharomyces Pombe	100–10 000 µg/plate	±	Negative	F. DUBINI 1984
Chromosome aberration assays				
Micronucleus test (male/female mice)	2 × 2000 or 2 × 4000 mg/kg	−	Negative	B. HERBOLD 1981
Dominant lethal test (male mice)	1 × 8000 mg/kg	−	Negative	B. HERBOLD 1983
DNA damage assays				
1. Test compound				
Saccharomyces cerevisiae D4 *ADE2* locus	100–10 000 µg/ml	±	Negative	F. DUBINI 1984
S. cerevisiae TRP5 locus	100–10 000 µg/ml	±	Negative	F. DUBINI 1984
Host-mediated assay				
(*S. cerevisiae* D4), mice p.o.	2 × 10–1000 mg/kg	−	Negative	F. DUBINI 1984
HeLa cell/DNA repair	10–1000 µg/ml	±	Negative	F. DUBINI 1984
2. Urine from Swiss mice				
S. cerevisiae D4 *ADE2* locus	2 × 10–1000 mg/kg	−	Negative	F. DUBINI 1984
S. cerevisiae TRP5 locus	2 × 10–1000 mg/kg	−	Negative	F. DUBINI 1984

[a] All data unpublished, Bayer AG.

incidence of clear cell foci (all treated groups) were seen. The incidence and/or severity of senile nephropathy was increased at 120 ppm and above, and there was an increased incidence of cortical cysts in males at 1000 ppm. The mineralization in kidneys seen in the chronic feeding study (see above) was not recorded here possibly due to the masking effect of senile nephropathy. The incidence of hyperplasia in the parathyroids was increased in males at 120 ppm and above. The incidence of osteodystrophia fibrosa in femur was increased in males at 120 ppm and above. There was no effect on the number of animals bearing tumours nor on the type and incidence of tumours at all locations. Altered resorption patterns in the intestine with resulting ionic imbalance could explain the findings in the kidneys, parathyroids and femur (R. Eiben 1993, unpublished data, Bayer).

2. Mouse

A 91-week feeding study in NMRI mice ($n = 50$) with animals being put to death at an interim stage ($n = 10$) after 52 weeks was conducted at dose levels of 0, 100, 300 or 900 ppm. Body weights of males receiving 900 ppm were significantly lower (up to 7%) than controls. In this group food consumption was slightly increased. An increased number of males at 900 pm scheduled for death at the end of the study showed an increase in hepatocytic glycogen retention. There was no indication of a carcinogenic potential (E. Bomhard 1994, unpublished data, Bayer).

V. Toxicokinetics

Data on toxicokinetic investigations are not yet available. Pharmacokinetic data are presented by Krause and Ahr (see Chap. 19, this volume).

VI. Special Studies

Hepatic and muscular glycogen concentrations were investigated in the overnight-fasted female Wistar rat ($n = 8$) after 3, 7 and 28 days treatment with single daily doses of 5, 50 or 500 mg/kg. Ten animals treated with the vehicle served as controls. Hepatic glycogen concentrations were dose dependently increased at 50 and 500 mg/kg at all time points. This increase was due to lysosomal storage of glycogen only. At these dose levels there was also a dose-dependent tendency towards higher lysosomal glycogen concentrations in the soleus muscle. In fed rats, 500 mg/kg administered over 7 days by gavage decreased postprandial hepatic total glycogen concentrations (Lembcke et al. 1991).

C. Miglitol

I. General Toxicology

1. Acute Toxicity

The results of acute toxicity experiments after oral or intravenous administration of miglitol to various species are summarized in Table 13. Miglitol has proved to be virtually non-toxic. While oral administration produced no symptoms in mice, diarrhoea occurred in the other species up to 48 h after administration. After intravenous administration, exaggerated breathing and decreased mobility were the predominant symptoms particularly at the higher doses. In dogs, diarrhoea also occurred after intravenous administration of 5000 mg/kg (P. MÜRMANN 1980, unpublished data, Bayer).

2. Subacute Toxicity

a) Rat

Groups of ten male and ten female Wistar rats received doses of 0, 50, 200 or 800 mg/kg once daily for 4 weeks by intraperitoneal injection. In male rats at 800 mg/kg, growth was slightly retarded with a simultaneous small reduction in feed consumption. There were no other effects in this and the other groups (E. BOMHARD and C. RÜHL 1984, unpublished data, Bayer).

b) Dog

Groups of two male and two female Beagle dogs received 0, 10, 32 or 100 mg/kg once daily over a 3-week period by intravenous injection. Dose levels up to 32 mg/kg were tolerated with no damage. The only effects at 100 mg/kg were a drop in serum Fe values, a rise in free iron-binding

Table 13. Acute toxicity of miglitol

Species	Strain	Sex	Route	LD_{50} (mg/kg)	Reference[a]
Mouse	CF_1	M	Oral	>10 000	P. MÜRMANN 1980
Rat	Wistar	M + F	Oral	>10 000	P. MÜRMANN 1980
Rabbit	Chinchilla	F	Oral	ca. 10 000	P. MÜRMANN 1980
Dog	Beagle	M + F	Oral	>10 000	P. MÜRMANN 1980
Mouse	CF_1	M	i.v.	>10 000	P. MÜRMANN 1980
Rat	Wistar	M	i.v.	>8000<10 000	P. MÜRMANN 1980
Rat	Wistar	F	i.v.	7095*	P. MÜRMANN 1980
Rabbit	Chinchilla	F	i.v.	ca. 7000	P. MÜRMANN 1980
Dog	Beagle	M + F	i.v.	>5000	P. MÜRMANN 1980

* Confidence interval for $P \leq 0.05$, 6093–7868 mg/kg.
[a] All data unpublished, Bayer AG.

capacity and a marked increase in plasma aspartate aminotransferase. A slight increase in cases of soft faeces or diarrhoea in the initial week in all treated groups can be explained pharmacodynamically (K. Detzer and O. Vogel 1984, unpublished data, Bayer).

3. Subchronic Toxicity

a) Rat

Groups of 15 male and 15 female Wistar rats were each administered doses of 0, 100, 330 and 1000 mg/kg by gavage once a day for 3 months. There were no adverse effects in any of the groups. There was the functional change of an increase in water consumption in male rats at 1000 mg/kg (E. Bomhard and P. Gröning 1980, unpublished data, Bayer).

b) Dog

Groups of three male and three female beagle dogs received single daily oral administrations of 0, 50, 150 and 450 mg/kg in gelatin capsules for 3 months. Diarrhoea together with some retardation of body weight gain occurred in animals at all dose levels. An increase in AST occurred in some animals in the mid-dose group, and in all animals in the high-dose group, which was associated with several instances of abnormal ECGs (repolarization disturbances in the form of distinct depressions of the ST segment), which might suggest myocardial damage. However, no specific myocardial lesions could be identified at necropsy and histopathology. The drug was otherwise well tolerated at 50 mg/kg (G. Schlüter and B. Schilde 1981, unpublished data, Bayer).

4. Chronic Toxicity

a) Rat

Groups of 20 male and 20 female Wistar rats were given miglitol at concentrations of 0, 250, 1000 and 4000 ppm in their feed for 12 months. Diarrhoea, increased water consumption and marked retardation of growth were observed at 4000 ppm. Growth was temporarily retarded in 1000-ppm males (H. Suberg and C. Wood 1987, unpublished data, Bayer).

b) Dog

Groups of four male and four female beagle dogs received 0, 20, 60 and 180 mg/kg. All these animals were fed a low-carbohydrate diet to avoid excessive diarrhoea. An additional control group was provided with standard food. The only effect was a significant increase in AST activity in plasma at 180 mg/kg (K. Detzer and C. Rühl 1987, unpublished data, Beyer).

II. Reproduction Toxicology

1. Fertility and General Reproductive Performance in Rats

In a combined male and female fertility study, groups of 24 male and 60 female Wistar rats each received single oral daily doses of 0, 30, 100 and 300 mg/kg by gavage. Males were treated for 10 weeks prior to mating and throughout a 3-week mating period. Females were dosed for 3 weeks prior to mating, during mating and until day 7 of gestation. About half of the F_0 generation dams were subjected to Caesarean section on day 20 of gestation; the other half was allowed to deliver and rear their young for 3 weeks. Selected male and female neonates from each litter of the control and high-dose groups were mated after reaching maturity. None of the reproductive parameters investigated was affected in the dose range investigated (M. RENHOF 1986, unpublished data, Bayer).

2. Embryotoxicity/Teratogenicity in Rats

Groups of 25 inseminated BAY:FB30 rats were treated once daily from day 6 to day 15 of pregnancy at doses of 0, 50, 150 and 450 mg/kg by gavage. On day 20 of pregnancy the animals were subjected to Caesarean section and the fetuses examined. There were no adverse effects on dams. Embryo development was not impaired up to 150 mg/kg. Lower average fetal weights were noted at 450 mg/kg, indicating an embryotoxic effect. No teratogenic effects were noted at any dose (G. SCHLÜTER and B. SCHILDE 1981, unpublished data, Bayer).

3. Peri-/Postnatal Study in Rats

Groups of 50 inseminated female rats were treated by gavage from the 16th day of gestation in daily doses of 0, 30, 100 or 300 mg/kg. The young of approximately 50% of the females were delivered by Caesarean section on the 20th day of gestation, while the remaining dams were allowed to litter naturally and to rear their pups for 3 weeks, during which time the treatment was continued. At least one male and one female from each litter in the control and the highest dose group were reared to sexual maturity to allow a check on the reproductive performance of the F_1 generation. Miglitol induced no late sequelae in the animals of the first or the second generation (M. RENHOF 1985, unpublished data, Bayer).

4. Embryotoxicity/Teratogenicity in Rabbits

Groups of 15 inseminated Himalayan rabbits were given daily oral doses of 0, 30, 100 or 300 mg/kg on days 6–18 of gestation. Body weight gain was reduced in dams at 300 mg/kg, both throughout the entire gestation period and during the administration period. At this maternally toxic dose there were signs of secondary embryotoxicity (reduced number of fetuses, increased number of runts and resorptions). No indications of teratogenicity were detected up to the maximum dose tested (M. Renhof 1985, unpublished data, Bayer).

III. Genotoxicity/Mutagenicity

The series of assays performed covers all three end points. The test systems used, dose/concentration ranges tested and the results are compiled in Table 14. In summary, in none of these tests were there any indications of a genotoxic/mutagenic potential of miglitol.

IV. Carcinogenicity

1. Rat

Groups of 50 male and 50 female Wistar rats received concentrations of 0, 120, 360 and 1000 ppm miglitol in the feed over 24 months. Ten animals of each sex from each dose group were put to death after 1 year for an interim evaluation. Diarrhoea occurred transiently in five males at 1000 ppm during the first 29 weeks. After 1000 ppm, growth was retarded throughout the study in the males and during some periods in the females. According to type, number and localization of tumours found, there was no indication of carcinogenic potential (E. Bomhard 1989, unpublished data, Bayer).

2. Mouse

Groups of 50 male and 50 female NMRI mice were administered concentrations of 0, 200, 600 and 1800 ppm miglitol in the feed over 21 months. Ten animals of each sex from each dose group were put to death after 1 year. After administration of 1800 ppm, mortality was slightly increased during the first weeks and body weights were lower than in controls. There was no indication of a carcinogenic effect in the dose range investigated (E. Bomhard 1988, unpublished data, Bayer).

V. Toxicokinetics

Several studies have been performed which have copied the conditions of the main toxicological and carcinogenicity studies with respect to species, strain, route, dose levels and mode of administration. From these experiments an assessment of the systemic exposure for the chronic studies in rats, mice and dogs is possible by consideration of peak concentrations in plasma and AUC (Table 15). In principle the maximum systemic exposure to miglitol achievable in the different species was limited with respect to peak concentration but also with respect to AUC due to a saturation of absorption at higher doses.

Table 14. Mutagenicity/genotoxicity studies with miglitol

Test system	Dose/concentration	S9 mix	Result	Reference[a]
Point mutation assays				
Salmonella typhimurium TA 98	20–12 500 µg/plate	±	Negative	B. HERBOLD 1980
TA 100	20–12 500 µg/plate	±	Negative	B. HERBOLD 1980
TA 1535	20–12 500 µg/plate	±	Negative	B. HERBOLD 1980
TA 1535	20–12 580 µg/plate	±	Negative	B. HERBOLD 1980
TA 1537	20–12 500 µg/plate	±	Negative	B. HERBOLD 1980
Schizosaccharomyces Pombe	100–10 000 µg/plate	±	Negative	F. DUBINI 1984
CHO/HPRT	1–2500 µg/ml	±	Negative	H. LEHN 1990
Chromosome aberration assays				
Micronucleus test (male/female mice)	2 × 4000 or 2 × 8000 mg/kg	–	Negative	B. HERBOLD 1982
Dominant lethal test (male mice)	1 × 2000 mg/kg	–	Negative	B. HERBOLD 1984
DNA damage assays				
1. Test compound				
Saccharomyces cerevisiae D4 *ADE2* locus	100–10 000 µg/ml	±	Negative	F. DUBINI 1984
S. cerevisiae *TRP5* locus	100–10 000 µg/ml	±	Negative	F. DUBINI 1984
Host-mediated assay (*S. cerevisiae* D4), mice p.o.	2 × 10–1000 mg/kg	–	Negative	F. DUBINI 1984
Rat primary hepatocyte culture/DNA repair	375–3 000 µg/ml	–	Negative	H. LEHN 1990
HeLa cell/DNA repair	10–1000 µg/ml	±	Negative	F. DUBINI 1984
2. Urine from Swiss mice				
S. cerevisiae D4 *ADE2* locus	2 × 10–1 000 mg/kg	–	Negative	F. DUBINI 1984
S. cerevisiae D4 *TRP5* locus	2 × 10–1 000 mg/kg	–	Negative	F. DUBINI 1984

[a] All data unpublished, Bayer AG.

Table 15. Miglitol: pharmacokinetic parameters characterizing the systemic exposure in different species. Data are obtained from pharmacokinetic studies which have copied the conditions of the main toxicological studies

Species	Dosing in main studies	Doses [mg/kg]		C_{max} [mg/l]		AUC 0–24 h [mg·h/l]	
		NOEL	High	NOEL	High	NOEL	High
Rat	Food, 1 year (250–4000 ppm)	89	374	6.3	5.1[a]	58	122[a]
Rat	Food, 2 years (120–1000 ppm)	89	89	6.3	6.3	58	58
Rat	Intraperitoneal 4 weeks (50–800 mg/kg)	200	800	135	570[a]	172	765[a]
Mouse	Food (200–1800 ppm)	31	106	2.6	5.9	35	101
Dog	Gavage, 3 months (50–450 mg/kg)	50	450	34	106	176	580
Dog	Gavage, 1 year (20–180 mg/kg)	60	180	41	52	211	274
Human	Tablet (3 × 100 mg)	4.3		Approx. 2		33	

NOEL, no adverse effect level.
[a] Estimated values.

Following oral administration to rats by gavage, saturation was achieved at approximately 25 mg/kg (H.J. Ahr et al. 1994, unpublished data, Bayer). By administration with feed as used in the chronic toxicological studies, saturation was reached at approximately 5000 ppm miglitol admixed to feed (J. Petersen-von Gehr and E. Brendel 1994, unpublished data, Bayer). Under these conditions, peak concentrations of approximately 6 mg/l represented the maximum plasma level achievable in rats after oral administration. Due to the mode, of administration, the plasma concentration was relatively constant during the dosing interval of 24 h, resulting in a relatively high AUC value (58 mg × h/l at 1000 ppm, approximately 120 mg × h/l at 4000 ppm). Higher systemic exposure could be achieved in rats only by parenteral administration, for instance, via the intraperitoneal route.

In NMRI mice dosed with miglitol admixed to feed (60–5400 ppm), peak concentrations of 12 mg/l at 5400 ppm were reached (Petersen-von Gehr and Brendel 1994, unpublished data, Bayer). At this dose, distinct signs of saturation of absorption were found. In the mice carcinogenicity study (200–1800 ppm), maximum peak concentrations of approximately 6 mg/l and AUC values (0–24 h) of 101 mg × h/l at 1800 ppm can be assumed.

The highest systemic exposure to miglitol has been reached in dogs in a 3-month study (50–450 mg/kg (C_{max}, 106 mg/l; AUC, 580 mg × h/l). In a chronic study for 1 year, peak concentrations of 52 mg/l and an AUC of 274 mg × h/l were reached in the high-dose group (180 mg/kg). In dogs, too, starting from 60 mg/kg a tendency for saturation of absorption was observed (H.J. Ahr et al. 1993, unpublished data, Bayer).

In the chronic and subchronic toxicological studies the achieved systemic exposure to miglitol was, at least in the high-dose group, distinctly higher than exposure in humans at 100 mg tid. With respect to plasma concentrations, peak concentrations in rats and mice are limited by saturation of absorption (multiples to maximum human exposure of approximately 3). However, in dogs multiples of 25 (180 mg/kg) and 52 (450 mg/kg) have been achieved (Table 15). With respect to AUC, multiples of 2–4 in rats (limited by absorption), 3 in mice and 17.6 (3 months) or 8.3 (1 year) in dogs have been achieved.

VI. Special Studies

Dose levels of 0, 5, 50 and 500 mg/kg were administered to groups of eight female Wistar rats for 3, 7 and 28 days once daily by gavage. Hepatic and muscular glycogen concentrations were investigated in overnight-fasted animals. Miglitol led to a significant storage of hepatic glycogen after 3, 7 or 28 days at the highest dose only while lysosomal storage was insignificant. In the soleus muscle glycogen concentrations were significantly elevated after 500 mg/kg at days 7 and 28 (Lembcke et al. 1991).

Groups of eight female Wistar rats were administered dose levels of 0 and 500 mg/kg once daily for 7 days by gavage. Hepatic and muscular glycogen concentrations were studied in fed animals. Postprandial hepatic glycogen concentrations were significantly decreased but were unchanged in the soleus muscle (Lembcke et al. 1991).

Groups of 15 female Sprague-Dawley rats were administered single daily doses of 400 mg/kg by gavage or 200 mg/kg by intraperitoneal injection or were fed continuously 4000 ppm in the diet. A group of 15 untreated females served as controls. Before necropsy the animals were fasted for 24 h. Only a few animals showed very slight lysosomal glycogen storage in the liver essentially after parenteral administration (E. Bomhard and A. Popp 1993, unpublished data, Bayer).

Miglitol was not irritating to the skin when applied to three female rabbits under semiocclusive conditions for 4 h. Likewise, no irritant effect to the eye was observed in three female rabbits when 100 μl was administered into the conjunctival sac of one eye for an exposure period of 24 h (T. Märtins 1990, unpublished data, Bayer).

Intraperitoneal injection of 400 mg/kg to Sprague-Dawley rats did not induce anaphylactoid reactions (Doherty and Beaver 1990).

D. Voglibose

I. General Toxicology

1. Acute Toxicity

In Table 16 the results of the acute toxicity experiments with oral and intravenous administration to mice, rats and dogs are shown. At dose levels of 10000 (mice) or 14700 (rats) mg/kg by gavage the following symptoms were noted: reduced motility, dyspnoea, ataxia, muscular hypotonia (mice), myosis (mice), abdominal position, tonic convulsions, liquid faeces (rats), ptosis (rats) and piloerection (rats). Intravenous injection of 3830 (mice) and 3160 (rats) or more mg/kg resulted in reduced motility, dyspnoea, ataxia, tonoclonic convulsions, exophthalmus (mice), mydriasis and lateral and abdominal position (rats). At necropsy the livers of diseased animals were pale, as were the kidneys after intravenous treatment of mice. In the dogs diarrhoea (at 500 mg/kg and above) and decreased body weight gain (at 1000 and 2000 mg/kg) were seen. On the day after dosing a slight and transient increase in serum AST activity occurred (Anon 1992, unpublished data, Takeda Chemical).

2. Subacute Toxicity

a) Rat, Oral

Groups of ten male and female Wistar rats were administered dose levels of 0, 30, 100 or 300 mg/kg once daily for 5 weeks by gavage. Diarrhoea or soft faeces were observed in all treated groups. Abdominal distension was seen in males receiving 100 or 300 mg/kg. Food consumption was slightly decreased in males on 300 mg/kg during the 1st week. Plasma AST was slightly

Table 16. Acute toxicity of voglibose

Species	Strain	Sex	Route	LD_{50} (mg/kg)	95% Confidence limits	Reference
Mouse	NMRI	M + F	Oral	>14700 <21500	–	Anonymous 1992
Rat	Sprague-	M	Oral	20500	–	Anonymous 1992
	Dawley	F	Oral	22500	–	Anonymous 1992
Dog	Beagle	M	Oral	>2000	–	Anonymous 1992
Mouse	CF_1	M + F	i.v.	7820	7288–8391	Anonymous 1992
Rat	Sprague-	M	i.v.	6300	5836–6800	Anonymous 1992
	Dawley	F	i.v.	6580	5847–7405	Anonymous 1992

[a] All data unpublished, Takeda Chemical Industries, Ltd.

increased in the 300-mg/kg group. At necropsy, swelling of the jejunum, ileum, caecum and colon was noted, and caecum weight was significantly increased in all treated groups. Slight discoloration of the kidneys was observed in males receiving 100 and 300 mg/kg and in females on 300 mg/kg. In histopathology hypertrophy of the epithelium in the small intestine was seen at 100 and 300 mg/kg, and slight hypertrophy of the gob let cells in the large intestine and caecum at 300 mg/kg. Slight vacuolization of the proximal tubular epithelium in the kidney was observed in 300-mg/kg males and females at 100 and 300 mg/kg. The number of hyaline droplets was increased in 300-mg/kg males. Slight single cell necrosis of hepatocytes was seen in one male and one female receiving 300 mg/kg. The no-adverse-effect level was considered to be 30 mg/kg (ANON 1992, unpublished data, Takeda Chemical).

b) Dog, Oral

Groups of three male and three female beagle dogs received doses of 0, 30, 100, 300 and 1000 mg/kg once daily for 30 days by gavage. An additional one male and one female of the 1000-mg/kg group and the control were allocated for a 28-day recovery period. Soft faeces were occasionally seen in each group including the control with a relatively high incidence in animals of treated groups. Body weight of one male at 1000 mg/kg was decreased at the end of the treatment period but returned to normal during the recovery period. At 1000 mg/kg, AST activity and potassium were increased and sodium and β-globulin decreased in the serum. Serum calcium was decreased in animals receiving 300 or 1000 mg/kg. Since these changes were very small, the possibility of incidental findings was discussed. No effects were seen at the end of the recovery period (ANON 1992, unpublished data, Takeda Chemical).

c) Rat, Intravenous

Voglibose was administered intravenously to male Wistar rats for 4 weeks at daily dose levels of 0, 3, 10 and 30 mg/kg. At necropsy, the caecum weight including the contents was increased in each treated group as was the liver and spleen weight at 30 mg/kg. Slight focal vacuolization of the proximal tubular epithelium in the kidney at 10 and 30 mg/kg was diagnosed on histopathology (ANON 1992, unpublished data, Takeda Chemical).

3. Subchronic Toxicity

a) Rat

Groups of ten male and ten female Wistar rats were administered dose levels of 0, 10, 30, or 100 mg/kg once daily for 90 days by gavage. Diarrhoea was observed in most rats at 100 mg/kg during the whole experimental period, while swollen belly, emaciation, partly closed eyes, weakness, pilo-erection and/or dirty fur were generally transient. Some of these clinical signs were also observed in some of the rats at 30 mg/kg in the first 2 weeks of the study.

Body weights were decreased markedly in 100-mg/kg males (up to 20%) and less marked in 100-mg/kg females (up to 10%) throughout the test period and slightly in males at 30 mg/kg (up to about 8%) in the first 2 weeks of the study. Food intake was decreased in 100-mg/kg males in the first few weeks (up to 25%), whereas, at the end of the study, it was increased in males (up to 25%) and females (up to 15%) at 100 mg/kg. At 100 mg/kg, water intake in males was 2.5-fold, in females 1.3-fold, higher than that of controls. Thrombocyte count was slightly decreased in females and there was a shift in the lymphocyte/neutrophil ratio (increase in neutro-phils at the expense of lymphocytes) in males both at 100 mg/kg.

At 100 mg/kg, decreased total protein (both sexes), chloride (females), glucose (males) and albumin concentrations (females), increased ALT (males), decreased GGT (males) and LDH (both sexes) activities, increased calcium (both sexes), bilirubin (females) and inorganic phosphate concen-trations (both sexes), and decreased bilirubin (females) were measured in blood. Urinary volume was increased while density and pH were decreased in both sexes.

The weight of the filled and empty caecum showed a dose-related increase in all test groups. There was a dose-related enlargement of the caecum in males and females from 30 mg/kg. Upon microscopy, epithelial hyperplasia in caecum, colon and small intestines was noticed at 100 mg/kg. Signs of tubular nephrotoxicity were observed in the mid- and top-dose group. Increased cortico-/medullary mineralization and pelvic urothelial hyperplasia were noticed in males of all test groups and in females from 30 mg/kg, accompanied by intraluminal calcareous deposits in the urinary bladder at 100 mg/kg. It was concluded that the intestinal and renal changes observed were not considered to represent direct effects of the test com-pound, but rather secondary changes known to occur upon increased micro-bial fermentation of carbohydrates in the large intestine (Lina et al. 1991).

b) Dog

Groups of four males and four females each were administered 0, 30, 100 or 300 mg/kg once daily by gavage for a period of 3 months. All test dogs showed severe diarrhoea during the whole experimental period. A few dogs in different test groups showed a decline in physical condition as indicated by emaciation, decreased skin turgor and piloerection. Body weights were decreased (up to 20%) in all test groups, the effect being more pronounced in males than in females. Mean food intake was increased (22%–23% on average in males, 13%–19% on average in females) in the test groups than in the controls.

Haemoglobin content, packed cell volume and red cell blood count were decreased in all test groups and both sexes. Clinical chemistry of blood plasma revealed no marked changes. Slightly decreased sodium concentrations were occasionally observed in all dose groups. Urinary pH decreased in all test groups after 37 and 79 days. Upon microscopic examination, a higher number of livers with a decreased lobular pattern was observed in all test groups in both sexes.

Diarrhoea, growth retardation, decreased urinary pH and changes in plasma sodium levels were not considered to represent direct effects of the test substance, but rather secondary changes which are known to occur upon increased microbial fermentation of undigested carbohydrates due to the pharmacodynamic effect of the test substance. The slight alterations in the liver and red blood cell picture might also be due to the decreased utilization of carbohydrates (TIL et al. 1991a).

4. Chronic Toxicity

a) Rat

Groups of 30 male and 30 female Sprague-Dawley rats received dose levels of 0, 3, 10, 30 or 100 mg/kg once daily for 1 year by gavage. A transient increase in the incidence of soft stool was observed in the 30- and 100-m/kg groups at the early dosing period (up to week 3). Body weights in males were comparable between control and treatment groups. Females treated with 30 or 100 mg/kg showed a slight retardation (less than 10%) in body weight gain. Starting with week 2, food consumption was increased over controls up to 21% at 30 and 100 mg/kg, suggesting that the animals in these groups were not utilizing the feed as efficiently as the control animals. Medullary hyperplasia of the adrenal gland occurred with a higher incidence in 30- and 100-mg/kg males. In the medulla and pelvic cavity of the kidneys, a dose-related increase in mineral deposition was noted in the 10- , 30- and 100-mg/kg males with respect to incidence and severity. Based on these results, the no-effect level was considered to be 3 mg/kg (ATKINSON et al. 1991a).

b) Dog

Groups of four male and four female beagle dogs received dose levels of 0, 1, 3, 10 or 30 mg/kg once daily for 1 year by gavage. Body weight gain was lower in 30-mg/kg males and females than in control animals throughout the study. Activated partial thromboplastin time was increased over controls by 15%–18% in 30-mg/kg males at 6 and 12 months. The no-effect level was considered to be 10 mg/kg (ATKINSON et al. 1991c).

II. Reproduction Toxicology

1. Fertility and General Reproductive Performance in Rats

Voglibose was administered daily by gavage to Crl:CD Br rats at doses of 0, 10, 30 and 100 mg/kg per day; 26 males/group were treated for 9 weeks prior to mating and through delivery of the pups; 26 females/group were treated for 2 weeks prior to mating and through gestation day 12 (approximately 13/group) or lactation day 21.

Two females in the 100-mg/kg group were found dead during lactation. Clinical observations which appeared in a dose-related pattern included soft faeces in both sexes and urine stains in the females prior to and during the mating period as well ad ataxia, rough haircoat, soft faeces, swollen abdomen, thin appearance and urine stains during lactation. There was no significant effect on body weight gain and food consumption in males and in females during premating and gestation periods. From day 14 of lactation, body weights were lower the 100-mg/kg group females.

Reproductive performance was not affected and there was no indication of embryolethality in the data obtained from the gestation 13-day uterine examinations. The viability index (pup survival day 0–4) was similar for all groups; however, the weaning index (survival day 4 postcull to weaning) was significantly lower in the 100-mg/kg group. In addition, mean pup body weight values at birth were similar for all groups; however, there was a dose-related retardation in body weight gain during lactation. Despite rather large differences in the mean body weight values, especially in the 100-mg/kg group (about only half of the mean body weights of controls at day 22), only slight delays in teeth eruption and eye opening were evident.

Gross necropsy of F_0 male and female animals revealed increased caecum weights in all treated groups. Necropsies of F_1 pups revealed no congenital malformations or toxicity-related findings.

F_1 skeletal evaluations indicated a slight increase in delayed ossification of some centres in the 100-mg/kg group pups, but no treatment-related skeletal malformations. In conclusion, up to 100 mg/kg, voglibose appears to have no effect on reproductive performance or intrauterine growth. However, when voglibose was administered to the F_0 females during lactation at dosage levels as low as 10 mg/kg, the growth of the F_1 pups was impaired. Physical and functional development appeared unaffected (Morseth and Nakatsu 1991a).

2. Embryotoxicity/Teratogenicity in Rats

Groups of 32–36 pregnant Crl:CD BR rats were administered by gavage doses of 0, 100, 300 and 900 mg/kg per day from day 6 to day 17 of gestation. From each group 21–23 animals were assigned to section on day 20, the rest for natural delivery. Dose-related clinical observations during gestation included soft faeces, alopecia and urine-stained fur. Mean maternal body weight gain was significantly retarded in the 300- and 900-mg/kg groups during the treatment interval. Mean maternal food consumption was significantly reduced part of the time in the 300- and 900-mg/kg groups during the early phase of gestation but increased in a dose-related fashion during the second half of gestation.

Mean implantation efficiency and fetal viability were similar in all groups. Significantly reduced mean fetal weights (male and female) were seen in the 300- and 900-mg/kg groups. No evidence of teratogenicity was noted in F1 fetuses. However, a dose-related increase in the fetal incidence of wavy/bent ribs was seen. In addition, skeletal (300 and 900 mg/kg) and visceral (900 mg/kg) variations were consistent with small for gestational age fetuses and included delayed kidney development, slightly dilated lateral ventricles and delayed bone ossification. The gestation index (litters with live born pups) was 100% in the control, 100- and 300-mg/kg groups and 91% in the 900-mg/kg group. Viability index was 96%, 99%, 98% and 77% in the control, 100-, 300- and 900-mg/kg groups, respectively. The weaning index was 99% or 100% in all groups. There were no effects on postnatal development including reproductive performance. The authors concluded that the no-adverse-effect level was 100 mg/kg for dams, fetuses and F_1 animals (MORSETH and NAKATSU 1991b).

3. Peri-/Postnatal Study in Rats

Groups of 22–23 pregnant Crl:CD BR rats were administered by gavage doses of 0, 30, 100 and 300 mg/kg once daily from day 15 of gestation to day 21 of lactation, and effects on late fetal development, delivery, lactation, neonatal viability, growth and reproductive ability of the F_1 generation were assessed. A dose-related increase in the number of females with soft faeces was evident throughout the dosing period in the 100- and 300-mg/kg groups. Body weight gain was retarded in 100- and 300-mg/kg group animals. Mean maternal food consumption was sometimes reduced during late gestation in the 100- and 300-mg/kg group. During lactation, total litter death occurred in 1, 2, 3 and 14 litters in the control, 30-, 100- and 300-mg/kg groups, respectively.

General indications of lack of maternal care were evident during lactation at 300 mg/kg. Empty stomach was a common finding at necropsy in the 300-mg/kg F_1 pups found dead or put to death in a moribund condition. Mean pup weights during lactation were significantly less than the control at 30 mg/kg (day 22), at 100 mg/kg (days 4, 7, 14, 22) and at 300 mg/kg (days 0, 4, 7, 14 and 22). Indications of small-for-age pups or delayed development were seen in the mean achieved day of pinna detachment, generalized hair growth, eye opening and negative geotaxis in the 100- and 300-mg/kg groups.

F_1 skeletal evaluations at day 22 revealed no significant differences between groups. During the F_1 growth phase significantly lower mean body weight values were noted in all treatment groups. There were no clear treatment-related differences in the open field or water T-maze measurements. There were no dose-related differences in mean gestation length, implantation data, pregnancy rates, gestation or viability index. The weaning index was significantly reduced from 100% in the control and 30-mg/kg groups to 76% at 100 mg/kg and 30% at 300 mg/kg. Body weights of the F_2 pups were below those of the control group at 100 and 300 mg/kg.

In conclusion, adverse effects of treatment in the 100- and 300-mg/kg groups included evidence of an effect on the dam during treatment and the offspring during lactation. There was no evidence of treatment-related effects in F_1 pup autopsy findings and skeletal examinations. The reproductive performance of the survivors of the F_1 generation was uncompromised by exposure in utero (Morseth and Nakatsu 1991c).

4. Embryotoxicity/Teratogenicity in Rabbits

Groups of 13–17 pregnant female HRA:NZW rabbits were administered by gavage doses of 0, 100, 300 and 1000 mg/kg once daily from day 6 to day 18 of gestation. In the 100-mg/kg group four females died, five aborted, one was put to death in a moribund condition, one gave birth on day 29 of gestation and one had only resorptions. In the 300-mg/kg group, three females aborted compared with one in the control group. A higher incidence of anorexia was noted in the 300- and 1000-mg/kg groups. During the treatment period soft faeces (300- and 1000-mg/kg), urine stains and languid behaviour (1000 mg/kg) were observed. Mean body weight values in the 1000-mg/kg group were less than the control values during the treatment and posttreatment periods. Mean food consumption was reduced at 300 and 1000 mg/kg during the treatment period. Maternal gross pathology findings included distension of the caecum at 1000 mg/kg. Mean implantation efficiency values did not indicate a treatment effect. Incidence of resorptions was 3%, 13%, 26% and 50% for the control, 100-, 300- and 1000-mg/kg groups, respectively. The slight decrease in mean fetal viability at 100 mg/kg was caused by one litter with a high percentage of resorptions and therefore was not attributed to treatment. Mean fetal body weights were slightly lower in the 300-mg/kg animals. Neither the incidence nor the type of fetal malformations indicated a treatment-related effect. In conclusion, the no-adverse-effect level was 100 mg/kg for dams and doses up to 300 mg/kg had no teratogenic effect on the fetuses (Morseth and Nakatsu 1991d).

5. Peri-/Postnatal Study in Rats Fed on a Glucose Diet

Groups of 14–15 female Crl:CD rats were administered doses of 0 and 300 mg/kg once daily from day 15 of gestation to day 21 of lactation by gavage and were fed either a commercial diet or a glucose diet (glucose, 66.1%). When a commercial diet was given, the dams had several clinical signs (soft faeces, diarrhoea, emaciation, swollen abdomen), reduced body weight gain and food consumption, blood chemistry changes (e.g., increased calcium, urea nitrogen and inorganic phosphorous concentrations as well as increased AST and ALT activities, lower albumin concentrations and A/G ratio) and some organ weight changes (e.g., caecum weight increases). In the F_1 pups the viability and the weaning index were decreased and growth was retarded. None of these effects were seen when a glucose diet was given (ANON 1992, unpublished data, Takeda Chemical).

III. Genotoxicity/Mutagenicity

Three bacterial point mutation studies have been reported, using the following strains: *E. coli* WP2uvrA and *S. typhimurium* TA98, TA100. TA1535 and TA1537. Irrespective of whether or not metabolic activation was applied, the experiments were conducted with doses of 313 – 5000 μg/plate for all strains. In one additional experiment, the doses were 2500–10000 μg/plate without adding S9 mix. Voglibose caused no increases in the number of revertant colonies.

Chinese hamster lung (CHL) cells were used in two chromosomal aberration tests, giving 24 and 48 h treatment without metabolic activation, and one experiment with 6 h treatment with and without S9 mix metabolic activation. The concentration range investigated was 0.625–10 mM. No significant increases in the incidence of cells showing chromosomal aberrations were seen. In the micronucleus test (C3HxSWV), F_1 male mice were given single oral doses of 1250, 2500 or 5000 mg/kg, or four doses of 1250 mg/kg at 24-h intervals. The incidence of micronucleated polychromatic erythrocytes in femoral bone marrow was not increased in any of the groups (SAKAMOTO et al. 1991).

IV. Carcinogenicity

1. Rat

Groups of 60 male and 60 female F344/DuCrj rats were administered by gavage dosage levels of 3, 10 or 30 mg/kg once daily for 24 months. Control animals (120 per sex) received the vehicle (distilled water). Soft faeces were observed beginning in week 1 in the 3- (males), 10- and 30-mg/kg (male and female) groups, but the number of animals affected decreased daily. Body weight gain was slightly (less than 10% difference in body weight compared with controls) retarded in males from 10 mg/kg and in females of all treatment groups. In the 30-mg/kg group, food consumption was decreased during week 1. At necropsy, distension of the caecum was observed in the 30-mg/kg group and in the 10-mg/kg males. Amongst the neoplastic lesions found, some statistically significant differences were identified in the number of males with testicular interstitial cell tumours and females with uterine tumours. A selection of tumours from organs with particular interest is shown in Table 17. The incidence of interstitial cell tumours in F344 rats is

Table 17. Incidence of selected tumours in a 24-month study with gavage administration of voglibose to F344 rats

Organ/tissue Tumour type	Dose group (mg/kg)							
	Males				Females			
	0	3	10	30	0	3	10	30
Pancreas	(120)	(15)	(13)	(60)	(120)	(14)	(15)	(60)
Islet cell adenoma	10	1	2	6	3	0	0	0
Pituitary	(120)	(32)	(31)	(60)	(119)	(41)	(37)	(60)
Adenoma	51	9	15	11	48	18	16	25
Adenocarcinoma	2	2	2	0	0	2	1	1
Adrenal medulla	(120)	(60)	(60)	(60)	(120)	(15)	(14)	(60)
Phaeochromocytoma (B)	9	5	7	6	3	1	0	4
Phaeochromocytoma (M)	2	1	2	3	0	0	0	0
Mammary gland	(120)	(60)	(60)	(60)	(120)	(60)	(60)	(60)
Adenoma	0	0	0	0	1	1	0	3
Fibroadenoma	1	1	1	1	8	5	8	10
Subcutaneous tissue	(120)	(60)	(60)	(60)	(120)	(60)	(60)	(60)
Fibroma	6	0	0	2	2	0	0	1
Lipoma	9	6	2	0	1	1	0	0
Fibrosarcoma	2	2	0	1	0	0	0	0
Uterus	–	–	–	–	(119)	(60)	(60)	(59)
Adenoma	–	–	–	–	3	2	2	2
Adenocarcinoma	–	–	–	–	3	3	1	6
Testes	(120)	(60)	(60)	(60)	–	–	–	–
Leydig cell tumour (B)	101	55	55	58	–	–	–	–

( ), number of animals examined microscopically.
B, benign; M, malignant.

known to reach nearly 100%. The significant increase in this tumour was therefore not considered to be indicative of oncogenicity. When taking both hyperplasia regarded as precancerous lesions and uterine adenomas into account no differences were found. In addition, the incidence of uterine adenomas was within the range of variation of historical controls; it was thus not considered to be treatment related. Non-neoplastic lesions observed at a slightly higher incidence in the 30-mg/kg group were calcification of the renal pelvis, hyperplasia of the adrenal medulla (males), vacuolated focus in the adrenal cortex (females) and cysts in the uterus. However, the progress of spontaneous nephropathy was suppressed in the 30-mg/kg group. It was concluded that voglibose was not oncogenic under the conditions of this study (Nonoyama et al. 1991).

2. Mouse

Voglibose was administered once daily by gavage to CD-1 mice (60/sex/group) at dosage levels of 0, 15, 50 and 150 mg/kg for up to 2 years. Dose levels were selected based on the results of a 3-month dosefinding study in mice (0, 150, 300, 600 mg/kg): at 300 and 600 mg/kg death occurred and moribund killings were necessary, and decreased carcass weight was observed in all treated groups. After 83 weeks of treatment, mortality had reached 75% in the low-dose group males, and all male groups were terminated prior to schedule. Females in the 50- and 150-mg/kg groups were put to death during weeks 97 and 101, respectively, also due to the mortality rate. The low-dose and control females remained in the study for the entire 2-year period (104 weeks). The mortality/moribundity was due to chronic nephropathy, which was not considered to be treatment related since this finding was also evident in control animals. In the 50- and 150-mg/kg per day groups, group mean body weight was occasionally significantly lower than that in the control group. Voglibose under the conditions of this study was not on-cogenic (ATKINSON et al. 1991b).

V. Toxicokinetics

Data on toxicokinetic investigations are only available from the study published by SUZUKI et al. (1991) (see below). Plasma levels of voglibose were determined before and 0.5, 1.0, 2.0 and 4.0 h after dosing on days 38 and 80 of the 13-week study with Wistar rats. Daily administration of 300 mg/kg by gavage resulted in peak plasma levels of up to about 12 μg/ml. Females appeared to have slightly higer peak levels. Simultaneous feeding with glucose had no obvious effect on plasma concentrations. Pharmacokinetic data have been published (MAESHIBA et al. 1991) and are dealt with in Chap. 19, this volume.

VI. Special Studies

1. Antigenicity in Guinea Pigs and Mice

Active systemic anaphylaxis was not elicited by injection of voglibose in guinea pigs which has been treated with voglibose orally or with Freund's complete adjuvant (FCA) emulsion of voglibose intradermally. Homologous passive cutaneous anaphylaxis (PCA) using sera of these animals was not elicited by voglibose. The rat 24-h PCA reaction indicated that no hapten-specific antibodies were produced in mice treated with voglibose orally or with FCA emulsion of voglibose subcutaneously. These results indicated that voglibose was non-antigenic under the conditions of this study (NAKAI et al. 1991).

2. Effect of Dietary Carbohydrates on the Oral Toxicity in Rats

Groups of ten male and ten female Jcl:Wistar rats were treated once daily for 13 weeks by gavage at dosage levels of 0 or 300 mg/kg and were fed either a commercial diet or a semisynthetic glucose diet (glucose content, 66.1%). In voglibose-treated rats fed a commercial diet, soft faeces, increases in food consumption and water intake, elevated urinary calcium and plasma urea nitrogen, ALP, AST and ALT, increased liver and caecum weights, caecum distension and reduced adipose tissue in the abdominal cavity were observed. No treatment-related changes were observed in voglibose-treated rats fed a glucose diet. Thus, the treatment-related changes in rats fed a commercial diet were considered to have been caused by excessive manifestation of the pharmacological activity (inhibition of disaccharidase) of voglibose (Suzuki et al. 1991).

3. Effect of Low-Protein Diet on Plasma Transaminases in Rats

Groups of 12 male Jcl:Wistar rats were fed ad libitum a low-protein diet (6% casein, 78% starch), a low-protein glucose diet (6% casein, 78% glucose) and a commercial diet (21% protein) for 15 days. During the terminal 8 days half of the rats in each group were given daily doses of 300 mg/kg by gavage. Plasma levels of ALT and AST were increased or tended to be increased in the voglibose-treated rats fed a low-protein diet but not when fed a low-protein glucose or a commercial diet. The activities of hepatic ALT and AST were markedly increased in voglibose-treated rats fed a low-protein diet but not on the low-protein glucose or the commercial diet. Small hepatocytes were seen in voglibose-treated rats fed a low-protein diet. These results may indicate that plasma transaminases were increased as a result of enhanced catabolism, i.e. gluconeogenesis, in the liver associated with increased liver transaminase activities that might have occurred to compensate sustained malabsorption of glucose induced by voglibose treatment (Anon 1992, unpublished data, Takeda Chemical).

E. Castanospermine

Castanospermine is an alkaloid found in the seeds of the Australian tree *Castanospermum australe*, which inhibits a number of glycohydrolases. It is no longer under development as an oral antidiabetic but for comparative purposes it seemed worthwhile to include at least some of the published toxicology data.

I. General Toxicology

1. Acute/Subacute Toxicity

An LD_{50} value >500 mg/kg in mice has been published but with no details regarding application route, sex, strain of mice, clinical or postmortem observations (Sunkara et al. 1987).

Indirect evidence for a rather low acute toxicity also in rats comes from a study where young Sprague-Dawley rats were injected intraperitoneally doses of up to 2 g/kg daily over a 3-day period. Some animals given 2 g/kg developed diarrhoea and had high bacterial counts in their intestines when fed normal rat chow. Using a special diet devoid of sucrose and starch, animals exhibited no gastrointestinal symptoms and did not have diarrhoea. Since castanospermine inhibited intestinal sucrase and maltase, the gastrointestinal symptoms suggested that these animals might be suffering from a maltose intolerance (SAUL et al. 1985).

II. Reproduction Toxicology

Treatment of Syrian hamsters with dose levels up to 660 mg/kg per os did not result in fetal or maternal toxicity. The only effect was decreased fetal body weight (MACGREGOR et al. 1989).

III. Genotoxicity/Mutagenicity

No data are available regarding genotoxicity and mutagenicity.

IV. Carcinogenicity

No data are available regarding carcinogenicity.

V. Toxicokinetics

No data are available regarding toxicokinetics.

VI. Special Studies

Young Sprague-Dawley rats were injected intraperitoneally with doses ranging from 0.1 to 0.2 g/kg daily over a 3-day period and put to death on the 4th day. Dose levels of 500 mg/kg or higher decreased hepatic α-glucosidase activity to 40% of control values, whereas α-glucosidase in brain was reduced to 25% and that in spleen and kidney to about 40%. In liver, both the neutral and the acidic α-glucosidase activities were inhibited, the neutral being more susceptible. In treated animals fed normal rat chow, the hepatocytes were smaller in size and simplified in structure, whereas feeding a high-glucose diet lessened these alterations. At dose levels of 1 g/kg or higher there was a marked decrease in the amount of glycogen in the cytoplasm, while in lysosomes it was largely increased. Glycogen levels in liver were somewhat depressed at 1.5 and 2.0 g/kg. Castanospermine also inhibited intestinal glycosidases (i.e., maltase and sucrase). Since the normal rat diet contains large amounts of sucrose and starch, and these sugars presumably cannot be metabolized in treated animals, the animals exhibit symptoms like those of individuals with lactose intolerance (i.e., high intestinal bacterial flora, diarrhoea, growth retardation, etc.). On a different diet in which sucrose and starch were replaced with glucose, these symptoms were overcome. But glycogen was still found in lysosomes, which indicated that glucosidase activities in these animals were still inhibited (Saul et al. 1985).

Male Sprague-Dawley rats were treated orally by gavage or by i.p. or i.v. injections once at dose levels ranging from 0.001 to 1000 mg/kg or with daily oral doses of 0.1 or 1.0 mg/kg for 12 days. Rats were killed after fasting for 17–19 h at various time points after treatment. Liver lysosomal glycogen accumulation positively correlated with lysosomal acid α-glucosidase inhibition when inhibition was about 50% or greater. Glycogen did not accumulate when inhibition was less than 50%. The route of administration had little effect on the amount of inhibition observed. In rats killed 17 h after administration, the doses estimated to cause 50% inhibition were 0.77, 0.11 and 0.22 mg/kg for i.p., i.v. and oral administration, respectively. Accumulated glycogen disappeared as lysosomal acid α-glucosidase activity recovered. Surprisingly, 12 daily doses of 1 mg/kg had only a small cumulative effect on inhibition and did not cause more glycogen accumulation than a single dose (Rhinehart et al. 1991).

Single intraperitoneal injections of 10–400 mg/kg as well as 400 mg/kg orally by gavage caused an anaphylactoid reaction similar to that induced by dextran, i.e., erythema and oedema of the snout, ears and paws for several hours after administration. Susceptible rat strains were Sprague-Dawley and Lewis (breeder, Charles River), Sprague-Dawley rats (breeder, Harlan), Fischer 344, Wistar-Kyoto, spontaneously hypertensive rats (breeder, Charles River) as well as Hartley guinea pigs and CD-1 mice showed no such responses. Rats which responded to castanospermine showed marked, but transient, tachyphylaxis to a second dose of castanospermine or dextran. Rats maintained on a complex carbohydrate-free diet also responded to castanospermine, excluding the possibility that the effect was due to absorption of dextran-like, dietary complex carbohydrates (Doherty and Beaver 1990).

F. Common Toxicological Characteristics of Glucosidase Inhibitors

The four α-glucosidase inhibitors which are either on the market or are under development as antidiabetics underwent the whole range of toxicological testing usually required for registration of compounds intended for long-term human therapy. In addition, and due to unexpected findings as well as those which could be anticipated due to the inherent enzyme inhibition property, an array of additional studies have been performed. Although, in some respect, the picture sometimes looks heterogeneous, there were basically three primary effects observed which are all related to the pharmacodynamic activity. These are:

1. Reduced glucose utilization
2. Lysosomal glycogen storage
3. Accumulation of undigested carbohydrates in the large intestine

Each of these three effects can but must not necessarily result in a cascade of further functional and/or morphological changes. Some of them seem to be species specific, while others are more or less dependent on certain experimental conditions. Direct comparison is often difficult due to different dose levels, strain of animals used, mode of administration or other factors.

I. Reduced Glucose Utilization

When administered shortly before or simultaneously with feed, the α-glucosidase inhibitors by their intrinsic properties reduce glucose utilization to an extent where the isocaloric state can be lost. Severe body weight retardation is one of the consequences especially seen in rats, dogs and rabbits. Mice and hamsters (tested with acarbose only) seem to be more resistant. Emaciation and cachexia have occurred in dogs and rabbits and some of these animals died in the experiments performed with acarbose, emiglitate and voglibose. In the case of pregnant animals, doses close to borderline of maternal "toxicity" seemed to secondarily affect especially the rabbit embryo, where a slight increase in resorptions and consequently a reduced number of fetuses are the main effects. In long-term rat studies, on the other hand, this "caloric restriction" leads to an increased longevity as well as decreased tumour incidences especially in those organs which are hormone dependent. Both effects are well known from experiments with dietary restriction or low-calorie feeding.

On the organ level (liver, kidney), signs of increased gluconeogenesis from glucogenic amino acids occurred. In the liver this results in an induction of enzymes involved in the amino acid/glucose metabolism such as AST and ALT. Increased activities of both enzymes in serum or plasma can best be explained in this way, which have been observed at high oral doses of acarbose, emiglitate and voglibose in the rat (without obvious strain specificity).

Interestingly, in dogs, only moderately increased AST activities occurred after emiglitate and miglitol treatment. In the dog, as a carnivorous species, gluconeogenesis from amino acids is normally much higher than in the rat, which would explain why the dog was less responsive with regard to the induction of these enzymes. The lack of a consistent effect on these enzymes even after subacute high-dose parenteral treatment with acarbose, emiglitate and miglitol in rats and dogs argues against a compound-related hepatotoxic potential. In addition, as has been demonstrated by W. Puls et al. (1981, unpublished data, Bayer), serum ALT activity increased nearly to the same levels as after acarbose treatment when male and female Sprague-Dawley rats were fed a low-carbohydrate diet. Finally, feeding of high glucose levels (66.1%) to voglibose-treated Wistar rats (300 mg/kg daily for 12 weeks by gavage) completely prevented the rise in AST and ALT plasma activity (Suzuki et al. 1991). It is well known that, e.g. aspartate, pyruvate, glutamate and alanine, make a major contribution to gluconeogenesis in rats under conditions such as high-protein/low-carbohydrate intake, starvation or glucocorticoid excess. The activity of enzymes involved in the metabolism of these amino acids is thereby increased in the liver as well as in peripheral blood (Heard et al. 1977; Fafournoux et al. 1993; Otto 1965; Rowsell et al. 1969, 1973; Yeung and Oliver 1967).

In acarbose-treated hamsters as well as in emiglitate- and miglitol-treated mice, no comparable effects on these enzymes were to be seen. There is also the possibility that the morphological changes seen in the liver of emiglitate-treated rats and dogs and castanospermine-treated rats can be connected to this increased gluconeogenesis.

In the kidney, at least for acarbose, it was shown that gluconeogenesis is increased in the proximal tubules. An increase in glucose transporter type I in epithelial cells of the collecting ducts, on the other hand, suggests an increased glucose uptake by these cells, leading to a locally increased glucose availability and possibly to an increased incidence of renal cell tumours (Enzmann et al. 1994).

II. Lysosomal Glycogen Storage

α-Glucosidase inhibition at the lysosomal level leads to glycogen storage within these cellular organelles. This effect is not restricted to a specific organ. Consequently, systemically available α-glucosidase inhibitors as well as metabolites or absorbed intestinal breakdown products with inhibitory activity can induce lysosomal glycogen storage. All compounds which have been adequately tested exert this effect. Voglibose has not yet been tested in this respect. For compounds with relatively low bioavailability such as acarbose, parenteral administration or very high oral dose levels are necessary to demonstrate this accumulation. The difference in potency between emiglitate and miglitol can be explained by their difference in lipophilicity. Emiglitate with its lipophilic properties reaches the lysosomes much more easily than miglitol and is therefore much more potent.

Although the morphological picture resembles that of human glycogen storage disease (the so-called Pompe's disease), deleterious effects which could be attributed to this glycogen storage have not been seen so far even after chronic high-dose administration. As mentioned before, increased activity of enzymes such as ALT and AST, generally regarded as liver disease markers, is most likely due to increased gluconeogenesis.

III. Accumulation of Undigested Carbohydrates in the Large Intestine

Administration of α-glucosidase inhibitors shortly before or during food consumption also leads to several changes in the physiology of digestion which may have secondary consequences in several other organ functions. The inhibition of intestinal α-glucosidase leads to an accumulation of undigested carbohydrates in the large intestine, thereby leading to caecal enlargement and increased caecal weights. Soft faeces or even diarrhoea and increased food and water consumption are regularly observed. Changes in the activities and profiles of intestinal enzymes as well as the microbial population have been described under these conditions. The latter effect could be one of the reasons why mice did not tolerate administration of acarbose via feed; the former was demonstrated to contribute to the increased alkaline phosphatase activity in plasma.

Intestinal effects like those described are not specific to these α-glucosidase inhibitors but are rather common for a diversity of mostly poorly absorbed and/or osmotically active substances essentially of carbohydrate-like structures or sugar alcohols (LORD and NEWBERNE 1990). Epithelial hyperplasia has sometimes been reported in the small intestine, caecum and colon in these cases. The response of the caecum to this accumulation can be obviously quite different and can partly depend on the presence of compounds available at the caecum level. For example, LEMBCKE et al. (1987) demonstrated differences in the effects of acarbose, emiglitate and miglitol on caecal weight and total caecal carbohydrate content, suggesting that acarbose may partially inhibit bacterial carbohydrate degradation in the caecum. Several secondary consequences of these intestinal changes have been reported. These are:

1. Renal mineralization/nephrocalcinosis, in some cases with pelvic epithelial hyperplasia and cortical nephrosis
2. Adrenal medullary hyperplasia/neoplasia
3. Leydig cell hyperplasia/neoplasia
4. Osteodystrophia fibrosa
5. Hyperplasia in the parathyroids

All of these changes have been observed with different substances of unrelated chemical structure. They have been explained by the intestinal changes consisting of increased microbial fermentation, reduction of pH value of the intestinal content, increased absorption of fatty acids and lactic acid, aciduria, increased absorption and urinary excretion of calcium as well as proteinuria (DE GROOT 1987; GARDNER 1961). The effects on the kidneys have recently been reviewed by LORD and NEWBERNE (1990). The connections between calcium absorption and adrenal medullary proliferative diseases in the rats have been investigated previously (ROE and BÄR 1985).

The evidence between the occurrence of Leydig cell hyperplasia and neoplasia and changes in the intestinal physiology has been collated by Bär (1992) in the case of lactitol and lactose. Although not reported for a broader range of compounds in this connection, the parathyroid hyperplasia and osteodystrophia fibrosa seen after feeding of high doses of emiglitate to rats could easily be related to disturbances in the ionic imbalances and/or the consecutive changes in renal physiology. Parathyroid stimulation has been shown, e.g., in phosphate-induced nephrocalcinosis (Haase 1978).

There seems to be a number of factors finally involved, which determine which of these secondary effects become manifest. Systematic investigations into these factors are largely lacking. With the possible exception of renal mineralization, all other changes seem to be rat specific. This may indicate that the large intestine of the rat, especially the caecum, with its important digestive functions, is critical for these kinds of alterations.

Acknowledgements. The author thanks Dr. H. Bischoff and Prof. Dr. G. Schlüter for many valuable and stimulating discussions, Dr. H. J. Ahr for compiling the toxicokinetic data, Dr. A. Popp for providing the electron microscopic figure and Mrs. G. Kampermann for typing the manuscript. The supply of unpublished data by Dr. K. Kitazawa, Takeda Chemical Industries Ltd., is gratefully acknowledged.

References

Ahr HJ, Boberg M, Krause HP, Maul W, Müller FO, Ploschke HJ, Weber H, Wünsche C (1989) Pharmacokinetics of acarbose. Part I: absorption, concentration in plasma, metabolism and excretion after single administration of [^{14}C]acarbose to rats, dogs and man. Drug Res 39:1254–1260

Atkinson JE, Daly IW, Saulog TM, Suzuki T (1991a) One-year oral gavage toxicity study of AO-128 in rats. Jpn Pharmacol Ther 19:233–260

Atkinson JE, Daly IW, Saulog TM, Fukuda R, Tanakamaru Z, Nonoyama T (1991b) Oncogenicity study of AO-128 in mice. Jpn Pharmacol Ther 19:131–142

Atkinson JE, Daly IW, Saulog TM, Suzuki T (1991c) One-year oral toxicity study of AO-128 in beagle dogs. Jpn Pharmacol Ther 19:261–282

Bär A (1992) Significance of leydig cell neoplasia in rats fed lactitol or lactose. J Am Coll Toxicol 11:189–207

Balfour JA, McTavish D (1993) Acarbose. An update of its pharmacology and therapeutic use in diabetes mellitus. Drugs 46:1025–1054

Bomhard E (1992) Frequency of spontaneous tumours in wistar rats in 30-month studies. Exp Toxicol Pathol 44:381–392

Bomhard E, Karbe E, Löser E (1986) Spontaneous tumours of 2000 wistar TNO/W 70 rats in two-year carcinogenicity studies. J Environ Pathol Toxicol Oncol 7:35–52

Bomhard E, Rinke M (1994) Frequency of spontaneous tumours in wistar rats in 2-year studies. Exp Toxicol Pathol 46:17–29

De Groot AP (1987) Biological effects of low digestibility carbohydrates. In: Leegwater DC, Feron VJ, Hermus RJJ (eds) Low digestibility carbohydrates. Pudoc, Wageningen, pp 12–22

Doherty NS, Beaver TH (1990) Comparison of the anaphylactoid response induced in rats by castanospermine and dextran. Int Arch Allergy Appl Immunol 93:19–25

Fafournoux P, Rémésy C, Demigné C (1993) Control of alanine metabolism in rat liver by transport processes or cellular metabolism. Biochem J 210:645–652

Gardner KD (1961) The effect of pH on the filtration, reabsorption and excretion of protein by the rat kidney. J Clin Invest 40:525–535

Geddes R, Taylor JA (1985) Lysosomal glycogen storage induced by acarbose, a 1,4-α-glucosidase inhibitor. Biochem J 228:319–324

Gross L (1988) Inhibition of the development of tumours or leukemia in mice and rats after reduction of food intake. Cancer 62:1463–1465

Haase P (1978) Parathyroid stimulation in phosphate induced nephrocalcinosis. J Anat 125:299–311

Heard CR, Frangi SM, Wright PM, Mccartney PR (1977) Biochemical characteristics of different forms of protein-energy malnutrition: an experimental model using young rats. Br J Nutr 37:1–21

Kamikawa T, Yamashita S, Kobayashi N, Minaga T (1989a) Acute toxicity study of acarbose with a single oral or intravenous administration in rats (in Japanese). Clin Res 23(12):7–10

Kamikawa T, Yamashita S, Kobayashi N, Minaga T (1989b) Toxicity study in rats treated orally with acarbose for 13 weeks (in Japanese). Clin Res 23(12):11–34

Keenan K, Smith P, Ballam G, Soper K, Bokelman D (1992) The effect of diet and dietary optimisation (caloric restriction) on survival in carcinogenicity studies – an industry viewpoint. In: McAuslane JAN, Lumley CE, Walker SR (eds) The carcinogenicity debate. Quay Publishing, pp 77–102

Konishi Y, Hata Y, Fujimori K (1989) Formation of glycogenosomes in rat liver induced by injection of acarbose, an α-glucosidase inhibitor. Acta Histochem Cytochem 22:227–231

Konishi Y, Okawa Y, Hosokawa S, Fujimori K, Fuwa H (1990) Lysosomal glycogen accumulation in rat liver and its in vivo kinetics after a single intraperitoneal injection of acarbose, an α-glucosidase inhibitor. J Biochem 107:197–201

Lembcke B, Löser C, Frölsch UR, Wöhler J, Creutzfeld W (1987) Adaptive responses to pharmacological inhibition of small intestinal α-glucosidase in the rat. Gut 28:181–187

Lembcke B, Lamberts R, Wöhler J, Creutzfeld W (1991) Lysosomal storage of glycogen as a sequel of a α-glucosidase inhibition by the absorbed deoxynojirimycin derivative emiglitate (Bay o 1248). Res Exp Med (Berl) 191:389–404

Lina B, Dreef-Van Der Meulen HC, De Groot AP, Suzuki T (1991) Subchronic (90-day) oral toxicity study with AO-128 in rats. Jpn Pharmacol Ther 19:181–208

Lord GH, Newberne PM (1990) Renal mineralization – A ubiquitous lesion in chronic rat studies. Food Chem Toxicol 28:449–455

Lüllmann-Rauch R (1981) Lysosomal storage glycogen mimicking the cytological picture of Pompe's disease as induced in rats by injection of an α-glucosidase inhibitor. I. Alterations in liver. Virchows Arch 38:89–100

Lüllmann-Rauch R (1982) Lysosomal glycogen storage mimicking the cytological picture of Pompe's disease as induced in rats by injection of an α-glucosidase inhibitor. II. Alterations in kidney, adrenal gland, spleen and soleus muscle. Virchows Arch 39:187–202

Macgregor JT, Caldwell K, Crawford L (1989) Biological effects of plant secondary metabolites in mammals. Fedrip Database; National Technical Information Service (NTIS). Western Regional Res Center/Toxicology & Biological Eval/Research Unit. Albany, California 94710

Maeshiba Y, Kondo , Kobayashi T, Tanayama K (1991) Metabolic fate of AO-128, a new α-glucosidase inhibitor, in rats and dogs (in Japanese). Jpn Pharmacol Ther 19:247–257

Morseth SL, Nakatsu T (1991a) Reproduction study in rats with AO-128. Jpn Pharmacol Ther 19:29–44

Morseth SL, Nakatsu T (1991b) Teratological study in rats with AO-128. Jpn Pharmacol Ther 19:45–67

Morseth SL Nakatsu T (1991c) Perinatal and postnatal study in rats with AO-128. Jpn Pharmacol Ther 19:79–100

Morseth SL, Nakatsu T (1991d) Teratological study in rabbits with AO-128. Jpn Pharmacol Ther 19:69–78

Nakai Y, Doi T, Inoue S (1991) Antigenicity of AO-128 in guinea pigs and mice (in Japanese). Jpn Pharmacol Ther 19:101–105

Nonoyama T, Imai R, Tensha S, Nagai H, Sakura Y, Wada T (1991) Onocogenicity study of AO-128 in rats (in Japanese). Jpn Pharmacol Ther 19:119–129

Otto K (1965) Alanin-transaminase und gluconeogenese. Hoppe-Seylers Zschr Physiol Chem 341:99–104

Pickering RG, Pickering CE (1984) The effect of diet on the incidence of pituitary tumours in female wistar rats. Lab Anim 18:298–314

Rhinehart BL, Begovic ME, Robinson KM (1991) Quantitative relationship of lysosomal glycogen accumulation to lysosomal α-glucosidase inhibition in castanospermine treated rats. Biochem Pharmacol 41:223–228

Roe FJC, Bär A (1985) Enzootic and epizootic adrenal medullary proliferative disease of rats: influence of dietary factors which affect calcium absorption. Hum Toxicol 4:27–52

Rowsell EV, Snell K, Carnie JA, Al-Tai AH (1969) Liver L-alanine-glyoxylate and L-serine-pyruvate aminotransferase activities: an apparent association with gluconeogenesis. Biochem J 115:1071–1073

Rowsell EV, Al-Tai AH, Carnie JA (1973) Increased liver L-serine-pyruvate aminotransferase activity under gluconeogenic consitions. Biochem J 134:349–351

Sakamoto Y, Fujikawa K, Nakamura M, Hitotumachi S (1991) Mutagenicity studies of AO-128 (in Japanese). Jpn Pharmacol Ther 19:107–117

Saul R, Ghidoni JJ, Molyneux RJ, Elbein AD (1985) Castanospermine inhibits α-glucosidase activities and alters glycogen distribution in animals. Proc Natl Acad Sci 82:93–97

Saxton JA, Sperling GA, Barnes LL, Mccay CM (1948) The influence of nutrition upon the incidence of spontaneous tumours of the albino rat. Acta Unio Int Cancrum 6:423–431

Schlüter G (1982) Toxicology of acarbose. In: Creutzfeld W (ed) Proceedings of 1. international symposium of acarbose. Excerpta Medica Amsterdam 1982, International Congress Series 594, pp 49–54

Schlüter G (1988) Toxicology of acarbose, with special reference to long-term carcinogenicity studies. In: Creutzfeld W (ed) Acarbose for the treatment of diabetes mellitus. Springer, Berlin, pp 5–14

Sinkeldam EJ, Woutersen RA, Hollanders VMH, Til HP, Van Garderen-Hoetmer A, Bär A (1992) Subchronic and chronic toxicity/carcinogenicity feeding studies with lactitol in rats. J Am Coll Toxicol 11:165–188

Sunkara PS, Bowlin TL, Liu PS, Sjoerdsma A (1987) Antiretroviral activity of castanospermine and deoxynojirimycin, specific inhibitors of glycoprotein processing. Biochem Biophys Res Commun 148:206–210

Suzuki T, Nagayabu T, Sato S, Abe T, Miyajima H, Matsuo T (1991) Effect of carbohydrates on oral toxicity of AO-128 in rats (in Japanese). Jpn Pharmacol Ther 19:13–19

Til HP, Kuper CF, Falke He, De Groot AP, Suzuki T (1991) Subchronic (90-Day) oral toxicity study with AO-128 in dogs. Jpn Pharmacol Thet 19:209–231

Tucker MJ (1979) The effect of long-term food restriction on tumours in rodents. Int J Cancer 23:803–807

Yeung D, Oliver IT (1967) Gluconeogenesis from amino acids in neonatal rat liver. Biochem J 103:744–748

CHAPTER 21

Clinical Pharmacology of Glucosidase Inhibitors

E. Brendel and W. Wingender

A. Acarbose

I. Introduction

The clinical pharmacology of acarbose covers the findings obtained from clinical pharma cological studies performed in healthy and patient volunteers. Numerous controlled studies have confirmed the effects and mechanism of action of acarbose in man as observed in animal experiments.

II. Pharmacodynamics

1. Effects on Blood Sugar and Insulin

Acarbose has been shown in animal studies to be a competitive inhibitor of intestinal brush-border α-glucosidase enzymes. The inhibitory potency follows a rank order of glucoamylase > sucrase > maltase > isomaltase (Caspary and Graf 1979; Caspary et al. 1982; Goda et al. 1981). Acarbose has little or no effect on lactase and trehalase (Caspary and Graf 1979). Studies in healthy volunteers of normal weight and obese subjects demonstrated that acarbose reduced postprandial hyperglycemia and serum insulin concentrations (Hillebrand et al. 1979; Fölsch et al. 1981a; Sjöström and William-Olsson 1981; Dimitriadis et al. 1982a). Administration of acarbose (75–300 mg) with meals reduces dose-dependently blood glucose and serum insulin concentrations. Doses of 150 mg and 300 mg of the inhibitor are more effective than a dose of 75 mg (Hillebrand et al. 1979).

A clear dose-proportionality of single oral doses of 50, 100 and 200 mg acarbose was shown in respect to an increase in blood glucose and serum insulin after simultaneous ingestion of 75 g sucrose (Fig. 1; Azuma et al. 1990). The inhibition of the insulin response was identical following administration of 1×100 mg and 2×50 mg acarbose, thus demonstrating pharmacological bioequivalence. The reduction in postprandial insulin levels was also found to be dose dependent in obese subjects, in whom insulin resistance is often present. Furthermore, there was also a reduction in the insulinogenic index (insulin/glucose ratio), probably secondary to a reduction in insulin

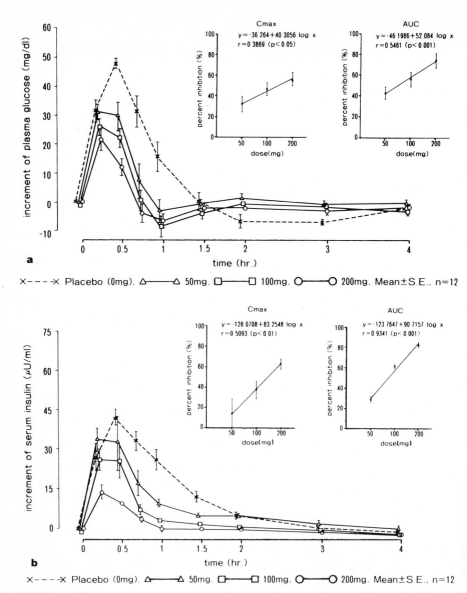

Fig. 1. a Effect of acarbose on plasma glucose increment after 75 g sucrose loading.
b Effect of acarbose on serum insulin increment after 75 g sucrose loading (ATUMA et al. 1990)

stimulation as a consequence of decreased absorption of glucose (HILLEBRAND et al. 1979).

Most carbohydrate in a normal meal is present in the form of starch rather than disaccharides. Since glucoamylase, sucrase, maltase and iso-maltase are intimately involved in the final digestion of starch and disaccharides, acarbose will have a major influence on the absorption of glucose derived from this carbohydrate source. It was thus important to investigate the specificity of the effects of acarbose with respect to different carbohydrate loads.

After administration of 200 mg acarbose, the area under the blood glucose curve was reduced by 89% after sucrose loading, by 80% after starch loading and by 19% after maltose loading (JENKINS et al. 1981). There was no effect following oral administration of lactose and glucose or after intravenous administration of glucose (DIMITRIADIS et al. 1982b). The retardation of carbohydrate digestion and absorption has also been confirmed by isotope tracer and slow marker perfusion studies (RADZIUK et al. 1984; HAGEL et al. 1985). With the aid of tritium-labeled glucose in the sucrose molecule, the reduced and delayed digestion of this disaccharide by acarbose was demonstrated by measuring the rate of glucose absorption.

As sucrose and starch together constitute approximately 90% of all dietary carbohydrate, the substrate specificity of acarbose is well adapted to the task of reducing postprandial hyperglycemia. Acarbose 100–300 mg with each main meal provides optimal inhibition of postprandial increases in glucose and insulin. However, metabolic studies using expiratory breath analysis found that higher doses of acarbose resulted in significant malabsorption of carbohydrate as estimated by increased breath hydrogen levels. Lower doses tended rather to delay and flatten the postprandial rise in the absence of a significant degree of malabsorption (FRITZ et al. 1985; CASPARY 1978; JENKINS et al. 1981). Other studies in healthy volunteers showed that 50 mg acarbose reduced postprandial blood glucose concentrations to the same extent as guar gum (14 g) and that a combination of low-dose acarbose and dietary fiber (guar gum and pectin) resulted in enhanced efficacy (JENKINS et al. 1988; FÖLSCH et al. 1982; TAYLOR et al. 1982a,b).

In summary, acarbose has been shown dose dependently to reduce postprandial increases of blood glucose and serum insulin.

2. Effects on Gastrointestinal Hormones and Exocrine Pancreatic Function

Because acarbose has been shown to alter some of the metabolic functions of the gastrointestinal mucosa, the question arose whether it also affects the release of pancreatic and gastrointestinal hormones. In healthy volunteers hormonal responses to a test meal were evaluated after administration of 200 mg t.i.d. acarbose for 7 days. Acarbose was shown to significantly reduce plasma concentrations of insulin (37% decrease in AUC), C peptide (35%), gastrin (71%) and pancreozymin (38%). In contrast there was a significant increase of somatostatin levels. Basal secretin concentrations were also lower after treatment with acarbose (RAPTIS et al. 1982).

Several investigators have studied the effect of acarbose on sucrose-induced release of gastric inhibitory polypeptide (GIP) (DIMITRIADIS et al. 1982a,b; FÖLSCH et al. 1981a; GROOP et al. 1986; UTTENTHAL et al. 1983,

1987). After single-dose administration of acarbose (400 mg), maximum inhibition of insulin and GIP responses to 31% and 28%, respectively, of control values were found. In the same study, total motilin response was slightly elevated while the enteroglucagon response was increased compared to control by a factor of 9 following a 200-mg dose. Postprandial responses of neurotensin, gastrin, cholecystokinins, somatostatin and pancreatic polypeptide were not significantly affected by acarbose (50–400 mg) (UTTENTHAL et al. 1987). While a single dose of acarbose (100 mg) did not affect glucose and insulin levels or the rate of glucose disappearance after administration of an intravenous glucose load, it did reduce postprandial glucose, insulin and GIP responses after a standard meal (DIMITRIADIS et al. 1982b). Acarbose 300 mg o.d. for 2 weeks significantly reduced the integrated responses of insulin and GIP without markedly affecting the integrated responses of glucose, C peptide or glucagon (GROOP et al. 1986).

The activity of acarbose on exocrine pancreatic function was investigated following administration of doses of 200–600 mg/day for 8 weeks. No effect on lipase, amylase or trypsin secretion was observed, nor was there any development of tolerance to the effects of acarbose on glucose or insulin levels (FÖLSCH et al. 1979).

In general, all changes in gastrointestinal hormones are consistent with the pattern of changes in glucose absorption induced by acarbose. Hormones secreted in stomach and upper small intestine tend to be reduced by a decrease in glucose absorption, whereas those secreted in the lower small intestine (enteroglucagon, somatostatin) are stimulated by the increased supply of glucose. A secondary effect could be mediated through the increased somatostatin levels, as this substance is known to inhibit the release of other gastrointestinal hormones (TAYLOR et al. 1982c). In any event, the changes in gut hormones would appear to be effects, rather than causes, of the favorable action of acarbose on glucose absorption.

3. Effects on Lipids and Lipoproteins

Studies in healthy volunteers have demonstrated that the main changes in lipid profile following treatment with acarbose have been reductions in serum triglyceride and very low density lipoprotein (VLDL) concentrations (HILLEBRAND et al. 1978; HOMMAS et al. 1982; NESTEL et al. 1985). Total cholesterol concentrations were not significantly altered (HOMMAS et al. 1982; NESTEL et al. 1985). Acarbose reduced the extent of carbohydrate-induced triglyceride overproduction in healthy volunteers. This effect may be mediated via a depression of the synthesis of VLDLs (NESTEL et al. 1985).

III. Pharmacokinetics

Acarbose was assayed in biological fluids using an enzyme inhibition test. Sucrase derived from porcine intestinal mucosa was incubated at 37°C with

4-nitrophenyl-maltoheptaoside (NPMH), from which 4-nitrophenol was liberated and photometrically quantified at 405 nm (K.D. RÄMSCH and H. HEUSCHEN 1989, unpublished data, Bayer). This enzymatic reaction was inhibited by acarbose in a concentration-dependent manner. Plasma was deproteinized either by precipitation with perchloric acid or by ultrafiltration prior to the enzyme inhibition test, while urine was purified by filtration through an extraction column with cyclohexyl-bonded silica. The limit of quantification was 5 µg/l for plasma and 10 µg/l for urine, with a working range up to 50 µg/l. In principle, this assay responds to all compounds with inhibitory activity for sucrase. Component 2, a metabolite of acarbose formed by microbial degradation in the large bowel, is the only compound derived from acarbose, which is known to have inhibitory activity besides the parent substance. However, the cross-reactivity is weak as its inhibitory activity is seven times less than that of acarbose.

In man, acarbose is practically not bound to plasma proteins (see Chap. 19, this volume). Since acarbose is a pseudotetrasaccharide, it is absorbed in only very small quantities as unchanged substance. After oral administration of 300 mg to ten healthy male volunteers, an absolute bioavailability of acarbose of 0.6% was observed when compared with an intravenous dose of 100 mg (PÜTTER 1980). An absolute bioavailability of 1.7% was reported in a study with six healthy male volunteers, who had been treated orally with 100 mg acarbose three time daily for 1 week, with 200 mg three times daily for a further 3 weeks and then with a single oral dose of 200 mg ^{14}C-labeled acarbose (Fig. 2; MÜLLER et al. 1988). A mean maximum concentration (C_{max}) of acarbose of 49 µg/l was observed at 2.1 h after administration. The plasma concentration time course of inhibitory activity corresponded well to the plasma concentration time profile of total radioactivity during the first 5 h, indicating that the drug-related material absorbed during this time mainly consisted of acarbose. Only at later stages acarbose metabolites not showing inhibitory activity preponderated in the systemic circulation. The measurement of total radioactivity indicated that 51.3% of the dose administered was excreted in the feces and 35.4% in the urine as drug-related material within 96 h. After intravenous administration of 100 mg acarbose (PÜTTER 1980) to healthy male volunteers, acarbose was nearly completely eliminated via the kidneys (89.8%). This means that acarbose is not metabolized substantially once it has reached the systemic circulation, but is nearly completely renally eliminated as unchanged compound. Nearly all of the drug-related material which was renally eliminated after oral administration of the ^{14}C-labeled acarbose (except for 1.7% consisting of unchanged acarbose and component 2) was therefore derived from prehepatic metabolization, most probably by bacterial degradation within the gut lumen. In fact, acarbose can be metabolized in vitro by incubation with human intestinal microflora (PFEFFER and SIEBERT 1986).

In man, systemically available acarbose is distributed with a half-life of 0.55 h (PÜTTER et al. 1982). The resultant volume of distribution at steady

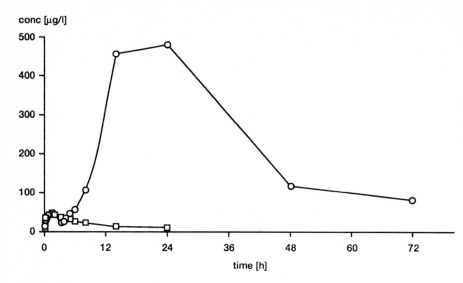

Fig. 2. Mean plasma concentration time profiles of inhibitory activity ($\square$, acarbose plus component 2) and total radioactivity ($\bigcirc$, given as acarbose equivalents) in six healthy male volunteers after oral administration of 200 mg ^{14}C-acarbose (MÜLLER et al. 1988)

state is 0.32 l/kg, which corresponds to the extracellular space. The terminal elimination half-life ranges from 2.7 h (PÜTTER et al. 1982) to 9.6 h (MÜLLER et al. 1988).

In a study investigating the efficacy of 100 mg acarbose t.i.d in patients with liver cirrhosis, the plasma concentration time course of acarbose was measured at steady state (H. HOBLER et al. 1990, unpublished data, Bayer). Elevated plasma concentrations could not be detected in any of subjects compared to healthy volunteers.

Acarbose, which is eliminated from the systemic circulation nearly exclusively via the kidneys, is likely to accumulate in patients with renal impairment. In fact, in a study comparing the pharmacokinetics of acarbose after oral doses of 300 mg t.i.d in patients with a creatinine clearance, CL_{cr}, of below 25 ml/min with those in patients with a CL_{cr} of 25–59 ml/min and a CL_{cr} >60 ml/min, the C_{max} of acarbose (plus component 2) at steady state was increased by a factor of 4 in the low-CL_{cr} group compared to the control group (J. LETTIERI 1991, unpublished data, Bayer). In contrast, C_{max} at steady state in the moderately renally impaired group was only twice as high as in the control group. AUC was increased by a similar factor. Consequently, acarbose should not be administered in patients with severe renal impairment (CL_{cr} <25 ml/min).

IV. Safety and Tolerability

In the clinical-pharmacological studies covered in this chapter the healthy volunteers received acarbose at doses ranging from 25 to 600 mg/day. The reduction in carbohydrate absorption induced by acarbose leads to an increase in the amount of carbohydrate available for fermentation by microorganisms in the lower bowel. The result is an increase in the formation of short-chain fatty acids, hydrogen and carbon dioxide, resulting in various gastrointestinal symptoms, such as flatulence, meteorism (flatulent defecation), soft stools and, less commonly, diarrhea which may be caused by osmotic factors. A proportion of the hydrogen present in the lower bowel is absorbed into the blood stream and excreted in expired air. Thus, the amount of hydrogen in expired air may be taken as an objective measure of the degree of carbohydrate fermentation in the colon. It should be noted, however, that breath hydrogen measurements do not necessarily correlate with the subjective symptoms experienced.

Following acarbose administration, the amount of hydrogen in expired air was observed in several studies to be dependent on the nature of the substrate and the dose of acarbose administered (CASPARY and KALISCH 1979; TAYLOR et al. 1982b). The symptoms showed a tendency to diminish with time (FÖLSCH et al. 1981a) but appeared earlier and were more intense following a sucrose load than after starch alone. A variety of substances was tested with the purpose of modifying the bacterially induced symptoms: antibiotics, an antifoaming agent, an antacid (to alter intestinal pH) and fibrous substances.

Neomycin (which itself may cause diarrhea) intensified the symptoms (LEMBCKE et al. 1980), as well as potentiating the metabolic effects (FÖLSCH et al. 1980), whereas metronidazole slightly reduced the symptoms (LEMBCKE et al. 1982). Hydrotalcit had no effect in humans in contrast to results obtained in animal experiments (R. AUBELL and H. SCHMITZ 1980, unpublished data, Bayer). Addition of guar and pectin likewise had no effect (LEMBCKE et al. 1980a). The results with guar alone were variable. In its hydrolyzed form (taken as suspension), it had no effect at a dose of 10–14 g (JENKINS et al. 1988), whereas, in the form of a crispbread, the guar appeared to reduce the symptoms associated with acarbose, probably via an increase in transit time (JENKINS et al. 1979). Thus, none of the substances tested had a consistently favorable effect on the gastrointestinal symptoms, the most promising being guar in crispbread form.

The objective tolerability of acarbose, assessed from values of hematological and clinicochemical parameters in individual clinical-pharmacological studies, was uniformly good and did not prohibit clinical studies with acarbose in patients.

V. Interactions

In two studies the potential for an interaction of acarbose with digoxin and propranolol, respectively, was investigated (HILLEBRAND et al. 1981). In each study six healthy subjects received a single oral dose of 0.4 mg digoxin or 80 mg propranolol, respectively, followed by a 1-week treatment with 100 mg acarbose three times daily. Together with the last dose of acarbose, the doses of digoxin or propranolol were readministered. Neither digoxin nor propranolol had any effect on acarbose efficacy on lowering postprandial rises of blood glucose and serum insulin. Furthermore, acarbose had no effect on digoxin or propranolol pharmacokinetics.

Fölsch et al. (1981b) found an additive effect of acarbose and neomycin on lowering the postprandial rise of blood glucose. In a crossover design, seven healthy male volunteers were treated with oral doses of 200 mg acarbose three times daily, 1 g neomycin three times daily or the combination of both, each period lasting 9 days. Before the start and at the end of each period the subjects received an oral load of 75 g sucrose. Postprandial glucose rise was significantly decreased by neomycin alone and the corresponding effect of acarbose was enhanced by the combination of both drugs, whereas no effect of neomycin was detected on postprandial rise of serum insulin.

In a crossover study of similar design, 12 healthy volunteers received oral doses of 200 mg acarbose three times daily alone, 500 mg metronidazole three times daily alone or the combination of both drugs for 9 days (Lembcke et al. 1982). Metronidazole had no effect on postprandial blood glucose and serum insulin when 75 g sucrose was given nor did it influence the corresponding effect of acarbose. However, metronidazole coadministered with acarbose reduced flatulence and H_2-exhalation caused by acarbose. In addition, the potential of an interaction with nifedipine (Müller et al. 1988b), ranitidine (M. Seiberling et al. 1988, unpublished data, Bayer), glibenclamide (Gerard et al. 1984) and warfarin (F.O. Müller et al. 1990, unpublished data, Bayer) was investigated. In none of these studies was a clinically relevant interaction observed. Neither hydrotalcit nor dimeticone had a significant influence on acarbose efficacy, whereas 4 g colestyramine t.i.d coadministered with 100 mg acarbose t.i.d led to conflicting results. Neither treatment with acarbose alone nor treatment with the combination with colestyramine demonstrated an attenuation of the postprandial rise in blood glucose, while the effect on postprandial serum insulin was variable. This study therefore was inconclusive in respect to a potential interaction (F.O. Müller et al. 1988c, unpublished data, Bayer).

B. Deoxynojirimycin Derivatives (Miglitol, Emiglitate)

I. Introduction

Miglitol and emiglitate, two deoxynojirimycin derivatives, are competitive α-glucosidase inhibitors that undergo almost complete enteral absorption. They have an acarbose-like spectrum of activity and an even greater affinity for intestinal disaccharidases (Rämsch et al. 1985; Lembcke et al. 1985). These substances have their strongest competitive inhibitory effect on the glucoamylase, saccharase and maltase of the brush border in the mucosa of the small intestine, and thus inhibit not only the breakdown of the disaccharides, but also that of dextrins.

II. Pharmacodynamics

1. Effect on Blood Sugar and Insulin

Miglitol and emiglitate have been investigated in clinical-pharmacological studies in healthy volunteers in controlled trials after single and multiple doses. Miglitol was administered in single doses of 25, 50, 100 and 200 mg, and emiglitate in single doses of 10, 20 and 40 mg in the morning with the

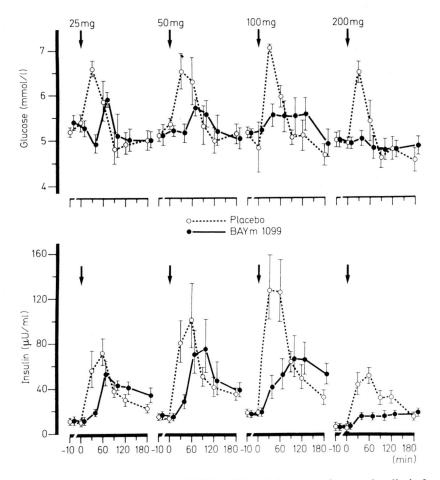

Fig. 3. Mean geometric values and SEM of blood glucose and serum insulin before and after breakfast with and without various dosages of miglitol in six healthy volunteers (HILLEBRAND et al. 1986)

first two mouthfuls of breakfast (Truscheit et al. 1988). Three mixed
meals were given on 1 day and were identical in all studies with emiglitate.
As shown in Figs. 3–4, there was a dose-dependent reduction in the postprandial rise in blood glucose and serum insulin. The highest doses of both
substances (200 mg miglitol and 40 mg emiglitate) completely blocked the
postprandial blood glucose elevation after breakfast. The effect on the
postprandial rise in blood glucose and insulin persisted for about 8 h after
administration of 10 mg emiglitate (Hillebrand et al. 1986).

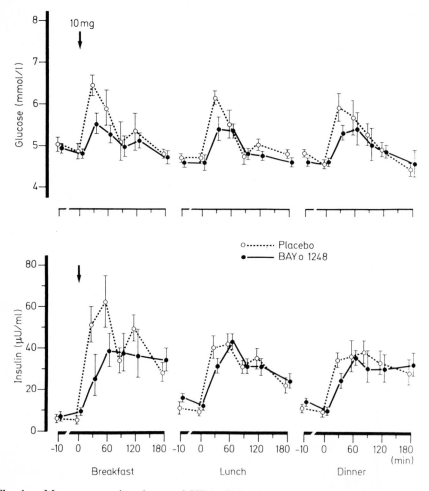

Fig. 4. a Mean geometric values and SEM of blood glucose and serum insulin before
and after breakfast, lunch and dinner with and without 10 mg emiglitate in healthy
volunteers. **b** Mean geometric values and SEM of blood glucose and serum insulin
before and after breakfast, lunch and dinner with and without 20 mg emiglitate in six
healthy volunteers. **c** Mean geometric values and SEM of blood glucose and serum
insulin before and after breakfast, lunch and dinner with and without 40 mg emiglitate
in six healthy volunteers (Hillebrand et al. 1986)

An intraindividual dose comparison between miglitol (25, 50, 100 mg) and emiglitate (10, 20, 40 mg) after sucrose loading showed miglitol only to produce a dose-dependent reduction in blood glucose and serum insulin concentrations in the morning, while the dose-dependent effect of emiglitate was still fully present at noon and less intensely present in the evening (FÖLSCH et al. 1984).

After loading with starch, emiglitate was found to exert the same persistent effect on blood glucose and serum insulin concentrations as after sucrose loading, whereas the effect of miglitol was much more pronounced after starch loading when administered in comparable doses (FÖLSCH et al. 1985).

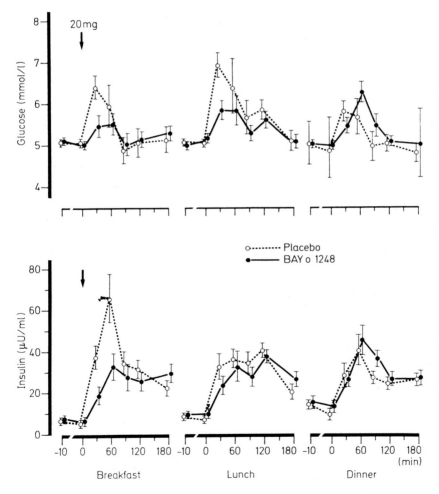

Fig. 4. b

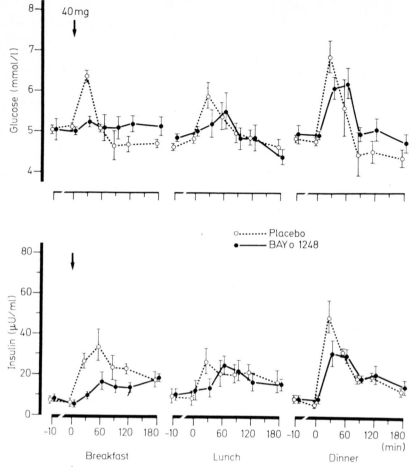

Fig. 4. c

The difference in substrate specificity between miglitol and emiglitate has been investigated by their effects on the absorption of starch, maltose and sucrose test meals. Both miglitol (50 mg) and emiglitate (20 mg) caused almost total sucrose malabsorption and reduced glycemic rises significantly. A high reduction in blood glucose peak occurred with miglitol but not with emiglitate after maltose loading. Both inhibitors reduced post-starch glycemic peaks significantly (TAYLOR et al. 1985).

The mechanisms by which miglitol and emiglitate lower glycemia following sucrose ingestion in man were investigated by isotope tracer studies. ^{3}H-glucose was infused i.v. to monitor changing metabolic clearance rates of glucose. Sucrose labeled with ^{14}C-glucose without or with varying doses of miglitol or emiglitate was ingested. Simultaneously, the gastric contents were monitored by using ^{99m}Tc-sulphur colloid mixed with the drink. With miglitol, sucrose digestion was significantly decreased. The gastric emptying pattern did not change. Emiglitate reduced sucrose digestion significantly and induced in addition a markedly slowed gastric emptying rate. Thus, slowed gastric emptying can contribute to the decrease in absorption rates and glycemia when the digestion of complex sugars is inhibited (PYE et al. 1987). The effect of miglitol after a glucose load in healthy volunteers was unexpected. Although one would expect an α-glucosidase inhibitor to have no effect on blood glucose concentrations after glucose load, miglitol produced a significant decrease in blood glucose concentrations after the absorption peak. This could be due to enhancement of insulin effects or to depression of anti-insulin factors (JOUBERT et al. 1987). Multiple daily administration of miglitol (50 mg t.i.d.) together with standardized meals caused a reduction in the postprandial rise in blood glucose and serum insulin, particularly in the morning and afternoon. The effect was less marked after evening treatment (LAUBE et al. 1984). Treatment over several days with emiglitate in increasing doses to healthy volunteers resulted in a significant reduction in the postprandial blood glucose and insulin increments compared to the values on placebo. The effect of 20 mg emiglitate persisted for 12 h (HILLEBRAND et al. 1984).

In summary, emiglitate appears to be more effective than miglitol in healthy volunteers on a milligram to milligram basis and its action persists for a longer time period.

2. Effects on Gastrointestinal Hormones

Postprandial gut hormone profile is determined by meal composition, volume and other factors. Consequently changes are measured in response to a standard meal given with intestinal glucosidase inhibitors. In addition to the reduction of blood glucose and serum insulin concentrations, miglitol and emiglitate inhibit dose-dependently GIP increases after oral administration of starch (LEMBCKE et al. 1985). Administration of miglitol (50 mg) and emiglitate (20 mg) after a breakfast containing a high proportion of carbohydrates was followed by a rise in gastrin and enteroglucagon, whereas pancreatic glucagon showed no change (BUCHANAN et al. 1985).

In general the measured effects on gastrointestinal hormones are the result of a slower digestion, reducing insulin and GIP release, but unabsorbed nutrients stimulate enteroglucagon distally. Slow absorption in contrast does not affect gastrin and pancreatic glucagon.

3. Effects on Lipids

Triglyceride levels were measured over 12 h in healthy volunteers after ingestion ·of standard breakfast, lunch and dinner. Neither miglitol nor emiglitate produced any significant difference in plasma triglyceride responses (KENNEDY et al. 1985, 1987). Significant reductions of plasma free fatty acids after sucrose loading were achieved with miglitol (50 mg) and emiglitate (10 mg), indicating delays in lipid oxidation (CAUDERAY et al. 1986).

III. Pharmacokinetics

In order to quantify miglitol in biological matrices, an enzymatic inhibition test was applied similar to that for acarbose, the difference between both being the use of 4-nitrophenyl-α-D-glucopyranoside (NPG) as substrate instead of NPMH (K.D. RÄMSCH et al. 1990, unpublished data, Bayer). By changing the substrate, the assay was adapted to the higher plasma concentrations of miglitol compared to acarbose by shifting the working range to 0.05–2.0 mg/l for plasma and to 0.1–2.0 mg/l for urine. Prior to the enzyme inhibition test, plasma was deproteinized by ultrafiltration and urine was purified by filtration through an extraction column with carbon-18-bonded silica.

As acarbose, miglitol is not bound to human plasma proteins (see Chap. 19, this volume). In contrast to acarbose, miglitol is nearly completely absorbed from the gastrointestinal tract at low doses in man. After absorption, miglitol is not metabolized (RÄMSCH et al. 1985), but almost completely renally eliminated as unchanged compound. Following an intravenous dose of 100 mg [^{3}H]miglitol to six healthy male volunteers, 96% of the dose was found in the urine excreted as unchanged drug within 96 h (AHR et al. 1994). Only 0.92% was found in the feces. The volume of distribution at steady state was 0.28 l/kg. As acarbose, miglitol is therefore distributed mainly in the extracellular space. The total clearance of 0.103 l/h per kilogram corresponds to the glomerular filtration rate. After oral administration of 25 mg and 50 mg, 95.7% and 90.1% of the dose was found in the urine as parent compound (BENEKE et al. 1991). Increasing the dose to 100 mg and 200 mg dose normalized C_{max} and AUC decreased, indicating nonlinear pharmacokinetics with a saturable absorption. In parallel, the urinary recovery decreased to 75.5% and 59.7%, respectively. Peak concentrations of miglitol of 0.78 mg/l (25 mg), 1.22 mg/l (50 mg), 1.86 mg/l (100 mg) and 2.61 mg/l (200 mg) were observed at 2.1–3.3 h after administration. Terminal elimination half-life ranged from 1.8 to 2.9 h.

Emiglitate and its acidic metabolite were determined in biological matrices using a high-performance liquid chromatography (HPLC) assay with UV detection at 254 nm after solid-phase extraction on carbon-18 extraction columns. The limit of quantification was 15 μg/l for both compounds. After an oral dose of 40 mg emiglitate to six healthy volunteers, a mean maximum plasma concentration of unchanged drug of 80 μg/l was observed, with a t_{max} of 45 min (RÄMSCH et al. 1985). Emiglitate was eliminated with a terminal half-life of 6.7 h, but only 2.6% of the dose was found in the urine as unchanged compound. Emiglitate was mainly metabolized to its free carboxylic acid. Its peak plasma concentration of $160 \mu g/l$ occurred at 14 h after the 40-mg dose. The metabolite was eliminated with a terminal half-life of 4.5 h and 25% of the dose was recovered from the urine.

IV. Safety and Tolerability

In general, miglitol and emiglitate were well tolerated in all clinical pharmacological studies in healthy volunteers. Reported side-effects were those associated with fermentation of malabsorbed carbohydrates as flatulence, meteorism, loose stools and, less commonly, osmotic diarrhea.

In order to obtain objective data on intestinal symptoms, hydrogen was determined in expired air in various studies. Sucrose, maltose or starch were given after oral doses of miglitol or emiglitate to healthy volunteers. The breath hydrogen level was dose-dependently increased by miglitol and emiglitate with sucrose indicating substrate malabsorption. There was no effect on breath hydrogen level by both inhibitors after maltose. A small late rise in the breath hydrogen level occurred after miglitol but not after emiglitate administration with starch (TAYLOR et al. 1986).

Almost similar results were obtained with both inhibitors after sucrose loading. The H_2-exhalation of volunteers was found to undergo an increase of equal magnitude up to 150 ppm at all dosages after administration of 10, 20 and 40 mg emiglitate. Under identical conditions, a dose-dependent rise in H_2-exhalation was observed on miglitol. Miglitol (100 mg) showed an increase in H_2 up to 150 ppm, 50 mg miglitol produced a slight increase (up to 60 ppm) and 25 mg miglitol showed no effect (LEMBCKE et al. 1984).

In a controlled study over an 8-week period in healthy volunteers with miglitol (100 mg t.i.d.), the stimulation of breath hydrogen exhalation proving carbohydrate malabsorption with starch and sucrose was investigated. Comparing breath hydrogen exhalation, responses were more pronounced after sucrose than after starch loading tests but remained unchanged over time. Symptoms (bloating, flatulence, diarrhea, cramps) were merely noticeable with starch as substrate, but clearly present after sucrose. These symptoms were substantially curtailed during continuous drug intake. It is concluded that symptoms of gaseousness due to carbohydrate malabsorption may undergo habituation (LEMBCKE et al. 1990).

Despite a shortening of the mouth-to-cecum transit time (LADAS et al. 1992), miglitol did not affect fecal weight, calories, protein, fat or fiber content (HOLT et al. 1985). Single doses of emiglitate were studied in healthy subjects after repeated oral sucrose or maize starch loading. Emiglitate increased breath hydrogen exhalation and induced sucrose malabsorption dose-dependently. After administration of emiglitate (10 mg), meteorism was the main symptom, but with emiglitate (20 mg) flatulence was more noticeable, and diarrhea became the main symptom with emiglitate

(40 mg). Emiglitate dose-dependently increased breath hydrogen after starch and was well tolerated with mild symptoms of carbohydrate malabsorption (Lembcke et al. 1991).

Objective safety assessed on the basis of clinicochemical and hematological parameters was good in all studies and was comparable to that of acarbose.

V. Interactions

After a loading dose of 0.6 mg β-acetyldigoxin once daily for 3 days, ten healthy male volunteers received 0.3 mg β-acetyldigoxin once daily for a further 19 days (Weber et al. 1989). On days 9–15, 50 mg miglitol t.i.d. was coadministered followed by 100 mg miglitol t.i.d. on days 16–22. At coadministration, a slight decrease in steady state plasma concentration of digoxin from 0.81 ng/ml (β-acetyldigoxin alone) to 0.66 ng/ml (plus 50 mg miglitol) and 0.59 ng/ml (plus 100 mg miglitol) was observed. The investigators considered this small change as clinically not relevant.

C. Other Glucosidase Inhibitors

I. Introduction

Research in α-glucosidase inhibitors has been promoted in the past 20 years, especially by west European and Japanese pharmaceutical companies. The inhibitors, produced mainly by microorganisms, belong to quite different classes of substances: proteins, glycoproteins and carbohydrates of varying structures (Truscheit et al. 1988). Only few inhibitors reached the stage of clinical development where most of them were discontinued early in phase I. With the exception of acarbose and miglitol, the most advanced α-glucosidase inhibitor is voglibose, whose efficacy has been confirmed in clinical trials.

II. Pharmacodynamics

An α-amylase inhibitor, BAY d 7791, from wheat flour proved to be relatively potent against human salivary and pancreatic amylases (Truscheit et al. 1981b). If administered in starch loading tests to healthy volunteers, this inhibitor reduced postprandial hyperglycemia and hyperinsulinemia (Puls and Keup 1973). Because of its proteinous nature, this substance was inactivated very rapidly by the proteases of the digestive tract and therefore did not meet the criteria for further clinical development.

BAY e 4609, an α-amylase inhibitor of microbial origin, was characterized to be of carbohydrate nature and, therefore, showed a broad stability against the proteases of the digestive tract. The specific activity of BAY e 4609 in inhibiting α-amylases was tenfold higher than that of the inhibitor from wheat. Because of its resistance to acids and proteases, the passage of the substance through the stomach caused no inactivation problems.

In clinical-pharmacological studies in healthy volunteers, the inhibitor showed the expected effect (reduction of postprandial hyperglycemia and hyperinsulinemia) but only with diets deficient, or very low, in sucrose. The rapid cleavage of sucrose to glucose and fructose by the small intestinal brush-border enzyme sucrase, which is not inhibited by BAY e 4609, overrides to a large extent the retardation of the liberation of glucose from starch, effected by the inhibition of α-amylase by BAY e 4609 (SCHMIDT et al. 1982).

Tendamistate, a soluble polypeptide isolated from *Streptomyces tendae*, binds irreversibly with α-amylase, thereby inhibiting gastrointestinal absorption of starch (VÉRTESY et al. 1982).

In controlled studies in healthy volunteers, tendamistate was characterized to inhibit starch absorption dose-dependently (MEYER et al. 1983a,b). Administration of tendamistate significantly attenuated postprandial glycemia; it had no effect on postprandial triglyceride levels and was associated with a reduction in free fatty acid levels (MEYER et al. 1984). Because of its allergic effects (apparently due to its protein structure), the further development of tendamistate was stopped.

Trestatin, a complex oligosaccharide, was shown to be a powerful inhibitor of pancreatic α-amylase. There was a marked reduction of hyperglycemia and hyperinsulinemia following a starch load in healthy subjects. Trestatin was observed to have only few effects on plasma glucose and insulin levels following sucrose ingestion. This suggests that its main site of action is at the α-amylase level (TAPPY et al. 1986). MDL 25637, a homonojirimycin glucoside, is a competitive inhibitor of intestinal glycohydrolase enzymes. The effects of this inhibitor were studied on carbohydrate metabolism in healthy volunteers. The inhibitor was shown to effectively reduce postprandial glucose and insulin after a sucrose challenge (CRAMER et al. 1991).

Voglibose, a synthetic valiolamine derivative, is a potent inhibitor of α-glucosidases. In healthy subjects. 1 or 2 mg/day voglibose orally administered before meals inhibited increases in postprandial blood glucose and serum insulin, dose-dependently (MATSUZAWA et al. 1986). After treatment with voglibose (0.6 mg/day) for 1 week in healthy male volunteers, increases in plasma glucose and insulin levels were attenuated and the increase in plasma GIP levels was diminished, while the increase in enteroglucagon levels was sustained much longer. It is concluded that GIP secretion is stimulated by glucose absorption and enteroglucagon secretion by the presence of sucrose in the gut (FUKASE et al. 1992).

References

Ahr HJ, Boberg M, Brendel E, Krause HP, Müller FO, Steinke W (1995) Pharmacokinetics of miglitol. Absorption, distribution, metabolism and excretion following administration of tritium or carbon-14 labelled miglitol to rats, dogs and man. (to be published)

Azuma J, Awata N, Mochizuki N, Doukai A, Ryski T, Tawara K, Hikima Y (1990) Bioequivalency study of acarbose, an intestinal tract α-glucosidase inhibitor; pharmacological equivalency in insulin and glucose response against sucrose loading. In: Kuhlmann J, Wingender W (eds) Dose-response relationship of drugs. Zuckschwerdt, Munich

Beneke PC, Müller FO, Van Dyk M, Luus HG, Groenewoud G, Hundt HKL (1991) The pharmacokinetics of miglitol over a therapeutic dose range. South Afr Med J 80(11–12):622

Buchanan KD, Gill DJ, Raju HS (1985) Alteration in release of carbohydrate dependent gastrointestinal hormones by alpha-glucosidase inhibitors in man. Diabetes Res Clin Pract [Suppl 1]: p 75, abstract 190

Caspary WF (1978) Sucrose malabsorption in man after ingestion of alpha-glucoside-hydrolase inhibitor. Lancet I (8076):1231–1233

Caspary WF, Graf S (1979) Inhibition of human intestinal α-glucoside hydrolases by a new complex oligosaccharide. Res Exp Med 175:1–6

Caspary WF, Kalisch H (1987) Effect of alpha-glucoside hydrolase inhibition on intestinal absorption of sucrose, water and sodium in man. Gut 20:750–755

Caspary WF, Lembcke B, Creutzfeldt W (1982) Inhibition of human intestinal α-glucoside hydrolase activity by acarbose and clinical consequences. In Creutzfeldt W (ed) Proceedings First International Symposium on Acarbose, Montreux, October 1981. Excerpta Medica, Amsterdam, pp 27–37

Cauderay M, Tappy L, Temler E, Jequier E, Hillebrand I, Felber JP (1986) Effect of alpha-glycohydrolase inhibitors BAY m 1099 and BAY o 1248 on sucrose metabolism in normal men. Metabolism 35:472–477

Cramer MB, Heck K, Taylor CH, McGees BD (1991) Effects of a homonojirimycin glucoside on carbohydrate digestion in normal males. Clin Pharmacol Ther 49:148

Dimitriadis G, Tessari P, Go V, Gerich J (1982a) Effects of the disaccharidase inhibitor acarbose on meal and intravenous glucose tolerance in normal man. Metabolism 31:841–843

Dimitriadis G, Tessari P, Go V, Gerich J (1982b) Effects of acarbose on metabolic and hormonal responses to meal ingestion and intravenous glucose in normal man. In: Creutzfeldt W (ed) Proceedings First International Symposium Acarbose, Montreux, October 1981. Excerpta Medica, Amsterdam, pp 216–222

Fölsch UR, Schwamen E van, Graf S, Caspary WF, Creutzfeldt W (1979) Influence of long-term intake of a glycoside hydrolase-inhibitor (BAY g 5421) on pancreatic exocrine secretion and contents of intestinal brush-border enzymes. Gastroent Clin Biol 3:286–287

Fölsch UR, Lembcke B, Caspary WF, Ebert R, Creutzfeldt W (1980) Influence of neomycin on postprandial metabolic changes and side effects of an α-glycoside-hydrolase inhibitor (BAY g 5421), II. Effects on sucrose tolerance, serum IRI and GIP. In: Creutzfeldt W (ed) Front hormone research. (The entero-insular axis, vol. 7, pp 295–296) Karger, Basel

Fölsch UR, Ebert R, Creutzfeldt W (1981a) Response of serum levels of gastric inhibitory polypeptide and insulin to sucrose ingestion during long-term application of acarbose. Scand J Gastroent 16:629–632

Fölsch UR, Lembcke B, Caspary WF, Ebert R, Creutzfeld W (1981b) Influence of neomycin on postprandial metabolic change of a glycoside α-hydrolase inhibitor (acarbose). In: Berchtold P et al. (eds) Regulators of intestinal absorption in obesity, diabetes and nutrition. Proceedings of satellite symposium, no 7, 3rd

International Congress on Obesity, October 1980. Rome; Societa Editrice Universo, vol 2, pp 249–258

Fölsch UR, Lembcke B, Ebert R, Caspary WF, Creutzfeldt W (1982) Effect of acarbose and fiber on glucose and hormone response to sucrose load or mixed meal. In: Creutzfeldt W (ed) Proceedings First International Symposium Acarbose, Montreux, October 1981. Excerpta Medica, Amsterdam, pp 97–106

Fölsch UR, Lembcke B, Jücke B, Hillebrand I, Ebert R, Creutzfeldt W (1984) The effect of semi-synthetic glucosidase inhibitors (BAY o 1248 and BAY m 1099) on glucose, GIP and insulin in serum after oral administration of saccharose. Akt Endokrin 5:87

Fölsch UR, Lembcke B, Gatzemeier W, Hillebrand I, Ebert R, Creutzfeldt W (1985) Effect of the glucosidase inhibitors BAY o 1248 and BAY m 1099 on hormone response and H_2-exhalation to repeated sucrose and starch loading. Diabetes Res Clin Pract [Suppl 1]:171–172

Fritz M, Kaspar H, Schrezenmeier J, Siebert G (1985) Effect of acarbose on the production of hydrogen and methane and on hormonal parameters in young-adults under standardized low-fiber mixed diet. Z Ernährungswiss 24:1–18

Fukase N, Takahashi H, Manaka H, Igarashi M, Yamatani K, Daimon M, Fugiyama K, Tominaga M, Sasaki H (1992) Differences in glucagon-like peptide-1 and GIP responses following sucrose ingestion. Diabetes Res Clin Pract 15:187–195

Gérard J, Levebvre PJ, Luyckx AS (1984) Glibenclamide pharmacokinetics in acarbose-treated type 2 diabetics. Eur J Clin Pharmacol 27:233–236

Goda T, Yamada K, Howya N, Moriuchi S (1981) Effects of α-glucosidase inhibitor BAY g 5421 on rat intestinal disaccharidases. Food Nutr 34:139–143

Groop P-H, Groop L, Toetterman KJ, Fyhrquist F (1986) Effects of acarbose on the relationship between changes in GIP and insulin responses to meals in normal subjects. Acta Endocr 112:361–366

Hagel J, Pickl J, Ruppin H, Feuerbach W, Bloom S, Domschke W (1985) Fate and effects of alpha-glucosidase inhibitors, BAY g 5421 acarbose, on the small intestine – a perfusion study in humans. Z Gastroenterol 23:439

Hillebrand I, Boehme K, Frank G, Fink H, Berchtold P (1978) Effects of the glycoside-hydrolase-inhibitor BAY g 5421 on postprandial blood-glucose, serum insulin and triglyceride levels: dose-time relationships in man. Diabetologia 15:239

Hillebrand I, Boehme K, Frank G, Fink H, Berchtold P (1979) Effects of the glycoside hydrolase inhibitor (BAY g 5421) on postprandial blood glucose, serum insulin and triglyceride levels in man. In: Creutzfeldt W (ed) Frontiers of hormone research, the entero-insular axis, vol 7. Karger, Basel, pp 290–291

Hillebrand I, Graefe KH, Bischoff H, Frank G, Rämsch KD, Berchtold P (1981) Serum digoxin and propranolol levels during acarbose treatment. Diabetologia 21:282–283

Hillebrand I, Skipton EA, Boehme K, Schulz H, Wehling K, Müller O, Fink H (1984) Clinical pharmacological studies over several days on healthy volunteers with the new glucosidase inhibitors BAY m 1099 and BAY o 1248. Akt Endokrin 5:92

Hillebrand I, Boehme K, Graefe KH, Wehling K (1986) The effect of new alpha-glucosidase inhibitors (BAY m 1099 and BAY o 1248) on meal-stimulated increases in glucose and insulin levels in man. Klin Wochenschr 64:393–396

Holt PR, Thea D, Kotler DP, Yang M (1985) Effective pharmacologic retardation of dietary carbohydrate absorption without fecal caloric wastage. Clin Res 33 (2PT2):705A

Homma Y, Irie N, Yaw Y, Nakaya N (1982) Changes in plasma lipoprotein levels during medication with a glucoside hydrolase inhibitor acarbose. Tokai J Exp Clin Med 7:393–396

Jenkins DJ, Taylor RH, Nineham R, Bloom SR, Larson D, George K, Alberti KG (1979) Combined use of guar and acarbose in reduction of postprandial glycemia. Lancet II (8149):924–927

Jenkins DJ, Taylor RH, Goff DV, Fielden H, Misiewicz JJ, Larson DL, Bloom SR, Alberti KG (1981) Scope and specificity in slowing carbohydrate absorption in man. Diabetes 30:951–954

Jenkins DJ, Taylor RH, Nineham R, Goff DV, Bloom SR (1988) Manipulation of gut hormone response to food by soluble fiber and alpha-glucosidase inhibition. Am J Gastroenterol 83:393–397

Joubert PH, Foukaridis GN, Bopape ML (1987) Miglitol may have a blood glucose lowering effect unrelated to inhibition of alpha-glucosidase. Eur J Clin Pharmacol 31:723–724

Kennedy FP, Heiling V, Kluge K, Lund T, Kahl J (1985) Effect of a new alpha-glucosidase inhibitor BAY m 1099 on postprandial plasma glucose, insulin, and triglyceride levels in normal volunteers. Diabetes 34 [Suppl 1]:199A, abstract 755

Kennedy FP, Miles JM, Heiling V, Gerich JE (1987) The effect of two new alpha-glucosidase inhibitors on metabolic responses to a mixed meal in normal volunteers. Clin Exp Pharmacol Physiol 14:633–640

Ladas SD, Foydas A, Papadopoulos A, Raptis SA (1992) Effects of alpha-glucosidase inhibitors on mouth to coecum transit time in humans. Gut 33:1246–1248

Laube H, Mehlburger L, Federlin K, Hillebrand I (1984) BAY m 1099, a new enzyme inhibitor, effecting a lowering of blood sugar levels. Akt Endokrin 5:103

Lembcke B, Caspary WF, Hönig M, Frerichs H, Ebert R, Creutzfeld W (1980a) Effect of carbohydrate gelling agents in addition to an α-glucoside-hydrolase inhibitor (α-GHI) on postprandial blood glucose, IRI, GIP and intestinal H_2 production. In: Creutzfeld W (ed) Front hormone research. Karger, Basel (The entero-insular axis, vol 7. pp 293–294)

Lembcke B, Caspary WF, Fölsch UR, Creutzfeldt W (1980b) Influence of neomycin on postprandial metabolic changes and side effects of an α-glucoside hydrolase inhibitor (BAY g 5421). I. Effects on intestinal hydrogen gas production and flatulence. In: Creutzfeldt W (ed) Front hormone research. Karger, Basel (The entero-insular axis, vol 7. pp 294–295)

Lembcke B, Fölsch UR, Caspary WF, Creutzfeldt W (1982) Influence of metronidazole on intestinal side-effects of acarbose. In: Creutzfeldt W (ed) Proceedings first international symposium acarbose, Montreux, October 1981. Excerpta Medica, Amsterdam, pp 236–238

Lembcke B, Fölsch UR, Klausgrete E, Hillebrand I, Creutzfeldt W (1984) Deoxynojirimycin derivative (BAY o 1248/BAY m 1099). Which dosage leads to carbohydrate malabsorption. Akt Endokrin 5:104

Lembcke B, Fölsch UR, Gatzemeier W, Hillebrand I, Creutzfeldt W (1985) Effect of semi-synthetic alpha-glucosidase inhibitors (BAY m 1099 and BAY o 1248) on blood glucose, insulin and GIP in serum as well as H_2-exhalation after repeated starch load. Akt Endokrin 6:97

Lembcke B, Fölsch UR, Creutzfeldt W (1985) Effect of 1-desoxynojirimycin derivatives on small intestinal disaccharidase activities and on active transport in vitro. Digestion 31:120–127

Lembcke B, Diederich M, Fölsch UR, Creutzfeldt W (1990) Postprandial glycemic control, hormonal effects and carbohydrate malabsorption during long-term administration of the alpha-glucosidase inhibitor miglitol. Digestion 47:47–55

Lembcke B, Fölsch UR, Gatzemeier W, Lücke B, Ebert R, Siegel E (1991) Inhibition of sucrose- and starch-induced glycaemic and hormonal responses by the alpha-glucosidase inhibitor emiglitate (BAY o 1248) in healthy volunteers. Eur J Clin Pharmacol 41:561–567

Matsuzawa Y, Tokunaga K, Fujioka S, Kameda Y, Nakajima T, Yamashita S, Sho N, Ueyama Y, Kihara S, Tarui S, Shutoh H (1986) Effects of AO-128, a novel alpha-glucosidase inhibitor on postprandial blood glucose and on secretion of insulin. 7th Conference of Japan Society for the Study of Obesity

Meyer BH, Müller FO, Clur BK, Grigoleit HG (1983a) Effects of tendamistate (α-amylase inactivator) on starch metabolism. Br J Clin Pharmacol 16:145–148

Meyer BH, Müller FO, Grigoleit HG, Clur BK (1983b) Inhibition of starch absorption by tendamistate (an alpha-amylase inactivator). S Afr Med J 64:284–285

Meyer BH, Müller FO, Grigoleit HG, Esterhuysen AJ, Clur BK (1984) Effects of tendamistate on postprandial plasma glucose, free fatty acid and triglyceride levels. S Afr Med J 66:224–225

Müller FO, Hundt HKL, Luus HG, Van Dyk M, Groenewoud G, Hillebrand I (1988) The disposition and pharmacokinetics of acarbose in man. In: Creutzfeld W (ed) Acarbose for the treatment of diabetes mellitus. Second international symposium on acarbose, November 12–14, 1987. Springer, Berlin Heidelberg New York, pp 17–26

Nestel PJ, Bazelmans, J, Reardon MF, Boston RC (1985) Lower triglyceride production during carbohydrate-rich diets through acarbose, a glucoside hydrolase inhibitor. Diabetes Metab 11:316–317

Pfeffer M, Siebert G (1986) Prefeeding-dependent anaerobic metabolization of xenobiotics by intestinal bacteria-methods for acarbose metabolites in an artificial colon. Z Ernährungswiss 25:189–195

Puls W, Keup U (1973) Influence of an α-amylase inhibitor (BAY d 7791) on blood glucose, serum insulin and NEFA in starch loading tests in rats, dogs and man. Diabetologia 9:97

Pütter J (1980) Studies on the pharmacokinetics of acarbose in humans. In: Brodbeck U (ed) Enzyme inhibitors, proceedings of a meeting March 20 and 21, 1980. Verlag Chemie, Weinheim, pp 139–151

Pütter J, Keup U, Krause HP, Müller L, Weber H (1982) Pharmacokinetics of acarbose. In: Creutzfeldt W (ed) Proceedings of the first international symposium on acarbose, October 8–10, 1981. Excerpta Medica, Amsterdam, pp 38–48

Pye S, Hillebrand I, Radziuk J (1987) Mechanisms by which alpha-glucosidase inhibitors lower glycaemia following sucrose ingestion in man. Diabetologia 30:572A, abstract 454

Radziuk J, Kemmer F, Morishima T, Berchtold P, Vranic M (1984) The effects of an alpha-glucoside hydrolase inhibitor on glycemia and the absorption of sucrose in man determined using a tracer method. Diabetes 33:207–213

Rämsch KD, Wetzelsberger N, Pütter J, Maul W (1985) Pharmacokinetics and metabolism of the desoxynojirimycin derivatives BAY m 1099 and BAY o 1248. Diabetes Res Clin Pract [Suppl 1]:461(A)

Raptis S, Dimitriadis G, Etzrodt H, Karaiskos C, Hadjidakis D, Rosenthal J, Zoupas C, Diamantopoulos E (1982) The effects of acarbose treatment on release of pancreatic and gastrointestinal hormones in man. In: Creutzfeldt W (ed) Proceedings first international symposium acarbose. Montreux, October 1981. Excerpta Medica, Amsterdam, pp 210–215

Schmidt DD, Frommer W, Junge B, Müller L, Wingender W, Truscheit E (1982) α-Glucosidase inhibitors of microbial origin. In: Creutzfeldt W (ed) Proceedings first international symposium acarbose in Montreux, October 1981. Excerpta Medica, Amsterdam, pp 5–15

Sjöström K, William-Olsson T (1981) The effect of a new glycoside hydrolase inhibitor on glucose and insulin levels during sucrose loads in obese subjects. Curr Ther Res 30:351–366

Tappy L, Buckert A, Griessen M, Golay A, Jequier E, Felber J (1986) Effect of trestatin, a new inhibitor of pancreatic alpha-amylase on starch metabolism in man. Int J Obesity 10:185–192

Taylor RH, Jenkins DJA, Goff DV, Fielden H, Misiewicz JJ, Alberti KG (1982a) Scope and specificity of acarbose in slowing carbohydrate-absorption. In:

Creutzfeldt W (ed) Proceedings first international symposium acarbose, Montreux, October 1981. Excerpta Medica, Amsterdam, pp 199–202

Taylor RH, Jenkins DJA, Barber HM, Lee DA, Allen HB, MacDonald G (1982b) Effect of acarbose on carbohydrate-absorption and 24-hour glycaemic profile. In: Creutzfeldt W (ed) Proceedings first international symposium acarbose. Montreux, October 1981. Excerpta Medica, Amsterdam, pp 203–205

Taylor RH, Jenkins DJA, Goff DV, Bloom SR, Larson DL, Misiewicz JJ, Alberti KG (1982c) Gut hormone response to carbohydrate with acarbose and guar. In: Creutzfeldt W (ed) Proceedings first international symposium acarbose. Montreux, October 1981. Excerpta Medica, Amsterdam, pp 206–209

Taylor RH, Barker HM, Bowey EA, Canfield JE (1985) Regulation of intestinal carbohydrate absorption in man by two new selective enzyme inhibitors. Gut 26:A549–A550

Taylor RH, Barker HM, Bowey EA, Canfield JE (1986) Regulation of the absorption of dietary carbohydrate in man by two new glucosidase inhibitors. Gut 27: 1471–1478

Truscheit E, Frommer W, Junge B, Müller L, Schmidt DD, Wingender W (1981a) Chemistry and biochemistry of microbial alpha-glucosidases inhibitors. Angew Chem Int English Edn. 20:744–761

Truscheit E, Schmidt DD, Arens A (1981b) Further characterization of new α-amylase inhibitors from wheat flour. In: Regulators of intestinal absorption in obesity, diabetes and nutrition, vol 2. Societa Editrice Universo, Rome, pp 157–179

Truscheit E, Hillebrand I, Junge B, Puls W, Schmidt DD (1988) Microbial α-glucosidase inhibitors: chemistry, biochemistry and therapeutic potential. In: Progress in clinical biochemistry and medicine, vol 7. Springer, Berlin Heidelberg New York, pp 19–99

Uttenthal LO, Ukpoumwan OO, Ghatel MA, Bloom SR (1983) Acute and short term effects of intestinal alpha-glucosidase inhibiton on gut hormone responses in man. Gut 24:A461–A462

Uttenthal LO, Ukpoumwan OO, Ghiglione M, Bloom SR (1987) Acute and short term effects of intestinal alpha-glucosidase inhibition on gut hormone responses in man. Dig Dis Sci 32:139–144

Vértesy L, Oeding V, Bender R, Nesemann G, Jukatsch D, Zepf KH (1982) Chemistry and biochemistry of a new type α-amylase-inactivator HOE 467 from Streptomyces tendae. In: 13th International Symposium Chem Nat Prod Aug 2–6, Pretoria, Abstract L17

Weber H, Horstmann R, Rämsch KD, Wingender W, Schmitz H, Kuhlmann J (1989) Influence of the α-glucosidase-inhibitor miglitol on the steady state pharmacokinetics of digoxin in healthy volunteers. Eur J Clin Pharmacol 36 [Suppl]:Abstract pp 11.15

CHAPTER 22

Clinical Evaluation of Glucosidase Inhibitors

R.H. Taylor and E.M. Bardolph

A. Introduction

Glucosidase inhibitors act by reducing activity of the digestive enzymes of the small intestinal brush border. Most of those which have shown any clinical promise are reversible, competitive inhibitors of these enzymes with reasonably high specificity.

I. Carbohydrate Digestion

Dietary carbohydrate is digested in the gut in a series of stages of enzymatic hydrolysis prior to absorption through the surface of the enterocyte, mainly in the form of monosaccharides and much smaller amounts of oligosaccharides. In most Western diets and in diabetic diets as currently recommended, the staple carbohydrate source is starch, which may provide up to 55% of dietary energy intake. It occurs naturally in two forms which are present in variable proportions in natural foods. Amylose consists of long chains of glucose residues joined by α 1,4-linkages, whereas amylopectin has similar α 1,4-linkages in addition to α 1,6-linkages, which give it a multiply branched structure. Cellulose and related complex polysaccharides are joined by β-linkages, which cannot be digested by human digestive enzymes specific for the α-configuration, though other animal species and also bacteria can digest them.

Starch is digested by α-amylase from the salivary glands and the pancreas and is broken down to maltose, maltotriose and α-limit dextrins within the lumen of the upper gastrointestinal tract. The rate of digestion is dependent on pH, particle size, presence of protein, fat and fibre and the complexity of the natural matrix structure of foods. These products of digestion join the dietary disaccharides, sucrose, lactose, maltose and trehalose, for the final stage of digestion at the brush border of the small intestinal enterocyte.

There are four α-glucosidases and one β-galactosidase in the brush border which complete the digestion of oligosaccharides to the monosaccharides, glucose, fructose and galactose, which are then absorbed by the cellular transport mechanisms. Glucoamylase hydrolyses α 1,4-linkages, releasing glucose residues. Sucrase-α-dextrinase is a hybrid enzyme with two independent active sites, sucrase, which splits sucrose to glucose, and fructose and α-dextrinase, which hydrolyses the α 1,6-linkages of the α-limit dextrins. Trehalase digests the disaccharide trehalose, which occurs only in mushrooms. The β-galactosidase is lactase, which hydrolyses lactose to glucose and galactose. Its activity declines in many people after infancy.

The monosaccharide products of intraluminal and brush border digestion of carbohydrate and any dietary monosaccharides are absorbed rapidly through the enterocyte and into the circulation. Glucose and galactose enter by a sodium-dependent carrier, a second non-sodium-dependent carrier and by passive diffusion, and fructose has a separate active carrier.

II. Diet and Glucosidase Inhibitors

All normal dietary carbohydrates, except for the relatively small amounts of lactose and monosaccharides, are digested by the intestinal α-glucosidases in saliva, pancreatic secretions and the enterocyte brush border. Inhibition of the action of these enzymes by specific inhibitors results in a reduction or even an abolition of hydrolytic activity. This is of great therapeutic importance for three reasons. Firstly the carbohydrate composition of the diet eaten by patients taking glucosidase inhibitors determines the substrate load for the various enzymes. This matters because the degree of inhibition of the various α-glucosidases is different for particular inhibitors, and the relationship between the inhibitory profile and the balance of carbohydrate sources in the diet is critical. This must be considered in assessing clinical response to treatment and particularly in evaluating the results of clinical trials. The second crucial point is that, if given in excessively high dosage, glucosidase inhibitors could result in malabsorption of some carbohydrate through inhibition of its digestion. As a consequence, an unknown amount of the dietary carbohydrate intake may be absorbed, making diabetic control less predictable. Secondary to carbohydrate malabsorption there may also be malabsorption of other macro- and micronutrients due to intestinal hurry. Malabsorbed nutrients in any quantity will provoke an osmotic diarrhoea and cause associated symptoms. The third important point is that, in a state of severe malabsorption induced by a glucosidase inhibitor, the only carbohydrate that can be absorbed through the intestine must be in a form in which no enzymatic digestion is required, that is monosaccharides and in particular glucose. This has important implications for resuscitation in hypoglycaemia induced by other agents given with a glucosidase inhibitor.

Any use of a glucosidase inhibitor to improve diabetic control, either as the main agent or as an adjunct, depends greatly on the composition of the diet taken. All dietetic guidelines in diabetes now recommend a diet which is high in starch and fibre and low in fats and free sugars. In using these inhibitors it is essential that consistent dietary guidelines are used which should be in keeping with the general dietary recommendations. The inhibitory specificity profile must be matched to the diet and not the other way round. This can be achieved (JENKINS and TAYLOR 1982; TAYLOR et al. 1982), and the dietary recommendations for patients on glucosidase inhibitors should be identical to those given to all (TOELLER 1992).

III. Role of Glucosidase Inhibitors in Diabetes

Regulation of the rate of carbohydrate digestion and thus its absorption can result in better control of the rate of delivery of monosaccharides to the circulation and to the tissues of the body. In those patients with some endogenous insulin production, this may achieve a better match between the rate of carbohydrate absorption and that of insulin release from the impaired endocrine pancreas. This should stabilise blood glucose levels and relieve pressure on alternative metabolic pathways. Therefore in diet-controlled patients and those taking oral hypoglycaemic agents, glucosidase inhibition may slow and smooth the rate of carbohydrate delivery, allowing the endogenous insulin output to match it and reducing swings in blood glucose levels.

In type I diabetes, the availability of insulin can be regulated with precision using modern delivery systems and the newer insulin preparations to match postprandial glycaemic rises and regulate them to achieve good diabetic control. However, glucosidase inhibitors may still help to smooth glycaemic swings.

B. Acarbose

Acarbose (Bayer, BAY g5421, "Glucobay") is the glucosidase inhibitor which has been studied in most detail and was the first one to become available for clinical use. This pseudotetrasaccharide is a potent inhibitor of glucoamylase and, in descending order of potency, of sucrase, maltase and isomaltase with little effect on amylase activity (TRUSCHEIT et al. 1981).

HILLEBRAND et al. (1988) summarised the clinical experience of the use of acarbose in treatment of diabetes by pooling data from all trials to date. A total of 2952 (811 type I, median age 52 years, and 2141 type II, median age 63 years) diabetic patients had been treated with acarbose 150–300 mg daily, of whom 54% had been treated for more than 3 months. Postprandial blood glucose levels were reduced overall and it appeared that diabetic control was helped by acarbose. The female patients were overweight but the males had normal Broca indices and acarbose had no effect on body weight. Of the total, 58% experienced some side effects and 53% had gastrointestinal side effects such as flatulence, meteorism or diarrhoea; 85% of these were reported early in treatment. Because these pooled data came from a diversity of trial sources, some unpublished, with widely varied protocols, meta-analysis is impossible and the conclusions to be drawn must be very general.

Tolerance has been established and the side effect profile defined. Safety data is now available from 8800 patients in trials worldwide (HOLLANDER 1992). The only side effects that occur more commonly on acarbose than on placebo are gastrointestinal and a consequence of its mode of action. A few minor changes in laboratory measures have been noted more commonly during acarbose treatment than placebo. An unexplained, reversible elevation of liver transaminases was found in tolerance studies in which subjects were taking acarbose 200 mg t.d.s. or more, but this has not been noted in clinical studies using lower doses (BAYER, personal communication). In late 1993, there were 9211 patients in the data pool who had taken acarbose and a further 10 462 in the German postmarketing surveillance study. Data on these patients has not revealed any new or unexpected information (BAYER, personal communication).

Many of the studies published before 1985 were small and their contribution is limited. There are now considerably more pooled data on file, though some remains unpublished. Evidence of efficacy is seen more clearly in the published reports of individual trials which are reviewed here.

I. Studies in Type I Diabetes

Studies of acarbose in type I diabetes are limited and the numbers reported are small. VIVIANI et al. (1987) studied 30 type I diabetic patients in an 8-week double-blind, crossover study in which the patients were stabilised initially on a very low carbohydrate diet (15% of calories). The diet was then changed to an isocaloric 30% carbohydrate diet and acarbose 100 mg or placebo given for 4 weeks followed by the other treatment with no change in insulin throughout. Blood glucose profiles were improved on acarbose and predictably worse without. Four patients dropped out and a number experienced abdominal side effects of flatulence or loose stools or became hypoglycaemic whilst on acarbose. The low carbohydrate intake and the inflexible insulin dose make it difficult to interpret some of the results. In a 6-month open study of 14 patients, DIMITRIADIS et al. (1986) found a 20%–30% reduction in postprandial glycaemia and were able to reduce insulin doses by 40%. Gastrointestinal symptoms were reported but had disappeared by the end of the 1st month. MARENA et al. (1991) conducted a detailed, double-blind, crossover 12-week study in 14 poorly controlled type I diabetic patients comparing acarbose 100 mg t.d.s. with placebo. On the last day of each limb of the study the patients were kept euglycaemic on a Biostator. Acarbose resulted in significant reductions in mean blood glucose, HbAlc and daily insulin requirement on the glycaemic clamp.

AUSTENAT (1991) reported a smoothing of the daily blood glucose profile in 11 brittle diabetics taking insulin and acarbose and a significant reduction in their triglyceride and total cholesterol levels. In a 12-week study in 32 overweight type I diabetics taking acarbose 100 mg t.d.s., she was able to reduce insulin dosages by 0%–15% and there was an overall fall in glycosylated haemoglobin levels. In a related double-blind crossover study in 23 patients, she found that acarbose caused a reduction in "excessive" appetite during treatment. GERARD et al. (1981) did an open comparison in 28 patients treated for 2 months each with acarbose 100 mg t.d.s. or placebo and found a sustained effect.

In a crossover study in eight patients whose glycaemic control was maintained for 2 days by feedback using a Biostator, ITO et al. (1989) found a significant reduction in insulin requirement in the 3 h following meals when the subjects were taking acarbose. They also reported a significant delay in the postprandial glycaemic peaks after acarbose. LECAVALIER et al. (1986) studied patients controlled on a Biostator after a 75-g sucrose load with acarbose 0, 50, 100 or 200 mg in four patients, or after 50, 75 or 100 g sucrose with or without acarbose 100 mg in five patients, and found a significant acarbose dose-related reduction in insulin requirement. However, in such a dosage acarbose results in gross malabsorption of sucrose (JENKINS et al. 1981), which would inevitably reduce the insulin requirement.

These rather limited studies suggest that acarbose can have a beneficial effect in stabilising type I diabetic patients, improving their control and reducing their insulin requirement, though this may be due in part to their malabsorbing carbohydrate.

II. Studies in Type II Diabetes

1. Clinical Trials

One of the largest trials in type II diabetic patients is the Canadian multicentre study reported initially by Ross et al. (1992). This was a randomised, placebo-controlled, double-blind trial of 1 year of treatment in 354 patients who were also treated with diet alone, or diet and sulphonylurea, insulin or metformin. Fasting blood glucose levels were unchanged, but HbAlc fell significantly in all except the insulin group with acarbose, as did the area under the postprandial glycaemic response curve. Digestive side effects were reported but tolerated.

Low-dose treatment with acarbose can improve metabolic control without causing side effects (JENKINS and TAYLOR 1982). In a double-blind, crossover trial in six patients, acarbose 25 mg t.d.s. was compared with placebo over 3-month treatment periods (JENNY et al. 1993). There was significant improvement in almost all measures except insulin production, suggesting that there is a therapeutic dose which avoids side effects. SANTEUSANIO et al. (1993) compared the effects of acarbose 50 mg t.d.s., 100 mg t.d.s. or placebo in three groups of diet-controlled, type II patients for 16 weeks and found that good control could be obtained on the lower dosage with fewer side effects.

In an open study of 27 patients, poorly controlled on diet alone, who took acarbose for 12 weeks, SAKAMOTO et al. (1989) found no change in the fasting blood glucose level and a slight but insignificant decline in the HbAlc. There was a significant fall in the postprandial blood glucose. The same group did a randomised, placebo-controlled 24- to 28-week trial in similar patients and then found a significant improvement in postprandial blood glucose and in HbAlc (Sakamoto et al. 1990). Similar results have been reported by Hotta et al. (1993a).

van Gaal et al. (1991) did a double-blind, crossover study in 24 patients with 10 weeks each on acarbose or placebo in random order. They found the expected improvement in diabetic control but also no changes in electrolytes, iron, vitamin B_{12} and folic acid levels, indicating the absence of any evidence of nutritional consequences of slowed absorption over this period of time.

Patients who were poorly controlled on oral hypoglycaemic agents were studied in a 4-week crossover study of acarbose compared with placebo (Scott et al. 1984). There was no difference in fasting glucose, HbAlc or glucose tolerance on treatment with acarbose in the 18 patients. Treatment was continued for 6 months for 12 of them, with no evidence of further benefit. These patients were difficult to control and the benefits of acarbose were insufficient to compensate for their inadequate insulin production.

Longer term studies indicate that the therapeutic benefits are maintained. Toto et al. (1989) treated 96 patients for 6–12 months in an open study. Diabetic control improved, triglyceride levels fell but changes in total and HDL cholesterol were insignificant. Side effects reduced with time. Aubell et al. (1983) reported a multicentre 1-year study on a mixed group of 245 type I and type II patients, of whom 152 were evaluable at the end of the study. Overall diabetic control improved and there was a significant fall in fasting and postprandial blood glucose levels. Control reverted to previous levels by 3 months after the cessation of treatment. Preliminary reports on other longer term studies have appeared (Foelsch et al. 1990). Zimmerman (1992) argues that these long-term metabolic benefits of acarbose, alone or in combination, may be valuable in reducing lipid levels and thereby preventing or delaying complications of diabetes.

2. Effect on Lipids

Healthy volunteer studies have shown that acarbose has a clear lipid-lowering effect, particularly on triglyceride and very low density lipoprotein (VLDL) cholesterol (Homma et al. 1982). Interpretation of data on the effects of acarbose on lipid levels in diabetic patients is more difficult, because most of the data has been obtained as a secondary objective or as a chance finding in trials addressing the quality of diabetic control. In some studies lipid levels were not raised in all patients at the start and comparisons are of limited value.

In a carefully controlled and detailed metabolic study on eight type II diabetic patients treated with acarbose 100 mg t.d.s. for 1 week, Baron et al. (1987) reported no change in fasting blood glucose but a reduction in postprandial values as in other studies. They also found a significant fall in triglycerides and in total cholesterol, with a small rise in high-density lipoprotein (HDL) cholesterol and a beneficial improvement in the ratio. They measured a reduction in adipose tissue lipoprotein lipase activity and an increase in sensitivity to insulin-mediated suppression of hepatic glucose output.

A larger randomised, double-blind, placebo-controlled trial of the effect of acarbose on lipids was done by LEONHARDT et al. (1991). After pretreatment on diet for at least 3 months, 94 patients were allocated to treatment with acarbose 100 mg t.d.s. or placebo for 24 weeks. Blood glucose and HbAlc improved and all but the lowest cholesterol levels fell significantly, compared with placebo treatment. HDL cholesterol rose and fasting triglycerides fell significantly in both groups, indicating the benefits of intensive supervision and dietary control. However, there was a highly significant fall in postprandial triglycerides which only occurred in the acarbose group. This suggests that acarbose can be beneficial in correcting lipid abnormalities in diabetic patients. However, this result was contradicted in a similar study in 59 type II diabetic patients who were controlled on insulin for 24 weeks and randomly allocated to treatment with acarbose or placebo (RYBKA et al. 1990). Diabetic control improved and the insulin dosage required fell. There was no significant change in total cholesterol or triglycerides in either group over the treatment period. HOTTA et al. (1993b) reported a similar study using insulin in nine patients and found a reduced requirement for insulin with improved control but no changes in lipid levels.

Overall it seems that, in type II diabetic patients with poor control of their diabetes and raised lipid levels, there may be an additional benefit of lowering lipid levels whilst improving glycaemic control. This effect may be directly due to improved diabetic control rather than evidence of a separate, independent action of acarbose.

3. Studies Comparing Effects with Other Antidiabetic Agents

There have been a number of studies comparing the effects of acarbose with that of oral hypoglycaemic agents, though because of the differences in mode of action such comparisons must be viewed with caution. Some of these studies were small or had some methodological limitations, which limits their significance, but the results are noted for completeness.

BUCHANAN et al. (1988) studied 29 poorly controlled patients taking oral hypoglycaemic agents which were withdrawn at the start of the study, which resulted in a worsening in their control, and the patients allocated randomly to acarbose or placebo in increasing dosage over 16 weeks. Only 20 patients completed the study (nine on acarbose, most of whom experienced side effects). Glycaemic control deteriorated in all and there was no significant difference between the groups on completion. The authors conclude that acarbose was less effective than sulphonylureas for these patients.

Overweight diabetic patients with poor control on sulphonylureas were randomly allocated to additional treatment with either acarbose 100 mg t.d.s. or placebo for 24 weeks in a study reported by Rao and Spathis (1990). Control in the 20 patients was monitored using 2 h postprandial blood glucose, HbAlc and 24 h urinary glucose excretion. Blood glucose and urinary glucose loss both fell significantly in the acarbose group, but were not significantly different when compared with the placebo group at the end of treatment. Glucose tolerance test results did not change significantly for either group during or at the end of treatment. The authors concluded that the effect of acarbose was not sufficient to overcome the severe insulin insufficiency found in these patients.

By contrast, Reaven et al. (1990), in an open study of 3 months treatment of 12 patients poorly controlled on sulphonylureas with added acarbose, reported significant improvements in glycaemia and reductions in HbAlc and triglycerides. There was no change in insulin response to meals or insulin-stimulated glucose uptake, which adds emphasis to Rao and Spathis's observation about the inadequacy of insulin production in these patients. There were clear clinical benefits despite that.

Hillebrand et al. (1987) compared efficacy and tolerance of acarbose or miglitol with glibenclamide in a double-blind, crossover trial with 12-week treatment periods for the 18 patients. Glibenclamide reduced the fasting blood glucose more effectively but acarbose reduced the postprandial level more than glibenclamide did. Overall control was similar for the two agents and there was no change in any other measures. Side effects were reported in 15 of 18 patients on acarbose 200 mg b.d.

In an open, randomised, multicentre trial comparing acarbose and glibenclamide in increasing doses with placebo in 96 patients, there was no difference in the quality of diabetic control achieved in terms of blood glucose, glycosuria or HbAlc over the 24-week treatment period, but both treatments were better than placebo (Hoffmann and Spengler, to be published). However, doses were increased to achieve control, and side effects were experienced in both groups. On acarbose these were more common and were gastrointestinal, whereas nine patients became hypoglycaemic on glibenclamide, which did not occur with acarbose.

In a double-blind crossover study in ten patients taking acarbose 300 mg or metformin 500 mg daily for 4-week periods, there was no significant difference between the two treatments in terms of postprandial blood glucose, HbAlc or frequency or severity of side effects, though they were different in character (Johansen 1984).

Pagano et al. (1986) compared the effects of acarbose and phenformin on diabetic control in two groups of 16 patients controlled with glibenclamide. Acarbose 300 mg daily had a similar effect to phenformin 75 mg.

Lebovitz (1992) in a review recommends α-glucosidase inhibitors as first-line treatment on theoretical grounds of mechanism of action for patients with mild to moderate hyperglycaemia who might be at risk of hypoglycaemia or lactic acidosis, in preference to sulphonylureas or big-uanides. He also recommends use in combination to improve control.

HANEFELD et al. (1991) evaluated acarbose, since first-line treatment in 94 patients after intensive dietary pretreatment for at least 3 months previously had not achieved satisfactory control. Patients were randomly allocated in a double-blind trial to treatment with acarbose 100 mg t.d.s. or placebo for 24 weeks. There was a significant reduction in fasting and postprandial blood glucose and HbAlc with acarbose. C peptide and fasting insulin levels were unchanged but postprandial insulin release fell by 30% compared with placebo. Gastrointestinal symptoms became less troublesome with time.

Acarbose has established a clear role in the treatment of type II diabetes, particularly in combination with other agents. In lower dosage it is well tolerated and most side effects, which have been mainly gastrointestinal, have been experienced at higher dosage levels (AUBELL et al. 1983). Its additional lipid-lowering effect, possibly secondary to improved diabetic control, is an extra benefit for patients with hyperlipidaemia.

III. Other Clinical Studies

Though the prime therapeutic indication for acarbose is in the metabolic control of diabetes, there have been some other studies. These have investigated its effects on blood glucose regulation in other circumstances in which it may be impaired.

ZILLIKENS et al. (1989) did a placebo-controlled study of 11 patients with alcoholic cirrhosis using acarbose 100 mg with test meals to regulate metabolism. After a test breakfast the blood glucose rise in the first 90 min was less after acarbose but it was significantly higher at 150 and 180 min. Insulin release and breath hydrogen were unchanged but glucagon levels were higher at 60 min. After an evening test meal with acarbose, fasting β-hydroxybutyrate levels were significantly lower the following morning than after placebo. Insulin and glucagon levels were unchanged though free fatty acid levels were lower.

Hopes of achieving weight loss in obesity by using acarbose, or other glucosidase inhibitors, to induce a pharmacological state of malabsorption have not been fulfilled. Weight loss did not occur in the diabetic studies, and in the limited other trials in obesity the results were totally disappointing. This use has been reviewed by BERGER (1992).

C. Deoxynojirimycin Derivatives

A new family of α-glucosidase inhibitors has been synthesised, derived from nojirimycin and its reduced form 1-desoxynojirimycin. They have a variety of different inhibitory profiles and a range of durations of action. Miglitol

(Bayer, BAY m1099) is a potent short-acting inhibitor with its greatest effect on sucrase and glucoamylase. Emiglitate (Bayer, BAY o1248) is longer acting and has a similar inhibitory profile to acarbose but is ten times more potent (Taylor et al. 1986; Holt et al. 1988; Lembke et al. 1990; Lembke et al. 1991). These two have been developed and researched extensively. A number of other compounds have been entered into animal studies but none have reached clinical evaluation.

I. Studies in Type I Diabetes

Serrano-rios et al. (1988) did a double-blind, short-term, placebo-controlled study in nine type I diabetic patients comparing the effects of miglitol and emiglitate on blood glucose control when the patients were maintained on a Biostator. Miglitol reduced postprandial glycaemia compared to placebo after breakfast and dinner, whereas emiglitate had a similar effect after breakfast and lunch. Insulin requirement fell after both compounds compared to placebo. Other gut hormones were not affected and there were no side effects. The differences reflect the different durations of action and inhibitory profiles.

A longer comparison was made in a double-blind, placebo-controlled, crossover study comparing the two drugs given for 7-day periods with a 7-day placebo interval between (Gerard et al. 1987). The seven patients had them in random order. Metabolic and hormonal studies were done at the end of each period. There was no change in insulin requirement or HbAlc on either compound but both reduced the post-breakfast glycaemic rise. Free fatty acid levels were unchanged. There were occasional gastrointestinal side effects.

In a larger scale study, Dimitriadis et al. (1988) administered miglitol (100 mg b.d.) and emiglitate (40 mg o.m.) for 1 month each in a double-blind, crossover fashion to 17 type I patients. On the last day of each study period they had metabolic studies for 24 h on a Biostator while taking test meals. Emiglitate reduced the insulin requirement for breakfast and lunch and reduced the glycaemic rise compared with placebo, while miglitol had a similar effect for breakfast and dinner. Both drugs caused a modest rise in breath hydrogen, indicating mild malabsorption.

A number of small studies in patients have examined the effects of miglitol on the response to test meals under controlled conditions. Kennedy and Gerich (1987) studied nine patients on short-term treatment and found a reduction in insulin requirement, a reduced postprandial glucose rise after all three meals on the test day but no change in postprandial triglyceride levels on miglitol compared with placebo. Side effects reported were flatulence in four, diarrhoea in four and abdominal discomfort in three patients. Wing et al. (1990) in a similar acute study of 11 patients found the same order of improvement in glycaemic control but fewer and milder side effects were reported. HILLMAN et al. (1989) also reported similar results in 13 patients.

DIMITRIADIS et al. (1991) examined the impact of the timing of insulin administration, with miglitol compared with placebo, on the postprandial response. They found that, with miglitol, diabetic control could be as good with insulin injected at the time of the meal, as when it was given 30 or 60 min before when placebo had been taken, thereby simplifying management for the patient. KENNEDY and GERICH (1988) came to the same conclusion from their study of nine patients.

II. Studies in Type II Diabetes

Two studies have compared the effects of miglitol and emiglitate in type II diabetic patients. FEDERLIN et al. (1987) evaluated the two in healthy volunteers first and then treated a group of ten patients in a randomised, controlled trial for 1-week periods with miglitol 25 mg b.d., emiglitate 15 mg o.m. or placebo. Both improved glycaemic control but produced side effects which the authors suggested could be reduced if lower dosages were used. WILLMS et al. (1985) studied 12 patients who were controlled on sulphonylureas, in a triple double-blind, crossover protocol with 4-day treatment periods separated by 2 days washout. Postprandial glycaemic rises were reduced and side effects were few. The preferred treatment was emiglitate 40 mg.

SCOTT and TATTERSALL (1988) carried out a dose response study with miglitol in 20 patients giving a single dose with a test breakfast and established 50 mg as a therapeutic dose with few side effects. In the second part of their study, they treated 13 poorly controlled patients who were taking sulphonylureas with miglitol 50 mg t.d.s. for 4 weeks compared with placebo. Postprandial blood glucoses were reduced but there was no change in fasting values or in fructosamine or HbAlc. They noted considerable individual variation in response and suggested higher doses may be required for some patients.

HILLEBRAND et al. (1987) compared the effects of acarbose 200 mg b.d. and miglitol 200 mg b.d. with glibenclamide 7 mg over 3-month periods in 18 glibenclamide-treated patients in double-blind, crossover format. Glibenclamide was stopped before starting the study and the three treatments produced very similar levels of control.

KATSILAMBROS et al. (1986) did a double-blind 1-day dosing study in 12 type II diabetic patients with detailed metabolic monitoring. Postprandial glycaemic rises were reduced after test meals but there were no changes in any other parameters. Other studies in small numbers of patients have confirmed the improved glycaemic control in glibenclamide-treated patients (HEINZ et al. 1989; ARENDS and WILLMS 1986).

There are fewer long-term studies on miglitol in type II diabetes. RAPTIS et al. (1992) conducted a 24-week double-blind, randomised, placebo-controlled study in 120 patients. Those on miglitol had improved diabetic control compared with the placebo group and there were few side effects reported on a dose of 100 mg t.d.s. Samad et al. (1988) were more cautious in their report of a 4-week randomised, crossover study in 12 patients in whom they studied their response to test meals based on starch, sucrose or starch and sucrose together in the last week of each treatment period. They found miglitol 50 mg t.d.s. was effective in diminishing and delaying the postprandial rises of glucose, lactate and pyruvate after all of the meals but overall diabetic control was unchanged. Schnack et al. (1989) did a detailed study in 15 patients in a placebo-controlled, crossover study with 8-week treatment periods, using miglitol 300 mg daily or placebo, separated by a 4-week washout. They found a continuing benefit throughout treatment and this was greatest in type II patients taking insulin.

The specificity of glucosidase inhibitors for particular enzymes makes the balance of carbohydrate substrates in the diet particularly important. Kingma et al. (1992) studied the effects of varying the percentage of starch in the carbohydrate component of test meals in a group of 36 type II patients taking miglitol 100 mg or placebo with the test meals. They found the reduction of postprandial response was independent of the starch percentage and suggested that miglitol may have an extra-intestinal effect on glucose disposition or insulin regulation. This has also been postulated by Joubert et al. (1987; 1990). Schneider et al. (1987) did test meal studies in ten patients using rapidly or slowly absorbed carbohydrate-based meals with miglitol 50 mg or placebo. They found a similar reduction in the postprandial blood glucose response to both meals taken with miglitol as compared with the different response with placebo. The delaying effect on digestion caused by miglitol was effectively masking the rather smaller natural difference in digestibility. Viscous fibres slow digestion and absorption and can have a synergistic effect with glucosidase inhibitors (Jenkins et al. 1979). Requejo et al. (1990) examined the effect of the combinations of miglitol and guar granules on glycaemic control in 12 sulphonylurea-treated patients in a randomised, double-blind, placebo-controlled trial. They found differences in the gut hormone responses to the different combinations but that the combined treatment gave no further reduction in postprandial responses than guar or miglitol did alone. This is obviously a dose-related phenomenon in which there may have been synergism at lower dosages.

These deoxynojirimycin-derived glucosidase inhibitors have shown themselves to be effective therapeutic agents with potential in the treatment of diabetes.

D. Other Glucosidase Inhibitors

A number of other glucosidase inhibitors have been investigated and some have gone into further investigation but none have reached full clinical evaluation for the treatment of diabetes.

AO-128 (Voglibose, Takeda) is one of a large group of related compounds developed as anti-obesity agents, which are potent inhibitors of starch and sucrose digestion and also have lipid-lowering properties. In limited studies they have been shown to have a therapeutic effect in healthy subjects and in type II diabetic patients (HORII et al. 1986).

Castanospermine (Merrell Dow) is a plant alkaloid which is a potent inhibitor of glucosidases in vitro especially sucrase and it reduces the glycaemic response to sucrose in man. In 1987 it was noted to have some anti-HIV properties, and subsequent development has been in that direction with no clinical studies in diabetes.

MDL 25637 (Merrell Dow) is a potent inhibitor of all human brush border glucosidases but a relatively weak inhibitor of amylase and lactase. It has not yet been used in clinical trials in diabetes.

Trestatin complex (Hoffmann-La Roche, Ro 9-0154) is a family of amino sugar derivatives, with varying degrees of α-amylase-inhibiting properties, which slows starch digestion. This results in a dose-related, reduction in post-starch glycaemia and insulin release in healthy volunteers and in type II diabetic patients when compared with the effects of placebo. The effect was maintained over a 4-week dosing period in the diabetic group (EICHLER et al. 1984).

Tendamistat (Hoechst, HOE 467-A) is an extensively investigated polypeptide which is a specific inhibitor of α-amylase. It proved ineffectual in modifying postprandial glucose or insulin responses after sucrose, maltose or glucose. It has not entered clinical trials.

E. Conclusions

Patients with diabetes produce insufficient insulin to satisfy their body's requirement. Use of α-glucosidase inhibitors does nothing to alter that deficiency, but can help to regulate the arrival of carbohydrate substrates in the tissues to a more manageable rate. The evidence from clinical evaluation of α-glucosidase inhibitors, much of which is based on acarbose, shows that they can be effective therapeutic agents in both type I and particularly in type II diabetes. Recently diagnosed type II patients, for whom diet and sulphonylureas have not provided adequate control, appear to benefit most from the addition of a glucosidase inhibitor to their therapy. In all cases they need to be used in conjunction with appropriate dietary guidelines.

They have little effect on fasting blood glucose values but reduce postprandial peaks and HbAlc. In some studies they appear to have a beneficial secondary effect in lowering lipid levels, especially total cholesterol and triglycerides, and in raising HDL cholesterol.

Slowing digestion by the use of doses in excess of the therapeutic range can result in malabsorption of carbohydrate and sometimes other nutrients, which is the source of most of the adverse side effects. These can be reduced by careful dose adjustment whilst maintaining therapeutic benefits. Clinical experience suggests that dosage adjusted to individual requirements, rather than at fixed levels in therapeutic trials, reduces adverse effects appreciably.

References

Arends J, Willms BHL (1986) Smoothing effect of a new alpha-glucosidase inhibitor Bay-m1099 on blood glucose profiles of sulfonylurea treated type II diabetic patients. Horm Metab Res 18:761–764

Aubell R, Boehme K, Berchtold P (1983) Blood glucose concentrations and glycosuria during and after one year of acarbose therapy. Arzneimittelforschung 33:1314–1318

Austenat E (1991) Neuere Therapiestudien mit acarbose bei typ-I-Diabetikern. Akt Endokr Stoffw 12:19–24

Baron AD, Eckel RH, Schmeiser L, Kolterman OG (1987) The effect of short term alpha glucosidase inhibition on carbohydrate and lipid metabolism in type II (non-insulin-dependent) diabetics. Metabolism 35:409–415

Berger M (1992) Pharmacological treatment of obesity: digestion and absorption – clinical perspective. Am J Clin Nutr 55:318–319

Buchanan DR, Collier A, Rodrigues E, Millar AM, Gray RS, Clarke BF (1988) Effectiveness of acarbose, an alpha-glucosidase inhibitor, in uncontrolled non-obese non-insulin dependent diabetes. Eur J Clin Pharmocol 34:51–53

Dimitriadis G, Karaiskos C, Raptis S (1986) Effects of prolonged (6 months) alpha-glucosidase inhibition on blood glucose control and insulin requirements in patients with insulin-dependent diabetes mellitus. Horm Metab Res 18:253–255

Dimitriadis G, Hatziagellaki E, Ladas S, Linos A, Hillebrand I (1988) Effects of prolonged administration of two new alpha-glucosidase inhibitors on blood glucose control, insulin requirements and breath hydrogen excretion in patients with insulin-dependent diabetes mellitus. Eur J Clin Invest 18:33–38

Dimitriadis G, Hatziagellaki E, Alexopoulos E, Kordonouri O, Komesidou V (1991) Effects of alpha-glucosidase inhibition on meal glucose-tolerance and timing of insulin administration in patients with type-I diabetes-mellitus. Diabetes Care 14:393–398

Eichler HG, Korn A, Gasic S, Pirson W, Businger J (1984) The effect of a new specific α-amylase inhibitor on post-prandial glucose and insulin excursions in normal subjects and type 2 (non-insulin-dependent) diabetic patients. Diabetologia 26:278–281

Federlin KF, Mehlburger L, Hillebrand I, Laube H (1987) The effect of two new glucosidase inhibitors and blood glucose in healthy volunteers and in type II diabetics. Acta Diabetol Lat 24:213–221

Foelsch UR, Spengler M, Boehme K, Sommerauer B (1990) Efficacy of glucosidase inhibitors compared to sulphonylureas in the treatment and metabolic control of diet treated type II diabetic subjects: two long-term comparative studies. Diabetes Nutr Metab 3:63–68

van Gaal L, Nobels F, Leeuw IDE (1991) Effects of acarbose on carbohydrate metabolism, electrolytes, minerals and vitamins in fairly well-controlled non-insulin-dependent diabetes mellitus. Z Gastroenterol 29:642–644

Gerard J, Luyckx A, Lefebvre PJ (1981) Improvement of metabolic control in insulin dependent diabetics treated with the alpha-glucosidase inhibitor acarbose for two months. Diabetologia 21:446–451

Gerard J, Hillebrand I, Lefebvre P (1987) Assessment of the clinical efficacy and tolerance of two new alpha-glucosidase inhibitors in insulin-treated diabetics. Int J Clin Pharmacol Ther Toxicol 25:483–488

Hanefeld M, Fischer S, Schulze J, Spengler M, Wargenau M, Scholler K, Fuecker K (1991) Therapeutic potentials of acarbose as first-line drug in NIDDM insufficiently treated with diet alone. Diabetes Care 14:732–737

Heinz G, Komjati M, Korn A, Waldhaeusl W (1989) Reduction of postprandial blood-glucose by the alpha-glucosidase inhibitor miglitol (Bay-m1099) in type II diabetes. Eur J Clin Pharmacol 37:33–36

Hillebrand I, Englert R, Boehme K (1987) Wirksamkeit und vertraeglichkeit einer jeweils 12woechigen behandlung mit acarbose (Bay-g5421), miglitol (Bay-m1099) und glibenclamid (the efficacy and tolerance of 12week periods of treatment by acarbose (Bay-g5421), miglitol (Bay-m1099) and glibenclamide). Klin Wochenschr 65:196–197

Hillebrand I, Cagatay M, Schulz H, Streicher-Saied U (1988). Efficacy and tolerability of the glucosidase inhibitor acarbose (Bay-g5421) evaluated by clinical data pool. Therapie 43:153

Hillman RJ, Scott M, Gray RS (1989) Effect of alpha-glucosidase inhibition on acarbose profiles in insulin-dependent diabetes. Diabetes Res 10:81–84

Hoffmann J, Spengler M (in press) Efficacy of 24 weeks monotherapy with acarbose, glibenclamide or placebo in NIDDM patients: the Essen study. Diabetes Care

Hollander P (1992) Safety profile of acarbose, an alpha-glucosidase inhibitor. Drugs 44:47–53

Holt PR, Thea D, Yang M-Y, Kotler DP (1988) Intestinal and metabolic responses to an alpha-glucosidase inhibitor in normal volunteers. Metabolism 37:1163–1170

Homma Y, Irie N, Yano Y, Nakaya N, Goto Y (1982) Changes in plasma lipoprotein levels during medication with a glucoside hydrolase inhibitor acarbose. Tokai J Exp Clin Med 7(3):393–396

Horii S, Fukase H, Matsuo T, Kameda Y, Asano N, Matsui K (1986) Synthesis and α-D-glucosidase inhibitory activity of N-substituted valiolamine derivatives as potential oral antidiabetic agents. J Med Chem 29:1038–1046

Hotta N, Kakuta H, Sano T, Matsumae H, Yamada H, Kitazawa S, Sakamoto N (1993a) Long-term effect of acarbose on glycaemic control in non-insulin-dependent diabetes mellitus: a placebo-controlled double-blind study. Diabet Med 10:134–138

Hotta N, Kakuta H, Koh N, Sakakibara F, Haga T, Sano T, Okuyama M, Sakamoto N (1993b) The effect of acarbose on blood glucose profiles of type II diabetic patients receiving insulin therapy. Diabet Med 10:355–358

Ito K, Mimura A, Tsuruoka A, Sasaki T, Saito S, Kageyama S, Ikeda Y (1989) Effects of acarbose on glycaemic control in insulin-dependent diabetes with an artificial endocrine pancreas [Biostator (registered trade mark)]. J Jpn Diabetes Soc/Tonyobyo 32:573–577

Jenkins DJA, Taylor RH (1982) Acarbose: dosage and its interaction with sugars, starch and fibre. In: Creutzfeldt W (ed) Proceedings of first international symposium on acarbose. Excerpta Medica, Amsterdam, pp 86–96

Jenkins DJA, Taylor RH, Nineham R, Goff DV, Bloom SR, Sarson D, Alberti KGMM (1979) Combined use of guar and acarbose in reduction of postprandial glycaemia. Lancet II (8149):924–927

Jenkins DJA, Taylor RH, Goff DV, Fielden H, Misiewicz JJ, Sarson DL, Bloom SR, Alberti KGMM (1981) Scope and specificity of acarbose in slowing carbohydrate absorption in man. Diabetes 30:951–954

Jenny A, Proietto J, O'Dea K, Nankervis A, Trainedes K, D'Embden H (1993) Low dose acarbose improves glycaemic control in NIDDM patients without changes in insulin sensitivity. Diabetes Care 16:499–502

Johansen K (1984) Acarbose treatment of sulfonylurea-treated non-insulin dependent diabetics. A double blind cross-over comparison of an alpha-glucosidase inhibitor with metformin. Diabete Metabol 10:219–223

Joubert PH, Foukaridis GN, Bopape ML (1987) Miglitol may have a blood glucose lowering effect unrelated to inhibition of alpha glucosidase. Eur J Clin Pharmacol 31:723–724

Joubert PH, Venter HL, Foukaridis GN (1990) The effect of miglitol and acarbose after an oral glucose load – a novel hypoglycaemic mechanism. Br J Clin Pharmacol 30:391–396

Katsilambros N, Philippedes P, Toskas A et al. (1986) A double-blind study on the efficacy and tolerance of a new alpha-glucosidase inhibitor in type II diabetics. Arzneimittelforschung 36:1136–1138

Kennedy FP, Gerich JE (1987) A new alpha-glucosidase inhibitor (Bay-m1099) reduces insulin requirements with meals in insulin-dependent diabetes mellitus. Clin Pharmacol Ther 42:455–458

Kennedy FP, Gerich JE (1988) Alpha-glucosidase inhibition and timing of pre-prandial insulin in patients with insulin-dependent diabetes mellitus (IDDM). Diabetes Res Clin Pract 4:309–312

Kingma PJ, Menheere PP, Sels JP, Kruseman AC (1992) Alpha-glucosidase inhibition by miglitol in NIDDM patients. Diabetes Care 15:478–483

Lebovitz HE (1992) Oral antidiabetic agents: the emergence of alpha-glucosidase inhibitors. Drugs 44:21–28

Lecavalier L, Hamet P, Chiasson JL (1986) The effects of sucrose meal on insulin requirement in IDDM and its modulation by acarbose. Diabete Metab 12:156–161

Lembcke B, Diederich M, Folsch UR, Creutzfeldt W (1990) Postprandial glycaemic control, hormonal effects and carbohydrate malabsorption during long-term administration of the alpha-glucosidase inhibitor miglitol. Digestion 47:47–55

Lembcke B, Foelsch UR, Gatzemeir W, Ebert R, Siegel E, Creutzfeldt W (1991) Inhibition of glycaemic and hormonal responses after repetitive sucrose and starch loads by different doses of the alpha-glucosidase inhibitor miglitol (Bay-m1099) in man. Pharmacology 43:318–328

Leonhardt W, Hanefeld M, Fischer S, Schulze J, Spengler M (1991) Beneficial effects on serum lipids in non-insulin dependent diabetics by acarbose treatment. Arzneimittelforschung 41:735–738

Marena S, Tagliaferro V, Cavallero G, Pagani A, Montegrosso G, Bianchi W, Zaccarini P, Pagano G (1991) Double-blind crossover study in type I diabetic patients. Diabet Med 8:674–678

Pagano G, Cassader M, Cavallo-Perin P, Dal Molin V, Salvini P et al. (1986) Comparison of acarbose and phenformin treatment in glibenclamide-treated non-insulin dependent diabetics. Curr Ther Res Clin Exp 39:143–148

Rao RH, Spathis GS (1990) Alpha-glucosidase inhibitor therapy does not improve glycaemic control in overweight diabetics poorly controlled on sulfonylureas. Diabetes Nutr Metab Clin Exp 3:17–22

Raptis SA, Hadjidakis D, Tountas N, Bauer RJ, Schulz H (1992) Long-term effectiveness of a new alpha-glucosidase inhibitor (Bay-m1099 – miglitol) in insulin treated type II diabetes mellitus. Diabetes 42:193A

Reaven GM, Lardinois CK, Greenfield MS, Schwartz HC, Vreman HJ (1990) Effect of acarbose on carbohydrate and lipid metabolism in NIDDM patients poorly controlled by sulfonylureas. Diabetes Care 13:32–26

Requejo F, Uttenthal LO, Bloom SR (1990) Effects of alpha-glucosidase inhibition and viscous fibre on diabetic control and postprandial gut hormone responses. Diabet Med 7:515–520

Ross S, Hunt J, Josse R, Mukherjee J, Palmason C, Rodger W (1992) Acarbose significantly improves glucose control in non-insulin dependent diabetes mellitus subjects (NIDDM): results of a multicenter Canadian trial. Diabetes 41:193A

Rybka J, Gregorova A, Zmydlena A, Jaron P (1990) Clinical study of acarbose. Drug Invest 2:264–267

Sakamoto N, Shibata M, Hotta N, Tomita A, Tsuchida I, Nagashima M, Katsumata K (1989) Clinical study on Bay-g5421 (acarbose) for non-insulin-dependent diabetes mellitus (NIDDM). Jpn Pharmacol Ther/Yakuri Chiro 17:285–301

Sakamoto N, Hotta N, Kakuta H, Sano T, Yamada H, Matsumae H, Kitazawa T (1990) Usefulness of long-term administration of Bay-g5421 on the patients with NIDDM. Jpn J Clin Exp Med/Rinsho Kenkyu 67:219–233

Samad AH, Willing TS, Alberti KGMM, Taylor R (1988) Effects of Bay 1099, new alpha-glucosidase inhibitor, on acute metabolic responses and metabolic control in NIDDM over 1 month. Diabetes Care 11:337–344

Santeusanio F, Ventura MM, Contadini S, Compagnucci P, Moriconi V, Zaccarini P, Marra G, Amigoni S, Bianchi W, Brunetti P (1993) Efficacy and safety of two different dosages of acarbose in non-insulin-dependent diabetic patients treated by diet alone. Diabet Nutr Metab 6:147–154

Schnack C, Prager RJ, Winkler J, Klauser RM, Schneider BG (1989) Effects of 8-week alpha-glucosidase inhibition on metabolic control. C-peptide secretion, hepatic glucose output, and peripheral insulin sensitivity in poorly controlled type II diabetics. Diabetes Care 12:537–543

Schneider J, Schauberger G, Nichting M, Niklas L, Otto H (1987) Wirkung des Glukosidasehemmers miglitol (Bay 1099) auf die postprandiale Hyperglykaemie nach schnell und langsam resorbierbaren Kohlehydraten. Akt Endokrin Stoffw 8:110–111

Scott AR, Tattersall RB (1988) Alpha glucosidase inhibition in the treatment of non-insulin dependent diabetes mellitus. Diabet Med 5:42–46

Scott RS, Knowles RL, Beaven DW (1984) Treatment of poorly controlled non-insulin-dependent diabetic patients with acarbose. Aust NZ J Med 14:649–654

Serrano-Rios M, Saban J, Navascues I, Canizo JF, Hillebrand I (1988) Effect of two new alpha-glucosidase inhibitors in insulin-dependent diabetic patients. Diabetes Res Clin Pract 4:111–116

Taylor RH, Jenkins DJA, Barker HM, Fielden H, Goff DV, Misiewicz JJ, Lee DA, Allen B, McDonald G, Wallrabe H (1982) Effect of acarbose on the 24-hour blood glucose profile and pattern of carbohydrate absorption. Diabetes Care 5:92–96

Taylor RH, Barker HM, Bowey EA, Canfield JE (1986) Regulation of the absorption of dietary carbohydrate in man by two new glucosidase inhibitors. Gut 27:1471–1478

Toeller M (1992) Nutritional recommendations for diabetic patients and treatment with alpha-glucosidase inhibitors. Drugs 44:13–20

Toto Y, Toyota T, Oikawa S, Maruhama Y, Tamura M, Komatsu K, Fujitani H, Abe R, Kikuchi H (1989) Long term treatment of Bay 5421 (acarbose) in non-insulin dependent diabetes mellitus. Jpn Pharmacol Ther/Yakuri Chiro 17: 4115–4129

Truscheit E, Frommer W, Junge B, Muller L, Schmidt DD, Wingender W (1981) Chemistry and biochemistry of microbial α-glucosidase inhibitors. Angew Chem 20:744–761

Viviani GL, Camogliano L, Borgoglio MG, Ohnmeiss H, Adezati L (1987) Acarbose treatment in insulin-dependent diabetics – a double-blind crossover study. Curr Ther Res Clin Exp 42:1–11

Willms B, Schumann E, Arends J (1985) Vergleich zweier alpha-Glucosidasehemmer (Bay-m1099 und Bay-o1248) in klinischen Studien bei Sulfonylharnstoff (sh-) behandelten typ II-Diabetikern. Akt Endokrin Stoffw 6:114

Wing J, Kalk WJ, Berzin M, Diamond TH, Griffiths RF, Smit AM, Osler CE (1990) The acute effects of glucosidase inhibition on post-meal glucose increments in insulin-dependent diabetics. S Afr Med J 77:286–288

Zillikens MC, Swart GR, Vandenberg JW, Wilson JH (1989) Effects of the glucosidase inhibitor acarbose in patients with liver cirrhosis. Aliment Pharmacol Ther 3:453–457

Zimmerman BR (1992) Preventing long-term complications. Implications for combination therapy with acarbose. Drugs 44:54–60

CHAPTER 23

Oral Antidiabetic Drugs in Research and Development

H. Bischoff and H.E. Lebovitz

A. Introduction

Current oral antidiabetic therapy has numerous pharmacologically efficacious drugs at its disposal – the recognized sulfonylurea and biguanide derivatives – and, more recently, the α-glucosidase inhibitors. However, a review of the experience with the well-established sulfonylurea and biguanide drugs bears out that the ultimate goal of antidiabetic drug therapy, the prevention of diabetic complications, cannot as yet be met satisfactorily.

The results of preclinical experiments and the observation of clinical data indicate that elevated blood glucose levels contribute to the development of microvascular diseases. It is undisputed that, in non-insulin-dependent diabetes mellitus (NIDDM) patients, metabolic control, in terms of blood glucose control, correlates strongly with the development of the diabetic complications neuropathy, retinopathy, and nephropathy (Pirart 1978). Therefore, treatment is aimed to achieve glycemic control, and this was shown convincingly for insulin-dependent diabetes mellitus (IDDM)-patients by the results of the Diabetes Control and Complications Trial (DCCT) (DCCT 1993). All epidemiological results clearly indicate the strong medical need for new therapeutic strategies and approaches to restore euglycemia and additionally to prevent diabetic complications in diabetic patients directly. This review will provide an overview of new pharmacological approaches that were studied in the past years or are in current research for oral treatment of diabetes.

Because of the genetic heterogeneity of diabetes, new pharmacological approaches should consider a variety of underlying metabolic defects. It is widely accepted that in NIDDM, insulin secretion and the sensitivity of liver and peripheral tissues to insulin is out of balance. The impairment of insulin response is overcome transiently by increased insulin secretion of the pancreas, resulting in postprandial and postabsorptive hyperinsulinemia. In the insulin-resistant state, glucose uptake and metabolism by peripheral skeletal muscle tissue are reduced, whereas hepatic gluconeogenesis and glucose output are normal or even increased despite elevated fasting plasma insulin concentrations (DeFronzo 1988; Gerich 1991). As a consequence of inadequate insulin action, fasting hyperglycemia is sustained due to insufficient suppression of hepatic gluconeogenesis and disposal of glucose by extrahepatic tissues. Moreover, an excess of glucagon, substrate availability, and fatty acid oxidation contribute to overactivity of hepatic glucose production. Independently of the ongoing controversy regarding which tissue contributes more to the development of the disease (DeFronzo 1992) – the liver, due to increased hepatic glucose output (Gerich 1991; Consoli 1992), or the skeletal muscle, due to impaired glucose utilization (Häring and Mehnert 1993; Beck-Nielsen et al. 1994) – new pharmacological approaches could help enhance the action of insulin in the cells. This is why, over the past 10 years, antidiabetic pharmaceutical research has been impressed by the discovery of the "insulin sensitizers"; due to lack of sufficient therapeutic success, however, this research will probably still be ongoing for a longer period of time. Oral drugs which could completely mimic insulin's action should be superior to compounds that "only" correct or improve its action. However, this extraordinary goal of antidiabetic pharmacotherapy is still far from being realized. Therefore, pharmacological research aiming to prevent the deleterious effects caused by glucose – i.e., the diabetes-related complications neuropathy, nephropathy, and retinopathy – is still justified. This field of research concerns the inhibition of the polyol pathway by aldose reductase inhibitors to prevent sorbitol accumulation in insulin-independent tissues, and the inhibition of protein glycation to prevent the formation of advanced glycation end products.

B. Stimulation and Modulation of Insulin Secretion

Today, at the end of the twentieth century, the most important oral antidiabetic drugs are still the insulin secretagogues, all of which belong to one class of chemical compounds, the sulfonylureas. The mode of action by which the sulfonylureas stimulate insulin secretion – closure of the ATP-sensitive K^+ channel through binding at a receptor-linked binding site – is a very efficacious one. The inhibition of K^+ permeability leads to depolarization of the plasma membrane and subsequently to activation of the voltage-dependent Ca^{2+} channel to promote Ca^{2+} influx. However, this is not the only site of action for compounds which potentially increase insulin secretion. Potential sites of action for orally active compounds could also be an intracellular site of the B-cell or an extracellular site at the plasma membrane. The intracellular sites of action are aimed at B-cell energy metabolism as is particularly known for glucose and other insulin-releasing nutrients (Sener and Malaisse 1984). Another topic of great interest consists in the mechanisms regulating the insulin secretory process under the control of Ca^{2+} ions, including enzyme systems like adenylate cyclase (Malaisse 1990), protein kinases (Tamagawa et al. 1985), lipoxygenase, and cyclooxygenase (Robertson 1986; Turk et al. 1987). However, the extracellular sites of action at the B-cell membrane have been elucidated to a much greater extent than have the intracellular mechanisms for the regulation of insulin release:

- Uptake of cationic amino acids such as arginine also depolarizes the plasma membrane (HENQUIN and MEISSNER 1981) with subsequent Ca^{2+} influx.
- Modulation (prolongation) of the open time of the voltage-dependent L-type Ca^{2+} channel (MALAISSE-LAGAE et al. 1984; RORSMAN and TRUBE 1986).
- Activation of stimulatory hormone receptors (e.g., glucagon, glucagon-like peptide 1, glucose-dependent insulinotropic peptide, cholecystokinin, acetylcholine) via adenylate cyclase, phospholipases, or ionic channels.
- Antagonization of inhibitory hormone receptors (α_2-adrenoreceptors, somatostatin, or galanin receptors).

This shows that distinctly different sites of action exist which could theoretically be a target for pharmacological intervention. However, only a few are actual subjects for drug research, since in all probability not all mechanisms are suitable:

- The uptake of cationic amino acids is probably not very efficacious (HENQUIN and MEISSNER 1984; PANTEN 1987).
- The activation of stimulatory hormone receptors should be possible; however, to our knowledge, specific research has not been conducted.
- The major drawback of the intracellular sites of action is that control and regulation of intracellular B-cell Ca^{2+} metabolism are still a "black box."

This explains why direct membrane depolarization via blockage of the ATP-sensitive K^+ channel with subsequent stimulation of Ca^{2+} influx are still the major pharmacological targets for stimulating insulin release. The only other two pharmacological targets currently under investigation affect (a) the inhibitory hormone receptor sites, specifically the α_2-adrenoceptor which, after activation, inhibits insulin release via inhibition of adenylate cyclase (ROBERTSON and PORTE 1973), and (b) the modulation of voltage-dependent Ca^{2+} channels (MALAISSE-LAGAE et al. 1984; KOMATSU et al. 1989).

I. Depolarization of the B-Cell Membrane

1. Inhibitors of the ATP-Sensitive K^+ Channel

a) Sulfonylurea Derivatives

Although sulfonylurea research came to an end in the 1970s, and this area of pharmacological research brought into focus new antidiabetic mechanisms, pharmaceutical research has engaged in further efforts over the last two decades to increase, prolong, or shorten the action of sulfonylureas, and above all to accelerate their onset of action. Glimepiride, a structure closely related to glibenclamide (glyburide) (Fig. 1), in which the phenylsubstitutent of glibenclamide was exchanged with pyrroline, is characterized by a faster

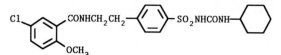

HOE 419
glibenclamide

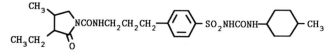

HOE 490
glimepiride

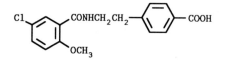

HB 699
meglitinide

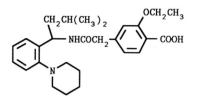

AG-EE 388 ZW (racemate)

AG-EE 623, repaglinide
(S-enantiomer)

Fig. 1. Sulfonylurea and benzoic acid derivatives

onset and longer-lasting effect in blood glucose-lowering activity in several animal species (Geisen 1988). In vitro studies utilizing perifused pancreatic islets from euglycemic rats, showed that both glibenclamide and glimepiride exhibited similar behavior on insulin secretion and did not affect intracellular glucose metabolism (Malaisse et al. 1993).

The more rapid increase in serum insulin and decrease in blood glucose concentrations in rats and dogs after oral administration of glimepiride compared with glibenclamide has been explained by quicker absorption of glimepiride (Geisen 1988). In rats a sex difference in the pharmacokinetics of glimepiride was observed. Female rats showed a higher AUC and longer half-life of the unchanged substance compared to male animals due to sex-dependent cytochrome P-450 metabolism in liver microsomes (YAMAZAKI and TABATA 1993). Like other sulfonylureas, glimepiride was also described to be more or less nontoxic in rats in acute, subchronic, and chronic toxicity studies. In acute studies 10000 mg/kg after oral and 3950 mg/kg after intraperitoneal administration were tolerated (DONAUBAUER and MAYER 1993).

In abstract form it was reported that, in NIDDM-patients, 3 mg glimepiride once daily was as potent as 10.5 mg glibenclamide per day administered in two doses (SCHERRER et al. 1988). In healthy subjects the apparent serum half-life of glimepiride ($t_{1/2}$ = 2.7 +/−1 h) resembled that for glipizide ($t_{1/2}$ = 3–4 h) rather than that of glibenclamide ($t_{1/2}$ = 1.4–3 h) (RATHEISER et al. 1993). Compared with glibenclamide, glimepiride, being in phase III of clinical development, showed longer-lasting hypoglycemic effects. Due to its pharmacokinetic profile the question may arise whether such a potent, long-acting insulin secretaguoge is well controllable. Possibly, substances having a rapid onset and shorter duration of action are more advantageous in controlling the blood glucose profile adequately and also in avoiding deleterious hypoglycemias.

b) Benzoic Acid Derivatives

Research on the structure–activity relationship of potent "second-generation" sulfonylureas like glibenclamide has revealed that the nonsulfonylurea moiety of glibenclamide (Fig. 1) also stimulated insulin secretion. The benzoic acid derivative meglitinide (HB 699) (Fig. 1) stimulated insulin release in vitro (FUSSGÄNGER and WOJCIKOWSKI 1977; HENQUIN et al. 1987) and in vivo (GEISEN et al. 1978; RIBES et al. 1981). In dogs, insulin secretion followed remarkable biphasic dynamics (GEISEN et al. 1978). The hypoglycemic activity of meglitinide (40 mg/kg i.v.) was three times lower compared to tolbutamide (12 mg/kg i.v.), but the duration of action was obviously markedly shorter (RIBES et al. 1981). In studies using the whole-cell configuration of the patch clamp technique, meglitinide showed a comparable half maximal inhibition of the ATP-dependent K^+ efflux (2.1 μmol/l) with tolbutamide (4.1 μmol/l), disclosing a distinctly lower potency compared to glibenclamide (4.0 nmol/l) (ZÜNKLER et al. 1988). In perifused mouse islets, meglitinide (3–100 μM) stimulated a steep increase in insulin release, and its secretory pattern was comparable to that of tolbutamide; however, it differed from glipizide (3–100 nM) and glibenclamide (3–100 nM) (PANTEN et al. 1989). Although the toxicological profile was probably clear (no detailed report was published), the very few published data revealed a difference from the corresponding sulfonylurea glibenclamide in terms of lower acute tolerability in mice (GEISEN et al. 1978). The clinical development of meglitinide was discontinued after phase II clinical trials.

The racemate AG-EE 388 ZW and its active (S)-enantiomer repaglinide (AG-EE 623 ZW) (Fig. 1), respectively, are benzoic acid derivatives with a remarkably higher potency. Repaglinide (10 nM) blocked the ATP-sensitive K^+ channel of rat B-cells within 1 min by 90% with a potency comparable to glibenclamide (20 nM) (Frøkjaer-Jensen et al. 1992). For another benzoic acid derivative of the same series, AZ-DF-265, sharing the same pharmacological properties, it could be demonstrated that this compound displaced [^{3}H]-glibenclamide from the sulfonylurea receptor-binding site (Ronner et al. 1992). In NIDDM patients, repaglinide seemed to be weakly active after low doses (1.5–4 mg), exhibiting blood glucose-lowering effects only on postprandial hyperglycemia (Wolffenbüttel et al. 1992). The racemate AG-EE 388 ZW (3 mg) was reported to be comparably active to glibenclamide (3.5 mg) in NIDDM patients concerning maximal plasma insulin levels (Profozić et al. 1993) with concomitant shorter duration of action. The elevation of plasma insulin levels lasted only 4 h after administration of AG-EE 388; however, it lasted 10 h after glibenclamide (Profozić et al. 1992). The pharmacokinetic behavior of both enantiomers seems to be of some interest. As described in a patent corporation treatment (PCT) application (Grell et al. 1993), the active (S)-enantiomer repaglinide (AG-EE 623 ZW) showed faster absorption, distinctly lower plasma concentrations, and a much shorter half-life compared with the inactive (R)-enantiomer AG-EE 624 ZW (Grell et al. 1993). Due to its pharmacokinetic behavior, the (S)-enantiomer suggested favorable pharmacological properties in terms of quick insulin release and short-acting blood glucose-lowering effects. Repaglinide is undergoing phase II clinical trial. No further detailed clinical or toxicological data were reported.

c) Guanidine Derivatives

Fully substituted guanidine derivatives represent another chemical class of nonsulfonylurea compounds which stimulate glucose-induced insulin secretion. The first agent also studied clinically was pirogliride (McN-3495) (Fig. 2), which induced elevated plasma insulin and lowered blood glucose concentrations in NIDDM patients (Johnson et al. 1980). Due to elevated hepatic transaminases, further development was discontinued (Johnson et al. 1980). As a follow-up the related compound linogliride was selected. Linogliride (McN-3935) (Fig. 2) reduced blood glucose concentrations in rats, mice, dogs, and monkeys (Tutwiler et al. 1986) dose-dependently. The minimum effective oral doses were between 1 and 5 mg/kg b.w. In starved nondiabetic rats linogliride brought about the hypoglycemic effect with an ED_{30} of 6–7 mg/kg. The compound was two times more potent than the parent substance pirogliride and 8 times more potent than tolbutamide (Tutwiler et al. 1986).

Studies to elucidate the mode of action showed a direct effect of linogliride on B-cells. Linogliride clearly stimulated the insulin secretion of

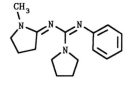

McN-3495
pirogliride

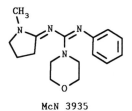

McN 3935
linogliride

Fig. 2. Guanidine derivatives

isolated rat islets (fivefold) in the presence of 5.5 mM glucose (ZAWALICH et al. 1987), and Ca^{2+} channel blockers inhibited linogliride-potentiated insulin secretion (TUMAN et al. 1990). It was recently shown that linogliride blocked the ATP-sensitive K^+ channel (RONNER et al. 1991) and decreased $^{86}Rb^+$ efflux from rat islets perifused with a glucose-free medium (TUMAN et al. 1990). Drug metabolism in animal studies and in healthy volunteers showed a rapid and extensive absorption. Linogliride was cleared primarily by renal excretion of unchanged drug (TUMAN et al. 1990). In clinical trials the daily effective oral doses were between 300 and 800 mg (TUMAN et al. 1990). The drug seemed to be effective and was well tolerated. However, due to preclinical safty issues related to direct CNS-toxic effects, linogliride was withdrawn from development (TUMAN et al. 1990).

d) 2-Substituted Imidazoline Derivatives

Midaglizole (DG-5128) (Fig. 3), an analogue of the adrenergic antagonist phentolamine (Fig. 3), moderately reduced blood glucose concentrations in various animal species in a dose-dependent manner (KAMEDA et al. 1982a) accompanied by elevated plasma insulin concentrations (KAMEDA et al. 1982b). In isolated islet cells, midaglizole reversed the clonidine- or adrenaline-induced inhibition of insulin release. In patients treated with diet alone, midaglizole lowered fasting blood glucose levels and improved oral glucose tolerance in doses of 150–250 mg tid due to an increase in plasma insulin concentrations (KAMAZU et al. 1987).

Midaglizole has been described to be a selective α_2-adrenoceptor antagonist (YAMANAKA et al. 1984), similar to other 2-substituted imidazoline

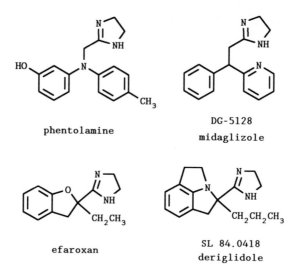

Fig. 3. 2-Substituted imidazoline derivatives

derivatives including phentolamine, idazoxan, and efaroxan. The observed effect that midaglizole antagonized the α_2-adrenergic-induced suppression of insulin secretion matched the concept that inhibitors of α_2-adrenoceptors should increase the insulin release of diabetic patients, since it is assumed that diabetic patients may exhibit a continuously elevated adrenergic tonus (Porte 1967; Robertson et al. 1976). These inhibitory effects on insulin release are mediated by stimulation of α-adrenoceptors located on the B-cell membrane (Porte 1967). On this basis the suggestion has been advanced that α_2-antagonists like midaglizole could be helpful in the managment of diabetes. However, in recent years, more and more evidence has emerged that adrenergic drugs of the imidazoline type can influence insulin release by mechanisms that are unrelated to adrenoceptor blockade.

Schulz and Hasselblatt (1988) found that specific α_2-antagonists like rauwolscine which are structurally unrelated to 2-imidazolines did not stimulate insulin release in contrast to phentolamine, which is structurally a 2-imidazoline derivative. After further studies using other 2-substituted imidazolines in the absence of adrenoceptor agonists (Schulz and Hasselblatt 1989), they suggested that the insulinotropic effect is due to the imidazoline structure rather than the α_2-adrenoceptor inhibitory activity by these compounds. Recent studies with phentolamine (Plant and Henquin 1990) confirmed these results, indicating that a change in membrane potassium permeability due to inhibition of the ATP-sensitive K^+ channel might be the cause for stimulated insulin release. Another α_2-adrenoceptor antagonist well characterized in vitro, efaroxan (Fig. 3) (Chapleo et al. 1984), also showed the capability to stimulate insulin secretion from pancreatic islets directly in the absence of any adrenoceptor agonist (Schulz and HASSELBLATT 1988; CHAN et al. 1988). Like midaglizole, efaroxan was developed for treatment of NIDDM as an insulin secretagogue due to its α_2-adrenoceptor antagonistic activity. However, corresponding to the results with phentolamine (PLANT and HENQUIN 1990), an interaction with the ATP-sensitive K^+ channel was also hypothesized for the 2-imidazoline efaroxan (CHAN and MORGAN 1990). Efaroxan reversed the diazoxide-induced inhibition of glucose-stimulated insulin secretion of isolated rat islets dose-dependently. This result was not expected since diazoxide is a sulfonamide and not an α_2-agonist (CHAN and MORGAN 1990). Further studies demonstrated that efaroxan did not affect cyclic adenosine monophosphate (cAMP) levels but reduced the diazoxide induced ^{86}Rb efflux from isolated rat islets. In isolated membrane patches the ATP-sensitive K^+ channel activity was inhibited with a K_i of $12 \mu M$ (MORGAN et al. 1991).

These studies provided evidence that these 2-substituted imidazoline derivatives, midaglizole and efaroxan, do not stimulate the insulin release as initially supposed due to its α_2-adrenoceptor antagonism. Similar to the sulfonylureas, they act by closing the ATP-sensitive K^+ channel and subsequently depolarizing the B-cell membrane. Recently it was suggested that imidazoline compounds and sulfonylureas bind to distinct sites on islet cells (BROWN et al. 1993), and that these sites can functionally interact to control ATP-sensitive K^+ channel activity and insulin secretion of islet cells. Although efaroxan showed a high selectivity for α_2-adrenoceptor binding sites compared with α_1-receptors, the compound increased diastolic blood pressure of animals dose-dependently and significantly (BERRIDGE et al. 1992) increased systolic blood pressure in humans after oral administration of 0.3 mg/kg. Higher doses (0.8 mg/kg) produced finger tremor and hot and cold flashes (SCRIP 1989). The development of efaroxan, which was in phase II clinical trials, was discontinued.

2. Compounds with Uncertain Sites of Action

A structurally new type of insulin secretagogues consists of amino acid compounds derived from phenylalanine. A-4166 (AY-4166) (Fig. 4) and related compounds showed stereospecific activity, and it was found that the R-configuration was essential for hypoglycemic activity (SHINKAI and SATO 1990).

It seems that the phenylmethyl moiety of the amino acid residue is essential. A variation in the amino acids such as glycine, alanine, phenylglycine, tyrosine, and others reduced the hypoglycemic activity (SHINKEY and SATO 1990). The pharmacological profile of A-4166 is characterized by a rapid onset of blood glucose decrease, which is the result of a quick rise in plasma insulin. In addition to the rapid onset of action, the duration of action seems to be markedly shorter than that of tolbutamide. In starved dogs the compound was three to four times more potent than tolbutamide (SHINKAI and SATO 1990). Oral administration of 5 mg/kg

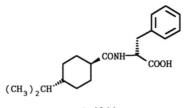

A-4166

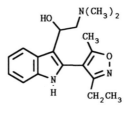

SaRI 59-801

Fig. 4. New structures of insulin secretagogues

lowered blood glucose concentrations by 36% (Sato et al. 1991). Recent in
vitro studies utilizing isolated perfused rat pancreas at glucose concentrations
higher than 5.5 mmol/l revealed that 3 and 30 μmol A-4166 stimulated
insulin secretion directly and dose-dependently. Unlike sulfonylurea compounds, A-4166 exerted little or no effect on glucagon secretion in vitro
(Hirose et al. 1994) and in vivo (Sato et al. 1991). To elucidate the
mechanism of action, the intracellular Ca^{2+} content was measured in single
rat B-cell preparations. In the presence of 2.8 mM glucose, 30 μM A-4166
induced a sustained increase in intracellular Ca^{2+} concentration comparable
to tolbutamide, but with a tenfold higher potency (Fujitani and Yada
1994). The Ca^{2+} increase could be blocked by the dihydropyridine derivative
nitrendipine, an inhibitor of the voltage-dependent L-type Ca^{2+} channel.
Nitrendipine also inhibited insulin release from perfused rat pancreases
stimulated by A-4166 (Fujitani and Yada 1994). The inhibition of intracellular metabolism by KCN and other inhibitors suppressed A-4166 response
only slightly at maximal concentrations. Although not yet determined, it has
been suggested that inhibition of the ATP-sensitive K^+ channel is at least
one of the modes of action (Yada and Fujitani 1992; Fujitani and Yada
1994); thus, it is possible that A-4166 stimulates insulin secretion by the
same mechanism as sulfonylureas. In rats the compound seemed to be
cleared and excreted mainly by the liver, not via the kidney (Shinkai and
Sato 1990). According to the few published toxicological animal data, the
oral LD_{50} in mice was more than 3 g/kg (Hirose et al. 1994), indicating good
tolerability. A-4166 is in phase II clinical trials. Published clinical results are
not yet available.

A completely different structure shows the isoxazolyl-indole derivative SaRI 59-801 (Fig. 4). This compound lowered blood glucose concentrations in mice, rats, and monkeys at doses ranging from 10 to 200 mg/kg. These hypoglycemic effects were correlated with increased plasma insulin concentrations in mice and rats (Ho et al. 1985). The rise in insulin levels could be confirmed by stimulation of insulin release from rat islets at SaRI 59-801 concentrations between 0.05 and 0.3 mM. These effects were comparable to that of tolbutamide (HANSON et al. 1985). Although the site of action for insulin stimulation is unknown, it may be supposed that this compound, too, exerted its activity by depolarization of the B-cell membrane, since it stimulated Ca^{2+} uptake by voltage-dependent Ca^{2+} channel rather than an intracellular increase in cAMP (HANSON and ISAACSON 1985). The clinical development of SaRI 59-801 was discontinued.

II. α_2-Adrenoceptor Antagonism

Based on the concept that an increased α-adrenergic tonus inhibits insulin release from B-cells (PORTE 1967; ROBERTSON and PORTE 1973; ROBERTSON et al. 1976), α_2-antagonists should elevate plasma insulin concentrations in diabetic patients. One of these α_2-antagonists (see also Sect. B.I.1.d) for which researchers are still claiming that specific mode of action, is under clinical development. Deriglidole (SL 84.0418) (Fig. 3) has been described as a selective α_2-adrenoceptor antagonist that binds like idazoxan to cerebroarterial membranes and inhibits binding of [^{3}H] idazoxan with a K_i value of 4.9 nM (ANGEL et al. 1992). In isolated rat islets, deriglidole did not affect glucose-induced insulin release but antagonized the inhibitory effect on insulin release of the α_2-adrenoceptor agonist UK 14304 (ANGEL et al. 1990). During a glucose infusion test in healthy volunteers, this compound (20–60 mg tid) facilitated insulin and C-peptide responses (BERLIN et al. 1991). During an oral glucose tolerance test, 50 and 100 mg deriglidole showed antihyperglycemic effects but did not affect plasma insulin and C-peptide response in healthy subjects (BERLIN et al. 1992). Since deriglidole is also a 2-substituted imidazoline derivative, the question arises whether the effect on insulin release is actually due to α_2-adrenergic atagonism. For further evaluation and to clarify this question, studies are necessary to rule out that deriglidole does not also act by closure of ATP-sensitive K^+ channels of B-cells. As long as data are not available showing the effect of deriglidole on the ATP-sensitive K^+-channel activity of B-cells, this question cannot be satisfactorily answered. Data that explain the antagonistic effects of deriglidole on epinephrine-induced hyperglycemia in mice by pancreatic α_2-adrenergic antagonism (ANGEL et al. 1993) are not sufficient. Rapid onset of hyperglycemia after 0.3 mg epinephrine/kg i.p. is primarily a consequence of hepatic glycogenolysis and glucose release rather than being due to inhibition of insulin secretion.

III. Ca²⁺ Channel Modulation

For nifedipine (Fig. 5) and other dihydropyridine Ca^{2+} antagonists it was shown that these compounds inhibited the insulin release of isolated islet cells (Malaisse and Sener 1981). A very similar compound of the same chemical class, BAY K 8644 (Fig. 5), produced the opposite effect. This Ca^{2+} channel modulator increased the Ca^{2+} influx and glucose-induced insulin release in vitro (Malaisse-Lagae et al. 1984).

These ionic channels are L-type, voltage-dependent Ca^{2+} channels; they are dihydropridine-sensitive, widely distributed in different tissues, and also located on the pancreatic B-cells (Rorsman and Trube 1986). These dihydropyridine derivatives modulate Ca^{2+} uptake (Schramm et al. 1983; Bechem et al. 1988) through direct action on the voltage-dependent Ca^{2+} channel (Schramm et al. 1983). Compounds like BAY K 8644 prolong the open state of the channel and lead immediately to an intracellular elevation of Ca^{2+} (Bechem et al. 1988), thereby increasing the insulin release of B-cells by a completely different mode of action than that of sulfonylurea drugs (Henquin et al. 1985) or compounds of other chemical classes that bind to the ATP-sensitive K^+ channel or to a closely related protein. Actually, in our own animal experiments we were able to demonstrate strong blood glucose-lowering effects with BAY K 8644. An oral dose of 10 mg/kg produced weak hypoglycemic effects in fed rats; however, a pronounced blood glucose reduction of 50% and more occurred in starved rats (Puls and Bischoff 1985). Surprisingly, concomitant plasma insulin concentrations were decreased in fed and unchanged in starved rats. Intravenous glucose and oral sucrose tolerance tests were markedly improved after 10 mg BAY K 8644/kg p.o. (AUC only 20% of untreated control) accompanied by a reduction in plasma insulin levels (AUC 35% of control). BAY K 8644 was also active in streptozotocin (STZ)-diabetic rats and normalized hyperglycemia in these animals dose-dependently. As a Ca^{2+} agonist this compound exhibited positive inotropic and strong vasoconstrictive properties; therefore, BAY K 8644 is not a candidate for antidiabetic treatment. However, both in vitro and in vivo results are interesting and

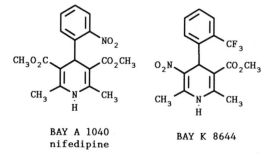

BAY A 1040
nifedipine

BAY K 8644

Fig. 5. Dihydropyridine derivatives

promising and may point the way to new antidiabetic mechanisms and the development of new blood glucose-lowering drugs, respectively.

C. Suppression of Hepatic Glucose Production

The liver represents an effective metabolic buffer system for blood glucose homeostasis and energy supply, which are controlled by insulin, counter-regulatory hormones, and substrate supply. In NIDDM this sensitive system is generally out of balance. Increased hepatic glucose output is thought to be a major factor contributing to fasting hyperglycemia (KOLTERMANN et al. 1981; CONSOLI et al. 1989; DEFRONZO et al. 1989; GERICH 1991). Although not yet defined, various mechanisms are very probably responsible for excessive hepatic glucose production: insulin insensitivity, oversecretion of glucagon and catecholamines, substrate availability, and fatty acid oxidation (CONSOLI 1992; DEFRONZO 1988, 1992). Because glucagon secretion leads to increased hepatic glucose release the obvious choice is to search for glucagon analogues that could block the binding and effects of glucagon in the liver (JOHNSON et al. 1981). In fact, glucagon as a polypeptide hormone offers opportunities for chemical derivation that allow binding to the receptor-binding site and competing with glucagon. However, such a polypeptide antagonist must not activate the glucagon receptor-mediated intracellular signaling cascade (UNSON et al. 1989) and would be useful only for parenteral administraion. For oral administration low molecular weight compounds must be used, and although research has been ongoing for years, there are currently no prospects of an oral drug with this mechanism. The only compound under preclinical development in that field is the benzamide derivative M&B 39890A (Fig. 6), which has been described as an inhibitor of glucagon secretion from islet cells (TADAYYON et al. 1987), but this compound also inhibited glucose- and arginine-stimulated insulin secretion from normal rat islets (YEN et al. 1991). There was no hypoglycemic effect observed in starved diabetic mice, normal mice, or STZ-diabetic rats; however, after subchronic administration with the diet, M&B 39890A reversed hyperglycemia in diabetic mice (YEN et al. 1991).

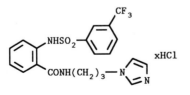

M&B 39890A

Fig. 6. Glucagon secretion inhibitor

Impairment of insulin-mediated suppression of plasma free fatty acid (FFA) concentrations, increased utilization of lipids, and oxidation of FFA have been reported in NIDDM (Chen et al. 1987; Consoli et al. 1989; Groop et al. 1989, 1991). These factors lead to a reduction in the consumption of glucose as an energy source. The relationship between fat and carbohydrate metabolism, the glucose–fatty acid cycle, was first described by Randle and coworkers in 1963 (Randle et al. 1963). This generally accepted hypothesis gave rise to ideas in the search for inhibitors of fatty acid oxidation. This approach should favor glucose instead of FFA as the energy substrate and should result in a lowering of blood glucose concentrations.

In the 1970s and 1980s, efforts in pharmacological research resulted in the discovery of a number of pharmacologically active substances of various chemical classes. These compounds, all of which inhibit FFA oxidation, lowered blood glucose concentrations in several animal species. Surprisingly, however, this was not due to increased glucose utilization as initially expected; rather, the observed hypoglycemic effects were the result of reduced hepatic gluconeogenesis.

I. Inhibition of Fatty Acid Oxidation

1. Inhibitors of Carnitine Acylcarnitine Translocase: Hydrazonopropionic Acid Derivatives

More than 30 years ago Weiss et al. (1959) reported that patients showed a marked reduction in their fasting blood glucose concentration 3–6 weeks after treatment with iproniazide, a monoamine oxidase inhibitor that belongs to the class of hydrazines. In the early 1960s phenelzine (Fig. 7), another antidepressant, was proposed as supplementary treatment in diabetic patients because its blood glucose-lowering effect was confirmed by several investigators (Van Praag and Leijnse 1963; Cooper and Keddie 1964; Wickström and Petterson 1964).

However, due to the adverse effects of such hydrazines (Goldberg 1964), clinical trials were discontinued, and the cause of hydrazine-related hypoglycemic activity was not further explored. Not until more than 10 years later was the mechanism of hypoglycemic action of phenelzine studied using a liver perfusion model by Haeckel and coworkers (Haeckel and Oellerich 1977). They found that a derivative of phenelzine, phenylethylhydrazonopropionate (PEHP) (Fig. 7), caused a strong hypoglycemic action, whereas the inhibition of monoamine oxidase (MAO) activity was reduced (Haeckel and Oellerich 1979). After that basic result, chemical synthesis and pharmacological research produced two hypoglycemic compounds with little or no MAO-inhibitory activity: 2-(3-methylcinnamylhydrazono)-propionate (MCHP; BM 42.304) and 2-(3-phenylpropoxyimido)-

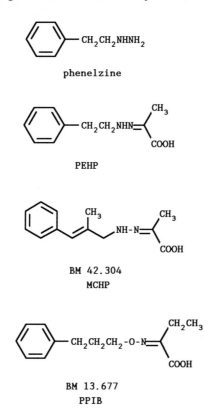

Fig. 7. Fatty acid oxidation inhibitors I: hydrazonopropionic acid derivatives and related substances

butyrate (PPIB; BM 13.677) (Fig. 7) (OELLERICH and HAECKEL 1980; KÜHNLE et al. 1984; WOLFF and KÜHNLE et al. 1985; KÜHNLE et al. 1990).

PPIB, unlike PEHP and MCHP, is not a hydrazonopropionate, but rather a compound with oxime structure (KÜHNLE et al. 1990). However, it is a compound structurally related to hydrazonopropionate showing similar pharmacological effects (HAECKEL et al. 1990).

a) Pharmacological Properties

Most animal studies using hydrazone and related derivatives have been performed with guinea pigs after intraperitoneal injection. In starved guinea pigs a single dose of 145 μmol PEHP/kg (ca. 30 mg/kg) reduced blood glucose concentration after i.p. injection by more than 50% (HAECKEL and OELLERICH 1979), and a similar effect was obtained by 15 mg MCHP/kg i.p. (KÜHNLE et al. 1984). Both compounds were less potent in rats, where a fourfold higher dose was necessary to achieve equivalent blood glucose-lowering effects. Hypoglycemic effects were also observed in diabetic mice after high-dose administration of PEHP (100 mg/kg), showing a more pronounced effect than in nondiabetic starved mice. MCHP significantly reduced ketone bodies in starved guinea pigs and increased plasma free fatty acids after i.p. injection (Deaciuc et al. 1983). PEHP and MCHP were both characterized by a pronounced species dependency. Blood glucose-lowering potency decreased in the following order: guinea pig > mouse > rat. Furthermore, these agents produced a marked hypoglycemia in *starved* animals only. In fed animals the blood glucose-lowering efficacy more or less disappeared. Interestingly, MCHP and other hydrazines inhibited intestinal glucose uptake (Haeckel et al. 1984), but this effect was not really a factor contributing to hypoglycemic activity since, particularly in fed animals, the blood glucose-lowering effects were negligible.

b) Effect on Gluconeogenesis and Mode of Action

Guinea pig liver perfusion studies have demonstrated that glucose production from pyruvate, propionate, lactate, alanine, and glutamine was inhibited by MCHP dose-dependently in the micromolar range (Deaciuc et al. 1983; Kühnle et al. 1984). Similar effects were also observed in isolated hepatocytes of starved rats. In accordance with the reported species dependency of MCHP in vivo, about tenfold higher concentrations of MCHP were again necessary for rat hepatocytes. The addition of octanoate to the perfusion system of guinea pig liver completely abolished the inhibitory effect of MCHP on hepatic glucose production, whereas the addition of oleate to the perfusate could not eliminate the inhibitory action of MCHP on hepatic glucose production (Deaciuc et al. 1983). The formation of acetyl-CoA from long-chain fatty acids in liver mitochondria was potently inhibited, and the intrahepatic acyl-CoA concentration was increased (Haeckel et al. 1985). These experiments indicated that the transport of long-chain fatty acids through the mitochondria membrane was inhibited by MCHP. This hypothesis was further corroborated by the observation that the oxygen consumption of isolated liver mitochondria was clearly reduced by MCHP when palmitoyl carnitine was added to the incubation. However, no effect of MCHP on mitochondrial oxygen uptake was noticed in the presence of succinate, glutamate, or octanoate (Deaciuc et al. 1983). It is well established that in the translocation process of medium-chain fatty acids from cytosol into mitochondria, carnitine is not required (Fritz 1959). Thus, it seemed very likely that the inhibition of long-chain fatty acid translocation into mitochondria was the primary mode of action of MCHP and the cause for its inhibition of fatty acid oxidation. Long-chain fatty acids provide the major energy source for the oxidation process in the mitochondria. Only the short- and medium-chain fatty acids, however, are translocated across the mitochondrial matrix (McGarry AND Foster 1980; Fritz 1959).

Long-chain fatty acids must be converted by cytosolic enzymes to long-chain fatty acyl-CoA esters. The acyl-CoA derivative, which cannot penetrate the mitochondrial membranes, is taken up by an enzyme complex in the outer mitochondrial membrane, the carnitine palmitoyl-transferase 1 (CPT 1), which substitutes carnitine for CoA. Only the long-chain acylcarnitine derivatives are translocated across the mitochondrial membranes. Another enzyme, carnitine acylcarnitine translocase, located in the mitochondrial inner membrane, facilitates the transport across the inner membrane. Once inside the mitochondria, carnitine palmitoyl-transferase 2 (CPT 2) of the inner membrane catalyzes the reverse reaction of CPT 1 in the presence of CoA, resulting in long-chain fatty acyl-CoA, which is subsequently oxidized to acetyl-CoA and ketone bodies.

To explore the site of action of MCHP, studies were conducted to show its possible inhibitory effect on a further enzyme system. It was demonstrated that MCHP did not affect either CPT 1 or CPT 2 (SCHMIDT et al. 1985). From these results it was proposed that inhibition of carnitine acylcarnitine translocase is possibly the site of action of MCHP. This suggestion was supported by a study, performed with isolated mitochondria, showing a concentration-dependent decrease in the rate of the translocase-mediated transport of carnitine (BENEKING et al. 1987).

Inhibition of fatty acid transport across the mitochondrial membrane reduces the oxidation of fatty acid, and consequently the production and supply of acetyl-CoA. For the initial step in gluconeogenesis, the reaction from pyruvate and CO_2 to oxaloacetate, acetyl-CoA is essential. The mitochondrial enzyme pyruvate carboxylase, which catalyzes this reaction, needs acetyl-CoA for allosteric activation (UTTER et al. 1975). Therefore, a lack of acetyl-CoA will indirectly affect the rate of new glucose production, and this causes the blood glucose-lowering action of MCHP, which is a result of decreased hepatic gluconeogenesis.

The oxime derivative PPIB (BM 13.677) (Fig. 7), which is structurally related to the hydrazonopropionic acid derivatives, may act in a similar manner, although some properties concerning the stimulation of substrate oxidation may differ from the hydrazones (HAECKEL et al. 1990; KÜHNLE et al. 1990).

The hypoglycemic activity of PPIB was less potent than with MCHP. However, it is remarkable that the same species dependency was seen as with MCHP. The potency of the drug decreased in the same order: guinea pig > mouse > rat (KÜHNLE et al. 1990). The most obvious difference between both drugs, besides potency, seemed to be another substrate dependency concerning the inhibitory effect on hepatic glucose production. Since glucose formation from propionate was only slightly inhibited by PPIB, it was speculated that PPIB possibly affects the pyruvate carboxylase reaction, because propionate is metabolized via succinyl-CoA, thereby circumventing the pyruvate carboxylase reaction. However, a final conclusion was not possible since the further development of PPIB has been suspended.

c) Clinical Studies

Of all hydrazones and related derivatives, only MCHP (BM 42.304) underwent phase II clinical trials. Only one study was reported; however, a blood glucose-lowering effect could not be demonstrated with doses of 200 mg MCHP in NIDDM patients (Paschke et al. 1989). Dosage and pharmacokinetic problems could both be the reasons for the drug failure. As recently communicated by the company, further development of MCHP has been discontinued.

2. Inhibitors of Carnitine Palmitoyltransferase I

a) Pharmacology and Effects on Gluconeogenesis

α) *2-Oxirane Carboxylic Acid Derivatives.* From the mechanism of long-chain fatty acid translocation across the mitochondrial membrane, it is obvious that in addition to translocase, other steps are also potential target sites of enzyme inhibition. Actually, the enzyme which is more frequently inhibited by various substances is the first step in the penetration process, the CPT 1. The most intensively studied class of compounds over the past decade was the substituted 2-oxirane carboxylic acids. These compounds themselves are not the inhibitors. They are prodrugs and must be converted to their corresponding CoA esters (Tutwiler and Dellevigne 1979). The CoA esters specifically inhibit CPT 1 due to covalent binding to the enzyme, causing irreversible inhibition. It has been suggested that they may interfere with the substrate binding site of CPT 1 (Bailey and Flatt 1989). Inhibition of CPT 1 seemed to be more effective than inhibition of translocase. One of the first compounds in that chemical class is methyl 2-tetradecylglycidate (MeTDGA, methyl palmoxirate, McN-3716) (Fig. 8). MeTDGA has been shown to be active after oral treatment in guinea pigs, as well as in normal and streptozotocin (STZ)-diabetic rats, diabetic (db/db) mice, and diabetic dogs (Bailey and Flatt 1990). In starved rats, 20 mg/kg po rapidly lowered the blood glucose concentration by 70% (Tutwiler et al. 1978). In the same doserange, the drug also effectively reduced the blood glucose concentration of STZ-diabetic rats by about 50%. As shown by in vitro studies, there is no doubt that the major mechanism of hypoglycemic action is the reduction of hepatic gluconeogenesis due to the inhibition of long-chain fatty acid oxidation (Tutwiler and Dellevigne 1979; Tutwiler et al. 1979; Ho et al. 1986). However, the inhibitory activity of this compound on long-chain fatty acid oxidation was not restricted to the liver but has also been demonstrated in peripheral tissues like the diaphragm, soleus muscle, heart, epididymal fat, and kidney cortex (Ho et al. 1986). Actually, the long-chain fatty acid oxidation was inhibited and not the gluconeogenic pathway itself; this was demonstrated by the addition of a short-chain fatty acid, octanoate, which almost completely reversed the inhibition of gluconeogenesis. As expected from its mode of action, MeTDGA also markedly reduced the number of

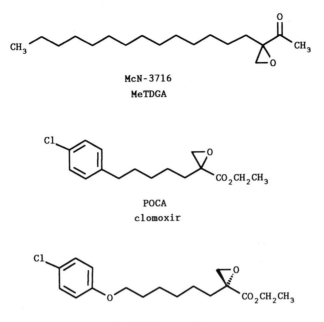

McN-3716
MeTDGA

POCA
clomoxir

(R)-(+)-etomoxir

Fig. 8. Fatty acid oxidation inhibitors II: 2-oxirane carboxylic acid derivatives

ketone bodies in animals. Compared to the hydrazone derivatives discussed above, the 2-oxirane derivatives seemed to be a more potent class of hypoglycemia-inducing substances. The stronger blood glucose-lowering activity in vivo corresponded to lower in vitro IC_{50} values for inhibition of gluconeogenesis. Compared with the hydrazones showing inhibitory activity in the micromolar range, MeTDGA was active in the submicromolar range (KIORPES et al. 1984).

This was also true for two other intensively studied 2-oxirane derivatives: 2-[5-(4-chlorophenyl)pentyl] oxirane-2-carboxylic acid (clomoxir, POCA) (Fig. 8) (WOLF et al. 1982; WOLF and ENGEL 1985) and ethyl-2-[6-(4-chlorophenoxy)hexyl] oxirane-2-carboxylate (etomoxir) (Fig. 8) (WOLF 1985; AGIUS and ALBERTI 1985).

In fact, the mode of action is the same: the irreversible inhibition of mitochondrial CPT 1 (TURNBULL et al. 1984; SELBY and SHERRATT 1989). It was demonstrated that the corresponding CoA esters are formed stereospecifically and only by the (R)-(+) enantiomers (WOLF 1990). Hypoglycemic efficacy after oral administration was demonstrated in rats, mice, guinea pigs, dogs, and pigs (EISTETTER and WOLF 1986; WOLF et al. 1982). In starved rats, 6.5 mg etomoxir/kg orally reduced the blood glucose concentration by 25%, as expected; the (+)-enantiomer was twice as effective as the racemate. In short-term studies using diabetic rats, clomoxir and etomoxir showed beneficial effects on the diabetic heart: the metabolic state and cardiac function were improved (Wolf 1990). Although under acute conditions the nonesterified fatty acids increased reversibly in starved animals, this was not the case under repeated administration. An important further effect besides the strong inhibition of fatty acid oxidation, ketogenesis, and gluconeogenesis was the observed decrease of serum cholesterol and triglyceride levels (Koundakjian et al. 1984). This corresponded to the additional hepatic inhibition of fatty acid and cholesterol synthesis that had been reported (Agius et al. 1985; Vaartjes et al. 1986). Recently it was shown that, in contrast to the stereospecific inhibition of fatty acid oxidation by only the (+)-enantiomer of etomoxir, fatty acid and cholesterol synthesis was inhibited by both the (+) and (−)-enantiomer of etomoxir (Agius et al. 1991).

Studies concerning the mRNA expression of enzymes involved in cholesterogenesis showed that etomoxir had no effect. It was suggested that the effects on lipid synthesis are also probably due to the reduction in acetyl-CoA supply (Asins et al. 1994).

β) Dioxolane Derivative. Scant information is available about a recently reported dioxolane derivative 2-[1,1 dimethylethyl]-2-[4-methylphenyl]-1,3-dioxolane (SDZ 51-641) (Young et al. 1990). SDZ 51-641 (Fig. 9) is an orally active blood glucose-lowering compound in normal and diabetic rats. Increased hepatic glucose production of STZ-diabetic rats was normalized by 100 mg SDZ 51-641/kg orally. Fatty acid oxidation was inhibited in isolated hepatocytes of starved rats. In contrast to oxirane derivatives, SDZ 51-641 had no effect on fatty acid oxidation in skeletal muscles. Therefore, it was suggested that SDZ 56-641 could be a liver-specific fatty acid oxidation inhibitor. This could theoretically be of potential benefit since fewer mechanism-related side effects should occur in the periphery due to lack of action there. In euglycemic clamp studies an effect on glucose utilization could not be confirmed (Young et al. 1990). All other metabolic effects are more or less comparable to the oxirane derivatives. Although its site of action has not been shown, SDZ 51-641 is assumed to be a liver-specific CPT 1-inhibitor (Foley 1992). Its structure suggests that the compound represents a prodrug: the methyl group can be oxidized to the carboxylic acid which in turn can be esterified with CoA, analogous to the oxirane derivatives. No safety and clinical data are available.

γ) β-Aminobetaine Structures. Emericedin (Fig. 9), a ß-aminobetaine structure, was isolated from fungi (*Emericella quadrilineata*) and found to inhibit specifically the oxidation of long-chain fatty acids in vitro. The IC_{50} for inhibiting carnitine-dependent palmitate oxidation in rat liver mitochondria was rather high at 8×10^{-3} M. The desacetyl derivative of emericedin, emeriamine (Fig. 9), exhibited stronger inhibitory activity with an IC_{50} of 3.2×10^{-6} M (Kanamaru et al. 1985). Studies concerning the effect on

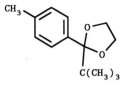

SDZ 51-641

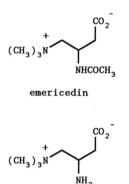

emericedin

emeriamine

Fig. 9. Fatty acid oxidation inhibitors III

L-carnitine-dependent transport of long-chain fatty acids indicated that emeriamine mainly inhibits the CPT 1 (KANAMARU et al. 1985).

The pharmacological effects of emeriamine on carbohydrate and lipid metabolism in animals are quite comparable to those of other CPT 1 inhibitors. Emeriamine showed hypoglycemic and antiketogenic effects only in starved animals. Blood glucose-lowering activity was induced at a dose higher than 3 mg/kg p.o. In fed animals, hypoglycemic effects could not be observed (KANAMARU et al. 1985). Clinical trials have not been reported with emeriamine and further development was discontinued.

b) Pharmacokinetics and Toxicology

Pharmacokinetic data were reported only for etomoxir. The bioavailability of etomoxir was 40% in male volunteers. Etomoxir was rapidly cleaved after absorption by ester hydrolysis to the free carboxylic acid. In human plasma two main metabolites were produced after hydrolysis of the epoxide ring and further via degradation of the alkyl chain. After single and multiple dosage of 150 mg, the mean maximum plasma concentration of the free acid ranged between 1.4 and 1.8 µg/ml. The elimination half-life of the free acid was 1.4 h and the substance was excreted via bile and kidney (WOLF 1990).

The toxicological profile of CPT 1 inhibitors and of 2-oxirane carboxylic acid derivatives is obviously not very clear. The accumulation of fatty acids and their metabolites has been associated with cardiac dysfunction and cell damage (Dhalla et al. 1992). Both MeTDGA and etomoxir have induced cardiac hypertrophy in rats and mice (Lee et al. 1985; Rupp and Jacob 1992). It was speculated that this phenomenon is not a "pathologic" hypertrophy but rather like a "physiologic" hypertrophy induced by exercise (Wolf 1990). In a clinical study, transaminases were reversibly increased, showing normalization 2 weeks after cessation of drug treatment (Ratheiser 1991).

c) Clinical Studies

Only a few clinical studies with CPT 1 inhibitors are available. Consistent results with various classes of compounds demonstrated lower hypoglycemic activity in clinical trials than in preclinical animal experiments. In abstract form only, one clinical study with MeTDGA, administered to insulin-treated IDDM patients, was reported (Verhaegen et al. 1984). A dosage of 50 mg MeTDGA per day (11 days) lowered blood glucose concentrations in 3 of 6 patients (11 days). In accordance with animal studies, ketonemia was already eliminated on the first day of treatment, indicating that there is also strong antiketogenic activity in humans due to this mechanism. Only a few further studies are available on etomoxir, which was administered in doses ranging between 25 and 200 mg/day. As with MeTDGA, etomoxir dosages were markedly lower than in pharmacological animal studies. This could be the reason for the smaller blood glucose-lowering effects in studies on humans. The reductions in ketone bodies and lipids, particularly in triglycerides, were obviously more pronounced (up to -60%) than the hypoglycemic effect (-20%) (Bliesath et al. 1987; Haupt et al. 1988). In healthy volunteers blood glucose levels were not affected after an 18-h fast, but they dropped after prolonged fasting for 36 h by 40% (Bliesath et al. 1987).

Interestingly, etomoxir increased insulin-mediated glucose uptake and oxidation by 33% in NIDDM patients as measured by the euglycemic clamp technique (Hübinger et al. 1993). However, it is still questionable whether etomoxir-related hypoglycemia is actually the result of improved insulin-mediated glucose uptake and utilization or more likely due to inhibition of gluconeogenesis, which both result from using the euglycemic clamp techniques (Ratheiser et al. 1991). In vitro and in vivo data, particularly from the studies on humans, have given rise to the interpretation that a shift to preferred glucose oxidation by inhibition of long-chain fatty acid oxidation is less important for hypoglycemic action than is the suppression of hepatic glucose production. The reason for this assumption is that reduced fatty acid oxidation leads to decreased ketogenesis and acetyl-CoA formation, since acetyl-CoA stimulates pyruvate carboxylase, which catalyzes the reaction to oxaloacetic acid as a precursor for gluocse synthesis in gluconeogenic tissues (liver, kidney) (Söling et al. 1968; Utter et al. 1975). Reduced acetyl-CoA formation thus directly effects a decreased gluconeogenesis and rate of glucose formation. Although this principle of inhibited hepatic glucose production seemed to be very promising in the early stages of research, none of these pharmacologically very potent inhibitors of fatty acid oxidation is still under clinical development for treatment of diabetes mellitus.

D. Enhancement of Insulin Action

In many populations and particularly in Western cultures, non-insulin-dependent diabetes mellitus (NIDDM) is frequently part of a cluster of metabolic abnormalities which include central obesity, hypertension, insulin resistance, hyperinsulinemia, dyslipidemia (increased plasma triglycerides and decreased HDL-cholesterol), hyperuricemia, increased plasminogen activator inhibitor levels, and increased prevalence of macrovascular disease (Lebovitz 1995). The relationship of these metabolic abnormalities to each other and the underlying pathogenetic mechanism responsible for them are the subject of considerable speculation and controversy (Stern 1994). One hypothesis is that insulin resistance and compensatory hyperinsulinemia not only precipitate NIDDM but are also the cause of the cluster which has in fact been named the insulin resistance syndrome (Reaven 1988). If this hypothesis is true, then agents which ameliorate insulin resistance and decrease hyperinsulinemia might be expected to lower blood pressure, improve dyslipidemia, and decrease macrovascular disease as well as improve glycemic control. Such agents would be ideal drugs for treatment or even prevention of NIDDM. Thiazolidinediones represent such a class of drugs (Fig. 10).

Ciglitazone was the first thiazolidinedione developed, and the results of initial studies were reported in 1983: Ciglitazone markedly reduced hyperglycemia and hyperinsulinemia in several rodent models of insulin-resistant diabetes while having little or no effect in normal rodents or those with insulin-deficient diabetes (Chang et al. 1983a; Fujita et al. 1983). Long-term therapy in animals caused lens cataracts and ciglitazone development was discontinued, although it continues to be used as a model for laboratory investigations into additional pharmacological properties and mechanism of action (Colca et al. 1988; Kraegen et al. 1989; Gill and Yen 1991).

Numerous analogues of ciglitazone have been synthesized. Three (pioglitazone, troglitazone, and englitazone) have been extensively studied in the laboratory, and two are currently in phase III clinical trials.

The in vivo models used to evaluate drugs that are thought to affect insulin action were those representing insulin-resistant states, and the results were compared to those obtained in insulin-deficient or normal animals. The insulin-resistant animals which have been used to study the thiazolidine-diones are Wistar or Zucker fatty rats, obese KK mice, ob/ob mice, db/db mice, high-fat-fed rats, and old rats. The streptozotocin (STZ)-diabetic rat

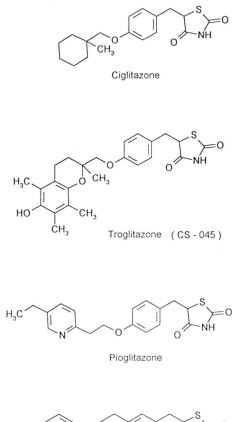

Ciglitazone

Troglitazone (CS - 045)

Pioglitazone

Englitazone

Fig. 10. 2,4-Thiazolidinedione derivatives

or mouse has been used most frequently to study the effects of these compounds on insulin-deficient states, although some of the earlier studies employed Chinese hamsters. Table 1 summarizes the results of thiazolidine-dione studies on in vivo models.

The magnitude of the in vivo effects varied considerably with the particular thiazolidinedione and the specific insulin-resistant animal model (Table 2); however, the qualitative actions were very similar.

All of these drugs caused a dramatic improvement in hyperglycemia, striking decreases in hyperinsulinemia, and significant reductions in plasma free fatty acids, triglycerides, and total cholesterol. These effects appear to be secondary to an increase in insulin action as measured by insulin tolerance

Table 1. Effects of thiazolidinediones on in vivo animal models

	Insulin-resistant diabetes[a–d]	Insulin-sensitive diabetes[a–d]	Normals[c]
Blood glucose	↓	0	0
Plasma insulin	↓	0	↓
Plasma triglycerides	↓	0	0
Plasma free fatty acids	↓	0	↓
Plasma cholesterol	↓	0	0
Insulin action	↑	0	↑
Obesity	0	0	0

[a] Troglitazone, 50–250 mg/kg per day (FUJIWARA et al. 1988).
[b] Pioglitazone, 5–10 mg/kg per day (IKEDA et al. 1990).
[c] Englitazone, 5–50 mg/kg per day (STEVENSON et al. 1990, 1991).
[d] Cliglitazone, 150–300 mg/kg per day (CHANG et al. 1983a,b; FUJITA et al. 1983).

Table 2. Reported effects of troglitazone (CS 045) in animal models of diabetes mellitus (LEBOVITZ and CHAIKEN 1994)

	Normal rats	Streptozotocin-diabetic rats	Insulin-resistant diabetic animals		
			KK mouse	ob/ob mouse	Zucker fatty rat
Plasma glucose	No effect	No effect	↓ −50%	↓ −30%	↓ <10%
Plasma insulin	No effect	No effect	↓ −60%	↓ −70%	↓ −60%
Plasma free fatty acids			↓ −58%		↓ −50%
Plasma triglycerides			↓ −45%		↓ −75%

[a] Data from FUJIWARA et al. (1988).

tests or euglycemic insulin clamps (FUJITA et al. 1983; GILL and YEN 1991; SUGIYAMA et al. 1990a,b).

The mechanisms by which tiazolidinediones improve insulin action have been investigated in cell cultures or in vitro studies with tissues removed from treated animals or placebo controls. Table 3 summarizes the salient findings.

Most studies show that thiazolidinediones do not alter insulin binding to its receptor. The exception is adipose tissue, where pioglitazone is reported not to alter insulin binding; however, both troglitazone and ciglitazone treatment of Zucker fatty rats, KK mice, or ob/ob mice have been shown to increase insulin binding to adipocytes. The increase in insulin binding is probably secondary to the marked reduction in hyperinsulinemia which would be expected to reverse downregulation of the insulin receptor.

In muscles, thiazolidinediones in the presence of insulin (1) increase autophosphorylation of the insulin recpetor and kinase activity to exogenous

Table 3. Effects of thiazolidinediones on isolated tissues

Cultured 3T3 cells
- Increased glucose transport (insulin-like action)
  (Kreutter et al. 1990)
- Enhanced insulin-induced differentiation to adipocytes
  (Kletzien et al. 1992)
- Increased GLUT 1 and GLUT 4 glucose transporter mRNA and protein
  (requires insulin)
  (Sandouk et al. 1993)

Adipose tissue
- Variable effect on insulin binding
  (Chang et al. 1983b; Fujiwara et al. 1988, Sugiyama et al. 1990a; Taketomi
  et al. 1988)
- No effect on insulin-mediated inhibition of lipolysis
  (Stevenson et al. 1990)
- Increased GLUT 4 glucose transporter mRNA and protein (requires insulin)
  (Hofmann et al. 1991)
- Increased glucose oxidation and lipogenesis
  (Sugiyama et al. 1990a; Stevens et al. 1990; Chang et al. 1983b)

Muscle
- No effect on insulin binding
  (Ciaraldi et al. 1990; Iwanishi and Kobayashi 1993; Kobayashi et al. 1992)
- Increased autophosphorylation of insulin receptor and kinase activity (requires
  insulin)
  (Iwanishi and Kobayashi 1993; Kobayashi et al. 1992)
- Increased GLUT 4 glucose transporter mRNA and protein (requires insulin)
  (Hofmann et al. 1991)
- Increased glycogen synthesis and glycolysis (requires insulin)
  (Ciaraldi et al. 1990; Sugiyama 1990a)

Liver
- No effect on GLUT 2 glucose transporter
  (Hofmann et al. 1992)
- Increased glucokinase
  (Sugiyama et al. 1990b)
- Decreased glucose-6-phosphatase
  (Sugiyama et al. 1990b)
- Decreased phosphoenolpyruvate carboxykinase
  (Hofmann et al. 1992)

substrates, (2) increase glucose transport by increasing GLUT 4 transporter levels, and (3) increase glycogen synthesis and glycolysis.

The effects of thiazolidinediones on liver probably represent increases in insulin sensitivity of the enzymes regulating hepatic glucose production. The thiazolidinediones not only have insulin-independent actions but also do not increase all insulin actions. For example, they have a direct insulin-like action in increasing glucose transport in cultured 3T3 cells. While they increase insulin-mediated glucose transport and lipid synthesis in adipose tissue, they have no effect on insulin-mediated antilipolysis. Most of the effects of the various thiazolidinediones require normal or elevated insulin levels. These compounds have no effect on insulin-deficient diabetic animals unless they are insulin-treated. Troglitazone administered at a dose of

70 mg/kg per day for 4 to 8 weeks in young obese Zucker rats decreased blood pressure, serum triglyceride, and plasma cholesterol levels (YOSHIOKA et al. 1993). Renal sodium excretion was increased as was the creatinine clearance. These data provide some support for a role of insulin resistance as a cause of hypertension in obese Zucker fatty rats.

Both pioglitazone and troglitazone are in phase II and phase III clinical trials (KUZUYA et al. 1991). Table 4 summarizes the results of several studies which have been published on the effects of troglitazone in NIDDM patients. Neither study included a placebo control group.

In contrast to the striking effects of thiazolidinediones in hyperglycemic hyperinsulinemic rodents, their effects in patients with NIDDM have been somewhat disappointing. Glycemia was decreased by 20% to 25%, as were plasma insulin levels. The effect on HbAlc was a modest decrease of 0.8%. The reduction in triglycerides and free fatty acids caused by darglitazone, an analogue of englitazone, were somewhat more striking than those obtained with troglitazone (LEBOVITZ and CHAIKEN 1994). The limited data in humans are compatible with the more extensive data in animals and suggest that the effects of thiazolidinediones in humans are most dramatic in hyper-insulinemics and much less so in individuals with normal or low insulin levels.

A recent study in obese nondiabetic subjects confirmed the speculation that troglitazone may be most effective in minimally hyperglycemic NIDDM individuals (most of whom are hyperinsulinemic) (NOLAN et al. 1994). In 12

Table 4. Effects of troglitazone (CS 045) in individuals with NIDDM (LEBOVITZ and CHAIKEN 1994)

	CS 045[a]	CS 045[b]	CS 045[c] + sulfonylurea
Number of patients	8	11	11
Duration of study (weeks)	12	6–12	12
Fasting plasma glucose (mM)	9.7 → 7.2 (↓26%)	12.5 → 10.7 (↓14%)	12.0 → 9.3 (↓22%)
Fasting plasma insulin (pM)	62 → 58 (↓6%)	180 → 96 (↓47%)	89 → 56 (↓37%)
Hemoglobin Alc	7.4 → 6.7		8.5 → 7.9
AUC of plasma glucose curve	↓26%	↓32%	↓16%
AUC of plasmas insulin curve	↓12%	↓35%	↓27%
AUC of plasma FFA curve		↓24%	
Glucose disposal to insulin infusion		↑59%	
Basal hepatic glucose output		↓17%	

[a] Data from IWAMOTO et al. (1991).
[b] Data from SUTER et al. (1992).
NIDDM, non-insulin-dependent diabetes mellitus; FFA, free fatty acid.

obese nondiabetic subjects, troglitazone treatment for 12 weeks decreased plasma insulin levels by 40%, improved insulin action by 28%, and lowered systolic blood pressure by 5 mmHg and diastolic blood pressure by 4 mmHg. Lipid levels did not change significantly nor did the fasting plasma glucose. However, the incremental plasma glucose rise after either a glucose load or a meal was decreased by 25%.

Sufficient data are available to indicate that thiazolidinediones reduce insulin resistance and hyperinsulinemia. They improve hyperglycemia in NIDDM patients and can ameliorate impaired glucose tolerance when it is present. Their ultimate clinical usefulness has yet to be determined.

Animal toxicity studies with several of the thiazolidinediones suggest that in rodents they may cause myocardial hypertrophy and/or significant anemia. Cardiac hypertrophy has not been observed in humans in phase III studies using echocardiographic techniques. Neither has a significant decrease in hemoglobin been reported.

E. Mimicking of Insulin Action

Insulin action occurs through a complex series of events that enable the binding of insulin on the outside of the cell to trigger a variety of actions within the cell. The initial event is the attachment of insulin to a specific binding site on the extracellular region of the A subunit of the insulin receptor. Following transduction of the signal through the membrane portion of the insulin receptor, the tyrosine-specific kinase on the B subunit is activated; through a series of tyrosine and subsequent serine phosphorylations of different substrates, the various actions of insulin occur inside the cell (Kahn 1995). This cascade of biochemical events provides the potential to develop drugs that can enter the cell and directly stimulate, at a variety of sites, some or all of these insulin actions. Not enough is known about the individual intracellular effector systems to model such agents at the present time.

Observations with naturally occurring substances could provide considerable insight into this area of drug development. Extensive data on potential insulinomimetic mechanisms are available with the trace element vanadium (Shechter 1990). In the late 1970s and early 1980s various vanadium preparations were shown to have in vitro insulin-like actions on glucose transport and metabolism in skeletal muscles and adipocytes. Table 5 lists the documented insulin-like actions of vanadate in vitro. Those observations led to in vivo studies in 1985 which demonstrated that administration of vanadate to STZ-diabetic rats decreased hyperglycemia (Heyliger et al. 1985).

Studies using insulin-deficient STZ-diabetic rats (Blondel et al. 1989; Cam et al. 1993; Dai et al. 1994), partially pancreatectomized rats (Rossetti and Laughlin 1989), and insulin-resistant ob/ob mice (Brichard et al. 1990) have shown that vanadium salts decrease hyperglycemia and reverse

Table 5. Insulin-like actions of vanadate demonstrated in in vitro systems (LEBOVITZ and CHAIKEN 1994)

Activity	In vitro systems
Intermediary metabolism	
Increased hexose transport	Adipose cells and muscle
Increased lipogenesis	Adipose cells
Increased glucose oxidation	Adipose cells
Decreased lipolysis	Adipose cells
Increased glycogen synthase	Adipose cells, muscle, and hepatocytes
Growth and cell function	
Augmented mitogenic activity	Various culture cells
Augmented translocation of IGFII receptors	Adipose cells
Increased K^+ uptake	Cardiac muscle cells
Inhibited Ca^{2+}/Mg^{2+}-ATPase	Adipose cell membranes
Increased Ca^{2+} influx	Adipose cells
Elevated intracellular pH	A-431 cells

IGFII, insulin-like growth factor II.

many of the effects of diabetes. Specifically, vanadium salts given to diabetic rats or mice have been shown to improve peripheral glucose utilization to maximal or submaximal insulin infusion (CAM et al. 1993), normalize in vivo muscle glycogen repletion by increasing glycogen synthase activity, restore hepatic glycogen synthesis and glycogenolytic enzymes to near-normal levels (PUGAZHENTHI and KHANDELWAL 1990), and improve oral glucose tolerance. These effects appear to be independent of insulin secretion since they are observed with significantly reduced or unchanged plasma insulin levels. Vanadium salts have no effect on in vivo glucose metabolism in normal rats but decrease plasma insulin levels significantly (CAM et al. 1993; DAI et al. 1994). The effects of vanadate were seen with all vanadate preparations used (metavanadate, orthovanadate, and vanadyl sulfate).

Several mechanisms have been proposed for the insulin-like actions of vanadium salts. These include activation of the insulin receptor tyrosine kinase (ROSSETTI and LAUGHLIN 1989) either through direct stimulation of autophosphorylation or by virtue of its known effects of inhibiting tyrosine phosphatase. Other possibilities that have been suggested are effects on the exocytotic process by which the glucose transporter molecules are moved through the cell and incorporated into the cell plasma membrane (SHECHTER 1990). The major problem with vanadium treatment in animals has been its toxicity (DOMINGO et al. 1991a, 1991b, 1992). Vanadium salts have been reported to cause decreased food intake, diarrhea, weight loss extending even to emaciation, and hepatic and renal dysfunction. Vanadium salts accumulate in tissues, and chronic toxicity is potentially a serious problem. Controversy exists as to whether the toxicity is significantly less pronounced with some vanadium salts as compared to others.

While it is likely that vanadium salts have in vivo insulin-like effects, several studies suggest that some or all of the antidiabetic effects observed in diabetic rodents can be explained by decreased food intake and weight loss (Domingo et al. 1991a; Malabu et al. 1994). A weakness of many of the in vivo studies in rodents has been the method of estimating the dose of vanadium salt administered. The salt has been given by adding it to the water supply and the dose determined by calculating the estimated fluid intake times the known concentration of the salt. Two studies in which the vanadium salt was given by gavage and exact doses known failed to show any significant decrease in hyperglycemia (Domingo et al. 1991a; Malabu et al. 1994).

It has been claimed that the toxic effects of vanadium salts are less severe if they are administered with sodium chloride. It is unclear if this is so, and if so, what the mechanism might be. Because of the concern of chronic tissue accumulation, some investigators are recommending administration of chelating agents along with the vanadium salts (Domingo et al. 1992). One might suspect that the controversies concerning mechanism of effects and toxicity would preclude human trials for many years to come. Interestingly, several clinical trials of the effects of vanadium salts in patients with diabetes mellitus have been funded and are currently being carried out in the U.S. Thus we should learn much about insulinomimetic drug treatment in humans soon.

F. Prevention of Deleterious Effects Caused by Glucose

An alternative approach to preventing the complications of diabetes mellitus through normoglycemic regulation is via blockade of the specific biochemical processes through which hyperglycemia causes its chronic complications. This has been the dream of diabetologists and pharmacologists for the last decade or two.

Obviously, this approach requires elucidation of the mechanisms by which hyperglycemia causes microvascular and neuropathic diseases and accelerates macrovascular disease. Research during the last two decades has shown that there are two major pathways which are likely to explain the development of chronic diabetic complications, and drugs have been developed to block these pathways. Clinical evidence of the efficacy of these pharmacological approaches is being actively pursued. The two mechanisms consist of (a) increased flux through the polyol pathway (Kirchain and Rendell 1990) and (b) glycation of proteins with the formation of advanced glycation end products (AGEs) (Vlassara 1994). The drugs which have been developed and are either in clinical use or in clinical investigations are aldose reductase inhibitors (Tomlinson et al. 1994) and aminoguanidine (Vlassara 1994), which is an inhibitor of AGE formation. The literature on both is voluminous and a complete review of the material is beyond our scope. We will attempt to summarize the more salient basic findings and focus on the results of clinical studies with these inhibitors.

I. Aldose Reductase Inhibitors

The polyol pathway converts glucose to fructose (KIRCHAIN and RENDELL 1990; TSAI and BURNAKIS 1993). It does so through two enzymes: aldose reductase, which converts glucose to sorbitol, and polyol dehydrogenase, which converts sorbitol to fructose. The first conversion is rapid while the second is slow. These enzymes, while present in most tissues, vary considerably in their concentrations. When intracellular glucose levels rise, the intracellular concentration of sorbitol becomes excessive because of its lack of permeability across the cell membrane and its slow conversion to fructose. Those tissues with high concentrations of aldose reductase enzymes will accumulate the highest quantities of sorbitol. Associated with an increase in polyol pathway activity is a decrease in cellular uptake of myoinositol, a decrease in Na^+/K^+ ATPase activity, and an increase in the intracellular nicotinamide adenine dinucleotide (NADH/NAD) ratio (KIRCHAIN and RENDELL 1990; YUE and BROOKS 1993; ZENON et al. 1990). An increase in polyol pathway activity due to hyperglycemia could therefore damage cells by increasing their intracellular osmolality if very high sorbitol concentrations occur (lens), by decreasing myoinositol levels or altering Na^+/K^+ ATPase activity (nerves), or by altering the redox potential of cells.

Drugs which block aldose reductase activity include carboxylic acid derivatives (tolrestat and ponalrestat), spirohydantions (sorbinil), and flavonoids (KADOR et al. 1985). The effect of various aldose reductase inhibitors will vary in different tissues based on the tissue uptake of the drug and the characteristics of the various tissue aldose reductase isozymes. Figure 11 shows the structures of the three aldose reductase inhibitors which have been studied most extensively.

Studies in experimental diabetic animals have yielded both positive and negative results as regards the benefits of aldose reductase inhibitors in preventing or treating chronic diabetic complications. Aldose reductase inhibitors uniformly prevented diabetic cataracts and improved neurologic dysfunction in experimental diabetic animals (KADOR et al. 1985). However, they did not reduce diabetic retinopathy in diabetic dogs (KERN and ENGERMAN 1991).

Extensive clinical investigations have been carried out over the last 10 years with sorbinil, tolrestat, and ponalrestat. These studies have focused on the treatment of retinopathy, neuropathy, and nephropathy. While some positive results have been obtained, the observed benefits are of small magnitude and questionable clinical significance. It is not clear whether this lack of clear-cut benefits is the result of a limited role of the polyol pathway in the pathogenesis of the complications, or whether it is due to poor design of the clinical intervention studies or a lack of effective tissue penetration

Sorbinil

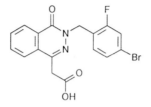

Statil

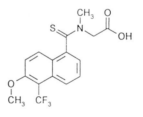

Tolrestat

Fig. 11. Aldose reductase inhibitors

of the drugs used. The toxic effects of sorbinil proved to be severe and frequent and its development has been discontinued. Ponalrestat has generally been ineffective in clincial trials. Tolrestat is relatively nontoxic, has some beneficial effects in neuropathy, and is being marketed in a few countries. Several newer aldose reductase inhibitors are in various stages of development.

The clinical trials which have assessed the effect of aldose reductase inhibitors on diabetic retinopathy have had disappointing results. Administration of sorbinil in a randomized placebo-controlled trial of 497 type-1 diabetic patients for a median of 41 months failed to show any significant benefit (SORBINIL RETINOPATHY TRIAL 1990). About 7% of the patients assigned to sorbinil developed a hypersensitivity reaction in the first 3 months. Smaller studies carried out over a shorter time interval with ponalrestat also showed no benefit (TROMP et al. 1991). A large multicenter trial with tolrestat on diabetic retinopathy has been underway for several years and results are still not available.

The effect of aldose reductase inhibitors on glomerular filtration rate and urinary albumin excretion have been carried out short term in a small number of patients. Sorbinil and tolrestat have both been reported to decrease urinary albumin excretion in type-1 diabetic patients (JENNINGS et al. 1990; PASSARIELLO et al. 1993) while ponalrestat was found to have no effect (RANGANATHAN et al. 1993). The hyperfiltration of IDDM was reduced significantly during 6 months of treatment with both tolrestat (PASSARIELLO et al. 1993) and ponalrestat (PEDERSEN 1991). No long-term studies of the effect of aldose reductase inhibitors on renal function in patients with established diabetic nephropathy have been reported.

Most clinical studies with aldose reductase inhibitors have focused on the treatment of diabetic neuropathy. The large sorbinil trial which investigated effects on retinopathy had an equally extensive evaluation of changes in neuropathy. Overall, sorbinil had no benefit in reducing the clinical signs and symptoms of diabetic neuropathy (SORBINIL RETINOPATHY TRIAL 1993). Changes in nerve conduction velocities were improved by sorbinil for the peroneal nerve but not for the median motor or sensory nerves. Several clinical trials with ponalrestat have failed to show significant beneficial effects on peripheral polyneuropathy or autonomic neuropathy over treatment periods as long as 18 months (FLORKOWSKI et al. 1991; KRENTZ et al. 1992; ZIEGLER et al. 1991; SUNDKUIST et al. 1992). The most extensive investigations of aldose reductase inhibition in diabetic neuropathy have been carried out with tolrestat (TSAI and BURNAKIS 1993; GIUGLIANO et al. 1993). Two large multicenter placebo-controlled trials (BOULTON et al. 1990; MACLEOD et al. 1992) failed to show any beneficial effect of tolrestat as compared to placebo on painful symptoms; there were conflicting data on paresthesias (one trial showed greater benefit from tolrestat and the other from placebo), and improvement was seen in nerve conduction in some motor nerves. The overall improvement seen in patients on tolrestat treatment can be summarized as minimal. The most positive data with tolrestat have been obtained on a double-blind, placebo-controlled study in which tolrestat was replaced by placebo in half of 372 patients who had been on tolrestat for an average of 4.2 years (SANTIAGO et al. 1993). The patients were followed up for 52 weeks. Patients assigned to placebo had a significant deterioration in motor nerve conduction velocity while those remaining on tolrestat did not. Symptoms (pain and sensation) were said to be better in those maintained on tolrestat. The clinical endpoints for improvement of neuropathy have been so difficult to quantify that current studies are using sural nerve biopsies to determine whether morphologic evidence of nerve regeneration can be used to demonstrate beneficial effects of tolrestat (Sima et al. 1993). Such a study involving 600 patients has been conducted, and the data are currently being evaluated. The only significant side effects noted with tolrestat treatment have been elevations of liver enzymes which are reversible on discontinuing the drug.

II. Aminoguanidine: An Inhibitor of Advanced Glycation End Product Formation

Studies over the last 15 years have provided an enormous amount of data indicating that one of the major mechanism by which glucose causes pathologic changes is through the formation of advanced glycation end products (Brownlee 1994; Vlassara 1994). These end products are formed by chemical rearrangements of highly reactive carbonyl compounds generated from Amadori rearrangement of the Schiff bases formed during the interaction of glucose and amino groups. These end products are generated over long periods of time; involve proteins, lipids, and nucleic acids; are irreversible, resistant to proteolytic digestion, and highly reactive. They alter function wherever they are formed. They have been implicated in the pathogenesis of microvascular, neuropathic, and macrovascular diseases in diabetic animals. Aminoguanidine, a small hydrazine-like compound (Fig. 12), appears to prevent the formation of glycation end products by preventing the late rearrangement reaction (Brownlee et al. 1986). Consequently, aminoguanidine has been used to prevent or treat the early complications of diabetes mellitus in experimental animals (Table 6).

Human studies of aminoguanidine's effects on diabetic complications are just beginning. Several preliminary studies suggest that this drug will be effective in decreasing end products. A 30-day treatment of 18 diabetic patients lowered end product hemoglobin levels by 30%. A similar decrease in plasma low-density lipoprotein (LDL) cholesterol occurred in patients with diabetes given aminoguanidine for 28 days (Bucala et al. 1994). Whether such changes will result in decreases in the chronic microvascular, neuropathic, and macrovascular complications of diabetes has yet to be determined.

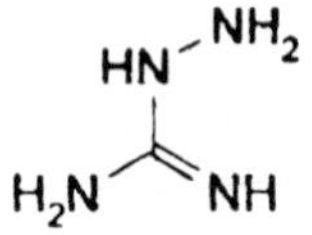

Fig. 12. The structure of aminoguanidine

Table 6. Effects of aminoguanidine on diabetic complications in animal models

Complication	Diabetic model	Duration of treatment (weeks)	Effect
Retinopathy			
HAMMES et al. 1991	STZ rats	75	Prevented background retinopathy
Neuropathy			
CAMERON et al. 1992	STZ rats	16	Improved motor nerve condution velocity Inhibited structural nerve changes
YAGIHASHO et al. 1992	STZ rats	8	Prevented decrease in motor nerve conduction velocity
Nephropathy			
SOULIS-LIPAROTA et al. 1991	STZ rats	32	Prevented mesangial expansion Attenuated albuminuria
EDELSTEIN and BROWNLEE 1992	STZ rats	24	Decreased albuminuria
YANG et al. 1994	Normal mice	4	Blocked glomerular hypertrophy and increases in extracellular matrix protein synthesis
Macrovascular disease			
BROWNLEE et al. 1986	Alloxan rats	16	Prevented arterial wall protein cross-linking
TILTON et al. 1993	STZ rats	4	Inhibited No synthase activity

STZ, streptozotocin; No, nitric oxide.

References

Agius L, Alberti KGMM (1985) Regulation of flux through pyruvate dehydrogenase and pyruvate carboxylase in rat hepatocytes. Effect of fatty acids and glucagon. Eur J Biochem 152:699–707

Agius L, Pillay D, Alberti KGMM, Sherratt HSA (1985) Effects of 2[5(4-chloro-phenylpentyl] oxirane-2-carboxylate on fatty acid synthesis and fatty acid oxidation in isolated rat hepatocytes. Biochem Pharmacol 34:2651–2654

Angel I, Schoemaker H, Duval N, Oblin A, Sevrin M, Langer SZ (1990) SL 84.0418: a new α_2-antagonist with anti-hyperglycemic properties. Eur J Pharmacol 183:990–991a

Angel I, Schoemaker H, Arbilla S, Galzin AM, Berry CN, Niddam R, Pimoule C, Sevrin M, Wick A, Langer SZ (1992) SL 84.0418: a novel, potent and selective alpha-2 adrenoceptor antagonist: I. In vitro pharmacological profile. J Pharmacol Exp Ther 263:1327

Angel I, Grosset A, Perault G, Schoemaker H, Langer SZ (1993) In vivo pharmacological profile of SL 84.0418, a new selective, peripherally active α_2-adrenoceptor antagonist. Eur J Pharmacol 234:137–145

Asins G, Serra D, Hegardt FG (1994) The effect of etomoxir on the mRNA levels of enzymes involved in ketogenesis and cholesterogenesis in rat liver. Biochem Pharmacol 47:1373–1379

Bailey CJ, Flatt PR (1989) Alkylglycidates. Diabetes 39:231–236

Bechem M, Hebisch S, Schramm M (1988) Ca^{2+} agonists: new, sensitive probes for Ca^{2+} channels. Trends Pharmacol Sci 9:257–261

Beck-Nielsen H, Hother-Nielsen O, Vaag A, Alford F (1994) Pathogenesis of type 2 (non-insulin-dependent) diabetes mellitus: the role of skeletal muscle glucose uptake and hepatic glucose production in the development of hyperglycaemia. A critical comment. Diabetologia 37:217–221

Beneking M, Oellerich M, Haeckel R, Binder L (1987) Inhibition of mitochondrial carnitine acylcarnitine translocase-mediated uptake of carnitine by 2-(3-methyl-cinnamyl-hydrazono)-propionate. J Clin Chem Clin Biochem 25:467–471

Berlin I, Rosenzweig P, Fusean E, Molinier P, Morselli P (1991) Effect of a new alpha-2 adrenoceptor blocking compound (SL 84.0418) on C-peptide and insulin responses to glucose infusion in healthy subjects. Diabetologia 34:A107

Berlin I, Rosenzweig P, Fusean E, Chalon S, Puech AJ (1992) The new alpha-2 adrenoceptor blocking compound (SL 84.0418) inhibits blood glucose increase following oral glucose tolerance test in healthy subjects. Diabetologia 35:A201

Berridge TL, Doxey JC, Roach AG, Smith CF (1992) Selectivity profile of the alpha-2 adrenoceptor antagonist efaroxan in relation to plasma glucose and insulin levels in the rat. Eur Pharmacol 213:205–212

Bliesath H, Haupt E, Lühmann R, Hoppe FU, Radtke HW (1987) First administration of etomoxir to type 2 (non-insulin-dependent) diabetic patients. Diabetologia 30:501A

Blondel O, Bailbe D, Portha B (1989) In vivo insulin resistance in streptozotocin-diabetic rats. Evidence for reversal following oral vanadate treatment. Diabetologia 32:185–190

Boulton AJM, Levin S, Comstock J (1990) A multicentre trial of the aldose-reductase inhibitor tolrestat in patients with symptomatic diabetic neuropathy. Diabetologia 33:431–437

Brichard SM, Bailey CJ, Henquin JC (1990) Marked improvement of glucose homeostasis in diabetic ob/ob mice given oral vanadate. Diabetes 39:1326–1332

Brown CA, Chan SLF, Stillings MR, Smith SA, Morgan NG (1993) Antagonism of the stimulatory effects of efaroxan and glibenclamide in rat pancreatic islets by the imidazoline, RX 801080. Br J Pharmacol 110:1017–1022

Brownlee M (1994) Glycation and diabetic complications. Diabetes 43:836–841

Brownlee M, Vlassara H, Kooney A, Ulrich P, Cerami A (1986) Aminoguanidine prevents diabetes–induced arterial wall protein cross-linking. Science 232:1629–1632

Bucala R, Makita Z, Vega G, Grundy S et al. (1994) Modification of low density lipoprotein by advanced glycation end products contributes to the dyslipidemia of diabetes and end stage renal insufficiency. Proc Natl Acad Sci USA 91:9441–9445

Cam MC, Pederson RA, Brownsey RW, McNeill JH (1993) Longterm effectiveness of oral vanadyl sulphate in streptozotocin–diabetic rats. Diabetologia 36:218–224

Cameron NE, Cotter MA, Dines K, Love A (1992) Effects of aminoguanidine on peripheral nerve function and polyol pathway metabolites in streptozotocin-diabetic rats. Diabetologia 35:946–950

Chan SLF, Morgan NG (1990) Stimulation of insulin secretion by efaroxan may involve interaction with potassium channels. Eur J Pharmacol 176:97–101

Chang AY, Wyse BM, Gilchrist BJ, Peterson T, Diani AR (1983a) Ciglitazone, a new hypoglycemic agent. I. Studies in ob/ob and db/db mice, diabetic Chinese hamsters and normal and streptozotocin-diabetic rats. Diabetes 32:830–838

Chang AY, Wyse BM, Gilchrist BJ (1983b) Ciglitazone, a new hypoglycemic agent. II. Effect on glucose and lipid metabolisms and insulin binding in the adipose tissue of C57BL/6J ob/ob and −+? mice. Diabetes 32:839–845

Chen YDI, Golay A, Swislock ALM, Reaven GM (1987) Resistance to insulin suppression of plasma free fatty acid concentrations and insulin stimulation of glucose uptake in noninsulin–dependent diabetes mellitus. J Clin Endocrinol Metab 64:17–21

Ciaraldi TP, Gilmore A, Olefsky JM, Goldberg M, Heidenreich RA (1990) In vitro studies on the action of CS 045, a new antidiabetic agent. Metabolism 39: 1056–1062

Colca JR, Wyse BM, Sawada G, Jodelis KS, Connell CL et al. (1988) Ciglitazone, a hypoglycemic agent: early effects on pancreatic islets of ob/ob mice. Metabolism 37:276–280

Consoli A (1992) Role of liver in pathophysiology of NIDDM. Diabetes Care 15:430–441

Consoli A, Nurjhan N, Capani F, Gerich J (1989) Predominant role of gluconeogenesis in increased hepatic glucose production in NIDDM. Diabetes 38: 550–557

Cooper AJ, Keddie KMG (1964) Hypotensive collapse and hypolycemia after mebanazine, a monoamine oxidase inhibitor. Lancet 1:1133–1135

Dai S, Thompson KH, McNeill JH (1994) One-year treatment of streptozotocin-induced diabetic rats with vanadyl sulphate. Pharmacol Toxicol 74:101–109

Deaciuc IV, Kühnle HF, Strauss KM, Schmidt FH (1983) Studies on the mechanism of action of the hypoglycemic agent, 2-(3)methylcinnamylhydrazono)-propionate(BM 42.304). Biochem Pharmacol 32:3405–3412

Dean PM, Matthews EK (1970) Glucose-induced electrical activity in pancreatic islet cells. J Physiol 210:255–64

DeFronzo RA (1988) The triumvirate: B-cell, muscle, liver. A collusion responsible for NIDDM. Diabetes 37:667–687

DeFronzo RA (1992) Pathogenesis of type 2 (non-insulin-dependent) diabetes mellitus: a balanced overview. Diabetologia 35:389–397

DeFronzo RA, Ferrannini E, Simonson D (1989) Fasting hyperglycemia in noninsulin-dependent diabetes mellitus: contributions of excessive hepatic glucose production and impaired tissue glucose uptake. Metabolism 38:387–395

Dhalla NS, Elimbam V, Rupp H (1992) Paradoxical role of lipid metabolism in heart function and dysfunction. Mol Cell Biochem 116:3–9

Diabetes Control and Complications Trial Research Group (1993) The effect of intensive treatment of diabetes on the development and progression of long-term complications in insulin-dependent diabetes mellitus. N Engl J Med 329: 977–986

Domingo JL, Ortega A, Llobet JM, Keen CL (1991a) No improvement of glucose homeostasis in diabetic rats by vanadate treatment when given by gavage. Trace Elem Med 8:181–186

Domingo JL, Comez M, Llobet JM, Corbella J, Keen CL (1991b) Oral vanadium administration to streptozotocin-diabetic rats has marked negative side effects which are independent of the form of vanadium used. Toxicology 66:279–287

Domingo JL, Gomez M, Sanchez DJ, Llobet JM, Keen CL (1992) Tiron administration minimizes the toxicity of vanadate but not its insulin mimetic properties. Life Sci 50:1311–1317

Donaubauer HH, Mayer D (1993) Acute, subchronic and chronic toxicity of the new sulfonylurea glimepiride in rats. Arzneim-Forsch/Drug Res 43:547–549

Edelstein D, Brownlee M (1992) Aminoguanidine ameliorates albuminuria in diabetic hypertensive rats. Diabetologia 35:96–97

Eistetter K, Wolf HPO (1982) Synthesis and hypoglycemic activity of phenylal-kyloxiranecarboxylic acid derivatives. J Med Chem 25:109–113

Eistetter K, Wolf HPO (1986) Etomoxir. Drugs of the Future 11:1034–1036

Florkowski CM, Rowe BR, Nightingale S, Harvey TC, Barnett AH (1991) Clinical and neurophysiological studies of aldose reductase inhibitor ponalrestat in chronic symptomatic diabetic peripheral neuropathy. Diabetes 40:129–133

Foley JE (1992) Rationale and application of fatty acid oxidation inhibitors in treatment of diabetes mellitus. Diabetes Care 15:773–784

Frøkjær-Jensen J, Kofod H, Godtfredsen SE (1992) Mechanism of action of AG–EE 623 ZW, a novel insulintropic agent. Diabetologia 35:A116

Fujita T, Sugiyama Y, Taketomi S, Sohda T et al. (1983) Reduction of insulin resistance in obese and/or diabetic animals by Ciglitazone, a new antidiabetic agent. Diabetes 32:804–810

Fujitani S, Yada T (1994) A novel D-phenylalanine-derivative hypoglycemic agent A-4166 increase cytosolic free Ca^{2+} in rat pancreatic beta-cells by stimulating Ca^{2+} influx. Endocrinology 134:1395–1400

Fujiwara T, Yoshioka S, Yoshioka T, Usiyama I, Horikoshi H (1988) Characterization of new oral antidiabetic agent CS 045. Studies in KK and ob/ob mice and Zucker fatty rats. Diabetes 37:1549–1558

Geisen K (1988) Special pharmacology of the new sulfonylurea glimepiride. Arzneim-Forsch/Drug Res 38:1120–1130

Geisen K, Hübner M, Hitzel V, Hrstka VE, Pfaff W, Bosies E, Regitz G, Kühnle HF, Schmidt FH, Weyer R (1978) Acylaminoalkyl-substituierte Benzoe- und Phenylalkansäuren mit blutglukosesenkender Wirkung. Arzneim-Forsch 28: 1081–1083

Gerich JE (1991) Is muscle the major site of insulin resistance in type 2 (non-insulin-dependent) diabetes mellitus? Diabetologia 34:607–610

Gill AM, Yen TT (1991) Effects of ciglitazone on endogenous plasma islet amyloid polypeptide and insulin sensitivity in obese-diabetic viable yellow mice. Life Sci 48:703–710

Gugliano D, Martella R, Quartraro A, Rosa N et al. (1993) Tolrestat for mild diabetic neuropathy. Ann Intern Med 118:7–11

Goldberg LI (1964) Monoamine oxidase inhibitors, adverse reactions and possible mechanisms. JAMA 190:456–462

Grell W, Greischel A, Zahn G, Mark M, Knorr H, Rupprecht E, Müller U (1993) (S)(+)-2-ethoxy-4-[N-[1-(2-piperidinophenyl)-3-methyl-1-butyl[aminocarbonylmethyl]-benzoic acid. PCT: WO 93/00337 (07.01.93)

Groop LC, Bonadonna RC, DelPrato S, Ratheiser K, Zyck K, Ferrannini E, DeFronzo RA (1989) Glucose and free fatty acid metabolism in noninsulin-dependent diabetes mellitus. Evidence for multiple sites of insulin resistance. J Clin Invest 84:205–213

Groop LC, Bonadonna RC, Shank M, Petrides AS, DeFronzo RA (1991) Role of free fatty acids and insulin in determining free fatty acid and lipid oxidation in man. J Clin Invest 87:83–89

Haeckel R, Oellerich M (1979) Hydrazonopropionic acids, a new class of hypoglycemic substances. 1. Hypoglycemic effect of 2-(phenylethylhydrazono)- and 2-(2-cyclohexylethylhydrazono)-propionic acid. Horm Metab Res 11:606–611

Haeckel R, Terlutter H, Schumann G, Oellerich M (1984) Hydrazonopropionic acids, a new class of hypoglycaemic substances. 3. Inhibition of jejunal glucose uptake in the rat and guinea pig. Horm Metab Res 16:423–427

Haeckel R, Oellerich M, Schumann G, Beneking M (1985) Hydrazonopropionic acids, a new class of hypoglycemic substances. 5. Inhibition of hepatic gluconeogenesis by 2-(3-methylcinnamyl-hydrazono)-propionate in the rat and guinea pig. Horm Metab Res 17:115–122

Hammes H-P, Martin S, Federlin K, Gersen K, Brownlee M (1991) Aminoguanidine treatment inhibits the development of experimental diabetic retinopathy. Proc Natl Acad Sci USA 88:11555–11558

Hanson RL, Isaacson CM (1985) Stimulation of insulin secretion from isolated rat islets by SaRI 59-801. Relation to cAMP concentration and Ca^{2+}-uptake. Diabetes 34:691–695

Hanson RL, Isaacson CM, Boyajy LD (1985) Stimulation of insulin secretion from isolated rat islets by SaRI 59-801. Diabetes 34:548–552

Häring HU, Mehnert H (1993) Pathogenesis of type 2 (non-insulin-dependent) diabetes mellitus: candidates for a signal transmitter defect causing insulin resistance of the skeletal muscle. Diabetologia 36:176–182

Haupt E, Bliesath H, Hoppe FU, Lühmann R, Wolf HPO, Radtke HW (1988) Treatment of type 2 diabetic patients (NIDDM) with different dosages of etomoxir: a placebo controlled drf-study. Diabetes Res Clin Pract 5:S615

Henquin JC, Meissner HP (1981) Effects of amino acids on membrane potential and ^{86}Rb fluxes in pancreatic B-cells. Am J Physiol 240:E245–E252

Henquin JC, Meissner HP (1984) Significance of ionic fluxes and changes in membrane potential for stimulus-secretion coupling in pancreatic B-cell. Experientia 40:1043–1052

Henquin JC, Schmeer W, Nenquin M, Meissner HP (1985) Effects of a calcium channel agonist on the electrical, ionic and secretory events in mouse pancreatic B-cells. Biochem Biophys Res Commun 131:980–986

Henquin JC, Garrino MG, Nenquin M (1987) Stimulation of insulin release by benzoic acid derivatives related to the non-sulphonylurea moiety of glibenclamide: structural requirements and cellular mechanisms. Eur J Pharmacol 141:243–251

Heyliger CE, Tahiliani AG, McNeill JH (1985) Effect of vanadate on elevated blood glucose and depressed cardiac performance of diabetic rats. Science 227:1474–1477

Hirose H, Maruyama H, Ito K, Seto Y, Koyama K, Dan K, Saruta T, Kato R (1994) Effects of N-[(trans-4-isopropylcyclohexyl)-carbonyl]-D-phenylalanine (A-4166) on insulin and glucagon secretion in isolated perfused rat pancreas. Pharmacology 48:205–210

Ho RS, Wiseberg JJ, Brand LJ, Nadelson J, Boyajy (1985) A novel, orally effective hypoglycemic agent, SaRI 59-801, in laboratory animals. Drug Dev Res 6:67–77

Ho W, Tutwiler GF, Cottrell SC, Morgans DJ, Tarhan D, Mohrbacher RJ (1986) Alkylglycidic acids: potential new hypoglycemic agents. J Med Chem 29:2184–2190

Hofmann CA, Colca JR (1992) New oral thiazolidinedione antidiabetic agents act as insulin sensitizers. Diabetes Care 35:1075–1078

Hofmann CA, Lorenz K, Colca JR (1991) Glucose transport deficiency in diabetic animals is corrected by treatment with the oral antihyperglycemic agent pioglitazone. Endocrinology 129:1915–1925

Hofmann CA, Edwards CW 3rd, Hillmann RM, Colca JR (1992) Treatment of insulin-resistant mice with the oral antidiabetic agent pioglitazone: evaluation of liver GLUT 2 and phosphoenolpyruvate carboxykinase expression. Endocrinology 130:735–740

Hübinger A, Knode O, Susanto F, Reinauer H, Gries FA (1993) Effects of etomoxir on insulin sensitivity, energy expenditure and substrate oxidation in NIDDM. Diabetologia 36:A75

Ikeda H, Taketomi S, Sugiyama Y, Shimura Y, Sodha T, Meguro K, Fujita T (1990) Effects of pioglitazone on glucose and lipid metabolism in normal and insulin-resistant animals. Arzneimittel-Forschung 40:156–162

Iwamoto Y, Kuzuya T, Matsuda A, Awata T, Kumakura S et al. (1991) Effect of new oral antidiabetic agent CS 045 on glucose tolerance and insulin secretion in patients with NIDDM. Diabetes Care 14:1083–1086

Iwanishi M, Kobayashi M (1993) Effect of pioglitazone on insulin receptors of skeletal muscles from high fat fed rats. Metabolism 42:1017–1021

Jennings PE, Nightingale S, Le Guew C, Lawson N et al. (1990) Prolonged aldose reductase inhibition in chronic peripheral diabetic neuropathy: effects on microangiopathy. Diabetic Medicine 7:63–68

Johnson DG, Dickerson J, Rutala P, Bressler R (1980) Clinical evaluation of pirogliride in patients with maturity-onset diabetes. Clin Res 28:81A

Johnson DG, Goebel CV, Hruby VJ, Bregman MD, Trivedi D (1981) Decrease in hyperglycemia of diabetic rats by a glucagon receptor antagonist. Science 215:1115–1116

Jonas JC, Plant TD, Henquin JC (1992) Imidazoline antagonists of α_2-adrenoceptors increase insulin release in vitro by inhibiting ATP-sensitive K^+ channels in pancreatic B-cells. Br J Pharmacol 107:8–14

Kador PF, Robison WG Jr, Kinoshita JH (1985) The pharmacology of aldose reductase inhibitors. Annu Rev Pharmacol Toxicol 25:691–714

Kahn CR (1995) The insulin receptor, insulin action and the mechanims of insulin resistance in diabetes and atherosclerosis. In: Schwartz CJ, Born GVR (eds) New horizons in diabetes mellitus and cardiovascular disease. Current Science, London, pp 46–56

Kameda K, Ono S, Abiko Y (1982a) Hypoglycemic action of 2-[2(4,5-dihydro-1H-imidazol-2-yl)-1-phenylethyl]pyridine dihydrochloride sesquihydrate (DG-5128), a new hypoglycemic agent. Arzneim-Forsch / Drug Res 32:39–44

Kameda K, Ono S, Koyama I, Abiko Y (1982b) Insulin releasing action of 2-[2(4,5-dihydro-1H-imidazol-2-yl)-1-phenylethyl]pyridine dihydrochloride sesquihydrate (DG-5128), a new, orally effective hypoglycaemic agent. Acta Endocrinol 99:410–415

Kanamaru T, Shinagawa S, Asai M, Okazaki H, Sugiyama Y, Fujita T, Iwatsuka H, Yoneda M (1985) Emeriamine, an antidiabetic β-aminobetaine derived from a novel fungal metabolite. Life Sci 37:217–223

Kawazu S, Suzuki M, Negishi K, Ishii J, Sando H, Katagiri H, Kanazawa Y, Yamanouchi S, Akanuma Y, Kajinuma H, Suzuki K, Watanabe K, Itoh T, Kobayashi T, Kosaka K (1987) Initial phase II clinical studies on midaglizole (DG-5128). A new hypoglycemic agent. Diabetes 36:221–226

Kern TS, Engerman RL (1991) Development of complication in diabetic dogs and galactosemic dogs: effect of aldose reductase inhibitors. In: Proceedings of a workshop on aldose reductase inhibitors. NIH, Bethesda, publication 81–3114

Kirchain WR, Rendell MS (1990) Aldose reductase inhibitors. Pharmacotherapy 10:326–336

Kletzien RF, Clarke SD, Ulrich PG (1992) Enhancement of adipocyte differentiation by an insulin-sensitizing agent. Mol Pharmacol 41:393–398

Kobayashi M, Iwanishi M, Egawa K, Shigeta Y (1992) Pioglitazone increases insulin sensitivity by activating insulin receptor kinase. Diabetes 41:476–483

Kolterman O, Gary R, Griffin J, Burstein P, Insel J, Scarlett J, Olefsky J (1981) Receptor and postreceptor defects contribute to the insulin resistance in non-insulin-dependent diabetes mellitus. J Clin Invest 68:957–969

Komatsu M, Rokokawa N, Takeda T, Nagasawa Y, Aizawa T, Yamada T (1989) Pharmacological characterization of the voltage-dependent calcium channel of pancreatic B-cell. Endocrinology 125:2008–2014

Koundakjian PP, Turnbull DM, Bone AJ, Rogers MP, Younan SIM, Sherratt HSA (1984) Metabolic changes in fed rats caused by chronic administration of ethyl 2-[5-(4-chlorophenyl)pentyl] oxirane-2-carboxylate, a new hypoglycaemic compound. Biochem Pharmacol 33:465–473

Kraegen EW, James DE, Jenkins AB, Chisholm DJ, Storlien LH (1989) A potent in vivo effect of ciglitazone on muscle insulin resistance induced by high fat feeding of rats. Metabolism 38:1089–1093

Krentz AJ, Honigsberger L, Ellis SH, Hardman M, Nattrass M (1992) A 12-month randomized controlled study of the aldose reductase inhibitor ponalrestat in patients with chronic symptomatic diabetic neuropathy. Diabet Med 9:463–468

Kreutter DK, Andrews KM, Gibbs EM, Hutson NJ, Stevenson RW (1990) Insulin-like activity of new antidiabetic agent CP 68722 in 3T3-L1 adipocytes. Diabetes 39:1414–1419

Kühnle HF, Schmidt FH, Deaciuc IV (1984) In vivo and in vitro effects of a new hypoglycemic agent, 2-(3-methylcinnamylhydrazono)-propionate (BM 42.304) on glucose metabolism in guinea pigs. Biochem Pharmacol 33:1437–1444

Kuzuya T, Iwamoto Y, Kosaka K, Takebe K, Yamanouchi T et al. (1991) A pilot clinical trial of a new oral hypoglycemic agent CS 045 in patients with non-insulin-dependent diabetes mellitus. Diabetes Res Clin Pract 11:147–153

Lebovitz HE (1995) The metabolic disease syndrome: lessons to be learned from radical and ethnic diversity. In: Schwartz CJ, Born GVR (eds) Diabetes mellitus and cardiovascular disease. Current Science, London, pp 75–80

Lebovitz HE, Chaiken RL (1994) Insulin-sensitizing and insulin-mimetic agents as potential treatment modalities for non-insulin-dependent diabetes mellitus. Av Diabetol 9:27–32

Lilly K, Chung C, Kerner J, Van Renterghem R, Bieber LL (1992) Effect of etomoxiryl-CoA on different carnitine acyltransferases. Biochem Pharmacol 43:353–361

Macleod AF, Boulton AJ, Owens DR, van Rooy P et al. (1992) A multicentre trial of the aldose-reductase inhibitor tolrestat in patients with symptomatic diabetic peripheral neuropathy. North European Tolrestat Study Group. Diabete Metab 18:14–20

Malabu UH, Dryden S, McCarthy HD, Kilpatrick A, Williams G (1994) Effect of chronic vanadate admimistration in the STZ-induced diabetic rat: the antihyper-glycemic action of vanadate is attributable entirely to its suppression of feeding. Diabetes 43:9–15

Malaisse WJ (1990) Regulation of insulin release by the intracellular mediators cyclic AMP, Ca^{2+}, inositol 1,4,5-trisphosphate, and diacylglycerol. In: Cuatrecasas P, Jacobs S (eds) Insulin. Springer, Berlin Heidelberg New York, pp 113–124 (Handbook of experimental pharmacology, vol 92)

Malaisse WJ, Sener A (1981) Calcium antagonists and islet function. XII. Comparison between nifedipine and chemically related drugs. Biochem Pharmacol 30: 1039–1041

Malaisse-Lagae F, Mathias PCF, Malaisse WJ (1984) Gating and blocking of calcium channels by dihydropyridines in the pancreatic B-cell. Biochem Biophys Res Commun 123:1062–1068

Malaisse WJ, Lebrun P, Sener A (1993) Modulation of the insulinotropic action of glibenclamide and glimepiride by nutrient secretagogues in pancreatic islets from normoglycemic and hyperglycemic rats. Biochem Pharmacol 45:1845–1849

McGarry JD, Foster DW (1980) Regulation of hepatic fatty acid oxidation and ketone body production. Annu Rev Biochem 49:395–420

Morgan NG, Chan SLF, Dumme MJ (1991) Stimulation of insulin secretion by α_2-antagonists is due to blockade of ATP-sensitive potassium channels. Diabetes 40 [Suppl 1]:79a

Nolan JJ, Ludvik B, Beerdsen P, Joyce M, Olefsky J (1994) Improvement in glucose tolerance and insulin resistance in obese subjects treated with troglitazone. N Engl J Med 331:1188–1193

Oellerich M, Haeckel R (1980) Hydrazonopropionic acids, a new class of hypo-glycemic substances. 2. Influence of 2-(phenylethyl-hydrazono)- and 2-(2-cyclohexyl-ethylhydrazono)-propionic acid on redox systems, acid base status and monoamino oxidase activity. Horm Metab Res 12:182–189

Oellerich M, Haeckel R, Wirries KH, Schumann G, Beneking M (1984) Hydrazono-propionic acids a new class of hypoglycemic substances. 4. Hypoglycemic effect of 2-(3-methyl-cinnamylhydrazono)-propionate in the rat and guinea pig. Horm Metab Res 16:619–625

Panten U (1987) Rapid control of insulin secretion from pancretic islets. ISI Atlas of Science: Pharmacology 1:307–310

Panten U, Burgfeld J, Goerke F, Rennicke M, Schwanstecher M, Wallasch A, Zünkler BJ, Lenzen S (1989) Control of insulin secretion by sulfonylureas, meglitinide and diazoxide in relation to their binding to the sulfonylurea receptor in pancreatic islets. Biochem Pharmacol 38:1217–1229

Passariello N, Sepe J, Marrazzo G, de Cicco A et al. (1993) Effect of aldose reductase inhibitor (tolrestat) on ordinary albumin excretion rate and glomerular filtration rate in IDDM subjects with nephropathy. Diabetes Care 16:789–795

Pedersen MM, Christiansen JS, Mogensen CE (1991) Reduction of glomerular hyperfiltration in normoalbuminuric IDDM patients by 6 mo of aldose reductase inhibition. Diabetes 40:527–531

Pirart J (1978) Diabetes mellitus and its degenerative complications: a prospective study of 4400 patients observed between 1947 and 1973. Diabetes Care 1:168–188

Plant TD, Henquin JC (1990) Phentolamine and yohimbine inhibit ATP-sensititve K channels in mouse pancreatic B-cells. Br J Pharmacol 101:115–120

Porte D Jr (1967) A receptor mechanism for the inhibition of insulin release by epinephrine in man. J Clin Invest 46:86–94

Profozić V, Babić D, Renar I, Rupprecht E, Škrabalo Z, Metelko Ž (1993) Benzoic acid derivative hypoglycemic activity in non-insulin dependent diabetic patients. Diabetologia 36:A183

Pugazhenthi S, Khandelwal RL (1990) Insulin-like effects of vanadate on hepatic glycogen metabolism in non-diabetic and streptozotocin-induced diabetic rats. Diabetes 39:821–827

Puls W, Bischoff H (1985) Hypoglycaemic effects of a novel dihydropyridine (DHP) analogue, Bay K 8644, in rats. Diabetes Res Clin Pract [Suppl 1]:457

Randle PJ, Hales CN, Garland PB, Newsholme EA (1963) The glucose-fatty acid cycle. Its role in insulin sensitivity and the metabolic disturbances of diabetes mellitus. Lancet i:785–789

Ranganathan S, Krempf M, Feraille E, Charbonnel B (1993) Short-term effect of an aldose reductase inhibitor on urinary albumin excretion rate (UAER) and glomerular filtration rate (GRF) in type 1 diabetic patients with incipient nephropathy. Diabete Metab 19:257–261

Ratheiser K, Schneeweiß B, Waldhäusl W, Fashcing P, Korn A, Nowotny P, Rohac M, Wolf HPO (1991) Inhibition by etomoxir of carnitine palmitoyltransferase 1 reduces hepatic glucose production and plasma lipids in non-insulin-dependent diabetes mellitus. Metabolism 40:1185–1190

Ratheiser K, Korn A, Waldhaeusl W, Komjati M, Vierhapper H, Badian M, Malerczyk V (1993) Dose relationship of stimulated insulin production following intravenous application of glimepiride in healthy man. Arzneim-Forsch 43:856–858

Reaven GM (1988) Role of insulin resistance in human disease. Diabetes 37:1595–1607

Ribes G, Trimble ER, Blayac JP, Wollheim CB, Puech R, Loubatières-Mariani MM (1981) Effect of a new hypoglycaemic agent (HB 699) on the in vivo secretion of pancreatic hormones in the dog. Diabetologia 20:501–505

Robertson RP (1986) Arachidonic acid metabolite regulation of insulin secretion. Diabetes Metab Rev 2:261–296

Robertson RP, Porte D (1973) Adrenergic modulation of basal insulin secretion in man. Diabetes 22:1–8

Robertson RP, Halter JB, Porte D (1976) A role for alpha-adrenergic receptors in abnormal insulin secretion in diabetes mellitus. J Clin Invest 57:791–795

Ronner P, Higgins TJ, Kimmich GA (1991) Inhibition of ATP-sensitive K^+-channels in pancreatic B-cells by nonsulfonylurea drug linogliride. Diabetes 40:885–892

Ronner P, Hang TL, Kraebber MJ, Higgins TJ (1992) Effect of the hypoglycaemic drug (−)-AZ-DF-265 on ATP-sensitive potassium channels in rat pancreatic B-cells. Br J Pharmacol 106:250–255

Rorsman P, Trube G (1986) Calcium and delayed potassium currents in mouse pancreatic B-cells under voltage-clamp conditions. J Physiol (Lond) 374:531–550

Rossetti L, Laughlin MR (1989) Correction of chronic hyperglycemia with vanadate, but not with phlorizin normalizes in vivo glycogen repletion and in vitro glycogen synthase activity in diabetic skeletal muscle. J Clin Invest 84:892–899

Sandouk T, Reda D, Hofmann C (1993) The antidiabetic agent pioglitazone increased expression of glucose transporters in 3T3-F442A cells by increasing messenger ribonucleic acid transcript stability. Endocrinology 133:352–359

Santiago JV, Sonksen PH, Boulton AJ, Macleod A et al. (1993) Withdrawal of the aldose reductase inhibitor tolrestat in patients with diabetic neuropathy: effect on nerve function. The Tolrestat Study Group. J Diabetes Complications 71: 170–178

Sato Y, Nishikawa M, Shinkai H, Sukegawa E (1991) Possibility of ideal blood glucose control by a new oral hypoglycemic agent, N-[(trans-4-iso-propylcyclohexyl)-carbonyl]-D-phenylalanine (A-4166), and its stimulatory effect on insulin secretion in animals. Diabetes Res Clin Pract 12:53–60

Scherrer R, Mürtz H, Schmeidl R, Draeger E, Usadel KH (1988) Glimepiride–a very potent new sulfonylurea for treatment of type II diabetes. Diabetes 37: 136A

Schmidt FH, Deaciuc IV, Kühnle HF (1985) A new inhibitor of the long-chain fatty acid transfer across the mitochondrial membrane: 2-(3-methylcinnamylhydrazono)-propionate (BM 42.304). Life Sci 36:63–67

Schramm M, Thomas G, Towart R, Franckowiak G (1983) Novel dihydropyridines with positive inotropic action through activation of Ca^{2+} channels. Nature 303:535–537

Schulz A, Hasselblatt A (1988) Phentolamine, a deceptive tool to investigate sympathetic nervous control of insulin release. Naunyn Schmiedebergs Arch Pharmacol 337:637–643

Schulz A, Hasselblatt A (1989) An insulin-releasing property of imidazoline derivatives is not limited to compounds that block α-adrenoceptors. Naunyn Schmiedbergs Arch Pharmacol 340:321–327

Selby PL, Sherratt HSA (1989) Substituted 2-oxiranecarboxylic acids: a new group of candidate hypoglycaemic drugs. TIPS 10:495–500

Sener A, Malaisse WJ (1984) Nutrient metabolism in islet cells. Experientia 40: 1026–1035

Shechter Y (1990) Insulin-mimetic effects of vanadate. Possible implications for future treatment of diabetes. Diabetes 39:1–5

Shinkai H, Sato Y (1990) Hypoglycaemic action of phenylalanine derivatives. In: Bailey CJ, Flatt PR (eds) New antidiabetic drugs. Smith Gordon, London, pp 249–254

Sima AA, Greene DA, Brown MB, Hohman TC et al. (1993) Effect of hyperglycemia and the aldose reductase inhibitor tolrestat on sural nerve biochemistry and morphometry in advanced diabetic peripheral polyneuropathy. J Diabetes Complications 7:157–169

Sorbinil Retinopathy Trial Research Group (1990) A randomized trial of sorbinil, an aldose reductase inhibitor in diabetic retinopathy. Arch Opthalmol 108: 1234–1244

Sorbinil Retinopathy Trial Research Group (1993) The sorbinil retinopathy trial: neuropathy results. Neurology 43:1141–1149

Soulis-Liparota T, Cooper M, Papazoglou D, Clarke B et al. (1991) Retardation by aminoguanidine of development of albuminuria, mesangial expansion and tissue fluorescence in streptozotocin-induced diabetic rat. Diabetes 40:1328–4324

Stern MP (1994) The insulin resistance syndrome. The controversy is dead, long live the controversy. Diabetologia 37:956–958

Stevenson RW, Hutson NJ, Krupp MN, Volmann RA, Holland GF et al. (1990) Actions of novel antidiabetic agent englitazone in hyperglycemic ob/ob mice. Diabetes 39:1218–1227

Stevenson RW, McPherson RK, Genereux PE, Danbury BH, Kreutter DK (1991) Antidiabetic agent englitazone enhances insulin action in nondiabetic rats without producing hyperglycemia. Metabolism 40:1268–1274

Sugiyama Y, Taketomi S, Shimura Y, Ikeda H, Fujita T (1990a) Effects of pioglitazone on glucose and lipid metabolism in Wistar fatty rats. Arzneim-Forsch 40:263–267

Sugiyama Y, Shimura Y, Ikeda H (1990b) Effects of pioglitazone on hepatic and peripheral insulin resistance in Wistar fatty rats. Arzneim-Forsch 40:436–440

Sundkuist G, Armstrong FM, Bradbury JE, Chaplin C et al. (1992) Peripheral and autonomic nerve function in 259 diabetic patients with peripheral neuropathy treated with ponalrestat or placebo for 18 months. J Diabetes Complications 6:123–130

Suter SL, Nolan JJ, Wallace P, Gumbiner B, Olefsky JM (1992) Metabolic effects of new oral hypoglycemic agent CS 04S in NIDDM subjects. Diabetes Care 15:193–203

Tadayyon M, Green I, Cook D, Pratt J (1987) Effect of a hypoglycemic agent M & B 29890 A on glucagon secretion in isolated rat islets of Langerhans. Diabetologia 30:41–43

Taketomi S, Fujita T, Yokono K (1988) Insulin receptor and postbinding defects in KK mouse adipocytes and improvement by Ciglitazone. Diabetes Res Clin Pract 5:125–134

Tamagawa T, Niki H, Niki A (1985) Insulin release independent of a rise in cytosolic free Ca^{2+} by forskolin and phorbol ester. FEBS Lett 183:430–432

Tilton RG, Chang K, Hasan KS et al. (1993) Prevention of diabetic vascular dysfunction by guanidines. Diabetes 42:221–232

Tomlinson DR, Stevens EJ, Diemel LT (1994) Aldose reductase inhibitors and their potential for treatment of diabetic complications. TIPS 15:293–297

Tromp A, Hooymans JM, Barendsen BC, van Doormaal JJ (1991) The effect of an aldose reductase inhibitor on the progression of diabetic retinopathy. Documenta Opthalmologica 78:153–159

Tsai SC, Burnakis TG (1993) Aldose reductase inhibitors: an update. Ann Pharmacother 27:751–754

Tuman RW, Tutwiler GF, Bowden CR (1990) Linogliride: a guanidine insulin secretagogue. In: Bailey CJ, Flatt PR (eds) New antidiabetic drugs. Smith Gordon, London, pp 163–169

Turk J, Wolf BA, McDaniel ML (1987) The role of phospholipid-derived mediators including arachidonic acid, its metabolites, and inositoltrisphosphate and of intracellular Ca^{2+} in glucose-induced insulin secretion by pancreatic islets. Prog Lipid Res 26:125–181

Tutwiler GF, Dellevigne P (1979) Action of the oral hypoglycemic agent 2-tetradecylglycidic acid on hepatic fatty acid oxidation and gluconeogenesis. J Biol Chem 254:2935–2941

Tutwiler GF, Kirsch T, Mohnbacher R, Ho W (1978) Pharmacologic profile of methyl2-tetradecylglycidate (McN3716)–an orally effective hypoglycemic agent. Metabolism 27:1539–1555

Tutwiler GF, Mohrbacher R, Ho W (1979) Methyl 2-tetradecylglycidate, an orally effective hypoglycemic agent that inhibits long chain fatty acid oxidation selectively. Diabetes 28:242–248

Tutwiler GF, Tuman RW, Joseph JM, Mihan BB, Fawthrop H, Brentzel HJ (1986) Pharmacologic profile and insulin secretagogue effect of linogliride (McN-3935), a new orally effective hypoglycemic agent. Drug Dev Res 9:273–292

Unson CG, Gurzenda EM, Merrifield RB (1989) Biological activities of des His [Glu^9] glucagon amide, a glucagon antagonist. Peptides 10:1171–77

Utter MF, Barden RE, Taylor BL (1975) Pyruvate carboxylase: an evaluation of the relationships between structure and mechanism and between structure and catalytic activity. Adv Enzymol Relat Areas Mol Biol 42:1–72

Vaartjes WJ, De Haas CGM, Haagsman HP (1986) Effects of sodium 2-5-(4-chlorophenyl)pentyl-oxirane-2-carboxylate (POCA) on intermediary metabolism in isolated rat-liver cells. Biochem Pharmacol 35:4267–4272

Van Praag HM, Leijnse B (1963) The influence of some antidepressant drugs of the hydrazine type on the glucose metabolism in depressed patients. Clin Chim Acta 8:466–475

Verhaegen J, Leempoels J, Brugmans J, Tutwiler GF (1984) Preliminary evaluation of methyl palmoxirate (Me-Palm) in type I diabetes. Diabetes 33 [Suppl 1]:180A

Vlassara H (1994) Recent progress on the biological and clinical significance of advanced glycosylation end products. J Lab Clin Med 124:19–30

Wickstrom I, Petterson K (1964) Treatment of diabetics with monoamine oxidase inhibitors. Lancet 2:995–997

Wolf HPO, Engel D (1985) Decrease of fatty acid oxidation, ketogenesis and gluconeogenesis in isolated perfused rat liver by phenylalkyl oxirane carboxylate (B 807-27) due to inhibition of CPT1 (EC 2.3.2.21). Eur J Biochem 146: 359–363

Wolff HP, Kühnle HF (1985) Synthesis and hypoglycemic activity of N-alkylated hydrazonopropionic acids. J Med Chem 28:1436–1440

Wolf HPO, Eistetter K, Ludwig G (1982) Phenylalkyloxirane carboxylic acids, a new class of hypoglycemic substances: hypoglycemic and hypoketonaemic effects of sodium 2-[5-(4-chlorophenyl)pentyl]-oxirane-2-carboxylate (B 807-27) in fasted animals. Diabetologia 22:456–463

Wolffenbüttel BHR, Nijst L, Sels JP, Menheere P, Muller PG, Nieuwenhuijzen-Kruseman AC (1992) Metabolic effect of a new oral hypoglycaemic agent, AG-EE 623 ZW, in sulphonylurea (SU) treated type 2 diabetic patients. Diabetologia 35:A200

Yada T, Fujitani S (1992) Action mechanisms of a novel hypoglycemic agent (A-4166) in pancreatic B-cells analyzed by measuring cytosolic Ca concentration. Diabetes 41:149A

Yagihashi S, Kamijo M, Baba M, Yagihashi N, Nagai K (1992) Effect of amino-guanidine on functional and structural abnormalities in peripheral nerve of STZ-induced diabetic rats. Diabetes 41:47–52

Yamanaka K, Kigoshi S, Muramatsu I (1984) The selectivity of DG-5128 as an α_2-adrenoceptor antagonist. Eur J Pharmacol 106:625–628

Yamazaki H, Tabata S (1993) Sex difference in pharmacokinetics of the novel sulfonylurea antidiabetic glimepiride in rats. Arzneim-Forsch/Drug Res 43: 1317–1321

Yang C-W, Vlassara H, Peten EP, He C-J, Striker GE et al. (1994) Advanced glycation end products up-regulate gene expression found in diabetic glomerular disease. Proc Natl Acad Sci USA 91:9436–9440

Yen TT, Schmiegel KK, Gold G, Williams GD, Dininger NB, Broderick CL, Gill AM (1991) Compound M & B 39890 A [N-(3-imidazol-1-ylpropyl)-2-(3-trifluoromethylbenzenesulfonamido)-benzamide hydrochloride], a glucagon and insulin secretion inhibitor, improves insulin sensitivity in viable yellow obese-diabetic mice. Arch Int Pharmacodyn Ther 310:162–174

Yoshioka S, Nishino H, Shiraki T, Ikeda K, Kolke H et al. (1993) Antihypertensive effects of CS 045 treatment in obese Zucker rats. Metabolism 42:75–80

Young DA, Ho RS, Bell PARH, Cohen DK, McIntosh RH, Navelson J, Foley JE (1990) Inhibition of hepatic glucose production by SDZ 51641. Diabetes 39: 1408–1413

Yue DK, Brooks B (1993) The role of aldose reductase inhibitors in the treatment of diabetic peripheral neuropathy. Med J Aust 159:76–78

Zawalich WS, Rasmussen G, Tuman RW, Tutwiler GF (1987) Influence of the oral hypoglycemic agent linogliride (McN-3935) on insulin secretion from isolated rat islets of Langerhans. Endocrinology 120:880–885

Zenon GJ, Abobo CY, Carter BL, Ball DW (1990) Potential use of aldose reductase inhibitors to prevent diabetic complications. Clin Pharmacy 9:446–456

Ziegler D, Mayer P, Rathmann W, Gries FA (1991) One-year treatment with the aldose reductase inhibitor ponalrestat in diabetic neuropathy. Diabetes Res Clin Prac 14:63–74

Zünkler BJ, Lenzen S, Männer K, Panten U, Trube G (1988) Concentration-dependent effects of tolbutamide, meglitinide, glipizide, glibenclamide and diazoxide on ATP-regulated K^+ currents in pancreatic B-cells. Naunyn Schmiedebergs Arch Pharmacol 337:225–230

CHAPTER 24

New Approaches for the Treatment of Diabetes Mellitus

J.M. AMATRUDA and M.L.McCALEB

A. Prevention

Non-insulin-dependent diabetes mellitus (NIDDM) is a major cause of cardiovascular mortality and morbidity and recent data indicate that the incidence of retinopathy and nephropathy in NIDDM is similar to that in insulin-dependent diabetes mellitus (IDDM). Furthermore, the incidence of neuropathy is very high in NIDDM and this, along with vascular disease, is a major etiologic factor in amputations. Because of these factors as well as epidemiologic data indicating that impaired glucose tolerance is a risk factor for cardiovascular disease, prevention of impaired glucose tolerance (IGT) and prevention of deterioration to NIDDM are important goals for new approaches to NIDDM.

I. Non-Insulin-Dependent Diabetes Mellitus

While it is not the purpose of this chapter to discuss diagnosis, any prevention strategy requires early diagnosis. With NIDDM, risk can be assessed with a knowledge of family history and ethnicity with some accuracy. Furthermore, preliminary trials are now being conducted to evaluate the effectiveness of α-glucosidase inhibitors in treating IGT. Assuming that α-glucosidase inhibitors are effective in treating IGT, they might also be effective in preventing deterioration to frank diabetes since the latter requires that the pancreas fail to secrete enough insulin to control plasma glucose (DEFRONZO et al. 1992). By preventing postprandial hyperglycemia, pancreatic failure could theoretically be prevented or delayed. Other methods of prevention might include biguanide therapy, agents that increase insulin sensitivity, sulfonylureas and, of course, weight loss and exercise. All of these therapies decrease the demand for insulin and could theoretically prevent or delay the decrease in insulin secretion characteristic of the appearance of hyperglycemia in NIDDM. Further support for this concept comes from the well-known fact that any treatment that lowers plasma glucose will improve insulin secretion in patients with NIDDM who have reduced insulin secretion, although it will not restore insulin secretion to normal (KOSAKA et al. 1980). Biguanides have the potential to prevent deterioration to frank diabetes through their inhibition of glucose absorption, decrease in hepatic glucose output or enhancement of peripheral glucose uptake.

Obesity is not only the most common cause of insulin resistance but also an extremely important risk factor for NIDDM. It is well known that weight loss can reverse the hyperglycemia of NIDDM if it is initiated early enough in the course of the disease, probably by improving insulin resistance. Weight loss can also lead to improvements in glucose tolerance in those patients with IGT but there are no long-term studies evaluating the effects of weight loss on the prevention of NIDDM. The reason for this is likely that there is no current strategy that is effective in maintaining weight loss and most patients regain the lost weight quickly. Furthermore, it is not known how much weight loss must be sustained to prevent or delay the onset of diabetes. It is clear that a relatively small weight loss can significntly lower plasma glucose in patients with NIDDM, but it is unknown how close to ideal body weight (IBW) a patient with IGT would have to be to prevent NIDDM.

II. Insulin-Dependent Diabetes Mellitus

With regard to IDDM, prevention again requires early diagnosis and the tools for early diagnosis are available although the population to be screened is less certain than for NIDDM (Bingley et al. 1992; Eisenbarth et al. 1993; Thai and Eisenbarth 1993). Several recent studies have demonstrated that decreases in the insulin response to an i.v. glucose load together with positive islet cell antibodies can be highly predictive of the development of diabetes within 3 years (Eisenbarth et al. 1993). Newer tests will include specific antigens such as glutamic acid decarboxylase (GAD) and possibly other pancreatic antigens such as 512 (Rabin et al. 1992) and others. It is anticipated that a panel of antigens will have higher sensitivity and specificity than existing tests and can be placed into an enzyme-linked immunoadsorbent assay (ELISA) format. This will allow for ease and standardization of testing and decreased cost compared to current tests for islet cell antibodies and GAD which are performed by immunofluorescence and immunoprecipitation, respectively. One of the factors that has limited enthusiasm for corporate involvement in commercializing tests for early diagnosis of IDDM is the lack of an established preventive therapy that shows efficacy and lacks intolerable side effects. Once such therapy is available, early diagnostic testing and preventive therapy will become the norm.

Future preventive therapy for IDDM will likely fall into three categories: immunosuppression, oral tolerance, and systemic tolerance and therapy (Skyler and Marks 1993; Kaufman et al. 1993; Tisch et al. 1993; Zhang et al. 1991). To date immunosuppression has been limited primarily to cyclosporin therapy while other immunosuppressive therapies have proven ineffective. The early cyclosporin trials were impressive; however, subsequent trials indicated that the remissions are not sustained even with continuing cyclosporin administration. There is a suggestion that shorter duration of diabetes prior to the initiation of immunosuppressive treatment and higher trough levels of cyclosporin may be determinants of a more prolonged insulin-free interval. There are serious concerns about cyclosporin therapy because of its general immunosuppressive properties, its renal toxicity and the potential for tumors with long-term immunosuppression. Consideration also needs to be given to the fact that only ~40% of patients with IDDM develop proliferative retinopathy and nephropathy. If immunosuppression is instituted prior to diagnosis or early after diagnosis, ~60% of individuals will be immunosuppressed only for the purpose of preventing insulin dependence. Even given the problems and lifestyle changes which insulin dependence engenders, chronic immunosuppression beginning for most in childhood does not seem to be a reasonable alternative for those who will not develop serious long-term complications.

While control of plasma glucose is certainly important when assessing risk within populations (DCCT REASEARCH GROUP 1993), some individuals with loose control do not develop complications while some with better control do. This, of course, raises the important question of identification of those at risk for devleoping sight-or life-threatening complications of diabetes. At the current time there are no known determinants of this at the time of or prior to diagnosis of IDDM. The only known risk factors such as blood glucose control, microalbuminuria, hyperfiltration and hypertension occur long after diagnosis when islet cell function has already deteriorated beyond repair. Certainly research into the genetic, metabolic and other factors that may be determinants of the development of complicaitons is important. Until we know more about these determinants, the results of the Diabetes Control and Complications Trial (DCCT) make it imperative that blood sugar control be as close to normal as possible. Certainly, prevention of IDDM would serve this purpose. In fact, even a delay in onset of IDDM would be very beneficial not only for quality of life considerations but in the delay in onset of complications.

Another possibility for immunotherapy for IDDM lies in specific tolerance. This necessitates knowledge of the inciting antigen, which is currently unknown in man. Recent data in the non-obese diabetic (NOD) mouse, however, indicate that a T-helper response to GAD develops in NOD mice at the same time as the onset of insulitis. Furhtermore, i.v. or intrathymic injection of GAD to NOD mice at 3–4 weeks of age either prevents or decreases the incidence of insulitis and diabetes (KAUFMAN et al. 1993; TISCH et al. 1993). These data are extremely exciting and, assuming that they are applicable to man, suggest that IDDM might be prevented by tolerization to GAD.

A new possibility of oral tolerance has recently arisen from the concept of "bystander suppression." This theory states that one can induce immunosuppression by administering a protein from the tissue which is participating in the immune response even though that protein is not the immunogen. Presumably, the antigen activates T cells in Peyer's patches and these T cells got to the tissue of interest, secrete (TGF-β) and induce immune suppresion.

Studies have indicated some efficacy of TGF β with regard to myelin administration to patients with multiple sclerosis, collagen administration to patients with rheumatoid arthritis and oral insulin for IDDM in NOD mice. In NOD mice, oral insulin therapy resulted in a 50% decrease in the incidence of IDDM over the 1-year study (ZHANG et al. 1991). These results need to be confirmed and other pancreatic antigens tested, since there is no evidence that insuln alone is the most efficacious antigen or data suggesting what antigen or combination of antigens might be more efficacious. Nevertheless, oral tolerance has shown substantial promise for the future for two major reasons: first, oral tolerance is expected to be nontoxic and safe. This makes it highly suitable for treatment of children (the average age of onset of IDDM is ~11 years old) and for long-term administration. The safety also negates the argument that not all patients with IDDM will develop life-threatening complications since prevention of IDDM by oral tolerance itself is safe. Secondly, the development of a panel of antibodies to diagnose those at risk for IDDM will undoubtedly not have 100% sensitivity and specificity. However, the oral administration of pancreatic antigens would not be contraindicated even in the few individuals who might not be at risk.

Another approach is systemic (subcutaneous) insulin administration at doses of 0.1–0.3 U/kg body weight. Preliminary human studies have suggested some efficacy of this approach without inducing hypoglycemia (KELLER et al. 1993). The principle is that by administering insulin one can "rest" the pancreas and preserve function (SKYLER and MARKS 1993). The reasoning is similar to the possible explanation for the decrease in insulin requirements early after initiating therapy for IDDM, i.e., the "honeymoon" period. This period suggests that there is functional reserve even after diagnosis of ketoacidosis. Confirmation of this rests in the cyclosporin trials which indicate that some patients can be maintained without insulin therapy if treated with cyclosporin early after diagnosis.

It is anticipated that future therapies for IDDM will incorporate early diagnosis of those at risk followed by specific immunosuppressive therapy, tolerization to the initiating target antigen or oral tolerance. Alternatively, if a benign immunization is feasible it is possible that the entire population could be "vaccinated" without testing or specific therapy for those at risk.

B. Obesity

Obesity is clearly a salient factor in the hyperglycemia of NIDDM. Over 80% of NIDDM patients are obese and reductions in body weight of many of these patients will result in clinically significant reductions in glycemia. Furthermore, excessive adiposity is linked to multiple other pathologies, such as cardiovascular diseases, which impact substantially on the mortality of NIDDM patients. As discussed above, successful treatment of obesity may be one of the most efficacious ways to treat NIDDM as well as retard the escalating increase in the number of NIDDM patients each decade. However, the prevalence of obesity has become epidemic in the industrialized countries and the long-term treatment of obesity must be an achievable therapeutic goal if we are going to provide complete care to patients.

I. Central Mechanisms

Drug treatment of excessive food consumption is currently limited to agents that alter either the catecholamine or serotonergic neurotransmitter systems in the brain. The traditonal anorectic drugs (i.e., phenylethylamines, amphetamines or mazindol) modulate the release of norepinephrine and/or dopamine. The other major class of drugs (i.e., fenfluramine or fluoxetine) enhance the release and inhibit the uptake of serotonin. All of these drugs have multiple side effects, such as depression, addiction and sedation, that are incompatible with long-term therapy. In addition, these anorectic agents do not provide a maintenance of reduced weight. There is a critical medical need for appetite suppressants that effectively prevent patients from regaining weight and help them comply with a restrictive diet.

Since modulation of the catecholaminergic or serotonergic systems have not proven to be clinically successful, future treatments will have to intervene at other neurochemical targets. We have only a fragmented understanding of the neurobiological control of appetite and minimal insight into its possible dysfunction during human obesity. It is hoped that, during the next decade, there will be substantial progress delineating the process by which the brain integrates the multiple signals that are involved in regulating body weight. Within the near future, success may be achieved by inhibiting the signaling pathways of two peptide neurotransmitters, galanin and neuropeptide Y (NPY). The injection of either peptide into the brain of rodents elicits a robust feeding response, which suggests that they are essential neurotransmitters for appetite. Furthermore, brain levels of these peptides are dependent upon nutritional and obesity conditions of animals. We do not know, yet, the importance of NPY and galanin in the regulation of appetite in humans.

A logial step is to identify antagonists for the respective neurotransmitters. Some peptide antagonists of galanin have been found to reduce food consumption in rodents. In the case of NPY, however, there are multiple receptor subtypes and it is not known which one is involved in appetite (MICHEL 1991). Subtypes of galanin receptor proteins are just recently being considered. Although the existence of multiple receptor subtypes provides additional hurdles to overcome initially, the receptor subtypes may allow pharmacologic specificity that could not otherwise be achieved. Both NPY and galanin are involved in other CNS functions, and a drug that completely antagonized all actions of either peptide would have unacceptable side effects.

Pharmacological intervention does not have to be limited to receptor blockade. The neurotransmitter signals could be antagonized during the intracellular signaling cascades if unique pathways can be identified. Both neurotransmitter receptors are part of the family of G-protein-coupled receptors that modulate adenyl cyclase activity. Because of the explosion of knowledge in this research area, multiple other pharmacologic targets may become evident during the next few years. Inhibitors of the synthesis or release of NPY and galanin may also become viable pharmacologic approaches. There will be a tremendous challenge, however, obtaining a means of achieving some degree of neuroanatomical specificity to limit the presumed side effects that would result from inhibition of such widely distributed neurotransmitters.

There is a particularly intriguing aspect of galanin that may prove to be the most relevant for clinical efficacy. Whereas the serotonergic and NPY neurochemical systems have been proposed to regulate the consumption of carbohydrates, galanin is thought to be intricately involved in the ingestion of fats (Leibowitz 1992). If the galanin neurotransmitter system is found to predominately control dietary fat intake, then it becomes a highly attractive and unique pharmacologic target. As discussed below, excessive dietary fat consumption by obese patients is a major contributor to their caloric imbalance.

The opioid neurochemical system may also be a viable target for pharmacologic intervention. Endorphins and dynorphins enhance feeding while the respective opioid antagonists diminish food consumption in rodents. Initial evaluation of the opioid antagonists naltrexone and naloxone in obese patients have been disappointing. However, there is significant interest in identifying specific antagonists of the mu, kappa or delta opioid receptors with the hope that more potent agents will provide proof of principle and perhaps even a clinically effective anorectic drug (Mitch et al. 1993).

Several peripherally derived substances may also provide some additional pharmacologic targets. Recently, there has been considerable interest to identify small molecular weight agonists of cholecystokinin (CCK) since the octapeptide of CCK reduces food consumption in a variety of species (Cooper and Dourish 1990). Two receptor subtypes of CCK have been demonstrated (Silvente-Poirot et al. 1993) and the anorectic activity of CCK appears to be elicited following binding to the CCK_A receptor subtype (Holladay et al. 1992), which is located predominately in the periphery. The clinical efficacy and safety of the CCK agonists have not been established. In animal studies the compounds are potent inhibitors of acute feeding, but it is also necessary that they provide a long-term reduction of body weight.

Another gut-derived protein, apolipoprotein A-IV (apo A-IV), has also been proposed to be a satiety factor. Although apo A-IV is a protein associated with lipoproteins, little is known about the role it has in lipid homeostasis, Recent experiments in rats have found that mesenteric lymph, collected following the consumption of a lipid meal, suppresses appetite.

Additional studies demonstrated that apo A-IV was responsible for the anorectic action of the lymph (FUJIMOTO et al. 1992). Interestingly, a micro-injection of apo A-IV directly into the brain of fasted rats suppressed food consumption (FUJIMOTO et al. 1993), which suggests that the lipoprotein is a centrally acting satiety factor.

The pentapeptide, enterostatin, has been found to suppppress fat, but not protein or carbohydrate, consumption by rats (ERLANSON-ALBERTSSON 1992). Enterostatin is a peptide fragment, derived from the cleavage of procolipase to colipase. The intestinal concentration of enterostatin in rats is increased by consumption of a high-fat diet, but predicted changes in blood levels have not been demonstrated yet. Enterostatin may also have a primary effect on the central nervous system since direct microinjection of the pentapeptide reduces fat consumption by rats (SHARGILL et al. 1991).

The initial, but provocative, results with apo A-IV and enterostatin will have to be investigated more thoroughly before we can assess the physiologic importance of these putative satiety factors. We must keep in mind, however, that the identification and development of small molecular weight drugs mimicking apo A-IV or enterostatin will entail substantial challenges, which will not be overcome easily. Therapeutic advances that augment the endogenous production or activity of these satiety signals may hold more promise of success.

II. Peripheral Mechanisms

1. Increased Energy Expenditure

It is becoming more accepted that obesity is due to increased energy intake rather than decreased energy expenditure as more data to support this conclusion emerge (AMATRUDA et al. 1993). The future role of agents that may increase energy expenditure is therefore small. Obese indivuduals already have increased energy expenditure. This is due to increased muscle mass, which increases basal energy expenditure, and the additional energy expenditure required for activities of daily living due to increased body weight. B3 agonists are currently being tested. The first agonists were not specific enough, caused side effects such as tremors and palpitations and were not highly efficacious, leading to an approximately 5-kg weight loss over control in an 18-week trial (CONNACHER et al. 1988). Newer agents, which are targeting the human B3 receptor, may be more specific but it is unlikely that increasing energy expenditure further will lead to significant weight loss without also controlling energy intake. Exercise, which can increase energy expenditure considerably, has never led to significant weight loss although it is highly efficacious in helping to keep weight off after weight loss (PAVLOR et al. 1989).

Whether B3 agonists are useful in the treatment of NIDDM is speculative, although recent data in animals are encouraging and studies in mice without brown adipose tissue suggest that a lack of brown adipose tissue may predispose the animals to obesity and diabetes (Lowell et al. 1993). Perhaps stimulating brown adipose tissue will have the opposite effect. Unfortunately, man has little brown adipose tissue although it is inducible.

2. Decreased Fat Absorption

Just as the ingestion of carbohydrates may worsen the control of blood glucose levels in diabetes, dietary fat has been proposed to contribute significantly to the enhanced lipid disposition in obesity. Humans have a limited capacity to synthesize fat from carbohydrates; therefore, almost all body fat is derived from dietary fat. Furthermore, each gram of fat can provide more than twice the amount of energy of an equivalent amount of carbohydrate or protein, making fat an efficient source of calories. Retarding the digestion and subsequent absorption of carbohydrates has proven to be an effective treatment modality for postprandial excursions of hyperglycemia. Similarly, it has been suggested that inhibition of fat absorption could therefore be a benign means of limiting excessive fat ingestion, and thus reducing body weight.

Significant progress has been made recently in the identification of a pharmacologic agent that can be used to prove or disprove the validity of this therapeutic hypothesis. Orlistat (Ro 18-0647) has been found to inhibit gastric and pancreatic lipases. This biochemical action prevents triglyceride hydrolysis and subsequent intestinal absorption (Hadvary et al. 1988). In the initial clinical evaluation of the drug, obese patients lost an average of 2 kg during the 12-week trial (Drent and van der Veen 1993). During this study and prior studies (Hauptman et al. 1992), intestinal side effects of the drug were observed. Unfortunately, the side effects may limit administration of this, or any other, lipase inhibitor, at dosages that reduce obesity.

There is also the possibility that patients may ingest larger quantities of fat and/or carbohydrate to compensate for the reduced absorption of fat. Multiple studies have demonstrated that replacement of dietary fat with nonabsorbable fat substitutes does not result in reductions of daily caloric intake. Experimental subjects exhibited a dramatic ability to compensate for the dietary dilution by increasing their daily intake of carbohydrates (Blundell et al. 1991). Additional studies will have to be completed before we can determine whether or not an inhibition of fat absorption proves to be an effective therapy for obesity.

C. Insulin Sensitizers

I. Rationale

There is substantial evidence linking insulin resistance to the pathogenesis of NIDDM (DeFronzo et al. 1992). First, in longitudinal studies insulin resistance and hyperinsulinemia are seen prior to the development of hyperglycemia and obvious B-cell secretory defects (DeFronzo et al. 1992; Lillioja et al. 1988). Hyperinsulinemia is a universal feature of early NIDDM, with insulin levels falling only after fasting hyperglycemia is clearly demonstrable. However, to develop NIDDM one must have a B-cell defect as well as an increased demand for insulin. Insulin resistance alone will not lead to NIDDM. All obese individuals are insulin resistant, yet only ~20% become diabetic. A similar argument can be made for pregnancy, glucocorticoid treatment, stress and other insulin-resistant states. Only susceptible individuals whose insulin secretion cannot keep up with the demand caused by insulin resistance develop diabetes. Other evidence linking insulin resistance to NIDDM are data from family studies indicating that family members of patients with NIDDM who are at risk for developing NIDDM are also insulin resistant (Lillioja et al. 1987; Warram et al. 1990). Also, prospective data in Pima Indians indicate that insulin resistance is a major risk factor for the development of NIDDM (Lillioja et al. 1993). These data, plus substantial data in animals linking insulin resistance to diabetes, have led to research efforts to develop drugs that will decrease insulin resistance or improve insulin sensitivity.

II. Specific Targets

The thiazolidinedione class has been most thoroughly tested. Many compounds in this class will improve or normalize blood glucose in animal models and some improve blood sugar in man. The precise molecular target of these compounds is unknown, although in vivo they appear to decrease hepatic glucose output. Serious problems with toxicity, however, are slowing the development of this class and may prevent their use.

New research programs are focusing on known molecular targets such as phosphotyrosine phosphatases that dephosphorylate the insulin receptor or substrates for tyrosine phosphorylation by the insulin receptor. Meaningful programs in drug discovery in these areas, however, will depend on identification of the specific phosphotyrosine phosphatases (PTPases) involved in insulin action since nonspecific inhibition of PTPases could activate growth factor receptors or oncogenes. Other targets which might improve sensitivity to insulin include increasing *Glut4* transcription or translocation. This approach, without also improving insulin signaling, might increase glucose transport but would not affect the routes by which the newly transported glucose is metabolized. This would likely lead to increased lactate production rather than increased glycogen, lipid and CO_2 production with the potential for resultant lactic acidosis. Another proposed target for regulation is hexokinase II in muscle; however, there is no clear evidence that increasing hexokinase II transcription or activity would lower plasma glucose, especially if glucose transport is rate limiting. The most fruitful targets for increasing insulin sensitivity will likely await a more complete definition of the insulin-signaling pathways responsible for the metabolic effects of insulin.

Other potential targets include glucagon receptor antagonists (RASKIN et al. 1978; BARON et al. 1987; JOHNSON et al. 1982; UNGER and FOSTER 1992) and glucokinase (VIONNET et al. 1992). Glucagon levels are elevated in patients with diabetes, and a substantial percentage of hepatic glucose output is driven by glucagon. Lowering glucagon levels in animals and man will reduce hepatic glucose output and ketogenesis in the presence and absence of insulin, respectively. Since patients with NIDDM frequently have substantial pancreatic reserve, lowering hepatic glucose output with bedtime insulin will frequently normalize fasting plasma glucose with minimal hyperglycemia during the day in response to meals. A synthetic oral glucagon antagonist would likely have the same effect and would also be useful for IDDM in conjunction with insulin to control hepatic glucose output (HGO) during the night and lower fasting plasma glucose. If postprandial hyperglycemia became excessive, the combination of a glucagon antagonist and an α-glucosidase inhibitor to reduce postprandial glucose would be an excellent combination. A glucagon antagonist would also be useful in IDDM to treat ketoacidosis since glucagon is necessary to make the liver "ketogenic," and neutralization of glucagon prevents and reverses ketoacidosis in animals and man.

Individuals with maturity onset diabetes of the young (MODY) and ~5% of patients with gestational diabetes have mutations in the glucokinase gene. This leads to decreased insulin secretion since glucokinase is an important component of the B-cell glucose sensor. It is unlikely that the more common forms of NIDDM have such defects, making glucokinase an unlikely future target for drug discovery. However, since the mechanism of B-cell failure is unknown, new targets in the B cell related to this pathophysiologic event will likely be identified. If effective weight loss measures and insulin sensitizers become available, however, it is likely that few patients would need a drug to increase insulin secretion.

III. Amylin and Amylin Blockers

For many years researchers have tried to identify blood factors responsible for insulin resistance associated with various diseases. If a single factor were found to be the cause of insulin resistance in NIDDM, then diabetic patients could be treated successfully by diminishing either the production or action of the resistance factor. Certainly, therapies focused on defined mechanisms are very attractive because they can be tremendously effective. However, the success of mechanism-based drug intervention depends largely upon the extent that the dysfunction is actually responsible for the pathology.

It has been suggested recently that the pancreatic polypeptide, amylin or islet amyloid polypeptide (IAPP), is responsible for the insulin resistance of NIDDM. Although this hypothesis has been a topic of substantial investigation and debate during the last few years, no firm conclusion has emerged (RINK et al. 1993; WESTERMARK et al. 1992). Multiple studies have shown that acute exposure of muscle tissue to amylin results in a reduction of insulin-stimulated glucose uptake and glycogen synthesis, two hallmarks of insulin resistance associated with NIDDM. The effects of amylin on hepatic insulin action have not been as consistent. In hyperinsulinemic glucose clamp studies, amylin partially diminished insulin's suppression of hepatic glucose production. However, many investigators have failed to demonstrate direct effects of amylin on isolated hepatocytes. It has been suggested that the augmentation of hepatic glucose production by amylin in the whole animal is a result of amylin's ability to elevate blood lactate levels by enhancing muscle glycogenolysis.

The pancreatic content of amylin is elevated considerably in both animal models of NIDDM as well as in diabetic patients. Similarly, the *relative* circulating levels of amylin are greater in NIDDM patients and rodents than nondiabetic subjects and animals. Because of the difficulties in quantifying plasma levels of amylin, there is considerable controversy concerning whether the blood concentrations required to induce insulin resistance are present in diabetic patients. Many investigators claim that concentrations of amylin necessary to impair insulin action are pharmacologic and do not reflect levels in diabetic patients. In a provocative new study, transgenic mice that overexpressed amylin did not exhibit any alterations in glucose or insulin homeostasis (Fox et al. 1993).

A major step in the ongoing evaluation of the putative pathologic role of amylin in NIDDM will be a determination of the antihyperglycemic activity of amylin receptor antagonists. One study has reported the ability of the peptide $^{8-37}$h-CGRP to antagonize the acute effects of an amylin infusion in anesthetized rats (Wang et al. 1991). More extensive in vivo experiments in animals, as well as possible clinical testing, will be required to adequately assess the potential ability of amylin antagonists to diminish diabetes-induced hyperglycemia.

D. Alternative Insulin Delivery

The normal B cell delivers insulin to the portal circulation in response to the absolute level and rate of change of plasma glucose. Studies in the 1970s with the first artificial pancreas demonstrated that, if the artificial pancreas was programmed to respond to the absolute level of glucose, patients with IDDM had impaired glucose tolerance. If the artificial pancreas was reprogrammed to respond to both the absolute level and rate of change of glucose, not only was glucose tolerance normal but the amount of insulin required was decreased substantially (Albisser et al. 1974a; Albisser et al. 1974b). This demonstrated the complexity of the B cell and the fact that, to control postprandial glucose reasonably, insulin must be injected prior to eating to allow absorption of the insulin prior to the carbohydrate. It also intensified the search for more physiologic insulin delivery systems, which is especially important since the publication of the DCCT trial indicating that the incidence of complications is related to the level of blood glucose control. Research has concentrated on nasal and oral insulin delivery, transplantation of the pancreas and islets, encapsulation of islets and the artificial pancreas as well as insulin analogues.

Nasal insulin appeared very promising because absorption is very rapid and the time course of effect brief, potentially making it very useful for preventing postprandial excursions in plasma glucose (Kimmerle et al. 1991). Unfortunately, two major problems quickly became apparent with this approach. First, to achieve significant absorption of insulin through the nasal mucosa it was necessary to use a surfactant, which causes nasal irritation. Second, the absorption through mucosal surfaces is inefficient and requires large amounts of insulin to be delivered. This raises significant cost considerations. Of less importance is the problem with absorption associated with nasal congestion. Most attempts to further develop this approach have been discontinued. There are much fewer data on oral insulin delivery. This approach suffers from the necessity to encapsulate the insulin in liposomes to avoid digestion in the intestine and the likelihood that large quantities of insulin will be required. Finally, physiologic delivery of insulin will be unlikely given the time course of absorption and the likely variability in absorption.

Pancreas transplantation is an established procedure in many major medical centers, but because of the necessity of immunosuppression is still undertaken primarily or exclusively in patients who are undergoing renal transplantation. Major breakthroughs in this area will come when advances are made in the area of transplantation tolerance through manipulation of the human leukocyte antigen (HLA) system or other manipulations. Islet cell transplantation has never met with significant success due to limitations of supply, the numbers of islets needed and the lack of preservation techniques. Many investigators have pursued the area of islet encapsulation in semipermeable membranes which are biocompatible (Lacy 1993; Lauza et al. 1992). Such membranes are available and encapsulated islets can normalize blood glucose in animals. However, significant issues remain with regard to the number of islets needed for people and the length of time that islets can be preserved functioning in vivo. Pig islets are most frequently mentioned as a source but the number of pigs which would have to be maintained under germ-free conditions make this source impractical. The major breakthrough in this area will likely come when a cell line is established which responds to physiologic concentrations of glucose, secretes significant amounts of insulin to render the number of cells practical, and is stable for long periods of time.

The artificial pancreas may be the closest solution to near normal blood glucose control in patients with IDDM or NIDDM. This device consists of an implantable glucose sensor connected to a small computer, which in turn drives an insulin pump. The sensor continuously monitors glucose levels (most likely subcutaneous) and the insulin infusion rate is adjusted based on the absolute glucose level and the rate of change of plasma glucose. Current implantable sensors last approximately 3–4 days and must be replaced (POITOUT et al. 1993). Improvements in the system would include noninvasive or i.v. glucose sensors; in the first case to avoid implantation and in the second to avoid the lag between changes in plasma glucose and subcutaneous glucose. An artificial pancreas would dramatically alter the lives of patients with IDDM, freeing them from multiple insulin injections, finger stick blood glucose measurements, dietary restrictions and planning and the concern and adjustments necessary during illness, exercise, travel and variable schedules.

E. Gene Therapy

The potential for gene therapy of any disease invokes considerable excitement because such treatment corrects the underlying causes of diseases rather than just reducing the disease symptoms. Recent advances in medicine and technology have made gene therapy a reality, and the Food and Drug Administration (FDA) has established a framework for the consideration and discussion of gene therapy protocols (KESSLER et al. 1993). However, a first step in the development of any gene therapy is the identification of the genetic lesion(s) that cause the disease. Elucidating the genetic components of NIDDM presents many challenges for the medical community.

The high concordance rate of NIDDM between monozygotic twins and increased incidence rate within families has clearly indicated that NIDDM is the result of genetic aberrations. However, defining the genetic lesions in a polygenic disease such as NIDDM is difficult. Patients with NIDDM possess some additional characteristics that will make this search even more arduous than usual. For example, the onset of the disease symptoms occurs late in life and therefore it is difficult to evaluate putative genetic markers in multiple generations. Furthermore, obesity, an associated pathology in most NIDDM patients and perhaps an additional etiologic factor itself, is also the result of polygenic aberrations (BOUCHARD et al. 1993).

Recent success in the delineation of glucokinase gene mutations in diabetic patients having MODY, a subtype of NIDDM that has unique clinical characteristics, has brought us into a new era of diabetes research (FROGUEL et al. 1993). Although mutations of the glucokinase gene are unlikely to be responsible for dysfunctional regulation of glucose homeostasis in most NIDDM patients, these exciting studies demonstrate the potential advancements that are to come in the future. There have been multiple attempts to identify mutations of the human gene for either the insulin receptor or insulin-sensitive glucose transporter (*Glut4*) in common-type NIDDM patients. It has been hypothesized that possible genetic alterations in either of these key proteins could be responsible for the insulin resistance of NIDDM. However, Kusari and colleagues (KUSARI et al. 1991) concluded that a majority of NIDDM patients do not have variations in the coding sequence of the β-subunit of the insulin receptor or the *Glut4* transporter genes. More recently, polymorphism of the glycogen synthase gene has been identified in a subgroup of NIDDM patients (Groop et al. 1993). Interestingly, however, the concentration of glycogen synthase in the muscle of these patients was normal. It is therefore unlikely that genetic variation of the glycogen synthase gene is responsible for alterations in function. Rather than trying to identify mutations in specific candidate genes, other investigators have utilized subtraction cDNA libraries to find genes expressed differentially between normal subjects and NIDDM patients. Using this strategy, Reynet and Kahn (1993) found a protein, called *Rad*, that was overexpressed in the skeletal muscle of NIDDM patients. They characterized the protein as a unique member of the *Ras* family of small GTP-binding proteins. How the overexpression of *Rad* could be responsible for alterations in insulin action, and whether insulin resistance could be corrected by inhibiting the protein, is unknown at this time.

Interesting new studies have begun to evaluate genetic factors that contribute to the development of diabetic complications (Krolewski et al. 1992). Genetic factors may be the reason a subset of diabetic patients are more susceptible to diabetic nephropathy and investigations are underway to evaluate the potential involvement of multiple candidate genes. It has been known that a predisposition to essential hypertension has genetic links and these genetic alterations may also contribute to the development of diabetic complications.

References

Albisser AM, Leibel BS, Ewart TG, Davidovac Z, Botz CK, Zingg W, Schipper H, Gander R (1974a) Clinical control of diabetes by the artificial pancreas. Diabetes 23:397–404

Albisser AM, Leibel BS, Ewart TG, Davidovac Z, Botz CK, Zingg W (1974b) An artificial endocrine pancreas. Diabetes 23:389–396

Amatruda JM, Statt MC, Welle SL (1993) Total and resting energy expenditure in obese women reduced to ideal body weight. J Clin Invest 92:1236–1242

Baron AD, Schaeffer L, Shragg P, Kolterman OG (1987) Role of hyperglucagonemia in maintenance of increased rates of hepatic glucose output in Type II diabetes. Diabetes 36:274–283

Bingley PJ, Bonifacio E, Gale EAM (1992) Can we really predict IDDM? Diabetes 42:213–220

Blundell JE, Burley VJ, Peters JC (1991) Dietary fat and human appetite: effects of non-absorbable fat on energy and nutrient intakes. In: Romsos DR, Himms-Hagen J, Suzuki M (eds) Obesity: dietary factors and control. Japan Scientific Societies Press/Karger, Tokyo, pp 3–13

Bouchard C, Despres JP, Mauriege P (1993) Genetic and nongenetic determinants of regional fat distribution. Endocr Rev 14:72–93

Connacher AA, Jung RT, Mitchell PEG (1988) Weight loss in obese subjects on a restricted diet given BRL 26830A. a new atypical β-adrenoreceptor agonist. Br Med J 296:1217

Cooper SJ, Dourish CT (1990) Multiple cholecystokinin (CCK) receptors and CCK-monoamine interactions are instrumental in the control of feeding. Physiol Behav 48:849–857

DeFronzo RA, Bonadonna RC, Ferrannini E (1992) Pathogenesis of NIDDM. Diabetes Care 15:318–368

Drent ML, Van der Veen EA (1993) Lipase inhibition: a novel concept in the treatment of obesity. Int J Obes 17:241–244

Eisenbarth GS, Verge CF, Allen H, Rewers MJ (1993) The design of trials for prevention of IDDM. Diabetes 42:941–947

Erlanson-Albertsson C (1992) Pancreatic colipase. Structural and physiological aspects. Biochim Biophys Acta 1125:1–7

Fox N, Schrementi J, Nishi M, Ohagi S, Chan SJ, Heisserman JA, Westermark GT, Leckstrom A, Westermark P, Steiner DF (1993) Human islet amyloid polypeptide transgenic mice as a model of non-insulin-dependent diabetes mellitus (NIDDM). FEBS Lett 323:40–44

Froguel P, Zouali H, Vionnet N, Velho G, Vaxillaire M, Sun F, Lesage S, Stoffell M, Takeda J, Passa P, Permutt MA, Beckman JS, Bell GI, Cohen D (1993) Familial hyperglycemia due to mutations in glucokinase. N Engl J Med 328:697–702

Fujimoto K, Cardelli JA, Tso P (1992) Increased apolipoprotein A-IV in rat mesenteric lymph afte lipid meal acts as a physiological signal for satiation. Am J Physiol 262:G1002–G1006

Fujimoto K, Fukagawa K, Sakata T, Tso P (1993) Suppression of food intake by apolipoprotein A-IV is mediated through the central nervous system in rats. J Clin Invest 91:1830–1833

Groop LC, Kankuri M, Schalin-Jäntti C, Ekstrand A, Nikula-Iläs P, Widén E, Kuismanen E, Eriksson J, Franssila-Kallunki A, Saloranta C, Koskimies (1993) Association between polymorphism of the glycogen synthase gene and non-insulin-dependent diabetes mellitus. N Engl J Med 328:10–14

Hadvary P, Lengsfeld H, Wolfer H (1988) Inhibition of pancreatic lipase in vitro by the covalent inhibitor tetrahydrolipstatin. Biochem J 256:357–361

Hauptman JB, Jeunet FS, Hartmann D (1992) Initial studies in humans with the novel gastrointestinal lipase inhibitor Ro 18-0647 (tetrahydrolipstatin). Am J Clin Nutr 55:309S–313S

Holladay MW, Bennett MJ, Tufano MD, Lin CW, Asin KE, Witte DG, Miller TR, Bianchi BR, Nikkel AL, Bednarz L, Nadzan AM (1992) Synthesis and biological activity of CCK heptapeptide analogues. Effects of conformational restraints and standard modifications on receptor subtype selectivity, functional activity in vitro, and appetite suppression in vivo. J Med Chem 35:2919–2928

Johnson DG, Goebel CV, Hruby VJ, Bregman MD, Trivedi D (1982) Hyperglycemia of diabetic rats decreased by a glucagon receptor antagonist. Science 215:1115–1116

Kaufman DL, Clare-Salzler M, Tian J, Forsthuber T, Ting GSP, Robinson P, Atkinson MA, Sercarz EE, Tobin AJ, Lehmann PV (1993) Spontaneous loss of T-cell tolerance to glutamic acid decarboxylase in murine insulin-dependent diabetes. Nature 366:69–72

Keller RJ, Eisenbarth GS, Jackson RA (1993) Insulin prophylaxis in individuals at high risk of Type I diabetes. Lancet 341:927–928

Kessler DA, Siegel JP, Noguchi PD, Zoon KC, Feiden KL, Woodcock J (1993) Regulation of somatic-cell therapy and gene therapy by the Food and Drug Administration. N Engl J Med 329:1169–1173

Kimmerle R, Griffing G, McCall A, Ruderman NB, Stoltz E, Melby JC (1991) Could intranasal insulin be useful in the treatment of non-insulin dependent diabetes mellitus. Diabetes Res Clin Pract 13:69–76

Kosaka K, Kuzuya T, Akanuma Y, Hagura R (1980) Increase in insulin response after treatment of overt maturity-onset diabetes is independent of the mode of treatment. Diabetologia 18:23–28

Krolewski AS, Doria A, Magre J, Warram JH, Housman D (1992) Molecular genetic approaches to the identification of genes involved in the development of nephropathy in insulin-dependent diabetes mellitus. J Am Soc Nephrol 3:S9–S17

Kusari J, Verma US, Buse JB, Henry RR, Olefsky JM (1991) Analysis of the gene sequences of the insulin receptor and the insulin-sensitive glucose transporter (GLUT-4) in patients with common-type non-insulin-dependent diabetes mellitus. J Clin Invest 88:1323–1330

Lacy PE (1993) Status of islet cell transplantation. Diabet Rev 1:76–92

Lauza RP, Sullivan SJ, Chick WL (1992) Islet transplantation with immunoisolation. Diabetes 41:1503–1510

Leibowitz SF (1992) Neurochemical-neuroendocrine systems in the brain controlling macronutrient intake and metabolism. Trends Neurosci 15:491–497

Lillioja S, Mott DM, Zawadzki JK, Young AA, Abbott GH, Knowler WC, Bennett PH, Moll P, Bogardus C (1987) In vivo insulin action is a familial characteristic in nondiabetic Pima Indians. Diabetes 36:1329–1335

Lillioja S, Mott DM, Howard BU, Bennett PH, Yki-Jarvinen H, Freymond D, Nyomba BL, Zurlo F, Swinburn B, Bogardus C (1988) Impaired glucose tolerance as a disorder of insulin action: longitudinal and cross-sectional studies in Pima Indians. N Engl J Med 318:1217–1225

Lillioja S, Mott MD, Spraul M, Ferraro R, Foley J, Ravussin E, Knowles WC, Bennett PH, Bogardus C (1993) Insulin resistance and insulin secretory dysfunction as precursors of non-insulin-dependent-diabetes mellitus. Prospective studies of Pima Indians. N Engl J Med 329:1988–1992

Lowell BB, S-Susalic V, Hamann A, Lawitts JA, Himms-Hagen J, Boyer BB, Kozak LP, Flier JS (1993) Development of obesity in transgenic mice after genetic ablation of brown adipose tissue. Nature 366:740–742

Michel MC (1991) Receptors for neuropeptide Y: multiple subtypes and multiple second messengers. Trends Pharmacol Sci 12:389–394

Mitch CH, Leander JD, Mendelsohn LG, Shaw WN, Wong DT, Cantrell BE, Johnson BG, Reel JK, Snoddy JD, Takemori AE, Zimmerman DM (1993) 3,4-Dimethyl-4-(3-hydroxyphenyl)piperidines: opioid antagonists with potent anorectant activity. J Med Chem 36:2842–2850

Pavlor KN, Krey S, Steffee WP (1989) Exercise as an adjunct to weight loss and maintenance in moderately obese subjects. Am J Clin Nutr 49:1115–1123

Poitout V, Moatti-Sirat D, Reach G, Zhang Y, Wilson GS, Lemonnier F, Klein JC (1993) A glucose monitoring system for on-line estimation in man of blood glucose concentration using a miniaturized glucose sensor implanted in the subcutaneous tissues, and a wearable control unit. Diabetologia 36:658–665

Rabin DV, Pleasic SM, Palmer-Crocker R, Shapiro JA (1992) Cloning and expression of IDDM-specific autoantigens. Diabetes 41:183–186

Raskin P, Unger RH (1978) Hyperglucagonemia and its suppression: importance in the metabolic control of diabetes. N Engl J Med 299:433–436

Reynet C, Kahn CR (1993) Rad: a member of the Ras family overexpressed in muscle of Type II diabetic humans. Science 262:1441–1443

Rink TJ, Beaumont K, Koda J, Young A (1993) Structure and biology of amylin. Trends Pharmacol Sci 14:113–118

Shargill NS, Tsujii S, Bray GA, Erlanson-Albertsson C (1991) Enterostatin suppresses food intake following injection into the third ventricle of rats. Brain Res 544:137–140

Silvente-Poirot S, Dufresne M, Vaysse N, Fourmy D (1993) The peripheral cholecystokinin receptors. Eur J Biochem 215:513–529

Skyler JS, Marks JB (1993) Immune intervention in type I diabetes mellitus. Diabet Rev 1:15–42

Thai AC, Eisenbarth GS (1993) Natural history of IDDM. Diabet Rev 1:1–14

The DCCT Research Group (1993) The effect of intensive insulin treatment of diabetes on the development and progression of long-term complications of insulin-dependent diabetes mellitus. N Engl J Med 329:977–986

Tisch R, Yang X-D, Singer SM, Liblan RS, Fugger L, McDevill HO (1993) Immune response to glutamic acid decarboxylase correlates with insulitis in non-obese diabetic mice. Nature 366:72–75

Unger RH, Foster DW (1992) Diabetes mellitus. In: Wilson JD, Foster DW (ed) Williams textbook of endocrinology. Saunders, Philadelphia, pp 1277–1281

Vionnet N, Stoffel M, Takeda J, Yakuda K, Bell GI, Zonali H, Lesage S, Velho G, Iris F, Passa P, Froquel P, Cohen D (1992) Nonsense mutation in the glucokinase gene causes early-onset non-insulin dependent diabetes mellitus. Nature 356: 721–722

Wang MW, Young AA, Rink TJ, Cooper GJS (1991) 8–37h-CGRP antagonizes actions of amylin on carbohydrate metabolism in vitro and in vivo. FEBS Lett 291:195–198

Warram JH, Martin BC, Krolewski AS, Soeldner JS, Khan CR (1990) Slow glucose removal rate and hyperinsulinemia precede the development of type II diabetes in the offspring of diabetic parents. Ann Intern Med 113:909–915

Westermark P, Johnson KH, O'Brien TD, Betsholtz C (1992) Islet amyloid polypeptide – a novel controversy in diabetes research. Diabetologia 35:297–303

Zhang ZJ, Davidson L, Eisenbarth G, Weiner HL (1991) Suppression of diabetes in nonobese diabetic mice by oral administration of porcine insulin. Proc Natl Acad Sci USA 88:10252–10256

Subject Index

Springer-Verlag
and the Environment

We at Springer-Verlag firmly believe that an international science publisher has a special obligation to the environment, and our corporate policies consistently reflect this conviction.

We also expect our business partners – paper mills, printers, packaging manufacturers, etc. – to commit themselves to using environmentally friendly materials and production processes.

The paper in this book is made from low- or no-chlorine pulp and is acid free, in conformance with international standards for paper permanency.

Printing: Mercedesdruck, Berlin
Binding: Buchbinderei Lüderitz & Bauer, Berlin